Ullmann's Polymers and Plastics

Excellence since 1914

Built from generations of expertise, for generations to come.

ULLMANN'S ENCYCLOPEDIA OF INDUSTRIAL CHEMISTRY

If you want to learn more about novel technologies in biochemistry or nanotechnology or discover unexpected new aspects of seemingly totally familiar processes in industrial chemistry – *ULLMANN'S Encyclopedia of Industrial Chemistry* is your first choice.

For over 100 years now, this reference provides top-notch information on the most diverse fields of industrial chemistry and chemical engineering.

When it comes to definite works on industrial chemistry, it has always been *ULLMANN'S*.

ULLMANN'S ENCYCLOPEDIA OF INDUSTRIAL CHEMISTRY ACADEMY – Kick-start your career – WILEY-VCH

Welcome to the ULLMANN'S ACADEMY

Kick-start your career!

Key topics in industrial chemistry explained by the *ULLMANN'S Encyclopedia* experts – for teaching and learning, or for simply refreshing your knowledge.

THE SMART ARTICLE – FASTER, SMARTER, CONTENT

What's new?

The Smart Article introduces new and enhanced article tools for chemistry content. It is now available within *ULLMANN'S*.

For further information on the features and functions available, go to wileyonlinelibrary.com/thesmartarticle

The information you need in the format you want.

DVD

- Released once a year.
- Fully networkable for up to 200 users.
- Time-limited access for up to 14 months (expires March 2016).
- 2015 Edition ISBN: 978-3-527-33754-5

Online

- Over 1,150 articles available online.
- Over 3,000 authors from over 30 countries have contributed.
- Offers flexible access 24/7 from your library, home, or on the road.
- Updated 4 times per year. ISBN: 978-3-527-30673-2

Print

- *Available as a comprehensive 40 Volume-Set.*
- The 7th Edition published in Aug 2011.
- ISBN: 978-3-527-32943-4

wileyonlinelibrary.com/ref/ullmanns

Only interested in a specific topic?

ULLMANN'S Energy
3 Volume Set • ISBN: 978-3-527-33370-7

ULLMANN'S Fine Chemicals
3 Volume Set • ISBN: 978-3-527-33477-3

ULLMANN'S Fibers
2 Volumes • ISBN: 978-3-527-31772-1

ULLMANN'S Modeling and Simulation
ISBN: 978-3-527-31605-2

ULLMANN'S Renewable Resources
ISBN: 978-3-527-33369-1

ULLMANN'S Agrochemicals
2 Volume Set • ISBN: 978-3-527-31604-5

ULLMANN'S Biotechnology and Biochemical Engineering
2 Volume Set • ISBN: 978-3-527-31603-8

ULLMANN'S Reaction Engineering
2 Volume Set • ISBN: 978-3-527-33371-4

ULLMANN'S Industrial Toxicology
2 Volume Set • ISBN: 978-3-527-31247-4

ULLMANN'S Chemical Engineering and Plant Design
2 Volume Set • ISBN: 978-3-527-31111-8

Visit our website to

Find sample chapters

Learn more about the history, the Editor-in-Chief and the Editorial Advisory Board

Discover the ULLMANN'S ACADEMY

and more...

Chemistry that delivers... Continuous product innovation

Create Innovate Inspire

WILEY-VCH

WILEY

Ullmann's Polymers and Plastics

Products and Processes

Volume 2

WILEY-VCH
Verlag GmbH & Co. KGaA

Editor in Chief:

Dr. Barbara Elvers, Hamburg, Germany

All books published by **Wiley-VCH** are carefully produced. Nevertheless, authors, editors, and publisher do not warrant the information contained in these books, including this book, to be free of errors. Readers are advised to keep in mind that statements, data, illustrations, procedural details or other items may inadvertently be inaccurate.

Library of Congress Card No.:
applied for

British Library Cataloguing-in-Publication Data
A catalogue record for this book is available from the British Library.

Bibliographic information published by the Deutsche Nationalbibliothek
The Deutsche Nationalbibliothek lists this publication in the Deutsche Nationalbibliografie; detailed bibliographic data are available on the Internet at <http://dnb.d-nb.de>.

© 2016 Wiley-VCH Verlag GmbH & Co. KGaA, Boschstr. 12, 69469 Weinheim, Germany

All rights reserved (including those of translation into other languages). No part of this book may be reproduced in any form – by photoprinting, microfilm, or any other means – nor transmitted or translated into a machine language without written permission from the publishers. Registered names, trademarks, etc. used in this book, even when not specifically marked as such, are not to be considered unprotected by law.

Print ISBN: 978-3-527-33823-8
ePDF ISBN: 978-3-527-68595-0
ePub ISBN: 978-3-527-68596-7
Mobi ISBN: 978-3-527-68597-4

Cover Design Grafik-Design Schulz, Fußgönheim, Germany
Typesetting Thomson Digital, Noida, India
Printing and Binding Markono Print Media Pte Ltd, Singapore

Printed on acid-free paper

Preface

This handbook features selected articles from the 7th edition of *ULLMANN'S Encyclopedia of Industrial Chemistry*, including newly written articles that have not been published in a printed edition before. True to the tradition of the ULLMANN'S Encyclopedia, polymers and plastics are addressed from an industrial perspective, including production figures, quality standards and patent protection issues where appropriate. Safety and environmental aspects which are a key concern for modern process industries are likewise considered.

More content on related topics can be found in the complete edition of the ULLMANN'S Encyclopedia.

About ULLMANN'S

ULLMANN'S Encyclopedia is the world's largest reference in applied chemistry, industrial chemistry, and chemical engineering. In its current edition, the Encyclopedia contains more than 30,000 pages, 15,000 tables, 25,000 figures, and innumerable literature sources and cross-references, offering a wealth of comprehensive and well-structured information on all facets of industrial chemistry.

1,100 major articles cover the following main areas:

- Agrochemicals
- Analytical Techniques
- Biochemistry and Biotechnology
- Chemical Reactions
- Dyes and Pigments
- Energy
- Environmental Protection and Industrial Safety
- Fat, Oil, Food and Feed, Cosmetics
- Inorganic Chemicals
- Materials
- Metals and Alloys
- Organic Chemicals
- Pharmaceuticals
- Polymers and Plastics
- Processes and Process Engineering
- Renewable Resources
- Special Topics

First published in 1914 by Professor Fritz Ullmann in Berlin, the *Enzyklopädie der Technischen Chemie* (as the German title read) quickly became the standard reference work in industrial chemistry. Generations of chemists have since relied on ULLMANN'S as their prime reference source. Three further German editions followed in 1928–1932, 1951–1970, and in 1972–1984. From 1985 to 1996, the 5th edition of ULLMANN'S Encyclopedia of Industrial Chemistry was the first edition to be published in English rather than German language. So far, two more complete English editions have been published in print; the 6th edition of 40 volumes in 2002, and the 7th edition in 2011, again comprising 40 volumes. In addition, a number of smaller topic-oriented editions have been published.

Since 1997, *ULLMANN'S Encyclopedia of Industrial Chemistry* has also been available in electronic format, first in a CD-ROM edition and, since 2000, in an enhanced online edition. Both electronic editions feature powerful search and navigation functions as well as regular content updates.

Contents

Volume 1 ....
Symbols and Units IX
Conversion Factors XI
Abbreviations XIII
Country Codes XVIII
Periodic Table of Elements XIX

Part 1: Fundamentals 1
Plastics, General Survey, 1. Definition,
    Molecular Structure and Properties 3
Plastics, General Survey, 2. Production of
    Polymers and Plastics 149
Plastics, General Survey, 3. Supermolecular
    Structures 187
Plastics, General Survey, 4. Polymer
    Composites 205
Plastics, General Survey, 5. Plastics and
    Sustainability 223
Plastics, Analysis 231
Polymerization Processes, 1. Fundamentals 265
Polymerization Processes, 2. Modeling of
    Processes and Reactors 315
Plastics, Processing, 1. Processing of
    Thermoplastics 367
Plastics, Processing, 2. Processing of
    Thermosets 407
Plastics Processing, 3. Machining, Bonding,
    Surface Treatment 439
Plastics, Properties and Testing 471
Plastics, Additives 527
Plasticizers 581

Volume 2 ....
Part 2: Organic Polymers 601
Fluoropolymers, Organic 603
Polyacrylamides and Poly(Acrylic Acids) 659
Polyacrylates 675
Polyamides 697
Polyaspartates and Polysuccinimide 733
Polybutenes 747
Polycarbonates 763
Polyester Resins, Unsaturated 781
Polyesters 791
Polyethylene 817
Polyimides 859
Polymethacrylates 885

Polyoxyalkylenes 899
Polyoxymethylenes 911
Poly(Phenylene Oxides) 927
Polypropylene 937

Volume 3 ....
Polystyrene and Styrene Copolymers 981
Polyureas 1029
Polyurethanes 1051
Poly(Vinyl Chloride) 1111
Polyvinyl Compounds, Others 1141
Poly(Vinyl Esters) 1165
Poly(Vinyl Ethers) 1175
Poly(Vinylidene Chloride) 1181
Polymer Blends 1197
Polymers, Biodegradable 1231
Polymers, Electrically Conducting 1261
Polymers, High-Temperature 1281
Reinforced Plastics 1325
Specialty Plastics 1343
Thermoplastic Elastomers 1365

Volume 4 ....
Part 3: Films, Fibers, Foams 1405
Films 1407
Fibers, 4. Polyamide Fibers 1435
Fibers, 5. Polyester Fibers 1453
Fibers, 6. Polyurethane Fibers 1487
Fibers, 7. Polyolefin Fibers 1495
Fibers, 8. Polyacrylonitrile Fibers 1513
Fibers, 9. Polyvinyl Fibers 1529
Fibers, 10. Polytetrafluoroethylene Fibers .. 1539
High-Performance Fibers 1541
Foamed Plastics 1563

Part 4: Resins 1595
Alkyd Resins 1597
Amino Resins 1615
Epoxy Resins 1643
Phenolic Resins 1733
Resins, Synthetic 1751

Part 5: Inorganic Polymers 1775
Inorganic Polymers 1777

Author Index 1817
Subject Index 1823

Symbols and Units

Symbols and units agree with SI standards (for conversion factors see page XI). The following list gives the most important symbols used in the encyclopedia. Articles with many specific units and symbols have a similar list as front matter.

Symbol	Unit	Physical Quantity
a_B		activity of substance B
A_r		relative atomic mass (atomic weight)
A	m^2	area
c_B	mol/m^3, mol/L (M)	concentration of substance B
C	C/V	electric capacity
c_p, c_v	$J\,kg^{-1}\,K^{-1}$	specific heat capacity
d	cm, m	diameter
d		relative density (ϱ/ϱ_{water})
D	m^2/s	diffusion coefficient
D	Gy (=J/kg)	absorbed dose
e	C	elementary charge
E	J	energy
E	V/m	electric field strength
E	V	electromotive force
E_A	J	activation energy
f		activity coefficient
F	C/mol	Faraday constant
F	N	force
g	m/s^2	acceleration due to gravity
G	J	Gibbs free energy
h	m	height
$\hbar$	$W \cdot s^2$	Planck constant
H	J	enthalpy
I	A	electric current
I	cd	luminous intensity
k	(variable)	rate constant of a chemical reaction
k	J/K	Boltzmann constant
K	(variable)	equilibrium constant
l	m	length
m	g, kg, t	mass
M_r		relative molecular mass (molecular weight)
n_D^{20}		refractive index (sodium D-line, 20 °C)
n	mol	amount of substance
N_A	mol^{-1}	Avogadro constant ($6.023 \times 10^{23}\,mol^{-1}$)
P	Pa, bar*	pressure
Q	J	quantity of heat
r	m	radius
R	$J\,K^{-1}\,mol^{-1}$	gas constant
R	Ω	electric resistance
S	J/K	entropy
t	s, min, h, d, month, a	time
t	°C	temperature
T	K	absolute temperature
u	m/s	velocity
U	V	electric potential

Symbols and Units (Continued from p. IX)

Symbol	Unit	Physical Quantity
U	J	internal energy
V	m^3, L, mL, μL	volume
w		mass fraction
W	J	work
x_B		mole fraction of substance B
Z		proton number, atomic number
α		cubic expansion coefficient
α	Wm^{-2}K^{-1}	heat-transfer coefficient (heat-transfer number)
α		degree of dissociation of electrolyte
$[\alpha]$	10^{-2}deg cm^{2}g^{-1}	specific rotation
η	Pa·s	dynamic viscosity
θ	°C	temperature
$\varkappa$		c_p/c_v
λ	Wm^{-1}K^{-1}	thermal conductivity
λ	nm, m	wavelength
μ		chemical potential
ν	Hz, s^{-1}	frequency
ν	m^2/s	kinematic viscosity (η/ϱ)
π	Pa	osmotic pressure
ϱ	g/cm^3	density
σ	N/m	surface tension
τ	Pa (N/m^2)	shear stress
φ		volume fraction
χ	Pa^{-1} (m^2/N)	compressibility

*The official unit of pressure is the pascal (Pa).

Conversion Factors

SI unit	Non-SI unit	From SI to non-SI multiply by
Mass		
kg	pound (avoirdupois)	2.205
kg	ton (long)	9.842×10^{-4}
kg	ton (short)	1.102×10^{-3}
Volume		
m^3	cubic inch	6.102×10^4
m^3	cubic foot	35.315
m^3	gallon (U.S., liquid)	2.642×10^2
m^3	gallon (Imperial)	2.200×10^2
Temperature		
°C	°F	°C $\times 1.8 + 32$
Force		
N	dyne	1.0×10^5
Energy, Work		
J	Btu (int.)	9.480×10^{-4}
J	cal (int.)	2.389×10^{-1}
J	eV	6.242×10^{18}
J	erg	1.0×10^7
J	kW·h	2.778×10^{-7}
J	kp·m	1.020×10^{-1}
Pressure		
MPa	at	10.20
MPa	atm	9.869
MPa	bar	10
kPa	mbar	10
kPa	mm Hg	7.502
kPa	psi	0.145
kPa	torr	7.502

Powers of Ten

E (exa)	10^{18}	d (deci)	10^{-1}
P (peta)	10^{15}	c (centi)	10^{-2}
T (tera)	10^{12}	m (milli)	10^{-3}
G (giga)	10^9	µ (micro)	10^{-6}
M (mega)	10^6	n (nano)	10^{-9}
k (kilo)	10^3	p (pico)	10^{-12}
h (hecto)	10^2	f (femto)	10^{-15}
da (deca)	10	a (atto)	10^{-18}

Abbreviations

The following is a list of the abbreviations used in the text. Common terms, the names of publications and institutions, and legal agreements are included along with their full identities. Other abbreviations will be defined wherever they first occur in an article. For further abbreviations, see page IX, Symbols and Units; page XVII, Frequently Cited Companies (Abbreviations), and page XVIII, Country Codes in patent references. The names of periodical publications are abbreviated exactly as done by Chemical Abstracts Service.

abs.	absolute	BGA	Bundesgesundheitsamt (Federal Republic of Germany)
a.c.	alternating current		
ACGIH	American Conference of Governmental Industrial Hygienists	BGBl.	Bundesgesetzblatt (Federal Republic of Germany)
ACS	American Chemical Society	BIOS	British Intelligence Objectives Subcommittee Report (see also FIAT)
ADI	acceptable daily intake		
ADN	accord européen relatif au transport international des marchandises dangereuses par voie de navigation interieure (European agreement concerning the international transportation of dangerous goods by inland waterways)	BOD	biological oxygen demand
		bp	boiling point
		B.P.	British Pharmacopeia
		BS	British Standard
		ca.	circa
		calcd.	calculated
ADNR	ADN par le Rhin (regulation concerning the transportation of dangerous goods on the Rhine and all national waterways of the countries concerned)	CAS	Chemical Abstracts Service
		cat.	catalyst, catalyzed
		CEN	Comité Européen de Normalisation
		cf.	compare
ADP	adenosine 5′-diphosphate	CFR	Code of Federal Regulations (United States)
ADR	accord européen relatif au transport international des marchandises dangereuses par route (European agreement concerning the international transportation of dangerous goods by road)		
		cfu	colony forming units
		Chap.	chapter
		ChemG	Chemikaliengesetz (Federal Republic of Germany)
AEC	Atomic Energy Commission (United States)	C.I.	Colour Index
		CIOS	Combined Intelligence Objectives Subcommittee Report (see also FIAT)
a.i.	active ingredient		
AIChE	American Institute of Chemical Engineers	CLP	Classification, Labelling and Packaging
		CNS	central nervous system
AIME	American Institute of Mining, Metallurgical, and Petroleum Engineers	Co.	Company
		COD	chemical oxygen demand
ANSI	American National Standards Institute	conc.	concentrated
AMP	adenosine 5′-monophosphate	const.	constant
APhA	American Pharmaceutical Association	Corp.	Corporation
API	American Petroleum Institute	crit.	critical
ASTM	American Society for Testing and Materials	CSA	Chemical Safety Assessment according to REACH
ATP	adenosine 5′-triphosphate	CSR	Chemical Safety Report according to REACH
BAM	Bundesanstalt für Materialprüfung (Federal Republic of Germany)		
		CTFA	The Cosmetic, Toiletry and Fragrance Association (United States)
BAT	Biologischer Arbeitsstofftoleranzwert (biological tolerance value for a working material, established by MAK Commission, see MAK)		
		DAB	Deutsches Arzneibuch, Deutscher Apotheker-Verlag, Stuttgart
		d.c.	direct current
		decomp.	decompose, decomposition
Beilstein	Beilstein's Handbook of Organic Chemistry, Springer, Berlin – Heidelberg – New York	DFG	Deutsche Forschungsgemeinschaft (German Science Foundation)
BET	Brunauer – Emmett – Teller	dil.	dilute, diluted

DIN	Deutsche Industrienorm (Federal Republic of Germany)		(regulation in the Federal Republic of Germany concerning the transportation of dangerous goods by rail)
DMF	dimethylformamide		
DNA	deoxyribonucleic acid	GGVS	Verordnung in der Bundesrepublik Deutschland über die Beförderung gefährlicher Güter auf der Straße (regulation in the Federal Republic of Germany concerning the transportation of dangerous goods by road)
DOE	Department of Energy (United States)		
DOT	Department of Transportation – Materials Transportation Bureau (United States)		
DTA	differential thermal analysis		
EC	effective concentration	GGVSee	Verordnung in der Bundesrepublik Deutschland über die Beförderung gefährlicher Güter mit Seeschiffen (regulation in the Federal Republic of Germany concerning the transportation of dangerous goods by sea-going vessels)
EC	European Community		
ed.	editor, edition, edited		
e.g.	for example		
emf	electromotive force		
EmS	Emergency Schedule		
EN	European Standard (European Community)		
		GHS	Globally Harmonised System of Chemicals (internationally agreed-upon system, created by theUN, designed to replace the various classification and labeling standards used in different countries by using consistent criteria for classification and labeling on a global level)
EPA	Environmental Protection Agency (United States)		
EPR	electron paramagnetic resonance		
Eq.	equation		
ESCA	electron spectroscopy for chemical analysis		
esp.	especially	GLC	gas-liquid chromatography
ESR	electron spin resonance	Gmelin	Gmelin's Handbook of Inorganic Chemistry, 8th ed., Springer, Berlin – Heidelberg –New York
Et	ethyl substituent $(-C_2H_5)$		
et al.	and others		
etc.	et cetera	GRAS	generally recognized as safe
EVO	Eisenbahnverkehrsordnung (Federal Republic of Germany)	Hal	halogen substituent $(-F, -Cl, -Br, -I)$
		Houben-Weyl	Methoden der organischen Chemie, 4th ed., Georg Thieme Verlag, Stuttgart
exp (. . .)	$e^{(\cdots)}$, mathematical exponent		
FAO	Food and Agriculture Organization (United Nations)		
		HPLC	high performance liquid chromatography
FDA	Food and Drug Administration (United States)		
		H statement	hazard statement in GHS
FD&C	Food, Drug and Cosmetic Act (United States)	IAEA	International Atomic Energy Agency
		IARC	International Agency for Research on Cancer, Lyon, France
FHSA	Federal Hazardous Substances Act (United States)		
		IATA-DGR	International Air Transport Association, Dangerous Goods Regulations
FIAT	Field Information Agency, Technical (United States reports on the chemical industry in Germany, 1945)		
		ICAO	International Civil Aviation Organization
Fig.	figure		
fp	freezing point	i.e.	that is
Friedländer	P. Friedländer, Fortschritte der Teerfarbenfabrikation und verwandter Industriezweige Vol. 1–25, Springer, Berlin 1888–1942	i.m.	intramuscular
		IMDG	International Maritime Dangerous Goods Code
		IMO	Inter-Governmental Maritime Consultive Organization (in the past: IMCO)
FT	Fourier transform		
(g)	gas, gaseous	Inst.	Institute
GC	gas chromatography	i.p.	intraperitoneal
GefStoffV	Gefahrstoffverordnung (regulations in the Federal Republic of Germany concerning hazardous substances)	IR	infrared
		ISO	International Organization for Standardization
GGVE	Verordnung in der Bundesrepublik Deutschland über die Beförderung gefährlicher Güter mit der Eisenbahn	IUPAC	International Union of Pure and Applied Chemistry
		i.v.	intravenous

Kirk-Othmer	Encyclopedia of Chemical Technology, 3rd ed., 1991–1998, 5th ed., 2004–2007, John Wiley & Sons, Hoboken	no.	number
		NOEL	no observed effect level
(l)	liquid	NRC	Nuclear Regulatory Commission (United States)
Landolt-Börnstein	Zahlenwerte u. Funktionen aus Physik, Chemie, Astronomie, Geophysik u. Technik, Springer, Heidelberg 1950–1980; Zahlenwerte und Funktionen aus Naturwissenschaften und Technik, Neue Serie, Springer, Heidelberg, since 1961	NRDC	National Research Development Corporation (United States)
		NSC	National Service Center (United States)
		NSF	National Science Foundation (United States)
		NTSB	National Transportation Safety Board (United States)
LC_{50}	lethal concentration for 50 % of the test animals	OECD	Organization for Economic Cooperation and Development
LCLo	lowest published lethal concentration	OSHA	Occupational Safety and Health Administration (United States)
LD_{50}	lethal dose for 50 % of the test animals		
LDLo	lowest published lethal dose	p., pp.	page, pages
ln	logarithm (base e)	Patty	G.D. Clayton, F.E. Clayton (eds.): Patty's Industrial Hygiene and Toxicology, 3rd ed., Wiley Interscience, New York
LNG	liquefied natural gas		
log	logarithm (base 10)		
LPG	liquefied petroleum gas		
M	mol/L	PB report	Publication Board Report (U.S. Department of Commerce, Scientific and Industrial Reports)
M	metal (in chemical formulas)		
MAK	Maximale Arbeitsplatzkonzentration (maximum concentration at the workplace in the Federal Republic of Germany); cf. Deutsche Forschungsgemeinschaft (ed.): Maximale Arbeitsplatzkonzentrationen (MAK) und Biologische Arbeitsstofftoleranzwerte (BAT), WILEY-VCH Verlag, Weinheim (published annually)		
		PEL	permitted exposure limit
		Ph	phenyl substituent ($-C_6H_5$)
		Ph. Eur.	European Pharmacopoeia, Council of Europe, Strasbourg
		phr	part per hundred rubber (resin)
		PNS	peripheral nervous system
		ppm	parts per million
		P statement	precautionary statement in GHS
max.	maximum	q.v.	which see (quod vide)
MCA	Manufacturing Chemists Association (United States)	REACH	Registration, Evaluation, Authorisation and Restriction of Chemicals (EU regulation addressing the production and use of chemical substances, and their potential impacts on both human health and the environment)
Me	methyl substituent ($-CH_3$)		
Methodicum Chimicum	Methodicum Chimicum, Georg Thieme Verlag, Stuttgart		
MFAG	Medical First Aid Guide for Use in Accidents Involving Dangerous Goods		
		ref.	refer, reference
MIK	maximale Immissionskonzentration (maximum immission concentration)	resp.	respectively
		R_f	retention factor (TLC)
min.	minimum	R.H.	relative humidity
mp	melting point	RID	réglement international concernant le transport des marchandises dangereuses par chemin de fer (international convention concerning the transportation of dangerous goods by rail)
MS	mass spectrum, mass spectrometry		
NAS	National Academy of Sciences (United States)		
NASA	National Aeronautics and Space Administration (United States)		
		RNA	ribonucleic acid
NBS	National Bureau of Standards (United States)	R phrase (R-Satz)	risk phrase according to ChemG and GefStoffV (Federal Republic of Germany)
NCTC	National Collection of Type Cultures (United States)		
		rpm	revolutions per minute
NIH	National Institutes of Health (United States)	RTECS	Registry of Toxic Effects of Chemical Substances, edited by the National Institute of Occupational Safety and Health (United States)
NIOSH	National Institute for Occupational Safety and Health (United States)		
NMR	nuclear magnetic resonance	(s)	solid

SAE	Society of Automotive Engineers (United States)		der Technischen Chemie, 4th ed., Verlag Chemie, Weinheim 1972–1984; 3rd ed., Urban und Schwarzenberg, München 1951–1970
SAICM	Strategic Approach on International Chemicals Management (international framework to foster the sound management of chemicals)	USAEC	United States Atomic Energy Commission
s.c.	subcutaneous	USAN	United States Adopted Names
SI	International System of Units	USD	United States Dispensatory
SIMS	secondary ion mass spectrometry	USDA	United States Department of Agriculture
S phrase (S-Satz)	safety phrase according to ChemG and GefStoffV (Federal Republic of Germany)	U.S.P.	United States Pharmacopeia
		UV	ultraviolet
STEL	Short Term Exposure Limit (see TLV)	UVV	Unfallverhütungsvorschriften der Berufsgenossenschaft (workplace safety regulations in the Federal Republic of Germany)
STP	standard temperature and pressure (0°C, 101.325 kPa)		
T_g	glass transition temperature	VbF	Verordnung in der Bundesrepublik Deutschland über die Errichtung und den Betrieb von Anlagen zur Lagerung, Abfüllung und Beförderung brennbarer Flüssigkeiten (regulation in the Federal Republic of Germany concerning the construction and operation of plants for storage, filling, and transportation of flammable liquids; classification according to the flash point of liquids, in accordance with the classification in the United States)
TA Luft	Technische Anleitung zur Reinhaltung der Luft (clean air regulation in Federal Republic of Germany)		
TA Lärm	Technische Anleitung zum Schutz gegen Lärm (low noise regulation in Federal Republic of Germany)		
TDLo	lowest published toxic dose		
THF	tetrahydrofuran		
TLC	thin layer chromatography		
TLV	Threshold Limit Value (TWA and STEL); published annually by the American Conference of Governmental Industrial Hygienists (ACGIH), Cincinnati, Ohio	VDE	Verband Deutscher Elektroingenieure (Federal Republic of Germany)
		VDI	Verein Deutscher Ingenieure (Federal Republic of Germany)
TOD	total oxygen demand		
TRK	Technische Richtkonzentration (lowest technically feasible level)	vol	volume
		vol.	volume (of a series of books)
TSCA	Toxic Substances Control Act (United States)	vs.	versus
		WGK	Wassergefährdungsklasse (water hazard class)
TÜV	Technischer Überwachungsverein (Technical Control Board of the Federal Republic of Germany)		
		WHO	World Health Organization (United Nations)
TWA	Time Weighted Average	Winnacker-Küchler	Chemische Technologie, 4th ed., Carl Hanser Verlag, München, 1982-1986; Winnacker-Küchler, Chemische Technik: Prozesse und Produkte, Wiley-VCH, Weinheim, 2003–2006
UBA	Umweltbundesamt (Federal Environmental Agency)		
Ullmann	Ullmann's Encyclopedia of Industrial Chemistry, 6th ed., Wiley-VCH, Weinheim 2002; Ullmann's Encyclopedia of Industrial Chemistry, 5th ed., VCH Verlagsgesellschaft, Weinheim 1985–1996; Ullmanns Encyklopädie		
		wt	weight
		$	U.S. dollar, unless otherwise stated

Frequently Cited Companies (Abbreviations)

Air Products	Air Products and Chemicals	IFP	Institut Français du Pétrole
Akzo	Algemene Koninklijke Zout Organon	INCO	International Nickel Company
		3M	Minnesota Mining and Manufacturing Company
Alcoa	Aluminum Company of America	Mitsubishi Chemical	Mitsubishi Chemical Industries
Allied	Allied Corporation		
Amer. Cyanamid	American Cyanamid Company	Monsanto	Monsanto Company
		Nippon Shokubai	Nippon Shokubai Kagaku Kogyo
BASF	BASF Aktiengesellschaft		
Bayer	Bayer AG	PCUK	Pechiney Ugine Kuhlmann
BP	British Petroleum Company	PPG	Pittsburg Plate Glass Industries
Celanese	Celanese Corporation	Searle	G.D. Searle & Company
Daicel	Daicel Chemical Industries	SKF	Smith Kline & French Laboratories
Dainippon	Dainippon Ink and Chemicals Inc.	SNAM	Societá Nazionale Metandotti
Dow Chemical	The Dow Chemical Company	Sohio	Standard Oil of Ohio
		Stauffer	Stauffer Chemical Company
DSM	Dutch Staats Mijnen	Sumitomo	Sumitomo Chemical Company
Du Pont	E.I. du Pont de Nemours & Company	Toray	Toray Industries Inc.
Exxon	Exxon Corporation	UCB	Union Chimique Belge
FMC	Food Machinery & Chemical Corporation	Union Carbide	Union Carbide Corporation
GAF	General Aniline & Film Corporation	UOP	Universal Oil Products Company
W.R. Grace	W.R. Grace & Company	VEBA	Vereinigte Elektrizitäts- und Bergwerks-AG
Hoechst	Hoechst Aktiengesellschaft	Wacker	Wacker Chemie GmbH
IBM	International Business Machines Corporation		
ICI	Imperial Chemical Industries		

Country Codes

The following list contains a selection of standard country codes used in the patent references.

AT	Austria	IL	Israel
AU	Australia	IT	Italy
BE	Belgium	JP	Japan*
BG	Bulgaria	LU	Luxembourg
BR	Brazil	MA	Morocco
CA	Canada	NL	Netherlands*
CH	Switzerland	NO	Norway
CS	Czechoslovakia	NZ	New Zealand
DD	German Democratic Republic	PL	Poland
DE	Federal Republic of Germany	PT	Portugal
	(and Germany before 1949)*	SE	Sweden
DK	Denmark	SU	Soviet Union
ES	Spain	US	United States of America
FI	Finland	YU	Yugoslavia
FR	France	ZA	South Africa
GB	United Kingdom	EP	European Patent Office*
GR	Greece	WO	World Intellectual Property
HU	Hungary		Organization
ID	Indonesia		

*For Europe, Federal Republic of Germany, Japan, and the Netherlands, the type of patent is specified: EP (patent), EP-A (application), DE (patent), DE-OS (Offenlegungsschrift), DE-AS (Auslegeschrift), JP (patent), JP-Kokai (Kokai tokkyo koho), NL (patent), and NL-A (application).

Periodic Table of Elements

element symbol, atomic number, and relative atomic mass (atomic weight)

- 1A "European" group designation and old IUPAC recommendation
- 1   group designation to 1986 IUPAC proposal
- IA "American" group designation, also used by the Chemical Abstracts Service until the end of 1986

1A 1 IA	2A 2 IIA	3A 3 IIIB	4A 4 IVB	5A 5 VB	6A 6 VIB	7A 7 VIIB	8 8 VIII	8 9 VIII	8 10 VIII	1B 11 IB	2B 12 IIB	3B 13 IIIA	4B 14 IVA	5B 15 VA	6B 16 VIA	7B 17 VIA	0 18 VIIIA
1 **H** 1.0079																	2 **He** 4.0026
3 **Li** 6.941	4 **Be** 9.0122											5 **B** 10.811	6 **C** 12.011	7 **N** 14.007	8 **O** 15.999	9 **F** 18.998	10 **Ne** 20.180
11 **Na** 22.990	12 **Mg** 24.305											13 **Al** 26.982	14 **Si** 28.086	15 **P** 30.974	16 **S** 32.066	17 **Cl** 35.453	18 **Ar** 39.948
19 **K** 39.098	20 **Ca** 40.078	21 **Sc** 44.956	22 **Ti** 47.867	23 **V** 50.942	24 **Cr** 51.996	25 **Mn** 54.938	26 **Fe** 55.845	27 **Co** 58.933	28 **Ni** 58.693	29 **Cu** 63.546	30 **Zn** 65.409	31 **Ga** 69.723	32 **Ge** 72.61	33 **As** 74.922	34 **Se** 78.96	35 **Br** 79.904	36 **Kr** 83.80
37 **Rb** 85.468	38 **Sr** 87.62	39 **Y** 88.906	40 **Zr** 91.224	41 **Nb** 92.906	42 **Mo** 95.94	43 **Tc*** 98.906	44 **Ru** 101.07	45 **Rh** 102.91	46 **Pd** 106.42	47 **Ag** 107.87	48 **Cd** 112.41	49 **In** 114.82	50 **Sn** 118.71	51 **Sb** 121.76	52 **Te** 127.60	53 **I** 126.90	54 **Xe** 131.29
55 **Cs** 132.91	56 **Ba** 137.33		72 **Hf** 178.49	73 **Ta** 180.95	74 **W** 183.84	75 **Re** 186.21	76 **Os** 190.23	77 **Ir** 192.22	78 **Pt** 195.08	79 **Au** 196.97	80 **Hg** 200.59	81 **Tl** 204.38	82 **Pb** 207.2	83 **Bi** 208.98	84 **Po*** 208.98	85 **At*** 209.99	86 **Rn*** 222.02
87 **Fr*** 223.02	88 **Ra*** 226.03		104 **Rf*** 261.11	105 **Db*** 262.11	106 **Sg**	107 **Bh**	108 **Hs**	109 **Mt**	110 **Ds**	111 **Rg**	112 **Cn**	113 **Uut**[a]	114 **Fl**	115 **Uup**[a]	116 **Lv**		118 **Uuo**[a]

[a] provisional IUPAC symbol

57 **La** 138.91	58 **Ce** 140.12	59 **Pr** 140.91	60 **Nd** 144.24	61 **Pm*** 146.92	62 **Sm** 150.36	63 **Eu** 151.97	64 **Gd** 157.25	65 **Tb** 158.93	66 **Dy** 162.50	67 **Ho** 164.93	68 **Er** 167.26	69 **Tm** 168.93	70 **Yb** 173.04	71 **Lu** 174.97
89 **Ac*** 227.03	90 **Th*** 232.04	91 **Pa*** 231.04	92 **U*** 238.03	93 **Np*** 237.05	94 **Pu*** 244.06	95 **Am*** 243.06	96 **Cm*** 247.07	97 **Bk*** 247.07	98 **Cf*** 251.08	99 **Es*** 252.08	100 **Fm*** 257.10	101 **Md*** 258.10	102 **No*** 259.10	103 **Lr*** 260.11

* radioactive element; mass of most important isotope given.

Part 2

Organic Polymers

Fluoropolymers, Organic

KLAUS HINTZER, 3M Dyneon, Burgkirchen, Germany

TILMAN ZIPPLIES, 3M Dyneon, Burgkirchen, Germany

D. PETER CARLSON, Du Pont de Nemours & Co., Polymer Products Dept., Experimental Station LaboratoryWilmington, United States

WALTER SCHMIEGEL, Du Pont de Nemours & Co., Polymer Products Dept., Experimental Station LaboratoryWilmington, United States

1.	Introduction	604
1.1.	History	604
1.2.	Monomers	604
1.3.	Polymers	604
1.4.	Production	605
1.5.	Properties	606
1.6.	Uses	606
2.	Fluoroplastics	606
2.1.	Introduction	606
2.2.	Polytetrafluoroethylene	607
2.2.1.	Production	608
2.2.2.	Properties	610
2.2.3.	Processing	613
2.2.4.	Uses	615
2.2.5.	New Materials—Modified Granular PTFE	615
2.3.	Tetrafluoroethylene–Hexafluoropropene Copolymers	617
2.3.1.	Production	617
2.3.2.	Properties	617
2.3.3.	Processing	619
2.3.4.	Uses	620
2.4.	Tetrafluoroethylene–Perfluoro (Alkyl Vinyl Ether) Copolymers	620
2.4.1.	Production	620
2.4.2.	Properties	621
2.4.3.	Processing	622
2.4.4.	Uses	624
2.5.	Tetrafluoroethylene–Ethylene Copolymers	624
2.5.1.	Production	624
2.5.2.	Properties	625
2.5.3.	Processing	628
2.5.4.	Uses	628
2.6.	Polychlorotrifluoroethylene	629
2.6.1.	Production	629
2.6.2.	Properties	629
2.6.3.	Processing	629
2.6.4.	Uses	630
2.7.	Chlorotrifluoroethylene–Ethylene Copolymers	630
2.7.1.	Production	630
2.7.2.	Properties	630
2.7.3.	Processing	631
2.7.4.	Uses	631
2.8.	Poly(Vinylidene Fluoride)	632
2.8.1.	Production	632
2.8.2.	Properties	632
2.8.3.	Processing	634
2.8.4.	Uses	634
2.9.	Poly(Vinyl Fluoride)	635
2.9.1.	Production	635
2.9.2.	Properties	635
2.9.3.	Processing	636
2.9.4.	Uses	636
2.10.	Amorphous Perfluoropolymers	636
2.10.1.	Production	636
2.10.2.	Properties	637
2.10.3.	Processing	638
2.10.4.	Uses	638
2.11.	Semicrystalline Tetrafluoroethylene-Hexafluoropropene-Vinylidene Fluoride Terpolymers	638
2.11.1.	Production	638
2.11.2.	Properties	638
2.11.3.	Processing	639
2.11.4.	Uses	640
3.	Fluoroelastomers (→ Rubber, 4. Emulsion Rubbers, Chap. 6)	640
3.1.	Introduction	640
3.2.	Elastomers Based on Vinylidene Fluoride (FKMs)	641
3.2.1.	Production	642
3.2.2.	Properties	643
3.2.3.	Processing	644

3.2.4.	Uses	648		3.5.	Fluoroelastomers containing Ethylene	651
3.3.	**Elastomers Based on Tetrafluoroethylene—Perfluoro (Methyl Vinyl Ether) Copolymers (FFKM)**	648		3.5.1.	TFE–E–PMVE Polymers	651
				3.5.2.	TFE–VDF–HFP–E Polymers	651
				3.6.	**Thermoplastic Elastomers**	651
3.3.1.	Production	648		3.7.	**Vinyl Ether Copolymers**	652
3.3.2.	Properties	649		4.	**Toxicology and Occupational Health**	653
3.3.3.	Processing and Uses	650				
3.4.	**Elastomers Based on Tetrafluoroethylene–Propene Copolymers**	650		4.1.	**Fluoroplastics**	653
				4.2.	**Fluoroelastomers**	653
				4.3.	**Environmental Aspects**	653
3.4.1.	Production	650		5.	**Economic Aspects**	654
3.4.2.	Properties	650			References	655
3.4.3.	Processing	651				
3.4.4.	Uses	651				

1. Introduction

1.1. History

Researchers of IG-Farbenindustrie in Hoechst/Frankfurt, Germany studied systematically the polymerizations of fluoroethylenes already in the early 1930s; polychlorotrifluoroethylene (PCTFE) and polytetrafluoroethylene (PTFE) including copolymers were prepared and the outstanding properties of these polymers were recognized. The first patent application for a fluoropolymer was filed in October 1934 by SCHLOFFER and SCHERER [7]. During the following years until 1948 there were not many activities anymore within this polymer field in Germany (due to World War II impacts).

Independently polytetrafluoroethylene was discovered by chance by PLUNKETT from DuPont while investigating refrigerants in 1938. The unique properties of PTFE were recognized during the Manhattan Project; there was an urgent need for a material withstanding the highly corrosive substance UF_6. PTFE apparently filled all the needs, spurring the development of processing and production methods for this unique polymer. In 1946, PTFE was commercialized by DuPont under the trademark Teflon. PTFE, PCTFE and all other fluoropolymers with their outstanding chemical, electrical, and surface properties gained immediate acceptance in various markets. During the decades, many fluoropolymers, including fluorothermoplastics and fluoroelastomers, were developed and commercialized. Fluoropolymers serve extremely high demanding applications in respect to performance, longevity, safety, etc., which no other polymer class can provide.

1.2. Monomers

There are only four key C_2-fluoroolefins, which have been polymerized to high molar mass homopolymers of commercial importance: tetrafluoroethylene (TFE), chlorotrifluoroethylene (CTFE), vinylidene fluoride (VDF), and vinyl fluoride (VF).

Most are prepared on industrial scale by pyrolysis of corresponding chlorinated or fluorinated hydrocarbons ($\rightarrow$ Fluorine Compounds, Organic). Special care has to be taken in producing and handling TFE due to its high tendency to self-decompose into carbon and tetrafluoromethane.

Other fluorinated or nonfluorinated olefins are used in combination with TFE to generate a large variety of crystalline or amorphous fluoropolymers. The most important monomers for commercial fluoropolymers are listed in Table 1.

1.3. Polymers

Fluoropolymers consist of carbon and fluorine in the polymer backbone. Fluorosilicones and fluoroacrylate polymers are not referred to as fluoropolymers and are not described in this article. Fluoropolymers can be divided into perfluorinated (e.g., PTFE, PFA, FEP, etc.)

Table 1. Monomers used in commercial fluoropolymers

Monomer	CAS registry number	Formula	Abbreviation	bp, °C	T_{crit}, °C	p_{crit}, MPa
Ethylene	[74-85-1]	$CH_2 = CH_2$	E	−104	79.6	75.1
Tetrafluoroethylene	[116-14-3]	$CF_2 = CF_2$	TFE	−75.6	33.3	3.7
Chlorotrifluoroethylene	[79-38-9]	$CF_2 = CClF$	CTFE	−28.4	105.8	3.9
Vinylidene fluoride	[79-38-7]	$CF_2 = CH_2$	VDF	−82.0	30.1	4.3
Vinyl fluoride	[75-02-5]	$CFH = CH_2$	VF	−72.2	54.7	5.4
Propene	[115-07-1]	$CH_3CH = CH_2$	P	−47	91.4	4.6
Hexafluoropropene	[116-15-4]	$CF_3CF = CF_2$	HFP	−29.4	86.1	2.8
Perfluoro(methyl vinyl ether)	[1187-93-5]	$CF_3OCF = CF_2$	PMVE	−21.8	96.2	3.4
Perfluoro(propyl vinyl ether)	[1623-05-8]	$CF_3CF_2CF_2OCF = CF_2$	PPVE	36	150.6	1.9

and partially fluorinated (e.g., ETFE, PVDF, etc.) types. A more convenient classification is to divide the fluoroplastics and amorphous fluoropolymers into four categories:

1. High-crystalline, nonmelt-processable PTFEs
2. Semicrystalline ($mp > 80°C$), melt-processable fluorothermoplastics
3. Amorphous fluoropolymers with $T_g > 70°C$
4. Amorphous, curable fluoroelastomers [9]

Table 2 provides an overview of the most important commercial fluoropolymers. A special group of TFE-copolymers containing sulfonic acids and/or carboxylic acids in side-chains are important polymers to make ion-exchange membranes for the chlorine industry and for fuel cell applications (→ Fluorine Compounds, Organic, Chap. 8.2 Fluorinated Alkanesulfonic Acids).

1.4. Production

The principal method for synthesizing fluoropolymers is free-radical polymerization. Due to the electrophilic nature of fluoroolefins, cationic polymerization catalysts are not very effective. Fluoroolefins and epoxides can be polymerized by anionic catalysts, but termination by fluoride ion elimination prevents formation of high molar masses. Coordination catalysts do not effect polymerization of fluoroolefins. The free radical polymerizations are mostly water-based, either as aqueous suspension polymerization or as aqueous emulsion polymerization in the presence of hydrocarbon emulsifiers but preferably in the presence of fluorinated emulsifiers. In previous times, the radical polymerization in fluorinated solvents (e.g., in R113) was widely spread, but due to the high emissions of the fluorinated solvents and due to the Montreal agreement, many polymerization processes were changed to water-based systems. The polymerization in supercritical media (e.g., CO_2), originally introduced as "green" polymerization technology, has not found widespread applicability [10, 11]. Other free-radical polymerization processes, e.g., in gas phase or ionic liquids are only of academic interest.

The polymerization and processing techniques described in this article are described in detail elsewhere (→ Polymerization Processes, 1. Fundamentals).

Table 2. Composition of important fluoropolymers

	Momomer(s)	mp, °C
Nonmelt processable PTFE		
PTFE	TFE	327
modified PTFE	TFE PVE (< 1 wt%)	326
Melt processable fluorothermoplastics		
PFA	TFE + PPVE	305
FEP	TFE + HFP	260
ETFE	TFE + E	270
THV	TFE + HFP + VDF	120–220
PVDF	VDF	170
PVF	VF	190
PCTFE	CTFE	210
ECTFE	CTFE + E	240
Amorphous fluoropolymer		T_g, °C
Teflon AF	PDD + TFE	160–240
Hyflon AD	TTD + TFE	90–125
Cytop	PBVE	108
Amorphous, curable fluoropolymer		
FKM	TFE + VDF + HFP + cure package	
TFEP	TFE + P + cure package	
FFKM	TFE + PMVE + funcitional momomer + cure package	

1.5. Properties

The properties of fluoropolymers vary with the fluorine content and with the distribution of fluorine atoms in the polymer chain. The unique properties of fluoropolymers are due to the strong carbon–carbon bond (ca. −340 kJ/mol) and the stable carbon–fluorine bond (ca. −490 kJ/mol; for comparison: C–H ca. −420 kJ/mol) of the polymer backbone. Substitution of fluorine for hydrogen improves three key physical properties of the polymer:

1. Increased service temperatures and reduced flammability
2. Low surface energy provides nonstick properties, anti-adhesiveness, low coefficient of friction, self-lubricating effects, and lower solubility in hydrocarbons
3. Excellent electrical and optical properties are seen in low high-frequency loss rates and in low refractive indices.

1.6. Uses

The outstanding properties enable applications in almost every industry (Table 3).

2. Fluoroplastics

2.1. Introduction

The family of fluoroplastic polymers (world consumption 2008: 1.8×10^5 t) [12] is headed by the nonmelt processable PTFEs (consumption 2008: 1.05×10^5 t/a). Melt processable fluorothermoplastics include [13]: TFE–HFP (FEP), TFE–PPVE copolymers (PFA), TFE–PMVE (MFA), TFE–ethylene (ETFE) TFE–HFP–VDF (THV), polychlorotrifluoroethylene (PCTFE), CTFE–ethylene (ECTFE), poly(vinylidene fluoride) (PVDF) including copolymers, and poly(vinyl fluoride) (PVF).

The progressing development of the various fluoropolymer grades was mostly triggered by

Table 3. Major applications of fluoropolymers

Industry or application area	Key properties	Typical uses
Chemical processing	chemical resistance good mechanical properties thermal stability cryogenic properties	gaskets, vessel liners, valve and pipe liners, tubing, coatings
Electrical and communications	low dielectric constant high volume and surface resistivity high dielectric breakdown voltage flame resistance, thermal stability low refractive indices	wire and cable insulation, connectors, optical fibers
Automotive and office equipment	low coefficient of friction good mechanical properties cryogenic properties chemical resistance low permeation properties	seals and rings in automotive power steering, transmission, air-conditioning, copier roller and food processing equipment covering, fuel management systems
Houseware	thermal stability low surface energy chemical resistance	cookware coatings
Medical	low surface energy stability excellent mechanical properties chemical resistance	cardiovascular grafts, heart patches, ligament replacement
Architectural fabrics/films	excellent weatherability flame resistance low surface energy	coated fabrics for buildings and roofs; Front/backside films for solar applications
Polymer processing additives	low coefficient of friction flame resistance	polyolefin processing to avoid surface defects, antidripping agents
Semiconductor industry	chemical resistance purity thermal stability	process surfaces, wafer carriers, tubing, valves, pumps, fittings

the trade-off of existing grades or by application needs. The high melt viscosity of PTFE resins prevents processing by conventional extrusion and molding techniques; they are processed by sintering techniques similar to those used in powder metallurgy. Although PTFE can be converted into practically any form (rods, billets, films, fibers, tubing, or coatings) by these techniques, the lack of melt processability is frequently a handicap. Research on melt-processable forms of PTFE began almost immediately after its discovery [14]. The first to be developed was a TFE–HFP copolymer called FEP (fluorinated ethylene–propene) copolymer [15]. The FEP resins, commercialized in 1960, share nearly all the properties of PTFE, but are also processable by extrusion and injection molding. However, the melting point of FEP is ca. 60–70°C lower than that of PTFE; its maximum use temperature is also lower. Further research on TFE copolymers led to the development of perfluoroalkoxy (PFA) resins [16] that were introduced commercially in 1972. These have the same high-temperature properties as PTFE resins and an even lower melt viscosity than FEP resins; they can be processed by extrusion and injection molding. In 1970, ETFE resins were commercialized. They exhibit improved mechanical properties and far better processability than other TFE copolymers. In addition, they can be cross-linked by high-energy radiation. The maximum use temperature, however, is much lower than that of PFA or FEP.

Among the fluoroplastics that contain monomers other than TFE, PCTFE was the first to be commercialized. Extensive development work was carried out on this polymer during World War II in conjunction with the Manhattan Project. Compared to TFE resins, PCTFE is harder, more resistant to creep, and less permeable. It has the lowest permeability to moisture of any plastic. Chlorotrifluoroethylene copolymerized with VDF provides an improved plastic for film manufacture. However, ECTFE is the most important CTFE copolymer [17]. These plastics have similar properties and uses as ETFE copolymers.

Poly(vinyl fluoride) contains the smallest amount of fluorine (41.3%) of any commercial fluoropolymer but possesses many of the properties of more highly fluorinated polymers [18].

It was commercialized in 1961 by DuPont under the trade name Tedlar. It is used mainly in coatings for metal, plastic, paper, and similar substrates to provide resistance to weather, chemicals, staining, and abrasion.

Poly(vinylidene fluoride) (PVDF) shares many of the characteristics of other fluoropolymers, such as thermal and oxidative stability, as well as outstanding weatherability [19]. However, the arrangement of alternate fluorine and hydrogen atoms leads to unusual polarity within the polymer chains, with a dramatic effect on dielectric properties and solubility. An unusual product made from PVDF is a film with piezoelectric properties. Copolymerization of VDF with small amounts (< 15%) of HFP reduces stiffness and improves processability for certain wire coating applications.

A comparison of manufacturing costs is provided in [20]; polymer designations and testing procedures for fluoroplastics are described in EN ISO 12086-1/2 (2006).

2.2. Polytetrafluoroethylene

Polytetrafluoroethylene [9002-84-0] (PTFE) is a straight-chain polymer of tetrafluoroethylene (TFE) of the general formula

$$-(CF_2CF_2-)_n$$

In 2008, the worldwide consumption of PTFE was 1.05×10^5 t/a, and its demand has grown by 2–3% per year over the decades.

Processed PTFE has a high crystalline melting point (327°C), a very high melt viscosity (10×10^9 to 100×10^9 Pa s at 380°C), and a high maximum use temperature (> 260°C). In addition, it exhibits unusual toughness down to very low temperatures (< −200°C); its molar mass is high (10^6–10^8 g/mol). It is insoluble in most known solvents and resists attack by most chemicals. Dielectric loss is low, whereas dielectric strength is high; antistick and antifriction properties are unusual. Although these properties give PTFE great commercial value, they also rule out processing by conventional thermoplastic techniques. Thus, PTFE is also called nonmelt-processable.

High molar mass polytetrafluoroethylene is produced in three different forms. *Granular PTFE* is prepared by aqueous suspension polymerization. The polymerized product is cut or milled to uniform particle sizes of 20–100 μm and processed by powder metallurgy techniques followed by machining. *Fine powder resins* are prepared by aqueous emulsion polymerization and coagulation of the resulting latices by high-speed agitation or by addition of electrolytes. The fine-powder resin agglomerates (average particle sizes 200–800 μm) consisting of the primary polymerization particles (average particle size 150–300 nm) are mixed with lubricants (e.g., kerosene, hydrocarbons) and processed by a cold extrusion technology called paste-extrusion, and usually after evaporating the lubricant the material is sintered. The third form of PTFE is an *aqueous dispersion*, containing about 40–60% solids and up to 10% nonionic surfactants. The dispersion products are processed by latex processing techniques (e.g., dip coating, spray coating, or roller coating) followed by baking or sintering at high temperatures that causes coalescence of the polymer particles into a continuous film.

Low molar mass ($< 10^6$ g/mol) PTFEs are offered as finely divided powders as so-called fluoro-additives or micropowders for various industries as lubricants or to improve wear-resistance, etc. Due to the reduced molar masses ($\sim 10^4$–16^6 g/mol; melt viscosity $\sim 10^5$ Pa s) these materials have no mechanical strength.

Manufacturers of PTFE include Asahi Glass (Fluon), Daikin (Polyflon), DuPont (Teflon), 3M (Dyneon), Solvay Solexis (Algoflon), and Halopolymer (Ftoroplast). Asian PTFE-producers established large PTFE capacities; the companies are Gujarat, Dongyue, 3F, Meilan, Chenguang, Fuxin, and Juhua.

2.2.1. Production

Polymerization. Tetrafluoroethylene must be stored, handled, and polymerized with great care. Even traces of oxygen can trigger polymerization; to prevent polymerization in storage vessels, inhibitors, such as terpenes or amines are added.

The inhibitors must be carefully removed prior TFE use by distillation, adsorption, e.g., on silica, or washing with sulfuric acid. The purity requirements for polymerization-grade TFE are very demanding (low ppm-range of impurities), because each chain-transfer active species in the TFE is detrimental to achieving the required high molar masses.

TFE also has a high tendency for self-decomposition into carbon and tetrafluoromethane:

$$F_2C = CF_2 \rightarrow C + CF_4 \quad \Delta H \sim -320 \text{ kJ/mol}$$

The generated power of this decomposition is close to that of explosives and can be initiated by so-called hot spots formed during polymerization. Due to the high heat of polymerization (~ -170 kJ/mol) and the excellent insulating properties of PTFE, hot spots can develop if the mixing or agitation and cooling of the aqueous suspension in the vessel are not adequate or if, in the case of aqueous emulsion polymerization, coagulation occurs during polymerization. In the latter case, the hot spots float to the water surface and ignite the TFE gas phase. Consequently, the PTFE manufactures try to polymerize at low temperatures and TFE pressures; it is also recommended to carry out polymerizations under inert gas blankets to avoid explosions [21]. Due to the risk of violent explosions, the whole polymerization equipment as well as transfer lines of TFE-monomers are protected by rupture disks and flame arrestors.

Tetrafluoroethylene is polymerized in aqueous medium by two technologies. Suspension polymerization is the method to produce high molar mass granular resins. In the suspension process, TFE is polymerized without or in the presence of small amounts of emulsifiers (e.g., nontelogenic fluorinated carboxylic acids). Preferred initiators are salts of persulfate, quite often in combination with redox systems (e.g., metal ions or bisulfite) to keep the polymerization temperatures at low levels; $KMnO_4$ is also a widely used low temperature initiator [22]. The vigorous agitation precipitates the initially formed polymer dispersion shortly after the polymerization starts, and most of the polymerization occurs then directly on the precipitated solid particles. Solid contents between 20–40% can be achieved. The irregular shaped polymer particles can show average particle sizes up to 5 mm.

Aqueous emulsion polymerization is used to produce high molar mass fine powder and dispersion PTFE. Significant amounts (up to 15 g/L water) of nontelogenic fluorinated emulsifiers (e.g., fluorinated carboxylic or sulfonic acids with up to 12 carbon atoms) are used. In contrast to suspension polymerization, very mild agitation conditions are applied. Because the PTFE latices are shear sensitive, horizontal reactors with paddle wheels are used. To improve latex stability and to avoid coagulations, paraffin wax is added, which is molten under polymerization conditions. This polymerization method is generating spherical, sometimes rodlike polymer particles with average particle sizes of 50–300 nm.

To fine tune end-use properties, seed polymerization and core-shell polymerizations are employed. The initiator is the same as with the suspension process. The obtained latices with solid contents from 20–40% can be either coagulated to generate fine powder PTFE or are further stabilized by addition of nonionic surfactants and concentrated to solid contents up to 65% PTFE dispersions.

Fluorinated emulsifiers (e.g., perfluorooctanoic acid) are used in aqueous emulsion polymerizations, which is a challenge to recover from downstream processes and products due to environmental and safety concerns.

Granular Resins. The dried as-polymerized PTFE cannot be used as such, because the irregular shaped particles possess a broad particle size distribution and show very high void contents in sintered articles.

During the sintering process, due to the high melt viscosity, fusion of the polymer particles takes place only at the upper surface where particles are in contact. Therefore, the as-polymerized particles must be cut or milled into more uniform and smaller particles. The coarse as polymerized PTFE is processed by wet or dry milling into materials of average particle sizes of 20–100 µm. These materials ensure that after sintering the products have a much lower void content. The polymer particles should also have a good deformability during molding, which can be achieved by polymerizing at lower temperatures. The milled products have poor flow properties and specific bulk densities of 300–500 g/L (specific surface area of 1–4 m^2/g); these materials are often called nonfree-flowing types and cannot be used in automatic or isostatic molding applications.

The fine cut PTFE resins are converted during an agglomeration process into soft agglomerates with good powder flow properties and much higher bulk densities (500–1 000 g/L). The agglomerates comprise of a number of fine cut particles and have average particle sizes of 200–800 µm. Agglomeration can be achieved by dry or wet technologies to get stable free-flowing PTFE resins. The wet process requires the use of mixtures of water and water-insoluble organic liquids (e.g., gasoline). Nonfree-flowing PTFE powder is mixed with such fluids in a stirred vessel to agglomerate the PTFE into free-flowing resins.

Presintered granular PTFE resins are also free-flowing powders of premelted PTFE, predominantly used for ram extrusion applications.

Fine Powder Resins. To modify the processing characteristics (e.g., extrusion pressure) as well as the final end-use properties, core-shell PTFE-polymers are made by aqueous emulsion polymerization. For example, the core of the polymer particle is of high molar mass and the shell is of much lower molar mass; this, for example, reduces the extrusion pressure. In addition, smaller amounts (< 1%) of modifiers, such as HFP, CTFE, or PPVE are used to tailor the properties of fine-powder PTFE. Numerous patent applications describe the benefits of all these modifications [23, 24].

The obtained PTFE-latices are subsequently coagulated to produce fine powders. Usually the latices are diluted and coagulation agents, such as water-soluble organic liquids (e.g., methanol, acetone) or electrolytes (e.g., ammonium carbonate, oxalic acid, hydrochloric acid) are added to cause coagulation under agitation. The primary as-polymerized PTFE-particles (average particle size 50–300 nm) form agglomerates (average particle sizes 300–800 µm, specific surface area 12 m^2/g), which are separated and carefully dried. Fine resin PTFE powder is quite sensitive to shear forces above the transition point (19°C) during work-up and handling processes due to the high tendency to fibrillate.

Aqueous Dispersions. Aqueous dispersions are made by the same polymerization process used to produce fine powder resins. The latices can be polymerized to give different average particle sizes (50–300 nm), which is important for the final applications; also bimodal particle sizes distributions have been recommended [25]. Usually commercial dispersions have average particle sizes of about 200–250 nm. The obtained latices are further stabilized by the addition of nonionic surfactants, such as Triton X-100 or other nonaromatic ethoxylated alcohols. These mixtures are concentrated from 20–40% to 60–65% solids by evaporation under vacuum, by decantation, or by ultrafiltration processes. The pH value, viscosity, etc., of the concentrated dispersion is adjusted to guarantee a high stability during storage, transportation, and in the final applications. The fluorinated emulsifier is recovered by anion exchange technology (→ see Section 4.3).

Filled Resins. A variety of fillers are utilized to improve some short-comings of unfilled PTFE [10], such as wear resistance, hardness, cold flow, or creep. These fillers include, for example, glass fibers, carbon, graphite, bronze, molybdenum sulfide, etc. Up to 40% by volume of filler can be added to the PTFE without complete loss of physical properties; the impact of fillers below 5% is low. About one third of granular PTFE is consumed in the form of filled compounds; nonfree-flowing as well as free flowing grades are offered. These PTFE-compounds can be produced by dry blending, alternatively by dry or wet agglomeration technologies of nonfree-flowing granular PTFE and filler.

Micropowders. PTFE micropowders are finely divided low molar mass (10^4–10^6 g/mol) PTFE-additives with small particle sizes (1–20 µm). Due to the low molar masses and reduced melt viscosities (10^2–10^5 Pa s), these materials have measurable melt flow indices, but usually no mechanical strength. Micropowders are almost entirely consumed by industries outside fluoropolymers. Major applications of micropowders include the addition into thermoplastics and elastomers, paints and coatings, printing inks, oils and greases [10]. Micropowders can be prepared by thermal or radiation degradation of high molar mass PTFE (e.g., using off-spec i. e., off-specification PTFE-material or scrap PTFE from processing) followed by milling steps [10]. Alternatively, micropowders can be produced by suspension polymerization or more conveniently by aqueous emulsion polymerization in the presence of chain-transfer agents.

2.2.2. Properties

Physical Properties. The structure of PTFE chains is unusual in that they are completely linear. The branching that occurs during the free-radical polymerization of ethylene cannot take place during the polymerization of TFE because an abstraction of the F-atom is not possible due to the strength of the C−F-bond. Furthermore, PTFE chains are stiffer than polyethylene chains because their fluorine atoms are larger than hydrogen atoms. Steric hindrance prevents a PTFE chain from assuming a planar zigzag structure; instead, it is forced to adopt a zigzag structure with a helical twist along the chain axis. Crystals of PTFE are made up of chains with a 180° twist every 13 to 15 carbon atoms, depending on the temperature. The high melting point of PTFE (327°C) is due to the small entropy change produced during melting, which in turn results from the stiffness of the chains [26].

The chemical bonding forces within the chains, between the chains, and the polymer surface are unusual.

```
  F   F   F   F   F   F
  |   |   |   |   |   |
- C - C - C - C - C - C -      ← C−F bond energy
  |   |   |   |   |   |          = 481.3 kJ/mol
  F   F   F   F   F   F

                                ← interchain energy
                                  = 3.2 kJ/mol

  F   F   F   F   F   F
  |   |   |   |   |   |
- C - C - C - C - C - C -
  |   |   |   |   |   |
  F   F   F   F   F   F        ← surface energy
                                  = 18.6 mN/m
```

The C−F bond energy is among the highest known. However, the interchain bonding forces are very weak, and the surface energy is very low. The combination of unusually strong and weak forces results in many unique properties.

The melting point of high molar mass, virgin unmelted PTFE is 342°C. Differential scanning calorimetry (DSC) indicates that the 342°C melting point is irreversible and the subsequent melting point is 327°C. Electron-microscopic studies of dispersion particles indicate that the rodlike crystals in virgin PTFE are fully extended with few defects [27]. The decrease of the melting temperature from 342°C to 327°C is due to the transition of fully extended chains to folded polymer chains. Virgin PTFE particles are 92–98% crystalline. The crystallinity of once-melted PTFE varies with molar mass and the rate of cooling from the melt. Processed commercial grades are usually 45–75% crystalline.

The molar mass of PTFE cannot be determined by the usual methods that require dissolution of the polymer because the solubility in fluorinated solvents is very limited for PTFE below 300°C. Above 300°C PTFE dissolves in fluorinated oligomers and light scattering measurements on lower molar mass PTFE are possible to determine the molar mass and distribution [11, 28]. The number-average molar mass can also be estimated from the determination measurements of end groups. Because the molar mass of PTFE is very high, the concentration of end groups is very low. A quantitative method must be sensitive to less than 1 ppm of end groups. An elegant method tags the end groups with radioactive sulfur in the initiator [29]; the sulfur from the initiator must be retained in the polymer with no loss by hydrolysis. When persulfate is used as initiator, the resulting sulfate end groups are hydrolyzed to carboxylic acid end groups and no radioactive sulfur is detected in the polymer. However, when an iron–bisulfite initiator is used, the sulfur is retained in the polymer because the initiating species (a bisulfite radical, $HOSO_2^{\bullet}$) gives rise to hydrolytically stable sulfonic acid end groups.

A number of PTFE-samples with molar masses of commercial interest where made using the radioactive iron–bisulfite-initiator system. These samples were used to calibrate other methods for measuring the molar mass of PTFE resins, such as the standard specific gravity (SSG) method [29]. Because the rate of crystallization varies inversely with molar mass, samples of high molar mass have a lower SSG than those of low molar mass; typical values for polymers of commercial interest range from 2.14 to 2.20. The SSG method is widely used with PTFE resins. The correlation between number-average molar mass $\overline{M}_n$ and the SSG is given by the following relationship:

$$SSG = 2.612 - 0.058 \log_{10} \overline{M}_n$$

Studies on the melting behavior of virgin PTFE by DSC [31] has led to another method for measuring molar mass. The following relationship was found between number-average molar mass and heat of crystallization (ΔH_c, J/g):

$$\overline{M}_n = 2.1 \times 10^{10} \times \Delta H_c^{-5.16}$$

In addition, rheological tools can be applied to determine molar mass and distribution [32, 33]. A pragmatic approach is to calculate the number-average molar mass of a polymerization batch as:

$$\overline{M}_n = \frac{2 \times M_{TFE}}{I} \times 100$$

where $\overline{M}_{TFE}$ is the number of moles of polymerized TFE, I is the number of moles of consumed initiator.

In addition to its crystalline melting point at 327°C, PTFE has several transition temperatures. Two crystalline transitions, at 19°C and 30°C, have a significant effect on product behavior. Below 19°C, the polymer crystallites are in an almost perfect three-dimensional array. Above 19°C, the triclinic unit cell changes to a hexagonal unit cell. Slight untwisting of the chains is observed in the cell from 13 CF_2 groups to 15 CF_2 groups per 180° twist. The transition at 30°C is only ca. one-tenth as large as that at 19°C. The hexagonal unit cell disappears above 30°C, but the rodlike hexagonal packing of the chains in the lateral direction is retained. Between 19°C and 30°C, the specific volume of PTFE expands by about 1.8%. PTFE has additional several first and second order transition temperatures ranging from −110°C to 140°C. The exact quantity of minor transitions is somewhat dependant on the methods used.

Mechanical Properties. The mechanical properties of PTFE are affected by processing variables, such as preform pressure, sintering

Table 4. Properties of PTFE resins

Property	Granular	Fine Powder	Standard
Standard specific gravity, g/cm^3	2.13–2.18	2.14–2.25	EN ISO 12086
Tensile strength, MPa	> 14	> 19	EN ISO 12086
Elongation, %	> 200	> 200	EN ISO 12086
Flexural modulus, MPa	350–630	280–630	ASTM D 747
Flexural strength, MPa	no break	no break	ASTM D 790
Impact strength, J/m	160		ASTM D 256
Compressive stress at 1% deformation, MPa	4.2		ASTM D 695
Hardness, Shore D	50–65	50–65	ASTM D 1706
Coefficient of thermal expansion, K^{-1}	12×10^{-5}	12×10^{-5}	ASTM D 696

temperature, and cooling rate, as well as by polymer variables, such as molar mass, particle size, and particle-size distribution. The resulting void content and the amorphous content determine some mechanical properties. Void content is determined by measuring gravimetrically the (apparent) density of a specimen and the intrinsic density via its IR-spectroscopically determined amorphous content. The required densities of the amorphous and crystalline phase are assumed to be 1.966 g/cm^3 and 2.340 g/cm^3 [34]. Properties affected are the flex life (the capability of a material to withstand repeated bending without fracture) tensile strength, elongation, permeability, stiffness, resilience, and impact strength. Properties, such as the coefficient of friction, low-temperature flexibility, and thermal stability are relatively unaffected. Typical values for the mechanical properties of granular and fine powder PTFE resins are listed in Table 4. Further data are given in [35].

The volume change caused by the crystalline transition at 19°C must be taken into account when precision parts made from PTFE are designed.

The coefficient of friction of PTFE is unusually low. Static friction decreases with increasing load and is less than the dynamic friction that frees PTFE from stick–slip problems. Because of the low surface energy, only liquids with low surface tension (e.g., fluorinated liquids) can wet a PTFE surface. Surfaces can be made wettable by treatment with alkali-metal compounds (e.g., sodium naphthenate) that, however, also increases the coefficient of friction.

The effect of fillers on the mechanical properties of PTFE is shown in Table 5; tensile strength, elongation are reduced, and the coefficient of friction is not significantly affected. In contrast, wear resistance is dramatically improved (e.g., the wear factor of a bronze-filled material can be reduced by orders of magnitudes compared to that of unfilled PTFE).

Electrical Properties. The very low dielectric constant of PTFE does not change over a wide range of temperature (−40 to 250°C) and frequency (5 Hz to 10 GHz). Such electrical behavior is attributed to PTFE's highly symmetrical structure composed solely of carbon–fluorine bonds, in which all electrical dipoles are exactly balanced. Similarly, the dissipation factor is extremely low and independent of temperature or frequency. Resistivity and dielectric strength, on the other hand, are very high, as is arc resistance; PTFE leaves no carbonized path. The electrical properties of PTFE are given in Table 6.

Table 5. Properties of unfilled and filled PTFE resins

Property	Unfilled resin	Filled resin	
		25% Glass fiber	60% Bronze
Standard specific gravity, g/cm^3	2.16	2.24	3.74
Tensile strength, MPa	28	18	14
Elongation, %	350	250	150
Impact strength, J/m	152	119	
Hardness, Shore D	51	57	70
Wear factor, mPa^{-1}	5 013	26.2	12
Coefficient of friction			
Static, 3.4 MPa load	0.08	0.13	0.10
Dynamic, at PV = 172 kPa m s^{-1}	0.01	0.17	0.15
Thermal conductivity, mW m^{-1} K^{-1}	0.24	0.45	0.46

Table 6. Electrical properties of PTFE resins

Property	Standard	Granular	Fine powder
Dielectric constant, 60 Hz to 2 GHz	ASTM D 150	2.1	2.1
Dissipation factor, 60 Hz to 2 GHz	ASTM D 150	0.0003	0.0001
Volume resistivity, Ω cm	ASTM D 257	$> 10^{18}$	$> 10^{18}$
Surface resistivity, Ω/m^2	ASTM D 257	$> 10^{16}$	$> 10^{16}$
Surface arc resistance, s	ASTM D 495	> 300	> 300
Dielectric strength (2 mm thickness), kV/mm	ASTM D 149	23.60	23.60

Electrical properties can be adversely affected by voids, which may reduce the dielectric strength and corona resistance of fabricated parts. Consequently, voids must be minimized to achieve optimum electrical insulation.

Chemical Resistance. Polytetrafluoroethylene is resistant to attack by most chemicals, including hot fuming nitric acid, hot caustic, gaseous chlorine, chlorosulfonic acid, organic esters, ketones, and alcohols. The only materials known to attack PTFE are molten alkali metals, chlorine trifluoride, and gaseous fluorine at elevated temperature and pressure. Some highly fluorinated oils can swell or dissolve PTFE near its melting point.

Flame Resistance. Polytetrafluoroethylene is one of the most flame-resistant polymers known and does not support combustion in air. It has an extremely high limiting oxygen index (LOI) of 96%, i.e., it burns only in almost pure (96 vol%) oxygen [36].

Weatherability. The weatherability of PTFE is remarkable; it is completely unaffected by all types of weather from dry desert heat to humid jungle conditions. Test films showed no change after 20–30 years of continuous exposure in Florida.

Radiation Resistance. On exposure to a high-energy electron beam or gamma radiation, PTFE undergoes degradation rather than cross-linkage. In the absence of oxygen, stable radicals are produced which slow down the rate of degradation. However, in the presence of air, the radicals react with oxygen that accelerates scission and degradation. On exposure to 10^4 Gy (1 Mrad) of radiation in air, PTFE loses 50% of its original tensile strength. Irradiation also affects its electrical properties: resistivity decreases, whereas the dielectric constant and dissipation factor increase.

Thermal Stability. In both air and nitrogen, PTFE has an extremely high thermal stability. Rates of decomposition are not measurable below ca. 440°C; decomposition rates are high at 540°C. PTFE can be pyrolyzed at about 600°C almost completely back into fluoroolefins, the TFE monomer being the main product [37]; the activation energy is 347 kJ/mol. The melt viscosity decreases during pyrolysis, which probably involves random chain cleavage followed by depolymerization (short kinetic chain length) and termination by disproportionation. In the presence of diverse metals, PTFE reacts violently [38].

2.2.3. Processing

Granular Resins. Moldings are produced from granular resins in three steps: First, the dry powder is placed in a mold and compressed at moderate temperature to produce a preform that is strong enough for handling without breaking. The preform is then placed in an oven at 380°C to allow the particles to coalesce. This operation is called sintering. The final step is controlled cooling to produce the desired crystallinity. Granular resins are available in a number of forms that are optimized for different types of molding. Free-flow resins are used in small and automatic moldings, also for isostatic molding methods. Finely divided resins are preferred for large billet moldings to which they impart superior properties. Presintered resins are easier to handle and are preferred in ram extrusion applications. Most PTFE manufacturers give detailed descriptions of molding equipment and procedures [10].

Automatically molded articles require no further finishing after molding. However, large billets and rods are subsequently skived or machined. Films and sheets are produced by skiving a billet on a lathe. Skiving and machining generate significant scrap rates, only on a minor portion of which are reused. Precision parts can be machined from ram-extruded rods. Another method, called coining, is used for articles that are too complicated to make by machining. In this operation, a sintered molding is heated to the melting point, quickly pressed into a mold cavity, and held under pressure until it resolidifies. Coined moldings are limited in their upper service temperature. After long periods at high temperature, they can return to their original precoining shape.

Fine Powder Resins. The paste-extrusion process for fine powder resins represented an important advance in PTFE processing. It permits the manufacture of continuous PTFE tubing and continuous PTFE coatings on wires.

In paste extrusion, 15–25 wt% of lubricant (a petroleum fraction, usually kerosene) is mixed with the fine powder resin. The wetted powder is then gently shaped into a preform at low pressure. (Fine powder resins are extremely sensitive to shear; great care must be taken to avoid shearing the powder before extrusion, otherwise processability may be lost.) The preform is forced through a die mounted in the extruder. The high shear exerted at the die fibrillates the powder and confers "green strength" to the extrudate. The lubricant is then evaporated and the extrudate is sintered at 380°C to coalesce the particles. Drying and sintering are performed consecutively by passing the extrudate through a multistage oven located directly after the extruder. Residence time in the oven can vary from a few seconds for thin-walled wire insulation to a few minutes for large-diameter tubing. Extrusion pressure depends on the reduction ratio, which is the ratio of the die diameter to the product diameter, the extrusion rate, the lubricant content, and the extruder characteristics.

Paste extrusion is also used to produce unsintered tape. Lubricant mixing and extrusion are the same as above. The product is extruded into rods, which are calendered on hot rolls to the desired dimensions.

Different fine powder resins have been developed for different applications. Powders that are suitable for high reduction-ratio applications, such as wire coatings, are usually not suitable for medium reduction-ratio applications, such as tubing. Thread sealant tapes are generally produced with other special grades. New types of fine powder resins are constantly developed, leading to both improved and new applications and processing techniques [7, 39].

W. L. GORE developed a unique expansion process to generate small pores in the structure of an article of PTFE; the trademark GoreTex is well known for lightweight and breathable fabrics, filtration membranes, medical implants, sealants, and high tensile fabrics [10, 39]. A good summary about the rheology and processing of PTFE fine powder is provided in [40].

Dispersion Resins. In general, PTFE dispersions are concentrated to 50–60% solids and stabilized by nonionic surfactants and sometimes with ionic surfactants. The dispersions are applied to various substrates (e.g., metal and hard surfaces in cookware applications, etc.) by spraying, flow or roll-coating, dipping, coagulating, or electrodepositing followed by drying and sintering. PTFE-films can be made by casting the aqueous dispersion on a supporting surface, drying, baking, and cooling. If the film is too thick, cracks develop; the maximum film thickness per application is ca. 40 µm. Thicker coatings are made by casting and baking a series of layers.

The most important application of PTFE dispersions is fabric coating. The fabric (e.g., fiberglass) is usually dipped into the dispersion and dried and sintered as in film formation. In some cases, the fiber is coated with the PTFE dispersion before weaving. Dispersions are also used in the production of PTFE fibers [41]. The dispersion is mixed with a matrix-forming medium, such as viscose and spun into fibers through a spinneret and coagulating bath. The fibers are heated to remove the matrix polymer, sintered, and drawn for spinning.

PTFE dispersions are added to other thermoplastic polymers to improve fire performance by suppressing dripping. Fibrillation of PTFE during processing and formation of a structure that retains high viscosity in molten state prevents dripping of the molten host polymer.

Another important application is the coagulation of PTFE-dispersions with fillers (e.g., graphite, bronze, etc.) to manufacture filled bearings.

Table 7. Uses of PTFE resins

Type of resin	Uses
Granular	
Agglomerates	gaskets, packing, seals, bearings, sheet, rod, heavy wall tubing, and tape
Coarse cut	tape, molded shapes for chemical, mechanical, electrical, and nonstick applications
Fine cut	molded sheets, tape, wire, wrapping, tubing, and gasketing
Presintered	rods and tubes
Fine powder	
High reduction ratio[*]	wire coating and thin-walled tubing
Medium reduction ratio[*]	tubing, pipe, overbraided hose, and spaghetti tubing
Low reduction ratio[*]	thread-sealant tape, pipe liners, tubing, and porous structures
Dispersion	
General purpose	impregnation, coating, and packing
Coating	film and coating

[*] Ratio of extrusion-die diameter to product diameter.

2.2.4. Uses

Overlapping end use designations (see Table 7) make it difficult to segment the markets for the three PTFE-classes. Taking the consumption in the United States as benchmark gives a good general overview [12].

- Granular PTFE is consumed by 45% for mechanical applications (e.g., seals, piston rings, bearings, and cylinder tubes), 35% by chemical processing industries (most PTFE for fluid handling parts as valve and pump linings, dip tubes, expansion bellows, nozzles, valve seats, laboratory apparatus, gaskets), and 20% by electrical applications (e.g., cable connectors, circuit breakers, stand-off insulators, coaxial cores, ribbon cables, skived insulating tapes).
- Fine powder is consumed by 30% in textile laminates (e.g., GoreTex-applications), 28% by the automotive industry (e.g., push-pull cables for throttles, clutches and brakes, hoses for various fluids), 16% by wire and cable (nonautomotive) applications (e.g., direct extrusion into wire and electrical tapes for coaxial cables, hookup wire, and computer wire), 15% by tubing applications (e.g., spaghetti tubing, liners for overbraided tubes, hoses and pipes in chemical processing), and about 11% for other applications including gaskets, pump packing, and thread seal tapes.
- Aqueous dispersions are widely used: about 40% to coat fabrics (e.g., glass-fiber fabrics) for architectural purposes, 15% in consumer and industrial coatings (e.g., cookware coating, manufacturing of conveyor belts, chute liners and rolls), 15% in PTFE-fiber markets (the fibers are used for fabrication of filters, pump packing, gaskets, and bearings), and about 30% in preparations of various items (e.g., metal and graphite parts in heat exchangers and bearings) and coagulations to prepare filled bearings.

2.2.5. New Materials—Modified Granular PTFE

PTFE as an engineering material suffers from three shortcomings (in particular granular PTFE-grades): Low creep resistance (cold flow), difficult weldability, and its insufficiently dense polymer structure (void content).

The reason for these weaknesses is the great tendency of the entirely linear polymer chains to crystallize from the melt. The crystalline phase typically amounts to 50–70% in end-use articles and exhibits a ductile behavior with practically no mechanical strength. Mechanical strength is provided by tie-molecules bridging adjacent lamellae. The entangled tie-molecules make up the amorphous phase. To achieve mechanical strength and toughness, the molar mass of commercial PTFE is required to be in the range of 10^7–10^8 g/mol to partially suppress crystallization and to provide sufficient high tie-chain concentrations. The high molar masses result in high melt viscosities (10^{10}–10^{12} Pa s), which also impairs perfect particle coalescence, and consequently, the processed PTFE shows significant void contents. The intrinsic shortcomings are largely overcome by reducing the molar mass and hence the melt viscosity. Simultaneously, enhanced crystallization is prevented by incorporating bulky side groups into the linear polymer chain [34].

Small amounts (0.01–0.2 mol%) of copolymerized perfluoro(alkyl vinyl ethers), e.g.,

PPVE, are sufficient to reduce the crystallinity of PTFE [42]. Such materials can be polymerized under the conditions described above; the chemical homogeneity and molar mass can be controlled by the mode of feeding the comonomer PPVE. Free-flowing as well as nonfree-flowing grades with different modifier contents are commercially available. Usually the PPVE-content of modified PTFE is in the range of 0.05 mol%; compared to commercial unmodified PTFE-grade, the melt viscosity of modified PTFE can be lower by an order of magnitude. Thus, more efficient particle coalescence during the sinter process is accomplished (also for agglomerated free-flowing grades). The denser polymer texture results in much lower void contents, lower permeability coefficient, higher dielectric strength (in particular for free-flowing grades), and reduced stretch void indices (Table 8).

Modified PTFE shows excellent weldability, revealing the successful establishment of effective chain entanglements, because of the lower melt viscosity and molar mass. The deformation under load is reduced by 50%, and the recovery enhanced by a factor of 2–3. These effects can be attributed to the smaller lamellae size and the increased concentration of tie-molecules that bridge adjacent lamellae.

The modification improves all technical important properties [43, 44]; the dielectrical loss is almost unaffected by the polarizable oxygen atoms of the modifier, and even the thermal stability is identical to unmodified PTFE.

Modified PTFE is processed with the same techniques as conventional PTFE. Due to its lower melt viscosity, modified PTFE is better suited for coined molding, blow molding, and other thermoforming technologies. Because of the reduced cold flow, modified PTFE is specially suited for gaskets.

Increasing the PPVE-modifier content further up to 0.7 mol%, melt processable PTFE-grades are generated with superior properties [9].

Table 8. Properties of modified granular PTFE and unmodified PTFE resins [34]

Property	Standard	Modified PTFE 0.1 wt% PPVE		Unmodified PTFE	
		Nonfree-flowing	Free-flowing	Nonfree-flowing	Free-flowing
Standard specific gravity, g/cm^3	EN ISO 12086	2.165	2.165	2.160	2.155
Void content^a, vol‰	EN ISO 12086^a	2.6	4.0	7.5	8.0
SVI value^b	EN ISO 12086^b	85	85	300	300
Dielectric strength (100 μm film), kV/mm	DIN 53481	105	100	105	80
Permeabililty, cm^3 mm m^{-2} d^{-1} bar^{-1}	DIN 53380, 25 °C				
SO$_2$		200		310	
HCl		210		300	
Cl$_2$		100		160	
O$_2$		210		240	
N$_2$		80		100	
CO$_2$		540		800	
Water vapor (100 °C), g mm m^{-2} d^{-1}	DIN 53122	2.6		3.8	
Tensile strength, MPa	EN ISO 12086	41	36	41	33
Elongation, %	100 μm film	630	600	450	400
Deformation under 15 MPa, %	ASTM D 621				
	24 h	8	10	16	16
Load	permanent	4	5	11	11
Flexural modulus, MPa	DIN 53457				
23 °C		650	630	600	555
100 °C		290	255	210	210
250 °C		60	60	40	40

^{a}Void content is determined by measuring gravimetrically the (apparent) density of a specimen and the intrinsic density via its IR spectroscopically determined amorphous content [42]. The required densities of the amorphous and crystalline phase are assumed to be 1.966 g/cm^3 and 2.340 g/cm^3, respectively.
^{b}The stretch void index (SVI) is calculated from the difference of gravimetrically determined densities of the unstretched and stretched specimen until break multiplied by 1 000 (see EN ISO 12086).

2.3. Tetrafluoroethylene–Hexafluoropropene Copolymers

Copolymers of tetrafluoroethylene (TFE) and hexafluoropropene (HFP) are commonly referred to as fluorinated ethylene–propene (FEP) resins [25067-11-2]. They are linear, perfluorinated copolymers:

$$-[(CF_2CF_2)_x CF_2 \underset{\underset{CF_3}{|}}{C}F]_y-$$

In 2008, the worldwide consumption in FEP copolymers was about 1.8×10^4 t [12]. There was a great demand for plenum cable and local area network applications during the 1990s. Currently the annual growth is below 2%.

The presence of HFP in the polymer chain reduces the tendency to crystallize. Thus, a copolymer with a much lower molar mass than the corresponding homopolymer PTFE retains toughness adequate for most applications. The molar mass of commercial FEP resins is only ca. 1/100 that of commercial PTFE resins. This results in a ca. 10^6-fold reduction in melt viscosity. Resins made of FEP have melt viscosities in the range of $1 \times 10^3 – 10 \times 10^3$ Pa s at 380°C, which enables standard plastics processing techniques, such as extrusion and injection molding, to be used. In spite of their much lower molar mass, the properties of FEP resins are largely comparable to those of PTFE resins. The FEP resins have outstanding chemical resistance, nonflammability, weatherability, and electrical, antiadhesive, and low-friction properties. The maximum use temperature of 200°C, however, is 60°C below that of PTFE resins.

Most commercial FEP grades have a melting point of $260 \pm 20°C$ and are offered with a wide range of melt viscosities (e.g., MFI 372°C/5 kg from < 1 to > 30 g/10 min) for the different applications and processing technologies. In addition, modified FEP-grades, e.g., modified with PPVE, are available. Aqueous dispersions of FEP with solid contents of 50–60% and nonionic surfactants are also on the market.

FEP resins are offered as melt-pelletized materials, as powders, or as aqueous dispersions by Daikin (Neoflon), DuPont (Teflon), and 3M (Dyneon). Other companies started also to commercialize FEP resins (e.g., Halopolymer, Dongyue, 3F, Juhua).

2.3.1. Production

Copolymerization takes place in aqueous or nonaqueous media. Because it is less reactive than TFE, HFP is present in larger proportions in the mixture. In aqueous media, fluorinated emulsifiers with carboxylic acid groups are normally employed. Persulfate is usually preferred as a free-radical initiator. Aqueous emulsion polymerization conditions are similar to those employed for PTFE fine powder. Standard polymerization equipment can be used; the polymerizations are usually carried out in a temperature range of 50–100°C and within a pressure range of $1.5 \times 10^6 – 3.0 \times 10^6$ Pa. PPVE-modifications (< 2 wt%) are also applied to improve e.g., stress crack resistance and processability.

The obtained latices have particle sizes of 100–250 nm and solid contents of about 30%. Coagulation by addition of electrolytes (e.g., acids) or applying high shear provides a very fine, nonfree-flowing powder, which usually needs to be agglomerated with organic solvents to get free-flowing properties. Persulfate-initiated polymers have many terminal COF- and COO$^-$-endgroups that are easily detected by FT-IR analysis. These carboxylic groups are thermally instable and must be removed, because they cause blistering, discoloration, and corrosion during processing. Various methods are applied to convert the unstable carboxylic groups in more stable CF_2H-group (e.g., by water-heat treatment) or into CF_3-groups by postfluorination [10]. Polymers synthesized in a nonaqueous medium, with fluorinated peroxides or azo compounds as initiators, do not contain unstable end groups [45] and no post-treatments are necessary.

Aqueous dispersions are prepared by adding nonionic surfactants and by removing the applied fluorinated emulsifiers (e.g., by an anionic ion-exchanger); further concentration to 50–60% solids is desirable.

2.3.2. Properties

Physical Properties. The FEP resins are random copolymers of TFE and HFP. The HFP content is usually expressed as an HFP index (HFPI), which is the ratio of IR absorbances at ca. 980 and 2 365 cm^{-1}. Commercial FEP resins have HFPIs between ca. 3 and 5.5 [15]. The actual HFP content of FEP resins can also be determined by a high-temperature ^{19}F-NMR technique [46]; a typical FEP resin contains about 9 mol% of HFP.

The crystallinity of as-polymerized FEP is lower than that of as-polymerized PTFE (70% vs. 98%). After processing, crystallinity depends on the molar mass and the rate of cooling from the melt. The crystallinity of FEP moldings is usually ca. 50%.

FEP resins are insoluble in most known solvents. Solubility is given in selected perfluorocarbons at higher temperatures [47] and in supercritical media (e.g., CO_2, CHF_3) [48]. This precludes the convenient determination of molar mass by the usual methods. Number-average molar masses have been estimated by measuring the concentration of acid end groups before finishing. Polymerization in aqueous media gives 0.05 to 0.20 mol% acid end groups. A convenient and reliable method to determine molar mass including distribution is based on dynamic rheometry data [49]. The number-average molar masses of commercial grades are between ca. 10^5 g/mol and 5×10^5 g/mol. However, molar mass is normally specified in terms of the melt-flow index, measured at 372°C under a load of 5 kg in a melt rheometer of the type used for polyethylene equipped with a corrosion-resistant barrel, piston, and die (ASTM D-2116-02). Melt-flow index values range from ca. < 1 for the high viscosity grade to ca. 35 for the low viscosity grades.

Typical FEP resins have a crystalline melting point from ca. 240 to 280°C, depending on their HFP content and do not exhibit any other crystalline transitions. (An amorphous transition is observed at 90–150°C, depending on frequency of measurement. A second amorphous transition occurs below −150°C [50].

The thermal stability of FEP resins is lower than that of PTFE resins, but is adequate for processing up to at least 380°C. HFP diads (2 adjacent HFP units [10]) are found to be a source for instability but in general, thermal decomposition occurs in two distinct stages. In the first, HFP is preferentially split out of the backbone at a rate that is about four times faster than that found in normal PTFE decomposition. The second stage involves the decomposition of the remaining PTFE backbone, at the same rate as that of PTFE decomposition.

Mechanical Properties. The mechanical properties of FEP are very similar to those of PTFE resins. The absence of the 19°C transition in FEP resins eliminates the volume change observed in PTFE resins at this temperature. The coefficient of friction is slightly higher in FEP resins than in PTFE. However, the static coefficient is lower than the dynamic coefficient, which avoids stick–slip. Typical values of mechanical properties are given in Table 9.

The temperature dependence of the tensile properties of FEP [51] and PTFE resins is similar from ca. −250 to 200°C. However, above 200°C, the FEP properties deteriorate rapidly, whereas the PTFE properties remain satisfactory up to at least 260°C. Thus, the maximum use temperature of FEP is 60°C below that of PTFE resins.

As with PTFE, fillers can be added to FEP resins to increase wear resistance and stiffness and to lower creep. The maximum loading of fillers is controlled by processing considerations. Available compositions include 30% bronze filled, 20% graphite filled, and 10% fiberglass filled. These all have high *PV* (pressure–velocity) values desirable for use as bearings.

The surfaces of FEP can be made wettable by corona discharge or by treatment with sodium naphthenate or with amines in a hot oxidizing atmosphere.

Electrical Properties. The electrical properties of FEP resins (Table 10) and PTFE resins

Table 9. Properties of FEP resins [2]

Property	MFI < 2	MFI 7	Standard
Specific gravity, g/cm^3	2.14–2.17	2.14–2.17	ASTM D 792
Tensile yield, MPa	12		ASTM D 638
Tensile strength, MPa	31	25	ASTM D 1708
Elongation, %	305	325	ASTM D 1708
Flexural modulus, MPa	590	620	ASTM D 790
Compressive strength, MPa	23	21	ASTM D 695
Hardness, Shore D	57	56	ASTM D 2240
Izod impact strength, J/m	no break	no break	ASTM D 256
Coefficient of thermal expansion (0–100 °C), K^{-1}	8×10^{-5}	14×10^{-5}	ASTM D 696

Table 10. Electrical properties of perfluoropolymers [34]

Property	Standard	PFA	MFA	FEP
Volume resistivity (from 23 to 150 °C), Ω cm	ASTM D 257	$> 10^{17}$	$> 10^{17}$	$> 10^{17}$
Surface resistivity, Ω	ASTM D 257	$> 10^{17}$	$> 10^{17}$	$> 10^{17}$
Arc resistance, s	ASTM D 495	> 200	> 200	> 200
Dielectric strength (1 mm thickness), kV/mm	ASTM D 149	30–32	30–32	26–30
Dielectric constant	ASTM D 150			
50 Hz, 23 °C		2.0–2.1	2.0–2.1	2.1
100 kHz, 23 °C		2.0	2.0	2.0
Dissipation factor	ASTM D 150			
50 Hz, 23 °C		3×10^{-4}	3×10^{-4}	3×10^{-4}
100 kHz, 23 °C		5×10^{-4}	5×10^{-4}	5×10^{-4}

(Table 6) are virtually identical. However, the effect of voids on the electrical properties that is observed in PTFE resins is absent in the void-free FEP resins. Because of the stability of FEP in dielectric measurements, the National Bureau of Standards has selected FEP resins as reference standards. The electrical properties of FEP are unaffected by prolonged exposure to water or a temperature of 200°C. Only corona ignition affects the dielectric life of FEP. If the voltage stress is too low to cause corona discharge, dielectric life is infinitely long at any frequency.

Chemical Resistance. FEP resins are inert to practically all chemicals and solvents, with the exception of fluorine at elevated temperature and pressure and molten alkali metals or sodium hydroxide. Adsorption of chemicals is usually less than 1% even after long exposure at an elevated temperature. Adsorption is completely reversible and has no effect on the polymer. Permeation of gas and vapor through FEP is slower than through most other plastics.

Radiation Resistance. The effect of gamma-ray or electron-beam irradiation on FEP and PTFE is largely degradative. In the absence of air, the radiation resistance of FEP is ten times greater than that of PTFE. In the presence of air, however, both degrade at about the same rate. In the absence of air, irradiation at low dose rates at a temperature above the alpha transition (> 150°C) results in cross-linking [52], which is, however, of no practical importance.

Weatherability. FEP resins exhibits an outstanding weatherability. After 15 years' exposure in Florida, the tensile properties and light transmission of a crystal clear FEP film remained unchanged.

Flame Resistance. The flame resistance of FEP is outstanding; it does not support combustion in air and has a LOI of 96%.

2.3.3. Processing

Extrusion and Injection Molding. Conventional methods, such as extrusion and injection molding are employed for processing FEP. However, FEP has a higher melt viscosity and lower critical shear rate than many other thermoplastic materials (e.g., PE, PP, PS), which makes processing more difficult. Molten FEP can corrode the metals used in extruders and injection molding machines. Thus, equipment parts exposed to molten FEP resin should be constructed from more corrosion-resistant materials (e.g., Hastelloy C or Inconel).

Critical shear rates at 380°C, above which melt fracture occurs, are low and range from 20 s^{-1} for FEP with MFI ~ 10 g/10 min to 2 s^{-1} for FEP with MFI < 2 g/10 min. The resulting low production rates preclude the use of pressure-extrusion techniques. Fortunately, FEP resins have a high melt strength and can tolerate high draw-down ratios without fracture. Melt-draw techniques make high production rates possible. In addition, FEP can be extruded into wire coating, tubing, rods, and film of various sizes. It can be extruded over such materials as silicone rubber, poly(vinyl chloride), glass braid, metal-shielded cables, twisted conductors, and parallel multiconductor cables. Concentrates of any desired color can be blended with standard resins.

Injection molding is best carried out with the low melt viscosity grades (e.g., FEPs with MFI ~ 20). Recommendations for molding equipment, designs, and operating conditions are available from the manufacturers. Bottles, flasks, beakers, and other laboratory ware are typical FEP products that are made by injection molding.

Dispersion Coating. Aqueous FEP dispersions are processed essentially as PTFE dispersions. They are usually coated on the substrate (e.g., glass cloth), and water is evaporated. The coating is then fused by heating above the melting point. Because of the lower melt viscosity of FEP, fusion is quicker than with PTFE resins. In addition, continuous films, free of pinholes, are more readily achieved. Sometimes FEP dispersions are used to provide a final coating on a PTFE coating.

2.3.4. Uses

More than 80% of all FEP resins are used in electrical applications, including hookup wires, computer and thermocouple wires, coaxial cables, etc. The leading market is the production of plenum wire and cables.

About 6% of FEP is used for film applications, e.g., release film, laminate with polyimide, in solar collectors, etc. Chemical processing industries use FEP-resins, e.g., for tubings in semiconductor industry, linings for fittings, pumps, bellows and valve components, and heat exchangers.

2.4. Tetrafluoroethylene–Perfluoro (Alkyl Vinyl Ether) Copolymers

Copolymers of TFE with perfluoro(propyl vinyl ether) (PPVE) are called propylfluoroalkoxy (PFA) resins [26655-00-5]. They have the following general formula:

$$-[(CF_2CF_2)_xCF_2CF]_y-$$
$$\qquad\qquad\qquad |$$
$$\qquad\qquad\quad OCF_2CF_2CF_3$$

Copolymers of TFE and perfluoro(methyl vinyl ether) (PMVE) are called methylfluoroalkoxy copolymer (MFA) and have the following structure [34]:

$$-[(CF_2CF_2)_xCF_2CF]_y$$
$$\qquad\qquad\qquad |$$
$$\qquad\qquad\quad OCF_3$$

The worldwide consumption was about 6 000 t in 2008 [12]. Growth rates (1–4%/a) depend on geographic regions.

PFA resins were developed after FEP resins and were designed to overcome the disadvantages of FEP resins in comparison with PTFE resins, i.e., lower thermal stability and poorer mechanical properties at high temperatures. The ether group in the side chain of PFA and MFA resins eliminates steric strain at the branching point, which is responsible for the reduced thermal stability of FEP resins [53]. In addition, the large perfluoropropoxy side group in PFA reduces the crystallinity much more than the trifluoromethyl side group in FEP; it is about five times more effective on a molar basis. The comonomer content of PFA resins is lower than that of FEP resins [34]. The melting point of PFA (305–310°C) is only ca. 20°C below that of PTFE. The thermal stability and maximum use temperature of PFA and PTFE resins are essentially the same. PFA and MFA resins have all the other advantages of PTFE resins: antistick performance, resistance to virtually all chemicals, low coefficient of friction, nonflammability, and excellent electrical properties.

Manufactures for fluoroalkoxy polymers are Daikin (Neoflon), DuPont (Teflon), 3M (Dyneon), Solvay Solexis (Hyflon PFA, MFA).

2.4.1. Production

Copolymerization. The copolymerization of TFE and PPVE is either carried out in aqueous [54] or nonaqueous media [55]. The aqueous process resembles the emulsion polymerization used for PTFE fine powders and FEP resins. However, only a small amount of PPVE is present in the PFA reaction mixture. This is because PPVE is ca. five times more reactive than HFP (copolymerization parameter for PPVE is about 5 and for HFP 24, respectively) and a much smaller amount of comonomer is incorporated into PFA compared to FEP.

To avoid chemical inhomogeneities, PPVE has to be replenished properly according to the copolymerization parameter; the polymerizations are usually done under constant TFE pressure in a range of $1.0 \times 10^6 - 2.5 \times 10^6$ Pa and at temperatures of 40–90°C. Water-soluble initiators, such as persulfates or potassium permanganate are preferred; chain-transfer agents are used to control the molar mass (e.g., H_2, R134, alkanes, alcohols).

A complication of TFE and PPVE copolymerization is the tendency of the growing PPVE radical to terminate by rearrangement:

$$-CF_2CF\cdot \atop | \atop O-C_3F_7 \longrightarrow -CF_2CF \atop \| \atop O + C_3F_7\cdot$$

To avoid this reaction, the polymerization conditions should be below 70°C. This rearrangement reaction occurs to a lesser extent with PMVE.

The polymerization rates can be increased by adding nontelogenic fluorinated solvents or by microemulsion technologies. The aqueous lattices show average particle sizes of about 150–200 nm and 30–40% solids.

Nonaqueous polymerizations in non- or low-telogenic fluorinated solvents use fluorinated peroxides (e.g., bisperfluoro alkanoic peroxides, perfluoro acyl hypofluorides) to generate stable end-groups. Chain transfer agents are used to control the molar mass. Usually high space–time yields are achieved with solid contents of 5–10%; higher solid contents have a gel-like behavior, and polymerizations are hard to control. The polymers are recovered by removing the solvent and unconverted initiator at higher temperatures; the polymer is then compacted for melt-extrusion.

The aqueous lattices are coagulated by shear or addition of electrolytes and are usually agglomerated with e.g., water–gasoline to obtain a free-flowing polymer powder. The polymer powders (in particular from aqueous polymerization) have some unstable end-groups (e.g., COO^-, COF) from initiators, chain transfer agent, or rearrangement reactions. Post-treatments with methanol or ammonia to stabilize the polymer have been recommended [56]. For demanding applications, all the unstable groups are removed by postfluorination with F_2–N_2-mixtures at temperatures up to 250°C [57, 58].

After addition of nonionic surfactants to the aqueous lattices, the fluorinated emulsifiers are removed and the lattices can be concentrated to 50–60% solids PFA-dispersions.

Polymerizations in supercritical CO_2 have been initiated, but did not find widespread applicability [10].

2.4.2. Properties

Physical Properties. PFA resins undergo a single first-order transition at their melting point (305–310°C). They do not display the other first-order transitions present in PTFE resins [59]. Torsion-pendulum measurements revealed three second-order transitions at −100, −30, and 90°C. The crystallinity of as-polymerized dispersion polymer was 65–75%. Samples slowly cooled from the melt had 58% crystallinity.

The number-average molar mass $\overline{M}_n$ for PFA and MFA can be deduced by quantification of the end-groups and weight-average molar mass $\overline{M}_w$ from the extrapolated plateau modulus of the complex viscosity via dynamic melt viscosity. The dynamic rheometry allows also the determination of the molar mass distribution (MWD) [33, 60]. The molar mass is usually in the range of $1-5 \times 10^5$ g/mol.

The amount of incorporated vinyl ethers, the number of polar end-groups, and the amorphous content can be determined by FT-IR-spectroscopy [34]. The melting point is lowered by about 10°C/PPVE mol% (HFP: ∼8°C/HFP mol%).

The thermal stability of PFA resins is comparable to that of PTFE resins. Although they can be processed at temperatures up to 425°C [49], degradation occurs on prolonged heating at or above 425°C. Below its melting point, PFA does not decompose. Samples of PFA aged in an air oven at 285°C for years showed no deterioration; in fact, tensile strength, ultimate elongation, and flex life increased.

Some general properties of PFA and MFA are given in Table 11.

Table 11. General properties of PFA polymers and MFA polymers [34]

Property	Standard	PFA	MFA
Specific gravity, g/cm^3	ASTM D 792	2.12–2.17	2.12–2.17
Melting temperature, °C	ASTM D 2116	300–310	280–290
Coefficient of linear thermal expansion, K^{-1}	ASTM E 831	$(12-20) \times 10^{-5}$	$(12-20) \times 10^{-5}$
Specific heat, kJ kg^{-1} K^{-1}		1.0	1.1
Flammability	(UL 94)	V-O	V-O
Oxygen index, %	ASTM D 2863	> 95	> 95
Hardness, Shore D	ASTM D 2240	55–60	55–60
Friction coefficient (on steel)		0.2	0.2
Water absorption, %	ASTM D 570	< 0.03	< 0.03
Upper working temperature, °C		260	230
Refractive index		1.35	1.35

Mechanical Properties. The mechanical properties of PFA resins resemble those of PTFE and FEP resins at room temperature, but above 200°C, PFA resins are superior to FEP resins. The high-temperature mechanical properties of PFA resins are equal or superior to those of PTFE resins. Both PFA and PTFE resins have a maximum use temperature of 260°C. The low-temperature mechanical properties of the two resins are equivalent. Some properties of PFA compared to PTFE are shown in Table 12.

The wear and friction properties of PFA, MFA, and FEP resins are similar. The static coefficient of friction is lower than the dynamic coefficient; consequently, the polymer is free of stick–slip behavior. Fillers, such as fiberglass or bronze powder improve the wear resistance of PFA, FEP, and PTFE resins. Among the partially and perfluorinated thermoplastics PFA, MFA, and FEP exhibit (up to 100°C) the lowest creep resistance; they show much better creep properties above 150°C compared to PVDF, ECTFE, and ETFE [34].

Electrical Properties. The electrical properties of PFA, MFA, and FEP resins are essentially the same [34] (see Table 10).

Flame Resistance. Perfluoroalkoxy resins have the same nonflammability characteristics as PTFE resins. LOI is over 95%. According to Underwriters Laboratories (UL 94), PFA resins are self-extinguishing and classified 94 V-O.

Radiation Resistance. High-energy irradiation affects PFA, MFA, FEP, and PTFE resins similarly. The PFA and MFA resins have about the same rate of degradation as PTFE resins in air, but resist radiation better under vacuum. The main effect of radiation is the reduction of tensile strength and elongation.

Chemical Resistance. Perfluoroalkoxy resins have outstanding resistance to chemicals and solvents, and are unaffected by strong oxidizing acids, bases, halogens, organic acids, aromatic and aliphatic hydrocarbons, alcohols, aldehydes, ketones, ethers, amines, esters, chlorinated compounds, and polymer solvents. Perfluoroalkoxy resins can be attacked by alkali metals and elemental fluorine at elevated temperature and pressure.

Optical Properties. Perfluoroalkoxy films have good optical properties; they are colorless and transparent in thin sections, and are translucent in thicker sections. Light transmission is best in the IR region (87–93%) followed by the visible (80–87%) and UV regions (55–80%). Refractive index is about 1.35.

2.4.3. Processing

High-temperature PFA and MFA melts can be corrosive to conventional stainless steels; more corrosion-resistant metals, such as Hastelloy-C or Inconel are recommended for equipment parts that are exposed to the melts. Processing temperatures of ca. 330–425°C are recommended. Residence times at the upper temperature limit should be kept short (< 10 min) to avoid polymer degradation. PFA and MFA are processed e.g., by extrusion or injection molding techniques; temperatures up to 425°C are required. For these processing techniques, PFA

Table 12. Properties of PFA (3.5 wt% PPVE) and granular nonfree-flowing PTFE resin [34]

Property	Standard	PFA	PTFE
Permeability, cm^3 mm m^{-2} d^{-1} bar^{-1}			
SO_2, 25 °C		190	310
SO_2, 100 °C		1 400	1 600
Cl_2, 25 °C		110	160
Cl_2, 100 °C		900	1 000
O_2, 25 °C	DIN 53380	240	240
O_2, 100 °C		1 400	1 400
N_2, 25 °C		75	100
N_2, 100 °C		700	700
CO_2, 25 °C		600	800
CO_2, 100 °C		2 100	2 100
HCl, 25 °C		180	300
HCl, 100 °C		1 800	1 800
Water vapor, g mm m^{-2} d^{-1}	DIN 53122		
25 °C		0.02	0.03
100 °C		4.5	3.8
Flexural modulus, MPa	DIN 53457		
25 °C		670	600
100 °C		200	210
150 °C		120	80
250 °C		30	40
Deformation (under 15 MPa load), %	ASTM D 621		
24 h		6.7	16
100 h		7.0	17.5
Permanent	permanent	1.6	11
Dielectric strength (100 µm film), kV/mm	DIN 53481	125	105
SVI value	ISO 12086	30	300

and MFA grades of higher MFI values (lower molar masses) are used.

Although perfluoroalkoxy-resins are stable resins, thermal degradation is unavoidable. This is a function of temperature, residence time, and shear rate. Thermal degradation occurs mainly from the end-groups. Chain scission becomes evident at temperatures above 400°C depending on shear rates. As with many other thermoplastics, melt fracture occurs above critical shear rates leading to rough extrudate surfaces [61]. The allowable flow rates are constrained at the low end by the thermal degradation due to longer residence times and at the high end by the melt fracture and chain scissions. Higher extrusion rates can also be achieved by broadening the molar mass distribution.

Transfer molding uses temperatures of 350–380°C and lower shear rates. At these conditions chain scission does not occur. Usually PFA and MFA grades with lower MFI (higher molar masses) are preferred. Transfer molding techniques are typically used for valve and liner applications.

Because of their higher inherent thermal stability, PFA and MFA resins can be processed at a higher temperature than FEP. This results in a further increase in melt-flow index and critical shear rate for PFA and MFA resins. Thus, PFA resins can also be processed at significantly higher rates than FEP (Table 13).

Perfluoroalkoxy resins have sufficient thermal stability for rotocasting and powder coating applications. For both applications, the resins are applied as powder. Rotocasting is used for coating large structures e.g., drums and tanks. During rotocasting operations, the heating cycles

Table 13. Melt-flow index and shear rate of FEP and PFA resins [2]

Teflon grade		MFI (372 °C/5 kg), g/10 min	Critical shear rate, s^{-1}
FEP	PFA		
100		6	20
	340	14	50
140		3	8
	310	6	16
160		1	2
	350	2	6

may range from 90 to 180 min at 350–370°C. During powder coating applications, finely divided powders are used to produce thin layers of PFA or MFA and various substrates. The surfaces to be coated are heated above the melting point of the resins and the powder is sprayed on. The powder melts on the surface, which is then cooled.

Aqueous dispersions are available for coating metals, glass fabrics, and similar substrates. The dispersion can be processed by the techniques used with PTFE dispersions.

2.4.4. Uses

PFA and MFA resins are used to large extent (~ 80%) in chemical processing and semiconductor industry e.g., for wafer carriers, pumps, valves, bellows, pipe and fittings, filtration systems, and tubing; also the lining of storage tanks and processing kettles is an important segment. PFA and MFA films are also produced; these films can be oriented to provide specialized effects. About 20% of the perfluoroalkoxy resins are used in wire and cable applications e.g., in heat trace cable, insulations of hook-up wire cables as well as in appliance, computer, and aerospace wire. MFA has displaced PFA in many wire and cable applications.

2.5. Tetrafluoroethylene–Ethylene Copolymers

Tetrafluoroethylene and ethylene have a strong tendency to alternate during polymerization. Thus, TFE–ethylene copolymers [25038-71-5] (ETFE resins) are composed mainly of alternating sequences of the two monomers and have the following structure:

$-(CF_2CF_2CH_2CH_2-)_n$

The world consumption of ETFE resins was about 6 000 t in 2008 [12], with expected growth rates of 1–2%. The alternating comonomer units are responsible for the unique properties of ETFE resins; they have a high tensile strength, moderate stiffness, high flex life, outstanding impact strength, good abrasion resistance, and very high cut-through resistance. Their electrical properties are similar to those of other TFE polymers. The ETFE resins have low dielectric constant, high dielectric strength, excellent resistivity, and a low dissipation factor; they also display good thermal stability and excellent chemical resistance. Unlike other TFE copolymers, ETFE resins have a high resistance to high-energy radiation. They are cross-linked by low doses of electron beam or gamma ray radiation. The cross-linked products have improved high-temperature properties, including resistance to cut-through by a hot soldering iron. The processability of ETFE resins is superior to that of other TFE copolymers (the critical shear rate of ETFE resin is at least a magnitude higher than that of FEP resin). They are particularly well suited to extrusion and injection molding.

Unmodified TFE–ethylene copolymers have poor thermal stress–crack resistance, which severely limits their utility. This problem has been largely overcome by incorporating 1–10 mol% of additional comonomers (modifiers), such as perfluoro(alkyl vinyl ethers), perfluoroalkylethylenes, and HFP in the polymer backbone [62–64]. Commercial ETFE resins are all modified in this way. The thermal stress–crack resistance of modified ETFE resins is a function of the modifier content and the melt viscosity of the resin. Processability (extrusion rate), on the other hand, is mainly a function of the melt viscosity. Lowering the melt viscosity improves processability, but reduces stress–crack resistance. Commercially available modified ETFE resins provide different compromises between processability and thermal stress–crack resistance to meet a variety of use and processing requirements.

ETFE resins are produced by Asahi Glass (Aflon COP), Daikin (Neoflon EP), DuPont (Tefzel), 3M (Dyneon ET), and Solvay Solexis (Halon ET).

2.5.1. Production

Tetrafluoroethylene and ethylene monomer mixtures can undergo thermal decomposition reactions. The released energy can be even greater than for decomposition of pure TFE. Therefore, great care must be exercised during the copolymerization of these monomers.

Copolymers of TFE and ethylene can be prepared in aqueous, nonaqueous, or mixed systems. Nonaqueous suspension polymerizations are performed in non or low telogenic hydrofluorocarbon solvents (e.g., $CF_3-(CF_2)_{1-8}-H$, $CF_3-(CF_2)_{1-4}-CH_2-CH_3$) [140]. Fluorinated or hydrogenated organic peroxides are used as initiators; in addition, hydrogen or fluorine containing chain-transfer agents are used to control molar mass. The solid contents are usually less than 10 wt% because the viscosity of the polymer slurry gets quite high.

Mixtures of water and fluorinated solvents may also be used [205]; in this case, the polymerization occurs in the organic phase, while the water serves to lower the viscosity of the mixture and to remove the heat of polymerization.

Aqueous emulsion polymerizations use mainly fluorinated emulsifiers (e.g., ammonium perfluoro octanoate). Organic peroxides or manganic salts are used as initiators; inorganic peroxides, such as APS are less preferred because stable sulfate end-groups are generated that might cause discoloration during workup and processing. Solid contents between 20–35% can be achieved. The polymerization temperature is usually in the range of 30°C to 80°C; polymerization pressure is below 2.0×10^6 Pa.

The following four propagation reactions take place in TFE–ethylene copolymerizations:

1) $\sim\!\!\sim CF_2CF_2\cdot + CF_2=CF_2 \xrightarrow{k_{11}} \sim\!\!\sim CF_2CF_2CF_2CF_2\cdot$
2) $\sim\!\!\sim CF_2CF_2\cdot + CH_2=CH_2 \xrightarrow{k_{12}} \sim\!\!\sim CF_2CF_2CH_2CH_2\cdot$
3) $\sim\!\!\sim CH_2CH_2\cdot + CF_2=CF_2 \xrightarrow{k_{21}} \sim\!\!\sim CH_2CH_2CF_2CF_2\cdot$
4) $\sim\!\!\sim CH_2CH_2\cdot + CH_2=CH_2 \xrightarrow{k_{22}} \sim\!\!\sim CH_2CH_2CH_2CH_2\cdot$

k = rate constant; reactivity ratio r_1 (C_2F_4) = k_{11}/k_{12}; reactivity ratio r_2 (C_2H_4) = k_{22}/k_{21}.

The reactivity ratios are at 65°C: r_1 = 0.045, r_2 = 0.14; at $-30°C$: r_1 = 0.014, r_2 = 0.010. These values demonstrate the strong alternation tendencies that decrease with increasing temperature. Calculation show that a 1 : 1 ETFE obtained at $-30°C$ and at 65°C have a 97% and 93% of alternating sequences [2].

Copolymers of ETFE have poor stress crack resistance, even at higher molar masses. To improve stress crack resistance, commercial ETFE grades are modified with 1–10 mol% of additional comonomers, such as perfluoro (vinyl ether) (PPVE, PMVE), HFP, or perfluoro(alkylethylenes).

The copolymers are isolated from the aqueous polymerization medium by coagulating, filtering, washing, and drying. The product is recovered from nonaqueous polymerization by evaporating the fluorinated solvent. Finishing usually includes a melt-compaction step in which the dried powders are converted into extruded pellets.

2.5.2. Properties

Physical Properties. The crystal structure of equimolar ETFE copolymer is determined by X-ray analysis. The unit cell is orthorhombic or monoclinic (a = 0.96 nm, b = 0.925 nm, c = 0.5 nm, μ = 96°), and the calculated crystalline density is 1.9 g/cm^3. The carbon chain is in a planar zigzag orientation.

Adjacent $-CF_2$ groups are on opposite sides of the chain, followed by a similar unit for adjacent $-CH_2$ units. Interpenetration of adjacent chains occurs when electronic interactions cause bulky $-CF_2$ groups of one chain to nestle into the space above smaller $-CH_2$ groups of an adjacent chain. Thus, each molecule has four nearest neighbors, creating an orthorhombic lattice. Therefore, ETFE has exceptionally low creep, high tensile strength, and high modulus among fluoropolymers. Interchain forces hold this matrix until the alpha transition occurs at about 110°C, where the physical properties of ETFE begin to decline and more closely resemble perfluoropolymer properties at the same temperature. Other transitions occur at $-120°C$ (γ) and about $-25°C$ (β).

As the degree of monomer alternation is reduced, the ability of chains to interlock is decreased. Thus, as the ethylene content is either increased or decreased from the 50 : 50 ratio, the crystalline content is reduced. Melting points of equimolar ETFE are higher than predicted based on a linear relationship between polyethylene and PTFE melting points. This positive melting point deviation occurs from 35 : 65 to 65 : 35 mole ratios and has its maximum at 50 : 50 alternating copolymer, which melts about 285°C. Melting points are lowered by incorporating modifiers, but the overall shape of a curve with positive melting point deviation is unaltered. Figure 1A provides a graphical representation of this phenomenon. Most copolymers of TFE with two or three-carbon alkenes show a linear relationship or a negative deviation when melting point is plotted against polymer composition. Ethylene is the only common monomer that shows a positive deviation. Normally produced ETFE has about 88% alternating sequences, and the melting point is reduced to about 270°C [65–67].

A melting point minimum is reached at about 65–70 mol% TFE. At higher TFE levels, melt processability is lowered and the copolymer behaves increasingly like PTFE as TFE–TFE sequences dominate the properties. Below about 45 mol% TFE, ethylene–ethylene sequences increase exponentially. Properties, such as chemical and oxidation resistance are degraded. Most commercial ETFE grades fall between 50 and 65 mol% TFE. Crystallinity ranges from 40–60%, depending on quench rates.

In addition to the crystalline melting point, ETFE resins exhibit three second-order transitions [68]. At 1 Hz, α-, β-, and γ-relaxations were observed at 110, −25, and −120°C, respectively, in a quenched sample. The α- and γ-relaxations are attributed to motions of long and short segments, respectively, in the amorphous regions. The β-transition appears to occur in the crystalline regions. These transition temperatures depend on frequency. The temperature of the α-transition is 145°C when measured at 100 Hz, and 170°C when measured at 10^5 Hz. The γ-transition occurs at −35°C when measured at 10^5 Hz.

Under ambient conditions, ETFE resins are insoluble in many known solvent. This precludes the measurement of molar mass by the usual methods. However, diisobutyl adipate dissolves ETFE resins above 230°C. A high-temperature laser light-scattering apparatus is used to measure the weight-average molar mass $\overline{M}_w$ of ETFE resin in diisobutyl adipate solution at 240°C [69]. Weight-average molar masses, $\overline{M}_w$, are obtained ranging from 5.6×10^5 to 1.1×10^6, with a polydispersity index $(\overline{M}_w/\overline{M}_n)$ of ca. 1.4. The correlation between melt viscosity (in Pa s) and $\overline{M}_w$ was shown to be

$$\eta(280°) = 1.99 \times 10^{-16} \times \overline{M}_w^{3.4}$$

The molar mass of ETFE resins is normally specified in terms of melt-flow index (MFI),

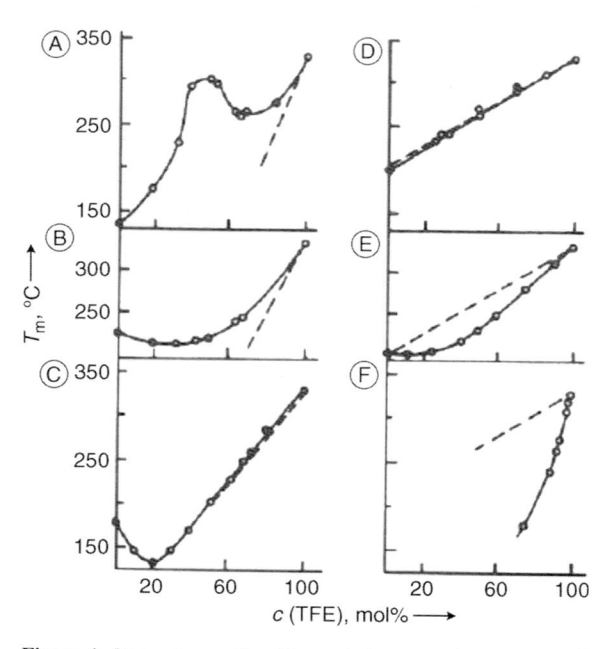

Figure 1. Dependence of melting endotherm maximum on tetrafluoroethylene copolymer composition [34, 67] A) $CH_2 = CH_2$; B) $CH_2 = CHF$; C) $CH_2 = CF_2$; D) $CF_2 = CHF$; E) $CF_2 = CF_2Cl$; F) $CF_2 = CFCF_3$

Table 14. Properties of ETFE-polymers, FEP, and PTFE [70]

Property	Standard	PTFE	FEP	ETFE[*]	ETFE (25 wt% glass fiber)[*]
Specific gravity, g/cm^3	ASTM D 792	2.16	2.14–2.17	1.70	1.86
Melting point, °C	ASTM D 2116	327	260	270	270
Coefficient of linear thermal expansion, K^{-1}	ASTM E 831	12×10^{-5}	9×10^{-5}	9×10^{-5}	3×10^{-5}
Specific heat capacity, kJ kg^{-1} K^{-1}		1.0	1.1	1.9–2.0	
Thermal conductivity at 23 °C, W K^{-1} m^{-1}	ASTM D 696	0.24	0.24	0.24	
Flammability	UL 94	V-O	V-O	V-O	V-O
Oxygen index, %	ASTM D 2863	> 95	> 95	30–32	
Hardness					
Rockwell R	ASTM D 785			R 50	R 74
Durometer D	ASTM D 2240	D 50–65	D 55–57	D 75	
Friction coefficient		0.05–0.08	0.27	0.4	0.3
Water absorption, %	ASTM D 570	< 0.005	< 0.01	0.029	0.022
Upper working temperature, °C		260	200	160	200
Refractive index		1.38	1.344	1.403	

[*] Values based on ETFE Tefzel grades [2].

measured at 297°C/5 kg in a melt rheometer; values are between 2 and 40 g/10 min (ASTM D 3159-06). General properties are listed in Table 14 [70].

Mechanical Properties. Modified ETFE resins are tougher and stiffer and have higher tensile strength than other TFE polymers, such as PTFE, FEP, or PFA. Typical properties are given in Table 14. As wire insulation, ETFE resins exhibit excellent cut-through and abrasion resistance. It has also excellent impact resistance down to −100°C.

The addition of fillers enhances creep and wear resistance. Unlike other TFE polymers, ETFE can be reinforced by glass fibers. The properties of a 25% glass-reinforced Tefzel ETFE resin are shown in Table 15; it has about twice the tensile strength and heat distortion temperature of the unreinforced resin, and its flexural modulus is also much higher. Its ultimate elongation is low, which is characteristic of reinforced compositions. However, its high toughness is indicated by the high Izod impact values at room temperature as well as at −54°C.

Physical properties, such as modulus, tensile strength, and high-temperature elongation are strongly influenced by the degree of alternation of the comonomers. Room temperature modulus and tensile strength decrease as the ratio of TFE–

Table 15. Properties of ETFE-resins [2]

Property	Standard	ETFE	ETFE (25 wt% glass fiber)
Tensile strength, MPa	ASTM D 638	45	83
Elongation, %	ASTM D 887	200	8
Flexural modulus, MPa	ASTM D 790	1380	6500
Tensile modulus, MPa	ASTM D 638	830	8300
Compressive strength, MPa	ASTM D 695	49	69
Shear strength, MPa		41	45
Izod impact strength, J/m	ASTM D 256		
−54 °C		> 1 070	370
23 °C		no break	490
Deformation under load (13.7 MPa at 50 °C), %		4.1	0.7
Heat deflection temperature, °C	ASTM D 648		
0.45 MPa		104	265
1.8 MPa		74	210
Dielectric constant, kHz	ASTM D 150	2.6	
Dielectric strength (3 mm thickness), kV/mm	ASTM D 149	16–20	
Dissipation factor (1 kHz)	ASTM D 150	8×10^{-4}	
Volume resistivity at 23 °C, Ω cm	ASTM D 256	> 10^{16}	

ethylene is increased. High-temperature elongation is improved; the drop off in tensile strength with increasing temperature is not as great in samples with higher TFE fractions. The operating temperature range of ETFE resins is broad (ca. $-100°C$ to $\geq 150°C$). This upper limit is a conservative estimate based on extensive high-temperature testing. At least half of the original properties are retained after 20 000 h exposure at this temperature. Cross-linking by high-energy radiation or cross-linking agents further enhances its high-temperature mechanical properties [71]. Some radiation-cross-linked ETFE wire constructions have been rated for continuous service at 200°C. Short-term excursions to 240°C are possible for highly cross-linked resins.

Electrical Properties. The ETFE resins serve as excellent insulating materials. Their dielectric constant is low (2.6) and is essentially independent of the frequency. The dielectric strength ranges from 16–20 kV/mm, when measured on 3-mm thick specimens, to 160–200 kV/mm on films 25–75 μm thick [2].

Radiation Resistance. Resistance to gamma and electron beam radiation is excellent. Properties are still useful after exposure to more than 10^6 Gy (100 Mrad). Radiation is more harmful in air than in nitrogen.

Chemical Resistance. Resistance to chemicals and solvents is also excellent [2]. Strong acids and bases, aromatic and aliphatic hydrocarbons, alcohols, ketones, and esters have no effect on ETFE resins. Strong oxidizing acids, organic bases, and sulfonic acids attack ETFE resins to varying degrees at elevated temperatures. Certain high molar mass esters, such as diisobutyl adipate are solvents for ETFE resins above 200°C.

Weatherability. Like other TFE polymers, ETFE resins have outstanding weatherability. Extended exposure in Florida and Arizona had no effect.

Flame Resistance. The ETFE resins are nonflammable in air. Wires insulated with ETFE resins are rated 94V-O by the Underwriters Laboratories, with a LOI of 31%.

2.5.3. Processing

Common techniques such as injection molding, compression molding, blow molding, transfer molding, rotational molding, extrusion, and coating techniques are used for processing ETFE resins. Processing temperatures are usually 280–340°C. Molten ETFE can corrode some steels, and corrosion-resistant steel such as Hastelloy C or Inconel is recommended for equipment parts that come in prolonged contact with molten ETFE resins.

High extrusion rates without melt fracture can be achieved [34]. The critical shear rates for ETFE resins are at about two orders of magnitude greater than for perfluorinated copolymers. Typical values for some ETFE Tefzel grades are reported in Table 16.

ETFE powders can be sprayed electrostatically, followed by baking to coalesce the particles into coatings.

Parts produced from ETFE are easily welded by melt-bonding, such as spin welding, ultrasonic welding, and conventional butt welding using a flame and an ETFE rod. ETFE also bonds well to untreated metal surfaces.

2.5.4. Uses

The largest market for ETFE is for wire and cable applications including automotive wire, computer back panels and back planes, lighting and instrument wiring in mass transit vehicles, heat trace cable, and down hole cables for data transmission during oil exploration. About 30% of ETFE resins are used in chemical processing and architectural applications, including tank, pump, valve and pipe linings, and extruded or molded items, such as tower packings, valve seats, mist eliminators, fittings, pipes, and tubes. Other applications include heat-shrink tubing, wire tie wraps, sockets, connectors, insulators, lithium battery cases, seal glands, and pipe plugs as well as low-permeability multilayer tubing for fuel management systems. Films for architectural applications is a fast growing segment, due to the transparency to UV-light it

Table 16. Melt-flow index and critical shear rate for some ETFE Tefzel grades

Product grade	Melt-flow index (297 °C/5 kg), g/10 min	Critical shear rate, s^{-1}
Tefzel 210	45	12 000
Tefzel 200	8	4 800
Tefzel 280	3	4 500

is used in greenhouses, but also in membrane structures for building.

2.6. Polychlorotrifluoroethylene

Polychlorotrifluoroethylene [9002-83-9] (PCTFE) is a linear, high molar mass polymer with the general formula:

$$-(CFCF_2)_n-$$
$$|$$
$$Cl$$

The worldwide consumption of PCTFE including VDF modified polymers was about 5 400 t in 2008. PCTFE is semicrystalline (*mp* ca. 210°C) and can be processed by conventional methods. It has good mechanical properties, chemical inertness, radiation resistance, and low vapor permeability. The introduction of chlorine atom in the polymer structure interrupts the crystallization ability of the polymer chain and results in lower crystallinity, lower melting temperature, better intermolecular interaction, and better mechanical properties compared to PTFE. Manufacturers for PCTFE are Arkema (Voltalef), Daikin (Neoflon), Honeywell (Aclon); Honeywell also produces PCTFE-VDF-copolymers (Aflon).

2.6.1. Production

Polychlorotrifluoroethylene can be polymerized to high molar masses in bulk [72], solution [73], suspension [11], or emulsion systems [74] using free-radical initiators. The emulsion process is carried out at low temperature with a persulfate–bisulfite redox catalyst or organic peroxides. The polymer emulsion is coagulated by freezing or by the addition of salts. The isolated polymer is washed to remove initiator residues and dried.

The dried powders can be stabilized by the addition of phenols or salts of organic acids to help prevent discoloration and bubbling during processing. The tendency for PCTFE to become brittle during use can be reduced by incorporating a small amount (< 5%) of vinylidene fluoride during polymerization; this modification improves also processability.

2.6.2. Properties

The crystallinity of PCTFE ranges from 45% (quick-quenched) to 65% (slow-cooled). The molar mass of PCTFE resin ranges from 1×10^5 g/mol to 5×10^5 g/mol. The useful temperature range is between −200 and 180°C. Other properties are shown in [75–77] and below:

mp, °C	211–216
Second-order transition (amorphous), °C	71–99
Coefficient of thermal expansion, K^{-1}	$(4-7) \times 10^{-5}$
Specific heat, kJ kg^{-1} K^{-1}	0.92
Specific gravity, g/cm^3	2.10–2.18
Specific gravity (amorphous), g/cm^3	2.077
Specific gravity (crystalline), g/cm^3	2.187
Tensile strength at 23°C, MPa	30–40
Elongation at 23°C, %	100–200
Flexural modulus, MPa	1 000–2 000
Hardness, Shore D	80
Dielectric constant at 100 Hz to 1 MHz	2.3–2.7
Dissipation factor at 1 kHz	0.023
Dielectric strength (0.1 mm film), kV/mm	120–150

The α-, β-, μ-relaxations have been determined at various frequencies. Resistance to attack by most chemicals is good. However, chlorinated fluorinated compounds, ketones, and aromatic compounds, such as benzene, toluene, and xylene swell PCTFE resins. They are impermeable to water vapor as well as other gases. They are nonflammable (LOI = 95%) and exhibit outstanding stability to UV and gamma radiation. The coefficient of friction is low; antistick properties are excellent.

CTFE–VDF copolymers have the lowest water vapor transmission of any available transparent plastic. PCTFE displays excellent embrittlement-resistance properties in contact with liquid gases. It also exhibits superior outgasing (barrier) characteristics and is impervious to UV-light and γ-radiation. In addition, the PCTFE-resins show excellent creep resistance at cryogenic temperatures.

2.6.3. Processing

Polychlorotrifluoroethylene resins can be processed by the usual methods for thermoplastic materials, such as extrusion, injection molding, and transfer molding. As with other fluoropolymers, corrosion-resistant materials should be used for equipment parts in contact with the melt. Processing temperatures are in the range of 280–350°C. However, degradation begins at ca. 280°C and thus always accompanies the processing of PCTFE resins. Partial degradation does not seem to affect properties greatly, but severely degraded PCTFE becomes highly crystalline and its mechanical properties suffer. In injection molding, the mold should be heated to 70–130°C. Corrosion-resistant coatings of PCTFE resins can be applied from aqueous dispersions (30–60% solids) by dipping or spraying followed by drying and sintering. Finely divided PCTFE powders (particle size < 150 μm) are used for powder coating.

2.6.4. Uses

Due to the outstanding barrier properties, PCTE resins are widely used as packaging films for drugs and medical products (blister packs for drugs that require oxygen-free storage). Electronics is another large market segment, e.g., for chemical resistant electrical insulation, semiconductor-processing equipment, computer chip production uses PCTFE because of the low moisture absorption properties. PCTFE's toughness in harsh cryogenic environment (e.g., liquid oxygen, nitrogen) for seals and globes is considered useful in the military sector. Low molar mass oils and greases are used as sealants and lubricants.

2.7. Chlorotrifluoroethylene–Ethylene Copolymers

Copolymers of CTFE and ethylene (ECTFE resins) [25101-45-5] are linear, semicrystalline polymers composed of alternating monomer units:

$$-(CF_2CFCH_2CH_2)_n-$$
$$|$$
$$Cl$$

The worldwide consumption was about 2 000 t in 2008. These resins have high tensile strength, moderate stiffness, and good flex life and impact strength. Electrical properties, chemical resistance, weatherability, radiation resistance, and nonflammability are good. The usual processing methods, such as extrusion, injection molding, and transfer molding, are employed. The resins are readily converted into fibers, filaments, films, sheets, and wire and cable insulation. Powder forms are available for the preparation of void-free, low-permeability coatings on metals by electrostatic and fluidized-bed processes. Solvay Solexis is the sole world producer of ECTFE (Halar).

2.7.1. Production

Ethylene and CTFE can be copolymerized by the techniques used for other fluoroolefins. However, to prepare commercially useful products, the copolymerization must be carried out at low temperature (preferably 10°C or below) to reduce the amount of ethylene blocks in the polymer backbone that are susceptible to thermal degradation. Commercial resins with a overall 1 : 1 CTFE : ethylene ratio contain less than 10 mol% of ethylene and CTFE-blocks. A typical polymerization employs a mixture of CTFE and ethylene in water at 10°C, with trichloroacetyl peroxide as catalyst and chloroform as chain-transfer agent [78]. Under these conditions, the copolymerization takes place in the condensed CTFE phase. Water acts as a dispersant and heat-transfer medium. Copolymerization has been carried out at even lower temperatures (between −40 and −80°C), using oxygen-activated trialkylborons as initiators [79]. Modified copolymers with improved thermal stress–crack resistance are prepared similarly [34]. Modifiers include perfluoro(alkyl vinyl ethers), perfluoroalkylethylenes [80], and 2-trifluoromethyl-3,3,3-trifluoroprop-1-ene. Over the course of polymerization, the polymer can agglomerate into spherical beads. Low concentrations of stabilizer additives are added to the dried resins to improve thermal stability. The stabilized powders are melt-pelletized or ground and screened into powder coating grades.

2.7.2. Properties

Physical Properties. The crystallinity of commercial ECTFE resins is in the 50–60% range. X-ray diffraction studies indicate that the unit cell of the alternating copolymer is hexagonal, with a chain repeat distance of $c = 0.50$ nm [78]. The molecular conformation is an extended zigzag, with an ethylene unit on one chain adjacent to a CTFE unit on another chain.

The melting point of an ECTFE copolymer is a function of the molar ratio and the degree of alternation of the two monomers. Commercial ECTFE resins are 1 : 1 copolymers of CTFE and ethylene, 80–90% alternating. They have melting points in the range of 235–245°C. A 1 : 1 copolymer prepared at −80°C, essentially 100% alternating, melts at 264°C [79]; second-order transitions are observed at 140°C, 90°C, and −65°C. The α- and γ-transitions are associated with the motion of chain segments in the crystalline regions, whereas the β-transition seems to be associated with the amorphous region.

The number-average molar mass $(\overline{M}_n)$ of ECTFE resins, determined by osmometry at 150°C, ranges from 1×10^5 g/mol to 5×10^5 g/mol; other properties are shown below:

mp, °C	236–246
Coefficent of thermal expansion, K^{-1}	8×10^{-5}
Specific gravity, g/cm^3	1.63 – 1.73
Tensile strength, MPa	45–55
Elongation, %	200–260
Flexural modulus, MPa	1 800–2 500
Hardness, Shore D	78
Izod impact strength, J/m	
23°C	no break
−40°C	120–60
Dielectric constant (60 Hz to 1 kHz)	2.55–2.60
Dissipation factor (ASTM D 150)	
60 Hz	< 0.005
1 MHz	0.0024
Dielectric strength (0.2 mm film), kV/mm	80–90
Volume resistivity, Ω cm	$> 10^{15}$
Maximum service temperature, °C	150

Mechanical Properties. The ECTFE resins are tough, moderately stiff, and creep resistant. They are usable over a broad temperature range (−100 to 150°C). Abrasion resistance and the heat distortion temperature are good and can be enhanced with fillers. ECTFE-film is the most abrasion resistant fluoropolymer film and has the highest tensile strength.

Chemical Resistance. Both ECTFE and PCTFE resins are similar in chemical resistance. They exhibit good stability toward acids, bases, and halogens. There are no known solvents for these polymers below 120°C, but amines, esters, and ketones can attack ECTFE resins, especially when warm. These resins have excellent barrier properties to water vapor and other gases. In this respect, they are superior to the perfluorinated plastics.

Weatherability and Radiation Resistance. Like most fluoropolymers, ECTFE resins exhibit outstanding weathering resistance. They also resist high-energy gamma and beta emission. The polymer retains useful properties after absorbing more than 10^6 Gy (100 Mrads) of radiation. Cross-linking by exposure to low doses 50–150 kGy (5–15 Mrads) enhances the high-temperature mechanical properties.

Flame Resistance. The ECTFE resins are nonflammable; they burn when exposed to a flame, but are self-extinguishing when the flame is removed. They are nondripping and form a char; LOI is 64%.

2.7.3. Processing

The standard processing methods for thermoplastics, such as extrusion, injection molding, and transfer molding, are employed to process ECTFE. Extrusion is used to manufacture tubing, hose, and wire and cable insulation. Temperatures from 260 to 300°C can be used, but should be kept as low as possible to prevent degradation. Extrusion of ECTFE resins is possible at high rates without degradation. Molten ECTFE resins can be corrosive. A corrosion-resistant steel, such as Hastelloy-C or Inconel is recommended for metal surfaces in contact with the molten polymer. In injection molding of ECTFE resins, the melt temperature should be from 260 to 300°C; for best results, the mold should be heated to 60–100°C. Metals are coated by powders of average particle size below 150 μm using an electrostatic or fluidized-bed process. Coatings of 1 mm or thicker can be obtained free of voids and cracks.

2.7.4. Uses

The principal use of ECTFE resin is in flame-resistant wire and cable insulation [81]. Monofilament made from ECTFE resin is used in the manufacture of chemical-resistant filters and screens. Other applications include coatings and linings for chemical process equipment. Rotomolded tanks and containers made from ECTFE resins are used to store corrosive chemicals, e.g., nitric and hydrochloric acids. Extruded sheet is used for thermoforming various products, such as battery cases for heart pacemakers. Tubing made from ECTFE resin is used in laboratory and some automotive applications.

The smooth surface of ECTFE reduces the tendency for adhesion of dirt and microorganisms, which is essential for semiconductor and water applications [34].

2.8. Poly(Vinylidene Fluoride)

Poly(vinylidene fluoride) [9002-85-1], poly (1,1-difluoroethylene) (PVDF) is a linear, high molar mass, semicrystalline polymer that can exist in three distinct crystalline forms and has the following structure:

$$-(CH_2CF_2-)_n$$

The worldwide consumption including copolymers was about 2.9×10^4 t in 2008. The alternating CH_2 and CF_2 groups create a dipole that confers solubility in highly polar solvents, such as dimethylformamide and dimethylacetamide. In addition, PVDF has a high dielectric constant and exhibits piezoelectric properties under certain conditions. It has excellent resistance to abrasion, stress fatigue, and cold flow. Like PTFE and TFE copolymers, it has outstanding weatherability. It also exhibits excellent resistance to gamma and beta radiation. The usable temperature range is from ca. −40 to +150°C. The resins are processable by extrusion, injection molding, and other molding techniques. Copolymerization with HFP (< 15%) improves processability, flexibility, and impact strength.

Uses for PVDF resins include durable finishes for exterior metal siding, electrical insulation, and chemical process equipment. Manufacturers for PVDF and copolymers are Arkema (Kynar), Daikin (Neoflon), 3M (Dyneon), Kureha (KF Polymer), and Solvay Solexis (Solef). Chinese fluoropolymer producers have also started PVDF production.

2.8.1. Production

The most common methods for producing PVDF homopolymers and copolymers are emulsion and suspension polymerization. In suspension polymerization, organic peroxides are used as initiators including chain-transfer agents to control molar mass; colloidal dispersants (e.g., cellulose derivatives, polyvinyl alcohol) are often employed. The product is a slurry of suspended particles, which are recovered by filtration. Inorganic peroxides (e.g., persulfates) give polymers with lower thermostability and which are prone to discoloration. In emulsion polymerizations organic peroxides, chain-transfer agents, and mostly fluorinated emulsifiers are also used; due to environmental concerns, nonfluorinated surfactants are nowadays recommended [82–85]. The polymerizations are typically carried out at 70–150°C at pressures up to $7–10^6$ Pa.

HFP and CTFE are most commonly employed to prepare PVDF-copolymers (commercial thermoplastic resins contain less than 15 mol% comonomers). To obtain constant compositions, the comonomers have to be adjusted according to their reactivity ratios [34]:

VDF : CTFE $\quad r_1 = 0.13 \quad r_2 = 3.73$
VDF : HFP $\quad r_1 = 2.45 \quad r_2 = 0.00$

2.8.2. Properties

Physical Properties. Poly(vinylidene fluoride) has a highly regular structure, with most VDF units joined head-to-tail and with ca. 5% of the monomer units joined head-to-head. Crystallinity is between 40% and 60%; three different crystalline forms exist.

In phase I (β-form), the polymer crystals are orthorhombic, and the molecules adopt a planar zigzag conformation. In phase II (α-form), the crystals are monoclinic, and the molecules adopt a gauche-trans-gauche-trans conformation. In phase III (γ-form), the crystals are believed to be monoclinic, but the molecules adopt a planar zigzag conformation. Slow or rapid cooling of polymer from the melt yields only phase II. Conditions for converting one form to another have been established [86].

Because of its polymorphism, the melting point of PVDF is not precisely defined (154–184°C). The melting point of the more

Table 17. Properties of PVDF homopolymers and PVDF copolymers [34]

Property	Standard	PVDF homopolymer	PVDF copolymers
Specific gravity, g/cm^3	ASTM D 792	1.75–1.80	1.75–1.79
Melting temperature, °C	ASTM D 34518	160–180	125–170
Refractive index n_D^{25}	ASTM D 542	1.41–1.42	1.40–1.42
Water absorption, %	ASTM D 570	0.01–0.04	0.03–0.09
Tensile strength, MPa	ASTM D 638		
yield		31–57	14–42
break		27–52	15–42
Elongation at break, %	ASTM D 638	50–250	200–500
Tensile modulus, MPa	ASTM D 882	1 030–2 410	480–1040
Flexural modulus, MPa	ASTM D 790	1 130–2 280	240–1180
Izod impact strength (25 °C), J/m	ASTM D 256		
notched		110–300	320–NB[*]
unnotched		800–4 500	3 600–NB[*]
Hardness, Shore D	ASTM D 2240	75–80	55–75
Abrasion resistance, mg/1 000 cycles	CS-17, 1 000 g	5–9	6–20
Friction coefficient (to steel)		0.14–0.17	
Heat deflection temperature, °C	ASTM D 648		
0.46 MPa		119–148	
1.82 MPa		84–118	
Thermal conductivity, W K^{-1} m^{-1}	ASTM D 433	0.17–0.19	0.16–0.18
Coefficient of linear expansion (~ RT), K^{-1}	ASTM D 696	0.7–1.5 × 10^{-4}	1.0–1.5 × 10^{-4}
Limiting oxygen index, %	ASTM D 2863	42–80	42–95

[*] NB = no break.

common phase II form is ca. 170°C; a glass transition temperature is observed at −40°C. Properties of PVDF resins are shown in Table 17.

PVDF has one of the lowest melting points of commercial fluorothermoplastics, but has the highest heat deflection temperature under load (see Table 18).

PVDF and ETFE are isomeric, but PVDF has a melting point in the range of 150–180°C compared to 260–275°C for ETFE. In both polymers, the alternating units can crystallize with larger CF$_2$ groups adjacent to smaller CH$_2$ units of adjacent chains. This penetration increases the moduli for ETFE and PVDF. Perfluoropolymer chains do not have such penetration; thus, they slide past one another, leading to higher creep and lower modulus values.

The high crystallinity and surface tension properties of PVDF provide very low permeation values (see Table 19) compared to other fluoropolymers.

Mechanical Properties. A comparison of mechanical properties of PVDF to those of ETFE and ECTFE at room temperature is provided in Table 20. The tensile modulus is similar but impact strength is lower. Modification with HFP or CTFE lowers the modulus, but increases elongation and impact strength (see Table 17).

Electrical Properties [10]. A high dielectric constant and dissipation factor impair the utility of PVDF as a primary wire and cable insulation. However, these "poor" electrical properties allow the production of PVDF films with their piezoelectric and pyroelectric properties. These films are prepared from extruded films in the phase I (β-form) conformation, obtained by stretching the film as it is extruded; ultrapure polymer is required. Both surfaces of the film are metallized, and it is then subjected to a high voltage, which leaves it permanently polarized. Such films generate a voltage when stretched or compressed (piezoelectricity) or heated (pyroelectricity).

Table 18. Comparison of deflection temperature and mp of fluoropolymers (ASTM D 648) [34, 87]

	Deflection temperature, °C		mp, °C
	0.5 MPa	1.8 MPa	
PVDF	148	113	178
PCTFE	126	75	218
PTFE	121	56	327
ECTFE	116	77	240
ETFE	104	74	270
PFA	73	48	310
FEP	70	51	270

Table 19. Gas permeability of fluoropolymers[a] [34, 88]

	PTFE	PFA	FEP	ETFE	PCTFE	ECTFE	PVDF	PVF
Water vapor, g m^{-2} d^{-1} bar^{-1}	5	8	1	2	1	2	2	7
Air, cm^3 m^{-2} d^{-1} bar^{-1}	2000	1150	600	175	NA[**]	40	7	50
Oxygen, cm^3 m^{-2} d^{-1} bar^{-1}	1500	NA[**]	2900	350	60	100	20	12
Nitrogen, cm^3 m^{-2} d^{-1} bar^{-1}	500	NA[**]	1200	120	10	40	30	1
Helium, cm^3 m^{-2} d^{-1} bar^{-1}	3500	17 000	18 000	3 700	NA[**]	3 500	600	300
Carbon dioxide, cm^3 m^{-2} d^{-1} bar^{-1}	15 000	7000	47 00	1 300	150	400	100	60

[a] Based on 100 µm film thickness at 23 °C. Gases were measured according to ASTM D 1434-82; Water permeation was measured according to DIN 53122.
[**] NA = not available.

Chemical Resistance. Resistance of PVDF to halogens, most inorganic acids and bases, alcohols, halogenated hydrocarbons, and aromatics is good. PVDF is, however, soluble in polar solvents, such as acetone and ethyl, butyl, and amyl acetate, amines and amides. It is also soluble in supercritical media (e.g., CO_2, CHF_3, CH_2F_2, etc.) [89].

Weather and Radiation Resistance. Resistance of PVDF to UV light is excellent (no effect after 5 000 h exposure in Weather-O-Meter). Tolerance to ionizing radiation is very high. Low radiation doses of 20–150 kGy (2–15 Mrads) cause cross-linking, which enhances the high-temperature modulus of PVDF.

Flame Resistance. When exposed to flame, PVDF is nonflammable and nondripping; it is self-extinguishing after the flame is removed. The LOI is 44%.

2.8.3. Processing

Poly(vinylidene fluoride) can be processed by the usual techniques for thermoplastics, such as extrusion and transfer, injection and compression molding. Metal coatings are applied from organisols (mixtures of resin and solvent) in a process called coil coating. Typically, the organisol is applied to coil stock of steel or aluminum. The film is formed on the metal surface as the solvent evaporates. In extrusion, attention should be paid to the elimination of dead spots where degradation of the polymer can start. The temperature should be between 215 and 270°C. Similarly, in injection molding, the melt temperature should be between 200 and 270°C.

2.8.4. Uses

The largest application for PVDF is within architectural coatings. Typically, PVDF is

Table 20. Typical physical properties of PVDF, ETFE, and ECTFE

Property	Standard	ETFE	PVDF	ECTFE
Melting temperature, °C	ASTM D 3418	250–275	154–184	236–246
Specific gravity, g/cm^3	ASTM D 792			
Melt		1.3		
Solid		1.72	1.75–1.80	1.7
Tensile strength (23 °C), MPa	ASTM D 638	38–48	36–56	45–60
Elongation (23 °C), %	ASTM D 638	100–350	25–500	150–250
Tensile modulus (23 °C), MPa	ASTM D 638	830	1 340–2 000	
Izod impact strength (23 °C), J/m	ASTM D 256	no break	160–530	no break
Coefficient of thermal expansion	ASTM D 696	9×10^{-5}	$\sim 10^{-4}$	5×10^{-5}
Critical shear rate, s^{-1}		200–10000		
Processing temperature range, °C		300–345	200–300	260–300
Dielectric constant (1 kHz)	ASTM D 150	2.6	7.5–13.2	2.6
Dielectric strength, kV/mm	ASTM D 149	59	260–950	80–90
Dissipation factor (1 kHz)	ASTM D 150	0.0008	0.013–0.019	0.0024
Limiting oxygen index, %	ASTM D 2863	30	44	64

sold as powder for coating metal coils for the production of metallic roofing, window frames, panel siding, and other construction units. It is also used as wire insulation in the aircraft and electronics industries. Heat-shrinkable tubing made from PVDF resin forms solder sleeves that are used in electronic, aerospace, and aircraft industries. It is used on a large scale for solid PVDF pipe and fittings and as a liner for pipes and valves. Piezoelectric PVDF films are very sensitive transducers with various applications [90].

2.9. Poly(Vinyl Fluoride)

Poly(1-fluoroethylene) [24981-14-4], poly(vinyl fluoride) (PVF) is a high molar mass, semicrystalline polymer with a melting point of 180–210°C and the following structure:

$$-(CH_2CH)_n-$$
$$|$$
$$F$$

It is considered a thermoplastic but cannot be processed by conventional thermoplastic techniques because it is unstable above its melting point. It can be fabricated into films by mixing with latent solvents (i.e., solvents that alone do not dissolve the polymer, but behaves as an active solvent when used in conjunction with an active solvent) and extruding at a temperature below the melting point. (Dissolution in latent solvents takes place at an elevated temperature below the melting point of the PVF.) Films made from PVF have exceptional resistance to sunlight, chemical attack, and water adsorption. They are also extremely tough and have good abrasion resistance. These unusual properties make PVF highly suitable for outdoor coatings or laminates. Films of various tensile modifications and thicknesses are available with adhereable and nonadhereable surfaces. These films are manufactured by DuPont under the Tedlar trade name. DuPont is the sole producer for PVF; the worldwide consumption was about 3 000 t in 2008.

2.9.1. Production

Vinyl fluoride is polymerized by free-radical processes [10]. It is, however, more difficult to polymerize than TFE or VDF and requires higher pressures [91]. Polymerization is carried out in aqueous media between 50°C and 150°C at a pressure of 3.4×10^6–34.4×10^6 Pa. The usual catalysts are peroxides and azo compounds. A continuous process employs water, VF, and catalyst at 100°C and 27.5×10^6 Pa [92]. Polymers have also been prepared at lower temperatures (-30 to $55°C$) using oxygen-activated trialkylboron catalysts [93]. The polymerization temperature affects the crystallinity and the melting point of the product; higher temperatures also increase branching.

2.9.2. Properties

Physical and Mechanical Properties. The monomer units in PVF are predominantly head-to-tail, with 12–18% head-to-head [94]. The proportion of head-to-head units decreases with decreasing polymerization temperature. Conversely, randomness increases with increasing polymerization temperature.

The number-average molar mass of PVF resin ranges from 7.6×10^4 g/mol to 2.3×10^5 g/mol, the weight-average molar mass ranges from 1.43×10^5 g/mol to 6.54×10^5 g/mol. The ratio of weight-average to number-average molar mass is 2.5 : 5.6 [95].

The melting point of PVF varies with polymerization temperature but is in the range of 185–210°C. Torsion-pendulum studies indicate that PVF has second-order transitions at 41°C and $-20°C$. The temperature range for continuous use is -70 to $110°C$. The mechanical (ASTM D 882 and 1004) and physical properties of PVF film are shown in [94]:

mp, °C	185–190
Specific gravity, g/cm^3	1.38–1.72
Tensile strength, MPa	~50
Elongation, %	115–250
Tensile modulus, MPa	1 700–2 600
Impact strength, kJ/m	10–22
Tear strength, kJ/m	174–239
Coefficient of thermal expansion, K^{-1}	5×10^{-5}
Dielectric constant, 1 MHz	6.5
Dielectric strength (0.1 mm film), kV/mm	120–140

Chemical Resistance. Poly(vinyl fluoride) is insoluble in ordinary solvents below 100°C. At higher temperature, it is soluble in polar solvents, such as amides, ketones, tetramethylene sulfone, and tetramethylurea.

Films show good resistance to acids and bases at room temperature. The resistance to aliphatic (petroleum), aromatic, and alcoholic liquids is also very good. Warm acetone attacks PVF.

Weather and Radiation Resistance. PVF films have excellent light and weather resistance, showing no change after 10 years outdoor exposure. They are transparent to both UV and visible light. A 25-μm thick film has over 90% transmission between 400 and 1400 nm and over 80% between 230 and 400 nm.

In addition, PVF films show good resistance to high-energy radiation. Usable properties are retained up to 1 MJ/kg exposure to electron beam radiation.

2.9.3. Processing

The polymer is dissolved in a latent solvent. Pigments, stabilizers, and plasticizers can be added to the PVF organosol before extrusion. The organosol is extruded into film, and the solvent is evaporated and recovered. The film can be oriented in varying degrees to achieve different levels of tensile and tear strength. Exposure to such treatments as corona discharge or alkali metal in liquid ammonia solution imparts adhesiveness.

2.9.4. Uses

Poly(vinyl fluoride) films are laminated to sheets of metal with special adhesives. The laminates are used for exterior siding on industrial and residential buildings. Pigmented films can be made in any color. Films can also be laminated to hardboard, paper, polystyrene, polyurethane, and other substrates. These laminates are used for wall coverings, aircraft cabin interiors, pipe covering, duct liners, as well as other applications. A weather-resistant PVF barrier is also resistant to abrasion, chemicals, solvents, and staining. Another application for PVF film is as a release sheeting in the molding of phenolic, epoxide, and polyester plastics, and as a release film for bag molding. It does not become brittle and strips easily from the mold. Toughness is maintained at elevated temperatures at all degrees of humidity and in contact with common solvents. The film has been used for photovoltaic applications because of its high transparency and resistance to solar radiation.

2.10. Amorphous Perfluoropolymers

Before 1980, all commercial fluoropolymers were semicrystalline materials, having shortcomings, such as low optical transparency, high creep, or poor solubility. Since mid-1980s, amorphous perfluoropolymers with high glass transition temperatures have been developed and commercialized. These polymers are linear, high molar mass polymers with the general formula:

$$\left(CF-CF\right)_n\left(CF_2-CF_2\right)_m$$
$$\underset{F_3C\quad CF_3}{O\diagdown C\diagup O} \quad \mathbf{1}$$

$$\left(CF_2\underset{CF-CF}{\diagdown\diagup}CF_2\right)_n$$
$$\underset{CF_2}{O\diagdown\quad\diagup CF_2}\quad \mathbf{2}$$

$$\left(CF-\underset{|}{\overset{OCF_3}{C}}\right)_n\left(CF_2-CF_2\right)_m$$
$$\underset{CF_2}{O\diagdown\quad\diagup O}\quad \mathbf{3}$$

These polymers retain outstanding chemical, thermal, and surface properties of perfluorinated polymers in addition of having excellent electrical and optical properties. They are also soluble at ambient temperatures in fluorinated solvents.

These resins are manufactured by Asahi Glass (Cytop, **2**), DuPont (Teflon AF, **1**), and Solvay Solexis (Hyflon AD, **3**). Current consumption of amorphous fluoropolymers is still very small.

2.10.1. Production

The commercial grades of Teflon AF [37626-13-4] and Hyflon AD [161611-79-6] are copolymers of tetrafluoroethylene and the corresponding dioxole monomers. They are produced by free-radical initiators in nonaqueous media or in aqueous media with a fluorinated emulsifier [96]. Both dioxole monomers form amorphous homopolymers with glass transition temperatures of 335°C for the Teflon AF dioxole-monomer (PDD = perfluoro-2,2-dimethyl-1,3-dioxole) and of 162°C for the Hyflon AD dioxole-monomer (TTD = 2,2,4-trifluoro-5-trifluoromethoxy-1,3-dioxole) [97]. The PDD-homopolymer is difficult to melt-process because of the narrow processing window below its decomposition temperature. TFE-copolymers with less than 20 mol% PDD are no longer amorphous. At 20 mol% PDD, the T_g of the copolymer is about 80°C. During aqueous polymerization, small amounts of acid fluoride or carboxylic acid groups may be produced by ring opening. These unstable groups are removed and converted to perfluorinated groups by first treating the polymer with ammonia followed by post-fluorination [98].

Cytop [101-182-89-2] is a homopolymer that is synthesized via a unique cyclopolymerization of perfluoro-3-(butenyl vinyl ether) (PBVE) [99, 34].

The glass transition temperature is about 108°C.

2.10.2. Properties

Amorphous perfluoropolymers exhibits good thermal stability with small mass losses at temperatures up to 400°C. In addition, they have excellent chemical resistance, low surface energy, and high LOI.

The mechanical properties differ from those of semicrystalline polymers. Below the glass transition temperature, the tensile module is higher and the creep is generally less and shows less variation with temperature.

These amorphous polymers are soluble at ambient temperature in a variety of fluorinated solvents [34]. The optical properties are outstanding; the optical transmission is greater than 90% throughout most of the UV, visible, and near-IR-spectrum. Teflon AF grades show the lowest refractive indices of any known polymer.

Investigations of Teflon AF grades reveal that the structure is characterized by microvoids; these voids result in a lower than expected polymer specific gravity, low dielectric constant, low refractive index, high gas permeability, and low thermal conductivity. Most likely, the origin of these microvoids is loose chain packing caused by the high energy for rotation and reorientation of the dioxole ring containing polymer chains. The dielectric constant of Teflon AF is the lowest of any know polymer, on the other hand, the gas permeabilities are more than two orders of magnitude higher than, for example, for PTFE [34].

The microstructure of Cytop was also analyzed by positron annihilation lifetime spectroscopy [100]. General properties of amorphous polymers are given in Table 21.

Table 21. Properties of amorphous perfluoropolymers, PFA and PMMA [34]

Property	Cytop	Teflon AF	PFA	PMMA	Remarks
Light transmittance, %	95	95	opaque	93	visible region
Refractive index	1.34	1.29–1.31	1.35	1.49	Abbe's refractometer
Abbe's number	90	92–113		55	Abbe's refractometer
Dielectric constant	2.1 ~ 2.2	1.9	2.1	4	60 Hz–1 MHz
Dissipation factor	7×10^{-4}	12×10^{-4}	$<2 \times 10^{-4}$	0.04	60 Hz
Volume resistivity, Ω cm	$>10^{17}$		$>10^{18}$	$>10^{16}$	RT in air
Dielectric strength, kV/0.1 mm	11	1.9–2.1 [2]	12	2	RT in air
T_g, °C	108	160–240	75	105–120	DSC
mp, °C	not observed	not observed	310	160 (isotactic)	DSC
Specific gravity, g/cm³	2.03	1.7–1.8	2.12–2.17	1.20	25 °C
Contact angle of water, °C	110	104	115	80	25 °C
Critical surface tension, N/cm	19×10^{-5}	16×10^{-5}	18×10^{-5}	39×10^{-5}	25 °C
Water absorption, %	< 0.01	< 0.01	< 0.01	0.3	60 °C, H$_2$O
Tensile strength, MPa	39	27	> 28	65–73	
Elongation at break, %	150	> 20	> 280	3–5	
Yield strength, MPa	40		11–15		
Tensile modulus, MPa	1 200	1 600	580	3 000	

2.10.3. Processing

The commercial amorphous perfluoropolymer can be melt-processed by all conventional techniques, such as extrusion, compression, or injection molding. Solutions of amorphous polymers can be used for spin coating, dip coating, spraying, or casting. Spin coating is used to produce very thin and uniform coatings on flat substrates. Dip coating is suitable for nonplanar surfaces. Laser ablation and vacuum pyrolysis techniques are also used.

2.10.4. Uses

Because of their low refractive indexes and excellent optical clarity, amorphous fluoropolymers are used in fiber optics cladding and core applications. Other applications include photolithography, lens covers, antireflective coatings, passivation, and protective coatings for various microwave, radar, optical, and optoelectronic devices in medical, military, aerospace, and industrial markets. Electronic applications include dielectric layers for high density and hybrid integrated circuits, passivation layers, and encapsulation of hybrid and sandwich integrated circuit packaging.

2.11. Semicrystalline Tetrafluoroethylene-Hexafluoropropene-Vinylidene Fluoride Terpolymers

Terpolymers of variable amounts of tetrafluoroethylene (TFE), hexafluoropropene (HFP) and vinylidene fluoride (VDF) [THV resins] are linear, semicrystalline polymers with the following structure:

$$-(CF_2CF)_x(CF_2CF_2)_y(CH_2CF_2)_z-$$
$$|$$
$$CF_3$$

THV polymers were invented by Hoechst during the late 1980s and commercialized by 3M in 1995. By combining these three monomers, unique properties for THV resins can be achieved, for example, low processing temperatures, bondability, high flexibility, excellent clarity, and low refractive indexes as well as efficient E-beam cross-linking. THV resins are manufactured by 3M; the worldwide consumption was about 1 000 t in 2008.

2.11.1. Production

Based on the Q/e reactivity ratios of the three monomers, most information to design a wide range of THV terpolymer are available; known Q/e value are scattered, but show a common trend: TFE ($Q \sim 0.032$, $e \sim 1.63$), HFP ($Q \sim 0.005$; $e \sim 1.50$), VDF ($Q \sim 0.015$; $e \sim 0.50$) [101]. THV polymers are polymerized in aqueous emulsion in the presence of fluorinated emulsifiers. The preferred initiator is the $KMnO_4$-system for temperature ranges from 30°C up to 80°C; the pressure can go up to 2.0×10^6 Pa. Chain-transfer agents are hydrogen, C_1–C_5-alkanes, or dimethyl ether [102].

The obtained latices have solid contents of about 35 wt%. $KMnO_4$-initiators provide carboxylic and $-CH_2OH$ endgroups, which do not lead to discoloration, whereas persulfate-initiated polymerizations have significant amounts of sulfate endgroups, which cause discoloration during certain processing methods. Hydrocarbon surfactants can also be used to polymerize THV polymers [103]. THV resins are isolated from the aqueous latices by applying high shear forces; agglomerated powders as well as melt-pelletized granules are commercially available. Concentrated dispersions with solid contents up to 50 wt% and stabilized with hydrocarbon surfactants are also offered.

2.11.2. Properties

Depending on the ratio of the three monomers, THV resins can be tailored to have melting points from 120°C up to 230°C (see Fig. 2). Melt viscosity is adjustable over a wide range. THV polymers are well characterized by GPC and rheological measurements. The polymers have usually a linear topography, and MFI values (265°C/5 kg) correlate well with molar mass ($1/\mathrm{MFI} \sim \eta_0 \sim \overline{M}_w^{3.8}$). Molar mass distribution is usually quite narrow ($\overline{M}_w/\overline{M}_n \sim 1.7$) [104]. Long chain branched (LCB) THV polymers are prepared by the addition of small amounts of reactive modifiers (e.g., $CF_2=CHBr$, diolefins) during polymerization; LCB-

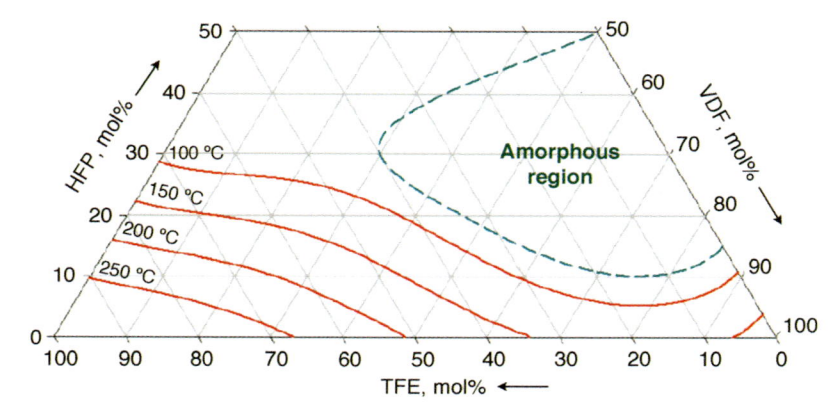

Figure 2. THV melting points versus chemical composition

THVs combine good mechanical properties with high structural viscosity [105, 106].

The low melting grade THV 200 is soluble in common organic solvents. THV 200 and THV 500 have relatively low processing temperatures (220–250°C), which offer new options for coprocessing. Both grades have also low flexural moduli between 8×10^7 Pa and 2.5×10^8 Pa. THV polymers are the most flexible of the melt-processable semicrystalline fluoropolymers. THVs are transparent to a broad band of light energy (UV to IR); the refractive indices are also very low (e.g., 1.355). All THVs retain properties common to other semicrystalline fluoropolymers, such as chemical resistance, weatherability, and low flammability. General properties are provided in Table 22.

2.11.3. Processing

Most THV resins (e.g., THV 200 and THV 500) extrude at lower temperatures (220–250°C) than other semicrystalline fluoropolymers. This provides two benefits:

1. In many cases corrosion-protected equipment is not required to extrude THV
2. It is easier to coprocess THVs with hydrocarbon plastics and elastomers

The low processing temperatures of THV combined with the ability to develop strong chemical bonds to many other materials make THV an ideal candidate for multilayer constructions. For instance, coextruded tubes of THV on

Table 22. Properties of THV-polymers

Property	Standard	THV 200	THV 500	THV 815
Specific gravity, g/cm³	ASTM D 792	1.95	1.98	2.06
Melting temperature, °C	ASTM D 3418	115–125	160–170	220–230
Thermal decomposition in air[*], °C		420	440	460
Limiting oxygen index, %	ASTM D 2863	65	75	80
Tensile strength at break, MPa	ASTM D 638[a]	29	28	30
Elongation at break, %	ASTM D 638[a]	600	500	400
Flexural modulus, MPa	ASTM D 790	80	210	520
Harness, Shore D	ASTM D 2240	44	54	59
Dielectric constant at 23 °C				
100 kHz	ASTM D 149	6.6	5.6	5.1
10 MHz	ASTM D 149	4.6	3.9	3.8
MFI (265°C/5 kg), g/10 min	ASTM D 1238	20	10	12
E-beam cured high temperature resistance, °C		> 150		

[*] Measured by thermogravimetric analysis.

the inside (for chemical and permeation resistance) with a less expensive plastic on the outside (for strength and protection) can be easily made, or the reverse can be done [34].

THV can also be processed as pure material by the usual techniques for thermoplastics, such as extrusion, blow-, and injection-molding. THV 200 can be applied for solutions in solvents; the aqueous dispersion grades (e.g., THV 340) can be processed like any other fluoropolymer dispersion.

2.11.4. Uses

Most applications of THV are multilayer constructions, where a thin layer of THV provides enhanced barrier properties to other layers. Fuel hoses and fuel management systems are major applications due to the excellent permeation and chemical resistance including high flexibility and ease of processing. Further applications are wire, cable, and protective coatings. THV can also serve as flexible liner for chemical holding tanks. Emerging applications include speciality films for architectural segment solar energy (solar panels) and as safety glass compound.

3. Fluoroelastomers (→ Rubber, 4. Emulsion Rubbers, Chap. 6)

3.1. Introduction

Fluorine-containing elastomers are designed to maintain rubberlike elasticity in extremely severe environments at high and low temperatures as well as in contact with various chemicals. Their primary use is in seals in automotive e.g., as O-rings or shaft seals, where they function to prevent leakage after prolonged compression. The chemical composition determines the balance between e.g., oil resistance and low-temperature characteristics. As fluoroelastomers show compared to hydrocarbon elastomers a rather slow relaxation and recovery from strain, they are rarely subjected to tension and then only at modest elongations. Preferably, they are used in static rather than dynamic applications.

A second application for uncured fluoroelastomers makes use of their ability to enhance processability of polyolefins. As a processing additive, they reduce extrusion related gels, increase throughput, and reduce die build-up [107–109].

Technologically important fluoroelastomers can be divided into organic polymers with carbon–carbon linkages in the backbone and fluoroelastomers with inorganic backbones, such as fluorosilicones, which were developed in the 1950s and the fluoroalkoxyphosphazines introduced in 1975, which represent only a small share of the fluoroelastomer market.

Elastomers have to be flexible and recover from deformation at service temperatures above 0°C. Noncross-linked elastomers are highly viscous, incompressible liquids. To prevent permanent flow deformation under an imposed force and to develop reversible elastic properties, these flexible macromolecules must be cross-linked. The restoring force in a deformed elastomeric network is based on the lower entropy of the extended and hence oriented chains. In general, cross-linking is performed chemically by vulcanization to produce tie points between chains to minimize irreversible flow under stress. However, physical cross-linking by crystallization of hard segments in segmented or grafted copolymers can also provide a network structure in socalled thermoplastic elastomers [110]. The few known thermoplastic fluoroelastomers have a much lower high-temperature service ceiling than thermoset elastomers; nevertheless, they possess certain advantages in processability because they can be reversibly cross-linked. Furthermore, they generally yield less extractable material, because they contain no curatives.

Fluorocarbon elastomers are copolymers made up of two or more major monomer units. One or more monomers give straight chain segments, which tend to crystallize if long enough. A second monomer with a bulky side group is incorporated at intervals to break up the crystallization tendency and produce a substantially amorphous elastomer.

The characteristics of a fluoroelastomer can be tailored by the chemical composition. The monomers VDF, TFE, and E increase the crystallinity, whereas HFP, PMVE, and P have bulky side-groups that hinder crystallization [111]. High solvent resistance and thermal stability in fluorocarbon polymers are important properties of fluoroelastomers but low-temperature flexibility and high fluorine content are also desirable. These requirements conflict to some extent. TFE increases the chemical resistance, but the crystallinity is also increased, which reduces flexibility.

Copolymerization of VDF and HFP in the molar ratio of 3.5 : 1.0 produces a fluoroelastomer [9011-17-0] with no crystallinity and a T_g of ca. $-20°C$. This composition (60 wt% VDF) is the basis of the fluorocarbon elastomer industry:

$$-[(CH_2CF_2)_{3.5}CF_2CF]_n-$$
$$\qquad\qquad\qquad\quad |$$
$$\qquad\qquad\qquad\ CF_3$$

Other combinations of VDF with HFP and TFE, chlorotrifluoroethylene (CTFE), PMVE, or PMVE–TFE also produce substantially amorphous copolymers with glass transition temperatures low enough for practical use. PMVE–VDF-based elastomers have the lowest T_g ($-40°C$) and are important for the use in aircrafts.

The service temperature range and fluorine and hydrogen content of fluoroelastomers are given in Table 23.

The combinations of TFE with PMVE or other perfluoro(alkoxy alkyl vinyl ethers) with fluorinated cure site monomers lead to perfluoroelastomers with the highest attainable fluid and thermal resistance.

The low-temperature properties of fluorosilicones and fluoroalkoxyphosphazenes ($\rightarrow$ Inorganic Polymers, Chap. 4) are superior to those elastomers with carbon backbones substituted by hydrogen, fluorine, chlorine, fluoroalkyl groups, or simple fluoroalkoxy groups. Their outstanding chain flexibility is a result of the alternating oxygen or nitrogen atoms in the backbone. However, fluorosilicones and fluoroalkoxyphosphazenes contain less fluorine than fluorocarbon elastomers; therefore, their solvent resistance and high-temperature stability are inferior.

Commercial fluorosilicone elastomers are polydimethylsiloxane homopolymers in which one methyl substituent of each silicon atom is replaced by a 3,3,3-trifluoropropyl group:

$$\qquad\ CH_3$$
$$-(Si-O)_n-$$
$$\quad\ |$$
$$\ CH_2CH_2CF_3$$
[25791-89-3]

Fluorosilicone processing is the same as for standard silicone elastomers and quite different from fluoroelastomer processing.

Fluoroalkoxyphosphazene elastomers have been produced by replacing the chlorine in dichlorophosphazene polymers by fluorocarbon alkoxides [112] but have been withdrawn from the market:

$$\qquad\ OCH_2CF_3$$
$$-(P=N)_n-$$
$$\quad\ |$$
$$\ OCH_2CF_2CF_3$$
[37002-15-6] ($T_g = -77\ °C$)

3.2. Elastomers Based on Vinylidene Fluoride (FKMs)

Copolymers of 1,1-difluoroethylene and chlorotrifluoroethylene prepared by M. W. KELLOGG were first described in 1955 [113–115]. After vulcanization, this fluoroelastomer (Kel-F) exhibited previously unattainable resistance to heat and solvents. However, it was soon surpassed by an even more stable fluoroelastomer, a copolymer of VDF and HFP, that was described in 1956 and commercialized by DuPont in 1958 as Viton A [116–118]. In 1960, DuPont introduced a TFE-containing terpolymer [25190-89-0] (Viton B) that was even more thermally stable and solvent resistant than Viton A [119, 120]. 3M, who had acquired the fluorocarbon polymer assets of M. W. KELLOGG, introduced a VDF–HFP copolymer named Fluorel in 1958 and continued its manufacture after 1962 under license from DuPont. These products are now sold as Dyneon.

Table 23. Service temperature, fluorine content, and hydrogen content of fluoroelastomers

Type	Service temperature*, °C	Fluorine content, %	Hydrogen content, %
VDF–HFP	−18–210	66	1.9
VDF–HFP–TFE	−12–230	66–69.5	1.1–1.9
VDF–PMVE–TFE	−27–230	64–66.5	1.1–1.7
TFE-P	5–200	54	4.3
Perfluorocarbon	0–260	73	0

*Temperature of elastomeric side of glass transition to maximum temperature for continuous service.

Elastomeric copolymers of VDF and 1-hydropentafluoropropene [32552-63-9] and a TFE-containing terpolymer [29830-35-1] were introduced by Montecatini–Edison during the 1960s [121–123]. However, these polymers, sold as Tecnoflon, lacked the stability of similar HFP compositions, and were replaced by HFP analogs as early patents for poly(VDF–HFP) and poly(VDF–HFP–TFE) expired. In the meanwhile, these products are produced by Solvay-Solexis. Daikin began production of VDF-based fluoroelastomers (Dai-el) under license from DuPont in 1970.

In Russia produced fluoroelastomers are commercialized as Levatherm F by Lanxess. Four Chinese manufacturers (Meilan Shanghai, 3F, Juhua, Chenguang) are also on the market. Trade designations and manufacturers of commercial VDF-based fluoroelastomers are shown in Table 24.

3.2.1. Production

1,1-Difluoroethylene-based elastomers are usually prepared by continuous or semibatch polymerization in aqueous emulsion with free-radical initiators. The monomers are derived from fluorocarbon refrigerants and their intermediates. Since the Montreal Protocol, the raw material base is smaller, due to the phase-out of ozone depleting chlorofluorocarbon compounds.

Fluoroelastomers are produced in an aqueous emulsion polymerization, resulting in polymer latex with particles sizes in the range of 100–1 000 nm. Typical polymerization conditions are at $1 \times 10^6 - 3 \times 10^6$ Pa and 50–120°C for semibatch and $2 \times 10^6 - 7 \times 10^6$ Pa and 60–130°C. Commonly used initiators are inorganic peroxides, such as ammonium persulfate. However, organic peroxides and redox systems are also used, especially at low temperatures where the thermal decomposition of persulfate is slow. Emulsifying agents must be compatible with the highly reactive fluorocarbon radicals of growing polymer chains; salts of fluorocarbon acids are used to prevent chain transfer. Molar mass is regulated by controlling the rate of initiation, rate of polymerization, and termination by chain-transfer agents. Such agents generally have reactive carbon–hydrogen bonds; diethyl malonate [105-53-3] and dimethyl ether [115-10-6] are typically used. Buffers control the pH of the polymerization and stabilize the emulsification system. Molar mass (M_w) ranges from 10^5 to 10^6 g/mol.

In a new approach [124, 125] perfluorinated diiodo compounds are used to increase the iodine content and the architecture of the polymer. In a pseudo-living polymerization, long branches of the polymer can be controlled by using bifunctional molecules that can be very effectively cured by peroxides. The use of perfluoropolyether microemulsions yields high molar mass polymers at good reaction rates.

In semibatch polymerizations, monomers can be added continuously under pressure at rates that correspond to their consumption, providing better control of the copolymer composition.

In a continuous process, water, monomers, initiator, and other components are fed to the reactor while the polymer latex is removed at a corresponding rate. Composition and molar mass can be precisely controlled. Monomers and other reactants of widely ranging reactivities can be used without undue variation in the polymer structure. Unreacted monomers are evaporated from the latex and recycled.

In either process, the latex from the reactor is coagulated by the addition of salts or acids. The polymer crumb is filtered or centrifuged and washed to remove coagulant and water-soluble residues. After being dried, it may be mixed with curing (cross-linking) agents. The finished

Table 24. Commercial VDF-based elastomers

Copolymer	Trade name	Manufacturer
VDF–HFP	Viton	DuPont
	Dyneon	3M
	Tecnoflon	Solvay-Solexis
	Dai-el	Daikin
	Levatherm F	Lanxess
VDF–HFP–TFE	Viton	DuPont
	Dyneon	3M
	Tecnoflon	Solvay-Solexis
	Dai-el	Daikin
VDF–PMVE–TFE	Viton	DuPont
	Dyneon	3M
	Tecnoflon	Solvay-Solexis
	Dai-el	Daikin

Table 25. Physical properties of VDF-based fluoroelastomers

Property	Range	ASTM
Specific gravity, g/cm^3 of raw polymer	1.77–1.91	
Tensile properties		
100% modulus, MPa	2–15	
Tensile strength, MPa	5–20	
Elongation at break, %	50–500	
Hardness, durometer A	60–95	
Compression set*, %		D-395
70 h at 25 °C	10–15	
70 h at 200 °C	10–40	
1 000 h at 200 °C	50–70	
Low-temperature properties		
Clash–Berg torsional-stiffness temperature (69 MPa), °C	−31 to −5	
Retraction temperature (TR-10), °C	−31 to −6	D-1329
Brittle point, °C	−51 to −34	D-746

* Method B, 24.5 × 3.5 mm O-rings.

product is packaged as extruded pellets and lumps or in milled sheets.

3.2.2. Properties

Heat Resistance. 1,1-Difluoroethylene-based copolymers are the largest group of fluoroelastomers in terms of production volume and variety. Their properties are summarized in Table 25. 1,1-Difluoroethylene-based elastomers can be used continuously at temperatures as high as 230°C, depending on polymer type and formulation. Copolymers retain over 50% of their tensile strength for more than one year at 200°C or two months at 260°C. Functionalized hydrocarbon elastomers that are considered resistant to oil, such as nitrile or acrylate rubber, lose 50% of their elongation aging at 150°C for 10 and 25 days, respectively. In comparison, VDF-based fluoroelastomers retain over 95% of their elongation after one year at 200°C [126].

Only VDF–HFP at 60 : 40 wt% is of commercial importance. A wide variety of viscosities and formulations of this composition is sold for different applications. Most of these products are cured with bisphenol.

Fluid and Chemical Resistance. For improved fluid resistance terpolymers with TFE are used. They show higher stabilities with a 30% VDF content without affecting low temperature flexibility. Higher fluorine content increases the chemical resistance.

The resistance of VDF–HFP–TFE elastomers to petroleum-based fuels and oils is excellent. Certain lubricants, however, contain additives that disperse sludge and inhibit metal corrosion but are highly basic or decompose to basic products that attack the elastomer. Therefore, the performance of the fluoroelastomer must be evaluated in contact with the particular fluid containing proprietary additives. Formulators of oil additives should examine not only lubricant stabilization and metal protection [127–129] but also the behavior of seals. Fluoroelastomers to be used with hydrocarbon–alcohol fuels (gasohol) must be chosen with care [130]. A VDF–HFP–TFE polymer of 69.5% fluorine content is used in methanol blends to prevent excessive swelling. This polymer is also recommended for certain phosphate ester-based hydraulic fluids. Fluid resistances guides are available [131–133].

In contact with aromatic solvents, polymers with 68% or 69.5% fluorine content give the best performance. Alcohols other than methanol do not greatly swell VDF-based elastomers despite their polar nature, but blends of alcohols and other fluids may act differently. Polar solvents that are only hydrogen bond acceptors, tend to have a high affinity for these elastomers and may dissolve raw polymer or destructively swell vulcanizates. Examples of such solvents are low-boiling ketones and esters (acetone, methyl ethyl ketone, ethyl acetate), amides (N,N-dimethylformamide, N,N-dimethylacetamide), and ethers (tetrahydrofuran, 2-methoxyethyl ether). Mixtures that contain substantial amounts of these solvents cannot be used with this group of fluoroelastomers.

In addition to the destructive action of highly swelling solvents, the vulcanizate may be attacked chemically. Dehydrofluorination by primary and secondary aliphatic amines occurs readily and is the basis of amine cross-linking. The resulting unsaturated bonds are susceptible to nucleophilic attack, which can lead to cross-linking and embrittlement. With amine-based curing systems, which are seldom used, dehydrofluorination is a controlled process; however, uncontrolled exposure to soluble aliphatic amines and other soluble strong organic bases degrades these vulcanizates and must be

Table 26. Fluid and chemical resistance of VDF-based fluoroelastomers

Outstanding resistance	Good to excellent resistance	Poor resistance
Hydrocarbon solvents	Low-polarity solvents	Strong alkaline solutions
Automotive fuels (free of methanol)	Oxidative environments	Ammonia and amines
Engine oils (free of amine additives)	Aqueous acids	Polar solvents (ketones, methanol)
Nonpolar chlorinated solvents	Dilute aqueous alkaline solutions	Organic acids
Oils and greases	Aromatic solvents	
Halogen oxidizers e.g., chlorine	Water and salt solutions	

avoided. Resistance to aqueous alkali is far greater, but causes surface cracks and failure. Phase-transfer catalysts, such as quaternary ammonium ions accelerate degradation by aqueous alkali. Improvement in base resistance can be achieved by using PMVE instead of HFP. Although these polymers are also attacked by amines, the resulting double bonds are not reactive for further cross-linking. Thus, embrittlement does not occur and service life is significantly extended. Resistance to aqueous acids is generally good, especially if acid acceptors, such as calcium hydroxide, magnesium oxide, or calcium oxide are not present in large amounts. Table 26 gives an overview on fluid resistances of FKMs.

Other Types of Resistance. Ozone resistance and weatherability are excellent [134] and permeability to gases is low. Vulcanizates can be formulated to be self-extinguishing or nonflammable. Abrasion resistance varies. Resistance to β-radiation is moderate and independent of VDF copolymer composition. Radiation doses are cumulative and a total exposure to 10^4–10^5 J/kg produces moderate to severe effects on physical properties. Radiation failure occurs by embrittlement rather than chain cleavage [135].

3.2.3. Processing

Formulation. Elastomer processing generally includes formulation. Optimum conditions must also be selected to provide the properties desired after vulcanization (cross-linking). The composition (in parts per hundred of rubber, phr) of a typical fluorocarbon elastomer is:

Ingredients	Composition, phr
Fluorocarbon elastomer	100
Acid acceptor (inorganic base)	3–15
Filler	0–50
Cross-linker	1–3
Accelerator	0.2–1
Activator	4–6
Processing aids	0–3

Typical acid acceptors for high steam and acid resistance are MgO, ZnO, or PbO. Carbon black, silica, or barium sulfate are often used as filler, and carnauba wax as processing aid acts as a flow lubricant or mold release aid. $Ca(OH)_2$ is used as accelerator, which activates dehydrofluorination. The fluorocarbon elastomer is chosen according to the intended use of the end product.

In O-ring formulations (Table 27), compression-set resistance is of critical importance because sealing can be maintained only if the O-ring can retain some tendency to regain its uncompressed dimensions. Higher molar mass and thus higher viscosity usually display better

Table 27. Fluoroelastomer molding and O-ring compounds

	General-purpose molding	O-ring
material		
Formulation		
Poly(VDF–HFP) [a]	100	
Poly(VDF–HFP) [b]		100
Benzyltriphenylphosphonium chloride	0.25	0.60
Hexafluoroisopropylidenediphenol	1.0	2.0
Calcium hydroxide	6.0	6.0
Magnesium oxide [c]	3.0	3.0
MT carbon black (N-908)	30	30
Compound viscosity [d]	27	35
Vulcanizate properties [e]		
Tensile, 25 °C		
100% Modulus, MPa	2.2	5.3
Tensile strength, MPa	11.2	12.4
Elongation at break, %	360	190
Compression set [f], %	47	16
Hardness, durometer A	66	72

[a] Viton E-45 (DuPont), 100 °C Mooney viscosity, ML-10 = 45.
[b] Viton E-60 (DuPont), 100 °C Mooney viscosity, ML-10 = 100.
[c] Maglite D (Merck & Co).
[d] Mooney scorch minimum viscosity, MS at 121 °C for 30 min.
[e] Press-cured 10 min at 177 °C, postcured 24 h at 232 °C.
[f] ASTM D-395, Method B (O-rings), 70 h at 200 °C.

compression-set resistance. The choice of curing systems is also important because the stability of the cross-links limits the stability of the elastomer network. Bisphenol curing system offers greater compression-set resistance than peroxide curing system and is generally preferred for O-ring and similar sealing applications. However, peroxide-cured stocks usually have better resistance to steam and acids and are often specified in these applications [136].

In formulations for molded articles (Table 27), compression-set resistance is usually sacrificed in favor of elongation. A lower cure state and greater elongation are achieved by using less cross-linking agent or a lower molar mass polymer; this reduces viscosity and increases mold flow. A minimum elongation is required, because many mold designs require substantial elongation of the hot vulcanizate to facilitate removal from the mold. If these processing requirements for elongation have been satisfied, molded articles are not usually used in tension but rather under slight compression although some elongation may occur during installation.

1,1-Difluoroethylene-based elastomers are available in a range of Mooney viscosity values of 0–160 (ML-10, 121°C) (Mooney viscosity is a measure of polymer plasticity), that correspond to molar masses of 1×10^4–1.5×10^5 g/mol. The choice of fillers and acid acceptor in fluoroelastomers has a major effect on viscosity and, therefore, influences processability and properties after vulcanization [137, 138].

Specialized polymers can be compounded for exceptional resistance to fluids and low-temperature flexibility (Table 28). The PMVE polymer exhibits the best low-temperature properties of any commercial fluorocarbon elastomer. Excellent resistance to methanol and methanol–fuel blends is characteristic of compounds based on a polymer containing 69.5% fluorine [138]. Higher fluorine containing elastomers based on TFE and special vinyl ethers guarantee low solvent swelling combined with lowest T_g (−40°C) [141, 142].

Mixing. Standard rubber compounding equipment and techniques are used to mix VDF-based fluoroelastomer stocks. Rubber mills are suitable for most viscosity grades, although some reduction in molar mass may occur when high-viscosity polymers are cold milled, and the clearance between rolls is small. Banbury and other internal mixers are also suited for preparing precompounds that contain fillers, inorganic acid acceptors, and processing aids. Detailed mixing procedures are given in a processing guide for fluoroelastomers [143]. Producers commonly supply precompounds that are fully compounded by a custom-mixing firm or by the molder.

Molding. Compression molding is the oldest and simplest way of vulcanization. In this process, a piece of uncured rubber is placed in the mold cavity and cured under pressure. Compression molding has several advantages: Loss of expensive material is minimized, flexibility for small series, and low equipment cost. Disadvantages are the high labor cost and variability of quality. Usually products with high molar mass are processed by compression molding, which give parts with excellent mechanical properties and environmental resistance [144].

In transfer molding, lower viscosities are required, because the compound needs to flow easily. Bisphenol-curable compounds are suitable for transfer molding because premature vulcanization during processing is low.

Injection molding requires a higher capital investment but is less labor intensive. Its use has increased and is justified in large-scale production where labor savings offset equipment costs and increased waste from mold filling channels. Injection molding provides the maximum product consistency and shortest cycle times compared to other molding processes. Compound formulation is critical for injection molding, because high flow rates and compound safety are required. Injection molding of thermoset elastomers is much more demanding than of thermoplastics or thermoplastic elastomers, because the former case entails irreversible chemical reactions that increase viscosity and creates cross-links during hot flow; in the latter case, the polymer flows easily in the melt and is simply cooled or crystallized in the mold.

Molding cycles are usually 1–5 min at 160–200°C and depend on the article size

Table 28. Specialty compounds for increased fluid resistance and low-temperature flexibility

	Terpolymers			Standard copolymer
	Methanol resistant	Standard	Low-temperature resistant	
Formulation				
Poly(VDF–HFP–TFE)[a], 69.5% F	100			
Poly(VDF–HFP–TFE)[b], 68.5% F		100		
Poly(VDF–PMVE–TFE)[c]			100	
Poly(VDF–HFP)[d], 65% F				100
Triallylisocyanurate	3		4	
Organic peroxide[e]	3		4	
Sublimed litharge	3			
MT black (N-908)	30	30	30	30
Calcium hydroxide		3	4	6
Magnesium oxide[f]		3		3
Benzyltriphenylphosphonium chloride		1.0		0.6
Hexafluoroisopropylidenediphenol		2.0		2.0
Compound viscosity[g]	40	72	32	35
Vulcanizate properties [h]				
100% modulus at 25 °C, MPa	5.4	7.0	5.3	5.3
Tensile strength, MPa				
25 °C	17.6	15.0	13.8	12.4
−30 °C			37.4	45.5
−40 °C			39.9	28.4
Elongation at break, %				
25 °C	210	210	200	190
−30 °C			160	60
−40 °C			60	20
Brittle point (1.91 mm thick)[i], °C	−45	−42	−51	−34
Clash–Berg torsional-stiffness temperature (69 MPa), °C	−5	−12	−31	−17
Compression set[j], %	38	30	30	16
Hardness, durometer A	76	80	67	72
Volume swell, 24 °C for 7 d, %				
Toluene	4	13		
Reference fuel C[k]	3	6		
Methanol	3	22		
Reference fuel C/methanol (85/15)	9	20		
Phosphate ester hydraulic fluid[l], 121 °C	45	127		
Reference fuel C/ethanol (85/15), 100 °C	18	24		
Permeability to automotive fuel[m], g				
Reference fuel C	1	3		6
Reference fuel C/ethanol (85/15)	11	19		28
Reference fuel C/methanol (85/15)	17	57		125

[a]Viton GF, contains peroxide-sensitive cure site (DuPont).
[b]Viton B (Du Pont).
[c]Viton GLT, contains peroxide-sensitive cure site (DuPont).
[d]Viton E-60 (DuPont).
[e]Luperco 101-XL (Pennwalt Corp.).
[f]Maglite D (Merck & Co.).
[g]Mooney scorch minimum viscosity, MS at 121 °C for 30 min.
[h]Press-cured 10 min at 177 °C, postcured 24 h at 232 °C.
[i]ASTM D-2137.
[j]ASTM D-395, Method B (O-rings), 70 h at 200 °C.
[k]50% toluene, 50% isooctane.
[l]Skydrol 500-B (Monsanto Co.).
[m]ASTM E-96, 0.45 mm membrane, 24 °C.

due to heat transfer limitations. In injection molding, however, frictional or shear heating, produced by high flow rates through small mold orifices, raises the temperature in the mold significantly. In large moldings produced by compression or injection molding, the heat of reaction of vulcanization must also be considered.

Extrusion. Fluorocarbon elastomers can be shaped by extrusion. Elastomer fed to the extruder is moved through the barrel by a screw to a die to get the desired extrudate cross-section. Because elastomers are essentially amorphous, viscous liquids, melting is not required. Extrusion of compounds must be carried out at temperatures below 120°C to avoid premature curing. The continuous extrudate, which may be hose or tubing, rod stock for spliced O-rings, or profile stock for cut gaskets, is rolled up or otherwise placed in a steam autoclave, where curing takes place at ca. 160°C. Extruded stock is also used to cut preforms for subsequent curing by compression or transfer molding.

Vulcanization Chemistry. Bisphenol curing (Fig. 3) is the most widely used curing system for VDF–HFP and VDF–HFP–TFE fluoroelastomers. It has the advantage of excellent processing safety, fast cures, excellent final properties, and high temperature compression set resistance in seals. Preferred cross-linker is bisphenol AF.

The chemical reactions that result in formation of networks in bisphenol-cured fluoroelastomers involve dehydrofluorination of HFP–VDF–HFP or HFP–VDF–TFE sites forming $-C(CF_3)=CH-$ double bonds. Fluoride ions catalyze a shift of the double bond to form $-CH=CF-$. The resulting double bonds are susceptible to nucleophilic attack because of their electron-deficient character. Addition of a nucleophile and subsequent elimination of a fluoride leads to a new double bond, which in turn is attacked by an intermediate phenolate accelerated by quaternary phosphonium or ammonium phenolate [136, 145, 146]. Because fluoride ions are formed in the curing reaction, a fluoride scavenger is needed.

Peroxide-initiated cross-linking of bromine or iodine containing fluoroelastomers in the presence of radical traps, such as triallyl isocyanurate (TAIC) [1025-15-6] starts by the attack of a radical at the allyl functions of TAIC. The propagating TAIC radicals abstract bromine or iodine atoms from the fluoroelastomer, and the resulting fluorocarbon radicals attack intact TAIC double bonds. The driving force for this reaction is the addition of an electron-deficient free radical to an electron-rich double bond. Propagation reactions may involve redistribution of the free-radical intermediates from fluoroelastomer radical addition to a double bond of the partially polymerized radical trap [136, 147] (Fig. 4).

Diamine cure is nowadays restricted to applications where FDA compliance is needed.

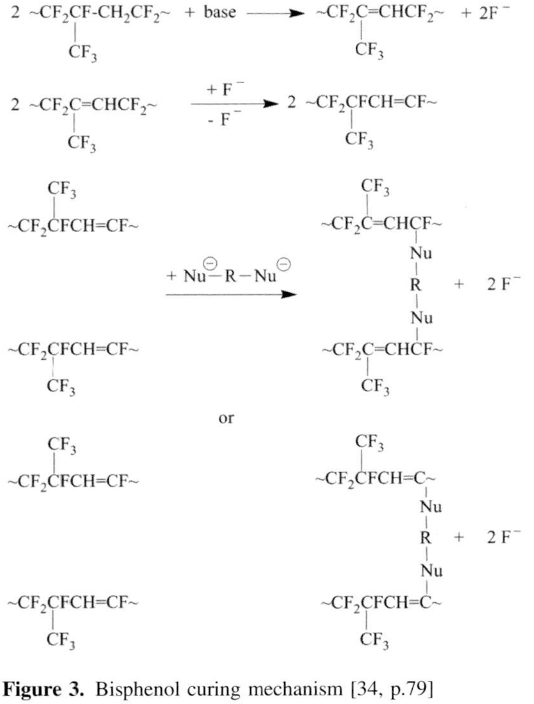

Figure 3. Bisphenol curing mechanism [34, p.79]

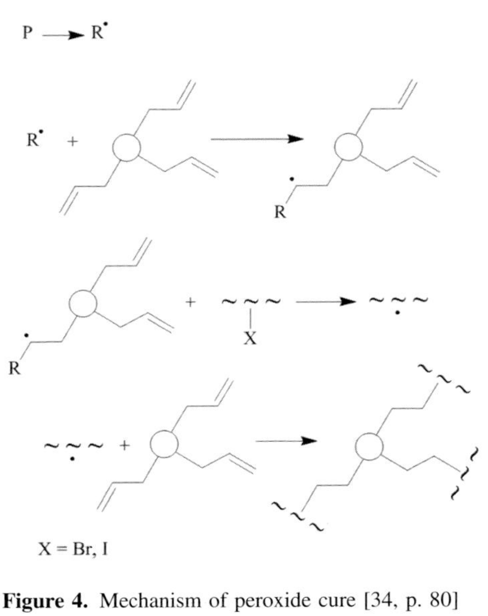

Figure 4. Mechanism of peroxide cure [34, p. 80]

Postcuring. The curing chemistry has to be designed such that during the processing of the elastomer, at temperatures of 120–130°C, no change in the viscosity occurs, but during the curing step at 170–180°C a high cure state is reached within 5 min. To develop optimum physical properties, such as compression-set resistance and tensile strength, fluorocarbon elastomers are usually postcured [148]. The postcure cycle subjects a press-cured article to circulating air at atmospheric pressure at 200–260°C for 12–24 h. Small amounts of volatile substances are removed, and slight shrinkage occurs. Typical bisphenol-curable stocks undergo a 0.3% mass loss on press curing and a further 1.2% loss on postcuring [149, 150]. Postcuring is believed to complete the vulcanization and deactivate remaining intermediates. The conditions for a postcure cycle are established arbitrarily, because each elastomer-curing system has a point at which heating a vulcanizate in air gives optimum properties, beyond which further heating or heating at a higher temperature affects physical properties undesirably with the onset of heat aging.

3.2.4. Uses

1,1-Difluoroethylene-based elastomers are used widely in vehicle construction because of their excellent resistance to heat and oil. Important fluoroelastomer parts are engine oil seals of both the rotating and reciprocating types, drive-train seals, and fuel-handling components, such as gaskets, O-rings, valve seals, and hoses. Other applications are hydraulic and pneumatic seals, hoses and seals for chemical industry, flue-duct expansion joints for power plants, pollution control equipment, and fuser rolls for copying machines. The aircraft industry uses more fluorosilicone than fluorocarbon elastomers because of the extreme low-temperature requirements of high-altitude flight [151].

3.3. Elastomers Based on Tetrafluoroethylene—Perfluoro(Methyl Vinyl Ether) Copolymers (FFKM)

Despite the vast improvement that VDF-based elastomers offer over conventional hydrocarbon elastomers in thermooxidative stability and oil resistance, even more stable elastomers are required for critical industrial and military applications. In response to this need, perfluorinated materials were developed at DuPont in the 1960s. The goal was an elastomer with the inertness and solubility characteristics of PTFE or FEP.

To create such a stable elastomer, a comonomer was required that interrupted TFE crystallinity and imparted good low-temperature properties. At this time, the epoxidation of HFP was developed and the resulting hexafluoropropene oxide (2,2,3-trifluoro-3-(trifluoromethyl) oxirane) [428-59-1] used to produce perfluoro (methyl vinyl ether) (PMVE) (→ Fluorine Compounds, Organic, Section 6.2). Perfluorovinyl Ethers).

Copolymerization of PMVE with TFE results in an elastomer [26425-79-6] whose thermal, oxidative, and chemical stability before cross-linking approach that of perfluorocarbon plastics [152]. However, the inertness of this base elastomer and the absence of cure sites prevented cross-linking, and most of the effort to develop a practical elastomer went into cure site monomers [153–158]. Several types of curing systems were developed, and articles made from them were commercialized by DuPont in the mid-1970s as Kalrez Perfluorocarbon Elastomer Part. Due to the great difficulties in processing and curing, only finished Kalrez parts are sold. Raw perfluoroelastomer gum, based on triazine or peroxide curing, is sold by Daikin, 3M, and Solvay Solexis [159].

3.3.1. Production

Copolymerization of TFE–PMVE and cure-site monomers is conducted in an aqueous emulsion like that used with VDF–HFP-based systems (see Section 3.2.1). Polymerization rates in continuous reactors are about one-tenth of those of VDF–HFP-based systems at temperatures that give useful molar masses. An important difference between the polymerization of VDF–HFP and TFE–PMVE is that the latter requires a surfactant. Hydrocarbon-based surfactants cannot be used because of chain-transfer reactions with the highly reactive perfluorocarbon chain radicals; salts of C_8–C_{12} perfluorocarbon acids are satisfactory. For the different curing systems (peroxide, bisphenol, or nitrile), different cure site monomers have to be incorporated (Table 29).

Table 29. Heat aging* and fluid swell of perfluoro-vulcanizates [163]

Cure site	-R_fCN	-H	-R_fBr
Curative	Ph_4Sn	bisphenol	TAIC
Probable cure site monomer	8-CNVE	VDF	BTFB
Heat aging, %			
225 °C		−3	−11
250 °C		−24	−54
275 °C		−45	−80
290 °C	24	−36	
Fluid resistance (swell after 70 h exposure), %			
Concentrated nitric acid, 85 °C	3	8	9
Glacial acetic acid, 100 °C	3	32	16
Butyraldehyde, 70 °C	13	10	14
Methyl ethyl ketone, 70 °C	6	4	4
Toluene, 100 °C	7	7	7
Water, 225 °C	10	60	7

*Change in T_B (tensile strength at break) after aging ten days in air.

The polymer latex is coagulated by addition of polyvalent metal salt solutions. The polymer crumb is filtered or centrifuged, washed to remove coagulant, initiator, and surfactant, and dried [156].

3.3.2. Properties

Heat Resistance. The mass loss of a TFE–PMVE copolymer in air is 6.5% at 316°C over seven days; an uncured terpolymer that contains the proprietary perfluoro(phenoxy ether)-based cure-site monomer loses 12% under the same conditions [156]. The higher loss of a perfluoroelastomer is because the cure site has a more complex structure than the backbone and is more prone to degradation under these extreme conditions. Degradation of the cure site leads to some local degradation of the main chain.

Cured perfluoroelastomer specimens are also exceptionally stable and are differentiated by the stability of their cross-links. A perfluoroelastomer with a perfluorocarbon nitrile group as cure site retains physical properties best under extreme heat conditions. The triazine cross-links are formed by trimerization of the nitrile functions, catalyzed by tetraphenyltin and are sufficiently stable to prevent any change from the original postcured properties with respect to 100% modulus, tensile strength, and elongation at break in air at 288°C over 18 days. Even at 316°C, useful properties are maintained: 100% modulus is reduced by less than 20%, tensile strength by less than 40%, and elongation at break is unchanged. Other cross-link reactions, avoiding tetraphenyltin and resulting in comparable stabilities, have been developed by 3M [160, 161].

Although perfluoroelastomers cross-linked by perfluorophenoxy groups are less stable to heat, they are still far more stable than hydrofluoroelastomers: After 20 days at 288°C in air, 100% modulus is reduced by less than 25% and tensile strength by less than 15%; elongation at break is increased by 120%. Table 29 gives some properties of FFKM for heat aging and fluid swell.

Fluid and Chemical Resistance. The chemical resistance of perfluoroelastomers depends on the cure system [162–164]. Uncured TFE–PMVE-based perfluoroelastomers are insoluble in most solvents except certain perfluorocarbons and chlorofluorocarbons.

Typical physical properties of a K_2–AF-cured vulcanizate loaded with carbon black follow [157]:

100% modulus, MPa	9
Tensile strength, MPa	16
Elongation at break, %	160
Hardness, Shore A	75
Compression set B (O-rings), %	
70 h at 121°C	23
70 h at 288°C	45
Clash–Berg torsional-stiffness temperature (66.9 MPa), °C	−2
Brittle point, °C	−38
Retraction temperature, °C	
TR-10	−1
TR-50	+8
Specific gravity, g/cm^3	2.05

3.3.3. Processing and Uses

Perfluoroelastomers based on TFE–PMVE are difficult to process because of the stiffness of the fully fluorinated chains. Because the polymer is also difficult to cross-link and expensive to produce, these materials are available as finished articles (Kalrez Perfluorocarbon Elastomer Parts). A substantial market was found for O-rings, flat and lathe-cut gaskets, rotating and reciprocating shaft seals, and diaphragms. Other products include coated fabrics, hose linings, tubing, industrial gloves, and extruded articles, such as autoclave and oven seals. The high cost of these products is justified in applications where no other material is equally resistant to oxidative, thermal, fluid, and chemical exposure. They are used in the manufacturing, aerospace, and automobile industries.

3.4. Elastomers Based on Tetrafluoroethylene–Propene Copolymers

In the 1960s, DuPont [165] and Asahi Glass [166–168] established independently that TFE and propene (P) undergo a nearly alternating polymerization. The resultant TFE–P fluoroelastomer [27029-05-6] exhibits very good thermal and chemical resistance, but has a high glass transition temperature ($-2°C$) despite its low fluorine content (54%). Although resistance to swelling by aromatic hydrocarbons is inadequate, these elastomers offer excellent resistance to certain polar solvents. An unusual property of TFE–P fluoroelastomers is their exceptional resistance to dehydrofluorination by organic bases.

Due to the high stability against dehydrofluorination, high temperatures (300–360°C, 2–4 h) are needed to introduce double bonds, which are accessible for peroxide cure [169]. Alternatively, a specific cure site is used [170, 171].

Asahi Glass introduced TFE–P polymers in the mid-1970s as Aflas. Compared to VDF–HFP–TFE fluoroelastomers, TFE–P vulcanizates have poorer low-temperature characteristics and resistance to compression set. Because of the excellent electrical properties, TFE–P has been used for wire and cable coating applications.

To make TFE–P polymers susceptible to bisphenol curing TFE–P–VDF polymers have been developed [172–175]. These polymers show significantly better resistance to lubricating oil than VDF–HFP containing polymers and show better processing and curing characteristics than TFE–P copolymers, but base resistance is significantly compromised by the presence of a large fraction of VDF units. In the late 1990s, workers at DuPont Dow Elastomers reinvestigated the TFE–P elastomer family to develop fully base resistant products with better processing and curing characteristics. Certain cure-site monomers, such as trifluoropropene incorporated into TFE–P makes the elastomer susceptible to bisphenol curing [176]. These terpolymers have better base resistance than VDF-containing TFE–P elastomers and have better hydrocarbon fluid resistance than TFE–P copolymers [177].

3.4.1. Production

Tetrafluoroethylene–propene elastomers are prepared in aqueous emulsion with a perfluorocarbon surfactant. Batch copolymerization at 70–80°C, using a conventional ammonium persulfate–sodium sulfite redox catalyst, produces elastomers of relatively low molar mass ($\overline{M}_w = 70\,000$ g/mol) due to chain transfer. To minimize transfer reactions, more active redox initiators were developed [139]. With these catalysts, combinations of ammonium persulfate, ferrous sulfate, ethylenediaminetetraacetic acid, and sodium hydroxymethanesulfinate, efficient batch polymerization to a $\overline{M}_w$ of 150 000 g/mol can be carried out at 25–35°C.

The polymer latex is coagulated by addition of solutions of polyvalent salts, and the coagulated crumb is washed and dried.

3.4.2. Properties

Typical properties of a TFE–P elastomer vulcanized by a peroxide–radical trap curing system (press-cured at 160°C for 30 min, postcured at 200°C for 2 h) [168] are given below:

Raw polymer	
Specific gravity, g/cm³	1.55
Mooney viscosity (ML-10, 100°C)	85
Appearance	dark brown
Formulation, phr	
Poly(TFE–P)	100
α,α′-Bis(*t*-butylperoxy)diisopropylbenzene	2
Triallylisocyanurate	3
MT carbon black (N-908)	30
Vulcanizate properties	
Specific gravity, g/cm³	1.60
Tensile properties at 25°C	
100% modulus, MPa	3
Tensile strength, MPa	20
Elongation at break, %	300
Compression set at 200°C, %	
1 d	40
30 d	65
Hardness, Durometer A	72
Brittle point, °C	−40
Retraction temperature (TR-10), °C	3
Volume increase after immersion, %	
95% H_2SO_4, 100°C, 3 d	4.4
Water, 150°C, 3 d	8.7
Steam, 160°C, 3 d	4.6
Fuel oil B, 25°C, 7 d	59
Benzene, 25°C, 7 d	40
Methanol, 25°C, 7 d	0.2
Ethyl acetate, 25°C, 7 d	88

3.4.3. Processing

The isolated tetrafluoroethylene–propene copolymer is subjected to a heat treatment to generate enough unsaturation to allow peroxide curing [176]. Dialkyl peroxides are used together with radical traps, such as polyallyl or polyvinyl compounds or *N,N'-m*-phenenedimaleimide for curing. An inorganic acid acceptor, such as calcium hydroxide or magnesium oxide is also present. Fillers, such as silicas can be used, but carbon blacks are usually chosen for reinforcement [168].

VDF-containing TFE–P elastomers show a similar curing behavior with bisphenol as the standard VDF–HFP polymers [178].

Tetrafluoroethylene–propene elastomers process less smoothly on conventional rubber processing equipment than VDF–HFP-based fluoroelastomers. Processing aids, such as carnauba wax or sodium stearate are widely employed. Similarly, longer press-cure cycles are used to develop optimum properties: Press-cured vulcanizates are usually postcured at 200°C for 16 h [179, 180].

3.4.4. Uses

The TFE–P fluoroelastomers are used where high temperatures and destructive media are encountered. Applications are similar to those described for VDF–HFP-based polymers (Section 3.2.4), but the choice may depend on specific requirements. Thus, TFE–P is recommended for exposure to strong organic bases and certain highly polar solvents, but not for maximum oil resistance and low-temperature properties.

3.5. Fluoroelastomers containing Ethylene

3.5.1. TFE–E–PMVE Polymers

In 1986, DuPont introduced fluoroelastomers based on TFE, E, and PMVE. Peroxide curing is possible by using bromine cure sites [181]. They can be used in severe environments, such as oilfield applications, where the seals are exposed to high temperatures, H_2S, and amine based corrosion inhibitors. They show better low temperature properties than TFE–PMVE and TFE–P elastomers and have better fluid resistance against hydrocarbons than the latter.

3.5.2. TFE–VDF–HFP–E Polymers

Solvay-Solexis developed a polymer based on TFE–VDF–HFP–E [182]. The polymer contains about 5% ethylene, which shields the vinylidene fluoride hydrogens, making the polymer less susceptible to base attack. Peroxide cure is possible due to bromine containing cure site monomers. Polymerization rate can be increased by a special microemulsion polymerization technique [183].

3.6. Thermoplastic Elastomers (→Thermoplastic Elastomers)

Thermoplastic elastomers combine elastomeric properties with standard thermoplastic processing. They are either polymer blends of a thermoplastic polymer and an elastomer or A-B-A block copolymers of hard (A) and soft (B) (elastomeric) segments. During processing, the polymer is completely molten, when cooled, the hard segments cocrystallize, thus forming physical cross-links between the soft segments. This gives the polymer elastomeric properties [34].

Daikin offers fluorinated TPEs commercially as Dai-el. To tailor block copolymers the "living" polymerization technology is used. In a first step, the elastomeric block is polymerized, consisting of VDF–HFP–TFE that has two iodine end-groups. In a second step, a thermoplastic polymer consisting of TFE–E–HFP is polymerized starting from the iodine end-groups to produce the A–B–A structure. The melting point is about 230°C; dimensional stability is good up to 120°C. The high temperature properties can be improved by either chemical or radiation curing. Typical applications include tubing, sheet, O-rings, and molded parts.

3.7. Vinyl Ether Copolymers

A new class of fluorinated coatings was developed by Asahi Glass Co. Ltd. under the trade name Lumiflon [184, 185].

The base polymer is a fluoroethylene-vinyl ether (FE–VE) copolymer, which comprises an alternating sequence of a fluorinated ethylene, normally CTFE, and different vinyl ether units.

FE–VE polymers are the first solvent-soluble fluoropolymers for coatings, which can be cured at room temperature conditions. The polymers are completely amorphous. The polymers combine various characteristics, such as solubility in organic solvents, film-forming ability, and transparency of the resulting film, which make them good base polymers for paints. FE–VE based coatings maintain excellent appearances (gloss, color) of buildings and bridges, where on-site coating is required, for more than twenty years. The coatings protect steel or concrete from sunshine, wind, rain, and corrosion, thereby reducing the total maintenance cost, such as repainting or cleaning.

In addition, FE–VE polymers that can be applied as emulsion or powder have been developed [186].

The molecular design of FE–VE gives some possibilities to tailor the resin applicable to different coating purposes (Fig. 5).

(R_1 - R_4 : alkyl, cycloalkyl, alkylene, etc.)

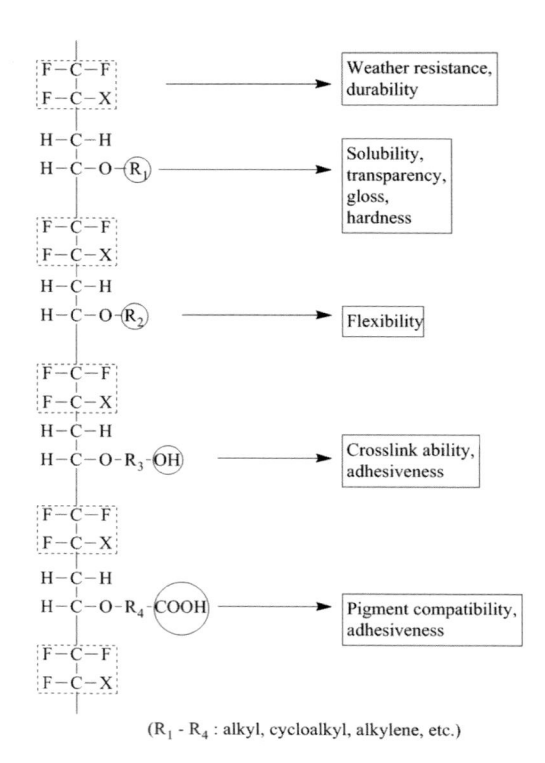

Figure 5. Molecular design of FE-VE polymers [34, p. 560]

CTFE as the fluorinated olefin improves the solubility of the polymer compared to TFE. The FE and VE components are highly alternating, which protects the rather unstable VE units in the backbone, adding excellent weather resistance to the resin. Several different VE compounds are used in the polymerization. Alkyl, cycloalkyl, or alkylene substituted VE are responsible for the solubility and the glass transition temperature. Introducing hydroxyl groups in the side chain enables the molecule to be cured with isocyanate to urethane polymers. Compatibility with pigments is achieved by partial carboxylation of hydroxyl groups. Copolymers with a higher COOH content are soluble in an aqueous medium.

The great advantage of FE–VE paints is their ability to cure at room temperature. PVDF, which is also used for protective coating, has to be baked at > 250°C, to form a continuous film. Due to the fluorinated backbone, FE–VE based paints show a much higher weathering stability, compared to fluorinated acrylic based coatings.

Some fundamental properties of a FE–VE polymer [34, 187] are:

Fluorine content, wt%	25–30
OH value, mg KOH/g	47–52
COOH value, mg KOH/g	0–5
Number-average molar mass $\overline{M}_n$, g/mol	0.8–6×10^4
Weight-average molar mass $\overline{M}_w$, g/mol	1–15×10^4
Specific gravity, g/cm^3	1.4–1.5
T_g, °C	20–70
Decomposition temperature, °C	240–250
Solubility parameter	8.8 (calc.)

4. Toxicology and Occupational Health

All commercial fluoropolymers are low in toxicity and safe under normal conditions. They are physiologically inert and some have been approved by the FDA for use in contact with food or for human implants. Incineration of fluoropolymers should be avoided because the combustion products include hydrogen fluoride, a toxic and corrosive gas.

4.1. Fluoroplastics

Polytetrafluoroethylene resins are subjected to prolonged heating at 380°C during sintering. Because a small quantity of toxic fumes can be emitted during this process, adequate ventilation should be provided to prevent inhalation of the fumes, which can cause a condition known as polymer fume fever. This temporary flulike disorder occurs several hours after exposure and lasts for 36–48 h. The effect is not cumulative. It is often caused by smoking tobacco contaminated with PTFE powder. Therefore, smoking and tobacco should be forbidden in areas where PTFE is processed. Safe handling practice must be observed during processing [188]. Enormous quantities of PTFE resins have been manufactured, processed, and used without causing any known cases of serious illness or death.

The same precautions should be followed with all TFE-containing polymers. In the extrusion or molding of TFE copolymers, the resin is heated above 300°C (to 425°C in the case of PFA resins). Exhaust hoods should be used over the die and feed hopper to remove any toxic fumes that may be evolved.

The main hazard associated with the processing of PVDF, ECTFE, and PVF resins is the evolution of hydrogen fluoride; this is prevented by avoiding overheating. Exhaust hoods should be used over the feed and die sections of the extruder or molding machine to remove hydrogen fluoride.

4.2. Fluoroelastomers

Fluorocarbon elastomers are nontoxic and non-irritating when handled in accordance with manufacturers' recommendations. Detailed literature on safety is available from the manufacturers; the following general principles should be observed:

1. Adequate ventilation of compounding, extrusion, curing, and postcuring areas should be provided
2. Dispersion of finely divided active metals or high concentrations of active dehydrofluorinating agents (e.g., aliphatic amines and combinations of phase-transfer catalysts with metal oxides or hydroxides) should be avoided
3. Fluoroelastomers should not be heated to or above their decomposition temperature
4. Incineration should not be used as a waste disposal method because of the attendant evolution of hydrogen fluoride.

Certain compounding ingredients may pose specific hazards, and manufacturers' or suppliers' recommendations should be followed.

Certain fluoroelastomers have been approved for contact with foods under the Code of Federal Regulations, 21, Food and Drugs, Part 177.2600.

4.3. Environmental Aspects

Presently, no overall and sustainable concepts for the recycling of fluoropolymers from processors or postconsumer waste are available. Only limited amounts from processor waste streams get recycled, the rest including postconsumer waste is uncontrolled incinerated or landfilled. In future, legislation will change this situation, e.g., forcing manufacturer and processors to develop concepts to reuse or recycle fluoropolymer waste, e.g., pyrolyzing perfluorinated polymers to TFE, HFP or to incinerate fluoropolymer in a controlled way [187, 189–191]. Europe and Japan will certainly be the front runners to handle scrap and waste streams of fluoropolymers in a sustainable manner.

In the late 1990s, attention became focused on ammonium perfluorooctanoate (APFO), a perfluorinated emulsifier used in aqueous emulsion polymerizations of fluoropolymers [12]. Use of a nontelogenic emulsifier in aqueous emulsion polymerization is essential, both to stabilize the final fluoropolymer latices and as process safety measure to control the exothermic polymerizations. APFO has been widely used by all fluoropolymer manufactures since the 1950s. However, evidence of wide spread distribution and persistence of APFO in the environment, its biopersistence in humans, and its toxicity in laboratory animal studies have raised serious concerns about its continued use [192–194]. The fluoropolymer industry developed and implemented containment technologies to recover and reuse APFO from off-gas stream, from aqueous wastestreams, and from products (e.g., aqueous dispersions) and closed the manufacturing loops [195–197].

Nevertheless, in 2006 the U.S. Environmental Protection Agency (EPA) announced a stewardship program challenging fluoropolymer manufacturers to eliminated APFO and related chemicals from emissions and products by 2015. As a result, considerable efforts have been made to reduce the use of APFO and to develop safer alternatives. Meanwhile a variety of nontelogenic, highly fluorinated, oxygen-containing carboxylic acid [198] have been developed as APFO substitute; these alternatives have superior EHS profiles compared to APFO [199]. Containment technologies, developed for APFO can be applied for the new generation of fluorinated emulsifiers to recover and reuse these materials [200–203].

5. Economic Aspects

Fluoroplastics. In 2008, world consumption of fluoropolymers reached 182 000 t and exceeded 3.0×10^9 $ [12]. World production capacity was more than 260 000 t/a. Annual growth of consumption is expected to be in the range of 3–6%. China has added huge PTFE production capacities and is the world's largest producer of PTFE with 30%. The landscape of fluoropolymer producers has changed by mergers and acquisitions since the late 1990s. Major producers include Asahi Glass (JP), Arkema (EU), Daikin (JP), DuPont (U. S.), 3M (U.S.), Honeywell (U.S.), Solvay Solexis (EU), Halopolymer (Russia), Gujarat (IN), 3F (CN), Meilan (CN), Dongyue (CN), Juhua (CN), Chenguang (CN), Fuxin (CN), and Kureha (JP).

Fluoroelastomers. In 2009, world consumption of fluorocarbon elastomers (FKM incl. FFKM) reached 19 500 t; the global demand for fluorosilicones (FSR) was about 1 700 t; the total market value is estimated to about 1.1×10^9 $ [204]. Based on sales volumes, the United States accounted for 26% of world consumption, Europe for 29%, and Asia for 42% in 2009. Total demand for fluoroelastomers (FKM, FFKM, FSR) is forecast to grow at about $5 - 6\%/a$. Included in FKM demand is their use as polymer processing acids (PPA) e. g., for the polyolefin industry. Currently there is a large overcapacity of FKM in China, which has a production capacity of 10 000 t. The landscape of producers changed during the last decade; major FKM- and FFKM-producers include: Asahi Glass (JP), Daikin (JP), DuPont (U.S.), 3M (U.S.), Halopolymer (Russia), Solvay Solexis (EU), Meilan (CN), Shanghai 3F (CN), Juhua (CN), Chenguang (CN), Dongyue (CN).

Abbreviations used in this article:

APFO ammonium perfluorooctanoate
CTFE chlorotrifluoroethylene
DSC differential scanning calorimetry
ECTFE ethylene–chlorotrifluoroethylene copolymer
E ethylene
ETFE ethylene–tetrafluoroethylene copolymer
FE fluoroethylene
FEP fluorinated ethylene–propene copolymer

FFKM	perfluoroelastomer
FKM	fluorohydrocarbon elastomer
FSR	fluorosilicones
HFP	hexafluoropropene
HFPI	hexafluoropropene index
IR	infrared
LCB	long chain branched
LOI	limiting oxygen index
MFA	methylfluoroalkoxy copolymer
MFI	melt-flow index
MWD	molar mass distribution
P	propene
PBVE	perfluoro-(3-butenyl vinyl ether)
PCTFE	polychlorotrifluoroethylene
PDD	perfluoro-2,2-dimethyl-1,3-dioxole
PE	polyethylene
PFA	perfluoroalkoxy copolymer
phr	parts per hundred of rubber
PMVE	perfluoro(methyl vinyl ether)
PP	polypropylene
PPA	polymer processing acids
PPVE	perfluoro(propyl vinyl ether)
PS	polystyrene
PTFE	polytetrafluoroethylene
PV	pressure–velocity
PVC	poly(vinyl chloride)
PVDF	poly(vinylidene fluoride)
PVF	poly(vinyl fluoride)
SSG	standard specific gravity
SVI	stretch void index
TAIC	triallyl isocyanurate
TFE	tetrafluoroethylene
TFEP	tetrafluoroethylene–propene rubber
THV	tetrafluoroethylene–hexafluoropropene–vinylidene fluoride polymer
TTD	2,2,4-trifluoro-5-trifluoromethoxy-1,3-dioxole
VDF	vinylidene fluoride
VE	vinyl ether
VF	vinyl fluoride

References

General References

1 S. Ebnesajjad, P.R. Khaladkar (eds.): *Fluoropolymer Applications in Chemical Processing Industries*, Plastic Design Library, Norwich, NY, 2005.

2 S.V. Gangal in *Kirk–Othmer Encyclopedia of Chemical Technology*, 5th ed., vol. 18, John Wiley & Sons, New York 2007 pp. 288–353.

3 L.A. Wall: *Fluoropolymers*, vol. 25, Wiley-Interscience, New York 1972.

4 P. Eyerer, T. Hirth, P. Elsner (eds.): *Polymer Engineering: Technologien und Praxis*, Springer Verlag, Berlin 2008.

5 H.G. Elias (ed.): *Macromolecules*, vols 1-4, Wiley-VCH, Weinheim 2005.

6 K. Hintzer in *Houben Weyl – Methoden der organischen Chemie*, vol. E 20/2, G Thieme, Stuttgart 1987.

Specific References

7 IG-Farbenindustrie AG, DRP 677071, 1934 (Schloffer/Scherer).
8 W. Wetzel, *N.T.M.* **13** (2005) 79–91.
9 A. Frick et al., *Macromol. Mater. Eng.* **297** (2012) 329–341.
10 S. Ebnesajjad (ed.): *Fluoroplastics*, vols. 1 & 2, Plastic Design Library, Norwich, NY, 2000.
11 G. Hougham et al. (eds.): *Fluoropolymers*, vols. 1 & 2, Kluwer Academic/Plenum Publishers, New York 1999.
12 R.K. Will et al., *CEH Marketing Research Report: Fluoropolymers*, SRI Consulting, Menlo Park, CA 94025 2008.
13 H. Teng, *Appl. Sci.* **2** (2012) 496–512.
14 DuPont, US 2230654, 1941 (R.J. Plunkett).
15 DuPont, US 2946763, 1960 (M.I. Bro, B.W. Sandt).
16 DuPont, US 3180995, 1965 (J.F. Harris, D.I. McCane).
17 DuPont, US 2392378, 1946 (W.E. Hanford).
18 H.W. Starkweather, *J. Am. Chem. Soc.* **56** (1934) 1870.
19 DuPont, US 2435537, 1948 (T.A. Ford, W.E. Hanford).
20 S.L. Bell, *Fluoropolymers*, Report 166B, SRI Consulting, Menlo Park, CA 94025 2002.
21 Hoechst AG, DE 4022405, 1990 (B. Felix et al.).
22 Hoechst AG, DE 1720801, 1968 (R. Hartwimmer).
23 Daikin, EP 1816148, 2005 (Takahiro et al.).
24 Hoechst AG, EP 30663, 1980 (Kuhls et al.).
25 Dyneon, EP 657514, 1994 (H. Blaedel et. al).
26 C.A. Sperati, H.W. Starkweather, *Fortschr. Hochpolym. Forsch.* **2** (1961) 465–495.
27 H.D. Chanzy, P. Smith, *J. Polym. Sci. Polym. Lett. Ed.* **24** (1986) 557.
28 B. Chu et al., *Macromolecules* **22** (1989) 831–837.
29 R.C. Doban et al.: *130th meeting of the ACS*, Atlantic City, Sept. 1956.
30 ASTM D 4895-10, ASTM D 4894-07: Standard Specifications for polytetrafluoroethylene (PTFE).
31 T. Suwa, M. Takehisa, S. Machi, *J. App. Poly. Sci.* **17** (1973) 3253–3257.
32 G. Ajroldi et al., *J. Appl. Poly. Sci.* **14** (1970) 79–88.
33 W.H. Tuminello in *Encyclopedia of Fluid Mechanics, Polymer Flow Engineering*, vol. 9, Gulf Publishing Company, Houston 1990, chap. 6.
34 J. Scheirs (ed.): *Modern Fluoropolymers*, John Wiley & Sons, New York 1997.
35 J. Brandrup, E.H. Immergut, E.A. Grulke, *Polymer Handbook*, 4th ed., John Wiley & Sons, New York 1999.
36 S.C. Lin, B. Kent: "Flammability of Fluoropolymers" in C. Wilkie et al. (eds.): *Fire and Polymers*, ACS Symposium 2009, American Chemical Society, Washington, DC 2009, chap. 17.
37 J.C. Siegle et al., *J. Polym. Sci., Part A*, **2** (1964) 391.

38. E.-C. Koch (ed): *Metal-Fluorocarbon Based Energetic Materials*, Wiley-VCH, Weinheim 2012.
39. W.L. Gore,US 3962153, 1976 (R.W. Gore).
40. S.G. Hatzikiriakos in M. Kontopoulou (ed.): *Applied Polymer Rheology: Polymeric Fluids with Industrial Applications*, J. Wiley & Sons, New York 2012, chap. 11.
41. DuPont, US 3051545, 1962 (W. Steuber).
42. DuPont, US 3855191, 1974 (T.R. Doughty, C.A. Sperati).
43. M. Schlipf, *Verfahrenstechnik* **3** (1999).
44. M. Schlipf, M. Schuter, *Konstruieren mit PTFE*, Verlag Moderne Industrie, Munich 2007.
45. Daikin Kogyo, DE-A 2104077, 1971 (Y.T. Kometani et al.).
46. A.D. English, O.T. Garza, *Macromolecules* **12** (1979) 352.
47. W.H. Tuminello, *Int. J. Polymer Analysis & Characterization* **2** (1996) 141–149.
48. C.A. Mertdogan et al,*J. Appl. Polym. Sci.* **74** (1999) 2039–2045.
49. W.H. Tuminello, *Polymer Engineering and Science* **29** (1989) 645–653.
50. R.K. Eby, F.C. Wilson, *J. Appl. Phys.* **33** (1962) 2951.
51. Teflon *FEP Fluoropolymers Resin Properties Handbook*, Du Pont & Co., publication no. H-37052-3, June 1998, Wilmington DE.
52. R.Y.M. Huang, P.J.F. Kanitz, *Polym. Prepr.* **10** (1969)no. 2.
53. DuPont,*Du Pont Innovation* **5** (1973)no. 1, 16.
54. DuPont, US 3635926, 1972 (W.F. Gresham, A.F. Vogelpohl).
55. DuPont, US 3536733, 1970 (D.P. Carlson).
56. DuPont, US 3642742, 1972 (D.P. Carlson); Du Pont, US 4599386, 1986 (D.P. Carlson et al.).
57. DuPont, US 4743658, 1988 (J.F. Imbalzano, D.L. Kerbow).
58. Daikin, EP 457255, 1994 (K. Ihara et al.).
59. S.V. Gangal in *Encyclopedia of Polymer Science and Engineering*, vol. 16, Wiley-Interscience, New York 1985, p. 614.
60. M. Fleissner, *Macromol. Chem., Macromol. Symp.* **61** (1992) 324.
61. E.E. Rosenbaum et al., *Intern. Polymer Process.* **10** (1995) 204.
62. DuPont, US 3624250, 1971 (D.P. Carlson).
63. Asahi Glass Co., US 4123602, 1978 (H. Ukihashi, M. Yamabe).
64. Hoechst, US 4338287, 1982 (H. Sulzbach et al.).
65. F. Kostov, A. Nikolov, *J. Appl. Polym. Sci.* **57** (1995) 1545–1555.
66. N. Modena et al., *Polym. Letter* **10** (1972) 153.
67. R. Naberezhnykh et al.,*Vyskomol. Soedin* **19** (1977) 33–37.
68. H.W.Starkweather,Jr.,*J. Polym. Sci. Polym. Phys. Ed.***11**(1973)587.
69. B. Chu, C. Wu, *Macromolecules* **20** (1987) 93.
70. F. Garbassi, R. Po in *Kirk Othmer Encyclopedia of Chemical Technology*, 5th ed., vol. 10, J. Wiley & Sons, New York 2005.
71. DuPont, US 3738923, 1973 (D.P. Carlson, N.E. West).
72. DuPont, US 2626254, 1954 (W.T. Miller, J.T. Maynard).
73. Union Carbide, US 2700662, 1955 (D.M. Young, B. Thompson).
74. 3M, GB 840735, 1960.
75. H. Matsuo, *J. Polym. Sci.* **25** (1957) 234.
76. A. West in *Kirk Othmer Encyclopedia of Chemical Technology*, 3rd ed., vol. 11, J. Wiley & Sons, New York 1980, pp. 49–54.
77. S. Chandrasekaran in *Encyclopedia of Polymer Science and Engineering*, vol. 3, J. Wiley & Sons, New York 1985, pp. 463–480.
78. J.P. Sibilia, L. Roldin, S. Chandrasekaren, *J. Polym. Sci., Part B.* **10** (1972) 549.
79. Montecatini, US 3501446, 1970 (M. Ragazzini et al.).
80. H.K. Reimschuessel, F.J. Rahl, *J. Polym. Sci., Pt. A., Poly. Chem.* **25** (1987) 1871.
81. A.B. Robertson, *Appl. Polym. Symp.* **21** (1973) 89–100.
82. Arkema, WO 030784, 2012 (M. Durali, L. Hedhli).
83. Arkema, WO 073686, 2008 (R. Amin-Sanayei, et al.).
84. Arkema, US 8124699, 2012 (M. Durali, et al.).
85. Arkema, EP 1462461, 2004 (M. Durali et al.).
86. K.R. Hasegawa et al., *J. Polym. Sci. Polym. Phys. Ed.* **8** (1970) 1073.
87. R. Honselka et al., *Ultrapure Water 1987*,July/August 1987, p. 46–50.
88. H. Fitz *Kunststoffe* **70** (1980) 11–15.
89. T.P. Dinoia et al., *J. Polym. Sci., Part B Polym. Phys.* **38** (2000) 2832–2840.
90. M.G. Broadhorst, *J. Appl. Phys.* **49** (1978) 4992.
91. R.N. Haszeldine et al., *Polymer* **14** (1973) 221.
92. DuPont, US 3265678, 1966 (J.L. Hecht).
93. Montecatini, GB 1029635, 1966.
94. S. Ebnesajjad in *Kirk-Othmer, Encyclopedia of Chem. Technology*, 5th ed., vol. 20, Wiley-Interscience, New York 2006.
95. F.S. Cohen, N. Kraftin: *Encyclopedia of Polymer Science and Technology*, vol. 14, J. Wiley & Sons, New York 1971, pp. 522–540.
96. DuPont, US 3978030, 1976 (P.R. Resnick).
97. Solvay Solexis, US 5498682, 1996 (W. Navarini et al.).
98. DuPont, US 4946902, 1990 (P.G. Bekiarian et al.).
99. Asahi Glass, US 4897457, 1990 (Nakamura et al.).
100. G. Dlubek et al., *J. Polym. Sci., Part B Polymer Phys.* **45** (2007) 2519–2534.
101. B. Ameduri,B. Boutevin (eds.): *Well-architectured Fluoropolymers*, Elsevier, Amsterdam 2004.
102. 3M, WO 088203, 2002 (H. Kaspar et al.).
103. 3M, EP 1963380, 2008 (K. Hintzer et al.). 3M, WO 033721, 2008 (P. Bissinger et al.).
104. H. Kaspar et al., *Macromolecules* **40** (2007) 2409–2416.
105. H. Kaspar et al., *Macromolecules* **39** (2006) 2316–2324.
106. H. Kaspar, K. Hintzer, *Rheol. Acta* **50** (2011) 577–599. D. Moeller et al., *Rheol. Acta* **48** (2009) 509–516.
107. A. Rudin, J. Blacklock, S. Nam, A. Worm, *SPE ANTEC Tech. Papers* **54** (1986) 1154–1157.
108. S. Woods, S. Amos, *TAPPI P.L.C. Conf.* (1998) 675–685.
109. K. Focquet, G. Dewitte, J. Briers, L. Van den Bossche, O. Georjon, T. Donders, *Polyethylene World Congress*, Milano, (May 5–10) 1997.
110. P. Eyerer,T. Hirth,P. Elsner (eds.): *Polymer Engineering: Technologien und Praxis*, Springer Verlag, Berlin 2008.
111. A.L. Moore, *Fluoroelastomers Handbook*, William Andrew Publishing, Norwich, NY, 2006.
112. H.R. Allcock et al., *Inorg. Chem.* **5** (1966) 1709.
113. M.E. Conroy et al., *Rubber Age N.Y.* **76** (1955) 543.
114. C.B. Griffis, J.C. Montermoso, *Rubber Age N.Y.* **77** (1955) 559.
115. W.W. Jackson, D. Hale, *Rubber Age N.Y.* **77** (1955) 865.
116. S. Dixon et al., *Ind. Eng. Chem.* **49** (1957) 1687.
117. J.S. Rugg, A.C. Stevenson, *Rubber Age N.Y.* **82** (1957) 102.
118. DuPont, US 3051677, 1962 (D.R. Rexford).
119. DuPont, Report No. 59–4, 1959 (A.L. Moran).
120. DuPont, US 2968649, 1961 (J.R. Pailthorp, H.E. Schroeder).
121. Montecatini-Edison, US 3331823, 1967 (D. Sianesi et al.).
122. Montecatini-Edison, US 3335106, 1967 (D. Sianesi et al.).
123. A. Miglierina, G. Ceccato, *Fourth Int. Syn. Rubber Symp.* **2** (1969) 65.
124. Daikin Kogyo, US-A 4158678, 1977 (M. Tsuneo, T. Masayoshi).
125. M. Apostolo, G. Biressi, *Macromol. Symp.* **206** (2004) 347–360.
126. J.R. Dunn, H.A. Pfisterer, *Rubber Chem. Technol.* **48** (1976) 356.
127. A. Nersasian, *ASLE Trans.* **23** (1980) 343.
128. B. Frapin, *Rev. Gen. Caoutch. Plast.* **672** (1987) 125.
129. I.A. Abu-Isa, H.E. Trexler, *Rubber Chem. Technol.* **58** (1985) 326.

130. A. Nersasian, *Elastomerics* **112** (1980)no. 10, 26.
131. Chemical Resistance Guide,http://www.dupontelastomers.com/tech_info/chemical.asp Nov. 2011 (accessed: February 2013).
132. High Performance Fluoroelastomers, Product Comparison Guidehttp://multimedia.3m.com/mws/mediawebserver?mwsId=SSSSSufSevTsZxtUmx_elYtvevUqevTSevTSevTSeSSSSSS-&fn=Dyneon%20FKM%20Selection%20Guide.pdf (accessed: February 2013).
133. Fluoroelastomer DAIEL Selection Guide http://www.daikin.com/chm/products/pdfDown.php?url=pdf/select_elastomer_e.pdf (accessed: February 2013).
134. E.W. Bergstrom, *Elastomerics* **109** (1977)no. 2, 21.
135. R.L. Clough, K.T. Gillen, C.A. Quintana,SAND 83-2493 RV, Sandia National Labs., Albuquerque1984; *J. Polym. Sci. Polym. Chem. Ed.* **23** (1985) 359.
136. W.W. Schmiegel, A.L. Logothetis in J.C. Arthur (ed.): "Polymers for Fibers and Elastomers",ACS Symposium Series no. 260, American Chemical Society, Washington 1984, Chap. 10.
137. S. Bowers,Rubberchem '99, Antwerp (22.11.1999), Rapra Technology, Akron, OH 1999, pp. 6.1–6.8.
138. T. Dobel, C. Grant, *Fall 180th Technical Meeting of the Rubber Division*, American Chemical Society, Cleveland, OH,Oct. 11–13, pp. 798–814.
139. G. Kojima, M. Hisasue, *Makromol. Chem.* **182** (1981) 1429.
140. Asahi Glass, EP 638596, 1994 (A. Funaki et al.).
141. Ausimont, EP 1148072, 2001 (W. Navarrini).
142. 3M, WO 2004/024786, 2002 (H. Kaspar et al.).
143. Processing Guide, Viton Fluoroelastomer Technical Information Bulletin VTE-H90171-00-A0703, DuPont Dow Elastomers, 2003, http://www.rtvanderbilt.com/VTE-H90171-00-A0703.pdf (accessed: February 2013).
144. Molding Solutions, http://molders.com/ (accessed: February 2013).
145. W.W. Schmiegel, *Kautsch. Gummi Kunstst.* **31** (1978) 137.
146. V. Arcella, G. Chiodini, N. DelFanti, M. Pianca, *140th ACS Rubber Division Meeting*, Paper 57, Detroit, Mich., 1991
147. D. Apotheker et al., *Rubber Chem. Technol.* **55** (1982) 1004.
148. J.F. Smith, G.T. Perkins, *J. Appl. Polym. Sci.* **5** (1961) 460–467.
149. A.W. Fogiel et al., *Rubber Chem. Technol.* **49** (1976) 34.
150. J.D. MacLachlan, A.W. Fogiel, *Rubber Chem. Technol.* **49** (1976) 43.
151. R.H. Wehrenberg II, *Mater. Eng.* **94** (1981) no. 7, 44.
152. DuPont, US 3069401, 1962 (G.A. Gallagher).
153. DuPont, US 3467638, 1969; US 3876654, 1975 (D.B. Pattison).
154. DuPont, US 3 546 186, 1971 (E.K. Gladding, R. Sullivan).
155. DuPont, US 3 114 778, 1963 (A.L. Barney, W. Honsberg).
156. A.L. Barney et al., *J. Polym. Sci., Polym. Chem. Ed.* **8** (1970) 1091.
157. G.H. Kalb et al., *Adv. Chem. Ser.* **129** (1973) 13.
158. A.L. Barney et al., *Rubber Chem. Technol.* **44** (1971) 660.
159. G. Comino, S. Arrigoni, M. Aposotolo, M. Albano, V. Arcella, *Prog. Rubber Plast. Technol.* **17** (2001)no. 2, 101–111.
160. 3M, US 6844388, 2001 (K. Hintzer et al.).
161. 3M, US 6703068, 2001 (K. Hintzer et al.).
162. A.L. Logothetis, *International Rubber Conference*, Kyoto, Japan, 1985, p. 73.
163. A.L. Logothetis, *Centenary of the Discovery of Fluorine (Symposium)*, Paris, France, 1986, p. 226.
164. A.L. Logothetis, *Japan–U.S. Polymer Symposium*, Kyoto, Japan, 1985, p. 121.
165. DuPont, US 3467635, 1969 (W.R. Brasen, C.S. Cleaver).
166. Y. Tabata et al., *J. Polym. Sci. Part A* **2** (1964) 2235.
167. G. Kojima, Y. Tabata, *J. Macromol. Sci., Chem.* **A5** (1971) 1087.
168. G. Kojima et al., *Rubber Chem. Technol.* **50** (1977) 403.
169. G. Kojima, H. Wachi, *Rubber Chem. Technol.* **51** (1978) 940.
170. G. Kojima et al., *Rubber Chem. Technol.* **54** (1981) 779.
171. DuPont, US 4035565, 1977; US 4214060, 1980 (D. Apotheker, P.J. Krusic).
172. DuPont, US 3859259, 1975 (J. Harell, W. Schmiegel).
173. Asahi Glass, US 4645799, 1985 (H. Wachi, S. Kaya, G. Kojima).
174. Asahi Glass, US 4758618, 1988 (Y. Ito, H. Wachi).
175. 3M, US 4882390, 1989 (W. Grooteart, R. Kolb).
176. Asahi Glass, US 4148982, 1977 (M. Morozumi, G. Kojima, T. Abe).
177. DuPont Dow Elastomers, US 2003/0065132, 2002 (J. Bauerle, W. Schmiegel).
178. J.G. Bauerle, P.L. Tang, *SAE World Congress*, Detroit, Michigan, March 2002.
179. D.E. Hull, *Elastomerics* **115** (1983)no. 10, 40.
180. D.E. Hull, *Elastomerics* **114** (1982)no. 7, 27.
181. DuPont, US 4694045, 1986 (A. Moore)
182. Ausimont, US 5264509, 1993 (V. Arcella, M. Albano, G. Chlodini, A. Minutillo).
183. Ausimont, US 4864006, 1987 (E. Gianetti, M. Visca).
184. G. Kojima, M. Yamabe, *J. Synth. Org. Chem., Jpn.* **42** (1984) 841.
185. S. Munekata, *Prog. Org. Coat.* **16** (1988) 113.
186. M. Yamauchi et al., *European Coatings J.* **124** (1996).
187. Hoechst, US 5432259, 1995 (T. Schoettle et al.).
188. Guide for the Safe Handling of Fluoropolymer Resins, Plastics Europe, Brussels 2012.
189. E. Meissner et al.,*Polymer Degradation and Stability* **83** (2004) 163–172.
190. I.J. van der Walt, O.S.L. Bruinsma, *J. Appl. Polym. Sci.* **102** (2006) 2752–2759.
191. Asahi Glass, JP 231984, 2005 (T. Abe et al.).
192. L. Vierke et al., *Environmental Science Europe* **24** (2012) no. 1, 1–11.
193. G.L. Kennedy et al,*CRC Crit. Rev. Toxicol.* **34** (2004) 351–384.
194. C. Lan et al., *Toxicol. Sci.* **99** (2007) 366–394.
195. 3M/Dyneon, EP 731081, 1996 (R. Sulzbach et al.).
196. 3M/Dyneon, EP 1084097, EP 1093441, 1999 (B. Felix et al.).
197. 3M/Dyneon, EP 1155055, 1999 (H. Blaedel et al.).
198. 3M, EP 1963247, 2008 (K. Hintzer et al.).
199. S.C. Gordon, *Regul. Toxicol. Pharmacol.* **59** (2011) 64–80.
200. 3M, US 0015937, 2007 (K. Hintzer et al.).
201. 3M, US 7795332, 2010 (K. Hintzer et al.).
202. 3M, US 7671112, 2010 (K. Hintzer et al.).
203. 3M, US 0025902, 2007 (K. Hintzer et al.).
204. U. Löchner, Y. Inoguchi: *CEH Marketing Research Report: Fluoroelastomers*, SRI-Consulting, Menlo Park, CA 2010.
205. Daikin, US 4521575, 1985 (S. Nakagawa, K. Ihwa).
206. Asahi Glass, EP 638596, 1994 (■forename■ Funaki et al.).

Polyacrylamides and Poly(Acrylic Acids)

GREGOR HERTH, BASF SE, Ludwigshafen, Germany

GUNNAR SCHORNICK, BASF SE, Ludwigshafen, Germany

FREDRIC L. BUCHHOLZ, The Dow Chemical Company, Midland, Michigan 48667, United States

1.	Introduction	659
2.	Raw Materials.	659
3.	Production	661
3.1.	Free-Radical Polymerization	661
3.2.	Polymerization Processes	662
3.2.1.	Solution Polymerization	662
3.2.2.	Inverse Emulsion Polymerization . .	664
3.2.3.	Suspension Polymerization	664
3.2.4.	Precipitation Polymerization	665
3.2.5.	Polymer-Analogous Reactions.	665
3.2.6.	Residual Monomer Scavenging	665
4.	Properties	666
5.	Uses	668
5.1.	Poly(Acrylic Acids) and Their Salts	669
5.2.	Polyacrylamides and Hydrolyzed Polyacrylamides.	669
5.3.	Other Polymers	670
6.	Economic Aspects	671
7.	Toxicology, Occupational Health, and Environmental Aspects	671
	References.	672

1. Introduction

It is difficult to imagine an aspect of our modern world not touched by polyacrylamides and poly (acrylic acids). These hydrophilic polymers, their copolymers with other hydrophilic monomers, and their metallic salts are utilized in many applications. Polyacrylamides are used in applications, such as mining, water treatment, sewage treatment, papermaking, oil well drilling, oil production, and agriculture. Poly(acrylic acids) are used in boiler-water additives, superabsorbent diapers, paints, detergents, industrial coatings, cosmetics, ceramics, textiles, ion-exchange resins, industrial adhesives, cements, and drug delivery [1]. The term polyacrylamide is used for homopolymers as well as for copolymers of acrylamide. Common comonomers for acrylamide are acrylic acid and its salts ("anionic polyacrylamide") as well as cationic ester of acrylic acid ("cationic acrylamide").

The utility of these polymers is directly related to their chemical structure, functionality, and molecular mass. The acid and amide functions are highly solvated by water, and are responsible for the solubility of the polymers in aqueous solutions. When cross-linked, the polymers swell in water without dissolving. The properties of polymers containing acrylic or methacrylic acid are dependent on the pH of the aqueous solution in contact with the polymer. Because the metal ion salt interacts with more water molecules than the unneutralized acid group, higher viscosity or higher swelling is observed at elevated pH, which is useful in thickeners or absorbent polymers. The proximity of the carboxyl groups to one another in poly(acrylic acid) gives rise to chelating properties with di- and trivalent metal ions, which is exploited in boiler-scale control and detergent builders. The high polymerizability of the monomers allows the preparation of high molecular mass polymers, which are useful as flocculants and thickeners. When the molecular mass is reduced, the polymers can be used as dispersants and scale inhibitors. The reactivity of the amide and acid groups allows simple chemical modifications that change the polymer properties and amplify the number of applications.

2. Raw Materials

Some useful physical properties of the water-soluble monomers used in the production of polyacrylamides, poly(acrylic acid), and poly (methacrylic acid) are listed in Table 1. All the monomers are toxic and should be handled with great care. Contact with the skin, eyes, and mucous membranes should be avoided. Breathing of vapors should also be avoided; operations should be conducted in vessels vented to scrubbers or in laboratory fume hoods. The monomers are quite reactive; their contamination can lead to uncontrolled polymerization and must be avoided. Dedicated storage vessels are used for the monomer. Manufacturers' data should be consulted for the safe handling and use of the products, e.g., [2, 3]. The best sources are material safety data sheets of the raw material producers, which are updated if new information is available.

Acrylic acid is described in detail elsewhere (→ Acrylic Acid and Derivatives). It is manufactured via oxidation of propene and has the largest production volume of the hydrophilic monomers. Significant progress has been made towards bio-based acrylic acid [4]. 3-Hydroxypropionic acid (3-HP) is the most likely precursor in a fermentation process. Acrylic acid is very corrosive to the skin and eyes, and its vapors are irritating to the eyes and lungs. Personal protective equipment, such as rubber gloves, goggles, impervious body coverings, and an organic vapor respirator should be used in the case of possible contact with acrylic acid. Commercial acrylic acid is normally inhibited against polymerization with 4-methoxyphenol (methyl ether hydroquinone, MEHQ). As this inhibitor requires oxygen for long-term effectiveness, air should be added to storage containers. The monomer should not be allowed to freeze because this causes the inhibitor to separate from the monomer by fractional crystallization. The resulting uninhibited acrylic acid is hazardous due to the significantly increased potential for uncontrolled polymerization.

Runaway polymerizations have occurred. A rail car of acrylic acid exploded in Teesside (UK) in 1976 as a result of too rapid thawing of frozen acid. The highly exothermic, adiabatic vinyl polymerization can become so hot that decarboxylation of the polymer occurs, with concomitant production of large volumes of carbon dioxide. A less violent, unplanned, runaway polymerization has been described [5].

Table 1. Properties of monomers used in the production of polyacrylamides, poly(acrylic acids), and their copolymers

Name	CAS registry number	Molecular formula	M_r	mp, °C	bp, °C	Density, g/cm³	Flash point, °C	Heat of polymerization, kJ/mol	Oral (rat) LD$_{50}$, mg/kg
Acrylamide	[79-06-1]	C_3H_5NO	71.08	84.5	136 (3.33 kPa)	1.122 (30°C)		81.5	124
2-Acrylamido-2-methylpropanesulfonic acid	[15214-89-8]	$C_7H_{13}NSO_4$	207.21	185				92.0	1 410
Acrylic acid	[79-10-7]	$C_3H_4O_2$	72.06	13.5	141 (101.3 kPa)	1.040 (30°C)	160	77.5	340
N,N-Dimethylacrylamide	[2680-03-7]	C_5H_9NO	99.15	−40	81 (2.80 kPa)	0.926 (25°C)	50		700
N-Isopropylacrylamide	[2210-25-5]	$C_6H_{11}NO$	113.16	64			70		
Methacrylic acid	[79-41-4]	$C_4H_6O_2$	86.10	16	163	1.015		56.3	1 600 (mouse)
Methacrylamide	[79-39-0]	C_4H_7NO	85.12	110			76	56.0	459
N,N'-Methylene-bisacrylamide	[110-26-9]	$C_7H_{10}N_2O_2$	154.17	300		1.24 (30°C)		82.7	390

Here the major product of the vigorous reaction was polyester, which is formed in a Michael-type reaction between molecules of acrylic acid. In 2012 an explosion occurred at the Nippon Shokubai Himeji Plant. The explosion and subsequent fire in an acrylic acid intermediate tank killed one person and injured 36 [6].

Methacrylic acid is similar to acrylic acid in most regards and should be handled accordingly. For a detailed description of this compound, see →Methacrylic Acid and Derivatives.

Acrylamide is described elsewhere (→Acrylic Acid and Derivatives). It has toxic effects on the nervous system. A low-dose, long-term exposure study in rats showed an increased incidence of tumors. For a review on the toxicity of acrylamide see [7]. At higher doses, laboratory animals experienced reproductive deficiencies [2]. Contact with the monomer should be strictly avoided. Solid acrylamide sublimes readily so that contact with the vapor is a likely route for exposure. Therefore, acrylamide is supplied commercially as an aqueous solution to make handling simpler and to reduce potential exposure to vapors. Nowadays, most of the acrylamide monomer is produced by a biological process. Acrylonitrile is enzymatically hydrolyzed to acrylamide [8]. This process is cheaper and more selective than the copper-catalyzed hydrolysis of acrylonitrile. As with acrylic acid, oxygen is critical for inhibition of polymerization and dissolved oxygen should be present during storage because MEHQ is used as a free-radical scavenger. Acrylamide monomer is readily absorbed through the skin so that rubber gloves and impervious body clothing are required when contact with the aqueous solution is likely.

N,N-Dimethylacrylamide. At elevated temperatures and at high pH acrylamide can be hydrolyzed to acrylic acid. Therefore, in some special applications, acrylamide is replaced by the more stable N,N-dimethylacrylamide. However, this monomer is more expensive and used only in niche applications [9].

3. Production

3.1. Free-Radical Polymerization

Kinetics. An understanding of the polymerization kinetics of the monomers is essential for the control of polymer production. The molecular mass and molecular mass distribution of the polymers are key determinants of their physical properties and are influenced strongly by polymerization kinetics.

Advanced physical methods and modeling give a deeper insight to the kinetics, i.e., initiation, propagation, and termination [10]. Polymerization kinetics have been studied extensively for acrylic acid, its salts, and acrylamide. Less extensive data are available for the other monomers although some are available for methacrylic acid [11].

The kinetics of free-radical polymerization of acrylic acid, methacrylic acid, acrylamide, and methacrylamide have been investigated in water, dimethyl sulfoxide, tetrahydrofuran, and formamide at 30–60°C with azobisisobutyronitrile as initiator [12]. The polymerization rate is proportional to the monomer concentration and to the square root of the initiator concentration. The acrylics react faster than the methacrylics under identical conditions. The polymerization rate depends on the nature of the solvent; values of the kinetic constants for each monomer in each solvent were determined [12].

The polymerization kinetics for the acid monomers become complicated as the pH of the system changes. At constant overall monomer concentrations, the polymerization rate initially falls as the pH is increased from 2 to 7, then rises slowly from pH 7 to 12 [13, 14]. This has been attributed to a change in the propagation rate constant as the degree of ionization increases [15]. At pH 7 to 12, rates increase with an increase in the ionic strength of the medium [16]. This is accomplished either by adding simple monovalent salts or by increasing the ionized monomer concentration. Methacrylic acid shows a similar behavior [17]

Initiators play a complex role in polymerization kinetics, and initiator fragments can be important in the subsequent use of the polymer: For example, solution properties can deteriorate with time, depending on impurities [18]. Thermal initiators (e.g., azo compounds and

Table 2. Reactivity ratio as a function of pH [30]

pH	Acrylic acid	Acrylamide	Methacrylic acid	Acrylamide
	r_1	r_2	r_1	r_2
4	0.32 ± 0.046	0.57 ± 0.067	2.80 ± 0.44	0.20 ± 0.069
6	0.33 ± 0.20	0.85 ± 0.62	0.19 ± 0.08	0.55 ± 0.078
8	0.63 ± 0.004	0.12 ± 0.004	0.32 ± 0.004	0.53 ± 0.003

peroxides) produce free radicals in a homolytic scission reaction. The reaction, producing a pair of radicals, is characterized by the temperature at which the specific initiator is half consumed in 10 h. With those initiators the polymerization rate generally depends on the monomer concentration and the square root of the initiator concentration [19, 20]. The monomer may react with the initiator prior to the reaction of the free radical with monomer, and induce the decomposition of the initiator. In such cases (e.g., acrylic acid and sodium persulfate), the polymerization rate depends on the 1.5 power of monomer concentration and the square root of the initiator concentration [15, 16, 21]. Redox initiators produce free radicals in a reaction between a reducing agent and an oxidizing agent. Examples include the persulfate–thiosulfate reaction [22] and the manganic ion (Mn^{3+})-isobutyric acid reaction [23]. For redox initiators not involving metal ions, the polymerization rate usually depends on the first power of the monomer concentration. The rate also depends on the concentration of oxidant and reductant, the stoichiometry of their reaction, and the extent to which the initiators (usually the reductant) participate in radical termination reactions [22, 24–26]. When metal ions are involved in the redox reaction, the kinetics are more complex because the metal ion also participates in the termination step of the polymerization [23, 27, 28].

Copolymerization. See also→Polymerization Processes. The monomers listed in Table 1 are frequently copolymerized with one another and with other monomers. The relative reactivities of the monomers affect the microstructure and thus the physical properties of the polymer. When a comonomer is more reactive than its partner, it tends to be used up earlier in the reaction. The final product then has a distribution of composition as well as a distribution of molecular masses.

Reactivity ratios are derived from measurements of polymer composition as a function of conversion. Data from the entire conversion range of the copolymerization of acrylamide (AAm) and acrylic acid (AA) were utilized to calculate the following reactivity ratios [29]:

$r_1(AA) = 1.45 \pm 0.33$, $r_2(AAm) = 0.57 \pm 0.04$

Reactivity ratios for acrylic or methacrylic acid copolymerized with acrylamide have also been determined as a function of pH [30]. and the copolymerization has been studied [31]. Selected data are given in Table 2. Reactivity ratios for these and many other potential comonomers have been tabulated [32].

3.2. Polymerization Processes

The techniques described in this section are described in detail elsewhere (→Polymerization Processes).

3.2.1. Solution Polymerization

Batch Polymerization. The batch polymerization of the water-soluble monomers in solution is straightforward. The monomers are dissolved in water (or another solvent for both the monomer and polymer) at a desired concentration (usually 10–70%). Although acrylamide and acrylic acid have been polymerized in various organic solvents [12], water is used in industrial processes more or less exclusively. One of the very few exceptions is the polymerization of acrylic acid in 2-propanol, which introduces besides the carboxylic function additional functional groups into the molecule by acting as chain transfer agent [33].

The monomer solution is deoxygenated by bubbling an inert gas through the solution or by a series of evacuations and repressurization with an inert gas. The desired free-radical initiator is then added and the temperature, usually 40–80°C depending on the specific initiator used, is brought to the appropriate point for polymerization to begin.

The initial monomer concentration is limited because polymerization of the undiluted monomers is extremely dangerous due to the high heat of polymerization and rapid polymerization kinetics [34]. The concentration is also limited by the capacity of the equipment for removing the heat of polymerization and by the molecular mass of the desired polymer. If no heat can be removed from the reactor, the adiabatic temperature rise can be calculated for a given monomer concentration [35]. A suitable monomer concentration can then be calculated, based on the desired initial and final temperatures [36]. To have a better temperature control semibatch polymerizations are used: Part of the monomer solution is charged and the rest of the monomer solution is fed over a period of time [37]. The viscosity of the reaction mass is also important with regard to the equipment available for pumping the product. Higher molecular mass products generally dictate a lower monomer concentration because of their high viscosities, whereas low molecular mass products allow higher monomer concentrations.

Low molecular mass, water-soluble polymers are often desired, and several processes have been developed to produce them industrially. In one such process, sodium hydrogensulfite is added continuously to an aqueous monomer solution in a stirred tank while introducing fine air bubbles; polymerization occurs below 80°C [38]. The sulfite acts as a chain-transfer agent to lower the molecular mass. Sodium hypophosphite and 2-mercaptoethanol are commonly used for the same reason [39–41]. In another industrial process, poly(acrylic acid) of molecular mass 1 000–20 000 is prepared at 55 wt% concentration by polymerization with hydrogen peroxide and transition-metal activators, such as copper, iron, or manganese [42].

For some applications solid, dried products are preferred, e.g., powder detergents. The water may be evaporated using hot, rotating drum dryers or continuous belt, hot-air ovens, spray drying, or granulation devices. The dry flakes or lumps are then milled to the desired particle size and packaged in bags ranging in size from 22 to 1 000 kg. Depending on the process conditions, spray drying leads directly to powders or granules without further treatment [43]. Granules are more and more preferred over powders because of the much lower dust level.

Continuous Polymerization. Polymers can often be produced more efficiently by using a continuous polymerizer. An aqueous solution of monomers, initiators, and other additives is fed into one end of the polymerizer, and the polymeric product removed from the other. Low molecular mass copolymers of acrylic acid and acrylamide of controlled molecular mass (10 000 to 100 000) were produced in a packed-tube reactor operating at elevated pressure [44].

When a higher molecular mass is needed the aqueous monomer solution rapidly forms a gel after low conversions. This polymer gel is transformed by cutting, drying, and grinding into a powder. These powders contain up to 90 mass% of polymer. Such products can be used directly as powders (e.g., superabsorbents in diapers) or they are dissolved in water and used as diluted solutions (e.g., as flocculant in sewage plants). The production of superabsorbents is described elsewhere (→ Superabsorbents, Chap. 4). In principle the same processes can be used for the production of polyacrylamides [45]. Common are endless belt and tubelike reactors. If the molecular mass is less than 10^6 the gel is sticky after polymerization and not processable. Therefore, a minimum molecular mass is needed for gel polymerization.

A more flexible process is based on so-called smart scale production (SSP). It has a modular design and setup, and product switches can be handled in a shorter time frame [46]. In another process, 30–40 wt% aqueous monomer solution is pumped onto a continuous moving belt. Angled rollers at the edges of the belt cause a trough to be formed over a section of the belt. The trough holds the liquid monomers. As polymerization proceeds, a gel is formed from the reagents. The belt section flattens after it passes the final set of rollers, and the gel is released from the belt and is finally discharged from the end of the conveyor. Superabsorbent polyacrylates are prepared industrially in this manner [45]. Another process for superabsorbent polymers is based on kneader technology. The monomer solution is fed at one end into a twin shaft reactor and discharged at the other end [47]. A quite different process, the so-called drop polymerization is currently under development. A droplet generator at the top of a reaction chamber releases droplets of a monomer solution into a gaseous atmosphere [48].

3.2.2. Inverse Emulsion Polymerization

In inverse emulsion polymerization, tiny droplets of aqueous monomer solution are dispersed in a continuous organic phase (usually aromatic or aliphatic hydrocarbons) prior to polymerization. The droplets are stabilized by surfactants. Free-radical polymerization of the monomers is conducted similarly to that described for solution polymerization (Section 3.2.1) [49, 50].

An obvious reason for using this technique instead of solution polymerization is to simplify the handling of the polymer product. The low-viscosity dispersion of monomer and polymer droplets is easy to agitate and pump and allows more efficient heat removal during polymerization. In addition, the particle size of the product is more easily controlled because it depends on the nature and amount of the emulsifier used to form the droplets prior to polymerization. Droplet size is typically 0.1–10 µm. However, the process is more complicated; several more ingredients need to be controlled and recycled, especially the hydrocarbon phase (unless the product is sold in emulsion form). It is less efficient on a volumetric basis compared to a powder product, due to the presence of the continuous organic phase. However, it is possible to make such inverse emulsions with a high active content. For a cationic polyacrylamide it is possible to use up to 70 wt% of aqueous monomer phase that itself has a monomer concentration of 30–70 wt% [50]. The overall content of the polymer can reach nearly 50 mass%.

It is possible to increase the active content by removing the volatile byproducts and water [51]. Such products are more expensive due to the distillation step and are used for special applications, such as personal care and pharmaceutical formulations [52].

Modern inverse emulsion products are self-inverting. The monomer emulsion is made using a water-in-oil emulsifier (HLB 3.5–7.5). After polymerization a second emulsifier with higher HLB is added (9 to 14), which acts as an oil-in-water emulsifier [53]. If the product is added to a large amount of water, the oil is emulsified in the water, and the polymer is released. This process is very rapid, and the polymer is dissolved completely within a very short time so that inline dosage is feasible. This is an advantage of inverse emulsions over powder products.

Inverse emulsions can be converted into a powder by a spray-drying process [54]. However, this process causes additional costs and is, if at all, used for niche applications.

3.2.3. Suspension Polymerization

In suspension polymerization, small droplets of aqueous monomer solution are dispersed in a hydrocarbon phase, generally with the aid of water-in-oil emulsifier or other polymers [55, 56]. Poly(acrylic acid) (linear and cross-linked) and polyacrylamides can be made by this process. Free-radical polymerization of the monomers is then conducted in a manner similar to that described for emulsion systems (see Section 3.2.2). After the polymerization, the polymer is dried by boiling off the water. The beads are filtered from the organic phase using a centrifugal or other industrial filter. The residual organic phase is then evaporated from the beads using an explosion-proof dryer. The droplet size is controlled by the amount of dispersant used; more dispersant generally yields smaller particle size. Larger polymer particles can be obtained than with emulsion processes. Typically, 15–500 µm spherical beads are formed. Cellulose esters or cellulose ethers are used as protective colloids in the suspension polymerization of sodium acrylate, acrylamide, 2-acrylamido-2-methylpropanesulfonic acid, or sodium styrenesulfonate [57]. A mixed suspending agent for water-soluble monomers consisting of hydrophobic fumed silica and a copolymer of acrylic acid and lauryl methacrylate yields a controlled particle size distribution [58]. A copolymer of cetostearyl methacrylate and methacrylic acid is beneficial for the polymerization of cationic polyacrylamide [59].

3.2.4. Precipitation Polymerization

In precipitation polymerization, the monomer is polymerized in a medium which is a solvent for the monomer but a nonsolvent for the polymer. As the polymer forms, it precipitates from the monomer phase. With the proper choice of medium, polymer isolation and purification is simplified. For example, cross-linked polymers can be produced by copolymerizing acrylic or methacrylic acid with small amounts of an allyl phosphate or phosphite with free-radical initiators in a solvent, such as benzene, hexane, acetone, or chloroethane at ca. 20–90°C [60]. This technology is the basis for the process used industrially to produce Carbopol poly(acrylic acid) resins. The precipitating polymerization of poly(acrylic acid) has been accomplished in methyl ethyl ketone [61].

3.2.5. Polymer-Analogous Reactions

The term polymer-analogous reactions denotes reactions that are used to modify the functional groups of polymers. Such reactions are complex from a practical perspective because of the high viscosities of polymer solutions and the changing solubilities of the polymers as they are converted to a different functionality. Polymers of acrylate esters can be hydrolyzed to the corresponding acid polymers and their salts. Polyacrylate esters can also be reacted with amines to yield polyacrylamide or polymethacrylamide derivatives. Polyacrylonitrile can be saponified to a copolymer of acrylic acid salt and acrylamide.

Poly(acrylic acid) undergoes reactions typical of other organic carboxylic acids. It has been reacted with diazomethane to yield poly(methyl acrylate) and with thionyl chloride to yield the acid chloride, which has increased reactivity in acylation reactions, such as amidation. A direct amidation of polyacids is carried out in water at elevated temperature to obtain partially sulfonated polymers [62, 63]. Poly(acrylic acid) can also be dehydrated above 200°C to yield a polyanhydride. The polyacids are generally reacted with neutralizing agents, e.g., alkali-metal hydroxides, carbonates, or amines, to yield the salt. Esterifications and transesterifications have been carried out in organic media using enzymes as catalyst [64]. Polyacrylamides can also be chemically derivatized [65]. Polyacrylamide reacts with formaldehyde in aqueous solution to yield a partially hydroxymethylated polymer. It is also readily hydrolyzed with aqueous sodium hydroxide or sodium carbonate solution to give a polymer with a maximum of ca. 86% acrylic acid moieties [66]. Main issue for this process is the large amount of ammonia that is released. Amide groups are converted to amine groups via the Hofmann reaction [67]; the amide is reacted with sodium hydroxide and hypochlorite to yield the amine. Amine groups can also be introduced into polyacrylamide by the Mannich reaction. Polyacrylamide is reacted with formaldehyde and an amine to produce an N-aminomethylated polymer. Intramolecular imidization occurs when polyacrylamide is heated with a strong acid catalyst.

Due to the processing difficulties encountered in conducting polymer-analogous reactions, such reactions are not typically used to convert a polymer to one with different functional groups. Nevertheless, the technique is used widely for cross-linking polyacrylamides and poly(acrylic acids). Polyacrylamide is cross-linked by reaction of its amido groups with glyoxal (→ Glyoxal). Alternatively, a reactive monomer can be synthesized from acrylamide and glyoxylic acid, copolymerized with acrylamide, and used as a reactive site for subsequent cross-linking (→ Glyoxylic Acid). Poly(acrylic acid) is cross-linked by reaction of its carboxylic acid groups with glycols; this technique is used industrially to cross-link the surface of superabsorbent polymers (→ (Superabsorbents, Chap. 4)).

3.2.6. Residual Monomer Scavenging

An important part of the production of these polymers is the reduction of residual monomers, especially residual acrylamide as it is carcinogenic. Depending on the application, different types of polyacrylamides are available. For drinking water application, special grades are offered with very low residual acrylamide.

The residual monomer content can be reduced by sulfite solution or gaseous sulfur dioxide [68, 69]. The principle is the Michael addition of a scavenger to the double bound of the residual monomer [70]. Another option is the use of initiator combinations containing one component that reacts at later stage of the production process, e.g., a UV initiator is activated in the drying by irradiation [71].

For solution polymers and inverse emulsions a reshoot with additional polymerization initiator is possible. Simple distillation [72] and solvent extraction [73] have also been employed.

4. Properties

The properties of the polymers as solids and in solution (particularly aqueous solution) determine their applications. Their hydrophilic nature and molecular mass are key determinants of the usefulness of the polymers for specific applications. The molecular mass and the molecular mass distribution can be measured by light scattering, osmometry, and ultracentrifugation. Intrinsic viscosity is the most useful and widely used method of correlation between molecular mass and viscosity: Appropriate equations are given below for individual polymers. Molecular masses are also correlated to the radius of gyration of the molecules in solution by gel permeation chromatography (size-exclusion chromatography) or hydrodynamic chromatography. Very high molecular mass polymers are susceptible to degradation by high shear [74], oxidation [75], and conformational changes [76], which can be induced by impurities introduced as polymerization modifiers. The above methods are useful for measuring the extent of such changes.

Analysis of the polymers and determination of their content of comonomer residues are of practical importance. These can be determined by infrared spectroscopy [77] and nuclear magnetic resonance (NMR) spectrometry [78]. The tacticity of the polymers is also measured by NMR spectrometry [79]. Thermal transitions and polymer chemistry are investigated with differential scanning calorimetry, thermogravimetric analysis (TGA), and TGA-mass spectrometry.

Poly(acrylic acid) [9003-01-4] is soluble in many polar and usually hydrogen-bonded solvents such as water, methanol, ethanol, dioxane, dimethylformamide, 2-propanol, formamide, ethylene glycol, and acetic acid. The value of the glass transition temperature (T_g) is still in dispute because it is difficult to obtain a solventfree sample. T_g has been reported at 106°C [80], at 118°C [81], and at 126°C [82]. Mark–Houwink–Sakurada relationships for determining the weight-average molecular mass M_W from the intrinsic viscosity $[\eta]$ have been determined in many solvents but the following two are widely used:

$[\eta] = 0.085\, M_W^{0.5}$ mL/g   dioxane, 30°C [83]
$[\eta] = 0.1062\, M_W^{0.5}$ mL/g   0.2 mol/L HCl, 14°C [84]

The exponents of 0.5 indicate theta conditions for the polymer. Under theta conditions, a given polymer–solvent pair behaves as if it were an ideal solution. This is because the attraction between polymer chain segments just equals the loss of entropy brought about by interpenetration of polymer molecules. If conditions for solvation are made any poorer than those at the theta condition, the polymer precipitates. For poly(acrylic acid), lowering the temperature from the upper critical solution temperature, UCST, causes phase separation. Poly(acrylic acid) is stable in acid or base solution. The solid polymer slowly liberates water upon heating above 100°C, forming the anhydride. The polymer decomposes slowly above 250°C, liberating carbon dioxide.

The simple alkali-metal salts have been characterized extensively, especially poly(sodium acrylate) [9003-04-7]. Poly(potassium acrylate) [25608-12-2] and poly(ammonium acrylate) [9003-03-6] are also known. These polymers are soluble in water and in aqueous salt solutions. Small amounts of organic solvents dissolved in water are sometimes tolerated. The fully neutralized sodium salt has a glass transition temperature at about 250°C [81]. Obtaining an accurate value is difficult because of the hygroscopic nature of the polymer. Mark–Houwink–Sakurada relationships have been determined in many aqueous solutions; for example

$[\eta] = 1.24 \times 10^{-3} M_W^{0.5}$ mL/g 1.5 mol/L NaBr, 15°C [85]
$[\eta] = 0.146 \times 10^{-3} M_V^{0.80}$ mL/g 0.1 mol/L NaCl, 25°C [86]

Poly(acrylic acid) salts are more stable to heat than the acid polymer. Anhydride formation is not observed above 50 mol% neutralization of acid groups and up to 250°C, where carbon dioxide evolution is observed.

Polyacrylamide [9003-05-8] is soluble in polar, hydrogen-bonded solvents, such as water, aqueous salt solutions, ethylene glycol, formamide, ethanolamine, dimethyl sulfoxide, and morpholine. For cationic and anionic copolymers, water is the main solvent. The glass transition temperature is again difficult to determine, because it is difficult to prepare a solvent-free sample; T_g is reported as 170°C [87] and 188°C [82]. The density of the polymer is 1.302 g/cm^3 [88]. The diffusion coefficient D and the sedimentation coefficient S_0 have been determined as a function of molecular mass [89]:

$D = 1.24 \times 10^{-4} M_W^{-0.53}$ cm^2/s
$S_0 = 1.07 \times 10^{-15} M_W^{0.48}$ svedberg

Mark–Houwink–Sakurada relationships have been determined in many solvents [90]:

$[\eta] = 9.33 \times 10^{-3} M_W^{0.75}$ mL/g water
$[\eta] = 7.19 \times 10^{-3} M_W^{0.77}$ mL/g 0.5 mol/L NaCl
$[\eta] = 1.36 \times 10^{-1} M_W^{0.51}$ mL/g ethylene glycol

A theta condition has been established in a 2:3 ratio of methanol : water at 20°C [91], and has been estimated at −38°C in aqueous solution.

Solutions of polyacrylamide in strong acid gradually become turbid due to imide formation from amide groups. The amide groups hydrolyze to ammonia and carboxylic acid above pH 9. The solid polymer is stable below 210°C, but the amide groups decompose to ammonia above 210°C.

The determination of the molecular mass is quite difficult as most of the polymer have a molecular mass above the exclusion of GPC. Field-flow fractionation has been established as a method to evaluate molecular masses [91]. Very important for the characterization of polyacrylamides are rheological methods [92].

Poly(methacrylic acid) [25087-26-7] is soluble in ethanol, methanol, water, dimethylformamide, dioxane, and 2-ethoxyethanol. The polymer and its sodium salt [54193-36-1] have been characterized because of curiosity about the conformational change that occurs upon neutralization of the acid polymer to about 30 mol%. The glass transition temperature occurs at 185°C [93]. Mark–Houwink–Sakurada relationships are:

$[\eta] = 22.4 \times 10^{-3} M_W^{0.679}$ mL/g methanol, 25°C [94]
$[\eta] = 6.6 \times 10^{-3} M_W^{0.5}$ mL/g 0.002 mol/L HCl, 30°C [95]

The above theta condition at 30°C in dilute hydrochloric acid is questionable in light of experiments that show phase separation in poly(methacrylic acid) solutions when the temperature is raised to 56°C [84].

The stability of poly(methacrylic acid) and its salts to acids and bases is similar to that of poly(acrylic acid). The solid polymer evolves water above 180°C, forming an anhydride.

Polymethacrylamide [25014-12-4] is soluble in methanol, water, ethylene, glycol, and mixtures of acetone and water. A Mark–Houwink–Sakurada relationship has been established as

$[\eta] = 2.1 \times 10^{-3} M_W^{0.76}$ mL/g water [96]

Stability of polymethacrylamide, either in solution or as dry solid, has not been reported.

The major copolymers are those of acrylic acid and acrylamide, prepared either by partial hydrolysis of polyacrylamide [97] (hydrolyzed, sodium salt [25987-30-8] and hydrolyzed, potassium salt [31212-13-2]) or by copolymerization of the respective monomers [82]. The properties of the copolymers depend on the mole fractions of the two monomer residues [98] and on the distribution of molecular masses of the sample [99]. The intrinsic viscosity (and the second virial coefficient) of the copolymers shows a maximum at 33 mol% acrylic acid residues [98]. Flocculation rate increases with molecular mass; maximum activity was observed with a 70 mol% acrylate copolymer [100]. Solution characteristics of heterodisperse copolymers were determined at various compositions in salt solutions by viscometry, light scattering, and gel permeation chromatography. Mark–Houwink–Sakurada relationships have been determined over the entire composition range with corrections for heterodispersity [99]. Copolymers of methacrylic acid and acrylamide have been similarly characterized as a function of pH [101]. The sudden increase in intrinsic viscosity observed at pH 4–5.5 has been attributed to a sudden unfolding of the chains when the electrostatic repulsion of the ionic groups is just sufficient to overcome the hydrophobic bonding of the α-methyl groups.

Poly(N-isopropylacrylamide) [25189-55-3] is of interest because of a lower critical solution temperature (LCST) of just 31°C in aqueous solution [102]. Dissolved poly(N-isopropylacrylamide) precipitates and its cross-linked gels deswell when the temperature is raised above the LCST. It is soluble in water, methanol, and tetrahydrofuran. The glass transition temperature has been measured at 130°C [103]. A Mark–Houwink–Sakurada relationship has been determined:

$[\eta] = 0.112 \, M_W^{0.51}$ mL/g water, 20°C [104]

Poly(N,N-dimethylacrylamide) [26793-34-0] is soluble in water, methanol, 2-methoxyethanol, methyl ethyl ketone, diethyl ether, toluene, chloroform, and dioxane. It has a low glass transition temperature (89°C) [105]. The polymer has been extensively characterized only in methanol and water. The Mark–Houwink–Sakurada relationships are [106]:

$[\eta] = 17.5 \times 10^{-3} \, M_W^{0.68}$ mL/g methanol, 25°C
$[\eta] = 23.2 \times 10^{-3} \, M_W^{0.81}$ mL/g water, 25°C

Poly(2-acrylamido-2-methylpropanesulfonic acid) [27119-07-9] is soluble in water, aqueous salt solutions, and dimethylformamide. The sodium salt is soluble in aqueous solution. Mark–Houwink–Sakurada relationships for the homopolymer and its sodium salt are [107]:

$[\eta] = 2.11 \times 10^{-3} \, M_W^{0.8}$ mL/g H$^+$ form, 5 mol/L NaCl
$[\eta] = 1.95 \times 10^{-3} \, M_W^{0.85}$ mL/g Na$^+$ salt in 0.5 mol/L NaCl

Solution properties of carboxylated and sulfonated copolymers have been compared [108].

5. Uses

The molecular mass of a polymer is a key factor in determining its usefulness for a particular application. Polymers of molecular mass less than 20 000 are used as sequestrants and scale inhibitors. Polymers of molecular mass between 20 000 and 80 000 are useful as pigment dispersing agents. Polymers with molecular mass between 10^5 and 10^6 are used as textile finishing agents and retention aids for papermaking. Molecular masses exceeding 10^6 are necessary for flocculants and thickening agents [109]. High molecular mass, cross-linked polymers are used as fluid absorbents. Uses of functional polymers in general and in particular of poly(acrylic acids) and polyacrylamides have been described [1]. The ionic character of the polymer is another important factor in determining its applications. Choices range from nonionic polyacrylamides to partially hydrolyzed polyacrylamides or copolymers of acrylic acid and acrylamide and finally to the salts of poly(acrylic acid). The appropriate choice of ionic content depends on the pH, ionic strength, surface charge of any substrate with which the polymer interacts, and other conditions specific to the application.

The polymers are supplied as solutions, emulsions, and powders. Low molecular mass polymers are typically supplied as viscous, aqueous solutions containing 20–50 wt% polymer. Above a molecular mass of 2×10^5, the high viscosity of such solutions makes them too difficult to meter into processes and mix well; they are supplied as powders packaged in 22 kg bags. The powders are difficult to dissolve in water because the particles hydrate exceedingly fast, forming gelatinous clumps. Special mechanical dispersers are available from the suppliers to aid dissolution [110]. High molecular mass polymers are also supplied as inverse emulsions; micrometer-sized polymer particles are dispersed in a liquid hydrocarbon with the aid of surfactants. These emulsions are of relatively low viscosity and readily pumped and metered. When the emulsion is added to water the emulsion inverts and the small polymer particles quickly dissolve. Emulsion products are supplied in drums or tank trucks. Cross-linked polymers, such as superabsorbents are supplied as granules in 1-t sacks that empty from the bottom to make addition of the polymer into hoppers easier.

5.1. Poly(Acrylic Acids) and Their Salts

Water-soluble poly(acrylic acid) and its neutralized salts with a molecular mass of 2 000–5 000 are used as scale inhibitors, sludge (mud, silt, clay) dispersants in cooling water systems, and as dispersants for preparing $CaCO_3$ or clay slurries as fillers, pigments, or paper-coating materials [111].

Homo- or copolymers of acrylic acid and methacrylic acid and mixtures with up to 10 wt % alkyl acrylate comonomers prevent soil redeposition in liquid detergent formulations [112]. The polymer, neutralized or not, has a molecular mass of 10^5–10^6. Relatively small amounts of polymer prevent soil redeposition, improve cleaning performance and viscosity control.

Copolymers having small amounts of pendant hydrophobic groups are also useful as hydraulic fracturing fluids for oil well drilling [113]. The fluids are readily formulated to give an initial viscosity that is retained for long periods at high temperatures and under high shear rates; this is due to hydrophobic bonding between the pendant groups. The polymer molecular mass can be below 10^6 and yet provide high viscosity in solution because the associating hydrophobic groups create a high pseudomolecular mass.

Cross-linked poly(sodium acrylate) is used as an absorbent in baby diapers [114–116]. Other applications include adult incontinent products [117], feminine hygiene products [118], absorbents in transmission cables [119], and agricultural mulches [120]. Cross-linked polymers of acrylic or methacrylic acid, neutralized to greater than 50 mol%, may also be used to provide high water retention capacity and high viscosity in plaster [121]. A cross-linked, water-absorbent acrylic acid polymer can be used in formulations for sustained-release tablets for oral administration [122]. A blend of water-swellable particles of anionic polymer of acrylic acid or methacrylic acid and their soluble salts and cationic polymers of amino acrylate and aminoalkyl acrylamide is used in prepasted wall-covering adhesive [123] to reduce water absorption. Cross-linked poly(acrylic acid) is also used as a weakly acidic, cation-exchange resin ($\rightarrow$ Ion Exchangers).

5.2. Polyacrylamides and Hydrolyzed Polyacrylamides

Polyacrylamides are typically high molecular mass products ($>10^6$) made by gel polymerization or inverse emulsion polymerization. Less common products are inverse suspension polymers, which are used in some special applications.

Slightly hydrolyzed polyacrylamide with molecular mass $(1-5) \times 10^5$ is useful as a strengthening agent in finished paper. Polymers with molecular masses up to 10^6 add wet strength in paper production. Polyacrylamides with molecular masses of $(1-2) \times 10^6$ are used as retention aids in papermaking. The polymer binds the smaller fibers and any filler particles to the larger fibers.

Polyacrylamides and molecular masses of $(1-2) \times 10^6$ are used as retention aids in papermaking. The polymer binds the smaller fibers and any filler particles to the larger fibers.

Polyacrylamides and hydrolyzed polyacrylamides having molecular masses of $(2-20) \times 10^6$ are used in applications involving flocculation or to increase the viscosity of aqueous systems. Flocculation is the major use for polyacrylamides [124]($\rightarrow$Water, 4. Treatment by Flocculation and Filtration)) and ($\rightarrow$Acrylic Acid and Derivatives)). Flocculant applications include clarification of water used in industrial processes (e.g., as pretreatment of boiler water), in treatment of effluent from paper mills, and in sewage treatment. In sewage treatment, polyacrylamides increase the removal of solids in the primary settling step, improve clarification of the discharged water, aid in sludge thickening, improve clarity in the supernatant from digesters, and increase water removal during sludge dewatering [125]. Molecular mass and and mole fraction of ionic groups are optimized to obtain maximum performance. Copolymers of acrylamide and acrylic acid of different compositions were prepared and their flocculation activity toward kaolin clay suspensions studied [100]. Flocculation rate increased with molecular mass, maximum flocculation activity are observed with 70 mol% acrylate copolymer.

In the exploration and production of mineral oil, there are several applications for polyacrylamides. They are used as rheology modifier for aqueous drilling fluids (water-based muds) and shale inhibitors in drilling of boreholes [126]. The water-based mud is used to power and cool the drill bit and to stabilize the formation. Additionally, the water-based mud transports the cuttings to the surface. To be most efficient a complex rheology is necessary. Low molecular polyelectrolytes are used as fluid loss control agents. These products prevent the loss of water from the drilling into the formation. The same principle applies to the cementing of the borehole in which similar products are employed.

During the life-cycle of an oil field the enhancement of the gas or oil-flow might be necessary. This process is called stimulation. If the permeability of the reservoir rock is too low, a so-called fracturing might be applied. During this operation, aqueous solutions are pressed under very high pressure into the formation, which induces fractures. To minimize energy losses during pumping, so-called friction reducers (drag reducers) are used. Liquid inverse emulsions of anionic polyacrylamides are frequently used in this application. If the aquifer of the oil reservoir is too close to the producing well, high amounts of water are pumped with the oil. To reduce the amount of produced water, aqueous solutions of polyacrylamides are pumped with gelling agents through the production well into the formation. Common gelling agents are chromium and zirconium salts. This operation is called water shut-off.

With maturity of an oil field, the reservoir pressure is maintained by injection of water from the sides of the reservoir (secondary oil recovery). During this operation, water channels through the oil banks and thus increases the fraction of water in the produced fluid (mixture of oil and water). To reduce the channeling, the viscosity of the injection fluid is increased to that of the oil by the addition of polymers (tertiary oil recovery). Anionic polyacrylamides are used nearly exclusively for this application. For high salinity or high temperature oil fields sulfonated and/or hydrophobically modified polymers are used.

Some superabsorbents containing acrylamide as comonomer are used in special applications, such as soil conditioning or cable insulation (→ Superabsorbents, Chap. 6). Dust is controlled on roads, in mines, and particularly on surfaces of pulverized coal and mineral piles within open transit cars or trucks by applying oil- continuous emulsions of highly branched, cross-linked, water-swellable polymers of acrylamide and acrylic acid [127]. The polymers are in the form of microgelatinous particles ($\leq$ 1 mm diameter) that swell dramatically in water and then bind dust particles.

5.3. Other Polymers

The polymers described in this section are generally experimental products, some of which are available in developmental quantities.

Lightly cross-linked poly(N-isopropylacrylamide) hydrogels have a phase transition at 31°C. This property has been suggested to be useful in separations, such as concentration of soy protein from aqueous extracts [128] or in controlled drugs release [129].

The good balance of hydrophobic and hydrophilic groups in poly(N,N-dimethylacrylamide) and copolymers of N,N-dimethylacrylamide with other water-soluble monomers makes these materials soluble in a wide range of solvents. This suggests utility as thickeners in formulations with high concentrations of organic chemicals. The homopolymer is mutually soluble with poly(vinyl acetate), poly(methyl methacrylate), and polystyrene. Appropriate copolymers may serve as polymer compatibilizers [130]. The homopolymers of

2-acrylamido-2-methylpropanesulfonic acid and copolymers with acrylamide [131] and the corresponding alkali-metal salts have been used in applications where the chelation tendencies of carboxylate copolymers toward di- and trivalent ions are detrimental, and for thickening aqueous solutions of strong acids [132] (e.g., in the acidification of oil wells [133]).

6. Economic Aspects

The homo- and copolymers of acrylic acid, its salts, and acrylamide make up the majority of the production volume of these polymers. Manufacturers and trade names are listed in Table 3. In 2008, worldwide production of poly(acrylic acid) and polyacrylamide amounted to 1.58×10^9 kg and 750×10^6 kg, respectively.

The market for poly(acrylic acid) was divided as follows: 69.6% as superabsorbent polymers, 10.5% as detergent additives, 11.7% as industrial dispersants, 4.6% as water treatment chemicals (primarily sequestrants and scale inhibitors), and 3.8% as coatings components and drilling muds. The polyacrylamide market was divided as follows: 37% for water treatment (primarily flocculants, 22% for municipal and 15% for industrial water treatment), 25% for additives in pulp and paper, 20% for oil recovery, 10% for mineral processing, and 8% other. Geographically, 48.2% of the poly(acrylic acid) was used in the United States and Canada, 31.3% in Western Europe and 20.5% in Japan. For polyacrylamide the breakdown was 28% United States and Canada, 27% Europe, and 45% Asia.

Annual growth for all segments of the polyacrylamide market was expected to be 5–7% in 2008. The poly(acrylic acid) market is dominated by the superabsorbent polymers which is expected to grow by 3 to 4 % in 2006. For the other polymer a slightly lower growth is expected (around 3%)

7. Toxicology, Occupational Health, and Environmental Aspects

Polyacrylamides, hydrolyzed polyacrylamides, poly(acrylic acids) and their salts are relatively nontoxic materials. Care should be taken in the finishing of the polymers to strictly limit the content of any unreacted monomers because they are toxic [135]. Handling of the monomers is described in Chapter 2.

Poly(acrylic acid) showed low toxicity in acute screening tests indicated by an LD_{50} of 5 000 mg/kg (rat, oral) and 3 000 mg/kg (rabbit, dermal) [109, 136]. Screening tests for toxicity to aquatic wildlife showed that the LC_{50} was greater than 1 000 ppm for daphnia, sunfish, and trout [137]. Neutralized products of high alkalinity can be irritating to the eyes and skin.

The toxicity of polyacrylamide and hydrolyzed (anionic) polyacrylamide containing only 500–800 ppm residual monomers has been evaluated in rats, dogs, and fish [138]. Low toxicity is indicated by the single oral dose LD_{50} which is greater than 4 000 mg/kg in rats. The results from two-year feeding studies in rats showed no effect at 1% dietary concentration and a slight retardation in growth at the 5% dietary level. Dogs were unaffected by

Table 3. Manufacturers and trade names of poly(acrylic acids) and polyacrylamides

Company	Trade name	Type*
Poly(acrylic acids)		
Akzo Nobel	Alcosperse	b
BASF SE	Dispex	b
	Sokalan	b
	Hysor	a
BF Goodrich	Carbopol	d
Dow Chemical	Acusol	b
Evonik	Favor	a
Nippon Shokubai K.K.	Aqualic	a
Sanyo Chemical Industries	Sanwet	a
Polyacrylamides		
Alco Chemical	Aquatreat	b
BASF SE	Magnafloc, Zetag	c
	Percol	c, d
	Alcoflood	d
Beijing Hengiu	Hengfloc	c
Kemira	Superfloc	c
	Fennopol	e
Mitsubishi Rayon Co, Ltd.	Diafloc	c
Nalco	Ultimer	c, d, e
SNF Floerger	Aquasorb	a
	Flopam	c
	Floret	e
Solenis	Praestol	c, d
	Praestaret	e

*a = superabsorbent; b = dispersant, scale inhibitor; c = flocculant; d = thickener; e = retention aid.

feeding diets containing up to 6% of these polymers. Fathead minnows, bluegills, and rainbow trout were unaffected by up to 100 ppm of these polymers in the water for 90 days. Minnows were all killed by a solution of 2 500 ppm of hydrolyzed polyacrylamide; the fish had extreme difficulty swimming due to the high viscosity of this test solution. Cationic derivatives of polyacrylamides can kill fish and should not be allowed into natural waterways. Polyacrylamide, cationic- and anionic-modified polyacrylamides, and poly(acrylic acid) and its salts strongly adsorb to many solids. They are thereby effectively removed from sewage in a waste treatment plant as a component of the sludge. Polyacrylamides showe no evidence of toxicity at a dosage of 10^4 mg/kg in rabbits [65]. The fate of polyacrylamides in hydrosystems is reviewed in [139]

References

1. A. Goethlich, S. Koltzenburg, G. Schornick, *Chem. Unserer Zeit* **39** (2005) 262–273.
2. *Aqueous Acrylamide Monomer Technical Book*, Dow Chemical, Midland 1988.
3. L.S. Kirch, J.A. Kargol, J.W. Magee, W.S. Stuper, *Plant Oper. Prog.* **7** (1988) 270.
4. BASF SE, Cargill, Novozymes: Joint Press Release, September 15, 2014.
5. L.B. Levy, J.D. Penrod, *Plant Oper. Prog.* **8** (1989) 105–108.
6. Nippon Shokubai: Explosion and Fire at Acrylic Acid Production Facility – Investigation Report, March 2013.
7. A. Shipp, G. Lawrence, R. Gentry, T. McDonald, H. Bartow, J. Bounds, N. Macdonald, H. Clewell, B. Allen, and C. Van Landingham. *Critical Reviews in Toxicity*, **36** (2006) 481–608.
8. Dia-Nitrix, US0059349-A1, 2013 (K. Murao, M. Kato, Y. Hirata) and cited references.
9. BASF SE, WO120636-A1, 2013 (R. Reichenbach-Klinke, M. Wohlfahrt).
10. H. Kattner, M. Buback, *Macromol. Symp.* **333** (2013) 11–23; J. Barth, M. Buback, *Macromol. React. Eng.* **4** (2010) 288–301; M. Buback, F. Günzler, G.T. Russell, P. Vana, *Macromolecules* **42** (2009) 652–662; M. Buback, H. Frauendorf, O. Jansen, P. Vana, *J. Polym. Sci,: Part A: Polym. Chem.* **46** (2008) 6071–6081; S. Beuermann, M. Buback, P. Hesse, S. Kukuckova, I. Lacik, *Polym. Prepr.* **46** (2005) 929–930.
11. N.F.G. Wittenberg, M. Buback, R.A. Hutchison, *Macromol. React. Eng.* **7** (2013) 267–276; J. Barth, M. Buback, *Macromolecules* **44** (2011) 2474–2480; M. Buback et al., *Ind. Eng. Chem. Res.* **47** (2008) 8197–8204; S. Beuermann et al., *Macromolecules* **41** (2008) 3513–3520.
12. V.F. Gromov et al., *Eur. Polym. J.* **16** (1980) 529–535.
13. H. Ito, S. Shimizu, S. Suzuki, *Kogyo Kagaku Zasshi* **58** (1955) 194–196.
14. G. Blauer, *Trans. Faraday Soc.* **56** (1960) 606–612.
15. V.A. Kabanov, D.A. Topchiev, T.M. Karaputdze, *J. Polym. Sci. Polym. Symp.* **42** (1973) 173–183.
16. S.P. Manickam, K. Venkatarao, N.R. Subbarat-nam, *Eur. Polym. J.* **15** (1979) 483–487.
17. I. Lacík, L. Učňová, S. Kukučková, M. Buback, P. Hesse, S. Beuermann, *Macromolecules* **42** (2009) 7753–7761.
18. H. Kheradmand, J. Francois, V. Plazanet, *J. Appl. Polym. Sci.* **39** (1990) 1847–1857.
19. S.P. Manickam, K. Venkatarao, U.C. Singh, N.R. Subbaratnam, *J. Polym. Sci. Polym Chem. Ed.* **16** (1978) 2701–2702.
20. S.P. Manickam, N. R. Subbaratnam, *J. Polym. Sci. Polym. Chem. Ed.* **18** (1980) 1679–1684.
21. J.P. Riggs, F. Rodriguez, *J. Polym. Sci. Polym. Chem. Ed.* **5** (1967) 3151–3165.
22. J.P. Riggs, F. Rodriguez, *J. Polym. Sci. Polym. Chem. Ed.* **5** (1967) 3167–3181.
23. P. Elayaperumal, T. Balakrishnan, M. Santappa, R.W. Lenz, *J. Polym. Sci. Polym. Chem. Ed.* **18** (1980) 2471–2479.
24. S. Lenka, P.L. Nayak, S. Ray, *J. Polym. Sci. Polym. Chem. Ed.* **22** (1984) 959–965.
25. K. Behari, K. Gupta, *Colloid Polym. Sci.* **262** (1984) 677–682.
26. J.S. Shukla, R.K. Tiwari, *J. Polym. Sci. Polym. Chem. Ed.* **19** (1981) 1517–1524.
27. S. Lenka, P.L. Nayak, *J. Polym. Sci. Polym. Chem. Ed.* **25** (1987) 1563–1568.
28. R.K. Samal, P.L. Nayak, T.R. Mohanty, *Macromolecules* **10** (1977) 489–492.
29. S.M. Shawki, A.E. Hamielec, *J. Appl. Polym. Sci.* **23** (1979) 3155–3166.
30. S. Ponratnam, S.L. Kapur, *Makromol. Chem.* **178** (1977) 1029–1038.
31. M. Haque: A Kinetic Study of Acrylamide/Acrylic Acid Copolymerization, Ph.D. Thesis, University of Waterloo, Ontario, Canada, 2010.
32. R.Z. Greenley in J. Brandrup, E.H. Immergut, E. A. Grulke (eds.): *Polymer Handbook*, 4th ed., John Wiley & Sons, New York 2003.
33. BASF SE, EP1622838-B1, 2004 (P. Baum, E. Lutz, M. Guzmann, K.H. Büchner, G. Brodt).
34. A. Katchalsky, *J. Polym. Sci.* **2** (1947) 432.
35. R.A.M. Thomson, C.K. Ong, C.M. Rosser, J.M. Holt, *Makromol. Chem.* **184** (1983) 1885–1892.
36. American Cyanamid, AU8818380-A, 1989 (W.B. Davies).
37. T. Clarke-Pringle, J.F. McGregor, *Comput. Chem. Eng.* **21** (1997) 1395–1409.
38. Nippon Shokubai Kagaku, JP85024806-B, 1985 (T. Tsubakimoto, T. Shimomura, Y. Iric).
39. Rohm & Haas, US5216099-A, 1993 (K.A. Hughes, G. Swift).
40. BASF SE, US20120202937-A1, 2012 (B. Urtel, R. Wirschem, E. Heintz, D. Faul).
41. N.F.G. Wittenberg, M. Buback, M. Stach, I. Lacik, *Macromol. Chem. Phys. 2* **13** (2012) 2653–2658.
42. K. Hughes, G. Swift, *Adv. Polym. Ser.* **213** (1986) 145–151.
43. BASF SE, EP0192153-B1, 1988 (W. Denzinger, H. Hartmann, W. Trieselt, A. Hettche, R. Schneider, H.-J. Raubenheimer).
44. Nalco Chemical, US4143222, 1979 (L.A. Goretta, R.R. Otremba).
45. BASF SE, WO2011/015520 A1, 2011 (Th. Pfeuffer, R. Reichenbach-Klinke, St. Friedrich, M. Guzmann).
46. M. Kleiner, A. Brodhagen, *Chem. Ing. Tech.* **84** (2012) 1393–1402.
47. List AG, EP1546215-B1, 2005 (S. Belkhiria, P.A. Fleury, I.M. Al-Alim).
48. BASF SE, EP2297211-B1, 2011 (M. Krüger, S. Blei, W. Heide, M. Weismantel, U. Stueven).

49. Ondeo Nalco Company, EP1297039-B1, 2001 (W. Whipple, C. Maltesh, C. Johnson, T. Giuddendorf, A. Sivakumar, A. Zagala).
50. Cytec Technology Corp., US5945494, 1999 (R.E. Neff, J.J. Pellon, R.G. Ryles).
51. Ciba, WO2009/021849, 2008 (B. Loehner, M. Butter, D. Normington, C. Fleetwood).
52. Ciba, US6365656, 2002 (M. Green, E.B. Ridley, D.E. Gavin).
53. Ashland, EP2205643-B1, 2012 (J. Schulte, S. Bellmann, F. Moskwa).
54. Cytec, WO1997/48732, 1996 (W. Davies, J.E. Healy, G. Miller, J. Kozakiewicz, R. G. Ryles).
55. Elf Atochem, US5994419, 1999 (C. Collette, M. Hidalgo, J.-M. Corpart, A. Kowalik, P. Mallo).
56. Sumitumo Seika Chemicals, US 8084544, 2011 (S. Fukudome, J. Takatori, K. Matsuda, Y. Nawata).
57. Kao Soap, US4446261, 1984 (S. Harada, H. Yamasaki).
58. Dow Chemical, US4708997, 1988 (F.W. Stanley, Jr., J.C. Lamphere, Y. Chonde).
59. Allied Colloids, US4506062, 1985 (P. Flesher, A.S. Allen).
60. BF Goodrich, DE1595727-B, 1975.
61. P.J. Kay, F.E. Treloar, *Makromol. Chem.* **175** (1974) 3207.
62. BASF SE, WO2005/028527-A, 2003 (M. Guzmann, H. Becker, K. Michl, A.J. Kingma, S. Nied).
63. BASF SE, WO2005/044868-A, 2003 (M. Guzmann, K.H. Büchner, P. Baum, G. Brodt).
64. BASF SE, WO2006/063037-A, 2005 (T. Friedrich, A. Göthlich, D. Häring, B. Hauer, T.L. Herrera).
65. American Cyanamid, *Cyanamer, Polyacrylamides for the Processing Industries*, Wayne 1982.
66. V.F. Kurenkov, H.-G. Hartan, F.I. Lobanov, *Russ. J. Appl. Chem.* **74** (2001) 534–559.
67. S.P.C.M. SA, US2011/0263796, 2009 (R. Hund, C. Auriant).
68. Dow Chemical, US2960486, 1960 (D.J. Pye).
69. Dow Chemical, US3780006, 1973 (M. Zweigle).
70. Stockhausen, DE3724709-A1, 1987 (M. Chmelir).
71. Cytec, WO2001/25289, 2000 (D. Cywar, R. Ryles. M. Holdsworth).
72. American Cyanamid, US4132844, 1979 (M.W.C. Coville).
73. Casella, US4794166, 1988 (F. Engelhardt et al.).
74. W. Nagashiro, T. Tsunoda, *J. Appl. Polym. Sci.* **21** (1977) 1149.
75. R.D. Shupe, *JPT J. Pet. Technol.* **33** (1981) 1513.
76. W.-M. Kulicke, R. Kniewski, *Makromol. Chem.* **181** (1980) 823.
77. W.-M. Kulicke, H.-H. Hörl, *Colloid Polym. Sci.* **263** (1985) 530–540.
78. C. Chen, D.D. Muccio, T.S. Pierre, *Macromolecules* **18** (1985) 2154–2157.
79. K. Hatada, T. Kitayama, K. Ute, *Polym. Bull. (Berlin)* **9** (1983) 241–244.
80. L.J. Thompson Hughes, D.B. Fordyce, *J. Polym. Sci.* **22** (1956) 509.
81. G.S. Haldankar, H.G. Spencer, *J. Appl. Polym. Sci.* **37** (1989) 3137–3146.
82. J. Klein, R. Heitzmann, *Makromol. Chem.* **179** (1978) 1895–1904.
83. S. Newman, W.R. Krigbaum, C. Laugier, P.J. Flory, *J. Polym. Sci.* **14** (1954) 451.
84. A. Silberberg, J. Eliassaf, A. Katchalsky, *J. Polym. Sci.* **23** (1957) 259.
85. A. Takahashi, M. Nagasawa, *J. Am. Chem. Soc.* **86** (1964) 543.
86. T. Kato, T. Tokuya, T. Nozaki, A. Takahashi, *Polymer* **25** (1984) 218–224.
87. H.K. Yuen, E.P. Tam, J.W. Bulock, *Anal. Calorim.* **5** (1984) 13–24.
88. American Cyanamid, *The Chemistry of Acrylamide*, Wayne, NJ, 1969.
89. T. Schwartz, J. Francois, G. Weill, *Polymer* **21** (1980) 247.
90. W.-M. Kulicke, R. Kniewski, J. Klein, *Prog. Polym. Sci.* **8** (1982) 373–478.
91. M. Leeman, M.T. Islam, W.G. Haseltine, *J. Chromatogr. A* **1172** (2007) 194–203.
92. D.A.Z. Wever, F. Picchioni, A.A. Broekhuis, *Progr, Polym. Sci.* **36** (2011) 1558–1628 and references therein.
93. W.E. Fitzgerald, L.E. Nielsen, *Proc. R. Soc. London, A*: **282** (1964) 137.
94. N.M. Wiederhorn, A.R. Brown, *J. Polym. Sci.* **8** (1952) 651.
95. A. Katchalsky, *J. Polym. Sci.* **6** (1951) 145.
96. P. Molyneux: *Water Soluble Synthetic Polymers*, vol. **1**, CRC Press, Boca Raton, Florida 1984, p. 104.
97. K. Nagase, K. Sakaguchi, *J. Polym. Sci., Part A* **3** (1965) 2475–2482.
98. W.-M. Kulicke, H.-H. Hörl, *Colloid Polym. Sci.* **263** (1985) 530–540.
99. K.J. MacCarthy, C.W. Burkhardt, D.P. Parazak, *J. Appl. Polym. Sci.* **33** (1987) 1683–1698.
100. W.-M. Kulicke, R. Kniewske, H.-H. Hörl, *Angew. Makromol. Chem.* **87** (1980) 195–204.
101. H. Rios, L. Gargallo, D. Radic, *Makromol. Chem.* **182** (1981) 665–668.
102. M. Heskins, J.E. Guillet, *J. Macromol. Sci. Chem.* **A2** (1968) 1441–1455.
103. O. Smidsrod, J.E. Guillet, *Macromolecules* **2** (1969) 272.
104. K. Kubota, S. Fujishige, I. Ando, *Polym. J.* **22** (1990) 15–20.
105. S. Krause et al., *J. Polym. Sci. Part, A* **3** (1965) 3573.
106. L. Trossarelli, M. Meirone, *J. Polym. Sci.* **57** (1962) 445–452.
107. L.W. Fisher, A.R. Sochor, J.S. Tan, *Macromolecules* **10** (1979) 949–959.
108. C.L. McCormick, D.L. Elliot, *Macromolecules* **19** (1986) 542–547.
109. Röhm, *Acrylic Resins: Polyelectrolytes*, Darmstadt 1985.
110. Dow Chemical, *Pusher Mobility Control Polymers: Enhanced Oil Recovery*, Houston 1980.
111. Rohm & Haas, EP123482-B, 1986 (W.M. Hann, J. Dupre, J. Natoli).
112. BF Goodrich Co., EP213500-A, 1987 (M.K. Nagarajan, F.J. Wherley, J.W. Frimel).
113. Dow Chemical, US4541935, 1985 (V.G. Constein, M.T. King).
114. Dow Chemical, US3669103, 1972 (B.G. Harper, R.N. Bashaw, B.L. Atkins).
115. Procter & Gamble, US4673402, 1987 (D.A. Gellert, D.I. Houghton, P.T. Weisman).
116. Kao Soap Co., US4364992, 1982 (O. Ito, N. Kazunori).
117. Kimberly-Clark, US4675012, 1987 (J.A. Rooyakkers).
118. Kao Corp., US4519799, 1985 (A. Sakurai, H. Mizutani).
119. Dow Chemical, EP24631-A, 1981 (K.E. Bow).
120. Unilever NV, EP72214-B, 1986 (J.A. Bosly, R.B. Dehnel, S.A. Symien).
121. Nippon Shokubai Kagaku, JP88046724-B, 1988 (T. Tsubakimoto, T. Shimomura, Y. Iric, Y. Masuda).
122. BF Goodrich, US3330729, 1967 (C.H. Johnson Jr.).
123. Allied Colloids, EP77618-B, 1986 (J.B. Clarke, J.F. Firth).
124. M.B. Hocking, K.A. Klimuchuk, S. Lowen, *Rev. Macromol. Chem. Phys.* **C39** (1999) no. 2, 177–203.
125. B. Bolto, J. Gregory, *Water Res.* **41** (2007) 2301–2324.
126. R. Caenn, G.V. Chillingar, *J. Pet. Sci. Eng.* **14** (1996) 221–230.

127. Nalco Chemical, US4417992, 1983 (B.R. Bhattachar, W.J. Roe).
128. R.F.S. Freitas, E.L. Cussler, *Sep. Sci. Technol.* **22** (1987) 911.
129. A.S. Hoffman, A. Afrassiabi, L.C. Dong, *J. Controlled Release* **4** (1986) 213.
130. K. Ogino, *Yuki Gosei Kagaku Kyokaishi* **33** (1975) 504–506.
131. Dow Chemical, US4737541, 1988 (D.L. Stavenger, J.W. Sanner, P.L. Slaber).
132. General Mills Chemical, US3931089, 1976 (C.L. Karl).
133. Halliburton Co., US4107057, 1978 (W.R. Dill, E.A. Elphingstone).
134. M. Fox, T. Gibson, R. Mulach, T. Sasano: "Acrylic Acid and Esters", *Chemical Economics Handbook*, SRI International, Menlo Park, Feb. 1990.
135. C.B. Shaffer, US Environmental Protection Agency, Office of Toxic Substances, document no.878 211 672, Washington 1983.
136. Rohm and Haas, *Polymeric Additives for Aqueous Systems*, Philadelphia 1986.
137. Rohm and Haas, *Acrysol Polymers for Household Laundry Detergents*, Philadelphia 1986.
138. D.D. McCollister, C.L. Hake, S.E. Sadek, V.K. Rowe, *Toxicol. Appl. Pharmacol.* **7** (1965) 639–651.
139. A.G. Guezennee, C. Michel, K. Bru, S. Touzé, N. Desroche, I. Mnif, M. Motelica-Heino, *Environ. Sci. Pollut. Res.* (2014) DOI 10.1007/s11356-014-3556-6 (accessed 26 September 2014).

Further Reading

M. Chanda, S.K. Roy: *Industrial Polymers, Specialty Polymers, and Their Applications*, CRC Press, Boca Raton 2009.

C.-S. Chern: *Principles and Applications of Emulsion Polymerization*, Wiley, Hoboken, NJ 2008.

A.V. Herk (ed.): *Chemistry and Technology of Emulsion Polymerisation*, Wiley-Blackwell, Oxford 2005.

R.C. Klingender (ed.): *Handbook of Specialty Elastomers*, CRC Press, Boca Raton 2008.

K. Matyjaszewski, Y. Gnanou, L. Leibler (eds.): *Macromolecular Engineering*, Wiley-VCH, Weinheim 2007.

T. Meyer (ed.): *Handbook of Polymer Reaction Engineering*, Wiley-VCH, Weinheim 2005.

Polyacrylates

ERICH PENZEL, BASF Aktiengesellschaft, Ludwigshafen, Federal Republic of Germany

1.	Introduction	675	4.1.1.	Polymer Solutions	685
2.	Raw Materials	676	4.1.2.	Polymer Dispersions	685
3.	Production	676	4.2.	Properties of the Polymers	687
3.1.	Emulsion Polymerization	676	5.	Uses	689
3.1.1.	Starting Materials	680	5.1.	Paints and Coatings	689
3.1.2.	Production Processes	681	5.2.	Paper Industry	691
3.2.	Solution Polymerization	683	5.3.	Adhesives and Sealing Compounds	691
3.3.	Elimination of Residual Monomers	684	5.4.	Textile Industry	692
3.4.	Toxicology and Environmental Aspects	685	5.5.	Leather Industry	692
			5.6.	Miscellaneous Uses	692
4.	Properties	685	6.	Economic Aspects	692
4.1.	Polymer Solutions and Dispersions	685		References	693

1. Introduction

Acrylic acid and acrylate esters have been known since the middle of the nineteenth century. A systematic investigation of acrylate esters was published in 1901 by H. VON PECHMANN and O. RÖHM [12]. A process for the industrial production of acrylate esters was developed in 1928 by H. BAUER [13] of Röhm & Haas. Solution polymers have been produced from methyl methacrylate since 1927 by Röhm & Haas (Darmstadt, Germany). Emulsion polymers were first developed on an industrial scale in 1929/30 by H. FIKENTSCHER [14, 15], and were introduced onto the market by BASF as a polymer dispersion named "Corialgrund" for the surface finishing of leather.

The use of polyacrylates in many fields of application increased rapidly with the development of new methods for producing acrylic acid and acrylate esters (e.g., the process described by W. REPPE in 1953 [16]).

Acrylate esters are compounds of the general formula

$$CH_2=CH-\overset{O}{\underset{\|}{C}}-OR$$

where R is an alkyl group.

Methyl, ethyl, n-butyl, and 2-ethylhexyl acrylate are produced on a large scale. Other acrylate esters such as *tert*-butyl, isobutyl, or lauryl acrylate are also produced industrially, but have not yet become quantitatively important.

Homopolymers of acrylate esters are suitable only for a few areas of application, whereas their copolymers have numerous potential applications. Methacrylates (particularly methyl methacrylate), styrene, acrylonitrile, vinyl acetate, vinyl chloride, vinylidene chloride, and butadiene are used as comonomers to obtain the desired copolymer properties.

In addition, auxiliary monomers are frequently incorporated in polyacrylates to obtain specific technical properties in dispersions. Although these auxiliary monomers are only present in low concentration in the polymer, they have a substantial influence on the colloid chemistry and other properties of polymer dispersions. Commonly used auxiliary monomers include acrylic acid, methacrylic acid, acrylamide, methacrylamide, and other monomers with functional groups.

Polyacrylates can be produced by various processes, emulsion polymerization being the most important industrial method. Solution polymerization is also of practical importance. Bulk polymerization is only performed on a small scale, but will presumably become more important in the future; printing plates are produced by this process. Radiation-curable paints also undergo bulk polymerization. In contrast to the methacrylates, suspension polymerization is of no technical importance.

2. Raw Materials

The production and properties of acrylic acid and its esters are described elsewhere, → Acrylic Acid and Derivatives [17]. Acrylates produced industrially by esterification of acrylic acid are available with a purity of 99–99.5%. Special acrylates may have a somewhat lower purity. Acrylate esters are clear, colorless liquids, which are normally stabilized with 15 ± 5 ppm of hydroquinone monomethyl ether. These small amounts of stabilizer do not affect polymerization. Hydroquinone monomethyl ether does not cause discoloration during polymerization and is therefore preferred to hydroquinone. Since the stabilizer is effective only in the presence of oxygen, the monomers must be stored under air and not under an inert gas. In exceptional cases (e.g., overseas dispatch or in special acrylates), a higher amount of stabilizer is used (≥ 100 ppm).

Unstabilized monomeric acrylates can be stored for only a few weeks at temperatures below 10 °C. The stabilized monomers are generally stored below 20 °C. Acrylic acid and methacrylic acid should not, however, be stored below 18 °C because they crystallize and spontaneously polymerize at this temperature.

Some physical properties of acrylates, comonomers, and auxiliary monomers are summarized in Table 1. Further data on the industrially most important acrylate esters are listed in Table 2. Guidelines for analyzing acrylates to determine their purity, acid number, water content, and stabilizer content [17] can be obtained from the technical data sheets of the manufacturers [18].

Acrylates with appropriate substituents can be copolymerized to obtain copolymers with special properties such as low flammability, antistatic behavior, cross-linkability, elasticity, solubility or insolubility in certain solvents, hydrophilicity, or hydrophobicity [22]. Such substituents include halogens, hydroxyl, epoxy, or N-methylol groups. Hydrophilicity is achieved by incorporating acrylic acid, acrylamide, or hydroxy-functional monomers (e.g., hydroxypropyl or hydroxyethyl acrylate). Incorporation of relatively large amounts of acrylic acid can increase hydrophilicity to such an extent that the emulsion polymers dissolve completely in water as a result of salt formation, to form a clear solution at pH 6–8 [23].

Compatibility with cationic auxiliaries is achieved by incorporating cationic monomers such as dimethylaminoethyl methacrylate or diethylaminoethyl acrylate.

Physical properties of selected acrylates and methacrylates with reactive groups are listed in Table 3 [17].

3. Production

Acrylates can be polymerized extremely easily because their carbonyl groups are adjacent to a vinyl group. Polyacrylates are produced almost exclusively by radical polymerization; conventional radical formers (e.g., peroxides and other per compounds) or azo starters are used as initiators. Polymerization can also be initiated photochemically, by γ-rays, or by electron beams. Although ionic (particularly anionic) polymerization [24] is possible, this process is not used industrially. Ordered (i.e., cis or trans) polymers can be produced by this polymerization process. The heat of reaction in the exothermic polymerization of acrylates is ca. 60–80 kJ/mol, and must be removed if the process is to be controlled effectively [25, 26]. The kinetics of radical polymerization and copolymerization are described in detail elsewhere (→ Polymerization Processes, 1. Fundamentals, → Polymerization Processes). The reactivity ratios r_1 and r_2 for industrially important acrylates and comonomers are summarized in Table 4.

3.1. Emulsion Polymerization

Emulsion polymerization is the most important industrial process for producing polyacrylates and their copolymers [47–50]. Polymers with

Table 1. Physical properties of monomeric acrylates and other comonomers [18]

Monomers	CAS registry no.	M_r	fp, °C	bp (101.3 kPa), °C	d_4^{25}	n_D^{25}	Solubility in water, g/100 cm³	
							25 °C	80 °C
Acrylates								
Methyl acrylate	[96-33-3]	86	−74	80	0.951	1.4010	5.2	5.3
Ethyl acrylate	[140-88-5]	100	−69	99	0.917	1.4037	1.6	1.8
n-Propyl acrylate	[925-60-0]	114	−65	119	0.904	1.4100	1.5	1.6
n-Butyl acrylate	[141-32-2]	128	−61	148	0.895	1.4160	0.15	0.20
Isobutyl acrylate	[106-63-8]	128	−60	135	0.885	1.4122	0.18	0.22
sec-Butyl acrylate	[2998-08-5]	128	−63	131	0.887	1.4110	0.21	0.22
tert-Butyl acrylate	[1663-39-4]	128	−68	120	0.879	1.4080	0.15	0.15
n-Hexyl acrylate	[2499-95-8]	156	−46	89[a]	0.879	1.4247	0.04	0.08
2-Ethylhexyl acrylate	[103-11-7]	184		217	0.880	1.4332	0.04	0.04
Lauryl acrylate	[2156-97-0]	240	−8	129[b]	0.870	1.4437	<0.001	<0.001
Comonomers								
Methyl methacrylate	[80-62-6]	100	−48	101	0.939	1.4119	1.5	1.6
Ethyl methacrylate	[97-63-2]	114	−76	118	0.909	1.4138	0.54	0.70
n-Butyl methacrylate	[97-88-1]	142		163	0.889	1.4220	0.08	0.15
Isobutyl methacrylate	[97-86-9]	142	−35	155	0.879	1.4172	0.13	0.19
tert-Butyl methacrylate	[585-07-9]	142	−47	144	0.873	1.4120	0.10	0.16
n-Hexyl methacrylate	[142-09-6]	170	−65	207	0.894	1.4290	<0.03	0.03
Acrylonitrile	[107-13-1]	53	−82	78	0.805	1.3888	8.3	9.7
Methacrylonitrile	[126-98-7]	67	−36	90	0.795	1.3977	2.6	3.0
Styrene	[100-42-5]	104	−31	145	0.905	1.5455	0.02	0.06
1,3-Butadiene	[106-99-0]	54	−109	−4	0.615	1.4220[c]	0.081	
1,2-Butadiene	[590-19-2]	54	−136	19	0.645	1.4200		
Vinyl acetate	[108-05-4]	86	−97	73	0.933	1.3959	2.9	2.9
Vinyl propionate	[105-38-4]	100	−81	95	0.917	1.4042	0.67	0.68
Vinyl chloride	[75-01-4]	62	−154	−13	0.911	1.398[d]	0.11	
Vinylidene chloride	[75-35-4]	97	−123	32	1.208	1.4237	0.25	
Auxiliary monomers								
Acrylic acid	[79-10-7]	72	14	141	1.044	1.4185	∞	∞
Methacrylic acid	[79-41-4]	86	17	161	1.015	1.4288	∞	∞
Acrylamide	[79-06-1]	72	87	125[a]	1.122		76	∞
Methacrylamide	[79-39-0]	85	112				27	

[a] 2.5 kPa.
[b] 3.8 kPa.
[c] −6 °C.
[d] Calculated.

Table 2. Further properties of industrially important acrylates

Property	Methyl acrylate	Ethyl acrylate	Butyl acrylate	2-Ethylhexyl acrylate
Vapor pressure, kPa				
0 °C	4.2	1.2	0.14	
25 °C	13	4.9	0.68	0.03
50 °C	36	17	2.8	0.16
75 °C	83	44	8.2	0.63
100 °C			22	2.1
125 °C			52	5.8
Heat of vaporization at boiling point, kJ/g	0.39	0.35	0.28	0.26
Specific heat (liquid), J g^{-1} °C^{-1}	2.00	1.967	1.925	1.88
Heat of polymerization, kJ/g	0.915	0.777	0.605	0.424
Rate constant of chain growth, L mol^{-1} s^{-1}	1000 (50 °C) [19]	800 (50 °C) [20]	679 (30 °C) [21]	155 (50 °C) [20]
Rate constant of chain termination $k_t \times 10^{-6}$, L mol^{-1} s^{-1}	3.55 (50 °C) [19]	1.76 (50 °C) [20]	6.4 (30 °C) [21]	0.233 (50 °C) [20]

Table 3. Acrylates and methacrylates with functional groups

Monomers	Formula	CAS registry no.	M_r	fp, °C	bp, °C (kPa)[a]	d_4^{25}	n_D^{25}	Solubility in water at 25°C, g/100 cm^3
Carboxyl								
Fumaric acid	HOOCCH=CHCOOH	[110-17-8]	116	286	165 (0.17)	1.625		0.7
Maleic acid	HOOCCH=CHCOOH	[110-16-7]	116	130		1.609	1.544[b]	78.8
Crotonic acid	CH$_3$CH=CHCOOH	[3724-65-0]	86	16	169 (101)	1.026	1.442[c]	
Itaconic acid	CH$_2$=C(CH$_2$COOH)COOH	[97-65-4]	130	175	decomp.	1.573	1.512[b]	9.1
Di- and triacrylates								
Ethylene glycol diacrylate	CH$_2$=CHCOOCH$_2$CH$_2$OCOCH=CH$_2$	[2274-11-5]	170		70 (0.13)		1.453[c]	
Diethylene glycol diacrylate	H$_2$C=CHCOO(CH$_2$CH$_2$O)$_2$COCH=CH$_2$	[4074-88-8]	224		94 (0.03)		1.457[c]	
Triethylene glycol diacrylate	H$_2$C=CHCOO(CH$_2$CH$_2$O)$_3$COCH=CH$_2$	[1680-21-3]	258		120 (0.02)	1.108	1.462[c]	5.0
Tetraethylene glycol diacrylate	H$_2$C=CHCOO(CH$_2$CH$_2$O)$_4$COCH=CH$_2$	[17831-71-9]	302		328 (101)	1.078	1.463	9.1
Butanediol diacrylate	CH$_2$=CHCOO(CH$_2$)$_4$OOCCH=CH$_2$	[1070-70-8]	198	−53	120 (1.3)	1.049	1.455[c]	0.6
Hexanediol acrylate	CH$_2$=CHCOO(CH$_2$)$_6$OOCCH=CH$_2$	[13048-33-4]	226	−5	107 (0.03)	1.015	1.458	0.12
Neopentyl glycol diacrylate	CH$_2$=CHCOOCH$_2$C(CH$_3$)$_2$CH$_2$OOCCH=CH$_2$	[2223-82-7]	212	9	101 (0.4)	1.024	1.451	0.9
Trimethylolpropane triacrylate	(CH$_2$=CHCOOCH$_2$)$_3$CCH$_2$CH$_3$	[15625-89-5]	296	−22	126 (0.03)	1.102	1.474	0.09
Tripropylene glycol diacrylate	CH$_2$=CHCOO(CHCH$_2$O)$_3$OCCH=CH$_2$, CH$_3$	[42978-66-5]	300	−60	109 (0.03)	1.036		
Di-(meth)acrylamides								
N,N'-Methylene bisacrylamide	CH$_2$=CHCONHCH$_2$HNCOCH=CH$_2$	[110-26-9]	154	188				3.5
N,N'-Methylene bismethacrylamide	CH$_2$=C(CH$_3$)CONHCH$_2$HNCOC(CH$_3$)=CH$_2$	[2359-15-1]	182	164				
Hydroxyl								
2-Hydroxyethyl acrylate	CH$_2$=CHCOOCH$_2$CH$_2$OH	[818-61-1]	116	<−60	60 (0.13)	1.114[c]	1.451[c]	∞
2-Hydroxyethyl methacrylate	CH$_2$=CCOOCH$_2$CH$_2$OH, CH$_3$	[868-77-9]	130	−12	95 (1.3)	1.06[c]	1.450	∞
Hydroxypropyl acrylate	CH$_2$=CHCOO(CH$_2$)$_3$OH	[25584-83-2]	130		63 (0.13)	1.06[c]	1.448[c]	∞
Propylene glycol methacrylate	CH$_2$=CCOOCH$_2$CHCH$_3$, CH$_3$ OH	[2761-09-3]	144		96 (1.3)	1.024	1.446	13.0
Butanediol monoacrylate	CH$_2$=CHCOO(CH$_2$)$_4$OH	[2478-10-6]	144		130 (2.7)	1.039	1.453[c]	∞
Ethers								
Ethyldiglycol acrylate	CH$_2$=CHCO(OC$_2$H$_4$)$_2$OC$_2$H$_5$	[32002-24-7]	188	−66	99 (0.5)	1.006	1.438[c]	14.0
Amines								
Dimethylaminoethyl acrylate	CH$_2$=CHCOO(CH$_2$)$_2$N(CH$_3$)$_2$	[2439-35-2]	143	−92	70 (2.9)	0.936[c]	1.437[c]	∞
Dimethylaminoethyl methacrylate	CH$_2$=CCOO(CH$_2$)$_2$N(CH$_3$)$_2$, CH$_3$	[2867-47-2]	157	−30	68 (1.3)	0.921	1.438	∞

Name	Structure	CAS	bp, °C (kPa)[a]	mp, °C	d_4^{20}	n_D^{20}	Solubility in water, %

N-(3-Dimethylaminopropyl)-methacrylamide	$CH_2=CCONH(CH_2)_3N(CH_3)_2$
	      $|$
	     CH_3
		[5205-93-6]	182			1.479[c]	

Diethylaminoethyl acrylate	$CH_2=CHCOO(CH_2)_2N(C_2H_5)_2$	[2426-54-2]	171	−100	0.925[c]	1.443[c]	10.7

tert-Butylaminoethyl methacrylate	$CH_2=CCOO(CH_2)_2NHC(CH_3)_3$
	      $|$
	     CH_3
		[3775-90-4]	185	< −70	0.914	1.440	11.5

Halogen							
2-Chloroacrylonitrile	$CH_2=CClCN$	[920-37-6]	88	−65	1.085	1.428	1
N-Hydroxymethyl							
N-Hydroxymethyl acrylamide	$CH_2=CHCONHCH_2OH$	[924-42-5]	101	79			∞

N-Hydroxymethyl methacrylamide	$CH_2=CCONHCH_2OH$
	      $|$
	     CH_3
		[923-02-4]	115	54	1.145		∞

Sulfonic acid							

2-Sulfoethyl methacrylate	$CH_2=CCOOCH_2CH_2SO_3H$
	      $|$
	     CH_3
		[10595-80-9]	194		decomp.	1.325		∞

2-Acrylamido-2-methylpropanesulfonic acid	$CH_2=CHCONHC(CH_3)_2CH_2SO_3H$	[15214-89-8]	207	185	1.401	1.560[b]	150
Polyfunctional							

Glycidyl methacrylate	$CH_2=CCOOCH_2CH-CH_2$
	      $|$ $\backslash\ \ /$
	     CH_3 O
		[106-91-2]	142		75 (1.3)	1.073	1.448	3.0

Diacetone acrylamide	$CH_2=CHCONHC(CH_3)_2CH_2COCH_3$	[2873-97-4]	169	57			>100

Diacetone methacrylamide	$CH_2=CCONHC(CH_3)_2CH_2COCH_3$
	      $|$
	     CH_3
		[22029-67-0]	183		107 (0.03)
					138 (1.5)			

Acrylamidoglycolic acid	$CH_2=CHCONHCHCOOH$
	                $|$
	               OH
		[6737-24-2]	145	95			13

Methylacrylamidoglycol methyl ether	$CH_2=CHCONHCHCOOCH_3$
	                $|$
	               OCH_3
		[77402-03-0]	173	73			18

[a] To convert kPa to mm Hg, multiply by 7.5; to convert to mbar, multiply by 10. [b] Calculated from refractive index. [c] 20 °C.

Table 4. Copolymerization parameters r_1 and r_2* [27]

Monomer 1	Monomer 2	r_1	r_2	Reference
Methyl acrylate	acrylonitrile	0.844	1.540	[28]
$Q = 0.45$	butadiene	0.070	1.090	[29]
$e = 0.64$ [46]	methyl methacrylate	0.400	2.150	[30]
	styrene	0.168	0.722	[28]
	vinyl acetate	6.300	0.030	[31]
	vinyl chloride	4.400	0.093	[28]
	vinylidene chloride	1.000	1.000	[32]
Ethyl acrylate	acrylonitrile	0.810	1.160	[33]
$Q = 0.41$	methyl methacrylate	0.280	2.000	[34]
$e = 0.55$ [46]	styrene	0.152	0.787	[35]
	vinylidene chloride	0.720	0.580	[36]
Butyl acrylate	acrylonitrile	0.820	1.080	[37]
$Q = 0.38$	butadiene	0.074	1.040	[29]
$e = 0.85$ [46]	methylmethacrylate	0.430	1.880	[38]
	styrene	0.180	0.840	[39]
	vinyl acetate	3.480	0.018	[40]
	vinyl chloride	4.400	0.070	[41]
	vinylidene chloride	0.873	0.934	[42]
2-Ethylhexyl acrylate	styrene	0.310	0.960	[43]
$Q = 0.37$	vinyl acetate	7.500	0.040	[44]
$e = 0.24$ [46]	vinyl chloride	4.150	0.160	[45]

*For definitions, see → Polymerization Processes, 1. Fundamentals.

high molecular masses are obtained at high polymerization rates; polymerization times are consequently short. Since the viscosity is generally low, the heat of polymerization can easily be removed via the aqueous phase. Emulsion polymers with solids contents of up to 60 wt% and molecular masses of 10^5–10^6 can be produced. These represent advantages compared with solution polymerization, and in addition there is no safety hazard due to flammable solvents.

3.1.1. Starting Materials

The most important starting materials for emulsion polymerization are water, monomers, emulsifiers, initiators, and, if necessary, modifiers.

Water. The water should preferably be distilled or completely demineralized because electrolytes can affect emulsion stability.

Monomers. The monomer composition is governed by the properties required by the relevant area of application.

Emulsifiers. Anionic, nonionic, and, less often, cationic or amphoteric compounds are normally used for emulsification. A combination of ionic and nonionic emulsifiers can considerably improve the stability of the dispersions.

Anionic Emulsifiers include sodium, potassium, or ammonium salts of fatty acids and sulfonic acids; alkali salts of C_{12-16}-alkyl sulfates; ethoxylated and sulfated or sulfonated fatty alcohols; alkyl phenols; and sulfodicarboxylate esters. The emulsifiers are used at concentrations of 0.2–5 wt%, based on monomers.

Nonionic Emulsifiers include ethoxylated fatty alcohols and alkyl phenols with 2–150 ethylene oxide units per molecule.

Cationic Emulsifiers include ammonium, phosphonium, and sulfonium compounds with a hydrophobic moiety that contains at least one long aliphatic hydrocarbon chain.

Many companies offer very large ranges of emulsifiers and emulsifier mixtures that vary in terms of chemical structure. The emulsifier has several functions in emulsion polymerization. It is required for the preparation of the monomer emulsion, and later stabilizes the final dispersion. The particle size (diameter) of the dispersion can be adjusted in the range 50–1000 nm by the choice of emulsifier, its amount, and the mode of addition.

Protective colloids such as cellulose derivatives, starch, or poly(vinyl alcohol) can also be used to stabilize the dispersions. Protective colloids are, however, less suitable for initiating particle formation.

Initiators. Water-soluble peroxo compounds (e.g., alkali persulfates, ammonium persulfate, or hydrogen peroxide) are used as initiators with a polymerization temperature between 50 and 85 °C. The initiators are used in amounts of 0.05–1 wt% (usually 0.2–0.5 wt%) based on monomer. Higher initiator concentrations may adversely affect the stability of the dispersion (electrolyte content) or improve it (additional stabilization); they also lower the molecular mass of the product.

If redox systems are used for initiation [51–53], polymerization can be performed below 50 °C. Oxidizing agents include hydrogen peroxide and alkali persulfates. Iron(II) sulfate, sodium bisulfite, sodium thiosulfate, or sodium formaldehyde sulfoxylate can be used as reducing agent. Iron(II) salts can also be added in small amounts as catalysts. The optimal quantitative ratio of oxidizing agent to reducing agent has to be determined experimentally.

Alkali persulfate cannot be used as an initiator in the preparation of cationic dispersions due to salt formation; water-soluble organic peroxides or hyrogen peroxide are more suitable. Polymerization is generally performed in the pH range 2–7.

Modifiers. The molecular mass of the polymer can be reduced by adding modifiers with the result that the polymers become more tacky. Common modifiers include halogen-containing compounds (e.g., carbon tetrachloride, carbon tetrabromide, bromoform, benzyl bromide, bromotrichloromethane), or thiols (e.g., butyl or dodecyl thiol).

3.1.2. Production Processes

Emulsion polymers are produced from acrylates in batch and semibatch (continuous emulsion or monomer feed) processes. All processes should be performed under nitrogen since oxygen inhibits polymerization.

Batch Process. The monomer emulsion is added to the cooled reaction vessel (<15 °C). The redox initiator is then added, generally as an aqueous solution. Polymerization begins and the temperature rises over the course of 0.5–1 h and then falls due to constant cooling and removal of the heat of reaction. The polymerization rate can sometimes be controlled by appropriate metering of the initiator.

The advantage of the batch procedure is the larger space–time yield. Disadvantages include the poor utilization of the cooling capacity, as well as poor reproducibility of the procedure and product properties.

If cooling is insufficient to prevent the reaction going out of control, the batch can be split into two or more partial batches. If monomers and initiator solution are added in an infinite number of small steps, this is referred to as a continuous feed (semibatch) process [54]. Accurate temperature control is easier with the semibatch (continuous feed) process. Mixed batch and semibatch procedures are used in practice.

Semibatch Procedure. In the *semibatch procedure with emulsion feed* [55, 56], the reaction vessel contains some water and is heated to the desired reaction temperature. Part (5–10%) of the monomer emulsion and initiator solution is then added as an initial batch to the reactor.

Alternatively, some of the emulsifier and the water is added, the remaining emulsifier is then used to prepare the monomer emulsion. The initial batch is polymerized for 15–30 min. The number of particles and the particle size of the polymer dispersion are determined in this starting phase (batch).

Product properties can be modified by adjusting the proportion of emulsifier in the initial batch and the feed. Instead of an initial batch, a finely particulate dispersion can also be used as "seed". The monomer emulsion and initiator solution are then added from separate tanks over 2–3 h.

In the *semibatch procedure with monomer feed* the water and emulsifiers are first added to the reactor. The monomer mixture, initiator solution, and the solution of auxiliaries are then fed in continuously.

In semibatch processes specific product properties can be selectively adjusted by small

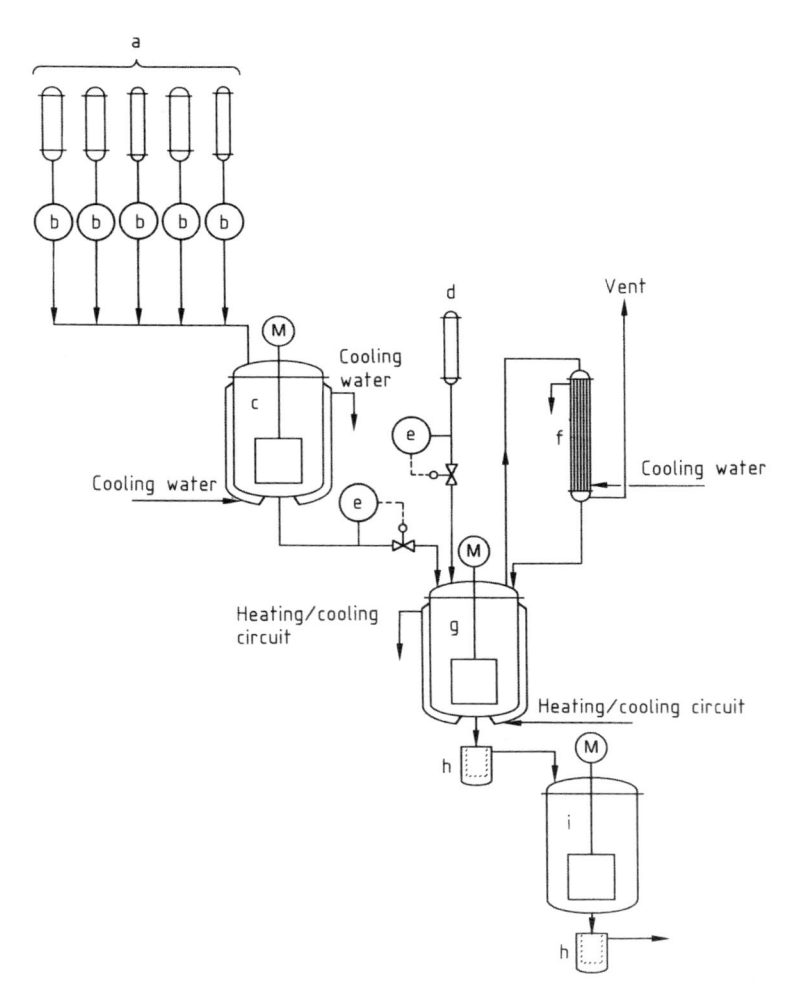

Figure 1. Production of polyacrylates by emulsion and solution polymerization
a) Storage tanks for monomers, distilled water, emulsifiers; b) Flow meter; c) Mixing tank; d) Storage tank for initiator; e) Flow registration control; f) Cooler; g) Polymerization reactor; h) Filter; i) Product storage tank

changes (e.g., of temperature, pH, and delayed addition of certain monomers) [57]. If polymerization is performed under specified conditions, the properties of the resulting dispersions remain virtually constant from batch to batch. These advantages have to be offset, however, by increased complexity, for example as regards temperature control and feed regulation.

Many "specialty" chemicals are produced by batch or semibatch processes. Continuous emulsion polymerization is mainly used on an industrial scale in the production of styrene–butadiene rubber [58–61]. No bulk polyacrylates have so far been produced by a continuous process.

Equipment. Figure 1 shows a production plant for batch and semibatch operation. Water, monomers, emulsifiers, and additives are contained in storage tanks (a) and fed via metering devices (b) to the mixing tank (c), in which the monomer emulsion is produced. The mixing tank is made of stainless steel (V2A, V4A) and equipped with a stirrer and cooling jacket. The shape of the stirrer and stirring rate are chosen to ensure thorough emulsification of the monomers. The mixing tank is, however, unnecessary if the individual components are accurately metered and an in-line mixer is used to prepare the monomer emulsion, or a monomer feed procedure is employed.

The monomer emulsion is usually added to the polymerization reaction batchwise or continuously over a period of minutes or hours. The reactor may be connected to further feed vessels for initiator and other additives, and usually has connections for flushing with inert gas. The internal reactor temperature can be regulated by wall cooling or evaporative cooling.

Vessels, stirrers, and all parts that come into contact with the product are made of stainless steel. The interior wall of the reactor should be as smooth as possible to prevent the deposition of wall scale. The polymerization reactor may have a capacity of 30 m^3 or more. The design, size, and power of the stirrer depend on the batch size and viscosity of the dispersion. High stirring rates (shear rate) may lead to coagulation, while low stirring rates may result in insufficient mixing.

After the end of the feed stage the dispersion is maintained for a further 1–2 h at the reaction temperature to reduce the residual monomer content before being cooled. The dispersion is discharged through a filter (h) and pumped to a storage tank (i). In the storage tank the solids content and pH value are often adjusted and formulations are prepared. Stabilizers and microbiocides are also added.

The dispersions are transported to the consumer or customer in polyethylene-lined metal drums, polyethylene vessels, or tank cars. The purchase of dispersions in barrels is being superseded by delivery in large-capacity containers for economic reasons.

3.2. Solution Polymerization

See also→ Polymerization Processes, 1. Fundamentals. Acrylates are polymerized as solutions in organic solvents [25, 62–64] if the user wishes to exploit specific properties of polymers in dissolved form (e.g., low molecular mass, good flow behavior, and homogeneous film formation after drying in paints or adhesives). As a result of growing environmental awareness (solvent disposal), solution polymerization of acrylates will probably not expand to the anticipated level.

Aromatic hydrocarbons such as benzene and toluene can be used as solvents for the polymerization of acrylates of long-chain alcohols; esters and ketones can be used for acrylates of short-chain alcohols.

Table 5. Transfer constants of various solvents

Monomer	Solvent	$k_u \times 10^4$	Temperature, °C	Reference
Methyl acrylate	acetone	0.622	80	[65]
	benzene	0.326	80	[65]
	2-butanone	3.238	80	[65]
	n-butanol	2.747	80	[65]
	carbon tetrachloride	1.25	80	[66]
	chloroform	2.10	80	[67]
	cyclohexane	0.027	80	[65]
	2-butanol	2.496	80	[65]
	toluene	1.775	80	[65]
Ethyl acrylate	acetone	1.10	80	[68]
	benzene	0.525	80	[68]
	n-butanol	5.85	80	[68]
	sec-butanol	22.20	80	[68]
	carbon tetrachloride	1.55	80	[68]
	chloroform	1.49	80	[68]
	cyclohexane	1.22	80	[68]
	ethyl acetate	0.89	80	[68]
	hexane	0.97	80	[68]
	2-propanol	21		[69]
	methanol	0.32	60	[70]
	toluene	2.60	80	[68]
Butyl acrylate	ethanol	4.28	80	[71]
	methanol	0.47	80	[71]
	propanol	3.78	80	[71]
	2-propanol	14.12	80	[71]
2-Ethylhexyl acrylate	toluene	2.13	80	[20]

If the solvent boils at the temperature used for polymerization, a large proportion of the heat of polymerization can be removed by evaporative cooling.

Solvents may act as modifiers in solution polymerization. This should be kept in mind when choosing the solvent [65–71]. Some transfer constants of solvents are summarized in Table 5. The lower the transfer constant, the higher will be the molecular mass of the polymer. The viscosity of the solution increases with the molecular mass of the polymer. This has an adverse effect on the mixing of the reactor contents and the removal of the heat of polymerization [72]; it its therefore essential to choose an optimum polymer concentration for a given molecular mass.

The molecular mass can be controlled by adding modifiers (e.g., halogenated aliphatic hydrocarbons and thiols [73]) or by adjusting the initiator concentration (increasing initiator concentrations lower the molecular mass).

Soluble azo compounds, peroxides, or hydroperoxides are used as initiators at a concentration of 0.01–2 wt% relative to the

Table 6. Half-lives of organic peroxides and azo compounds used as initiators (0.1 mol/L in benzene)

Compound	Temperature for a half-life of 1 h, °C	Half-life, h					
		20 °C	40 °C	60 °C	80 °C	100 °C	120 °C
Acetylcyclohexanesulfonyl peroxide	46	57	2.4	0.1			
Diisopropylperoxy dicarbonate	61	400	18	1.2			
tert-Butylperoxy pivalate	74		92	5.7	0.5		
Dilauroyl peroxide	80			12	1.0		
Diacetyl peroxide	87			35	2.4		
Dibenzoyl peroxide	91			48	3.8	0.4	
tert-Butylperoxy 2-ethylhexanoate	92				6	0.4	
tert-Butylperoxy isobutyrate	96				6.7	0.6	
tert-Butylperoxy benzoate	124					18	1.6
Dicumyl peroxide	136						5.7
Di-tert-butyl peroxide	146						18
tert-Butyl hydroperoxide	200						
2,2′-Azobis(2,4-dimethylvaleronitrile)	64		25	1.8	0.17	0.021	
Azobis(isobutyronitrile)	82			15	1.2	0.12	

monomer. Some common peroxides and azo compounds together with their half-lifes are given in Table 6 [74, 75].

An initiator should be chosen with a half-life that guarantees a sufficient level of initiator throughout the feed time. Radical formation may be influenced by the solvent [76]. Azo compounds, e.g., 2,2′-azobis(isobutyronitrile), are often used as initiators in polymerization in organic solvents; 0.01–2% of initiator is used depending on the desired product properties.

Solution polymerization can also be initiated with UV light ($\lambda < 360$ nm) or γ-radiation (see Section 5.1). However, gels are often formed due to cross-linking reactions.

As with emulsion polymerization, solution polymers can be produced by batch or semibatch processes (see Section 3.1.2).

Semibatch processes can be efficiently controlled as regards temperature. Part of the solvent and monomer mixture (or all of the solvent with part or all of the initiator) is charged to the reactor. The remaining monomer–solvent mixture (or monomers) is added continuously over the course of 1–4 h. The monomers can also be added in portions. The batch is fully polymerized after 2–24 h.

The apparatus used in solution polymerization is similar to that for emulsion polymerization (Fig. 1). The monomers and solvents are fed into a stirred mixing tank or directly into a polymerization reactor via an in-line mixer.

In addition to the main feed line, the polymerization reactor has inlets for the initiator and modifier solutions. It is equipped with a heating/cooling jacket and an evaporative cooler. Most of the heat is generally removed by evaporative cooling. The initial batch is first heated by the heating jacket. Cooling is generally only necessary to control the reaction in the event of disorders.

The polymer solution should not be subjected to large temperature changes during storage. At elevated temperature there is a danger of solvent loss, while at low temperature the viscosity rises sharply and the polymer solution can no longer be pumped. Storage facilities (e.g., tank cars) should therefore be temperature controlled.

3.3. Elimination of Residual Monomers

Although polymerization proceeds almost quantitatively, residual monomers sometimes have to be removed.

After emulsion polymerization is complete, the dispersion is heated for a short period at 95–98 °C, if necessary under reduced pressure, and the residual monomers are expelled. Foaming can be suppressed by adding antifoaming agents.

Continuous removal of the residual monomers is, however, more effective. Steam is blown into the dispersion, which is under vacuum and begins to foam strongly. The rising dispersion foam is destroyed mechanically and the dispersion returned to the reactor; the monomers are condensed in a cooling device [77]. This process is used on an industrial scale.

Table 7. Toxicity of the most important acrylate esters

Parameter	Methyl acrylate	Ethyl acrylate	Butyl acrylate	2-Ethylhexyl acrylate
OSHA				
PEL[a] (skin)	10 ppm or 35 mg/m^3	5 ppm or 20 mg/m^3	10 ppm or 55 mg/m^3	not regulated
STEL[b] (skin)		25 ppm or 100 mg/m^3		
Irritation data[78]				
rabbit eye	100 mg/24 h, moderate	1204 ppm/7 h	500 mg/24 h, mild	500 mg/24 h, mild
rabbit skin	10 mg/24 h open, mild	500 mg/open, mild	500 mg/open, mild	500 mg/open, mild
Toxicity data[78]				
LD$_{50}$ rat, oral, mg/kg	277	800	900	5660
LC$_{50}$ rat, inhalation, 4 h, ppm	1350	2180	2730	
LD$_{50}$ rabbit, dermal, mg/kg	1243	1834	2000	8480
MAK	5 ppm or 18 mg/m^3	5 ppm or 20 mg/m^3	10 ppm or 55 mg/m^3	
IARC Group[c]	3	2	3	

[a] Permissible environmental limit (8 h TWA).
[b] Short term exposure limit (15 min TWA).
[c] International Agency for Research on Cancer (1987).

In some cases the residual monomer content is lowered by adding small amounts of suitable initiators at the end of the polymerization.

3.4. Toxicology and Environmental Aspects

Polyacrylates, being high molecular mass compounds, are not toxic. Films and coatings produced from acrylate dispersions are permitted for food packaging in Germany [Recommendation XIV Plastics Dispersions (Empfehlung XIV Kunststoff Dispersionen) of the Federal Ministry of Health] and in the United States (21 CFR 176.189 Components of paper and paperboard in contact with dry food, FDA). Auxiliary substances also have to be taken into consideration in food legislation. The toxicology of polyacrylates is determined not by the polymers themselves, but by their auxiliaries (e.g., initiator residues, emulsifiers) and residual monomer content. Table 7 contains toxicity data for the most important acrylate esters. The residual monomer content of polymer dispersions is usually reduced to less than 300 ppm by the methods described in Section 3.3.

With solution polymers there are additional difficulties as regards toxicity, safety, and environmental pollution due to the use of organic solvents. This can be improved by choosing a suitable solvent. In most cases, however, solution polymers are not as environmentally safe as aqueous polymer dispersions; they are treated as combustible mixtures if the solvents are flammable.

4. Properties

4.1. Polymer Solutions and Dispersions

4.1.1. Polymer Solutions

The flow properties of polymer solutions and their formulations are of great importance as regards processability. The temperature dependence of these properties determines the temperature range in which the solution polymer can be handled.

The viscosity of polymer solutions depends on the molecular mass, molecular structure, temperature, and thermodynamic behavior of the solvent [72]. Information on polymer molecular mass, coil dimensions, interaction between solvents and polymers, and the thermodynamic behavior of a solvent are usually obtained from measurements on dilute solutions (light scattering, osmometry, sedimentation, viscosimetry), see also → Plastics, Analysis.

Interactions between the polymer molecules increase with rising concentration. Temporary polymer–polymer contacts are formed that finally lead to the formation of undesirable gels that do not flow and cannot be processed.

4.1.2. Polymer Dispersions

Dispersions consist of two phases, and are therefore more complicated systems than polymer solutions; they often also contain polyelectrolytes. Almost all investigations are carried out on dilute systems.

The most important properties of polymer dispersions are appearance, concentration, pH, surface tension [79], residual monomer content, particle size, particle size distribution, viscosity, and stability.

Residual Monomer Content. Residual monomers are determined by chromatography. The volatile monomers can be quantitatively determined by gas chromatography using calibration and addition of an internal standard (e.g., toluene or methyl ethyl ketone [80]). An accuracy of ca. 1 ppm can be achieved.

The content of water-soluble, nonvolatile monomers (e.g., acrylic acid) is determined by high-pressure liquid chromatography [81].

Particle Size and Particle Size Distribution. The particle size and particle size distribution of emulsion polymers influence other properties (e.g., the viscosity of dispersions with a high solids content). At the same concentration and with the same auxiliary system, coarse dispersions generally have a lower viscosity than fine dispersions due to electrostatic effects or, in the case of dispersions with protective colloids, to steric effects.

Particle size and particle size distribution can be determined by *electron microscopy* (measurement range: 1–1000 nm) [82].

A simple method for determining the mean particle size is to measure the *light transmission* of a highly dilute sample. The relationship between the turbidity and particle radius of spheres is described by Mie's theory [83]. The particle diameter is shown as a function of the light transmission in Figure 2. An important parameter is m, the ratio of the refractive index of the polymer to that of water. The approximate refractive index of copolymers can be calculated with the following formula:

$$n = w_1 n_1 + w_2 n_2 + w_3 n_3 + \ldots$$

where the mass fractions $w_1 + w_2 + w_3 \ldots = 1$ and n_1, n_2 are the refractive indices of the homopolymers.

A mean particle size can also be determined from *static light scattering* (measurement range: 300–1500 nm). Hydrodynamic diameters can be accurately measured by *quasielastic light scattering* (QELS); measurement range: 10–3000 nm) [84–88].

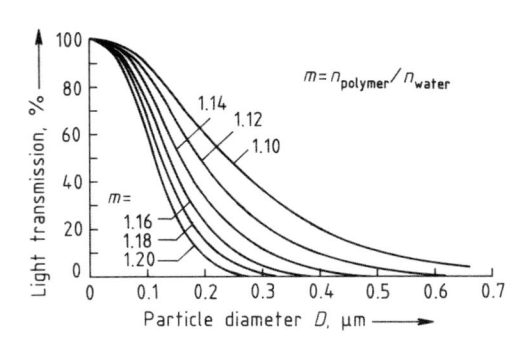

Figure 2. Relationship between light transmission and particle size for 0.01 wt% polymer dispersion (path length 2.5 cm, λ 546 nm)

The particle size distribution can be determined with the *ultracentrifuge* (measuring range: 10–20 000 nm in a gravity field of 20–200 000 g) [89]. The particles must be spherical.

The surface areas of latex (dispersion) particles are determined by *soap titration*, from which the particle size can be calculated [90–93]. Soap titration is simple and quick to perform, and yields reasonable values with homologous polymer dispersion series. Care should be taken, however, when evaluating latex surfaces of different hydrophilicity.

Viscosity. Dispersions with solids contents of 40–60 wt% exhibit a broad spectrum of rheological behavior [94], [95] that depends on the volume fraction, temperature, particle size, particle size distribution, and auxiliary system. Dispersions may be highly liquid to pasty, and may exhibit a Newtonian to viscoelastic flow behavior [96]. In dispersions structural viscosity ("shear thinning") is frequently observed; dilatancy is observed less often; thixotropy and rheopexy are rarely encountered.

Stability. Dispersions must have a certain degree of mechanical (shear) stability if coagulation is not to occur during pumping, stirring, or spraying. Mechanical stability is checked by determining the amount of coagulate after a stirring test; the stirrer, temperature, shear velocity, time, and other parameters should correspond to possible stresses occurring during use.

Thermal stability is checked by subjecting the dispersions to the anticipated service temperature conditions. A freeze–thaw test with five cycles is widely used. However, the results depend on the container size and temperature control so that the correlation between laboratory results and frost stability under relevant weathering conditions is generally unsatisfactory.

The polymer dispersion must be stable to electrolytes when diluted with hard water and when formulating products such as paints or adhesives. The stability is therefore tested by adding salts.

Minimum Film-Forming Temperature. Whether a dispersion forms a coherent film after evaporation of the water or whether a brittle, pulverizable layer is formed depends on the glass transition temperature of the polymer, residual water content, and temperature at which the film is formed. The lowest temperature at which a coherent film can still form from the copolymer dispersion is called the minimum film-forming temperature (MFT). The MFT is measured with a metal plate that has a temperature gradient [97, 98]. The temperature at which the film begins to crack or the so-called white point, at which the turbid film starts to become clear, is measured [99].

The MFT depends on the copolymer composition, as well as the particle size and polarity of the comonomers [100–102]. It is usually comparable to the glass transition temperature, but may be substantially lower, probably because the emulsifier and water act as a plasticizer.

4.2. Properties of the Polymers

Glass Transition Temperature. The glass transition temperatures (T_g) of polyacrylates and homopolymers of monomers that are frequently used in acrylate copolymers are summarized in Table 8. The values were determined by differential thermoanalysis on dispersion films (heating rate 20 °C/min, midpoint), the films being repeatedly heated until reproducible values were obtained. Other factors that influence the glass transition temperature (e.g., molecular mass, [103, 104]) are discussed elsewhere [105, 106].

The glass transition temperature of most copolymers can be accurately calculated with the nonlinear equation of GORDON and TAYLOR [107] given below. Other nonlinear equations

Table 8. Properties of homopolymers

Monomer	CAS registry no. of polymer	T_g^a, °C	d_4^{25}	n_D^{25}
Methyl acrylate	[9003-21-8]	22	1.20	1.479
Ethyl acrylate	[9003-32-1]	−8	1.13	1.464
n-Propyl acrylate	[24979-82-6]	−25	1.10	1.462[b]
n-Butyl acrylate	[9003-49-0]	−43	1.06	1.474
Isobutyl acrylate	[26335-74-0]	−17	1.05	1.464[b]
sec-Butyl acrylate	[30347-35-4]	−6	1.06	1.462[b]
tert-Butyl acrylate	[25232-27-3]	55	1.03	1.468
n-Hexyl acrylate	[27103-47-5]	−51	0.98	1.468[b]
2-Ethylhexyl acrylate	[9003-77-4]	−58	0.99	1.433[b]
Lauryl acrylate	[26246-92-4]	−17	0.99[c]	1.468[b]
Methyl methacrylate	[9011-14-7]	105	1.19	1.488
Ethyl methacrylate	[9003-42-3]	67	1.12	1.483
n-Butyl methacrylate	[9003-63-8]	32	1.06	1.483
Isobutyl methacrylate	[9011-15-8]	64	1.05	1.475
tert-Butyl methacrylate	[25189-00-8]	102	1.02	1.460
n-Hexyl methacrylate	[25087-17-6]	−14	1.01	1.479
Acrylonitrile	[25014-41-9]	105	1.18	1.519
Methacrylonitrile	[25067-61-2]	112	1.18	1.545[b]
Styrene	[9003-53-6]	107	1.05[d]	1.591[d]
trans-1,4-Butadiene	[40022-02-4]	−107	0.97	1.518
cis-1,4-Butadiene	[40022-03-5]	−102	1.01	1.52
Vinyl acetate	[9003-20-7]	42	1.19	1.463
Vinyl propionate	[25035-84-1]	8	1.14	1.465[b]
Vinyl chloride	[9002-86-2]	77[e]	1.40	1.545
Vinylidene chloride	[9002-85-1]	−18[f]	1.84[g]	1.618[b,g]
Acrylic acid	[9003-01-4]	130[h]	1.54[c]	1.544[b]
Methacrylic acid	[25087-26-7]	162[h]	1.45[c]	1.563[b]
Acrylamide	[9003-05-8]	220[h]	1.44[c]	1.603[b]
Methacrylamide	[25014-12-4]	243[h]	1.40[c]	1.616[b]

[a] Emulsion polymers or aqueous solution polymers, differential scanning calorimetry, 20 °C/min, midpoint.
[b] From the refractive increment.
[c] From the specific volume.
[d] Atactic.
[e] Polymerization temperature 90 °C.
[f] Amorphous.
[g] Crystalline.
[h] After repeated heating.

are discussed elsewhere [108].

$$T_g = \frac{T_{gA}w_A + cT_{gB}w_B}{w_A + cw_B}$$

where w_A and w_B are the mass fractions of the monomers A and B, and T_{gA} and T_{gB} are the absolute glass transition temperatures of the corresponding homopolymers ($T_{gA} < T_{gB}$). The constant c takes into account the expansion coefficients of the melt and glass state of the two homopolymers. Its value is generally between 0.5 and 2.0.

With the exception of *tert*-butyl acrylate, most acrylate homopolymers have extremely

Table 9. Determination of the molecular mass according to the Staudinger equation (intrinsic viscosity in cm^3/g)

Polymer	Tetrahydrofuran, 25 °C [109]		Acetone, 20 °C		Chloroform, 20 °C [110]	
	$K \times 10^3$, cm^3/g	a	$K \times 10^3$, cm^3/g	a	$K \times 10^3$, cm^3/g	a
Poly(methyl acrylate)	10.0	0.73	5.6 [110]	0.75	3.0	0.85
Poly(ethyl acrylate)	8.9	0.75	8.8 [110]	0.72	7.3	0.78
Poly(n-propyl acrylate)	11.9	0.72				
Poly(n-butyl acrylate)	7.4	0.75	5.2 [110]	0.73	13.1	0.69
Poly(isobutyl acrylate)	12.7	0.71				
Poly(tert-butyl acrylate)	43.4	0.60	4.7 (25 °C) [111]	0.75		
Poly(n-hexyl acrylate)	5.5	0.76				
Poly(2-ethylhexyl acrylate)	11.1	0.68				
Poly(n-lauryl acrylate)	27.3	0.58				

low glass transition temperatures. They are therefore too soft, too tacky, or have too high an elongation and insufficient strength for many areas of application. Polymers for specific applications can be produced by copolymerization with monomers whose homopolymers have high glass transition temperatures (e.g., styrene, acrylonitrile, methyl methacrylate, ethyl methacrylate).

Molecular Mass. See also → Plastics, Analysis, Chap. 5. The molecular mass can be calculated from the intrinsic viscosity [η] and a knowledge of the two constants K and a:

$$[\eta] = K \cdot M^a$$

Values for K and a in various solvents are listed in Table 9.

Molecular mass determination does not generally present any difficulties with solution polymers. With acrylate emulsion polymers, however, cross-linking of the dispersed particles may cause problems. In this case it is impossible to produce a molecular-disperse solution. Before the molecular mass is determined it must therefore be checked whether a true dilute solution exists.

Mechanical Properties of Polymer Films. The temperature dependence of mechanical moduli [112, 113] provides detailed information about the mechanical behavior and structure of polymer films. Stress–strain diagrams of polymer films [114] are plotted with conventional apparatus; rupture strength and elongation at break are thus obtained. The values exhibit a relatively high degree of scatter with dispersion films, however, due to film defects. It is therefore often more appropriate to use the force required for a specific elongation, or the elongation that occurs when a defined force is applied as a measure of the stress–strain behavior. Mechanical moduli can be accurately measured by performing tensile tests with small extensions.

Poly(methyl acrylate) films have a relatively low elongation at break and an extremely high rupture strength at room temperature. The elongation of poly(ethyl acrylate) films is greater and the rupture strength is considerably lower. Poly(ethyl acrylate) is elastic and almost rubberlike. The elongation of poly(butyl acrylate) is extremely high and the rupture strength very low; the films are not elastic, but exhibit viscous flow even at small elongations and cannot be freely handled. Poly(ethylhexyl acrylate) has an even higher elongation and the films have no strength.

Tackiness. Poly(methyl acrylate) is not tacky at room temperature, poly(ethyl acrylate) is only slightly tacky. Poly(butyl acrylate) and poly(ethylhexyl acrylate) are, however, extremely tacky.

Tackiness depends on the molecular mass: tacky films can be produced from poly(methyl acrylate) with a very low molecular mass, whereas tack-free films can be produced from poly(butyl acrylate) of very high molecular mass. Increasing chain length of the alcohol residue (>C_8) causes the tackiness to fall. The polyacrylates become increasingly "dry" and elongation at break decreases due to crystallization of the n-alkyl chain. For example, poly(n-hexadecyl acrylate) is waxy and rigid at room temperature.

Electrical and Optical Properties. The electrical and optical properties of acrylate homo- and copolymers are similar to those of the polymethacrylates. Polyacrylates hardly yellow under direct sunlight, because they only absorb UV radition below 290 nm. If UV stabilizers are added, their properties do not change at all under the action of light. This is an important advantage of polyacrylates.

Solubility. Polyacrylates with short side chains are relatively polar and thus soluble in polar solvents, aromatic hydrocarbons, and chlorinated hydrocarbons. Common solvents include tetrahydrofuran, dimethylformamide, acetone, butanone, ethyl acetate, and chloroform. Precipitating agents include water, methanol, ethanol, diethyl ether, and aliphatic hydrocarbons.

The solubility parameters for different polyacrylates calculated according to SMALL [115, 116] vary only slightly. The values for polymethyl, polyethyl, and polybutyl acrylate and for the corresponding polymethacrylates are 21.3–18 $(MPa)^{0.5}$ and 10.4–8.8 $(cal/cm^3)^{0.5}$, respectively. Solubility in a certain solvent can be expected if the solubility parameters of the solvent and substance to be dissolved are comparable [117].

Chemical Stability. Polyacrylates are extremely resistant to oxygen, and only decompose very slowly under extreme conditions such as high temperature and in an oxygen-rich atmosphere. When heated, polyacrylates depolymerize to monomers much less readily than the corresponding polymethacrylates. Poly (methyl acrylate) decomposes at 292–399 °C to produce mainly methanol and carbon dioxide [118]. Volatile decomposition products of poly (ethyl acrylate) at 300–500 °C are ethylene, ethanol, and carbon dioxide. Butene, butanol, and carbon dioxide are formed from poly(butyl acrylate) in the same temperature range. Volatile decomposition products of poly(2-ethylhexyl acrylate) are 2-ethylhexene, ethylhexanol, and carbon dioxide [119]. Polyacrylates are largely resistant to acid and alkaline hydrolysis. The longer the alkyl radical, the greater the resistance to hydrolysis [120].

Cross-Linking of Polyacrylate Films. Solubility and tackiness are reduced by cross-linking, the rupture strength increases, whereas the elongation at break decreases. Emulsion polymers and, in special cases, solution polymers are cross-linked by incorporation (copolymerization) of monomers with groups that react with one another under certain conditions (e.g., change in pH or temperature) or in the presence of additives. Such monomers include N-methylolacrylamide, N-methylolmethacrylamide, and the corresponding ethers. The methylol groups react in the acid pH range at elevated temperature to form methylene bisacrylamide linkages with elimination of formaldehyde. Reviews of cross-linking systems are given in [121–128]. Environmentally friendly systems are preferred that cross-link at room temperature and do not lead to the release of toxic substances such as formaldehyde [128, 129]; an example is cross-linking of acrylic acid via formation of metal salts.

Cross-linking and cross-linking density can be determined by measuring the solvent uptake of the polyacrylate film. Suitable solvents are tetrahydrofuran or dimethylformamide, in which the noncross-linked polyacrylates are completely soluble. The relationship between swelling and cross-linking density can be derived from a model used to describe rubber elasticity [130–134]. Measurement of the cross-linking density from the temperature dependence of the modulus of elasticity [135–138] or modulus of torsion is somewhat more time-consuming.

5. Uses

Many important physical properties of polyacrylates and their copolymers can be varied within wide limits by a suitable choice of comonomers, auxiliaries, and polymerization process to give a large number of products for a wide variety of applications. The most important areas of application are concerned with protection, binding, and adhesion.

5.1. Paints and Coatings

See also → Paints and Coatings, 1. Introduction.

Architectural Paints. Good pigmentability, processability, and film properties are important for the use of a polymer dispersion as a paint binder [139]. Vinyl ester copolymer dispersions mostly contain protective colloids, are coarsely particulate, and have favorable rheological properties for coating materials. However, vinyl acetate hydrolyzes on strongly basic substrates and the high water absorption of the dispersion films is also a disadvantage [140].

Acrylate copolymer dispersions usually contain emulsifiers and are finely particulate; they can therefore be highly extended with pigments [141, 142], but auxiliaries are necessary to adjust rheology. The low water absorption of the dispersion films is also an advantage. Emulsion paints based on these dispersions are stable to hydrolysis and are largely weather-resistant [143–145].

Copolymers with very high styrene contents tend to yellow in UV light and are therefore only used for interior paints. Coats with a matt, silk, or high-gloss finish can be formulated. Acrylate ester–methacrylate ester copolymers ("pure acrylates") have outstanding light stability, even in UV light, and are resistant to hydrolysis. They are therefore commonly used as binders for transparent and low-pigment coatings [146]. Specially selected pure acrylate dispersions with wet adhesion on alkyd substrates are used in gloss paints.

Vinyl ester, vinyl ester–acrylate ester, and acrylate ester–methacrylate ester copolymers are widely available in the United States. Styrene acrylics and pure acrylates are becoming increasingly important in Europe.

Trade names include Acronal, Propiofan, Luhydran (BASF); Uramul (DSM); Crilat (Montedison); Mowilith (Hoechst); Walpol (Reichhold Chemie); Primal, Rhoplex (Rohm and Haas); Revacryl (Revertex); Plextol (Röhm); Rhodopas (Rhône-Poulenc); Synthacryl (Synthopol); Ucecryl (UCB); Vinacryl and Vinamul (Vinyl Products); Vinnapas (Wacker).

Coatings and Lacquers. Metals are often coated with a waterborne priming coat and a solvent-containing topcoat. The trend is toward topcoats with high polymer concentrations to reduce solvent emissions. In cases where stringent requirements do not have to be satisfied as regards weather resistance and corrosion protection, the trend is toward waterborne topcoats [147–151].

In the automobile industry three coats are generally applied:

1. A waterborne priming coat, which is mainly responsible for corrosion protection
2. An intermediate coat (filler) which compensates for any unevenness of the substrate
3. A pigmented topcoat

In metallic-effect coatings the topcoat consists of an aluminum-pigmented, colored metallic basecoat and a clearcoat.

The primers may be waterborne polyacrylate systems. Waterborne or low-solvent systems are increasingly used in fillers and metallic basecoats. Acrylic resins with hydroxyl groups, which are reacted with melamine resins or isocyanates, are used in the clearcoat; solution polymers offer advantages as regards application, gloss, and weather resistance.

Trade names include Acronal, Larodur, Luhydran, Lumitol, Luprenal (BASF); Desmophen A (Bayer); Synocryl (Cray Valley); Degalan (Degussa); Elvacite, Lucite (Du Pont); Macrynal, Synthacryl (Hoechst); Joncryl (Johnson); Neocryl (Polyvinyl Chemie); Acryloid, Paraloid (Rohm and Haas); Plexigum (Röhm); Setalux (Synthex); Synthalat (Synthopol Chemie).

Radiation-Curable Systems. Coating methods in which monomers and oligomers are cured (hardened) by exposure to radiation (usually UV light but also electron beams) are experiencing considerable growth (see → Paints and Coatings, 1. Introduction). Advantages of these methods include low emission of volatile organic compounds (VOCs), energy savings, high production rates, and excellent gloss of the coating. However, radiation curing can only be employed on flat surfaces, and cannot therefore be used in the automobile industry. Unsaturated polyester resins (mainly acrylate and methacrylate esters of epoxides), polyesters, oligoethers, urethanes, and polyhydric alcohols are used as binders. Reactive species (radicals) are formed after exposure of photoinitiators to UV light [152–158]. In electron-beam curing, free radicals are generated by interaction of accelerated electrons with organic compounds.

The main areas of use include the coating of wood (doors, furniture, parquet floors), paper (printing lacquers for record sleeves, postcards), and plastics (PVC coverings).

5.2. Paper Industry [159]

High-quality papers are coated with pigments to improve their printability, appearance, gloss, and other properties [160, 161]. Natural products (e.g., starch and casein) were first used as binders. One of the first acrylate-based dispersions, Acronal 500 D, was introduced on the market in the 1950s. Acrylate-based substitution products (cobinders) also became available for casein and starch [162]. The first fully synthetic binders that could be used without cobinders were introduced in the 1960s [163]. Various dispersions with different polymer hardnesses are available for different types of paper and printing methods [164]. Vinyl esters, acrylonitrile, or styrene are used as comonomers for acrylates; styrene gives good adhesion.

In the United States, acrylate dispersions are of minor importance as binders in coating compositions for paper, butadiene copolymer dispersions are used instead. These products are becoming increasingly important in Europe.

Polyacrylate cationic dispersions [165], and dispersions containing protective colloids are used as sizing agents in paper production.

Trade names include Acronal, Acrosol, Basoplast, Styronal (BASF); Primal, Rhoplex (Rohm and Haas).

5.3. Adhesives and Sealing Compounds [166–169]

Laminating adhesives, pressure-sensitive adhesives, building construction adhesives, and sealing compounds are produced from polyacrylates; see also → Adhesives, 1. General. Homopolymers and copolymers with a low glass transition temperature (usually based on 2-ethylhexyl acrylate and butyl acrylate) are used to obtain the desired tackiness. The glass transition temperature can be adapted for individual applications by copolymerization with acrylonitrile, styrene, vinyl acetate, or methyl methacrylate. Products with a balanced adhesion–cohesion ratio can thus be obtained.

Tackiness is extremely important in emulsion adhesives and adhesives based on solution polymers. The higher the glass transition temperature, the lower has the molecular mass to be in order to obtain the necessary tackiness. The higher the molecular mass, the better is the cohesion of the film. Cohesion often has to be improved by additional cross-linking. Depending on the molecular mass, solution polymer concentrations of up to 50% can be produced. Solvent naphtha, aromatic compounds, ethyl acetate, and alcohols are used as solvents. However, solution polymers are being replaced by aqueous dispersions.

Dispersions and solution polymers are used as industrial laminating adhesives (e.g., high-gloss foil laminating, automobile industry). Laminating adhesives for food and cosmetic packaging must be solvent free. Good adhesion to the substrate is important in this area of application, and can be achieved by using monomers with reactive groups (e.g., carboxyl or amide) or special additives.

Polyacrylates are replacing natural products in pressure-sensitive adhesives (e.g., label adhesives, pressure-sensitive sheets, adhesive tapes, adhesive plasters); glyoxal derivatives are used as cross-linking agents (→ Glyoxal, Chap. 5.). The advantages of polyacrylates compared with natural rubber, for example, are their better resistance to aging and temperature. Both solution polymers and dispersions are used [170–172]; solution polymers have a better water resistance than dispersion adhesives [173].

Structural adhesives (adhesives for floor coverings and tiles) are mainly dispersions [174]. Adhesives for floor coverings are formulated with tack-imparting resins and incorporate organic solvents. Developments in this field include the use of solvent-free systems, i.e., addition of solvent-free resins (e.g., aqueous tackifier dispersions). Tile adhesives are solvent free and are generally mixed with cement; dry powder adhesives will be of particular interest in the future.

Sealing compounds are based on dispersions and must be tack-free in the dry state. One-component polyacrylate systems offer economic advantages compared with silicones. Disadvantages are their water sensitivity and poor adhesion to glass and ceramics.

Trade names include Coldac (Allied Colloids); Airflex, Flexbond, Flexcryl (Air Products); Acronal, Lutonal, Propiofan (BASF); Lacstar, Voncoat (Dai Nippon); Emultex

(Harco); Litex (Hüls); Revertex (Revertex); Rhodopas (Rhône–Poulenc); Texicryl (Scott Bader); Ucecryl, Ucefix (UCB); Vinnapas (Wacker).

5.4. Textile Industry

In the textile industry polyacrylates are used almost exclusively as emulsion polymers (e.g., as binders for pigment dyeing or printing) [175–180]. In almost all areas of application apart from bonding and laminating, soft, nontacky films are desired. The treated textiles generally have to be resistant to washing and drycleaning; this is achieved by cross-linking the polyacrylates after film formation. Drycleaning resistance can be improved by copolymerization with polar monomers such as acrylonitrile, but the handle and feel are then hard.

Polyacrylates are also used as binders for textile flocking to produce velvety surfaces. In addition, fabrics may be coated with water-resistant dispersions (e.g., wax cloth and umbrella material).

Polyacrylates are of minor importance in high-grade finishes [181] that are used to improve textile properties such as handle, feel, abrasion resistance, and crease resistance. A further field of application of polyacrylate dispersions is the consolidation of nonwoven fabrics [182–186] for uses such as interlinings [187], filter materials, and carrier materials for leather imitates [188]. Thermosensitive dispersions that coagulate at a specific temperature during the drying process are also used in the production of nonwovens (→ Nonwoven Fabrics, Section 2.2.).

Trade names include Acronal, Helizarinbinder, Perapret (BASF); Acramin (Bayer); Imperon (Hoechst); Primal (Rohm and Haas).

5.5. Leather Industry [189, 190]

The surface of leather is treated to make it hydrophobic while retaining its natural character, in particular appearance. Highly reversible stretching and good cohesion of the dispersion film are important for the use properties; the leather finish must not crack or tear after repeated folding and bending and should have a low tackiness [191–194]. See also → Leather, Chap. 9.. The first synthetic polymer dispersion for industrial use was "Corialgrund" developed by BASF [193].

Trade names include Corialgrund, Corial Binder (BASF); Euderm (Bayer); Hycar (Goodrich); Fondocryl (Henkel); Melio Resin A (K. J. Quinn); Revertex (Revertex); Primal (Rohm and Haas); Resin RA (Stahl Chemical Ind. B.V.).

5.6. Miscellaneous Uses

Polyacrylates, mainly emulsion polymers, are used in many minor areas: for example, as sizes in the production of textile and glass fibers [195–197]; as polishes for floor coverings, automobiles and shoes; and as additives for hydraulic binders. Polyacrylates are used as additives in mortar and concrete, as well as in agriculture for seed protection and spraying with fertilizers (→ Fertilizers, 3. Synthetic Soil Conditioners, Chap. 4.). They are also employed as matrices for ion exchangers cross-linked with divinylbenzene (→ Ion Exchangers, Section 2.1.).

6. Economic Aspects

Detailed production figures for acrylates are only available for the United States [198], and not for Western Europe and Japan. According to estimates for 1982 and 1984, the United States produced ca. 50%, Western Europe ca. 35%, and Japan ca. 15% of all acrylate esters. These figures have probably not changed substantially. Production figures for acrylate esters from 1969 to 1987 are given in Table 10. Total production has increased in the United States and presumably in Europe, whereas in Japan it has remained constant. The percentage proportions of methyl and 2-ethylhexyl acrylates have remained virtually unchanged. Butyl acrylate showed average annual growth rates of 7.3% between 1969 and 1984, while ethyl acrylate experienced a downturn of 0.6%. The consumption of acrylate esters for 1989 and various fields of application is given in Table 11 [198].

Most polyacrylates are produced by emulsion polymerization. In the United States the proportion of solution polymers is 12%; relevant figures are not available for Europe.

Table 10. Production of acrylates (10^3 t/a)[198]

Year	Methyl acrylate	Ethyl acrylate	n-Butyl acrylate	2-Ethylhexyl acrylate	Other acrylates	Total
United States						
1969	14	91	29	15	4	153
1975	20	109	81	15	9	234
1980	29	122	157	31	17	356
1984	28	139	193	39	27	426
1987		147	234			
Western Europe[a]						
1984	60–75	60–75	105	45–60	6	300
Japan[b]						
1975						67
1980						111
1984						106

[a] Chemical Economics Handbook (CEH) estimated values.
[b] Excluding Asahi Chemical.

Although the proportion of solution polymers (which are largely used for coatings and adhesives) may well fall, they will not be completely replaced by emulsion polymers. The use of polyacrylates in the adhesives and paper industries is substantially higher in Europe than in the United States. Methyl acrylate is used almost exclusively in the polymerization of acrylonitrile for fibers to improve dyeability. The mean annual growth rate in the consumption of polyacrylates for 1984–1989 is estimated to be 3.3% for the United States and 2.8% for Europe [198]; specialty fields have the highest growth rates.

Table 11. Estimated consumption of polyacrylates in 1989 expressed in 10^3 t/a (%) [198]

Area of use	United States	Western Europe	Japan[a]
Dispersions			
Surface coatings	124 (32)	103 (36)	38 (33.4)
Adhesives	27 (7)	48 (17)	
Textiles	68 (17.5)	33 (11)	37 (34.9)
Leather	7 (2)	33 (11)	37 (34.9)
Paper	18 (4.5)	49 (17)	
Polishes	13 (3)		
Miscellaneous uses	18 (4.5)		13 (11.9)
Other polymers			
Acrylonitrile fibers	14 (3.5)	26 (9)	15 (13.9)
Solution polymers for surface coatings	45 (12)	28 (10)	
Miscellaneous uses	53 (14)		7 (6.7)
Total	387 (100)	287 (100)	110 (100)

[a] 1984 excluding Asahi Chemicals.

References

General References

1 D. R. Bassett, A. E. Hamielec (eds.): "Emulsion Polymers and Emulsion Polymerization," *ACS Symp. Ser.* **165** (1981).
2 D. C. Blackley: *Emulsion Polymerisation, Theory and Practice*, Applied Sci. Publ., London 1975.
3 D. C. Blackley: *High Polymer Lactices. Their Science and Technology*, vols. **I and II**, Maclaren & Sons, London 1966.
4 F. A. Bovey, I. M. Kolthoff, A. I. Medalia, E. J. Meehan: "Emulsion Polymerization," in *High Polymers*, vol. IX, Interscience, New York 1955.
5 F. Hölscher: "Dispersionen synthetischer Hochpolymerer I: Eigenschaften, Herstellung und Prüfung," in *Chemie, Physik und Technologie der Kunststoffe*, vol. 13, Springer Verlag, Berlin 1969.
6 M. B. Horn: *Acrylic Resins*, Reinhold Publ., New York 1960.
7 I. Piirma, J. L. Gardon (eds.): "Emulsion Polymerization," *ACS Symp. Ser.* **24** (1976).
8 H. Rauch-Puntigam, T. Voelker: "Acryl- und Methacrylverbindungen," in *Chemie, Physik und Technologie der Kunststoffe*, vol. 9, Springer Verlag, Berlin 1967.
9 H. Reinhard: "Dispersionen synthetischer Hochpolymerer II: Anwendung," in *Chemie, Physik und Technologie der Kunststoffe*, vol. 14, Springer Verlag, Berlin 1969.
10 E. H. Riddle: *Monomeric Acrylic Esters*, Reinhold Publ., New York 1954.
11 H. Warson: *The Applications of Synthetic Resin Emulsions*, Ernest Benn, London 1972.

Specific References

12 H. v. Pechmann, O. Röhm, *Ber. Dtsch. Chem. Ges.* **34** (1901) 427.
13 Röhm & Haas, DE 571 123, 1928, US 1 829 **208** (W. Bauer).
14 IG Farbenind., DE 615 219, 1931, GB 387 **736** (E. Scharf, H. Fikentscher).
15 IG Farbenind., DE 654 989, 1930, GB 358 534 (H. Fikentscher, C. Heuck).
16 W. Reppe, *Justus Liebigs Ann. Chem.* **582** (1953) 1.
17 L. S. Luskin: "Vinyl and Diene Monomers," in C. Leonhard (ed.): *High Polymers*, vol. 24, part I, Wiley Interscience, New York 1970, p. 105.
18 *Acrylate Esters–A Guide to Safety and Handling* compiled by Badische Corporation, Celanese Chemical, Rohm and Haas, Union Carbide Corporation, Williamsburg. *The Handling and Storage of Acrylic Esters*, BASF, Ludwigshafen 1989.
19 Z. A. Sinitsyna, Kh. S. Bagdasaryan, *Zh. Fiz. Khim.* **32** (1958) 1319.
20 M. Rätzsch, I. Zschach, *Plaste und Kautsch.* **21** (1974) 345.
21 D. J. Liaw, K. C. Chung, *J. Chin. Inst. Chem. Eng.* **13** (1982) 145.
22 H. Spoor, *Angew. Makromol. Chem.* **4/5** (1968) 142.
23 H. Wesslau, *Makromol. Chem.* **69** (1963) 220.
24 H. Yuki, K. Hatada, K. Ohta, Y. Okamoto, *Appl. Polym. Symp.* **26** (1975) 39.
25 E. H. Riddle: *Monomeric Acrylic Esters*, Reinhold Publ., New York 1954.
26 E. Trommsdorff et al. in E. Müller (ed.): *Methoden der organischen Chemie*, vol. **XIV/1**, Thieme Verlag, Stuttgart 1961, p. 1010.

27. R. Z. Greenley in J. Brandrup,E. H. Immergut (eds.): *Polymer Handbook*, 3rd ed., Wiley-Interscience, New York 1989, p. II/153.
28. C. S. Marvel, R. Schwen, *J. Am. Chem. Soc.* **79** (1957) 6003.
29. C. Walling, J. A. Davison, *J. Am. Chem. Soc.* **73** (1951) 5736.
30. V. P. Zubov, L. I. Valuev, V. A. Kabanov, V. A. Kargin, *J. Polym. Sci. Polym. Chem.* **9** (1971) 833.
31. N. G. Kulkarni, N. Krishnamurti, P. C. Chatterjee, M. A. Sivasamban, *Makromol. Chem.* **139** (1970) 165.
32. F. R. Mayo, F. M. Lewis, C. Walling, *J. Am. Chem. Soc.* **70** (1948) 1529.
33. W. M. Ritchey, L. E. Ball, *J. Polym. Sci. Polym. Lett. Ed.* **4** (1966) 557.
34. G. Markert, *Makromol. Chem.* **103** (1967) 109.
35. A. Fehervari, T. Földes-Bereznich, F. Tüdös, *J. Macromol. Sci. Chem.* **A18** (1982) 337.
36. B. G. Elgood, B. J. Sauntson, *Chem. Ind. (London)* 1965, 1558.
37. T. Tamikado, Y. Iwakura, *J. Polym. Sci.* **36** (1959) 529.
38. N. Grassie, B. J. D. Torrance, J. D. Fortune, J. D. Gemmell, *Polymer* **6** (1965) 653.
39. J. H. Bradbury, H. W. Melville, *Proc. Royal Soc. London Ser. A* **222** (1954) 456.
40. A. F. Nikolaev et al.,*Vysokomol. Soedin Ser. A* **11** (1969) 2418.
41. G. V. Tkachenko, L. V. Stupen, L. P. Kofman, L. Z. Frolova, *Zh. Fiz. Khim.* **31** (1957) 2676.
42. E. F. Jordan, K. M. Doughty, W. S. Port, *J. Appl. Polym. Sci.* **4** (1960) 203.
43. J. T. Khamis, *Polymer* **6** (1965) 98.
44. K. Moser, R. Signer, H. U. Stuber, *Chimia* **23** (1969) 393.
45. G. Talamini, G. Vidotto, C. Garbuglio, *Chim. Ind. (Milan)* **47** (1965) 955.
46. R. Z. Greenley in J. Brandrup,E. H. Immergut (eds.): *Polymer Handbook*, 3rd ed., Wiley-Interscience, New York 1989, p. II/267.
47. H. Fikentscher, H. Gerrens, H. Schuller, *Angew. Chem.* **72** (1960) 856.
48. A. E. Alexander, D. H. Napper, *Prog. Polym. Sci.* **3** (1971) 145.
49. G. Markert, *Angew. Makromol. Chem.* **123/124** (1984) 285.
50. *Houben-Weyl*, 4th ed., **E20/2**, 1150.
51. W. Kern, *Makromol. Chem.* **1** (1947/48) 199.
52. W. Kern, *Angew. Chem.* **61** (1949) 471.
53. C. F. Fryling, A. E. Follett, *Polym. Sci.* **6** (1951) 59.
54. K. W. Min, W. H. Ray, *J. Macromol. Sci. Rev. Macromol. Chem.* **C11** (1974) 177.
55. J. SnuparekJr., *Acta Polym.* **32** (1981) 368.
56. J. SnuparekJr., *Macromol. Chem. Phys. Suppl.* **10/11** (1985) 129.
57. N. Suetterlin, *Macromol. Chem. Phys. Suppl.* **10/11** (1985) 403.
58. G. Ley, H. Gerrens, *Makromol. Chem.* **175** (1974) 563.
59. J. F. Mac Gregor, A. Penlidis, A. E. Hamielec: *IFAC Instrumentation and Automation in the Paper*, Rubber, Plastics and Polymerisation Industries, Antwerpen 1983, p. 375.
60. M. Nomura in K.-H. Reichert,W. Geiseler (eds.): *Polymer Reaction Engineering*, Hüthig and Wepf Verlag, Basel 1986, p. 41.
61. B. W. Brooks in K. H. Reichert,W. Geiseler (eds.): *Polymer Reaction Engineering*, Verlag Chemie, Weinheim, Germany 1989, p. 3.
62. D. H. Klein, *J. Paint Technol.* **42** (1970) 335.
63. W. H. Brown, T. J. Miranda, *J. Paint Technol.* 36, part 2, (1964) 92.
64. *Houben-Weyl*, 4th ed., **E20/2**, 1156.
65. J. N. Sen. U. Nandi, S. R. Palit, *J. Indian. Chem. Soc.* **40** (1963) 729.
66. S. D. Gadkary, S. L. Kapur, *Makromol. Chem.* **17** (1955/56) 29.
67. S. L. Kapur, S. D. Gadkary, *J. Sci. Ind. Res. Sect. B* **17** (1958) 152.
68. P. V. T. Raghuram, U. S. Nandi, *J. Polym. Sci. Polym. Chem. Ed.* **7** (1969) 2379.
69. I. M. Likhterova, E. M. Ludina, *Zh. Obshch. Khim.* **42** (1972) 194.
70. Y. Hachihama, H. Sumitomo, *Technol. Rep. Osaka Univ.* **5** (1955) 497.
71. U. S. Nandi, M. Singh, P. V. T. Raghuram, *Makromol. Chem.* **183** (1982) 1467.
72. H. U. Moritz, *Chem. Eng. Technol.* **12** (1989) 71.
73. I. L. O'Brien, F. Gornick, *J. Am. Chem. Soc.* **77** (1955) 4757.
74. Peroxidchemie, Organische Peroxide, company brochure.
75. Akzo, Initiators for Polymer Production, company brochure.
76. I. Czajlik, T. Földes-Bereznich, F. Tüdös, E. Vertes, *Eur. Polym. J.* **17** (1981) 131.
77. S. M. Englund, *Chem. Eng. Prog.* **77** (1981) no. 8, 55.
78. D. V. Sweet (ed.): *Registry of Toxic Effects of Chemical Substances*, U.S. Department of Health and Human Services, Washington 1988.
79. J. Hansmann, *Adhäsion* **21** (1977) 272.
80. G. Stoev, M. Angelova, *HRC CC J. High Resolut. Chromatogr. Chromatogr. Commun.* **10** (1987) 25.
81. T. R. Crompton: *Analysis of Polymers*, Pergamon Press, Oxford 1989, chap. 9.
82. E. B. Bradford, J. W. Vanderhoff, *J. Polym. Sci. Polym. Symp.* **3** (1963) 41.
83. M. Kerker: "The Scattering of Light and other Electromagnetic Radiation," in E. M. Loebl (ed.): *Physical Chemistry,* Academic Press, New York 1969.
84. R. Pecora (ed.): *Dynamic Light Scattering: Applications of Photon Correlation Spectroscopy*, Plenum Press, New York 1985.
85. R. S. Stock, W. H. Ray, *J. Polym. Sci. Polym. Phys. Ed.* **23** (1985) 1393.
86. P. G. Cummins, E. J. Staples, *Langmuir* **3** (1987) 1109.
87. S. E. Bott, *Polym. Mater. Sci. Eng.* **53** (1985) 68.
88. J. Wagner, *Chem. Ing. Tech.* **58** (1986) 578.
89. W. Mächtle, *Makromol. Chem.* **185** (1984) 1025.
90. T. R. Paxton, *J. Colloid Interface Sci.* **31** (1969) 19.
91. J. Kloubek, K. Friml, S. Petrikova, F. Krejci, *J. Polym. Sci. Polym. Phys. Ed.* **14** (1976) 1451.
92. J. G. Brodnyan, G. L. Brown, *J. Colloid Sci.* **15** (1960) 76.
93. H. Schuller, *Kolloid Z. Z. Polym.* **211** (1966) 113.
94. I. M. Krieger, *Adv. Colloid Interface Sci.* **3** (1972) 111.
95. J. Mewis, A. J. B. Spaull, *Adv. Colloid Interface Sci.* **6** (1976) 173.
96. F. B. Malihi, T. Provder, M. E. Koehler, *J. Coat. Technol.* **55** (1983) no. 702, 41.
97. T. F. Protzman, G. L. Brown, *J. Appl. Polym. Sci.* **4** (1960) 81.
98. G. Taschen, *Plaste Kautsch.* **24** (1977) 212.
99. DIN 53 787, 1974.
100. J. G. Brodnyan, T. Konen, *J. Appl. Polym. Sci.* **8** (1964) 687.
101. S. Eckersley, A. Rudin, *Polym. Mater. Sci. Eng.* **58** (1988) 1115.
102. DIN 53 765, 1991.
103. K. Ueberreiter, G. Kanig, *Z. Naturforsch.* **6A** (1951) 551.
104. K. Ueberreiter, G. Kanig, *J. Colloid Sci.* **7** (1952) 569.
105. M. C. Shen, A. Eisenberg, *Rubber Chem. Technol.* **43** (1970) 95.
106. A. Eisenberg, M. C. Shen, *Rubber Chem. Technol.* **43** (1970) 156.
107. M. Gordon, S. J. Taylor, *J. Appl. Chem.* **2** (1952) 493.

108 K. H. Illers, *Ber. Bunsenges. Phys. Chem.* **70** (1966) 353.
109 E. Penzel, N. Goetz, *Angew. Makromol. Chem.* **178** (1990) 191.
110 W. Wunderlich, *Angew. Makromol. Chem.* **11** (1970) 189.
111 R. Jerome, V. Desreux, *Eur. Polym. J.* **6** (1970) 411.
112 A. Zosel, *Farbe Lack* **94** (1988) 809.
113 D. J. Skrovanek, C. K. Schoff, *Prog. Org. Coat.* **16** (1988) 135.
114 A. Zosel, *Prog. Org. Coat.* **8** (1980) 47.
115 P. A. Small, *J. Appl. Chem.* **3** (1953) 71.
116 D. Mangaraj, S. Patra, S. B. Rath, *Makromol. Chem.* **67** (1963) 84.
117 H. Burrel, *Off. Digest* **27** (1955) part 2, 726.
118 S. L. Madorsky, *J. Polym. Sci.* **11** (1953) 491.
119 N. Grassie, J. G. Speakman, T. I. Davis, *J. Polym. Sci. Polym. Chem. Ed.* **9** (1971) 931.
120 R. F. B. Davies, G. E. J. Reynolds, *J. Appl. Polym. Sci.* **12** (1968) 47.
121 B. G. Bufkin, J. R. Grawe, *J. Coat. Technol.* **50** (1978) no. 641, 41.
122 J. R. Grawe, B. G. Bufkin, *J. Coat. Technol.* **50** (1978) no. 643, 67.
123 B. G. Bufkin, J. R. Grawe, *J. Coat. Technol.* **50** (1978) no. 644, 83.
124 J. R. Grawe, B. G. Bufkin, *J. Coat. Technol.* **50** (1978) no. 645, 70.
125 B. G. Bufkin, J. R. Grawe, *J. Coat. Technol.* **50** (1978) no. 647, 65.
126 J. R. Grawe, B. G. Bufkin, *J. Coat. Technol.* **51** (1979) no. 649, 34.
127 H. Warson, *Polym. Prepr. Am. Chem. Soc. Div. Polym. Chem.* **16** (1975) no. 1, 280.
128 H. R. Lucas, *J. Coat. Technol.* **57** (1985) no. 731, 49.
129 R. G. Lees, H. R. Lucas, G. A. Gelineau, *Polym. Mater. Sci. Eng.* **55** (1986) 315.
130 P. J. Flory, J. Rehner, *J. Chem. Phys.* **11** (1943) 512.
131 P. J. Flory, J. Rehner, *J. Chem. Phys.* **11** (1943) 521.
132 L. R. G. Treloar: *The Physics of Rubber Elasticity*, 3rd ed., Clarendon Press, Oxford 1975.
133 H. Gibmeier, K. Hummel, G. Kerrutt, *Kautsch. Gummi Kunstst.* **22** (1969) 537.
134 E. Penzel, A. Zosel, *Angew. Makromol. Chem.* **99** (1981) 23.
135 A. Zosel, *Farbe Lack* **82** (1976) 125.
136 A. Zosel, *Farbe Lack* **83** (1977) 804.
137 A. Zosel, *Prog. Org. Coat.* **8** (1980) 47.
138 A. Zosel, *Double Liaison Chim. Peint.* **34** (1987) no. 384, 305.
139 K. Weimann, *Farbe Lack* **93** (1987) 447.
140 E. Wistuba, *Congr. FATIPEC* **17** (1984) 209.
141 K. A. Safe, *J. Oil Colour Chem. Assoc.* **53** (1970) 599.
142 K. A. Safe, *Paint Manuf.* **30** (1960) 249.
143 E. V. Schmid, *Congr. FATIPEC* **14** (1978) 83.
144 R. Dhein, L. Fleiter, R. Küchenmeister, *Congr. FATIPEC* **14** (1978) 195.
145 L. A. Simpson, *Congr. FATIPEC* **14** (1978) 623.
146 K. Zimmerschied, *Congr. FATIPEC* **19** (1988) 105.
147 H. U. Schenck, H. Spoor, M. Marx, *Prog. Org. Coat.* **7** (1979) 1.
148 R. Zimmermann, *Farbe Lack* **82** (1976) 383.
149 W. Brushwell, *Farbe Lack* **86** (1980) 706.
150 W. Brushwell, *Farbe Lack* **90** (1984) 924.
151 G. Y. Tilak, *Prog. Org. Coat.* **13** (1985) 333.
152 G. A. Senich, R. E. Florin, *J. Macromol. Sci. Rev. Macromol. Chem. Phys.* **24** (1984) 239.
153 K. Nitzl, *Coating* **13** (1980) 274.
154 K. Nitzl, *Papier (Darmstadt)* **35** (1981) 10A, V 80.
155 C. Decker, *J. Coat. Technol.* **59** (1987) no. 751, 97.
156 W. Brushwell, *Farbe Lack* **91** (1985) 812.
157 S. P. Pappas, *J. Radiat. Curing* **14** (1987) July, 6.
158 A. Vrancken, *J. Oil Colour Chem. Assoc.* **67** (1984) 118.
159 J. Reinbold, *Papier (Darmstadt)* **40** (1986) 10A, V 119.
160 J. Watanabe, P. Lepoutre, *J. Appl. Polym. Sci.* **27** (1982) 4207.
161 H. Wilfinger, *Wochenbl. Papierfabr.* **108** (1980) 153.
162 G. Hirsch, *Papier (Darmstadt)* **32** (1978) 10A, V 66.
163 M. Aschwanden, *Wochenbl. Papierfabr.* **104** (1976) 493.
164 D. Meck, A. Stephan, *Papier (Darmstadt)* **37** (1983) 137.
165 F. Reichel, *Wochenbl. Papierfabr.* **112** (1984) 373.
166 W. C. Wake: *Adhesion and the Formulation of Adhesives*, Applied Science, London 1976.
167 K. Eisenträger, W. Druschke in I. Skeist (ed.): *Handbook of Adhesives*, 2nd ed., Van Nostrand Reinhold, London 1977, chap. 32.
168 A. Zosel, *Colloid Polym. Sci.* **263** (1985) 541.
169 V. E. Basin, *Prog. Org. Coat.* **12** (1984) 213.
170 H. Jaeger, *Adhäsion* **29** (1985) no. 9, 32.
171 W. Druschke, *Adhäsion* **31** (1987) no. 5, 29 and no. 6, 26.
172 H. W. J. Müller, *FINAT News* (1984) no. **1**, 25.
173 M. Toyama, T. Ito, *Polym. Plast. Technol. Eng.* **2** (1973) 161.
174 P. Fickeisen, R. Füßl, R. Hummerich, J. Neumann, *Adhäsion* **32** (1988) nos. 1 and 2, 24.
175 W. Schwindt, *Melliand Textilber.* **50** (1969) 670.
176 G. Frerker, *Melliand Textilber.* **50** (1969) 675.
177 U. Perkuhn, *Melliand Textilber.* **50** (1969) 678.
178 G. Faulhaber, *Melliand Textilber.* **50** (1969) 682.
179 W. Schwindt, G. Faulhaber, A. J. Moore, *Rev. Prog. Color. Relat. Top.* **2** (1970) 33.
180 D. Bechter, G. Kurz, *Melliand Textilber.* **58** (1977) 150.
181 *Ullmann*, 4th ed., **23**, 77.
182 A. Einwiller, *Allg. Vliesst. Rep.* 1976, 78.
183 H. Herlinger, D. Bechter, G. Kurz, *Chemiefasern Text. Ind.* **31/83** (1981) 936.
184 W. Loy, *Melliand Textilber. Int.* **69** (1988) 836.
185 K. Fischer, *Chemiefasern Text. Ind.* **36/88** (1986) 589.
186 D. Bechter, H. Herlinger, G. Kurz, *Adhäsion* **28** (1984) no. 9, 23.
187 H. C. Assent, *Melliand Textilber.* **65** (1984) 176.
188 R. Schmitt, *Chemiefasern Text. Ind.* **28/80** (1978) 911.
189 H. Kittel (ed.): *Lehrbuch der Lacke und Beschichtungen*, vol. 5, W. A. Colomb. Heenemann Verlagsges., Berlin 1977.
190 R. Schubert in H. Herfeld (ed.): *Bibliothek des Leders*, vol. 6, Umschau Verlag, Frankfurt/Main 1982.
191 L. Würtele, *Leder Häutemarkt Gerbereiwiss. + Prax.* **25** (1973) 224.
192 E. C. Cluthe, F. A. Desiderio, W. C. Prentiss, *J. Am. Leather Chem. Assoc.* **73** (1978) 22.
193 E. Penzel, G. Eckert, *Leder Häutemarkt Gerbereiwiss. + Prax.* **33** (1981) 418.
194 M. J. Osgood, *J. Oil Colour Chem. Assoc.* **70** (1987) 104.
195 *Ullmann*, 4th ed., **23**, 12.
196 P. Dürrbeck, H. Leitner, H. Schöpke, *Chemiefasern Text. Ind.* **37/89** (1987) 576.
197 H. Leitner, P. Dürrbeck, K. Stöhr, *Melliand Textilber. Int.* **70** (1989) 893.
198 *Chemical Economics Handbook*, SRI International, Menlo Park 1986.

Further Reading

M. Chanda, S. K. Roy: *Industrial Polymers, Specialty Polymers, and Their Applications*, CRC Press, Boca Raton 2009.
S. K. Ghosh (ed.): *Functional Coatings*, Wiley-VCH, Weinheim 2006.

R. C. Klingender (ed.): *Handbook of Specialty Elastomers*, CRC Press, Boca Raton 2008.

J. M. Margolis: *Engineering Plastics Handbook*, McGraw-Hill, New York 2006.

K. Matyjaszewski, Y. Gnanou, L. Leibler (eds.): *Macromolecular Engineering*, Wiley-VCH, Weinheim 2007.

T. Meyer (ed.): *Handbook of Polymer Reaction Engineering*, Wiley-VCH, Weinheim 2005.

E. M. Petrie: *Handbook of Adhesives and Sealants*, 2nd ed., McGraw-Hill, New York, NY 2007.

A. A. Tracton (ed.): *Coatings Technology*, CRC Press, Boca Raton 2007.

Polyamides

BEN HERZOG, INVISTA Intermediates, Wichita, Kansas, United States

MELVIN I. KOHAN, MIK Associates, Wilmington, Delaware, United States

STEVE A. MESTEMACHER, E. I. DuPont de Nemours & Co., Inc., Parkersburg, West Virginia, United States

ROLANDO U. PAGILAGAN, E. I. DuPont de Nemours & Co., Inc., Parkersburg, West Virginia, United States

KATE REDMOND, E. I. DuPont de Nemours & Co., Inc., Wilmington, Delaware, United States

1.	Introduction	697
2.	Nomenclature	698
3.	General Considerations in Polyamidation	701
3.1.	Molecular Mass	701
3.2.	Equilibrium and Rate Constants .	704
3.3.	Effects of Monomer Structure . . .	706
3.4.	Amide Interchange	707
3.5.	Structure-Property Considerations	707
4.	Other Polymerization Techniques .	708
4.1.	Variants of Hydrolytic Polymerization	708
4.2.	Interfacial and Solution Polymerization	708
4.3.	Ionic Polymerization	708
4.4.	Solid-Phase Polymerization	709
5.	Commercial Processes	710
5.1.	PA 46	710
5.2.	PA 66	711
5.2.1.	Batch Production	711
5.2.2.	Continuous Production	711
5.3.	Other AABB Polyamides	712
5.4.	PA 6	713
5.5.	PA 11	714
5.6.	PA 12	714
6.	Properties	715
6.1.	Properties of Unmodified Polyamides	715
6.2.	Copolymerization	718
6.3.	Modification by Additives	719
6.3.1.	Filled and Reinforced Polyamides . .	719
6.3.2.	Blends and Alloys	720
6.3.3.	Flame-Retardant Polyamides	721
6.3.4.	Conductive Polyamides	721
6.3.5.	Other Formulations	722
7.	Processing	722
8.	Uses	724
9.	Ecological Aspects and Toxicology	725
9.1.	Recycling	725
9.2.	Polyamide Monomers	727
9.3.	Other Aspects	727
10.	Economic Aspects	727
	References	728

1. Introduction

Polyamides are a versatile family of thermoplastics that have properties that vary broadly from relative flexibility to significant stiffness, strength, and toughness. The term "nylon" is generic and is equivalent to "polyamide". A key characteristic of nylons is their resistance to oils, greases, lubricants, hydrocarbons, and most chemicals. Phenols, strong mineral acids, and oxidizing agents attack them.

Polyamides are synthetic polymers that contain multiple amide –CONH– groups as a recurring part of the chain. The most important commercial nylons are the semicrystalline materials, which account for over 90 % of global usage. There is another significant class of nylons that are referred to as being "amorphous". These typically contain a ring structure. There is an almost unlimited range of nylon materials that can be made by using different monomers.

The first publications describing polycondensation were [1, 2]—the basic principles of nylon synthesis were in 1929. Synthesis of poly (hexamethylene adipamide), the original "nylon", was carried out by WALLACE H. CAROTHERS in a DuPont laboratory in 1935 and provided a material whose properties were judged to be appropriate for apparel use. This invention culminated in the first patent for the production of synthetic polyamides in 1937 [3], and their subsequent commercial introduction in 1938. The other principal commercially important polyamide, nylon 6, based on polycaprolactam, was first made at IG Farbenindustrie in Germany by P. SCHLACK in 1938, who was issued a patent in 1941 [4]. The history of these nylons and those made from 11-aminoundecanoic acid and dodecanolactam were reviewed in 1986 [5–7].

The intellectual achievement of creating nylon was a major breakthrough in the science of synthetic polymer chemistry. This development provided incontrovertible evidence for the Staudinger thesis [8] that polymers are high molecular mass species and not aggregated entities. Materials with new types of physical properties were achievable, and could be tailored to various end uses by adjusting parameters such as molecular mass, viscosity, backbone chemistry, additives, etc. CAROTHERS' classic 1931 review [9] of polymerization and his first definitions of the terminology of polycondensation are still valid today. Polycondensation polymerization is a type of step-growth or step polymerization [10] (see Chap. 3). The initial success of nylon led to the numerous scientific and engineering innovations that were required to commercially develop and extend polymer usage. Routes to starting materials were necessary, as was the development of polymerization equipment to manufacture the polymer, and analytical techniques to measure the materials.

Mechanical and physical properties, rate and equilibrium constants, thermal and oxidative stability, resistance to hydrolysis and solvents, and other chemical behavior depend on the specific nylon polymer, but the basic chemistry discussed herein applies to all of these polyamides. Even though nylon was initially used as a fiber, its application as a plastic for a variety of purposes such as brush filaments, wire coating, coil forms, gears, and a variety of parts occurred very soon thereafter. The focus of this article is on the polyamides used as plastics. The technology has evolved to the point where nylons, especially reinforced compositions, are widely used in place of metals, even in under-the-hood applications in automobiles (see Chap. 8). For a discussion of polyamides used as fibers, see → Fibers, 4. Polyamide Fibers.

2. Nomenclature

In current terminology the word "nylon" is a generic term reserved for aliphatic and semiaromatic polyamides. The fully aromatic polyamides are called "aramids" (see → High-Performance Fibers). The origin of the word nylon is uncertain but there is reasonable speculation that it is a mutation of the word "norun" from nylon's first use in hosiery [11, p. 3], [12]. Although nylons can be prepared from various reactive derivatives of carboxylic acids and amines, in general, especially in commercial manufacture, nylons are prepared directly from dicarboxylic acids and diamines [Eq. (1)], ω-aminoacids [Eq. (2)], or lactams [Eq. (3)]:

$$n\,H_2N(CH_2)_xNH_2 + n\,HOOC(CH_2)_{y-2}COOH \rightleftharpoons$$
$$H[NH(CH_2)_xNHCO(CH_2)_{y-2}CO]_nOH + (2n-1)H_2O \quad (1)$$

$$n\,H_2N(CH_2)_{x-1}COOH \rightleftharpoons$$
$$H[NH(CH_2)_{x-1}CO]_nOH + (n-1)H_2O \quad (2)$$

$$n\,HN(CH_2)_{x-1}C=O + H_2O \rightleftharpoons$$
$$H[NH(CH_2)_{x-1}CO]_nOH \quad (3)$$

Equation (3) represents the hydrolytic polymerization of a lactam. It is described in detail in Section 3.2. Two basic monomer types are possible.

1. AB type monomer; both the acid and amine functionalities are contained in the same molecule.
2. AABB type monomer; two monomers are required for polymerization, one monomer contains the amine functionality and the other the carboxylic moiety.

A and B stand for the functional groups NH_2 and COOH, respectively. Lactams are considered AB type monomers as these are derivatives of amino acids, which, by virtue of their structures, have a strong driving force for cyclization (lactam formation). In current practice, AABB and AB may describe products that do not involve polymerization via amine and acid ends but can be visualized as if so made. This is realistic because under conditions of fabrication and processing, moisture is always present.

Reaction of the polyamide with moisture under these conditions (hydrolysis), regardless of the starting monomers used in their preparation, would generate the corresponding amino acid or the acid and amine monomers. The AABB types are regarded as homopolymers because of the absolute requirement for alternation of both reactants to form a polymeric chain that includes both moieties in the repeating structure (see Chap. 3). This is consistent with IUPAC recommendations [13]. Copolymers result from simultaneous polymerization of AABB and AB types or use of more than one AA, BB, or AB type. They are common and were used as early as 1940 [6].

In the early development of nylons, the monomers used were most often linear aliphatic compounds. A simple system of nomenclature that is still used today is to designate these polyamides by the number of carbon atoms in the monomers. The polyamides (PA) or nylons are identified by numbers corresponding to the number of carbon atoms in the monomers (diamine first for the AABB type). Thus, two numbers represent an AABB polyamide; the first number refers to the diamine and the second number to the carboxyl monomer. An AB type is represented by a single number. Table 1 provides a list of the common names of the commercial nylon homopolymers, their numerical designations, their CAS registry numbers, and their CAS names. The monomers TMD in TMDT and PACM in PACMT are mixtures of isomers and TMDT and PACMT are in fact copolymers. The systematic IUPAC naming of polyamides [14] is awkward and is rarely encountered in commercial literature. The CAS names are largely in accord with IUPAC practice, but examination of Table 1 reveals some inconsistencies. Many of the IUPAC rules for nomenclature are conveniently assembled in a polymer handbook [15] wherein, however, older versions of the IUPAC names for PA 66 and 6 are used [16]: poly(iminoadipoyl-iminohexamethylene) and poly [imino(1-oxohexamethylene)].

Table 2 lists some of the most common monomers. As shown in the tables, numerical symbols are simple, convenient, and unambiguous only for linear aliphatic monomers. This system of nomenclature is inadequate and it is often necessary to use letters instead of numbers for

Table 1. Commercial polyamides or nylons

Common name	xy or x*	CAS name, CAS registry no.
Poly(tetramethylene adipamide)	46	poly[imino(1,4-dioxo-1,4-butanediyl)imino-1,6-hexanediyl] [24936-71-8]
Poly(hexamethylene adipamide)	66	poly[imino(1,6-dioxo-1,6-hexanediyl)imino-1,6-hexanediyl] [32131-17-2]
Poly(hexamethylene azelaamide)	69	poly[imino-1,6-hexanediyliminol(1,9-dioxo-1,9-nonanediyl)] [28757-63-3]
Poly(hexamethylene sebacamide)	610	poly[imino-1,6-hexanediylimino(1,10-dioxo-1,10-decanediyl)] [9008-66-6]
Poly(hexamethylene dodecanoamide)	612	poly[imino-1,6-hexanediylimino(1,12-dioxo-1,12-dodecanediyl)] [24936-74-1]
Poly(m-xylylene adipamide)	MXD 6	poly[iminomethylene-1,3-phenylene-methyleneimino(1,6-dioxo-1,6-hexanediyl)] [25805-74-7]
Poly(trimethylhexamethylene terephthalamide)	TMDT	1,4-benzenedicarboxylic acid, polymer with 2,2,4-trimethyl-1,6-hexanediamine and 2,4,4-trimethyl-1,6-hexanediamine [25497-66-9]
Poly(11-aminoundecanoamide)	11	poly[imino(1-oxo-1,11-undecanediyl)] [25035-04-5]
Polycaprolactam or polycaproamide	6	poly[imino(1-oxo-1,6-hexanediyl)] [25038-54-4]
Polydodecanolactam or polylaurolactam or polydodecanoamide	12	poly[imino(1-oxo-1,12-dodecanediyl)] [24937-16-4]
Hexamethyleneadipamide-hexamethyleneterephthalamide copolyamide	6T/66	1,4-benzenedicarboxylic acid, polymer with 1,6-hexanediamine and hexanedioic acid [25776-72-1]
-Caprolactam-hexamethyleneterephthalamide copolyamide	6T/6	1,4-benzenedicarboxylic acid, polymer with hexahydro-2H-axepine-2-one and 1,6-hexanediamine [25086-53-7]
Bis(4-aminocyclohexyl)methane-terephthalic acid copolymer	PACMT	1,4-Benzenedicarboxylic acid, polymer with 4,4′-methylenebis(cyclohexanamine) [117092-07-6]

* xy and x refer to the number of carbon atoms according to Equations (1)–(3).

Table 2. Monomers for polyamides

Common and (CAS) name	CAS registry no.	x or y [*]	Source	Properties and synthesis
Adipic acid (hexanedioic acid)	[124-04-9]	6	benzene, toluene	→ Adipic Acid
Azelaic acid (nonanedioic acid)	[123-99-9]	9	oleic acid	→ Dicarboxylic Acids, Aliphatic, Section 2.4.7.
Sebacic acid (decanedioic acid)	[111-20-6]	10	castor oil	→ Dicarboxylic Acids, Aliphatic, Section 2.4.8.
Dodecanedioic acid	[693-23-2]	12	butadiene	→ Dicarboxylic Acids, Aliphatic
Dimer acid (fatty acids, dimers)	[61788-89-4]	36	oleic and linoleic acids	→ Dicarboxylic Acids, Aliphatic, Section 3.4.2.
Isophthalic acid (1,3-benzenedicarboxylic acid)	[121-91-5]	I	m-xylene	→ Carboxylic Acids, Aromatic → Terephthalic Acid, Dimethyl Terephthalate, and Isophthalic Acid
Terephthalic acid (1,4-benzenedicarboxylic acid)	[100-21-0]	T	p-xylene	→ Terephthalic Acid, Dimethyl Terephthalate, and Isophthalic Acid
Tetramethylenediamine, 1,4-diaminobutane (1,4-butanediamine)	[110-60-1]	4	acrylonitrile and HCN propene	→ Amines, Aliphatic, Section 8.1.3.
Hexamethylenediamine, 1,6-diaminohexane (1,6-hexanediamine)	[124-09-4]	6	butadiene, propene	→ Hexamethylenediamine
2-Methylpentamethylenediamine (1,5-pentanediamine, 2-methyl)	[15520-10-2]	MPMD	butadiene	
4,4'-Diaminodicyclohexylmethane (cyclohexaneamine, 4,4'-methylenebis-)	[1761-71-3]	PACM	aniline and formaldehyde	
m-Xylylenediamine (1,3-benzenedimethaneamine)	[1477-55-0]	MXD	m-xylene	
2,2,4-Trimethylenehexamethylenediamine	[3236-53-1]	ND	acetone	
2,4,4-Trimethylenehexamethylenediamine	[3236-54-1]	IND	acetone	
Dodecamethylenediamine, 1,12-diaminododecane (1,12-dodecanediamine)	[2783-17-7]	12	butadiene	→ Amines, Aliphatic, Section 8.1.3.
11-Aminoundecanoic acid (undecanoic acid, 11-amino-)	[2432-99-7]	11	castor oil	
ϵ-Caprolactam (2H-azepin-2-one, hexahydro)	[105-60-2]	6	benzene, toluene	→ Caprolactam
Laurolactam, dodecanolactam, (azocyclotridecane-2-one)	[947-04-6]	12	butadiene	→ Cyclododecanol, Cyclododecanone, and Laurolactam

[*] $x, y =$ number of carbon atoms according to Equations (1)–(3).

monomers of more complex structures. The absence of unifying rules to deal with monomers other than linear, aliphatic monomers has led to different symbols for the same monomer. More recently ISO 1874-1 has been established and gives recommended symbols for some monomers; however, it fell short of establishing rules for assigning letter designation for complex monomers. Notice the new designation for the isomers of TMD wherein the two positional isomers are given designations of ND for the 2,2,4-trimethylhexamethylenediamine and IND for the 2,4,4-trimethylhexamethylenediamine. However, no distinct symbols were assigned to the *cis–cis, cis–trans*, and *trans–trans* isomers of PACM. Another aspect still not addressed by ISO 1874-1 is the configurational isomerism with polyamides of nonsymmetrical monomers such as ND, IND, and MPMD. These diamines can produce head-to-head, head-to-tail, and tail-to-tail isomerism in their polymers.

The ISO recommendation for polyamides is to use the symbol PA for polyamide followed by the monomer symbol. Pronunciation is in accord with the monomers so that the homopolymer from hexamethylenediamine and adipic acid, which is represented by PA 66, is pronounced PA six-six not sixty-six, PA 612 is PA six-twelve, PA 11 is PA eleven, etc. Often the space between PA and the monomer symbol is eliminated as in PA66, PA612, PA11. Copolymers are designated by "PA" followed by the polyamide symbols separated by a slash. For example, a copolymer of PA 6 and PA 66 is represented as PA 6/66, and a copolymer of PA 66 and PA 610 is PA 66/610. No provisions for polyamide blends have been established. KOHAN [11, p. 5] designates blends by separating the component polyamides with a slash. For example a blend of PA 6 and PA 66 is represented as PA 6/PA 66 to distinguish it from the copolymer PA 6/66.

3. General Considerations in Polyamidation

CAROTHERS [1] proposed the term "polycondensation" for polymerization reactions wherein two functional groups react to form a covalent bond with the elimination of a simple molecule as a byproduct in accordance with the definition of condensation reactions in organic chemistry. In this type of polymerization, chain growth involves the stepwise reaction of the functional groups. Step-growth polymerization, a broader term that encompasses all polycondensation polymerizations, was first used by LENZ [10]. Detailed coverage of step-growth polymerization is given in Polymerization Processes, 1. Fundamentals [17].

The basic concepts of equilibria and kinetics in polyamidation are in large measure owing to FLORY and are summarized in his classic text [18]. Derivations of the following important expressions are to be found in many places, e.g., [17–20], and will not be developed here.

3.1. Molecular Mass

The extent of reaction, p, is the fraction of possible reactions that has occurred. The number-average degree of polymerization, $\bar{P}_n$ (IUPAC recommends $\bar{X}_n$), is the number of monomer units in the average polymer chain and is $2n$ for the AABB type nylon and n for the AB type where n is as shown in Equations (1)–(3). The typical polymer consists of molecules of varying molecular masses, so it is necessary to talk of average values. The number-average molecular mass, $\bar{M}_n$, is the product of $\bar{P}_n$ and the average molecular mass of the units in the polymer (see also → Plastics, Analysis, Chap. 5.). This ignores the contribution of end groups, which is unimportant at usefully high molecular masses. $\bar{M}_n$ is, precisely, the sum over all species of the number fraction of each molecular mass species multiplied by its molecular mass. The relation between $\bar{P}_n$ and p is:

$$\bar{P}_n = \frac{1}{1-p} \qquad (4)$$

This equation applies strictly to equilibrium conditions but is well suited to commercial polyamides. For PA 6 or PA 66, the average mass per amide is 113. Thus, $\bar{P}_n$ is 100 for an $\bar{M}_n$ of 11 300 and p is 0.990. This is at the low end of the range of useful $\bar{M}_n$ values. For a more typical value of about 20 000, p is 0.994. The need for pure monomers and clean, essentially complete reactions is clear. $\bar{M}_n$ for most commercial nylons is in the range 11 000–34 000. Physical properties depend on $\bar{M}_n$ which can be conveniently determined by end group analysis.

The weight-average molecular mass, $\bar{M}_w$, is given by $\bar{M}_w = \sum M_i + \sum_i (M_e)_i$ i.e., the sum of the products of the weight fraction w_i of each of the molecular mass species i and its molecular mass M_i plus the contributions of end groups. Melt viscosity, important in processing, varies with $\bar{M}_w^{3.4}$ [21–23]. $\bar{M}_w$ is available from light scattering or equilibrium ultracentrifugation. Solution viscosity is often used to approximate molecular mass and corresponds more nearly to $\bar{M}_w$. Both $\bar{M}_n$ and $\bar{M}_w$ can be obtained by chromatographic techniques [24–27]. $\bar{M}_w$ relates to p as shown:

$$\bar{M}_w = (1+p)/(1-p) \tag{5}$$

The distribution of molecular masses is important. A much-used index of the distribution is the ratio, $\bar{M}_w/\bar{M}_n$. Dividing Equation 5 by Equation 4 yields

$$\bar{M}_w/\bar{M}_n = 1 + p \tag{6}$$

At useful molecular masses where p nearly equals 1, $\bar{M}_w/\bar{M}_n$ is almost exactly 2. This was defined by FLORY as "the most probable distribution" of molecular masses. In such a distribution, which applies closely to almost all commercial nylons, the number fraction of any species decreases with increasing molecular mass. The weight fraction versus molecular mass curve goes through a maximum, the position of which is equal to $\bar{P}_n$ [18, 28].

The most convenient analytical technique is solution viscometry, which delivers a so-called viscosity-average molecular mass, $\bar{M}_v$. Solution viscosity can be used to estimate both $\bar{M}_n$ and $\bar{M}_w$ as long as their ratio is known and remains the same. Formic acid is a convenient solvent for PA 6, PA 46, and PA 66, but m-cresol or sulfuric acid must be used for nylons with fewer amide groups. Details of the technique and expressions relating viscosity and molecular mass can be found in [19].

Added or adventitious reactants with more than two reacting groups per reactant molecule (or just two alike in the case of AB types) will lead to branching and, in turn, will change the equilibrium distribution of molecular masses [18, 29]:

$$\bar{M}_n = 2/[N_e - (f-2)b] \tag{7}$$

$$\bar{M}_w = 4/[N_e - f(f-2)b] \tag{8}$$

where

N_e = number of ends
f = functionality (number of reactive groups in a monomer molecule)
b = number of branch points

At onset of gelation, where $\bar{M}_w$ goes to infinity, the number of branch points is

$$b_{gel} = N_e/f(f-2) \tag{9}$$

Since the number of moles is $1/M_n$, b' the number of branch points per mole, is $b\bar{M}_n$. At the gel point

$$b' = 2 b_{gel}/[N_e - (f-2)b_{gel}] \tag{10}$$

For $f = 3$, $b_{gel} = N_e/3$, and $b'_{gel} = 2 b_{gel}/(N_e - b_{gel}) = 1$. That is, it takes only one branch point per polymer molecule to cause gelation if a trifunctional reactant such as trimesic acid [554-95-0] (1,3,5-benzenetricarboxylic acid), or bishexamethylenetriamine [143-23-7] [N-(6-aminohexyl)-1,6-hexanediamine] is present. The latter compound can be formed by intermolecular reaction of two amine ends of 1,6-hexamethylenediamine with loss of ammonia.

Star-type branching occurs when AB monomers react with a polyfunctional reactant that contains only A or B groups and not both kinds. In this special situation [30] gelation cannot occur. Broad distributions occur for low p values and narrow distributions for p approximating unity in which case $\bar{M}_w/\bar{M}_n = 1 + f^{-1}$. Branched polyamides were introduced commercially in the late 1990 [31–34] (see Section 6.2).

The effect of mixing two polymers of different molecular masses is often questioned. In the case of $(\bar{M}_w)_1/(\bar{M}_n)_1 = (\bar{M}_w)_2/(\bar{M}_n)_2 \approx 2$, which most often applies to nylons [19]:

$$(\bar{M}_w/\bar{M}_n)_{blend} = 2 + [2 F_1 F_2 (R-1)^2/R] \tag{11}$$

where F is the weight fraction of components 1 or 2 and R the ratio of number-average molecular masses of components 1 and 2. Thus, for $F_1 = F_2 = 0.5$ and $R = (\bar{M}_n)_2/(\bar{M}_n)_1 = 2$, an extreme case for nylons, $\bar{M}_w/\bar{M}_n = 2.25$. This type of blending, therefore, has little effect on the molecular mass ratio and the changes observed are due to the changes in $\bar{M}_w$ and $\bar{M}_n$.

The reaction of amine and acid to form an amide is described by the following equilibrium expression:

$$K_{eq} = \frac{[CONH][H_2O]}{[COOH][NH_2]} \tag{12}$$

At high molecular masses [CONH] is essentially constant and depends only on the average molecular mass per monomeric unit. For PA 6 or PA 66 this becomes 1/113. For numerical convenience the units of ends and other concentrations are often expressed as the number of moles (or equivalents) per 10^6 g so that [CONH] is 8850.

The product of end group concentrations, [COOH] [NH$_2$] = P, is determined by the temperature and moisture content. The molecular mass depends on the number of ends, which is minimized (and the molecular mass maximized) for a fixed P when the concentration of amine and acid ends is the same, i.e., balanced. An imbalance may be imposed on purpose to limit the molecular mass by adding an excess of one kind of monomer or a chain-terminating reactant that contains only one functional group.

Addition of AA or BB in the polymerization of AB yields a polymer that at high extents of reaction (p very close to unity) has essentially only one kind of end group and belongs to the class of star-type products mentioned above. In this case the distribution of molecular masses represented by $\bar{M}_w/\bar{M}_n$ no longer approximates to 2 and, because the functionality is 2, is about 1.5. The addition of monofunctional acids or amines in the polymerization of AB or AABB and the addition of excess diacid or diamine in the polymerization of AABB does not alter the distribution and can be treated in a general way. If n is the amount of additive expressed as the mole fraction of AB or either AA or BB, and not the sum of AA and BB, then $\bar{P}_n$ is simply $1/n$ in the case of the AB polymer and $2/n$ for the AABB polymer. The factor of 2 is due to the fact that reaction of a difunctional monomer yields two amide groups. Including the contribution of the additive requires adding 1 to $\bar{P}_n$.

A problem in AABB polymerization lies in the fact that the imbalance of ends often results from loss of diamine during polymerization because its volatility is significantly higher than that of the diacid. It cannot, therefore, be anticipated but must be determined by analysis of the product. Assume N_A and N_B are the numbers of amine and acid ends found in the product in convenient units such as equivalents per 10^6 g. Assume also that a monofunctional acid has been added in an amount equal to n defined as in the above paragraph. Assume further that the average of the molecular masses of AA and BB less the 2 mol of water eliminated by polymerization is M so that the concentration of amide functions in the polymer, in the same units, is $10^6/M$. The total number of ends, N_T, in the polymer is $N_A + N_B + n\ 10^6/2\ M$. Because the polymer is linear, the number of amide groups per polymer molecule, that is $\bar{P}_n$, is $2 \times 10^6/MN_T$. By way of example, if N_A, N_B, and n are 70, 30, and 0.01, respectively for PA 66, N_T is $100 + 0.01 \times 4425$ and $\bar{P}_n$ is $2 \times 10^6/(113 \times 144) = 123$.

A related question is the implication of imbalanced ends for the equilibrium moisture content where it is desired to achieve or maintain a given molecular mass, i.e., have the same sum of ends. $P_{bal} = N_e^2/4$ if the ends are balanced, i.e., the amine and acid ends are equal in number and add up to N_e. For imbalanced ends and the same N_e, let $P_{imb} = N_A N_B$ where N_A and N_B are the new concentrations of amine and acid ends. For excess acid ends, define $r = (N_B - N_A)/N_e$. Since $N_e = N_B + N_A$, it is readily shown that $1 - r^2$ is $4 N_A N_B/N_e^2$. Thus,

$$P_{imb} = (N_e^2/4)(1 - r^2) = P_{bal}(1 - r^2) \tag{13}$$

Because the equilibrium moisture content varies directly with the product of ends,

$$[H_2O]_{imb} = [H_2O]_{bal}(1 - r^2) \tag{14}$$

where $[H_2O]_{imb}$ is the equilibrium moisture content in the case of imbalanced ends. Thus, if $r = 0.5$, which corresponds to a ratio of unreacted amine ends to unreacted acid ends (or vice versa) of 3/1 and is rather large, $[H_2O]_{imb}$ is 0.75 $[H_2O]_{bal}$. Polymerization to a higher molecular mass or hydrolysis to a lower one occurs depending on whether the water content is above or below the equilibrium value. The significance for the drying of resin and melt processing is clear.

3.2. Equilibrium and Rate Constants

The equilibrium constant for the polycondensations corresponding to Equations (1) and (2) is defined in Equation 12. Few data exist for such polycondensations. Many equilibrium and rate studies have been concerned with the so-called hydrolytic polymerization of ϵ-caprolactam in the presence of water. A comprehensive coverage of lactam-based polyamides is given in [35]. This polymerization is not as simple as Equation (3) indicates. Three reactions in the hydrolytic polymerization of caprolactam are of prime concern:

1. Hydrolysis of the lactam by water to yield the amino acid (Eq. 3 with $n = 1$)
2. Condensation as in Equation (2)
3. Lactam addition to the amine end of the growing chain (Eq. 15):

$$n\ HN(CH_2)_5C=O + H[NH(CH_2)_5CO]_mOH \rightleftharpoons$$
$$H[NH(CH_2)_5CO]_{m+1}OH \quad (15)$$

Also contributing to the complexity is the formation of cyclic oligomers corresponding to the following formula [36]:

$$\left[\begin{array}{c} NH(CH_2)_5CO \\ [CO(CH_2)_5NH]_n \end{array} \right] \quad n \geq 1$$

Structures with n up to 7 have been reported [37]. The amount of lactam monomer ($n = 0$) at equilibrium depends on the temperature (Fig. 1), [19, 38, 39]; the monomer comprises about 75 % of the extractable material, i.e., monomer and cyclic oligomers [40]. Residual lactam monomer and its cyclic oligomers are relatively volatile and tend to vaporize and condense on cool surfaces during melt processing unless provision for their removal has been made. Cyclic

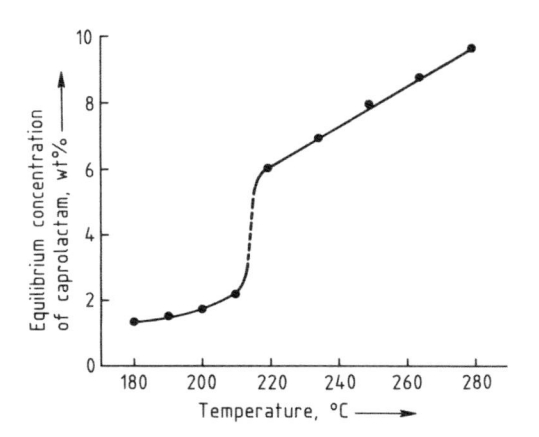

Figure 1. Dependence of monomer on temperature in polycaprolactam (data from [19, 38, 39])

oligomers exist not only in the higher AB-type nylons such as PA 11 and PA 12 but also in AABB types such as PA 66 and PA 610 [27, 28, 41–43]. However, the concentrations are much lower, and the extractable fractions are usually below 2 wt % [44–48] so that separation is not required as it is for PA 6.

There are further complications. The amine and acid end groups may exist as ammonium and carboxylate ions; this raises the possibility of alternative reactions involving condensation reactions between ions or between an unionized end and an ion. Degree of hydration is another question. Also a concern is the use of concentrations instead of the theoretically correct activity coefficients. The dependence of conventionally calculated equilibrium constants K_{eq} on water content, which can affect ionization, hydration, and activities, has been observed many times [49–53]; these K_{eq} values are thus apparent equilibrium constants. GIORI and HAYES [54] showed for PA 6 that the apparent equilibrium constant for polycondensation increased with water content up to about 2 wt% and decreased thereafter (Fig. 2). They used vapor–liquid equilibria, which showed a negative deviation from Raoult's law, to obtain activities [55]. K_{eq} (their K_{III}) showed much less deviation with percent water when using activity coefficients instead of concentrations (Fig. 3). Their K_{eq} at 2 wt % water (using concentrations) is about 600 at 240 °C and 450–500 at 270 °C. Values of 519 and 377

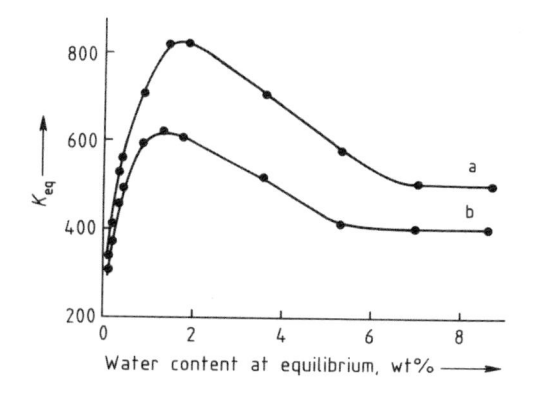

Figure 2. Apparent equilibrium constants for polycondensation versus equilibrium water content a) 240 °C; b) 260 °C
(reproduced with permission of John Wiley and Sons from Figure 1 in [54])

for comparable conditions can be calculated from the data of TAI and coworkers [56]. These variations are typical of the data available on the hydrolytic polymerization of caprolactam, most of which have been generated at water contents much higher than practical for melt processing.

Nylons are typically sold with about ≤ 0.2 wt% water. Extrapolation of data obtained at 0.2 wt% water at 240 and 260 °C by GIORI and HAYES [54] gives $K_{eq} = 280$ at 280 °C. TAI's data [56] yield $K_{eq} = 410$ at 280 °C at the lowest water content studied (0.42 mol/kg or 0.76 wt%). ZIMMERMAN [57] used a reported moisture content of 0.16 wt%, an end product of 3000,

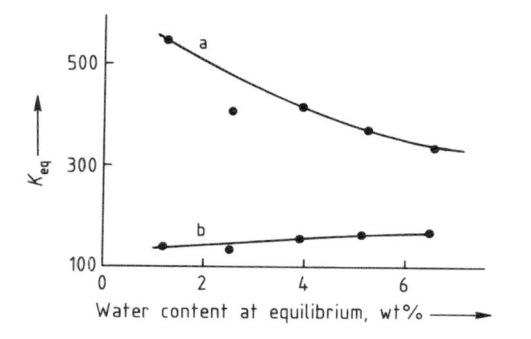

Figure 3. Equilibrium constants for nylon 6 polycondensation versus equilibrium water content at 270 °C a) Apparent equilibrium constants; b) Equilibrium constants corrected for the water activity coefficient
(reproduced with permission of John Wiley and Sons from Figure 8 in [54])

and an amide concentration of 8850 g equivalents/10^6 g at 101.3 kPa steam for PA 66 at 280 °C to calculate a K_{eq} of 260. The effect of moisture on processing requirements is best resolved by examining its significance for the stability of melt viscosity at the temperatures of concern as long as excessive holdup times and temperatures leading to thermal degradation are not involved [19].

Kinetic and thermodynamic constants for the polymerization of caprolactam have been summarized ([56], Table 3) and have been said to correlate with manufacturing experience with a deviation of less than 12 wt% [58]. The data in Table 3 are averaged for initial water contents of 0.82 and 1.18 mol/kg (1.48 and 2.12 wt%). Examination of the original data, which also included an initial water content of 0.42 mol/kg, indicates modest changes in the various constants except in the case of the temperature-independent constant, $A^c{}_2$ for catalyzed polycondensation which increased by a factor of 20.4 over the 2.8 increase in water content. Enhanced ionization and catalysis are thus indicated to be more important than reduced concentration of unionized species. Both second- and third-order kinetics (uncatalyzed and acid-catalyzed) for all three processes (Eqs. 2, 3 with $n = 1$, and 15) have been observed. For illustration the appropriate expressions for the polycondensation reactions are shown below:

$$-d[\text{end}]/dt = k_2^o[\text{NH}_2][\text{COOH}] \tag{16}$$

$$-d[\text{end}]/dt = k_2^c[\text{NH}_2][\text{COOH}]^2 \tag{17}$$

Amidation is reversible so that the contribution of hydrolysis cannot be ignored. The rate constant for hydrolysis is k/K_{eq} where K_{eq} is defined as in Equation 12 and k is the rate constant for amidation. At low water contents and high extents of reaction, as in the case of commercial nylons, third-order kinetics apply. The optimum water concentration for polymerization varies with conversion—high initially to promote opening of the ring and low in the later stages to promote high conversion. Figures 4 and 5 illustrate this effect of water content.

The amidation constants corresponding to Equations (1) and (2) are generally applicable because of the postulate that end group activity is independent of chain length [18], but overall

Table 3. Kinetic and thermodynamic constants in the hydrolytic polymerization of caprolactam[*]

i	A^0_i, kg mol^{-1} h^{-1}	E^0_i, cal/mol	A^c_i, kg^2 mol^{-2} h^{-1}	E^c_i, cal/mol	ΔS_i, e.u.	ΔH_i, cal/mol
1	5.9874×10^5	1.9880×10^4	4.3075×10^7	1.8806×10^4	−7.8846	1.9180×10^3
2	1.8942×10^{10}	2.3271×10^4	1.2114×10^{10}	2.0670×10^4	0.94374	−5.9458×10^3
3	2.8558×10^9	2.2845×10^4	1.6377×10^{10}	2.0107×10^4	−6.9457	−4.0438×10^3

[*] Average of data obtained with initial water contents of 0.82 and 1.18 mol/kg (1.48 and 2.12 wt%, respectively). A_i = pre-exponential factor and E_i = activation energy from $k_i = A_i \exp(-E_i/RT)$; ΔS_i = entropy; ΔH_i = enthalpy; $i = 1$, ring opening (hydrolysis of lactam to aminocaproic acid); $i = 2$, polycondensation (of aminocaproic acid); $i = 3$, polyaddition (addition polymerization of ε-caprolactam). Reproduced with permission of John Wiley and Sons from Table II in [56].

reaction rates and ring-chain equilibria vary because of differences in ring stability. Rings with 5, 6, 12, or 13 members are more stable than the 7-membered caprolactam. The larger rings are also, however, more difficult to form. This accounts for the use of catalysts in synthesizing PA 12 and for its lower equilibrium concentration of extractables. An excellent review of the properties of lactams has been provided by SEKIGUCHI [52].

Catalysts for amidation are not used for PA 6 or PA 66, although a small amount of aminocaproic acid or diammonium carboxylate such as the salt used to make PA 66 can be added to the caprolactam–water mix to eliminate the induction time required for lactam hydrolysis. Reported catalysts for amidation include metal oxides and carbonates or acidic materials [59], strong acids [60], lead monoxide [61], terephthalate esters [62], acid mixtures [63], titanium alkoxides or carboxylates [64], and phosphorus compounds [65–67].

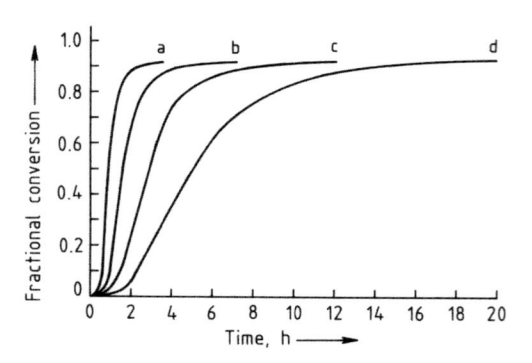

Figure 4. Effect of initial mass fraction of water (W_o) on rate and extent of conversion ($T = 265$ °C)
a) $W_o = 0.08$; b) $W_o = 0.04$; c) $W_o = 0.02$; d) $W_o = 0.01$
(reproduced with permission of John Wiley and Sons from Figure 8 in [51])

3.3. Effects of Monomer Structure

The ability of monomers to participate in polyamidation is markedly influenced by their tendency to form cyclic structures and the stability of those structures once formed. CAROTHERS [1] showed that it is difficult to form high molecular mass polymers when the formation of cyclic structures is favored. This principle affects the formation of both AB and AABB-type polymers, but in different ways. Lactams with five or six members are more stable than the seven-membered caprolactam rings, and their formation is favored over chain growth. The stability of the 5-membered ring lactam is so high that melt polymerization of pyrrolidone to PA 4 results only in low molecular mass polymers. Polymers of useful molecular mass have been obtained from these monomers only by using mild conditions such as solid-phase polymerization or anionic polymerization [68], and under

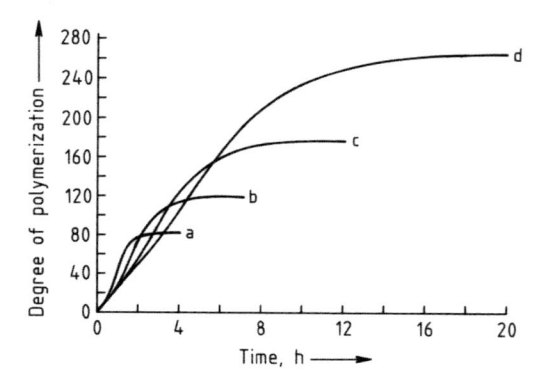

Figure 5. Effect of initial mass fraction of water (W_o) on rate and degree of polymerization
a) $W_o = 0.08$; b) $W_o = 0.04$; c) $W_o = 0.02$; d) $W_o = 0.01$
(reproduced with permission of John Wiley and Sons from Figure 10 in [51])

these conditions the monomer will be regenerated when the polymer is heated (Section 3.2). The 12- and 13- membered lactams are less reactive than caprolactam and require the use of catalysts for polymerization. Despite its enhanced reactivity, caprolactam has an equilibrium concentration of 8 % in the presence of its polymer. The equilibrium concentrations of 12- and 13-membered lactams in their polymerizations are less than 0.5 % because the reformation of these large rings is entropically unfavorable.

Ring formation can also lead to byproducts and chain-termination reactions in AABB polymerization. Succinic and glutaric acids, for example, produce only low molecular mass polymer under melt polyamidation conditions due to formation of the favored five- and six-membered cyclic imides derived from a reaction with the diamine comonomer. Although adipic acid has very little tendency to form its seven-membered cyclic imide, under typical polymerization conditions it does cyclize to the five-membered cyclopentanone by the elimination of carbon dioxide. The formation of pyrrolidine from 1,4-diaminobutane was reported by ROERDINK et al. [69, 70] in the manufacture of PA 46. The pyrrolidine formation and the instability of adipic acid preclude the attainment of high molecular mass in the melt polymerization of PA 46 (see Section 5.1).

3.4. Amide Interchange

Amide interchange (transamidation) is the interchange of amide groups between various polymer molecules or segments thereof. It is a factor when blending polyamides different molecular masses and/or different polymers. As noted above (Eq. 11), the effect of transamidation on molecular mass distribution is minor. Furthermore, there is rapid approach to an equilibrium value [53, 71]. The degree of interchange necessary for producing a random copolymer is slow enough, however, that block polymers are feasible [53, 72], although high conversion of the mix results in block lengths significantly smaller than the original chain lengths [73]. A study of the interchange reaction involving polymerization of sebacamide and N,N'-diacetylhexamethylenediamine suggested that the rate increased with water concentration and was proportional to $[COOH]^{1/2}$ [53]. MILLER studied the reaction of model compounds with each other and with PA 66 [73]. The primary reaction was acidolysis; aminolysis was acid catalyzed and did not occur in the absence of acid.

3.5. Structure-Property Considerations

Polyamides are valued as a plastic due to their good mechanical properties, excellent chemical resistance, and their ability to resist high temperatures. These qualities are due to their crystallinity and high melting points (see Chap. 6). Data that have been accumulated from the wide range of polyamides that have been studied have led to some generalizations on the effects of structure on their properties.

A large body of evidence has demonstrated that the underlying characteristic of polyamides that influences their properties is their ability to form strong hydrogen bonds. Any structural changes that affect the hydrogen bonding ability will have a very large effect on the properties. The dependence of the melting point, T_m, on the crystal heat of fusion, ΔH_u, and entropy of fusion, ΔS_u, is defined by:

$$T_m = \frac{\Delta H_u}{\Delta S_u}$$

The high melting points of polyamides are attributed to their low liquid-state entropy of fusion due to possible retention of hydrogen bonding in the melt and some stiffening of the amide group due to resonance [74]. The increase in melting point that goes with increased ability to form hydrogen bond is exemplified by PA 66 and PA 46. The latter has an increased ratio of CONH to CH_2 groups, which means more hydrogen bonding, and a increased melting point (see Chap. 6). On the other hand, lateral substituents on either the carbon chain or amide nitrogen or the presence of comonomers result in lower crystallinity and melting points. More drastic effects are exhibited when the substituent is bulky and/or on the nitrogen atom. The T_m for PA 66 is 265 °C but the polymer from N,N-dimethylhexamethylenediamine and adipic acid is amorphous and rubbery. A more detailed discussion of this topic is covered in [75].

4. Other Polymerization Techniques

4.1. Variants of Hydrolytic Polymerization

This category includes the use of acid derivatives or precursors instead of the acid itself [76] in high-temperature reactions. Examples are esters [77, 78]; acid halides [79]; amides [80]; and nitriles [81–85]. Another variant is the use of high-boiling, stable liquids that may or may not be a solvent for the nylon, e.g., selected phenols or white oil [86], glycols [87], and polar solvents such as N-methyl-2-pyrrolidone or sulfolane [88]. The use of amine derivatives is limited. The reaction of an aromatic diisocyanate with a dibasic acid is an example [89]; an appropriate choice of solvent and catalyst is necessary to avoid side reactions. A special case involves the use of the relatively reactive oxalate esters that react with diamines at ordinary temperatures.

4.2. Interfacial and Solution Polymerization

These are low-temperature methods for the preparation of polyamides whose stability and/or viscosity prevent melt polymerization. The reaction of an amine and acid chloride to yield aramid-type products is typically involved. The classic text on the subject is that of MORGAN [90]. Reference to the technique can also be found in [91]. (See also, → High-Performance Fibers, Section 2.2.1.. Aramid Fibers and → Polymerization Processes, 1. Fundamentals.)

4.3. Ionic Polymerization

Both anionic and cationic polymerization of anhydrous lactams have been demonstrated. Cationic methods have, however, failed to yield useful polymers because of side reactions that limit attainable molecular mass. Reviews of ionic polymerization have been provided [35, 51, 52].

The anhydrous, *anionic polymerization* of caprolactam has been of interest because the monomer is converted to polymer in minutes. This not only eliminates the need for costly polymerization facilities but also allows for use of inexpensive molds of complex design. It makes low-volume production feasible but is not economical for mass production of small parts. Unless modified, this method yields a polymer with properties somewhat different from hydrolytically polymerized PA 6 because of higher molecular mass and higher crystallinity [92].

Preferred techniques involve the use of the lactam anion, which is made in situ by reaction with a base (catalyst), and a lactam with an electrophilic substituent on the nitrogen atom (the cocatalyst). A very large number of publications and patents deal with the choice of catalyst and cocatalyst and their implication for advantages in process or properties. Examples of catalysts are metal hydrides, oxides, alkoxides, amides, carbonates, and Grignard reagents; cocatalysts are N-acyl caprolactams or compounds that form such derivatives in situ, e.g., anhydrides and isocyanates. Patents in the 1980s (e.g., [93–102]) have focused more on fillers, tougheners, applications, and process variants than on novel catalyst systems.

The mechanism of anionic polymerization is depicted below (M = metal, B = base, R = alkyl or aryl):

Anion formation:
$$HN(CH_2)_5C=O + MB \rightleftharpoons M^+ \ ^-N(CH_2)_5C=O + HB \quad (18)$$
$$\underbrace{}_{1} \qquad \underbrace{}_{2}$$

Initiation:
$$2 + RCON(CH_2)_5C=O \rightleftharpoons$$

$$RCON(CH_2)_5\overset{O^-}{\underset{|}{C}}-N(CH_2)_5C=O \quad (19)$$
$$\underbrace{}\underbrace{}_{3}$$

$$3 \rightleftharpoons RCON^-(CH_2)_5CON(CH_2)_5C=O \quad (20)$$
$$\underbrace{}_{4}$$

Anion regeneration:
$$1 + 4 \rightleftharpoons 2 + RCONH(CH_2)_5CON(CH_2)_5C=O \quad (21)$$
$$\underbrace{}_{5}$$

Propagation:
$$2 + 5 \rightleftharpoons \text{as in Equations (19) and (20)} \rightleftharpoons$$
$$RCONH(CH_2)_5CON^-(CH_2)_5CON(CH_2)_5C=O \quad (22)$$

SEKIGUCHI [52] prefers a mechanism that requires reaction of with to yield and . As

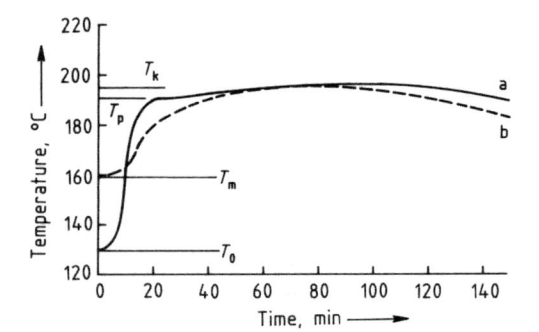

Figure 6. Temperature as a function of time in a typical cast polymerization of nylon 6 [92]
a) Temperature at center of large casting; b) Temperature of mold
T_o = initial temperature of polymerization (melt); T_p = final temperature of polymerization; T_k = maximum temperature; T_m = initial temperature of mold

he points out, there are many possible reaction pathways and side reactions, some of which arise from the increased probability of midchain ion formation as polymerization proceeds and the ratio of polymeric chain to lactam monomer increases.

Sample temperature is plotted versus time and density versus temperature for cocatalyzed, anionic polymerization of caprolactam in Figures 6 and 7, respectively. The polymer crystallizes during the polymerization since the maximum temperature of the system is lower than the melting temperature of the polymer. The maximum temperature of the system is caused by the heat of polymerization and the heat of crystallization, and is thus higher than the polymerization temperature which can be calculated from the heat of polymerization under adiabatic conditions. Anionically polymerized polycaprolactams are more crystalline than hydrolytically polymerized ones, and thus have higher densities (1.16 versus 1.14 g/cm^3).

A toughened, anionically polymerized caprolactam adaptable to reaction injection molding (RIM) technology was developed by Monsanto. Reaction of a bislactam with a soft polyether terminated by hydroxyl groups, e.g., a poly(propylene oxide), in the presence of caprolactam and a magnesium Grignard reagent yields a bisacylated prepolymer which, on further reaction with lactam, yields the polymeric product [103–105].

PA 12 has been made by cocatalyzed, anionic polymerization [106]. Other nylons (e.g., PA 4, PA 5, and PA 7) have been similarly prepared but have not been commercialized. Successful polymerization of PA 4 and PA 5 has, however, been of considerable theoretical interest because these nylons are not otherwise available. PA 4 has received a great deal of attention and has been commercialized in fiber applications. However, its thermal instability makes it unsatisfactory for molding and extrusion applications [35, Vol. II, pp. 74–117]. The best technique for the synthesis of PA 4 involves the use of carbon dioxide in place of an electrophilically substituted lactam, but it has been argued that this is not a cocatalyzed polymerization because the carbon dioxide cannot be found in the polymer [52].

4.4. Solid-Phase Polymerization

Solid-phase or solid-state polymerization (SPP or SSP) was first applied to low molecular mass polyamides by FLORY [107]. Applicability to a variety of polyamides with melting points ranging from about 190 to 290 °C was indicated. SPP of diamine–diacid salts has been discussed [108–110] but has not yet been commercialized. Highly crystalline PA 11 has been made in the laboratory by SPP of 11-aminoundecanoic acid [111]. MONROE [112] described a fluidized-bed process for increasing the molecular mass of polyamide particles not over 10 mm in diameter from an initial 1000 to 15 000 to over 15 000. BEATON [113] claimed a procedure which avoided agglomeration and permitted higher

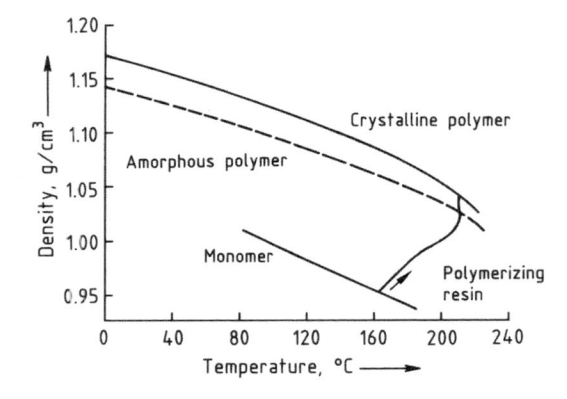

Figure 7. Density of caprolactam monomer and polymer as a function of temperature [92]

temperatures in a nonagitated bed of polymer. The rate of polymerization depends on particle size, temperature, and exposure to an inert gas maintained at a low relative humidity. As noted in [114], only the polycondensation reaction has been studied. SPP yielded higher molecular masses at equilibrium than expected for the water vapor pressure present [115]. This can be explained on the basis that the end groups and the absorbed water exist outside the crystalline regions. A broader than normal molecular mass distribution was also observed. Similarly, a transient distribution had been noted after the SPP of PA 12 wherein return to the equilibrium value ($\bar{M}_w/\bar{M}_n \approx 2$) occurred with a lowering of the weight-average molecular mass (but not of the degree of conversion) when the SPP product was melted [116]. The SPP of powdery 6-prepolymer was said to depend on the diffusion of acid ends and the initial molecular mass [117]. It is of interest that the number of bonds needed for the reported increases in molecular mass in both [112] and [117] is roughly proportional to the time required. Prior removal of oligomers facilitates SPP of PA 6 [118, 119].

The most significant application of SPP is in the synthesis of PA 46. A product with a narrow molecular mass distribution was obtained by SPP of 46-salt at 290 °C [120]. Prepolymerization in the presence of water followed by SPP yields a commercial quality product [121, 122]. Second-order kinetics are said to apply [121]. Special technique is required for PA 46 to control the tendency of the tetramethylenediamine to form a pyrrolidine ring which then acts as a monofunctional chain stopper. The utility of SPP in PA 46 is ascribed to the higher equilibrium constant for amidation at the lower temperature and to the higher activation energy for ring formation than for polymerization, 137 versus 83 kJ/mol [53, 69]. For a comprehensive coverage of SPP refer to [123].

5. Commercial Processes

The primary raw material sources of the monomers used to prepare commercial nylons have been listed in Table 2. Commentary on most monomer syntheses can be found in [5–7, 19]. The diamine for PA 46 is made via the addition of hydrogen cyanide to acrylonitrile [124]. For convenience, certain basic parameters of the commercial nylons have been assembled in Table 4.

5.1. PA 46

The 46-salt can be converted into prepolymer at 180–240 °C under steam pressure [122, 125]. SPP is carried out at 220–275 °C. Variations include a continuous polymerization method involving short exposure at 293 °C [125], water extraction of prepolymer to improve its color [126], a two-stage heat treatment to make prepolymer [127], control of temperature and steam pressure to yield high molecular mass [128], and polymerization with some of the adipic acid replaced by terephthalic acid to give a PA 46/4T copolymer [129].

Table 4. Basic parameters of commercial nylons

PA	$CH_2/CONH^a$	M_u^b	[CONH], $10^6/M_u^c$	T_m^d, °C	Density, g/cm^3
46	4	99.13	10 088	295	1.18
66	5	113.16	8 837	269	1.14
69	6.5	134.20	7 452	215	1.08
610	7	141.21	7 082	225	1.08
612	8	155.24	6 442	220	1.07
MXD 6		123.15	8 120	243	1.21
TMDT		144.20	6 935		1.12
11	10	183.30	5 456	190	1.05
6	5	113.16	8 837	228	1.14
12	11	197.30	5 068	180	1.02

^{a}Ratio of methylene to amide groups per repeating unit (or average monomer unit).
$^b M_u$ is the average molecular mass per amide group.
^{c}[CONH] is the concentration of amide groups.
^{d}The melting temperatures (points) are as determined by the disappearance of the last vestige of birefringence on a hot stage and correspond to the minimum fabricating temperature. Estimates were made where such data were not available.

5.2. PA 66

The first step is preparation of pure, balanced salt in aqueous solution. Stoichiometric equivalence is determined by pH measurement. Some excess diamine is normally added to compensate for losses due to its relative volatility. The equivalence pH for the aliphatic salts approximates 7.6 [130, 131]. Charcoal decolorization of the salt solution is required unless the diamine has been carefully refined and, in particular, is free of cis-1,2-diaminocyclohexane [132]. Purging a charcoal bed with inert gas increases its life [133]. Treatment with diacid followed by removal of acid with diamine before use with salt solution avoids initial losses in molecular mass [134]. A two-stage, recycling method of salt preparation is used in a continuous process [135]. Two-step addition to diamine with intermediate evaporation of water yields a concentrated salt solution [136]. Ways of mixing molten acid and diamine have been patented [137–139]. To avoid precipitation when stored at about 25 °C a 50% aqueous salt solution is normally prepared.

5.2.1. Batch Production

The stored salt solution is concentrated under pressure to 65–80% before charging to an autoclave. A typical cycle is shown in Figure 8 [140]. The essential features are heating to 210 °C under autogeneous pressure to reach a pressure of 1.75 MPa, gradually increasing the temperature to about 275 °C while releasing steam at a rate which maintains the pressure, reducing the pressure at a rate that avoids cooling and finally holding the batch at atmospheric or reduced pressure to obtain the target molecular mass before extruding the polymer under inert gas pressure. This procedure is designed to assure that there is enough water present to avoid freezing of the batch before the melting point has been reached. Water also minimizes excessive loss of diamine. Stirred autoclaves are used but are normally unnecessary. The extrudate is a wide ribbon that is quenched with water that is subsequently removed by jet blowers. The ribbon is cut into chips which are blended and packaged [141, 142]. Hermetically sealed containers are used to avoid absorption of moisture that can hydrolyze polymers to lower molecular masses during subsequent melt processing. The containers vary from 10 kg bags to 500 kg or larger cartons. Nylon molders and extruders have drying equipment to contend with situations where the moisture content of as-received or reprocessed material requires adjustment.

5.2.2. Continuous Production

The same concerns for control of the rate of removal of water and loss of diamine exist as in batch polymerization, but the situation is complicated by the needs of a continuous process. Typically, a first stage involves evaporation/reaction with controlled loss of water to form a prepolymer and minimize loss of diamine (Fig. 9). Further reaction occurs in subsequent stages with controlled evaporation in devices known as "separators" and "flashers". The desired molecular mass and water content are obtained in a "finisher". A large patent literature exists that claims improvement in each of these devices or combinations thereof for process simplification (see the review [140]). An early process that includes an additive port for delusterant is shown in Figure 9 [143]. A modification thereof (Fig. 10) includes a separator after the flash tubes to provide an essentially steam-free melt to the finisher [144]. Further improvements include the addition of a cage structure at the top of the screw section for better flow and reduced gel formation [145], insertion of static mixers in the flash tubes [146], putting holes in

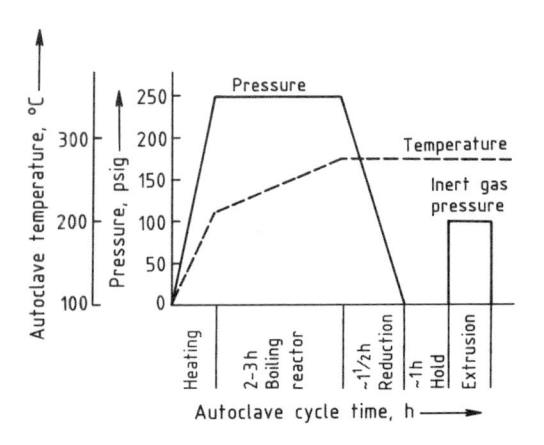

Figure 8. A typical autoclave cycle useful for batch preparation of nylon 66
(DuPont Technical Laboratory, Seaford, Delaware, reproduced with permission of John Wiley and Sons from Figure 8 in [140])

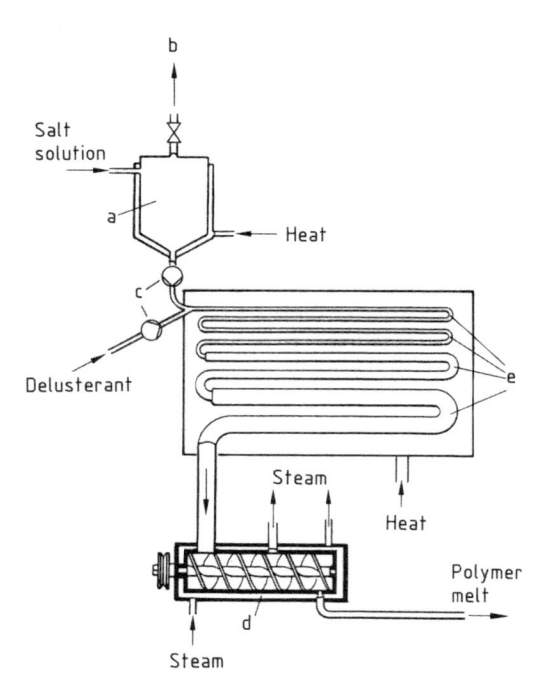

Figure 9. Continuous polymerizer for PA 66 [143]
a) Evaporator/reactor; b) Vent; c) Pump; d) Finisher; e) Flash tubes

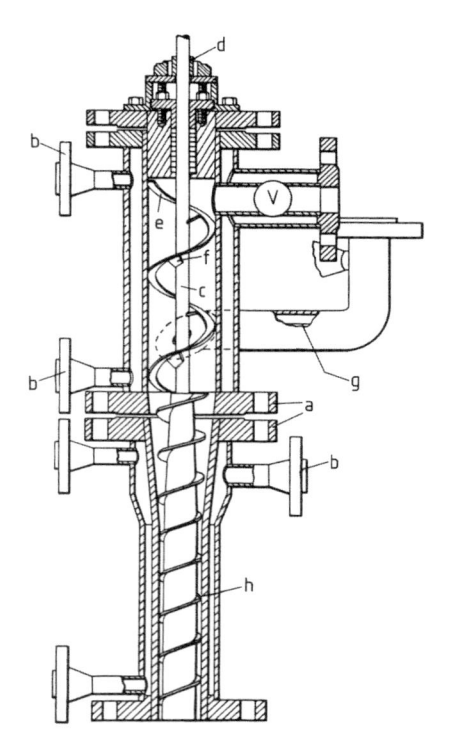

Figure 10. Two-part separator for polymerizer depicted in Figure 9 [144]
a) Flanges joining upper and lower members; b) Part of heating system; c) Shaft; d) Journal bearing; e) Ribbon flight; f) Spokes attaching ribbon to shaft; g) Feed from flash tubes; h) Transfer screw feeding melt to finisher

the ribbon flight in the separator [147], and providing flow splitters in the flash tubes [148]. A modification between the flash tubes and the separator significantly increases steam release and achievable molecular mass [149]. Dimensions, time, temperature, and pressure vary somewhat with specific design, but the same basic needs outlined above always apply in order to obtain a molten, high molecular mass product of low water content that can be fed to spinners or granulators. Severe constraints such as holdup spots and exposure to oxygen must always be avoided in continuously operated, high-temperature equipment such as polymerizers, extruders, or injection molding machines.

Another design (Fig. 11) involves essentially adiabatic evaporation in a second zone by directing the flow from an initial evaporator/reactor through a perforated ring nozzle against a heat exchanger with a surface of high thermal conductivity [150]. Another device emphasizing adiabatic evaporation makes use of the observation that the presence of low molecular mass polymer in salt solution decreases the tendency of the salt to precipitate [151], and another resorts to a multiplicity of zones [152]. Other procedures include rectification columns in the first zone, e. g., [153, 154]; a one-stage process with a long, narrow, coiled tube emptying into a stirred melt pool [155]; modified screw finishers [156–158]; designs to provide thin films of melt [159–161]; an apparatus that uses a dry salt charge [162]; and a dispersion technique for high-viscosity polymers involving cyclic monomers [163].

5.3. Other AABB Polyamides

The polymerizations yielding PA 69, PA 610 and PA 612 are subject to the same concerns as that yielding PA 66. These polyamides are made in much the same way with differences due to different melt viscosities and melting points. The choice between continuous or batch process depends mostly on production volume. Because

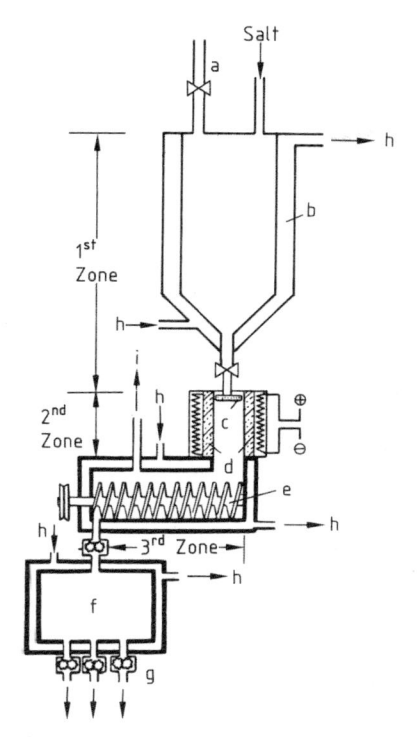

Figure 11. Polymerizer with essentially adiabatic evaporation in second zone [150]
a) Pressurized vent; b) Evaporator/reactor; c) Ring nozzle; d) Heat exchanger with surface of high thermal conductivity; e) Finisher; f) Heat exchanger; g) Gear pumps; h) Heat exchanger fluid; i) Exit tube to atmospheric vent or vacuum

of smaller sales volume, the batch method is favored for these specialty nylons.

The AABB principles described above can easily be extended to monomers containing an aromatic or cycloaliphatic moiety in the AA, the BB, or both portions of the molecule. Because of the presence of these structures these polyamides present special problems due in part to their high viscosity and/or stability at the high processing temperatures. An example is PA MXD6, a polyamide prepared from m-xylylenediamine (MXD) and adipic acid. Careful control of steam pressure in the early stage of polymerization of MXD6 is required to minimize decomposition of the aromatic diamine [164]. In the production of TMDT polymer, the salt is precipitated from an aqueous solution with isopropyl alcohol, partially condensed in a stirred reactor, and finally polymerized to a number-average molecular mass of about 20 000 in an extruder [165].

Polyphthalamides (PPAs), also known as high temperature nylons (HTNs) are high-melting semiaromatic nylons resulting from the addition of an aromatic ring to the polymer backbone. The high polymerization temperatures required for their manufacture necessitates either a short residence time in the melt by using an extruder or polymerization below their melting points (see Section 4.4) [166, 167].

5.4. PA 6

Continuous processes are used by the major manufacturers of PA 6. The so-called VK tube (Vereinfacht Kontinuierlich = simplified continuous) was developed in Germany. It is a vertical tube operated at atmospheric pressure wherein heating and prepolymerization take place in the upper part and polymer is formed in the lower section [168]. Problems of initial stoichiometry and ingredient volatilization, inherent with AABB polymerization as already discussed, do not apply; however, the need to free the product of its equilibrium content of monomer and oligomers presents a different kind of problem. A schematic of the BASF process [6] shows a VK tube feeding a pelletizer followed by a water extraction unit (Fig. 12). A process [169] involving vacuum extraction is shown in Figure 13. Claim is made that controlling initial conversion to 45 % yields a product with less than 2 % of cyclic oligomers with a vacuum finish that does not require prolonged heating at less than 665 Pa [170]. Improved processes include a baffled hydrolyzer operating in plug flow and at a carefully controlled temperature [171] and a two-stage hydrolyzer [172].

The simulation and optimization of the hydrolytic polymerization of caprolactam have been attempted [58, 173, 174]. Lack of detailed manufacturing data has been indicated to prevent confirmation of the analysis.

Polymerization in an extruder has also been discussed [175]. Numerous publications and patents continue to appear that deal with the anionic polymerization of caprolactam. No large-scale applications exist as yet although modified systems in reaction injection molding (RIM) for the production of block copolyesteramides and

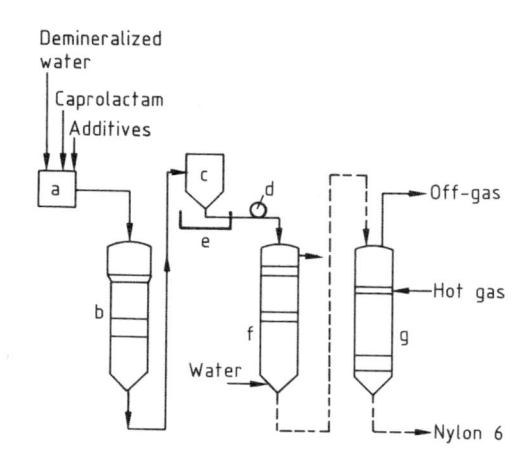

Figure 12. Production of nylon 6 by BASF
a) Feed tank; b) VK tube; c) Pourer; d) Pelletizer; e) Water bath; f) Extractor; g) Dryer
(reproduced with permission of Elsevier Science from Figure 5 in [6])

copolyetheramides have been developed by several companies [35]. Some details on recommended RIM procedures such as separation of feed streams, mold design, and temperature have been published [176, 177].

5.5. PA 11

PA 11 is made in a continuous process that is operated under inert gas at atmospheric pressure [178]. The monomer (11-aminoundecanoic acid, which is made from castor beans) is suspended in water and mixed with the desired additives (antioxidants, etc.) before charging to a three-zone, upright converter. The monomer is melted, water eliminated, and the polymerization begun in the top section. Near completion of the condensation occurs in the middle zone. Adjustment to the target molecular mass takes place in the bottom section which provides a melt pool for uniform discharge of product. The good thermal stability of the polymer and the low tendency of the monomer for cyclization simplify the process design.

Phosphoric acid catalyzes the polymerization as does hypophosphoric acid. The latter can, however, make the product insoluble if used in excess with prolonged heating [179].

5.6. PA 12

Polylaurolactam, PA 12, is made by heating dodecanolactam (which is made from butadiene) with an acid catalyst such as phosphoric acid in the presence of water to effect conversion to a prepolymer, then evaporating the water, and finally heating in an inert atmosphere to complete the polymerization [35, 180]. Temperatures up to 350 °C may be used to effect completion [181, 182]. Confirmation of the independence of reaction rate on molecular

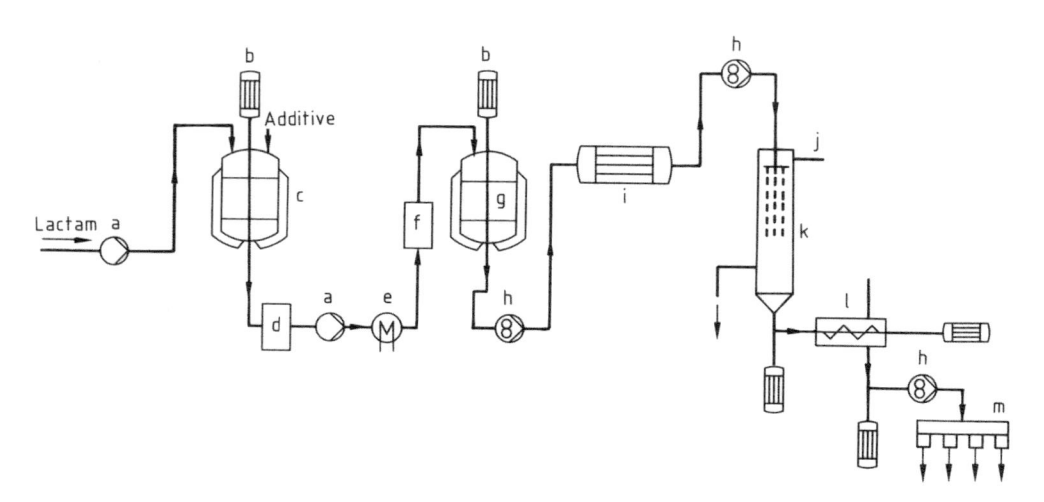

Figure 13. Production of nylon 6 by Allied Chemical [169]
a) Pump; b) Stirrer; c) Holding tank; d) Filter; e) Flowmeter; f) Preheater; g) Hydrolyzer; h) Metering pump; i) Polyaddition reactor; j) Vent; k) Vacuum flasher; l) Finisher; m) Spinning heads

mass and of a most probable molecular mass distribution together with an expression relating degree of polymerization to viscosity in *m*-cresol are provided in [183].

6. Properties

The enduring utility of nylons rests upon their combination of properties and upon their susceptibility to modification. Key properties are resistance to oils and solvents; toughness; fatigue and abrasion resistance; low friction and creep; high strength; stability at elevated temperatures; fire resistance; drawability; good appearance; and good processability.

Polyamides are readily modified in order to achieve desired changes in processing behavior or properties. This characteristic has allowed them to sustain their growth in spite of the advent of challenging new materials. However, modifications to enhance some specific properties can have negative effects on other properties. Basic modifications can be accomplished through changes in the polymer molecule itself and/or the incorporation of additives. Alternative techniques of modification include use of branching agents (see Section 6.2) and post-polymerization treatments such as annealing, conditioning to a specific moisture level,

dyeing, metallizing, painting, irradiation, or chemical reaction such as alkoxyalkylation.

6.1. Properties of Unmodified Polyamides

A problem common to most polymers is the lack of adequate and comparable data. Nylons are no exception; however, the increasing global acceptance of ISO protocols in measuring properties has alleviated the situation. A limited comparison of unmodified nylon resins is provided in Table 5. Even these data are not unambiguous because the molecular masses are not specified. The table is self-evident and reflects the differences in amide group concentration and crystallinity. Information on literally hundreds of nylon compositions can be obtained from various web sites maintained by the suppliers and in the Campus database [184]. Additional information is also available from the *Plastics Technology Handbook* and *Modern Plastics Encyclopedia*, which are both issued annually.

This introductory section, however, is concerned with those features of neat polyamides that determine their properties and account for differences between individual nylons. The most distinguishing feature is the amide group,

Table 5. Properties of unmodified polyamides[*]

Property	ASTM	66		6		612		11		12	
		Dry	50% R.H.	Dry	50% R.H.	Dry	50% R.H.	Dry	50% R.H.	Dry	50% R.H.
Tensile strength, MPa	D 638	83	77	81	69	61	61	57	54	49	47
Yield strength, MPa	D 638	83	59	81	44	61	51			49	40
Ultimate elongation, %	D 638	60	≥ 300	200	300	15	30	120	300	250	250
Elongation at yield, %	D 638	5	25	9		7	40			10	20
Flexural modulus, MPa	D 790	2830	1210	2700	970	2030	1240	1170	1030	1410	1030
Izod impact strength, J/m	D 256	53	112	58	215	53	75	40		58	64
Water absorption, %	D 570										
24 h			1.5		1.6		0.4		0.25		0.25
50% R.H.			2.5		2.7		1.3		0.8		0.7
100% R.H.			8.5		9.5		3.0		ca. 2.4		ca. 2.0
Heat deflection temperature, °C	D 638										
455 kPa			235		185		180		150		150
1820 kPa			90		65		90		55		52
Dielectric constant	D 150 (1 kHz)	3.9	7.0	3.8		4.0	5.3	3.7		3.6	
Flammability	UL 94	V 2		HB		HB		HB		HB	

[*] Data assembled from publications of various manufacturers for standard injection molding grades. Variations in molecular mass contribute to differences in properties.

which is responsible for strong hydrogen bonding between adjacent chains. Melting point increases with increasing ratio of CONH groups to CH_2 groups in the chain and the increased opportunity for hydrogen bonding, but not regularly because whether the number of CH_2 groups between the CONH groups is odd or even is important as well as the ratio of CH_2 to CONH. The odd number of CH_2 groups between the amide groups in PA 6 allows complete hydrogen bonding when the amides in adjacent chains have an opposed or antiparallel orientation but not when they have the same or parallel orientation (Fig. 14). Changing from a parallel to an antiparallel array requires inverting the entire molecular chain in this odd-numbered case, but only a one-segment lateral movement is needed if there is an even number of CH_2 groups, as in the case of PA 66 with its intervening number of 4 and 6 CH_2 groups. It is believed that this odd/even feature accounts for the lower melting point and percent crystallinity of PA 6 versus PA 66. It is the reason in general why PA-odd and PA-odd/even have lower melting points than comparable or similar PA-even/even, such as PA 6 versus PA 7.

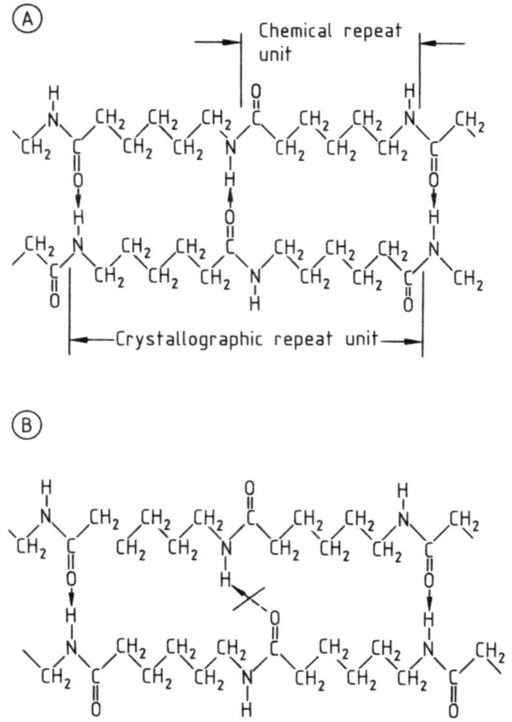

Increasing crystallinity results in higher stiffness, density, tensile and yield stress, chemical and abrasion resistance, and better dimensional stability. It decreases elongation, impact resistance, thermal expansion, and permeability. Crystallinity is responsible for the development of microscopically observable structures, known as spherulites, that scatter light and make nylons opalescent unless very thin. The transparent nylons are amorphous and are made so by appropriate copolymerization involving at least in part aromatic or cycloaliphatic AB, AA, or BB monomers in order to have acceptable stiffness.

Orientation [5, p. 300] contributes importantly to properties not only in film, fiber, and strapping but also in injection-molded articles. The resulting anisotropy means that properties vary with the direction of stress and must be considered in rationalizing resin behavior.

Water absorption is characteristic of nylons. Unless compensated for by increased crystallinity, a higher proportion of amide groups leads to higher water adsorption. Increased water content has an effect analogous to that of increased temperature, i.e., enhanced segmental mobility with, for example, concomitant loss in stiffness and tensile strength, gain in toughness, and growth in dimensions (elongation). At very low temperatures, however, water stiffens the nylon. Thus, the brittleness temperature (ASTM D 746) of PA 66 is − 80 °C if dry and −65 °C if conditioned to 50% relative humidity. Properties are frequently reported in the "dry", as-molded, condition corresponding to about 0.2% water or less, and after equilibration to a specified relative humidity such as 50 or 65%, and occasionally to 100%. The greatest change occurs in the vicinity of the glass transition temperature (T_g) so that a useful aid in understanding behavior is a knowledge of the effect of humidity on T_g. This is complicated because of variation with the method of calculation or measurement (Table 6), but it is generally true that nylons with fewer CONH groups and lower water adsorption have a lower dry T_g but show less change of T_g with relative humidity [19, 121, 185–190].

Figure 14. Schematic of a hydrogen bonded sheet of PA 6 with antiparallel (A) and parallel (B) orientation of amide groups

Table 6. Glass transition temperatures (T_g, °C) of polyamides at various relative humidities (R.H.)*

PA	Dry					50% R.H.		100% R.H.	
	A	B	C		D	C	D	C	D
46	102	78	78						
66	82	65	80, 78, 66		48	35	15	−15	−37	35
6 (extracted)	56	65	75, 65		41	20	3	−22	−32	20
610	56	46	67, 70		42	40	10			
612	52	40	60		45	40	20		20	
11	36	29	53		43					
12	29	25	54		42				42	
MXD 6	71		68							
TMDT			150							

* A) Estimated from melting temperatures using $T_g = (2/3) T_m$ [185]; B) Calculated from group contributions [186]; C) Dynamic measurements with torsion pendulum [121, 187–189]; D) Static measurements by differential thermal analysis or inflection point in curve of modulus of elasticity as a function of temperature [19, 190].

Little change in sensitivity to aliphatic hydrocarbons occurs with a change in polyamide type, but aromatic hydrocarbons are more strongly absorbed by polyamides with fewer amide groups. For example, PA 66 and PA 6 absorb 1 % toluene; PA 610, PA 3, and PA 11 absorb 6.8 %.

With this background now in place, a comparison of the properties of various nylons can be made, moving from the least aliphatic, to the most aliphatic, and then to semiaromatic nylons.

PA 46 is structurally similar to PA 66, and is produced from 1,4-diaminobutane and adipic acid. This nylon has the highest amide-to-methylene group ratio and the highest crystallinity of the commercial nylons. The high amide concentration accounts for the high moisture absorption of the polymer. Its melting point is 290 °C.

PA 66, which has one amide group for every 4 or 6 (with an average of 5) methylenes, has a high melting point of 269 °C, good strength and creep resistance, and excellent gas barrier properties. These qualities have led to its widespread use in automotive mechanical and electrical and electronic applications.

PA 6 also has an amide to methylene ratio of 1:5, but the odd number of methylene units retards crystallization and results in a more amorphous polymer with a lower melting point of 228 °C. These characteristics can be desirable for extruded film and wire and cable jacketing, but make PA 6 somewhat more sensitive to moisture and chemicals and limits its resistance to temperature spikes.

PA 69, 610 and 612 are all produced from hexamethylenediamine plus azelaic, sebacic, or dodecanedioic acids, respectively. PA 612 is intermediate between 6/66 and 11/12 with an average amide to methylene group ratio of 1:8, and provides an excellent compromise between PA 66 and higher nylons, with its excellent dimensional stability, reasonably high melting point (212 °C), strength, creep resistance, and barrier properties. This balance leads to widespread use in toothbrush filaments, battery components, and parts that are exposed to hot aqueous environments.

Higher nylons, like PA 11 and 12, have much lower amide to methylene group ratios (1:10 and 1:11 respectively), which results in more polyethylene-like materials that have excellent dimensional stability, low-temperature toughness, and stress-crack resistance at the expense of lower melting point and strength. They are readily plasticized for use in flexible, chemically-resistant tubing that has high burst pressures. The long aliphatic chains impart a low affinity for moisture. In addition, PA 11 is available in powder grades that may be applied by electrostatic spray or in fluidized beds for coating metallic parts.

Semiaromatic nylons are generally copolymers that are commonly made using terephthalic acid and/or isophthalic acid as comonomers often with an aliphatic or alicyclic acid. Common semiaromatic nylons will contain units derived from hexamethylenediamine and adipic acid or caprolactam. They typically have high melting points (295–325 °C for commercial grades) and exhibit a significant decrease in moisture absorption when compared to purely aliphatic nylons. The aromatic groups in the polymer backbone make them less mobile than linear aliphatic nylons and contribute to their high stiffness and good strength, creep resistance, and dimensional stability. Biaxially oriented films made from semiaromatic polymers provide excellent gas and oxygen barrier properties allowing for use in packaging materials and bottles.

Control of Rheological Properties. Control of melt rheology is very important for nylon processing (see Chap. 7). Low viscosity is desired for molding application for easy mold-filling. On the other hand, a high viscosity and melt strength are desired in extrusion and blow-molding applications. Control of molecular mass is the most convenient method for controlling rheological properties. Most polyamides have number-average molecular masses in the range of 11 000 to 34 000. Toughness increases with increasing molecular mass, but the effect is a moderate one. The major practical reason to change the molecular mass of a polyamide is to obtain a desired level of melt viscosity, which can range from less than 1000 Pa · s for injection molding to over 4000 Pa · s for shaped extrusion.

Another technique used to modify rheological properties is to change the molecular architecture. Introduction of low-level branching results in more non-Newtonian behavior, especially at low shear rates where an increase in melt strength and melt viscosity is obtained. Star nylons (see Section 3.1), although known for a number of years, are a new class of commercially available semicrystalline polyamides that exhibit a quasi-Newtonian rheological behavior and enhanced flow properties during processing [191]. The polymer is a modified PA 6 and has a star-like structure that is based on proprietary polymerization techniques. Polymerization catalysts and chemicals incorporated into the polymerization cycle produce a nylon with controlled branches. These materials can be compounded with large amounts of glass fibers or mineral fillers and molded with increased melt flow under normal processing conditions. The enhanced flow characteristics promote an excellent surface finish and aesthetic appearance. Commercial uses now include automotive engine covers, air-intake manifolds and trim parts, fishing reels, and appliance doors [192].

6.2. Copolymerization

Copolymerization to modify the properties of polyamides has long been practiced. This technique allows one to obtain varied properties without resorting to developing new polyamides from new monomers. Copolymerization increases the disorder in the polymer molecules and interferes with the ability of the molecules to pack and form hydrogen bonds. As a consequence, copolymers in general are characterized by a lower degree of crystallinity and a slower rate of crystallization, greater transparency, a lower modulus and strength, and greater solubility than the respective homopolymers. In a PA 66/6 copolyamide, for example, the melting point minimum is about 180 °C at 40 wt% of caprolactam units. PA 66/6 copolymers with about 40–65 wt% caprolactam units become soluble in hot methanol whereas the homopolymers PA 6 and PA 66 are insoluble in this solvent [19, 193, 194]. Terpolymers such as PA 66/6/610 are designed to further enhance solubility for use as binders [19, 195]. Transparency with good retention of properties is achieved by use of alicyclic or aromatic comonomers as has been noted for PA TMDT. A review of transparent polyamides that includes a comparison with other transparent engineering plastics is given in [196, 11, pp. 377–387].

There are quite a number of pairs of monomers, however, that have comparable distances between their functional groups and that yield copolymers in which the amide moieties can assume the proper orientation for hydrogen bonding. These molecular characteristics permit the comonomers to crystallize in the same morphology [197–199]. Instead of being depressed, the melting points of these isomorphous copolymers are weighted averages of the melting points of the component homopolymers. Some examples of isomorphous pairs of monomers are adipic and terephthalic acids; *p*-benzenedipropionic and sebacic acids; and *p*-xylylenediamine and *m*-xylylenediamine. For a more extensive discussion refer to [11, pp. 370–372].

6.3. Modification by Additives

Additives are materials used in small amounts (less than 5 wt% and usually less than 1 wt%) to affect various aspects of polymer processing, polymer properties, and/or appearance. Additives used to alter properties or appearance include antioxidants, antistatic agents, biodegradative agents, biopreservatives, blowing agents, colorants, fragrances, and stabilizers against hydrolysis, thermal degradation, or UV degradation. Processing additives include coloration inhibitors, lubricants, mold-release agents, nucleating agents, and viscosity thickeners or reducers. They have little effect on the physical properties except in the instance of nucleating agents, which not only increase the rate of crystallization and the degree of crystallinity, but impart reduced mold shrinkage and increased tensile and flexural properties. Enhanced crystallization may be desired to shorten molding cycles, and also to modify mechanical properties.

Modifiers are used in larger amounts, usually more than 5 wt%, in order to achieve desired changes in properties. Modifiers include mineral fillers, glass or carbon fibers, lubricants to improve wear and friction, plasticizers, fire retardants, electrically conductive materials, and/or blends with other polymers to toughen or otherwise affect the nylon. A broad discussion of additives is given in [11, pp. 361–448], [200].

6.3.1. Filled and Reinforced Polyamides

Fillers and reinforcing agents have contributed significantly to the sustained growth in the use of nylons. Particulate mineral fillers (e.g., aluminum or calcium silicates) always increase the modulus of elasticity and often the tensile strength whereas elongation always decreases. Impact strength as measured by the notched Izod test may or may not improve; it depends on the choice of filler and matrix. Temperature resistance is increased.

Reinforcement with glass and carbon fibers has similar but larger effects with consistently improved Izod impact strengths because of the fiber acicularity. Warpage can be a factor in some applications. This again relates to fiber acicularity and is, accordingly, less in the typical mineral-filled compositions. Mixtures of fillers and reinforcing agents are used, therefore, to cope with such situations. Important factors are the efficacy of the bonding agent used to promote adhesion between the filler or the reinforcing agent and the polyamide matrix and the length to diameter ratio of the fiber in the final product. The properties in Table 7 were chosen to illustrate these effects but because of the number of variables involved they should not be construed as limiting.

Long glass fibers are used in polyamide compositions with the attendant implications for properties and processing [201, 202]. Most of these compositions are produced by a pultrusion process [203, 204]. The continuous glass fibers are pulled through an impregnation line where they are coated with nylon resin. Upon cooling, the resin is chopped into long pellets that contain reinforcing fibers equal in length to the pellet. Traditional extrusion compounding of glass into nylon yields much shorter fibers, and hence less stiffness and worse transverse mechanical properties. The balance of properties makes long-fiber reinforced compositions ideal for the replacement of metals such as die-cast aluminum in many parts. In specialty applications, long fibers of carbon, aramids, and steel are used in addition to glass [205].

Table 7. Selected properties of dry, filled/reinforced PA 66[*]

Property	Neat	40 % Mineral[**]	33 % Glass fiber	Mineral and glass fiber	30 % Carbon fiber
Tensile strength, MPa (ASTM D 638)	83	89	186	130	241
Ultimate elongation, % (ASTM D 638)	60	17	3	12	
Flexural modulus, MPa (ASTM D 790)	2830	5240	8960	6900	20 000
Izod impact strength, J/m (ASTM D 256)	53	70	117	48	80
Heat deflection temperature, °C (ASTM D 638)					
at 455 kPa	235	230	260	257	
at 1820 kPa	90	185	255	239	257
Density, g/cm^3	1.14	1.51	1.38	1.45	1.28

[*] Data from publications of various manufacturers.
[**] "Mineral" as found in the nylon literature normally refers to various silicates and carbonates.

Nanocomposites are emerging with increased acceptance of the properties that they offer. They can improve strength, stiffness, heat deflection temperature, dimensional stability, gas barrier, flame retardancy, and electrical conductivity. Nanoparticle platelets are only 1 nm thick with an aspect ratio (L/D) sufficiently high that relatively low loadings (generally 2 to 6 wt%) provide significant property advantages. These clay-based additives have been used with PA 6, PA 612, and PA MXD 6 to improve stiffness heat resistance, oxygen barrier and gloss, but confer a disproportional loss of toughness. In 2000 these nylon compositions were first commercialized for container and packaging applications [206–208]. A more detailed discussion of the many aspects of nanotechnology is given in [209].

6.3.2. Blends and Alloys

Blends of nylon (and polymers in general) are mixtures that contain at least one other polymer or copolymer. The properties of the blends can be additive, averaged, synergistic, or antagonistic. An alloy is an immiscible polymer blend having a modified interface between both components and/or morphology. While positive synergism of the components is most desired, a weighted averaging of properties based on the volume component of resins is usually the rule. Nature rarely provides something for nothing so there are often tradeoffs of properties and for every advantage obtained by a resin blend, there are often disadvantages as well. The disadvantages are often so significant that the resulting material is unusable. The primary reason for the negative results is that most polymer pairs are immiscible because of high interfacial tension, which results in phase-separated components. A brief list of some common property tradeoffs is:

Stiffness, strength (e.g., glass, mineral)	vs.	toughness, ductility
Melt flow	vs.	toughness, elongation
Fire retardance	vs.	environmental considerations, toughness, melt stability
Solvent resistance (i.e. crystallinity)	vs.	mold shrinkage
Cycle time	vs.	shrinkage
Softness	vs.	creep, stiffness
Product complexity	vs.	compounding yield, speed, cost

The successful use of blending to tailor-make materials for specific end use applications depends on the type of resins used, their morphologies, the relative levels of the resins, and the property tradeoffs that are acceptable [210–212].

Some of the polymers, that have been successfully blended with polyamides include polypropylene (PP), high impact polystyrene (HIPS), styrene-acrylonitrile (SAN) and acrylonitrile-butadiene- styrene (ABS) copolymers, poly(methyl methacrylate) (PMMA), styrene-maleic anhydride (SMA) copolymer, EPDM elastomers, polycarbonate (PC), and poly(phenylene ether) (PPE) [213].

For all blends, it is essential to develop processes that control the morphology and the interfaces of the phase-separated components. Interfaces can be modified by the addition of reactive and nonreactive compatiblizers, as well as by reactive extrusion to assure a controlled chemical reaction between the components and thus compatibilize the blend in situ [214].

The broad utility of nylons was restricted for many years by their notch sensitivity and loss of toughness at very low temperatures. For example, the impact strength of PA 66 at 23 °C and 50% humidity is 37 J/m with a notch radius of 0.05 mm, 112 J/m with a 0.25 mm radius (the Izod specification of ASTM D 256), and 366 J/m with a 0.75 mm radius. At −40 °C with the standard radius of 0.25 mm, it is 27 J/m. Some improvements were made, but no major breakthrough occurred until suitably defined elastomeric modifiers were used [215]. This work yielded compositions that were ductile and exhibited Izod impact strengths over 534 J/m even when dry. Independence of notch radius down to 0.0025 mm; insensitivity to part thickness, orientation, proximity to the molding gate or rate of loading; and improvement down to about −20 °C were demonstrated [216]. One requirement in this work was a sufficiently small particle size of the dispersed elastomeric phase. Ductile behavior independent of part thickness has been shown for a PA 6–ABS blend (about 50/50 weight ratio) with a cocontinuous phase morphology [217]. Another variant of toughener involves a core – shell structure in the dispersed phase [218]. The effect of orientation, position in the molded part, rate of loading and temperature has not yet been shown for these alternative materials. Many studies on toughening nylons from both industrial laboratories and academia have appeared. Morphology, type of toughener, testing with sophisticated instrumentation, and analysis of the mechanism of toughening have been studied. It is, however, beyond the scope of this report to delve into this broad and complex subject.

The number of impacts to failure of toughened PA 66 increases, as would be expected, as the impact energy is lowered below that required for failure in a single blow [219]. Noteworthy, however, is the resistance to repeated impact of neat PA 66 [219, 220].

Mineral-filled and glass-reinforced nylons have also been toughened. For example, a dry, mineral-filled product with a tensile strength (TS) of 89 MPa, a flexural modulus (FM) of 5300 MPa, and a notched Izod impact strength (Iz) of 69 J/m was toughened to yield one with corresponding values of 79 MPa, 4600 MPa, and 130 J/m [221]. Similarly, a dry, glass-reinforced PA 66 with a TS of 200 MPa, a FM of 8960 MPa, and an Iz of 100 J/m was toughened to yield a TS of 140 MPa, a FM of 7000 MPa and an Iz of 250 J/m [221]. These examples illustrate the feasability of the wide spread practice combining various additives and modifiers to meet the demands of specific applications.

6.3.3. Flame-Retardant Polyamides

Many nylons are rated UL 94 V-2, and fire-retardant grades with a UL 94 rating of V-0 in thin sections are available. For a discussion of UL ratings, see →High-Performance Fibers. Common flame retardants are highly-halogenated organic compounds and polymers such as brominated polystyrenes, poly(dibromostyrenes), or dechlorane [222, 223]. These are often used in combination with a synergist such as an antimony oxide. Red phosphorus is also used as a flame retardant [224]. Concerns about the toxic fumes released when these materials are heated [225] and the hazards involved in handling red phosphorus have led to the development of halogen- and elemental-phosphorus-free flame retardants based on melamine cyanurates that function by generating molecular nitrogen upon decomposition, which smothers the flame [226, 227]. More recently, flame retardants comprising a mixture of phosphinates and melamine cyanurates or melamine phosphates have been introduced [228].

Flame-retardant polyamides have many automotive, electrical, and electronics applications. One survey discusses control of tracking (formation of a conducting path on the polymer surface) [229]. Another review addresses properties of glass-reinforced thermoplastics including PA 6, PA 66, and PA 610 [230]. Colorants can increase flammability [231]. Safety is increasingly stressed and is generating even greater concern for smoke suppression and the toxicity of gases formed during combustion as well as the inhibition of ignition and flame support [225].

6.3.4. Conductive Polyamides

The burgeoning market for electronics has led to a great demand for electrically conductive polymers. Electronic devices generate radio frequency radiation during operation, and to prevent the consequent electromagnetic interference (EMI) with other devices, it is necessary to shield the source with an electrically conductive material. Since polyamides are not intrinsically conductive, EMI shielding of nylon parts can be accomplished in either a two-step process that involves coating the plastic with a conductive material, or, preferably, filling it with a conductive additive, such as carbon fibers, metal-coated carbon or glass fibers, metal fibers, carbon black, or graphitic carbon fibrils [232, 233].

It should be noted that neat and glass-reinforced nylons have also been successfully used for some time in electrical power engineering and electronics [234]. A variety of plastics containing metal fillers or carbon to make them electrically conductive, including PA 66, were exposed to an aqueous detergent solution at 25 and 80 °C and to isopropyl alcohol; the effect on EMI shielding was assessed [235]. The compositions with carbon generally outperformed those with metal fillers. The only PA 66 evaluated contained carbon fibers, and its retention of shielding was equivalent to that of the other materials tested.

Unadulterated polyamides are *thermal insulators*, but when combined with heat-conductive additives such as graphite fibers or ceramics like boron nitride and aluminum nitride, they can be used in heat-removal applications that include heat sinks for electronics and electrical applications [236].

6.3.5. Other Formulations

The amount of polyamide used and the nature of its dispersion in a blend with other polymers may be such that it is properly defined as the modifier rather than the base resin. One commercial example is a blend of nylon and poly (2,6-dimethyl-*p*-phenylene oxide). Allusion has been made to a blend with poly(phenylene sulfide) as the major ingredient [237]. Another composition presumably involves mostly ABS and some nylon [238]. A developmental material is a glass-reinforced blend of polycarbonate and nylon [239]. The literature on blends is voluminous. Brief reviews of all commercial blends including some commentary on their processing are given in [237–239]. A recent text focuses specifically on commercial blends and alloys [240]. It may reflect success in finding a compatibilizing agent for essentially incompatible polymers.

A different kind of product in which nylon is the principal component arises from reaction during molding of pellets that differ in chemical constitution; at least one component is caused to form a cross-linked structure. In one commercial case a pseudo- or semi-interpenetrating network (SIPN) results which physically binds a silicone modifier to a nylon matrix [241].

With the appropriate additives, polyamides can be used in a variety of specialized applications. For example, polyamides can be combined with magnetic materials such as neodymium-iron-boron alloy powders, permitting the injection molding of magnetic parts [242].

High-density polyamide formulations can be obtained by compounding with metal fillers and powders such as aluminum, zinc, steel, tungsten, or lead. The densities of the materials can be fine-tuned by using metals of different densities and varying their loadings. Addition levels of up to 95 wt% have been achieved. However, the polymer can still comprise 40–70% of the volume, which permits these compositions to be processed using standard equipment. Wear and abrasion on extruder barrels, screws, and molds is an issue and hardened steel tooling is recommended. These polyamide metal materials can serve as substitutes for metal in both practical (such as for X-ray and radiation shielding applications, fishing-line sinkers, and bullets) and aesthetic applications. In the former, the substitution of moldable plastics for metal parts permits a greater degree of design freedom. In the latter, the added heft and metallic feel given to the plastics creates a perception of greater value [243–245].

Rigid parts made from polyamides can be given soft coatings by overmolding them with thermoplastic elastomers. Suitable elastomers include polypropylene-EPDM thermoplastic vulcanizates that contain small amounts of polyamide block copolymer to serve as a compatibilizer. Such "soft-touch" materials have a variety of applications in consumer goods [246, 247]. Alternatively, block copolymers of polyethers and polyamides have elastomeric properties and their hardness can be tuned over a wide range. These materials can be overmolded onto other materials for a variety of consumer, industrial, automotive, and adhesive applications [248] (→Thermoplastic Elastomers).

The amount of polyamide used and the nature of its dispersion in a blend with other polymers may be such that it is properly defined as the modifier rather than the base resin. One commercial example is a blend of nylon and poly (2, 6-dimethyl-p-phenylene oxide). Allusion has been made to a blend with poly(phenylene sulfide) as the major ingredient [194]. Another composition presumably involves mostly ABS and some nylon [195]. A developmental material is a glass-reinforced blend of polycarbonate and nylon [196]. The literature on blends is voluminous. Brief reviews of commercial blends including some commentary on their processing are given in [194–196]. A recent text focuses specifically on commercial blends and alloys [197]. Blends in which there is some level of bonding between matrix and modifier, whether due to likeness of solubility parameters, hydrogen bonding, or some other factor are sometimes termed alloys.

7. Processing

Nylons are processed into useful articles by a variety of techniques—injection molding, extrusion, blow molding, monomer casting, solution coating, fluidized-bed or electrostatic coating, or forming [19]. The major fraction of the annual polyamide volume is processed by injection molding, extrusion, and blow molding. Each of these melt processes requires different melt viscosities for optimum processing. Whereas injection molding requires the lowest melt viscosity, for extrusion and blow molding a high melt strength is needed, which by necessity requires a high melt viscosity. Molecular mass can be altered to meet the viscosity required for the desired process. In addition, modifying molecular architecture is sometimes used to alter the rheological properties (see Section 6.2).

Equation 12 governs the dynamics of the amidation/hydrolysis equilibrium in polyamides. The polyamide molecular mass will either increase or decrease, and consequently the melt viscosity, as well, depending on whether the moisture content is higher or lower, respectively, than the equilibrium moisture content. Thus, moisture control in polyamide melt processing is extremely important and because of the ubiquitous nature of moisture it is often the most difficult to control. Most problems in nylon processing could be attributed to poor control of moisture. These problems can manifest themselves in mechanical properties, appearance, and processibility.

Good flow and fast set-up make nylons relatively easy to process. Standard three-stage screws with or without mixing heads, corotating twin screws, and barrier screws with or without venting are all used. Most reciprocating screws in injection molding are of standard type. Moisture content should be somewhat lower for extrusion than for injection molding because the absence of high die pressure facilitates volatilization. Control of moisture to avoid hydrolysis or polymerization is always a concern. Processors use dehumidifiers to control the moisture present in molding materials especially when using regrind or while operating under hot, humid conditions. Failure to compensate for changes in molecular mass during measurement of melt viscosity because of failure to control moisture content adequately, e.g., by over drying has resulted in questionable estimates of the activation energy of viscous flow. Thermal degradation can also be a problem if the melt is held too long at too high a temperature. Thus, one study preferred to use a capillary mounted on an injection molding machine rather than an Instron capillary rheometer in part because of shorter hold-up times [249].

Values of 42–63 kJ/mol for the activation energy of viscous flow appear reasonable rather than those of 71–92 kJ/mol as have sometimes been reported [19, 250]. Master curves that are based on only the T_g and melt flow index (MFI) per ASTM D 1238 have been proposed for a number of polymers including nylons [251]. These curves plot the product of viscosity and MFI against shear rate divided by MFI. The sample curves in Figure 15, mostly at 280 °C, were taken from [252–255]. As shown in the figure, the apparent melt viscosity of the linear, aliphatic nylons starts to decrease rapidly at a shear stress of ca. 50 MPa.

Filled and reinforced nylons have higher melt viscosities than their neat analogues, but the differences become smaller as the level of shear increases [256, 257]. Processing variables affect the ultimate fiber size in the blend and can accordingly affect the properties of the molded product [258, 259]. Compounding studies indicated that the major fiber degradation occurs in the melting region [260]. A subsequent study came to the same conclusion and showed fiber fracture to be independent of fiber concentration and dispersion and to arise from fiber interaction with shear in the matrix [261].

Toughened nylons also tend to have higher melt viscosities than their neat counterparts. There is again diminution of the difference as shear increases. The lower modulus implies

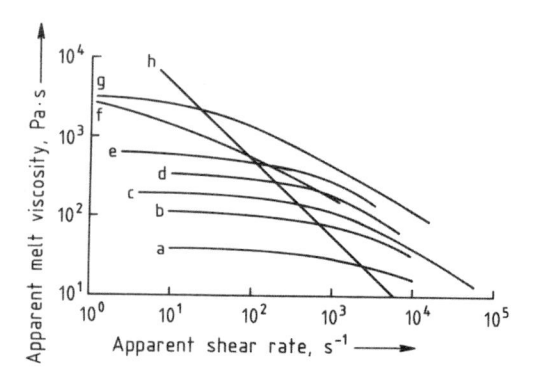

Figure 15. Dependence of apparent melt viscosity on apparent shear rate for various nylons at 280 °C (adapted from [19])

a) Nylon 610 [254]; b) Nylon copolymer [254]; c) Nylon 66, M_n ca. 18 000 [253, 254]; d) Nylon 6, M_n ca. 20 000 [253]; e) Nylon 6, M_n 19 100 at 203 °C [252]; f) TMDT at 250 °C [255]; g) Nylon 66, M_n ca. 34 000 [253, 254]; h) Line of constant shear stress of 50 MPa.

longer molding cycles, but in at least one case the toughener induced nucleation which made for fast molding [216].

8. Uses

Nylons are used in many and diverse ways. They are found in appliances, business equipment, consumer products, electrical/electronic devices, furniture, hardware, industrial/machinery, packaging, and automotive and truck. This diversity makes classification and analysis difficult as shown in Table 8, which shows the pattern of consumption in the major regions and globally [262].

Automotive and truck is the largest market for nylons. The good performance of nylon at high temperatures and pressures, as well as its chemical resistance and compatibility with many automotive fluids, along with the design flexibility available with plastics, paved the way for the adoption of polyamides in many under-the-hood applications. The introduction of high-performance mineral reinforced and glass reinforced versions of nylon greatly increased the ability to tailor properties for individual engine components. When government regulations required automobile manufactures to install pollution-control devices, designers turned to these proven materials for critical air and fuel system control components because of the polymer's performance and the relative ease of part manufacture. Utilizing nylon in the engine compartment reduces weight, which can improve fuel efficiency without sacrificing engine performance. Key developments were pioneered in Europe where smaller cars, higher gasoline costs, and emission control mandates drove innovation. Soon, commercial applications of nylons in air intake manifolds, brake fluid reservoirs, power steering reservoirs, radiator end tanks, and oil pans were introduced. These developments gained footholds in Europe and spread to other regions. Unreinforced resins are used in electrical connectors, wire jackets, windshield wiper and speedometer gears, and emission canisters. The softer nylons are used in fuel lines, air brake hoses, and spline shaft coatings. Glass reinforced nylons are used in engine fans, radiator headers and grilles, brake and power steering fluid reservoirs, valve covers, air intake manifolds, wheel caps, air brake contacts and head-rest shells. Applications for nylons combining tougheners and reinforcement include brackets, steering wheels, and accelerator and clutch pedals. Mineral-filled resins are used in wheel caps, radiator grilles, and mirror housings. Nylons containing both glass fibers and minerals are used in exterior parts such as fender extensions. Toughened nylons are found in stone shields and trim clips.

Electrical and electronic applications comprise a major market for nylons, albeit more so in Northeast Asia and Western Europe than in North America. Flame-retardant materials are particularly important in this area. Uses include color-coded components, plugs, connectors, coil forms, wiring devices, terminal blocks, antenna mounting devices, and harness ties. A review of nylon in electrical power engineering and electronics has already been cited [234]. Wire and cable jacketing is used mostly over primary insulation because of the solvent, wear, and abrasion resistance of nylons. Relays, fittings, and contact makers constitute a partial list of applications in telecommunications.

Industrial/machinery applications are attracted to the excellent fatigue resistance and repeated impact strength of nylons. Examples are hammer

Table 8. Pattern of nylon consumption globally and regionally in 2012; data sourced and used with permission of IHS Chemical [262]

Nylon Engineering Resin End-Use Demand	North America	South America	Greater Europe	Asia and rest of world	Global
Automotive and truck, %	49.9	58.0	36.9	29.1	35.9
Electrical and electronic, %	5.3	5.0	14.1	13.0	11.8
Appliances, %	2.5	4.0	6.2	11.1	7.9
Film, coating, cable, %	20.6	11.0	18.5	13.2	15.9
Consumer, %	11.5	5.0	8.3	10.7	9.8
Industrial/machinery, %	5.8	8.0	11.0	16.9	13.0
Others, %	4.5	9.0	5.1	6.1	5.6
Total, 10^3 t	399	100	810	1266	2575

handles and moving machine parts. The mechanical strength accounts for use in gears, bearings, antifriction parts, snap fits, and detents. Food- and textile-processing equipment, pumps, valves, agricultural and printing devices, and business and vending machines comprise a partial list of other industrial uses.

Consumer products exploit the toughness of nylons in ski boots, ice and roller skate supports, racquet equipment, and bicycle wheels. Kitchen utensils, toys, photographic equipment, brush bristles, fishing line, sewing thread, and lawn and garden equipment show that the breadth of utility of nylon also includes consumer products.

Appliances and power tools make use of the impact strength of nylons. Glass-reinforced resins, which combine stiffness at high temperatures with toughness and grease resistance, are used in handles, housings, and parts in contact with hot metal. Sewing machines, laundry equipment and dishwashers are examples of other appliances that have utilized nylons.

Film and coating applications have grown in importance because of the use of nylons in coextruded, composite films for food packaging. Polyamides are also used in cook-in bags and pouches. Another development involves dispersion of nylon platelets in a polyolefin matrix to combine both hydrocarbon and moisture barrier properties in blow molded containers [263, 264].

Other Applications. The soft polymer-modified, anionically polymerized caprolactam has been used via RIM to make machine housings, grain buckets, and fuel tanks [266]. Blow molding compositions [267], PA 46 and MXD 6, have extended the utility in the market [268–270]. Products with improved processing characteristics, high- and low-temperature performance, and new property profiles obtained via the use of new modifications and combinations thereof have been introduced [271, 272]. A PA 6 with reduced moisture sensitivity has been demonstrated [269], as has the use of block copolymers as thermoplastic elastomers [248].

The adaptability of nylon to be changed to even obtain diametrically opposite results is shown in Figure 16, which gives some examples of the applications involved. Nylons with qualified approval in the United States in 2002 for processing, handling, and/or packaging food are PA 66, 610, 66/610, 6/66, 11, 6, 6T/66, 612, 12, MXD6 and impact-modified MXD6, 12/AI (A = bis(4-amino-3-methylcyclohexyl)methane [6864-37-5]), 6I/6T, 6/69, 46, and TMDT [273].

9. Ecological Aspects and Toxicology

9.1. Recycling

Polyamides are quite stable, and are used primarily in durable applications that are not major contributors to the waste stream. Scrap and trim are generally reused during processing, further minimizing waste. Some consumer applications such as bristles, strapping, and composite film for packaging do involve routine disposal, but the total is well below 0.05 wt% of all collected refuse. As a result, commercial nylon recycling processes are not as well developed as they are for other thermoplastics. However, it is far more energy efficient to recycle nylon than to burn it [274], and European directives in 2001 that phase in a requirement that almost all plastics used in automobiles be recycled over the subsequent two decades will continue to spur development in the recycling of mixed thermoplastics. Several commercial initiatives to recycle readily-available PA 6 and PA 66 carpet fibers were conducted in the United States and Europe, but were only partly successful because of the low demand and the difficulties of efficiently and economically collecting raw materials [275–278].

Difficulties involved in recycling postconsumer polyamides include the fact that they are often used compounded with a wide variety of additives (see Section 6.3) that need to be removed and the fact that the properties of polyamides generally suffer strongly upon multiple reheatings, as the polymers tend to degrade and lose molecular mass. Most of the current commercial recycling processes involve the depolymerization of the polymer chains and isolation of the monomers, which are then used as feedstocks to make near-virgin polyamides. PA 6 can be depolymerized to caprolactam by heating the polymer in the presence of

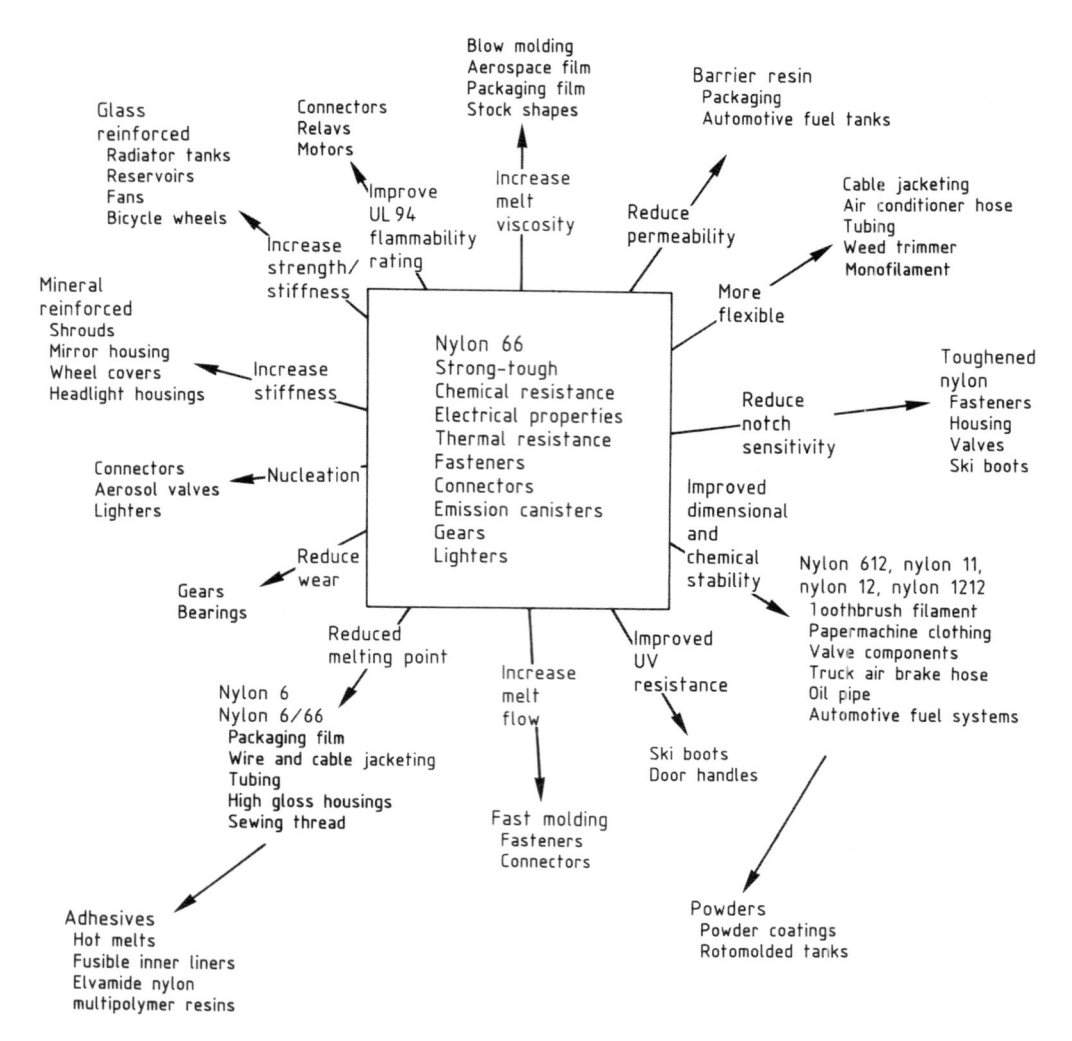

Figure 16. Nylon readily tailored to market needs

superheated steam and an acid catalyst such as phosphoric acid and collecting the monomer by distillation [279, 280]. PA 66 or PA 6 can be depolymerized by ammonolysis, which involves heating the polymer with ammonia at elevated pressures in the presence of a Lewis acid catalyst to yield hexamethylenediamine (HMD) or caprolactam and a series of nitriles and amides, some of which can be hydrogenated to yield HMD or caprolactam precursors [281, 282]. Other methods exist that convert PA 66 or PA 6 to adipic acid by treating the polymers with an aliphatic acid at elevated temperatures and oxidizing the resulting alkylamides and bisalkyamides [283].

Depolymerization followed by repolymerization is an expensive process, however, and mechanical recycling of polyamides, which involves grinding or crushing the resin, followed by pelletizing, is also used, although the resulting resins are of lower quality and have limited uses. A solution-based polyamide recovery process has been developed in which the polymer is dissolved in an anhydrous polyol or aliphatic carboxylic acid at elevated temperatures, insoluble fillers are removed by filtration, and more solvent at a sufficiently low temperature to cause the polyamide to precipitate is added quickly [284]. As a result of the rapid quenching, this process leads to relatively little degradation of the polymer, which can be then be reused. A disadvantage of this process, however, is that the resulting solutions are quite viscous, which makes filtration through fine filters, and hence the removal of submicron particles, such as titanium dioxide (which is often used in polyamide fiber applications) difficult. To address this problem, a process was developed to partially depolymerize a polyamide dissolved in methanol or ethanol to achieve a viscosity suitable for filtration, separate the insoluble material, and then repolymerize the recovered low molecular polymer by solid-phase or melt polymerization [285].

9.2. Polyamide Monomers

Most polyamide monomers come from petroleum-based feedstocks, although two commercially important monomers, sebacic acid (decanedioic acid) and 11-aminoundecanoic acid, are derived from castor beans. The processes for preparing many common monomers often use ingredients that can generate byproducts and thus must be handled carefully. For example, adipic acid is generally produced by oxidizing a mixture of cyclohexanone and cyclohexanol with nitric acid, generating nitrous oxide, which is believed to be a greenhouse gas, as a byproduct. These byproducts are contained under rigorous manufacturing controls and abated and not released into the environment. A variety of research projects have addressed these issues. A process has been developed that oxidizes benzene to phenol with nitrous oxide. The phenol is converted to cyclohexanone, which is oxidized to adipic acid with nitric oxide, and the resulting nitrous oxide byproduct is recycled back into the process [286, 287]. Recently a method has been reported to directly oxidize hexane to adipic acid in the presence of air using molecular sieves containing cobalt(III) ions [288]. A method has been developed to oxidize *p*-xylene to terephthalic acid using a heterogeneous catalyst in supercritical water [291].

9.3. Other Aspects

Polyamides degrade slowly in sunlight with the evolution of organic acids, aldehydes, and oxides of carbon. These are assimilated with other natural decomposition products. Resistance to attack by bacteria and fungi makes for very slow decomposition when buried in the soil. Nylons are normally inert, nonleaching, and well compacted in sanitary landfills so they pose no danger to subsurface water. Normal incineration in air at 800 °C yields the same gaseous products as do naturally occurring, proteinaceous materials such as wool, silk, hair, and leather, i.e., mainly carbon dioxide and carbon monoxide with small amounts of ammonia and hydrogen cyanide. Silk generates more of these products than does PA 66 under the indicated conditions of incineration.

Most polyamides are biologically inert and offer no toxicological problems except, possibly, in the instance of additives or some low molecular mass fragments. Caprolactam monomer has been reported to be a convulsant poison in sufficient doses although aminocaproic acid is not [290]. The same comment is applicable to the dermatological situation—no problem with the polymer but caution with respect to additives and caprolactam. Polyamides have been widely used as a suture material and many are approved for food uses. The use of wood fiber as a lightweight reinforcing agent in polyamides that comes from a renewable resource and leaves no ash upon incineration has also been investigated [292]. Polyamides can contribute to decreased emissions and fuel consumption when they replace heavier metals in automotive applications and can replace toxic metals in certain applications (see Section 6.3.5).

10. Economic Aspects

The following discussion is based on data sourced from and used with permission of IHS Chemical [262]. Global consumption of polyamides in engineering polymers applications has grown from 800 000 t/a in 1990 to 1 700 000 t/a in 2000 to 2 500 000 t/a in 2011. Even with the impact of the recession of 2008, the average annual compounded growth rate from 1990 to 2010 was 5.6%. Recovery to pre-2008 recession consumption levels is expected by 2012.

The largest polyamide consumption regions are Northeast Asia, Western Europe, and North

Table 9. Major producers of polyamides

Company	PA 6	PA 66	Other	Trade names
Asahi Kasei		x		Leona
Arkema	x	x	11, 12	Rilsan, Orgamid, Cristamid, Platamid
Ascend Performances Materials		x		Vydyne
BASF	x	x	610	Ultramid
Lanxess	x	x		Durethan
DSM	x	x	46, 610	Akulon, Stanyl
DuPont	x	x	612, phthalamides, amorphous	Zytel, Minlon
EMS-Grivory	x		12, amorphous	Grilon, Grilamid, Grivory
Evonik			12, amorphous	Vestamid, Trogamid
Invista		x		Torzen
Mitsubishi Gas			MXD6	Reny
Mitsubishi Petrochemicals			phthalamides	Arlen
Nilit		x		
Radici	x	x	610	
Rhodia	x	x	610	Technyl
PingMei Shenma		x		
Solvay			MXD 6, phthalamides	Ixef, Amodel
Toray	x	x	610, 12	Amilan
Ube	x	x	12	Ube
Unitika	x			Unitika

America with 38%, 29%, and 15% of global demand for polyamides respectively. Polyamide consumption in Northeast Asia grew from 100 000 t/a in 1990 to 900 000 t/a in 2010 with an average annual compounded growth rate from 1990 to 2010 of 12%. In the more developed markets, polyamide consumption in West Europe has grown from 275 000 t/a in 1990 to 690 000 t/a in 2010, which corresponds to an average annual compounded growth rate of about 3%. In North America, consumption of polyamides has grown from 290 000 t/a in 1990 to 360 000 t/a in 2010, corresponding to an average annual compounded growth rate of 1%. The impact of the recession in 2008 can be seen by comparing the polyamide consumption in North America of 430 000 t/a in 2006 to the 360 000 t/a consumption in 2010.

Major market segments worldwide for injection-molded polyamides are appliances, consumer products, electrical and electronic, and automotive and truck applications. Globally, automotive and truck and electrical and electronic applications are the fastest growing segments and are expected to grow at higher rates than the global demand in polyamides. In Northeast Asia the average annual compounded growth from 2006 to 2011 for the automotive and truck, electrical and electronic appliances, and film and coating segments was 7% for automotive and more than 3% for the others. In Western Europe and North America the average annual compounded growth from 2006 to 2011 dropped for most segments because of the recession in 2008.

Producers of polyamides with their products and trade names are listed in Table 9.

References

1. W.H. Carothers, *J. Am. Chem. Soc.* **51** (1929) 2548–2559.
2. W.H. Carothers, J.A. Arvin, [1], 2560–2570.
3. DuPont, US 2071250, 1937 (W.H. Carothers).
4. IG Farbenindustrie, US 2241321, 1941 (P. Schlack).
5. M.I. Kohan: The History and Development of Nylon-66 in R.B. Seymour, G.S. Kirshenbaum (eds.): *High Performance Polymers: Their Origin and Development*, Elsevier, New York 1986, pp. 19–37.
6. P. Matthies, W.F. Seydl: History and Development of Nylon 6, in [5] pp. 39–53.
7. G.B. Apgar, M.J. Koskoski: The History of Development of Nylons 11 and 12, in [5] pp. 55–65.
8. H. Staudinger, *Ber. Dtsch. Chem. Ges.* **53** (1920) 1073–1092.
9. W.H. Carothers, *Chem. Rev.* **8** (1931) 353–426.
10. R.W. Lenz: *Organic Chemistry of Synthetic High Polymers*, Interscience Publishers, New York 1967, pp. 1–9.
11. M.I. Kohan (ed.): *Nylon Plastics Handbook*, Hanser Verlag, Munich 1995, p. 3.
12. G.B. Kauffman, *Chem. Tech.* **18** (1988) 725–731.
13. Source-Based Nomenclature for Copolymers, IUPAC Macromolecular Division, Provisional Paper (1985) Rule 7.
14. Nomenclature of Regular Single-Strand Organic Polymers, IUPAC Macromolecular Division, Tentative, *Pol. Lett. Div.* **11** (1973) 389–414.

15. J. Brandrup: Nomenclature Rules in J. Brandrup, E.H. Immergut (eds.): *Polymer Handbook*, 3rd ed.Wiley-Interscience, New York 1989, pp. I/1–I/61.
16. R. Pfluger: "Physical Constants of Various Polyamides", in [15] p. V/109.
17. G.C. Eastmond, A. Ledwith, S. Russo, P. Sigwalt (eds.): Comprehensive Polymer Science, vol. 5, *Step Polymerization*, Pergamon Press, Oxford 1989.
18. P.J. Flory: *Principles of Polymer Chemistry*, Cornell Univ. Press, New York 1953.
19. M.I. Kohan (ed.): *Nylon Plastics*, Wiley-Interscience, New York 1973.
20. J. Zimmerman: *Polyamides in Encyclopedia of Polymer Science and Engineering*, 2nd ed., vol. 11, Wiley-Interscience, New York 1988, pp. 315–381.
21. F. Bueche, *J. Polym. Sci.* **43** (1960) 527–530.
22. T.G. Fox, S. Loshaek, *J. Appl. Phys.* **26** (1955) 1080–1082.
23. T.G. Fox, S. Gratch, S. Loshaek in F.R. Eirich (ed.): *Viscosity Relationships for Polymers in Bulk and in Concentrated Solution*, vol. **1**, Rheology-Academic Press, New York 1956, pp. 431–493.
24. M.A. Dudley, *J. Appl. Polym. Sci.* **16** (1972) 493–504.
25. E. Jacobi, H. Schuttenberg, R.C. Schultz, *Makromol. Chem. Rapid Commun.* **1** (1980) 397–402.
26. G. Pastuska, U. Just, H. August, *Angew. Makromol. Chem.* **107** (1982) 173–184.
27. E. Biagini, E. Gattiglia, E. Pedemonte, S. Russo, *Makromol. Chem.* **184** (1983) 1213–1222.
28. P.J. Flory, *J. Am. Chem. Soc.* **58** (1939) 1877–1885.
29. W.H. Stockmayer, *J. Chem. Phys.* **11** (1943) 45–55.
30. J.R. Schaefgen, P.J. Flory, *J. Am. Chem. Soc.* **70** (1948) 2709–2718.
31. DSM, N. V., WO 96/35739, 1996 (A. Nijenhuis, M. Serne, E. van den Berg, R. Aberson).
32. Nyltech, WO 97/24388, 1997 (A. Cucinella, G. Di Silvestro, C. Guaita, F. Speroni, H. Zhang).
33. Nyltech, WO 99/64496, 1999 (G. Di Silvestro, F. Speroni, C. Yuan, H. Zhang).
34. R. Puffr, Z. Tuzar, K. Mrkvickova, J. Sebenda, *Makromol. Chem.* **184** (1983) 1957.
35. R. Puffr, V. Kubanek: *Lactam-Based Polyamides*, Vols. **I & II**, CRC Press, Boca Raton 1991.
36. H. Zahn, G.B. Gleitsman, *Angew. Chem. Int. Ed. Engl.* **2** (1963) 410–420.
37. P. Kusch, H. Zahn, *Angew. Chem. Int. Ed. Engl.* **4** (1965) 696.
38. T.M. Cawthon, E.C. Smith, *Polym. Prepr. Am. Chem. Soc. Div. Polym. Chem.* **1** (1960) Sep., 98.
39. H.K. Reimschuessel, *J. Polym. Sci.* **41** (1959) 457–466.
40. H.K. Reimschuessel, *Macromol. Rev.* **12** (1977) 65–139.
41. P. Feldmann, R. Feinauer, *Angew. Makromol. Chem.* **34** (1973) 9–18.
42. C.J. Brown, H. Hill, P.V. Youle, *Nature (London)* **177** (1956) 128.
43. H. Zahn, P. Miro, F. Schmidt, *Chem. Ber.* **90** (1957) 411–418.
44. S. Schaaf, *Faserforsch. Textiltech.* **10** (1959) 328–330.
45. R. Aelion, *Ind. Eng. Chem.* **53** (1961) 826–828.
46. K. Dachs, E. Schwartz, *Angew. Chem. Int. Ed. Engl.* **1** (1962) 430–435.
47. M. Genas, *Angew. Chem.* **74** (1962) 535–540.
48. C.F. Horn, B.T. Freure, H. Vineyard, H.J. Decker, *J. Appl. Polym. Sci.* **7** (1963) 887–896.
49. H.K. Reimschuessel: Lactams, in K.C. Frisch, S.L. Reegen (eds.): *Ring-Opening Polymerizations*, Dekker, New York 1969, pp. 303–326.
50. D.C. Jones, T.R. White: Polyamides in D.H. Solomon (ed.): *Step-Growth Polymerization*, Dekker, New York 1972, pp. 41–94.
51. H.K. Reimschuessel, *Macromol. Rev.* **12** (1977) 65–139.
52. H. Sekiguchi: Lactams and Cyclic Imides in K.J. Ivin, T. Saegusa (eds.): *Ring-opening Polymerization*, vol. **2**, Elsevier, London 1984, pp. 809–918.
53. R.J. Gaymans, D.J. Sikkema: Aliphatic Polyamides in G.C. Eastmond, A. Ledwith, S. Russo, P. Sigwalt (eds.): *Comprehensive Polymer Science*, vol. **5**, Step Polymerization, Pergamon Press, Oxford 1989, pp. 357–373.
54. C. Giori, B.T. Hayes, *J. Polym. Sci. Polym. Chem. Ed.* **8** (1970) 335–349.
55. C. Giori, B.T. Hayes, *J. Polym. Sci. Polym. Chem. Ed.* **8** (1970) 351–358.
56. K. Tai, H. Teranishi, Y. Arai, T. Tagawa, *J. Appl. Poly. Sci.* **25** (1980) 77–87.
57. J. Zimmerman: *Polyamides in Encyclopedia of Polymer Science and Engineering*, 2nd ed., vol. **11**, Wiley-Interscience, New York 1988, pp. 315–381.
58. K. Tai, T. Tagawa, *Ind. Eng. Chem. Prod. Res. Dev.* **22** (1983) no. 2, 192–206.
59. DuPont, US 2130523, 1935 (W.H. Carothers).
60. DuPont, US 2244192, 1939 (P.J. Flory).
61. DuPont, US 2669556, 1953 (C.A. Sperati).
62. B.S. Khaitin et al., SU 775106, 1978; *Chem. Abstr.* **94** (1981) 48 033 f.
63. Firestone Tire and Rubber, US 3705811, 1969 (S.E. Schonfeld).
64. ICI, US 3705134, 1970 (D. James).
65. DuPont, WO 92/10526, 1992 (J.F. Buzinkai, M.R. DeWitt, R. C. Wheland).
66. DuPont, US 3173898, 1965 (W.M. Sum).
67. Allied Corp., US 4471081, 1984 (J.A. Bander, T.S. Brown, L. C. Lin, D.M. Vultaggio).
68. V.V. Korshak, T.M. Frunze: *Synthetic Hetero-chain Polyamides*, Jerusalem Program for Scientific Translations, Jerusalem 1964, pp. 17–21.
69. E. Roerdink, P.J. DeJong, J. Warnier, *Polym. Commun.* **25** (1984) 194–195.
70. E. Roerdink, J. M. M. Warnier, *Polymer* **26** (1985) 1582.
71. A.M. Kotliar, *J. Polym. Sci.: Polym. Chem. Ed.* **11** (1973) 1157–1165.
72. L.F. Beste, R.C. Houtz, *J. Polym. Sci.* **8** (1952) 395–407.
73. I.K. Miller, *J. Polym. Sci. Polym. Chem. Ed.* **14** (1976) 1403–1417.
74. F.W. Billmeyer, Jr.: *Textbook of Polymer Science*, 3rd ed., Wiley-Interscience, New York 1984, pp. 332–333.
75. J. Zimmerman: *Polyamides in Encyclopedia of Polymer Science and Engineering*, 2 ed., vol. **11**, Wiley-Interscience 1988, pp 340–353.
76. DuPont, US 2130523, 1935 (W.H. Carothers).
77. Chemische Werke Witten, US 3454536, 1964 (G. Schade, F. Blaschke).
78. General Electric, US 4567249, 1983 (D.W. Fox, S.J. Shafer).
79. DuPont, US 4009153, 1975 (H. Shin).
80. Standard Oil, US 4719285, 1986 (B.S. Curatolo, R.C. Sentman, G.P. Coffey).
81. Halcon International, US 3847876, 1972 (O.T. Onsager).
82. Standard Oil, US 4520190, 1983 (G.P. Coffey, B.S. Curatolo, R.L. Cepulla).
83. Standard Oil, US 4542205, 1984 (B.S. Curatolo, R.C. Sentman, G.P. Coffey).
84. Standard Oil, US 4603192, 1985 (G.P. Coffey, R.S. Sherrard, Jr., B.S. Curatolo, R.C. Sentman).

85 Standard Oil, US 4725666, 1986 (B.S. Curatolo, R.C. Seatman, G.P. Coffey).
86 DuPont, US 2190770, 1936 (W.H. Carothers).
87 RED-Gesellschaft, US 2918455, 1955 (E. Elod, R.-E. Dorr, H. Braun, J. Elian, M. Lepingle).
88 Sumitomo Chemical, US 4721772, 1982 (K. Ueno, K. Nagaoka, A. Miyashita).
89 W.J. Farrissey, B.K. Onder, P.S. Andrews, *Mod. Plast.* **58** (1981) March, 82,84.
90 P.W. Morgan: *Condensation Polymers: By Interfacial and Solution Methods*, Interscience, New York 1965.
91 L. Vollbracht: Aromatic Polyamides in G.C. Eastmond, A. Ledwith, S. Russo, P. Sigwalt (eds.): *Comprehensive Polymer Science*, vol. 5, Step Polymerization-Pergamon, Oxford 1989, pp. 375–386.
92 G.M. Carlyon: Monomer Casting in M.I. Kohan (ed.): *Nylon Plastics*, Wiley-Interscience, New York 1973, pp. 457–472.
93 Mitsubishi Monsanto, JP-Kokai 6303909, 1986 (T. Kato, H. Isekawa); *Chem. Abstr.* **108** (1988) 151 647 r.
94 Monsanto, US 4595746, 1984 (J.D. Gabbert, R.M. Hedrick).
95 BASF, DE-OS 3439461, 1984 (P. Horn, E. Baumann, M. Marx, P. Gerarts); *Chem. Abstr.* **105** (1988) 44 417 g.
96 Ube, JP-Kokai 60223822, 1984 (M. Ogasa, M. Matsumoto, A. Kawabata); *Chem. Abstr.* **104** (1986) 187 511 y.
97 PPG, US 4501821, 1983 (R.B. Hodeck, J.A. Seiner).
98 Istituto Chim, EP 130360, 1983 (G.C. Alfonso, S. Russo, E. Pedemonte, A. Turturro, C. Puglisi); *Chem. Abstr.* **102** (1985) 132 965 c.
99 Ube, JP-Kokai 5958030, 1982; *Chem. Abstr.* **101** (1984) 111 589 x.
100 Aisin Seiki, JP-Kokai 58210924, 1982; *Chem. Abstr.* **100** (1984) 140 442 d.
101 VEB Leuna-Werke "Walter Ulbricht", DD 202885, 1983 (H. Schade et al.); *Chem. Abstr.* **100** (1984) 140 337 y.
102 Ural Inst. Chem., SU 937475, 1980 (V.P. Begishev et al.); *Chem. Abstr.* **97** (1982) 199 022 q.
103 R.M. Hedrick, J.D. Gabbert, AJChE: *A New RIM System from Nylon 6 Block Copolymers: Chemistry and Structure*, National Mtg., Detroit 1981.
104 J.D. Gabbert, R.M. Hedrick, *Polym. Process Eng.* **4** (1986) 359–373.
105 J.A. Lefelar, K. Udipi, *Polym. Commun.* **30** (1989) Feb., 38–40.
106 Harwe, US 4223433, 1980 (W. Hartmann).
107 DuPont, US 2172374, 1937 (P.J. Flory).
108 BASF, GB 801733, 1955.
109 California Research, US 2987507, 1957 (I.E. Levine).
110 E.M. Kampouris, *Polymer* **17** (1976) 409–412.
111 E.M. Macchi, A.A. Giorgi, *Makromol. Chem.* **180** (1979) 1603–1605.
112 DuPont, US 3031433, 1958 (G.C. Monroe, Jr.).
113 DuPont, US 3821171, 1972 (D.H. Beaton).
114 F. Pilati: Solid-state Polymerization in G.C. Eastmond, A. Ledwith, S. Russo, P. Sigwalt (eds.): *Comprehensive Polymer Science*, vol. 5, Step Polymerization, Pergamon Press, Oxford 1989, pp. 201–216.
115 J. Zimmerman, *J. Polym. Sci. Part B* **2** (1964) 955–958.
116 R. Feldmann, R. Feinauer, *Angew. Makromol. Chem.* **34** (1973) 1–7.
117 R.J. Gaymans, J. Amirtharaj, H. Kamp, *J. Appl. Poly. Sci.* **27** (1982) 2513–2526.
118 DuPont, US 3015651, 1956 (E.W. Kjellmark, Jr.).
119 Unitika, JP 7251120, 1968 (Y. Nishida, K. Hirose); *Chem. Abstr.* **80** (1974) 15 507 w.
120 R.J. Gaymans, T. E. C. Van Utteren, J. W. A. Van Den Berg, J. Schuyer, *J. Polym. Sci. Polym. Chem. Ed.* **15** (1977) 537–545.
121 E. Roerding, J. M. M. Warnier, *Polymer* **26** (1985) 1582–1588.
122 Stamicarbon, NL-A 8001764, 1980 (R.J. Gaymans, E. H. J. P. Bour); *Chem. Abstr.* **96** (1982) 69 637 n.
123 J.M. Schultz, S. Fakirov: *Solid State Behavior of Linear Polyesters and Polyamides*, Prentice Hall, London 1990.
124 D. O'Sullivan, *Chem. Eng. News* (1984) May, 33–34.
125 Stamicarbon, EP-A 160337, 1984 (R.J. Gaymans, A. J. P. Bongers); *Chem. Abstr.* **104** (1986) 89 240 q.
126 Asahi, JP-Kokai 61188421, 1985 (K. Higami); *Chem. Abstr.* **106** (1987) 67 885 t.
127 Bayer, DE-OS 3526931, 1985 (W. Nielinger et al.); *Chem. Abstr.* **106** (1987) 120 440 x.
128 Stamicarbon, US 4722997, 1985 (E. Roerdink, J. M. M. Warnier).
129 Stamicarbon, JP-Kokai 62158722, 1986; *Chem. Abstr.* **107** (1987) 237 562 a.
130 DuPont, US 2130947, 1935 (W.H. Carothers).
131 DuPont, US 2163584, 1936 (W.H. Carothers, G.D. Graves).
132 V.D. Luedeke: Hexamethylenediamine, in J.J. Mcketta (ed.): *Encyclopedia of Chemical Processing and Design*, Dekker, New York 1987, pp. 222–237.
133 DuPont, US 3337612, 1964 (D.L. Sharps).
134 DuPont, US 3439025, 1965 (R. Gallay, S.P. LaPenta).
135 BASF, US 4233234, 1978 (R-H. Rotzoll et al.).
136 DuPont, US 4442260, 1983 (H.A. Larsen).
137 DuPont, US 2840547, 1955 (W.L. Stump).
138 Monsanto, CA 715476, 1962 (E.P. Brignac).
139 Monsanto, US 3300449, 1964 (E.P. Brignac).
140 D.B. Jacobs, J. Zimmerman: Preparation of 6, 6-Nylon and Related Polyamides in C.E. Schildknecht, I. Skeist (eds.): *Polymerization Processes, High Polymers*, vol. 29, Wiley-Interscience, New York 1977, pp. 424–467.
141 DuPont, US 2289774, 1939 (G. DeW. Graves).
142 DuPont, US 3491177, 1967 (W.G. Johnson).
143 DuPont, US 2689839, 1951 (W.W. Heckert).
144 DuPont, US 3113843, 1959 (W.H. Li).
145 DuPont, US 3361537, 1965 (N.A. Ferrante).
146 DuPont, US 3948862, 1974 (J.M. Iwasyk).
147 DuPont, US 4134736, 1978 (L.F. Hammond, Jr.).
148 DuPont, US 4299498, 1979 (R.D. Sauerbrunn).
149 DuPont, US 3789584, 1972 (J.M. Iwasyk, D.W. Rodeffer).
150 Vereinigte Glanzstoff-Fabriken, US 3027355, 1957 (H. Taul, F. Wiloth).
151 British Nylon Spinners, US 3185672, 1959 (P.C. Clemo, J.A. Briggs, W. Wilson, J.A. Carter).
152 BASF, US 4060517, 1974 (F. Mertes, H. Doerfel, E. Heil, C. Cordes).
153 Monsanto, US 3296217, 1964 (J.E. Tate).
154 DuPont of Canada, US 3900450, 1973 (I. Jaswal, K. Pugi).
155 British Nylon Spinners, US 3258313, 1962 (G.D. Griffiths).
156 DuPont, US 3118739, 1959 (W.G. Atkinson, J.L. Thomas, Jr.).
157 DuPont, US 3257173, 1960 (J.E. Parnell).
158 DuPont of Canada, US 3960820, 1974 (B. McD. Pinney).
159 American Enka, US 3044993, 1957 (H. G. N. Tiemersma).
160 Monsanto, US 3218297, 1960 (G.W. Sovereign).
161 Monsanto, US 3260703, 1964 (A.C. Coggeshall).
162 Vereinigte Glanzstoff-Fabriken, US 3130180, 1958 (F. Wiloth).
163 Amoco, US 4831108, 1987 (J.A. Richardson, W. Poppe, B.A. Bolton, E.E. Paschke).
164 California Res., US 2997463, 1957 (F.G. Lum).
165 G. Bier: "New Polycondensation Polymers," *Adv. Chem. Ser.* **91** (1969) 612–627.

166 BP Chemicals, US 5079307, 1992 (S.A. Taylor, M.B. Studholme, M.R. Orpin).
167 Mitsui Petrochemicals, US 6130312, 2000 (H. Murakami, S. Omari, K. Wakatsuru).
168 H. Jacobs, C. Schweigman: Mathematical Model for the Polymerization of Caprolactam to Nylon-6 in Chemical Reaction Engineering, *Proceedings of the Fifth European/Second International Symposium on Chemical Reaction Engineering*, Elsevier, Amsterdam 1972, pp. B 7–1 to 7–26.
169 Allied Chemical, US 3578640, 1968 (I.C. Twilley, G.J. Coll, Jr., D. W. H. Roth, Jr.).
170 Allied Chemical, CA 823290, 1967 (I.C. Twilley, D. W. H. Roth, Jr., R.A. Lofquist).
171 Allied Chemical, US 3813366, 1973 (W.H. Wright, A.J. Bingham, W.A. Fox).
172 Allied Chemical, EP-A 0038473, 1980 (S.L. Yates, C.J. Cole, A.H. Wiesner, J.W. Wagner).
173 A.K. Ray, S.K. Gupta, *Polym. Eng. Sci.* **26** (1986) no. 15, 1033–1044.
174 A. Kumar, S.K. Gupta, *J. Macromol. Sci. JMS-Rev. Macromol. Chem. Phys.* **C 26** (1986) no. 2, 183–247.
175 G. Menges, T. Bartilla, *Polym. Eng. Sci.* **27** (1987) 1216–1220.
176 P. Wagner, *Kunststoffe* **73** (1983) 588–590.
177 C.R. Dupre, J.D. Gabbert, R.M. Hedrick, *Am. Chem. Soc. Polym. Chem. Div. Preprints* **25** (1984) no. 2, 296–297.
178 L. Notarbartolo, *Ind. Plast. Mod.* **10** (1958) 44–52.
179 Societe-Organico, FR 951924, 1947 (M. Genas); *Chem. Abstr.* **46** (1952) 132 g.
180 N.V. Onderzoekingsinst. Res., BE 642232, 1963.
181 ICI, GB 680212, 1968 (H. McGrath).
182 BASF, DE-AS 1495149, 1963 (J.H. Kunde, J.H. Wilhelm, H. Metzger, H. Dorfel).
183 W. Griehl, D. Ruestem, *Ind. Eng. Chem.* **62** (1970) no. 3, 17–22.
184 www.campusplastics.com (accessed August 28, 2012).
185 R.G. Beaman, *J. Polym. Sci.* **9** (1952) 470–472.
186 H.G. Weyland, P.J. Hoftyzer, D. W. Van Krevelen, *Polymer* **11** (1970) 79–87.
187 K-H. Illers, *Makromol. Chem.* **38** (1960) 168–188.
188 A.M. Thomas, *Nature (London)* **179** (1957) 862.
189 U. Rhode-Liebenau, *Kunststoffe* **55** (1965) 302–305.
190 G.A. Gordon, *J. Polym. Sci. Polym. Phys. Ed.* **9** (1971) 1693–1702.
191 R. Leaversuch, L.M. Sherman, M.H. Naitove, *Plast. Technol.* **47** (2001) Sept., 65–67.
192 http://www.rhodia.com/en/binaries/technylStar_brochure_BD_PP.pdf (accessed November 12, 2012).
193 F. Stastny, *Kunststoffe* **40** (1950) 273–304.
194 E.D. Harvey, F.J. Hybart, *J. Appl. Polym. Sci.* **14** (1970) 2133–2143.
195 W.E. Catlin, E.P. Czerwin, R.H. Wiley, *J. Polym. Sci.* **2** (1947) 412–419.
196 H-G. Elias, *Polymer News* **11** (1985) 40–43.
197 Imperial Chemical Industries, GB Pat App. 604/9, 1949 (H. Plimmer).
198 Inventa AG, CH 280 367, 1949 (R. Gabler).
199 O.B. Edgar, R. Hill, *J. Pol. Sci.* **8** (1952), pp. 1–22.
200 R. Gächter, H. Müller: *Plastics Additive Handbook*, 4th ed., Hanser, Munich 1993.
201 J.M. Crosby, T.R. Drye, *Mod. Plast.* **63** (1986) Nov., 74–84.
202 D.A. Cianelli, J.E. Travis, R.S. Bailey, *Plast. Technol.* **34** (1988) Apr., 83–89.
203 Libbey-Owens-Ford Glass, US 2571717, 1951 (A.M. Howard, L.S. Meyer).
204 Glastrustions, US 2871911, 1959 (W.B. Goldsworthy, F. Landgraf).
205 A.H. Tullo, *Chem. Eng. News* **80**(4) (2002) Jan. 28, 21.
206 M. Holmes, *Plastics Additives & Compounding*, January 2002, pp. 30–33.
207 R. Leaversuch, *Plast. Technol.* **47** (2001) Oct., 64–69.
208 P.S. Wilson: Nanocomposite Market Opportunities in *Proceedings of Nanocomposites 2001*, Executive Conference Management, Chicago 2001.
209 Proceedings of Nanocomposites 2001, Executive Conference Management, Chicago 2001.
210 D.R. Paul, S. Newman: *Polymer Blends*, Academic Press, New York 1978.
211 M.J. Folkes: *Polymer Blends and Alloys*, Blackie, London 1993.
212 L.A. Utracki: *Polymer Alloys and Blends*, Hanser Verlag, Munich 1989.
213 L.A. Utracki: *Polymer Alloys and Blends*, Hanser Verlag, Munich 1989, pp. 248–270.
214 W.E. Baker, C. Scott, G.-H. Hu: *Reactive Polymer Blending*, Hanser Verlag, Munich 2001.
215 DuPont, US 4174358, 1977 (B.N. Epstein).
216 E.A. Flexman, Jr., *Polym. Eng. Sci.* **19** (1979) no. 8, 564–571.
217 R.E. Lavengood, F.M. Silver, *Soc. Plast. Eng. (Tech. Pap.)* **33** (1987) 1369–1374.
218 Rohm and Haas, US 3796771, 1972 (F.H. Owens, J.S. Clovis).
219 G.C. Adams: Impact Fatigue of Polymers Using an Instrumented Drop Tower Device in S.L. Kessler, G.C. Adams, S.B. Driscoll, D.R. Ireland (eds.): *Instrumented Impact Testing of Plastics and Composite Materials*, ASTM STP 936-Amer. Soc. Test. Matls., Philadelphia 1987, pp. 281–301.
220 J.R. Heater, E.M. Lacey, *Mod. Plast.* **41** (1964) May, 123–126, 180–182.
221 B. Epstein: Soc. Plast. Industry,Reinf. Plast./Composites Institute, *34th Ann. Tech. Conf.* (1979) 10-A, pp. 1–6.
222 R.C. Nametz, *Plast. Comp.* **2** (1979) Jan./Feb., 31–40; **7** (1984) July/Aug., 26–39.
223 N.A. Burditt, *Plast. Comp.* **11** (1988) Sep./Oct., 51–56.
224 M.C. McMurrer, *Plast. Comp.* **11** (1988) July/Aug., 60–61.
225 V. Wigotsky, *Plast. Eng.* **43** (1987) Oct., 25–30.
226 Ems-Inventa, US 4511684, 1982 (E. Schmid, M. Hoppe).
227 *Plast. Eng.* **44** (1988) Oct., 53.
228 Clariant, US 6255371, 2001 (E. Schlosser, B. Nass, W. Wanzke).
229 C.S. Hardo, J.J. Duffy, *Plast. Eng.* **41** (1985) Sep., 51–54.
230 J.E. Theberge, J.M. Crosby, K.L. Talley, *Plast. Eng.* **44** (1988) Aug., 47–52.
231 K.J. Nangrani, R. Wenger, P.G. Daugherty, *Plast. Comp.* **11** (1988) Mar./Apr., 28–31.
232 Asahi, US 6051307, 2000 (T. Kido, S. Hasegawa).
233 Hyperion, US 5651922, 1997 (P.R. Nahass, S.O. Friend, R.W. Hausslein).
234 E.M. Doll, M. Koetting, *Soc. Plast. Eng. (Tech. Pap.)* **33** (1987) 406–409.
235 D.M. Bigg, *Polym. Compos.* **8** (1987) 1–7.
236 L.M. Sherman, *Plast. Technol.* **47** (2001) June, 52–53, 55–57.
237 M.T. Shaw, *Polym. Eng. Sci.* **22** (1982) no. 2, 115–121.
238 V. Wigotsky, *Plast. Eng.* **42** (1986) July, 19–29.
239 P.A. Toensmeier, *Mod. Plast.* **65** (1988) Oct., 62–67.
240 Polymer Blends and Alloys-Technomic, Lancaster 1988.
241 B. Arkles, C. Carreno, *Soc. Plast. Eng. (Tech. Pap.)* **30** (1984) 486–487.
242 R. Leaversuch, *Plast. Technol.* **47** (2001) July, 43–45, 47.
243 R.D. Leaversuch, *Plast. Technol.* **47** (2001) June, 58–61, 63.

244. J.C. Gardner, P.J. Gardner, I.P. Oliver, T. Peake, EP-A 0641836, 1995.
245. "Ideas to Market",US 6048379, 2000 (A.V. Bray, B.A. Muskopf, M.L. Dingus).
246. J. De Gaspari, *Plast. Technol.* **44** (1998) June, 44–48.
247. J.A. Grande, *Mod. Plast.* **75** (1998) May, 74–79.
248. J.R. Flesher, Jr., *Mod. Plast.* **64** (1987) Sep., 100,105–106, 110.
249. H.W. Cox, C.C. Mentzer, R.C. Custer, *Polym. Eng. Sci.* **24** (1984) 501–510.
250. P. Perrini, D. Romanini, G.P. Righi, *Polymer* **17** (1976) 377–381.
251. A.V. Shenoy, D.R. Saini, V.M. Nadkarni, *Rheol. Acta* **22** (1983) 209–222.
252. G. Pezzin, G.B. Gechele, *J. Appl. Polym. Sci.* **8** (1964) 2195–2212.
253. R.F. Westover: Processing Properties in E.C. Bernhardt (ed.): *Processing of Thermoplastic Materials*, Reinhold Publ. Co., New York 1959, pp. 547–679.
254. Plastics Department, DuPont, *Tech. Bulletin: Molding DuPont Zytel^R Nylon Resins . . . a Handbook for the 70's*, 1970.
255. R. Gabler et al., *Chimia* **21** (1967) no. 2, 65–81.
256. J. C. L. Williams, D.W. Wood, I.F. Bodycot, B.N. Epstein, *Soc. Plast. Ind. Reinf. Plast. Div. Conf. Proceedings* **23** (1968) 2 C 1–2 C 8.
257. R.J. Crowson, M.J. Folkes, *Polym. Eng. Sci.* **20** (1980) 934–940.
258. W.C. Filbert, Jr., *SPE J.* **25** (1969) Jan., 65–69.
259. H. W. H. Yang, R. Farris, J. C. W. Chien, *J. Appl. Polym. Sci.* **23** (1979) 3375–3382.
260. J.M. Lunt, J.B. Shortall, *Plast. Rubber Process.* **5** (1980) June, 37–44.
261. R. v. Turkovich, L. Erwin, *Polym. Eng. Sci.* **23** (1983) 743–749.
262. IHS Chemical World Analysis: Nylon Engineering Resins 2012, www.ihs.com/products/chemical (accessed November 2012).
263. P.M. Subramanian, *Polym. Eng. Sci.* **25** (1985) June, 483–487.
264. P.M. Subramanian, V. Mehra, *Polym. Eng. Sci.* **27** (1987) May, 663–668.
265. R. Leaversuch, L.M. Sherman, M.H. Naitove, *Plast. Technol.* **47** (2001) Sept., 65–67.
266. L.K. English, *Mater. Eng. (Cleveland)* **106** (1989) Feb., 47–51.
267. W.B. Glenn, *Plast. Technol.* **31** (1985) June, 105–108.
268. R. Burns, *Plast. Technol.* **34** (1988) Feb., 60–65.
269. A.S. Wood, *Mod. Plast.* **66** (1989) Feb., 44–49.
270. M. Gabriele, *Plast. Technol.* **36** (1990) Mar., 55–57.
271. H. Frank, *Kunststoffe* **76** (1986) 853–855.
272. G. Blinne, E. Priebe, *Kunststoffe* **77** (1987) 988–993.
273. United States Code of Federal Regulations, Title 21, Part 177.1500(Apr., 2002) 265–271.
274. J.H. Schut, *Plast. Technol.* **39** (1993) April, 22–24.
275. J.H. Schut, *Plast. Technol.* **48** (2002) May, 37–39.
276. M. Reisch, *Chem. Eng. News* **79**(38) (2001) Sept. 17, 22.
277. A.H. Tullo, *Chem. Eng. News* **78**(4) (2000) Jan. 24, 23–24.
278. R.F. Jones, M.H. Baumann, *Mod. Plast.* **75** (1998) May, 95–96, 98.
279. BASF, US 5294707, 1994 (R. Kotek).
280. J.H. Schut, *Plast. Technol.* **48** (2002) May, 37–39.
281. DuPont, US 5302756, 1994 (R.J. McKinney).
282. DuPont, EP-B10740 648, 1998 (R.J. McKinney).
283. DuPont, EP-B10783 479, 1999 (E.F. Moran, Jr., R.J. McKinney).
284. DuPont, US 5430068, 1995 (P.M. Subramanian).
285. DuPont, WO-A2 01/94457, 2001 (M.S. McKinnon).
286. Solutia, WO 00/01654, 2000 (M.A. Rodkin et al.).
287. M. McCoy, *Chem. Eng. News* **78**(40) (2000) October 2, 32–34.
288. R. Raja, G. Sankar, J.M. Thomas, *Angew. Chem. Int. Ed.* **39** (2000), 2313–2316.
289. United States Code of Federal Regulations, Title 21, Part 177.1500(Apr., 1989) 240–245.
290. M.W. Goldblatt, M.E. Farquharson, G. Bennett, B.M. Askew, *Br. J. Ind. Med.* **11** (1954) 1–10.
291. S.K. Ritter, *Chem. Eng. News* **79**(29) (2001) July 16, 27–34.
292. P. Mapleston, *Mod. Plast.* **79** (2002) Jan., 36.

Further Reading

H.-J. Arpe: *Industrielle Organische Chemie*, 6th ed., Wiley-VCH, Weinheim 2007.

M. Biron: *Thermoplastics and Thermoplastic Composites*, Elsevier, Amsterdam 2006.

B.L. Deopura, R. Alagirusamy, M. Joshi, B. Gupta (eds.): *Polyesters and Polyamides*, CRC Press/Woodhead Publ., Cambridge, UK 2008.

P. Dubois, O. Coulembier, J.M. Raquez (eds.): *Handbook of Ring-Opening Polymerization*, Wiley-VCH, Weinheim 2009.

S. Fakirov (ed.): *Handbook of Condensation Thermoplastic Elastomers*, Wiley-VCH, Weinheim 2005.

W.B. Gratzer: *Giant Molecules*, Oxford University Press, Oxford, UK 2009.

C.D. Papaspyrides, S.N. Vouyiouka: *Solid State Polymerization*, Wiley, Hoboken, NJ 2009.

E.M. Petrie: *Handbook of Adhesives and Sealants*, 2nd ed., McGraw-Hill, New York, NY 2007.

D.W. van Krevelen, te Nijenhuis: *Properties of Polymers*, 4th ed., Elsevier, Amsterdam 2009.

Polyaspartates and Polysuccinimide

THOMAS KLEIN, Lanxess Deutschland GmbH, Leverkusen, Germany

RALF-JOHANN MORITZ, Lanxess Deutschland GmbH, Leverkusen, Germany

RENÉ GRAUPNER, Lanxess Deutschland GmbH, Leverkusen, Germany

1.	Introduction	733
2.	Synthesis and Production of Polyaspartates	734
3.	Analytical Characterization and Detection	736
3.1.	^{1}H NMR Spectroscopy	736
3.2.	^{13}C Spectroscopy	736
3.3.	Gel Permeation Chromatography	737
3.4.	Isotachophoresis	737
3.5.	Fluorescence Spectroscopy	737
3.6.	Precipitation Titration	737
3.7.	Polyelectrolyte Titration	738
4.	Toxicological Properties and Environmental Fate	738
5.	Function and Applications	738
5.1.	Functionality	738
5.1.1.	Dispersing Effect	738
5.1.2.	Threshold Effect	739
5.1.3.	Crystal Distortion	740
5.1.4.	Complexing Effect	740
5.1.5.	Corrosion Inhibition	740
5.2.	Applications and Uses of Polyaspartates	740
5.2.1.	Water Treatment	740
5.2.2.	Detergents and Cleaners	741
5.2.3.	Mining and Oil Recovery	741
5.2.4.	Other Uses	741
6.	Polysuccinimide	742
6.1.	Introduction	742
6.2.	Chemical Properties and Analytic Characterization	742
6.3.	Applications and Uses of Polysuccinimide	743
	References	743

Polysuccinimide is an intermediate in the production process of polyaspartates and may be easily isolated in a dry state. Tablets of polysuccinimide are formed which release traces of polyaspartate over a long period of time under neutral to slight alkaline pH conditions. Those tablets are mainly used for the treatment of waters. Applications of polysuccinimide are based on its hydrolyzation properties or on its reaction with nucleophiles such as amines and sulfides.

1. Introduction

Mineral formation and scaling, widely occurring problems in various industrial processes, are currently solved by the use of inhibitors and dispersants such as polycarboxylate homo- and copolymers. Most of these compounds are ultimately released to wastewater but they are usually poorly biologically degradable, if at all [1].

In nature, the control of inorganic crystallization is vital to almost all organisms. Inorganic crystals, for example, modified by association with organic components are constituents of skeletons, shells, and teeth. The study of such composite materials and the ways in which they are formed and modified is inherent in the more general field of biomineralization, which has been the subject of numerous publications [2–5].

A multitude of organic molecules from biological systems potentially interact with and modify inorganic crystallization. Key among these is a class of polyanionic proteins. The main constituents of these proteins are aspartic acid and phosphoserine, together accounting for roughly 60 mol % of the total composition of amino acids [6–8].

In aqueous solution, these polyanionic proteins can inhibit crystal nucleation and growth by altering the morphology of microscopically small crystals. Studies using modern analytical methods, e.g., atomic force microscopy [9], and molecular modeling methods [10] have been carried out in order to understand these interaction phenomena. In the 1980s, the first attempts were made to utilize this principle of action in industrial applications. Starting with natural proteins from, e.g., crustacea (biodesign), led to a new class of polypeptide-mimicking polyamides, the "polyaspartates". These can be used to solve industrial scale or dispersion problems, e.g., the stabilization of weakly soluble salts such as alkaline-earth carbonates, sulfates, or phosphates in aqueous solution, or the dispersion of inorganic and organic particles. These polymers, which are functional equivalents of the currently used inhibitors and dispersants are biodegradable and therefore might be a new and more environmentally acceptable alternative to the existing products (see → Disperse Systems and Dispersants, → Laundry Detergents).

Polyaspartates can be used in many applications such as laundry detergents, dishwasher chemicals, chemicals for the treatment of cooling and process water, dispersants for organic and inorganic solids, and as a basis for absorbent and gelling materials. (see Section 5.2)

Polyaspartates manufactured by using thermal processes (see Chapter 2) represent a group of acidic polyamides which have been known for a long time [12]. By the end of the nineteenth century, SCHIFF [13] already had polymerized aspartic acid. Fox [14], HARADA [15], and KOVACS [16] later continued this work during investigation of the formation of proteinoid structures under prebiotic conditions.

While the polyaspartic subunits in naturally occurring polyamino acids contain only α-moieties, poly(aspartic acids) obtained by thermal polymerization processes contain a mixture of α- and β-moieties:

2. Synthesis and Production of Polyaspartates

During the past decades, numerous methods for the synthesis of polyaspartates have been developed. It was found that the synthetic method used strongly influences the structure and properties of the resulting product. Because of their interesting properties, polyaspartates could be used on a large scale in different applications. Therefore, the production of polyaspartates has been the focus of extensive research.

Peptide Synthesis. Like other peptides, polyaspartates can be synthesized using an automated, solid-phase peptide synthesizer. A family of polyaspartates ranging in size from ASP_5 to ASP_{60} were prepared by this method and studied in the control of the crystallization of $CaCO_3$ [17].

Polymerization of Protected Aspartic Acids. In small scale, polyaspartates of well-defined structure can also be synthesized by

1. the polymerization of the *N*-carboxyanhydride of L-aspartic acid β-benzylester [12, 18, 19] or
2. the anionic polymerization of L-aspartic acid β-lactam benzylester [20].

After deprotection, poly(α-L-aspartic acids) or poly(β-L-aspartic acids) with molecular masses of 10 000–50 000 can be obtained:

Naturally occurring poly(aspartic acids)	Homopoly(aspartic acid) produced by thermal polymerization [a]
Structure (α-linked repeating units with OH groups)	Structure (mixed α,β-linked repeating units with OH groups)
Subunits: α-moieties	α,β-moieties, polydisperse
$n < 50$	depends on synthetic pathway

[a] Ideal structure.

[−NH−CH−CO−]$_n$
     |
     CH$_2$COOH

Poly(α-L-aspartic acid)

[−NH−CH−CH$_2$−CO−]$_m$
     |
     COOH

Poly(β-L-aspartic acid)

NMR studies (see Chapter 3) were done on the resulting compounds and it was found that the purity of the samples depends on the method of debenzylation [21–23]. During debenzylation of, e.g., poly(α-L-aspartic acid β-benzylester), a polysuccinimide-containing succinimide ring is formed which, after alkaline hydrolysis, yields mixtures of α- and β-linked aspartic acid units.

Solid-Phase Thermal Polycondensation. Commercial production would seem to be possible using the thermal polycondensation of aspartic acid in the solid phase, which has been known for a long time [14, 15, 17, 24–28]:

Aspartic acid → Polysuccinimide → α-Aspartate units + β-Aspartate units

The reaction can be carried out in different types of drying reactors. The generation of dust from the dry starting material and the reaction product is a problem, making the installation of efficient filters necessary.

Because of its commercial availability, L-aspartic acid, which is produced by enzymatic methods, is used for the polycondensation. The reaction is carried out at temperatures of 160–240 °C. Polysuccinimide is formed as an intermediate which is hydrolyzed to a 30:70 mixture of α- and β-linked aspartic acid units. It could be shown that increasing the temperature improves the yield and molecular mass of the product, but also reduces its optical purity, and may end in complete racemization [29]. The thermally synthesized polyaspartate with an average molecular mass of 6000 is biodegradable and, in a low-biomass sludge experiment (SCAS, modified Sturm test), yielded approximately 70% mineralization in 28 days [30] (see Chapter 4).

Thermal Polycondensation in the Presence of Catalysts. Another method for the synthesis of polyaspartates is the thermal polycondensation of L-aspartic acid in the presence of catalysts. Preferred catalysts are *ortho*-phosphoric acid and polyphosphoric acid [25]. The reaction yields a melt which is difficult to handle. In NMR and biodegradation studies, these polyaspartates showed less branching and improved biodegradability compared to those prepared without an acid catalyst [31–33] (see Chapter 4). In addition, these polyaspartates have a higher molecular mass of approximately 50 000.

Catalysts containing sulfur such as H_2SO_4, $NaHSO_4$, and NH_4HSO_4 were used to manufacture highly degradable, low-colored polyaspartates with molecular masses between 2000 and 20 000 [34] (see Chapter 4). Several acid catalysts were used in combination with organic solvents in which the polymer is dispersed to increase the molecular mass of polysuccinimide up to 64 000 [35]. To synthesize polyaspartates with special properties, biodegradable water-absorbent polymers were prepared by acid-catalyzed thermal polycondensation of L-aspartic acid in the presence of cyclic anhydrides [36]. In addition to organic and inorganic acids, transition metal salts, e.g., chlorides of Cu, Co, Ni, and Zn, were used as catalysts in the thermal polycondensation of L-aspartic acid and its ammonium salt. An increase in the molecular mass and preservation of the optical purity were achieved [37].

Thermal Polycondensation of Chemical Precursors of Aspartic Acid. All the aforementioned synthetic methods start with L-aspartic acid, global production of which is approx. 10 000 t/a. Worldwide demand for polycarboxylates (mostly non-biodegradable) is estimated at 100 000 t/a or more. Thus, a substitute should be available by a manufacture process which can match that market. L-Aspartic acid is therefore not an adequate source for the production of all polyaspartates. Especially in the case of laundry detergents, dishwasher chemicals, and water treatment agents, the breakthrough of polyaspartates can only be achieved if large-scale production of tens of thousands of tons is possible. This can be achieved by using chemical precursors of aspartic acid.

The thermal polycondensation of maleic acid ammonium salt was described long ago [12] and repeated in several investigations [15, 24] In addition, maleamic acid, D,L-aspartic ammonium salt and a mixture of asparagine and maleic acid were used [16]. Because of their good availability, derivatives of maleic acid anhydride with ammonia, e.g., maleic acid ammonium salts and maleamic acid and its ammonium salts, were of special interest [38, 39]. Fumaric acid ammonium salt is also used [40]. Maleic acid anhydride is produced in volumes of approx. 300 000 t/a (estimated) and thus represents a source for the large-scale production of thermal polyaspartates:

Maleic acid ammonium salt
R = H, NH_4

Polysuccinimide

Maleamic acid or Maleamic acid ammonium salt

In general, the thermal polycondensation of aspartic acid precursors which are prepared in the form of concentrated aqueous solutions or in a dry form is carried out in a reaction melt. Special reactors are necessary in which the reaction melt can be transformed to a dry product [41]. At reaction temperatures of 160 to 240 °C, the starting material is rapidly converted to polysuccinimide. The water from the aqueous solution and from the condensation reaction is distilled off from the reaction mixture until dry polysuccinimide is obtained. The polysuccinimide can be hydrolyzed, e.g., in aqueous sodium hydroxide, to poly(α,β-D,L-aspartic acid sodium salt) containing an $\alpha:\beta$ ratio of 30:70. The molecular masses of the resulting products are about 1200 (number average) and 2400–2500 (weight average), and the biodegradability measured according to OECD 301 E reaches 60–70% [41–43].

3. Analytical Characterization and Detection

The molecular structure of poly(aspartic acid) (PASP) has been investigated mainly by using spectroscopic methods [36]. Especially NMR spectroscopy was an important tool to obtain information about the structural composition of PASP. Other methods such as infrared spectroscopy were also useful to analyze PASP or its copolymers.

3.1. ^{1}H NMR Spectroscopy

Polyaspartic acid (PASP) can be identified by NMR spectra. The signals measured in alkaline solution are shown in Figure 1.

Due to the polymer structure, there is no observation of sharp peaks, and the signals appear in the form of broader multiplets. In a mixture, of which the resonance signals of polyaspartic acid and of other compounds do not overlap, it is possible to quantitatively analyze PASP by using integrals. It is also possible to identify possible copolymer units in the molecule. Matsuyama et. al [44] determined the ratio of α- and β- amide units in the main chain by integrating the separated methine signals in the ^{1}H NMR spectrum.

3.2. ^{13}C NMR Spectroscopy

^{13}C NMR-Spectroscopy has also been used to answer questions about the molecular structure

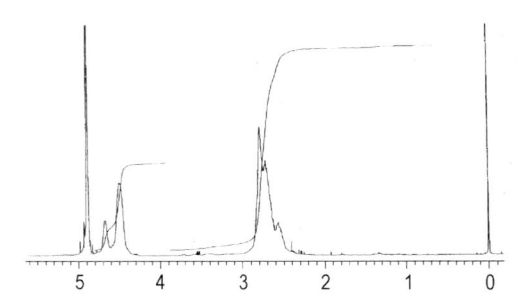

Figure 1. ^{1}H NMR spectrum of thermal PASP from aspartic acid in D_2O (pH 13, 400 MHz)

of PASP. Pivcova et al. [45] evaluated the ratio of α- and β- amide units using the methylene signals in the ^{13}C-NMR spectrum. They analyzed the amide bond sequence using the amide carbonyl signals and concluded that the distribution of the α- and β- bonds was random [47].

3.3. Gel Permeation Chromatography

The molecular mass was determined in aqueous solution by gel permeation chromatography (GPC). As a standard polymer, polyacrylates or polystyrene sulfonic acids can be used. There are also polyaspartic acids of a defined and unique chain length commercially available, which are suitable as standard materials. During the measurement it is important to keep the pH of the solution constant because the molecule extension is strongly dependent on the charge density which is influenced by the degree of dissociation. As a column filling, cross-linked polyhydroxyethyl methacrylate (poly-HEMA) can be used (see also →; Plastics, Analysis, Chap. 5.).

3.4. Isotachophoresis

Isotachophoresis (see → Electrophoresis, Chap. 9.) is a method which is low in cost because it can be carried out with a low amount of chemicals. Therefore, only a diluted aqueous buffer solution is necessary. It is very suitable for complex mixtures, also in extreme concentration properties. The method is based on the fact, that charged particles may be characterized by their migration behavior in the electric field. Different dissolved ions are sharply separated into defined fractions which are detected successively. Basic apparative device have been described by EVERAERTS [47] and KANIANSKY et al. [48]. This method allows a quantitative analysis of polyaspartic acid even in presence of a range of other ions.

The sample is injected between two electrolytes, from which one has the highest, and the other the lowest mobility of all ions present in the system (leading and terminating electrolyte). By applying an electric field, anions and cations are separated according to their different mobility, if these are between the mobility of the two electrolytes. After reaching the stationary state, all distinguished zones move through the solution with the same speed (therefore the name "isotacho") being in direct contact to each other to maintain the electric circuit. For detection, ultraviolet radiation or conductivity can be used.

The detection of PASP is mainly important in commercial applications where the user needs information about the actual product concentration in his process. Especially in water conditioning the process control must include a permanent feedback of what amount of PASP must be dosed to compensate product decreases.

3.5. Fluorescence Spectroscopy

Thermally synthesized polyaspartic acid can incorporate traces of chromophoric structures into the polymer chain, thus allowing a photometrical analysis. PASP is excited in the absorption maximum at 336 nm and emits fluorescence light with a maximum at 411 nm [49]. The intensity of the emitted radiation is directly transformed into a concentration, as it is proportional to the fluorescence intensity. This way of analysis is suitable for most applications, in which a fast and reliable online control of the concentration of polyaspartic acid is required. The actually measured signal determines the amount of product to be added, by this way a minimum concentration can be kept up without uneconomic overdose. Portable instruments for a mobile measurement station are commercially available. They can be specially adapted to purposes with a frequent local move of the analysis set, and under outdoor conditions, for example to screen the concentration PASP in drainage systems at different distances from the dosing equipment.

3.6. Precipitation Titration

A method to be carried out with common laboratory equipment is a precipitation titration with ferric trichloride. The sample solution is mixed with hydrochloric acid to adjust the pH to 8. A 0.05 M $FeCl_3$ solution is dosed, and the turbidity is detected photometrically. The collected data are evaluated by using calibration curves which are measured with defined concentrations of polymers with a similar distribution of their molecular masses.

3.7. Polyelectrolyte Titration

The basis of the determination of polyanions is the formation of a polysalt by reaction of the polyanion with a counter-polycation against which it is titrated [50]. The polysalt reaction meets the conditions for a titrimetric determination: it proceeds essentially quantitatively and is very rapid. In many cases it is a reaction in the sense of a 1:1 reaction [51–57]. The endpoint of the 1:1 reaction between a polyanion and a polycation is indicated by the formation of an associate between the exceeding polycation and anionic metachromic dye, e.g., eriochrome black T.

4. Toxicological Properties and Environmental Fate

Chapter 2, already discussed the fact that thermal polyaspartates are accessible through both catalyzed and noncatalyzed reactions. Polyaspartates synthesized by the first method with sulfur-containing catalysts were tested according to OECD 301 B. Here, the biodegradability was measured by the quantity of carbon dioxide produced by microorganisms during metabolism of the test substance. After 16 days, more than 90% had been degraded already and complete degradation was achieved after 28 days [34].

However, noncatalyzed reactions also yield polyaspartates [poly(α,β-D,L-aspartate) with 70% β-isomer] which show quite high degradation rates of 70% mineralization after 28 days, measured by the modified Sturm test (OECD 301 B) [30]. The good degradability of thermal polyaspartate was demonstrated also by measuring the biological oxygen demand (BOD) in a diluted secondary effluent from a wastewater treatment plant as a source of flora. The trials conducted with solutions containing 3 mg/L of polyaspartate showed only 40% of the theoretical BOD after three weeks [26].

Recent tests have shown that thermal polyaspartates synthesized using phosphoric acid as catalyst had degradation rates of more than 90% in the Zahn-Wellens test (OECD 302 B) [42]. In the same test series, the catalytically manufactured polyaspartate was fully degraded after 7 d in the SCAS test (OECD 302 A) simulating a semi-continuously operated wastewater treatment plant with the addition of samples and blowdown once a day.

The excellent degradation behavior and, moreover, the very favorable toxicological properties of this kind of poly(aspartic acids) are quoted in [41] and [43]:

a)	Acute toxicity (rat, LD_{50} oral):	$\geq$ 2000 mg/kg
b)	Fish toxicity (LC_0, 96h):	$\geq$ 3200 mg/kg
c)	Toxicity to daphnia magna (EC_0):	$\geq$ 2500 mg/L
d)	Bacteria (EC_0, OECD 209):	$\geq$ 10 000 mg/L
e)	Algal growth (LOEC):	395 mg/L
	(NOEC):	125 mg/L
	(E_rC_{50}):	1070 mg/L
f)	Biodegradation tests:	
	Modified screening test (OECD 301 E), DOC removal, 28 days:	60 to 67%
	Carbon dioxide evolution test (OECD 301 B)	50 to 75%
	Zahn–Wellens test (OECD 302 B), DOC removal:	77%

5. Function and Applications

Chemically, polyaspartates are assigned to the class of polycarboxylates, which refers to the repeating carboxylate units along the polymer chain (each monomer aspartate unit features one carboxylate group). Consequently, polyaspartates showing some similar properties, effects, and eventually applications as the most prominent representative of the polycarboxylates, namely the polyacrylates (see → Polyacrylates). A main difference, however, is the biodegradability of the polyaspartates, which can be attributed to the potentially cleavable amid bonds in the molecule. Thus, a more environmental friendly alternative for the non-biodegradable polyacrylates for a lot of applications is available.

5.1. Functionality

The applications of polyaspartates (see Section 5.2) are based mainly on the following effects, which are explained in the following.

5.1.1. Dispersing Effect

The dispersing effect describes the phenomenon that agglomeration and subsequent settling of solid particles present in aqueous suspensions is prevented. This is caused by the fact that polyaspartate, due to its polyanionic character, adsorbs onto the highly polar particle surfaces, resulting in equally (negatively) charged particles, which undergo an electrostatic repulsion.

Exemplarily, Figure 2 shows the dispersion effect of polyaspartate in comparison to polyacrylates of different molecular masses onto kaolin. High turbidity values mean high content of solids within the suspension after storage, i.e., reduced settling of kaolin.

The prevention of crystal agglomeration, due to electrostatic repulsion (and steric repulsion to a minor amount as well), is especially important within the process of mineral scale inhibition. Crystal agglomeration in oversaturated solutions is one cause for the formation of strongly adherent scale – a severe problem in water treatment and cleaning applications.

Besides the improvement of the suspension stability, the dispersing effect can cause a dramatic viscosity decrease of slurries (Fig. 3). Thus, a significant reduction of the water content for the production of, nevertheless easy-pumpable, high-solid-content slurries is achievable.

It seems to be that there is a dispersion effect of polyaspartate not only to inorganic solids but also to organic matter, because improved cleaning results in water/organic systems have been reported.

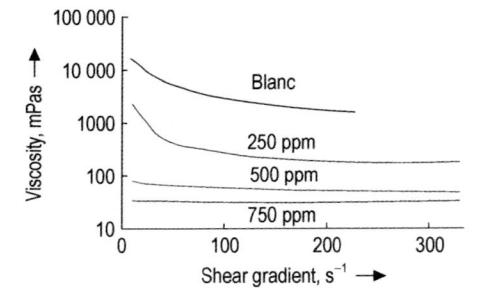

Figure 3. Viscosity of iron oxide slurries with polyaspartate as dispersing agent in various concentrations

5.1.2. Threshold Effect

The threshold effect describes the phenomenon that supersaturated solutions can be stabilized by inhibitors in far less than stoichiometric amounts related to the mineral ions causing supersaturation. Despite a considerable excess of the solubility product, the solutions remain clear in the presence of the inhibitor.

This threshold effect can be explained by an adsorption of the threshold inhibitor on the growth sites of the submicroscopic crystal nuclei which form in a supersaturated solution. This stops altogether, or at least delays, the growth of the crystal nuclei – such as added seed crystals and heterogeneous nuclei – for a long period of time. Thus, it can be concluded that the scale inhibition effect of threshold inhibitors is the result of kinetic rather than thermodynamic effects [58].

By stabilization of supersaturated solutions, threshold inhibitors are able to prevent the formation of mineral scale, a prevalent and severe problem in a lot of applications. The most important scale-forming minerals are carbonates and sulfates of the alkaline earth metals, in particular calcium and barium. Polyaspartate is a particular effective threshold inhibitor for alkaline earth metal sulfates (Fig. 4).

Polyaspartate is effective for calcium carbonate as well; however, it is to note that the phosphonates, which in contrast are not biodegradable, are more efficient with regard to this mineral.

It is obvious that scale inhibition is based on both, the threshold and the dispersing effect. However, it can be assumed that the threshold effect is the dominant effect in oversaturated

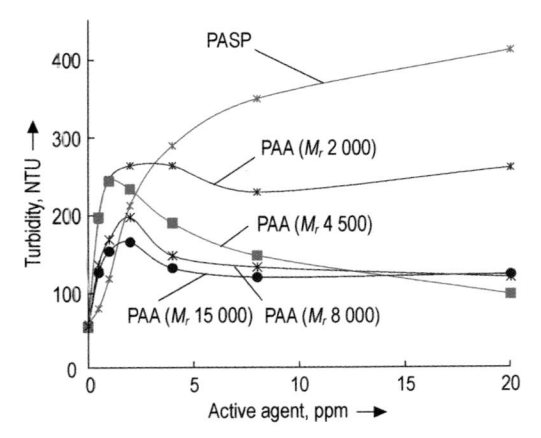

Figure 2. Comparison of polyacrylates (PAA) of different molecular masses and polyaspartate (PASP) on their dispersing activity to kaolin
NTU = national turpidity units

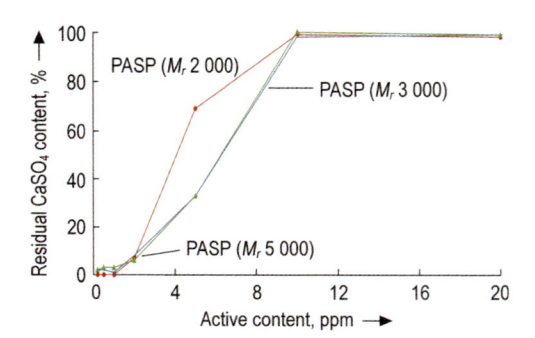

Figure 4. Calcium sulfate scale inhibition by polyaspartate Initial CaSO₄ content 10 000 ppm, pH 9, results after 24 h storage in a closed bottle at 60 °C (GPR 1391)

solutions in which no crystals (turbidity) have been formed (100% residual hardness or inhibition) because of the presence of an inhibitor.

5.1.3. Crystal Distortion

A contribution to scale inhibition is also crystal distortion. It means that the crystal shape is changed, which results in a loose, easy-removable precipitate. The mechanism of threshold inhibition through an adsorption of the inhibitor on the growth sites of the submicroscopic crystal nuclei also explains why the structure of the crystals which form when the inhibitor concentration is too low for complete inhibition is altered or distorted and incorporates a proportion of inhibitor.

5.1.4. Complexing Effect

In the case of the complexing effect, the active agent interacts with individual cations, i.e., in contrast to the dispersing and threshold effect mentioned above stoichiometric ratios are necessary. Polycarboxylates such as polyacrylates or polyaspartates have carboxylic groups or an amide nitrogen atom which may act as ligands towards metal ions. That means that these polycarboxylates may also show complexing activity. However, due to their polymeric structure the complexing activity is less prominent than the dispersant activity.

The higher carboxylate content per mass unit of polyacrylates compared to that of polyaspartates provokes a higher complexing activity especially towards calcium ions. On the other side, polyaspartates feature a more prominent dispersant character.

One exception may be the complexing activity of polyaspartates towards ferric irons. In the case of polyaspartate with a molecular mass of about 2000 (Baypure DS) a high complexing constant (log K) of 18.5 has been determined [60].

5.1.5. Corrosion Inhibition

The corrosion inhibiting effect of polyaspartate can be attributed to an adsorption of the polymer onto the metal surface, forming a protective film. It can be assumed that alkaline earth metals present in the aqueous solution are also incorporated into this film, as already known in the case of phosphonate corrosion inhibitors [60]. This could explain the better corrosion inhibition effectiveness at higher water hardness (Fig. 5). According to [65], polyaspartate is also effective in highly sour brine media.

5.2. Applications and Uses of Polyaspartates

5.2.1. Water Treatment

Industrial applications dealing with a large amount of water face the challenge of controlling corrosion and mineral scale formation onto functional equipment, e.g., piping and heat exchanger systems. Since some decades, mixtures mainly based of phosphonates and polycarboxylates have been used in the nowadays sophisticated

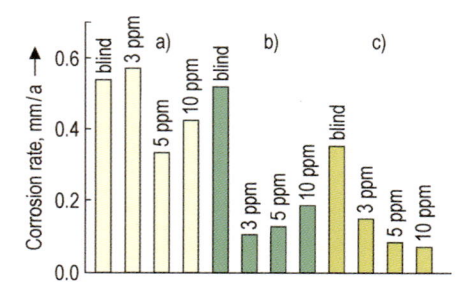

Figure 5. Corrosion inhibition effect of polyaspartate on carbon steel (Spinner test, St37) dGH = degrees of general hardness, dKH = deutsche Karbonathärte a) 17 dGH (300 ppm CaCO₃); 9 dKH (160 ppm CaCO₃), 3-d test period; b) 34 dGH (600 ppm CaCO₃); 18 dKH (320 ppm CaCO₃); 3-d test period; c) 54 dGH (900 ppm CaCO₃); 27 dKH (480 ppm CaCO₃); 7-d test period

cooling and process water treatment. Against the background of increasing interest in better biological degradable water treatment products, a lot of developments have been made in recent years. With regard to polyaspartate, they are based mainly on a substitution of the hitherto used non-biodegradable polycarboxylates (or in parts at least) by polyaspartates, which results in an improved overall biodegradability [62–66]. Due to the outstanding scale inhibition performance of phosphonates (e.g., PBTC) towards calcium carbonate, which in fact is the main scale forming species in cooling and process water applications, phosphonate substitution has taken place rather seldom. However, phosphorus-free products based on polyaspartate have been developed as well [67], which are of particular importance for applications/regions for which very strong legal constraints regarding environmental behavior exist. They are of interest also in the case that calcium sulfate is the predominant scale causer due to geological reasons.

Besides its good corrosion inhibiting effect, especially in combination with phosphonates or metal salts [68], an advantage of polyaspartate in water treatment is its outstanding calcium tolerance, i.e., no insoluble inhibitor salts are formed, which would reduce the active inhibitor content [62].

5.2.2. Detergents and Cleaners

Polycarboxylates – usually polyacrylate and poly(acrylate comaleinate) – have an outstanding function in the ternary builder system zeolith A–polycarboxylate–sodium carbonate in that they effectively prevent lime and soil depositions on laundry and washing machines (see → Laundry Detergents, → Zeolites, Chap. 7.). The mechanism of action is mainly based on the threshold effect [69, 70] (see Section 5.1.2). Environmentally friendly polyaspartates have been proposed as a substitute for non-biodegradable polyacrylates [71]. In addition, the detergency is improved [72]. This allows the formulation of a modern detergent with fully degradable structures. The efficiency of polyaspartate as a cobuilder in modern detergent formulations has been recently demonstrated [73]. Hydrophobic modified polyaspartates have been proposed for better detergency and improved whiteness of the fabric [74]. Sodium carbonate modified by polyaspartate can be used as carrier material in laundry detergents [75]. Low-temperature bleaching detergents containing (phthalimido)peroxyhexanoic acid are described with 3–12% polyaspartate as a cobuilder and stabilizer [76]. Polyaspartates are also suitable for the builder system of automatic dishwashing detergents with a good primary and secondary detergency [77], which may have also a good storage stability [78].

5.2.3. Mining and Oil Recovery

Due to its pronounced barium sulfate scale inhibition efficiency, polyaspartates find use in mining [79] and oil recovery applications. The biological degradability and the favorable ecotoxicological data are of a significant advantage because legal requirements are very high (especially within Europe) in this field of applications [80].

5.2.4. Other Uses

The following table gives a summary of other applications, the corresponding functions of polyaspartates and the resulting benefits.

Application	Function	Benefit
Paper	liquefaction of raw materials	clays and titanium dioxide are used as fillers for paper and must be liquefied before addition to the paper
Leather	dispersion of color pigments	improves the depth of color of the leather, thus reducing the amount of color pigments required
Textiles	dispersion of color pigments/ dyes	dispersion of colors in dyeing processes, especially in cellulosic fiber/reactive dye systems
Membrane cleaning	dispersion of dirt particles at the surface of the membrane	dispersion of dirt particles, cleaning and rinsing of scaled membrane surfaces, especially in the case of organic pigments
Agriculture	fertilizer/ nutrition enhancer	availability of components essential for nutrition to enhance the growth of the plants; no eutrophication because free from phosphorus; reduction of soil erosion
Pigments, colorants, coatings	dispersion of organic pigments	commonly available dispersants are not biologically degradable, e.g., lignine sulfonates; possible replacement by PASP as an environmentally friendly alternative

6. Polysuccinimide

6.1. Introduction

The synthesis and manufacture process was already described in Chapter 2. Usually, polysuccinimide is an intermediate in the production of polyaspartates, thus it may be easily isolated in a dry state.

6.2. Chemical Properties and Analytic Characterization

The chemical and physical properties of polysuccinimide are listed in the following:

Appearance:	slightly yellow- to brownish-colored powder, depending on the manufacturing process
Temperature of decomposition:	234–312 °C (exothermally)
Molar mass: [a]	4000 kg/mol
Solubility:	organic aprotic solvents such as dimethyformamide (DMF), dimethylacetamide (DMA), or dimethylsufoxide (DMSO); ethylene glycol (modest solubility) or di- or triethylenclycol.
Viscosity of a 30% solution in diethyleneglycol at 24 °C:	4800 mPas
Viscosity of a 30% solution in triethyleneglycol at 24 °C:	17 700 mPas
Heat of reaction of hydrolysis per monomer unit at 25 °C:	38.5 kJ/mol

[a] The PSI was obtained by a thermal, non-catalyzed process starting with maleic anhydride and ammonia.

Polysuccinimide may contain hydrolyzed succinimide rings which is equivalent to a content of polyaspartic acids or their salts. By means of IR spectroscopy the ratio of succinimide or amide carbonyl groups can be determined and thus the degree of hydrolysis in ratio to a given standard of polysuccinimide.

The hydrolysis of PSI to polyaspartate [81] depends on the following parameters:

pH
temperature
particle size/particle surface of the PSI

The hydrolysis can be monitored by the fluorescence technique (Fig. 6). To a water cycle

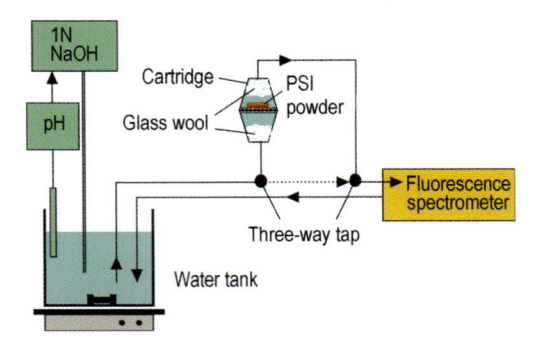

Figure 6. Hydrolysis of polysuccinimide in a closed water cycle Test parameters: $T = 22°C$; $V_{\text{deionized water}} = 2\,L$, pH 8, mass of polysuccinimide (PSI) = 0.5 g, flow rate = 2 mL/min

with a 2-L water reservoir, fluorescence detector and pH control a bypass (containing polysuccinimide powder enclosed in a cartridge) is connected through a tree-way tap. The latter allows the separation of the bypass from the cycle. The polysuccinimide dissolution as a function of water flow rate and pH was investigated. The separation of the bypass containing polysuccinimide during the dissolution process at pH 8 gave the surprising result that the fluorescence activity did not rise further after the bypass separation but the overall sodium hydroxide consumption (Fig. 7).

This can be explained by a stepwise hydrolysis of polysuccinimide via succinimide-containing intermediates [82]:

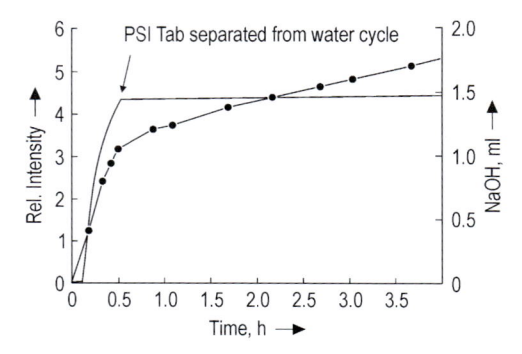

Figure 7. Monitoring the dissolution of polysuccinimide using the test and test conditions given in Figure 6

[Structures: Polysuccinimide → (H₂O) → Soluble partial hydrolysate → (H₂O) → Polyaspartic acid]

6.3. Applications and Uses of Polysuccinimide

Polysuccinimide (PSI) is an intermediate in the manufacture process of polyaspartates. Hydrolysis of PSI under alkaline conditions yields polyaspartates. The main application of PSI itself is to use it as a kind of depot for polyaspartate under more or less alkaline conditions. The slow hydrolysis at a neutral to slight alkaline pH allows the formulation of tablets with polysuccinimide as the main constituent which release traces of polyaspartate over a long period of time. These tablets can be used instead of a dosage system for polyaspartates in solution [83, 84]. The biodegradability and non-toxicity of the hydrolysis products as well as the long lifetime of the tablets in the range of several months make polysuccinimide the product of choice in the treatment of waters flowing through drainage systems in tunnels, disposal sites, dams, and power plants. The scaling is due to leaching of alkalinity and calcium ions from the cementous construction material [85].

The partial hydrolyzates occurring during hydrolysis of polysuccinimide are responsible for the erosion of calcite or other crystalline basic calcium salts. The deterioration of calcite can be demonstrated by a simple experiment: A tablet containing polysiccinimide is added on top of a column containing calcite (or marble) pieces. Tab water flows through the column from the top, so that the water washes around the tablet. After seven days of continuous flow, the calcite pieces have fragmented and are colored because of the adsorbed polyaspartate. The fragmentation can be explained by hydrolysis on the calcite surface [86]. Polyaspartate itself is capable of dissolving calcium carbonate at low pH; at high pH this is possible only at higher concentrations of 0.1 M [87]. The low pH may be obtained in the case of polysuccinimide application by hydrolysis of the adsorbed partial hydrolyzates.

If polysuccinimide is formulated with hydrolysis-accelerating substances (alkali metal hydroxide or carbonates, alkaline silicates or phosphates, solvents for polysuccinimide such as triethylene glycole), faster dissolving tabs can be made for cleaners or detergents. The advantage is a reduction of pH during the dissolution and the lower hygroscopicity of polysuccinimide compared to that of polyaspartate or other comparable polyanions such as polyacrylates [88].

The hydrolysis of polysuccinimide comes to an end at a pH of 3–4 [81]. A slurry is formed, which contains only small amounts of free polyaspartic acid together with most of the polysuccinimide intact is formed. This acidic dispersion can be used as a mild abrasive cleaner with a large depot for acidity in case of neutralization [88].

Polysuccinimide is a setting retarder for both cement [89] and gypsum [90]; in the latter case the redarding efficiency exceeds those of polyaspartate itself.

Polysuccinimide not only reacts with hydroxide but also with other nucleophiles as amines or sulfides. This can be used for odor-reducing formulations [91], also with zeolithes as a synergistic partner [92].

References

1. R. J. Larson *et al.*; *J. Env. Polymer Deg.* **5** (1997) 41.
2. L. Addadi, S. Weiner, *Angew. Chem., Int. Ed. Engl.* **31** (1992) 153–169.
3. S. J. Mann; *J. Chem. Soc., Dalton Trans.* (Inorganic Chemistry) (**1**)(1993) 1–9.
4. D. Allemand;J.-P. Cuif: "Fundamentals of biomineralization", *Proc. 7th Internat. Symp. on Biomineralization. Bull. Inst. oceanogr.* Monaco, special volume 14/1, 1–236
5. C. S. Sikes, A. Wierzbicki in S. Mann (ed.): *Biominetic Mater. Chem.*, VCH, New York 1996, 249. (Conference, General Review).
6. K. W. Rusenko, J. E. Donachy, A. P. Wheeler: Purification and characterization of a shell matrix phosphoprotein from the American oyster in C. S. Sikes, A. P. Wheeler (eds.): *Surface Reactive Peptides and Polymers: Discovery and Commercialization*, ACS Books, Washington DC 91, 107–121.
7. A. P. Wheeler, K. W. Rusenko, D. M. Swift, C. S. Sikes, *J. Mar. Biol.* **98** (1988) 71–80.
8. J. P. Gorski, *Calcif. Tissue Journal* **50** (1992) 391–396.
9. A. Wierzbicki, C. S. Sikes, J. D. Madura, B. Drake, *Calcif. Tissue Int.* **54** (1994) 133.
10. C. S. Sikes, A. Wierzbicki, *Corrosion* **94** (1994),No. 193; S. Fizwater, *ACS Symp. Ser.* **589** (1995) 316 (Conference, General Review).
11. M. Frankel, A. Berger, *J. Org. Chem.* **16** (1951) 1513–1518.
12. M. Dessaignes, *Comptes Rendues des Seances* **31** (1850) 432–433.
13. H. Schiff, *Chem. Ber.* **30** (1897) 2449.
14. K. Harada, Fox, *J. Am. Chem. Soc.* **80** (1958) 3361–1166.
15. K. Harada, *J. Org. Chem.* **24** (1959) 1662–1666.
16. J. Kovacs, H. N. Kovacs *et al.*, *J. Org. Chem.* **26** (1961) 1084–1091.
17. C. S. Sikes, A. P. Wheeler (eds.), Chemical Aspects of Regulation of Mineralization *Proc. Symp. Ind. Eng. Chem. ACS, University South Alabama*, Mobile, Alabama, (1988) 15–19.
18. R. H. Karlson, K. S. Norland, G. D. Fasman, E. R. Blout, *J. Am. Chem. Soc.* **82** (1960) 2268–2275
19. V. S. Rao, P. Lapointe, D. N. McGregor, *Makromol. Chem.* **194** (1993) 1095–1104
20. A. Rodriguez-Galan, S. Munoz-Guerra, J. A. Subirana, B. Chuong, H. Sekiguchi, *Makromol. Chem. Symp.* **6** (1986) 277–284.
21. H. Pivcova, V. Saudek, J. Drobnik, *Biopolymers* **20** (1981) 1615–1623.
22. H. Pivcova, V. Saudek, J. Drobnik, *Polymer* **23** (1982) 1237–1241.
23. H. Pivcova, V. Saudek, *Polymer* **26** (1985) 667–672.
24. S. W. Fox, K. Harada: *Analytical Methods of Protein Chemistry*, Pergamon Press, Oxford 1966, 127–135.
25. S. W. Fox, K. Harada, *Biosystems* **7** (1975) 213–21.
26. A. P. Wheeler, L. P. Koskan, *Mat. Res. Soc. Symp. Proc.* **292** (1993) 277–283.
27. L. P. Koskan, K. C. Low, *MSE* **69** (1993).
28. K. C. Low, A. P. Wheeler, L. P. Koskan, *Advances in Chemistry Series* (1996).
29. E. Kokufuta, K. Harada, S. Suzuki, *Bull. Chem. Soc. Japan* **51** (1978) 1555–1556.
30. D. D. Alford, A. P. Wheeler, C. A. Pettigrew, *J. Environ. Polym. Degrad.* **2** (1994)no. 4, 225–236.
31. H. Pivcova, V. Saudek, J. Drobnik, *Biopolymers* **20** (1981) 1605–14.
32. S. K. Wolk, G. Swift, Y. H. Paik, K. M. Yocom, R. L. Smith, E. S. Simon, *Macromolecules* **27** (1994) 7613–7620.
33. M. B. Freeman *et al.*, *Polymer Preprints*, (American Chemical Society, Division of Polymer Chemistry) **35**(2)(1994) 423–424.
34. R. J. Ross, D. A. Batzel, A. R. Meah, J. F. Kneller, *Macromol. Symp.* **123** (1997) 235–249.
35. T. Nakato, M. Kuramochi, K. Matsubara, M. Tomida, *Polym. Prepr. Am. Chem. Soc., Div. Polym. Chem.* **37** (1996) 555–556.
36. A. Kusunu *et al.*, *Polymer Preprints-America* **40** (1999) 113–114.
37. K. Okamoto, H. Ohata, T. Imano, J. Hidaka, *Viva Origino* **19** (1991) 133–146.
38. G. Boehmke, US 4,839,461, 1989.
39. Y. Fujimoto, US 3,846,380, 1972.
40. L. Wood, US 5,288,783, 1994.
41. M. Schwamborn, *Nachr. Chem. Tech. Lab.* **44** (1996) no. 12, 1167–1170.
42. M. Kroner, W. Bertleff, "Seifen, Öle, Fette, Wachse" *(Sepawa)* (1996).
43. M. Schwamborn, *Polym. Degrad. Stab.* **59** (1998) 39–45.
44. M. Matsuyama, E. Kokufuta, T. Kusumi, K. Harada: "On the Poly (β-DL-aspartic acid)", *Macromolecules* **13** (1980) 196–198.
45. H. Pivcova, V. Saudek, J. Drobnik, J. Vlasak, *J. Biopolymers* **20** (1981) 1605–1614.
46. H. Pivcova, V. Saudek, J. Drobnik, *J. Polymers* **23** (1982) 1237–1241.
47. F. M. Everaerts, T. P. Verheggen, F. E. Mikkers, *J. Chromatogr.* **169** (1979) 21.
48. D. Kaniansky, P. Havasi (eds.): *Trends in Analytical Chemistry* **2** (1983) 197–202.
49. Bayer AG, WO 2002018458, 2002 (T. Klein, T. Klausa, A. Elschner).
50. K. H. Wassmer, U. Schroeder, D. Horn, *Makromol. Chem.* **192** (1991) 553–565.
51. H. Terayama, *J. Polym. Sci.* **8** (1952) 243.
52. D. Horn: Polymeric Amines and Ammonium Salts, Pergamon Press, New York 1980, 333.
53. A. Domard, M. Rinaudo, *Macromolecules* **13** (1980) 898.
54. A. Domard, M. Rinaudo, *Macromolecules* **14** (1981) 620.
55. D. Horn, C. C. Heuck, *J. Biol. Chem.* **258** (1983) 1665.
56. B. Philipp, H. Dautzenberg, K. J. Linow, J. Kötz, W. Dawydoff, *Prog. Polym. Sci* **14** (1989) 91.
57. E. Tsuchida, K. Abe, *Adv. Polym. Sci.* (1982) 45.
58. T. A. Ring, *Powder Technology* **65** (1991) 195–206.
59. Lanxess, Baypure® product broshure, LXS-FCC17D, issue 2005-10
60. R. Ashcraft, G. Bohnsack, R. Holm, R. Kleinstück, S. Storp, NACE Corrosion Paper No. 328/1987.
61. G. Schmidt, A. Saleh: *Materials Performance* (2000) 62–65.
62. P. Kmec, D. Emerich, WO 0044677, 2000.
63. W. Hater, A. Friedrich, M. Schlag, WO 9739078, 1997.
64. R. Kleinstück, T. Groth, W. Joentgen, WO 0010928, 2000.
65. L. Zhang, *Hebeisheng Kexueyuan Xuebao* **22** (3)(2005) 52–54.
66. A. Graf, D. Frahne, EP 1350768, 2003.
67. M. Schweinsberg, W. Hater, J. Verdes, International Water Conference (2003) 64th, 250–262 CODEN:OIWECQ
68. Q. Xu, J. Gong, CN 1733622, 2006.
69. H. Upadeck, P. Krings, *Seifen Öle Fette Wachse* **117** (1991) 554–558.
70. P. Zini (ed.): *Polymeric additives for High Performing Detergents*, Technomic, Lancaster 1995, 179–212.
71. G. Wagner, *Waschmittel*, 3rd ed., Wiley-VCH, Weinheim 2005, 86–87.
72. M. Kroner, G. Schornick, R. Bauer, DE 4342316, 1995.

73. Lanxess Broshure: "Einsatzmöglichkeiten der Baypure®-Produkte in Waschmitteln," Lanxess Deutschland GmbH, D-51369 Leverkusen, ID-No. LXS-FCC18D (2005).
74. G. T. Jordan, E. P. Gosselink, WO 2003014193, 2003.
75. A. P. Chapple, R. J. Crawford, WO 2006081930, 2006.
76. R. Weber, DE 10020767, 2000.
77. A. Du Vosel, G. Muratori, EP 612842, 1994.
78. S. Maruyama, K. Yamaguchi, JP 2001003084, 2001.
79. W. Hater, Corrosion 98, Paper No. 213 (NACE International, Conference Division).
80. http://www.ospar.org/
81. J. Mosig; C. H. Gooding; A. P. Wheeler, *Ind. Eng. Chem. Res.* **36** (1997) 2163–2170.
82. T. Klein; A. Mitschker, R. Moritz: "Zur Hydrolyse von Polysuccinimid und zu dessen Wirkung gegen Tunnelversinterung, *Wasserwirtschaft* **94** (2004) no. 4, 32–37.
83. H. Sicius *et al.*, WO 2002016731.
84. T. Klein, F. Ebert, R. Moritz, EP 1702892, 2006.
85. A. Saxer, C. Draschitz, *Spritzbeton-Technologie* (2006) 87–103. (Berichtsband der 8. internationalen Fachtagung Alpbach)
86. G. Girmscheid, T. Gamisch: "Hardness stabilisation reduces maintenance costs", part 1 *Tunnel* (2005), 38–45; part 2 *Tunnel* (2005), 20–35.
87. K. Burns, Y.-T. Wu, C. S. Grant, *Langmuir* **19** (2003) 5669–5679.
88. T. Klein, J. Voss, H. Schmidt, F. Ebert, H.-G. Müller, EP 1260539, 2002.
89. C. C. Sykes, T. M. Vickers, US 5908885, 1999.
90. T. Staffel, T. Klein, G. Brix, F. Wahl, DE 10017133, 2000.
91. G. Calton, L. L. Wood, EP 909160, 1999.
92. G. Calton, J. B. Cook, WO 2000069478, 2000.

Further Reading

Y. Hamano (ed.): *Amino-Acid Homopolymers Occurring in Nature*, Springer, New York 2010.

M. Kjellin, I. Johansson (eds.): *Surfactants from Renewable Resources*, Wiley, Chichester 2010.

M. Showell (ed.): *Handbook of Detergents*, Taylor & Francis, Boca Raton, FL 2006.

Polybutenes

HARTMUT KOCH, Shell Chemical International, Shell Center, London, United Kingdom

RALPH L. MAWER, Shell Chemical International, Shell Center, London, United Kingdom

WOLFGANG IMMEL, BASF Aktiengesellschaft, Ludwigshafen, Federal Republic of Germany

1.	Poly(1-Butene)...............	747	2.2.2.	Low Molecular Mass Polyisobutylenes..............	755
1.1.	Production..................	747	2.2.3.	Medium Molecular Mass Polyisobutylenes..............	756
1.2.	Properties...................	748	2.2.4.	High Molecular Mass Polyisbutylenes.................	757
1.2.1.	Structure and Physical Properties....................	748	2.3.	Structure....................	758
1.2.2.	Application Properties..........	751	2.4.	Properties...................	758
1.3.	Processing...................	753	2.5.	Uses and Processing...........	759
1.4.	Applications.................	753	2.6.	Commercial Products..........	760
2.	Polyisobutylene...............	754	2.7.	Copolymers..................	760
2.1.	Introduction.................	754		References...................	760
2.2.	Production...................	755			
2.2.1.	Isobutene as Raw Material.......	755			

1. Poly(1-Butene)

1.1. Production

Introduction. Poly(1-butene) is a thermoplastic polymer produced by solution polymerization of 1-butene with a Ziegler – Natta catalyst system. Hüls developed a process during the 1960s and produced the polymer in a 12 000 t/a plant in the Federal Republic of Germany. Mobil started to develop a process in 1968 and, together with Witco, built a plant in Taft, Louisiana with a capacity of 30 000 t/a. Shell (USA) bought the plant in 1977 and is at present the only commercial producer.

The feedstock of pure 1-butene is supplied from the Shell Higher Olefins Plant in Geismar, Louisiana.

The poly(1-butene) plant produces grades specifically developed for pipe and speciality film applications.

Process. The current commercial poly(1-butene) production process consists of seven primary unit operations: (1) polymerization, (2) deashing, (3) polymer/monomer separation, (4) monomer purification, (5) extrusion, (6) pellet blending, and (7) packaging (Fig. 1).

In the polymerization step the catalyst system is fed into the reactor with the purified 1-butene recycle stream. Hydrogen is added as a chain terminator to control the melt flow index of the polymer. Ethylene is fed to the reactor when producing random 1-butene – ethylene copolymers.

Polymerization catalyst residues must be removed from the product to preserve its long-term properties. In the deashing step, the poly (1-butene) is mixed with dilute caustic soda to dissolve and extract the catalyst components. The cleaned hydrocarbon – polymer phase and the catalyst-containing aqueous phase are then separated.

During the polymer/monomer separation step, unreacted 1-butene is separated from the poly(1-butene) for recycle to the reactor. Separation is carried out by heating the polymer solution under pressure then flashing off the monomer by rapidly reducing the pressure. The remaining molten polymer feeds the extrusion step, while the vaporized monomer flows to the monomer purification section of the process.

The monomer purification step accomplishes separation of impurities both heavier and lighter than 1-butene. Water is removed by azeotropic

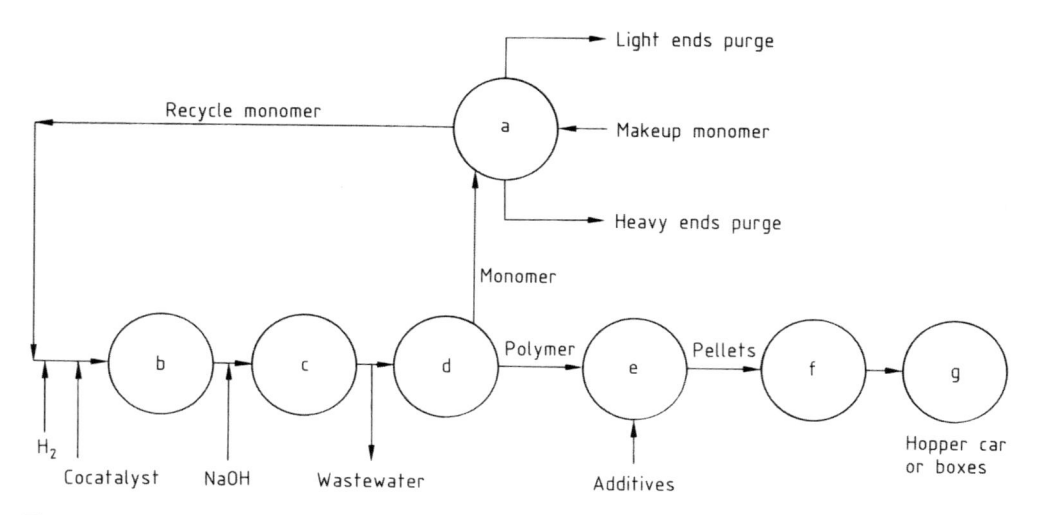

Figure 1. Flow sheet for the production of poly(1-butene)
a) Monomer purification; b) Polymerization; c) Deashing; d) Monomer/polymer separation; e) Extrusion and pelletizing; f) Blending and storage; g) Packaging

distillation. Fresh 1-butene is added to the process at this stage.

In the extrusion step further volatile hydrocarbons are removed, additives for color, product performance, processing, and stabilization are added, and the polymer is pelletized.

The pellets are pneumatically conveyed to the batch blending stage and then transferred to the packaging area. Pellets are typically sold in 450-kg boxes.

1.2. Properties

1.2.1. Structure and Physical Properties

The properties of a high-molecular material are essentially determined by the dimensions and structure of the molecular chain (fine structure) and by the thermomechanical history [1]. Since the fine structure of poly(1-butene) can be varied across a broad range by means of the polymerization process, a knowledge of this is essential for the spectrum of properties.

Molecular Structure. Polymerization with Ziegler–Natta catalysts produces both atactic and isotactic polymer. Atactic poly(1-butene) is virtually impossible to crystallize, while isotactic poly(1-butene) has a crystallinity of 40–55 % at normal cooling speeds.

The molecular chains in the crystal form helices. Depending on the crystallization conditions, poly(1-butene) forms 11_3, 3_1 or 10_3 helices, which differ in their helical dimensions [2–4] and give rise to different crystalline modifications (see below).

The isotactic content of poly(1-butene) is generally determined by the solubility in boiling diethyl ether ("ether extract"), whereby it is assumed that only the atactic components are soluble. However, infrared spectroscopy and X-ray crystallography have shown that the atactic extract contains crystallizable components, mostly of low molecular mass. A quicker and more reliable method of determining the atactic content is by evaluating the infrared band in the region of 920 cm^{-1} [5]. This leads to the following definition of the atactic content Y_a:

$$Y_a = \frac{\bar{v} - 918 \text{ cm}^{-1}}{30 \text{ cm}^{-1}}$$

where v is the band position of the unknown sample in cm^{-1}.

The advantage of this method is that it is independent of the molecular mass and of the morphology of the sample.

Molecular Mass. The average molecular mass may be determined either by direct methods such as light scattering or osmometry, or by indirect methods such as determination of the intrinsic viscosity [6, 7]. The molecular mass is calculated from the viscosity number using the Mark–Houwink equation [7]. Mark–Houwink

Table 1. Characteristic data for crystal modifications of poly(1-butene)

	Modification II (tetragonal)	Modification I (hexagonal)	Modification III (orthorhombic)	Modification I' (hexagonal)	Amorphous phase	Reference
Helix type	$11_3/40_{11}$	3_1	10_3	3_1		[3, 8, 10]
Helix length per monomer unit, nm	0.187	0.217		0.217		
Helix cross section, nm^2	0.055	0.044				
Specific volume at 25 °C, cm^3/g	1.1364	1.0453			1.1579	[11]
					1.16 for mod. II	
					1.14 for mod. I	[11]
Specific coefficient of volume expansion, $cm^3\,g^{-1}\,K^{-1}$	3.05×10^{-4}	3.13×10^{-4}			8.12×10^{-4}	[11]
Melting point, °C	ca. 125	ca. 140	93–98	100		[9, 12] [13] [14]
Heat of fusion, kJ/mol	ca. 3.1	ca. 6.7				[15] [16] [17]

equations for isotactic poly(1-butene), which give virtually identical molecular masses are as follows:

$[\eta]_{n\text{-nonane}} = 5.9 \times 10^{-5} \times M^{0.80}$
$[\eta]_{\text{dekalin,135°C}} = 6.2 \times 10^{-5} \times M^{0.79}$

These equations also apply to atactic poly(1-butene).

The molecular mass distribution can be determined easily by gel permeation chromatography in 1-chloronaphthalene as solvent. Industrial products have molecular nonuniformities $U = (M_w/M_n) - 1$ between 5 and 12 (M_w = weight-average value, M_n = number-average value).

Crystal Structure and Superstructure. Like most polymers isotactic poly(1-butene) can crystallize in various modifications. However, the polymorphism is more marked and of greater industrial significance in poly(1-butene) than in any other polymer. Four crystal forms of poly(1-butene) are known at present and are designated modification I, I', II, and III [2, 8]. Characteristic physical data of these modifications are listed in Table 1.

Modifications I' and III are formed on crystallization from solution. They are of no industrial importance. However, modifications II and I, which are obtained by crystallization from the melt, are industrially significant. All the crystal forms can be obtained from modification III in a single step [14], as shown in Figure 2. The transitions from one crystal form to another are monotropic. Modification I' crystallizes preferentially from solution at lower temperatures [8], while at high temperature the orthorhombic modification III is produced [18, 19]. Modification III is also formed by precipitation, e.g., by addition of methanol to a benzene solution of poly(1-butene) [20]. Modifications I' and III are also formed in solution or suspension polymerization [21].

The tetragonal modification II and the hexagonal form I are industrially important. As a melt cools, i.e., after most processing operations, form II crystallizes out first. This is thermodynamically unstable and in the absence of external mechanical stresses it changes monotropically into the stable hexagonal modification I after a period ranging from a few hours to several days, depending on the temperature and thermal history [2]. The transformation from II to I is associated with a change in various physical properties, e.g., an increase in melting point, density, [11] and modulus (Fig. 3).

These changes in properties are primarily due to the change in crystal modification. The

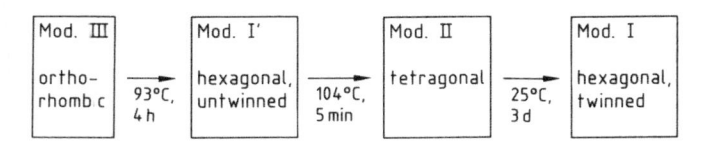

Mod. III orthorhombic → (93°C, 4h) → Mod. I' hexagonal, untwinned → (104°C, 5 min) → Mod. II tetragonal → (25°C, 3d) → Mod. I hexagonal, twinned

Figure 2. Monotropic interconversion of the four modifications of poly(1-butene)

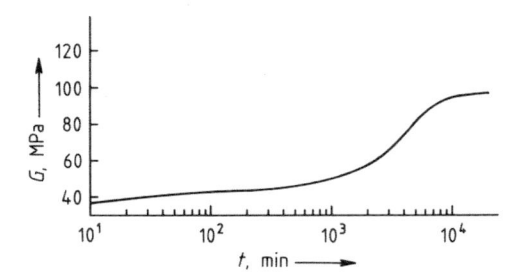

Figure 3. Time dependence of the shear modulus G of modification II of poly(1-butene)
Thermal history: 1 mm thick test bar cooled from the melt (160 °C) in air to the test temperature of 20 °C

degree of crystallization and the mean spherulite size remain virtually constant during the transition from II to I [11]. The crystallite size also remains unchanged [12]. This is a purely crystal – crystal transformation without experimentally detectable melting of modification II crystals.

The rate of conversion, as indicated for example by the half-value time $t_{1/2}$ depends on the thermal history (Fig. 4) and on the temperature (Fig. 4).

Mechanical stresses may significantly increase the conversion rate [22, 23]. Figure 6 shows the influence of constant tensile stress on the conversion rate, monitored by IR spectroscopy. Whereas these measurements show a stress-free sample to have a half-value time of 5500 min, the half value for a sample under a tensile load of 4.4 MPa is only ca. 10 min. Internal stresses and orientations have a similar effect to external stresses. The conversion of II

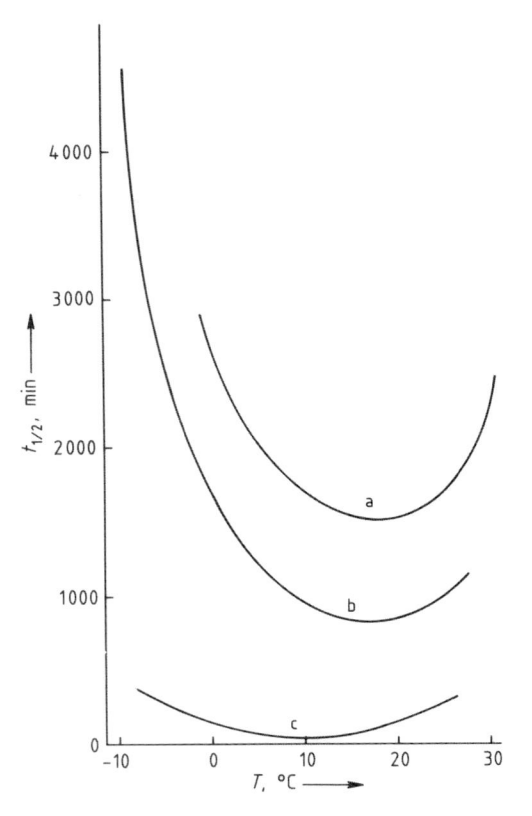

Figure 5. Temperature dependence of half-value time $t_{1/2}$ for the transition II → I
a) Pure poly(1-butene); b) Melt mixture of poly(1-butene) and polypropylene in the ratio of 1:1; c) Poly(1-butene) with 30% butyl stearate

to I is also accelerated by hydrostatic pressure [24]. The conversion rate can also be increased by certain additives, such as polypropylene or plasticizers (see Fig. 5) [8, 25, 26].

The molecular mass and the molecular mass distribution have virtually no effect on the conversion rate. However, the molecular mass does have significant influence on the rate of crystallization of modification II (Fig. 7). The crystallite and spherulite sizes of all crystal modifications can be varied over a broad range by modifying the cooling and crystallization conditions of the melt [16, 27, 28]. Surprisingly, the mean crystallite size and size distribution are not increased by subsequent tempering [27, 28]. This may be a reason for the low time dependence of certain mechanical properties (see Section 1.2.2).

The four modifications are distinguished analytically and characterized quantitatively

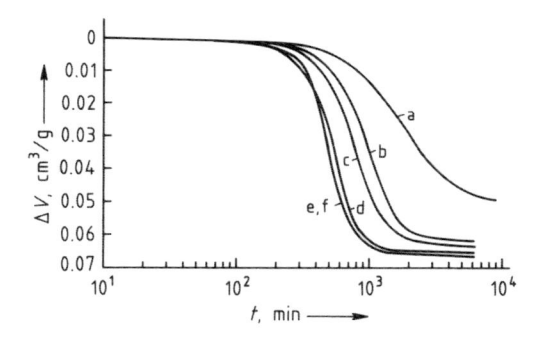

Figure 4. Time dependence of specific volume during transformation (II → I) of poly(1-butene) at 20 °C
a) Quenched; b) Crystallized for 200 min at 92 °C; c) Crystallized for 5700 min at 102 °C; d) Crystallized for 5600 min at 102 °C; e) Crystallized for 7500 min at 106 °C; f) Crystallized for 7200 min at 108 °C

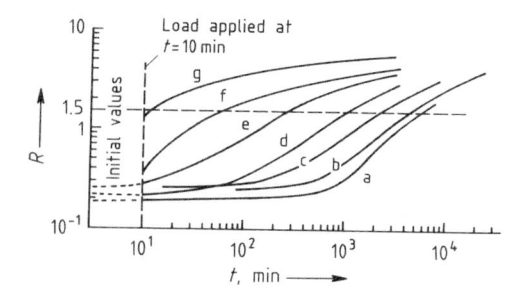

Figure 6. Poly(1-butene) conversion curves for modification II → I for various constant tensile stresses
R is the ratio of the IR extinction of modification I at 920 cm^{-1} to the extinction of modification II at 920 cm^{-1}. The half-value time $t_{1/2}$ (II → I) is reached when R has a value of ca. 1.5
a) 0 MPa; b) 0.12 MPa; c) 1.48 MPa; d) 1.91 MPa; e) 2.55 MPa; f) 2.95 MPa; g) 4.23 MPa

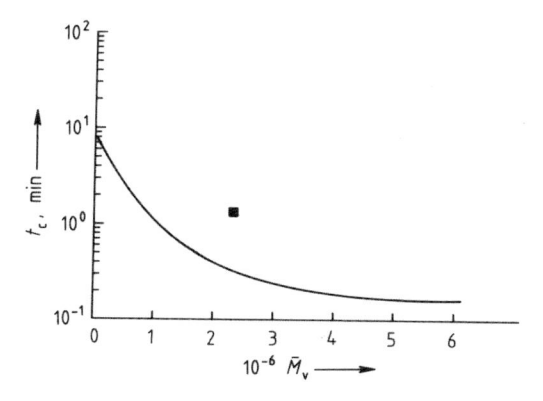

Figure 7. Crystallization time t_c of poly(1-butene) at 78 °C as a function of the viscosity-average molecular mass M_v
Curve: measurements on fractions, ■ unfractionated product

by a combination of X-ray, infrared, and differential thermal analysis [13–15, 29].

1.2.2. Application Properties

In normal processing operations, isotactic poly(1-butene) is heated to above its melting point. Therefore it is the unstable modification II which forms immediately after cooling, and is converted to the stable form I during storage or on application of mechanical stress. Since only the properties of form I are of interest from the point of view of applications, the statements below refer only to this modification.

In terms of density, melting point, and glass transition temperature, poly(1-butene) resembles polyethylene rather than polypropylene [30]. This similarity extends to cross-linked polyethylenes, although some of their properties naturally vary, depending on their cross-linking density [31]. Commercial material is around 40 % to 55 % crystalline. Since the crystallization rate is slow by comparison with polyethylene and polypropylene, superstructures which are of interest for applications can be obtained by the choice of cooling conditions [32]. Nucleators increase the rate of crystallization, leading in some conditions to a material that is somewhat more brittle at low temperature [32].

Mechanical Properties. Table 2 lists some mechanical properties of poly(1-butene), polyethylene, and polypropylene. According to the VDE hardness, poly(1-butene) is softer than polyethylene and polypropylene. However, the Shore hardness is virtually the same as for polyethylene. The surface gloss of components made from poly(1-butene) is similar to those made from polypropylene [33].

The rigidity of poly(1-butene) is relatively low. The flexural stress at conventional deflection and modulus of elasticity are lower than for polyethylene (Table 2).

Table 2. Hardness, rigidity and strength of low-pressure polyolefins [33]

Property	Test method	Polyethylene	Polypropylene	Poly(1-butene)
VDE hardness (60 s), MPa	VDE 0302	44 – 64	61 – 80	25 – 45
Shore D hardness	DIN 53 505	62 – 69	70 – 75	60 – 68
Flexural stress at conventional deflection, MPa	DIN 53 542	26 – 43	34 – 50	15 – 25
Modulus of elasticity, MPa	a	900 – 1200	1000 – 1500	500 – 900
Impact strength	DIN 53 453	no break	no break	no break
Notched impact strength, kJ/m^2	DIN 53 453	5 to no break	3 – 15	20 to no break
Brittle point, °C	ASTM D 746–57T	below – 70	+ 20 to – 5	– 10 to – 30

^{a} Measured in Hooke range by bending tests.

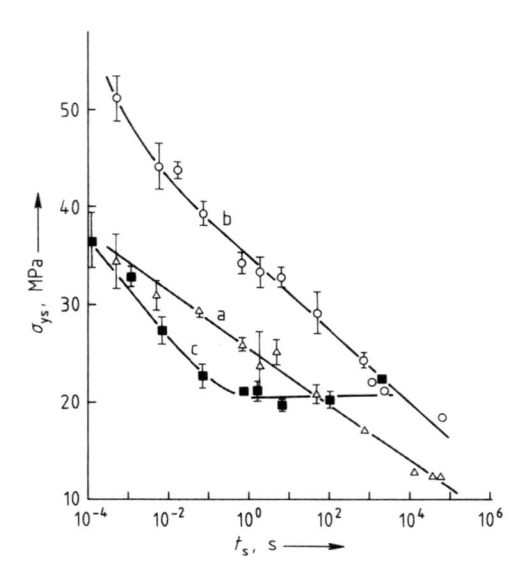

Figure 8. Yield strength σ_{ys}, measured in a tensile test at 22 °C, as a function of the stretching time t_s [34]
a) Polyethylene; b) Polypropylene; c) Poly(1-butene)

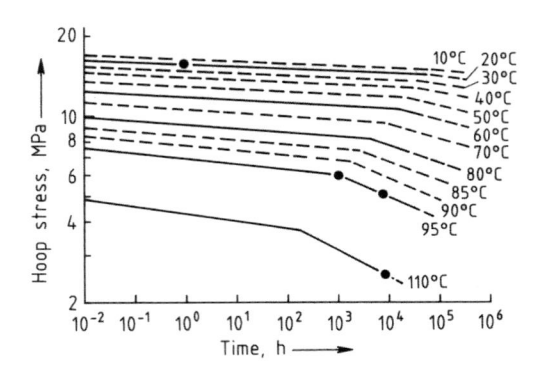

Figure 9. Burst hoop stress performance of poly(1-butene) pipes

The impact strength and notched impact strength are significantly higher than those of polypropylene. Above 0 °C the values are similar to those for high molecular mass polyethylene, but at lower temperature they are inferior [33]. Nevertheless the low-temperature strength, as characterized by the brittle point, is relatively good [22, 33].

The stress – strain behavior of modification I differs noticeably from that of other polyolefins. The differences are not only in the shape of the stress – strain curve, which for industrial material has virtually no maximum [33, 34], but also in the dependence on time and temperature [35]. Figure 8 shows the yield strength σ_{ys}, measured in an uniaxial tensile test as a function of the stretching time t_s. Whereas the yield strength of polyethylene and polypropylene declines sharply with increasing stretching time, the values for poly(1-butene) remain virtually constant above 0.1 s. The good creep resistance and resistance to stress cracking of poly (1-butene) pipes subject to internal pressure [30] is due to the low time dependence of the mechanical properties.

Since this low time dependence of the mechanical characteristics is maintained at higher temperatures, poly(1-butene) is particularly suitable for hot-water pressure pipes [31, 36]. Figure 9 shows the burst hoop stress performance of poly(1-butene) pipe under internal pressure according to ISO-SEM (DIS 9080) [37].

As well as the outstanding resistance to stress cracking, the unusually good creep behavior is significant. The tendency to creep, i.e., to undergo plastic deformation below the yield point, is significantly lower than for other polyolefins, even at temperatures up to 100 °C [31, 33, 38]. This is also an essential requirement for the use of poly(1-butene) as a film material, particularly for the packaging sector [38].

Electrical Properties. The electrical behavior of poly(1-butene) does not differ significantly from that of polyethylene and polypropylene [33]. Apart from the usual quantities of stabilizers, it only contains residues of polymerization catalysts. Because of its good resistance to chemicals, outstanding resistance to stress cracking, and flexibility, poly(1-butene) is an important material for special applications in the cable and electrical industry.

Resistance to Chemicals and Gas Permeability. Because of its paraffinic structure, poly(1-butene) is resistant to most common chemicals such as acids, alkalies, salts, alcohols, and ketones. However, it swells or dissolves in hot aromatic and chlorinated hydrocarbons. It is less inert to aliphatic hydrocarbons than polyethylene and polypropylene and more prone to permeation [33, 39, 40]. A detailed table on the resistance of poly(1-butene) to chemicals is given in [38].

Permeability to gases and vapors, in particular water and oxygen, is comparable to that of polyethylene of similar crystallinity. Permeability to carbon dioxide, however, is higher [32].

1.3. Processing

Processing poly(1-butene) on conventional machines presents no problems. Because of its low crystallite melting point and particularly because of the rapid decrease in melt viscosity with increasing temperature, poly(1-butene) can be processed by extrusion (pipes, profiles, sheets), extrusion blow molding (hollow bodies), and injection molding (engineering products, sheathing for metals) [33]. Sheets made by extrusion or compression molding can be shaped hot or cold. A special feature is cold shaping of the relatively soft modification II. Articles shaped from this modification then become increasingly rigid as the conversion to modification I progresses.

1.4. Applications

Pipe. The major outlet for poly(1-butene) is currently in high-performance pipes, which exploit its creep resistance and long lifetime at elevated service temperatures, flexibility, light weight, corrosion resistance, and its ability to be heat welded [31]. It competes in plumbing and radiator heating applications with copper, where progress is mainly limited by resistance to replacing traditional working practises. In the underfloor heating pipe market it competes with other plastics such as cross-linked polyethylene and polypropylene.

The advantage of poly(1-butene) over other polymers in both these applications is its better ability to retain its mechanical and performance properties at elevated temperature. For instance a comparison of tangent modulus, yield stress, ultimate stress, and elongation at break within and transverse to the extrusion direction of the pipes shows a more balanced performance for pipes made of poly(1-butene) [41]. In the last decade the long-term properties of poly(1-butene) pipes and their lifetime at temperatures up to 120 °C have been studied [42] and the methodology for such long-term tests has been developed for various plastics [43–45].

Extrapolation methods give estimated lifetimes of 50 years at 70 °C [41]. A recent evaluation of pipes based on polymers as well as traditional materials, such as copper, favored poly(1-butene) [46].

The problem of oxygen diffusion in plastic piping systems, which was given as one explanation of corrosion in metal radiators, tanks, and boilers in the past, has been resolved by the use of multilayer pipes containing an oxygen-barrier layer and the addition of inhibitors to the circulating hot water [46].

Packaging. In the packaging sector, poly(1-butene) has become established in speciality film applications which require resistance to cooking temperatures, and in heat-seal layers for oriented polypropylene, which exploit the high compatibility of poly(1-butene) with polypropylene. The incompatibility of poly(1-butene) in blends with various polyethylenes and ethylene copolymers, such as low-density polyethylene, linear low-density polyethylene, high-density polyethylene and ethylene – vinyl acetate copolymers, forms the basis of peelable seals (see Fig. 10).

Peel-seal technology provides a robust pack, yet allows the consumer to open it easily without ripping or tearing. The key attributes of systems based on poly(1-butene) are the broad heat sealing range and the cohesive failure mechanism, which results in an aesthetically attractive package after opening. This is because cohesive failure propagates through the peel-seal layer itself due to the weaker intermolecular forces between the dissimilar materials, and results in clean uniform separation. Additional advantages are good clarity and a clean peel with no stringing [47, 48]. The more conventional system relies on delamination between layers as a result of adhesive failure and is also more difficult to control.

A correlation of peel-seal strength with blend composition and temperature is shown in Figure 11.

Manufacture of packaging film or sheet incorporating a poly(1-butene) peel-seal blend is possible by all the conventional processes such as film blowing or casting, sheet extrusion

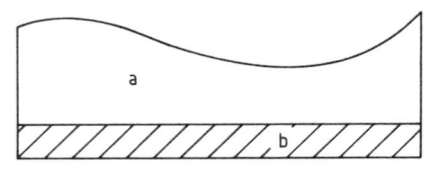

Figure 10. Cross section of a coextruded peel-seal layer on an LDPE substrate
a) LDPE substrate; b) Peel-seal blend LDPE in which incompatible poly(1-butene) particles are dispersed

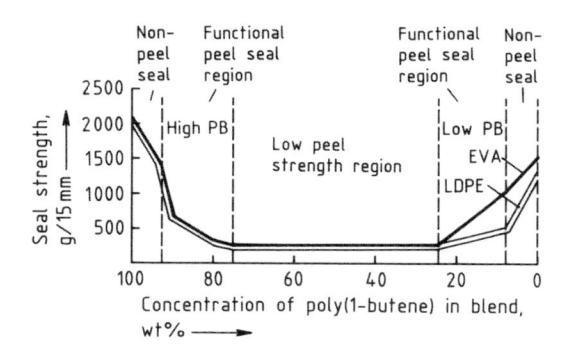

Figure 11. Dependence of peal-seal strength on blend composition

or coextrusion, extrusion or coextrusion coating, and lamination.

Poly(1-butene) peel seals are used in flexible packaging such as box liners for cereals, cake mixtures, and biscuits. In combination with more sophisticated barrier films, they are used for processed meat and vegetable packs. They are also used for packaging medical products.

An important growth area for peelable seals is in lids for rigid or semirigid containers. These range from the relatively simple yogurt, milk, and cream pots to individual-portion containers for microwave meals.

Two features of poly(1-butene) peel seals are considered to be of particular importance to the development of the next generation of high-performance packages. The first is their resistance to high-temperature cooking or sterilization. Often referred to as "retortability", this is an increasingly important feature both in the food and nonfood sectors. Poly(1-butene) peel seals can be made heat resistant by formulation with higher-temperature-resistant materials such as polypropylene or high-density polyethylene.

The second important feature is a distinct whitening effect when the seals are opened because of the characteristic cohesive failure mechanism. This gives a clear indication that the package has been opened or tampered with.

2. Polyisobutylene

2.1. Introduction

Industrially important polyisobutylenes can be divided into five groups:

1. Di-, tri-, and tetraisobutenes
2. Low molecular mass polyisobutylenes with $\bar{M}_n = 330 – 1600$
3. Medium molecular mass polyisobutylenes with $\bar{M}_n = 20\,000 – 45\,000$
4. High molecular mass polyisobutylenes with $\bar{M}_n = 75\,000 – 600\,000$
5. Ultra high molecular mass polyisobutylenes with $\bar{M}_n > 760\,000$

History. In 1925 LEBEDEW was the first to prepare polyisobutylene [49, 50]. He obtained products of varying chain length: liquids of oily to honey-like consistency; sticky, doughy materials; or solid, rubber-like materials.

Relatively uniform low molecular mass polyisobutylenes were first obtained by the company de Bataafse [51] by polymerizing isobutene at $+10$ to -10 °C in the presence of metal halides.

High molecular mass polyisobutylenes were first prepared by OTTO and MÜLLER-CUNRADI by polymerizing isobutene below -10 °C in presence of volatile inorganic halides [52].

Polymerization Mechanism. Even in the early studies, two important aspects of the polymerization of isobutene were recognized: (1) catalysis with initiators of the Friedel – Crafts type, and (2) the inverse dependence of the molecular mass on the temperature. Subsequent studies led to the first formulations of a reaction mechanism [53].

According to these ideas, which essentially remain valid today, the reaction consists of the following steps:

Initiation:

$$BF_3 + HX \longrightarrow H[BF_3X]$$

$$H[BF_3X] + \underset{CH_3}{\overset{CH_3}{C}}=CH_2 \longrightarrow \left[CH_3 - \underset{CH_3}{\overset{CH_3}{C}} \right]^+ [BF_3X]^-$$

Chain growth:

$$\left[CH_3 - \underset{CH_3}{\overset{CH_3}{C}} \right]^+ + n\,CH_2 = \underset{CH_3}{\overset{CH_3}{C}}$$

$$\longrightarrow \left[CH_3-\underset{\underset{CH_3}{|}}{\overset{\overset{CH_3}{|}}{C}}\left[-CH_2-\underset{\underset{CH_3}{|}}{\overset{\overset{CH_3}{|}}{C}}\right]_{n-1}-CH_2-\underset{\underset{CH_3}{|}}{\overset{\overset{CH_3}{|}}{C}} \right]^+$$

Chain termination:

$$R-\left[-CH_2-\underset{\underset{CH_3}{|}}{\overset{\overset{CH_3}{|}}{C}}\right]^+ \longrightarrow R-CH=\underset{\underset{CH_3}{|}}{\overset{\overset{CH_3}{|}}{C}} + H^+$$

$$\longrightarrow R-CH_2-\underset{\underset{CH_3}{|}}{C}=CH_2 + H^+$$

The initiator is a Lewis acid (BF_3, $AlCl_3$) in combination with a proton source, which is usually present as a trace impurity, e.g., water, hydrogen, fluoride, or hydrogen chloride.

The competition between the growth reaction and chain termination determines the chain length (i.e., the molecular mass). Since the activation energy of chain termination is higher, increasing the polymerization temperature lowers the polymer chain length ($\ln M \sim 1/T$).

Meanwhile [54], a series of studies have appeared which have revealed a considerably refined view of the course of the reaction; yet a comprehensive mechanism which is consistent with all experimental findings is still lacking. Other studies deal with new initiators (aluminum alkyls; inorganic initiators such as acid clays and molecular sieves; Ziegler catalysts; high-energy radiation as an initiator). However, none of these new catalyst systems are used industrially. Industrial processes for polyisobutylene production still use the conventional Friedel – Crafts catalysts, boron trifluoride and aluminum chloride.

2.2. Production

2.2.1. Isobutene as Raw Material

The production of isobutene is described in → Butenes, Chap. 4, → Butenes, Chap. 5.

The purity requirements for the isobutene differ according to the type of polyisobutylene produced.

For the production of low molecular mass polyisobutylenes, butadiene-free C_4 fractions can be used, which apart from isobutene contain n-butane, isobutane, and n-butenes; stricter requirements are set for the isobutene for high molecular mass and solid polyisobutylenes.

During isolation of isobutene from C_4 fractions, an isobutene must be obtained which is as free from n-butene as possible. Primarily the content of cis- and $trans$-2-butene must be minimized. Both are powerful polymerization poisons and chain terminators. On the other hand, 1-butene can be used as a regulator in the production of high molecular mass and solid polyisobutylenes.

2.2.2. Low Molecular Mass Polyisobutylenes

Figure 12 shows the layout of the Cosden Petroleum continuous process, which can be regarded as a kind of fluidized-bed process.

A liquefied refinery-gas mixture with $C_1 - C_5$ hydrocarbons and an isobutene content of 3 – 50 % is freed from pentenes in a preliminary distillation, enriched in isobutene in an absorption column, washed in a wash tower with ca. 20 % caustic soda solution to remove sulfur, cooled in a heat exchanger, dewatered in a water separator, dried in silica gel towers and, after further low-temperature cooling in a second heat exchanger, introduced into the lower part of the reaction vessel. The polymerization takes place here at temperatures between – 43 and +16 °C and pressures of 0.1 – 0.35 MPa, so that the liquid phase is maintained. The catalyst used is a mushy suspension of finely divided (50 – 100 mesh) aluminum chloride in dry, liquid polyisobutylene, which is prepared in a mixing tank and supplied to the reaction vessel by a gear pump.

The quantity of aluminum chloride is about 10 – 20 % of the total hydrocarbons in the reaction mixture. The activity of the aluminum chloride is increased by adding 0.08 – 0.12 % hydrogen chloride (relative to aluminum chloride) as activator and accelerator. Water or chloroform have the same effect. The diameter of the reaction zone (A) and the feed rate of the catalyst slurry are so matched to each other that the aluminum chloride particles do not settle. Zone (B) of the reaction vessel, with a diameter two to six times that of zone (A), reduces the flow rate by a factor of about three to four, and functions as a calming zone. It merges into the settling zone (C), in which the aluminum chloride particles settle and are pumped off as a suspension through the offtake pipe which extends into zone (B). This catalyst slurry

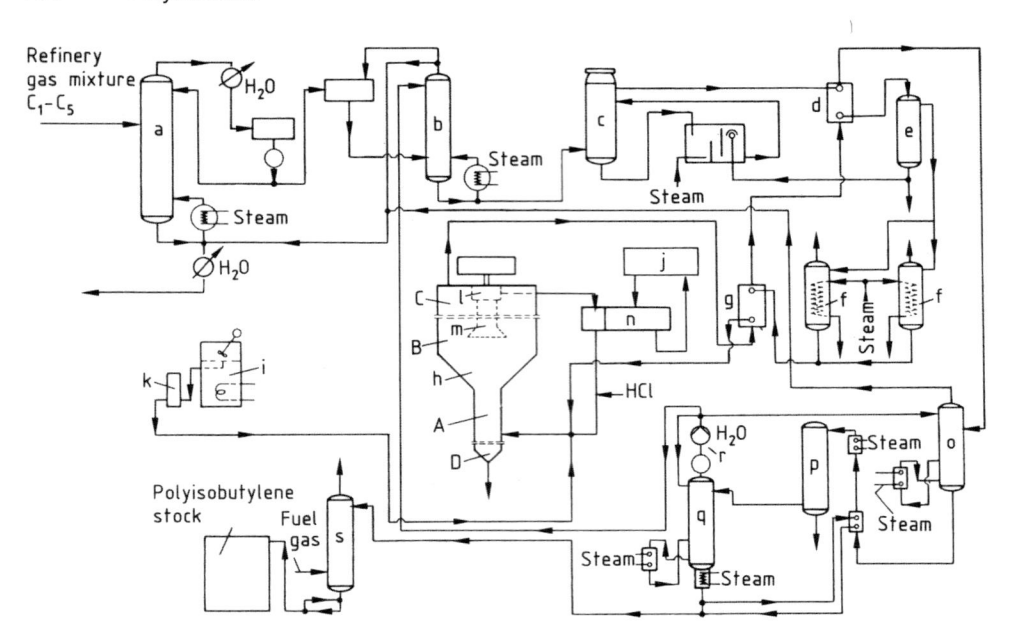

Figure 12. Cosden Petroleum process for producing low molecular mass polyisobutylene
a) Preliminary distillation; b) Absorption column; c) Caustic soda solution wash tower; d) Heat exchanger; e) Water separator; f) Silica gel towers; g) Heat exchanger; h) Reaction vessel; i) Mixing tank; j) Refrigeration unit; k) Gear pump; l) Pump; m) Offtake pipe; n) Cooler; o) Stripping column; p) Clay-filled tower; q) Distillation column; r) Water cooler; s) Pressure column; (For A to D, see text)

returns via a cooler into the reaction zone (A). Spent, aggregated catalyst is discharged at the bottom of the reaction zone.

A variant of the process uses separate vessels for polymerization and for catalyst settling.

From the settling zone flows a clear, light solution of liquid polyisobutylene in unreacted starting material. To improve the yield, the reaction mixture is pumped at least four times, but usually more than eight times, through the reaction zone (A). The clear polymer solution flows through the heat exchangers (g) and (d) to cool the isobutene and, after a water wash to remove acidic constituents, into a stripping column in which about one to two thirds of the inert, unreacted starting material is distilled off. After post-treatment of the concentrated polymer solution in a clay-filled tower, which may precede the stripping column, all volatile constituents of the starting material, especially unpolymerized isobutene, are vaporized in a distillation column and returned via a water cooler into the absorption column. In a further column, the last traces of volatile constituents are distilled from the liquid polyisobutylene at atmospheric pressure.

Aside from this process of the Cosden Petroleum Corp., others have been developed. For a description of the BASF and Exxon processes, see [55, pp. 94 – 97], [56].

2.2.3. Medium Molecular Mass Polyisobutylenes

The basic principle of the Exxon process for producing medium molecular mass polyisobutylenes (Fig. 13) [57] resembles the process for low molecular mass polyisobutylenes. However, a purer isobutene is used as starting material and pentane or hexane is added as solvent. The polymerization temperature must be lower to obtain a higher molecular mass. A mixture prepared from fresh isobutene, recycled isobutene, and hexane with 30 % isobutene and 70 % hexane is brought via a heat exchanger and a cooler to the polymerization temperature of $-40\ °C$ and fed into the reaction vessel. Simultaneously, a slurry of 5 % finely divided aluminum chloride in hexane at a temperature of below $-23\ °C$ flows into the reaction vessel.

The polymerization takes place with powerful stirring so that the aluminum chloride does

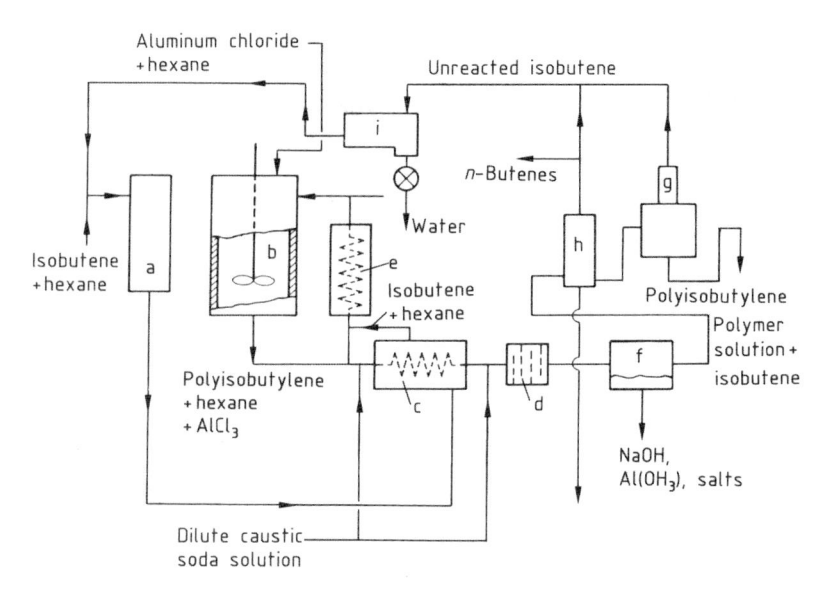

Figure 13. Exxon process for producing medium molecular mass polyisobutylenes
a) Mixing vessel; b) Reaction vessel; c) Heat exchanger; d) Nozzle mixer; e) Cooler; f) Settling tank; g) Degasser; h) Degassing vessel; i) Water separator

not settle, and external cooling. The throughput is 378.5 L/h isobutene – hexane mixture and 0.454 kg/h AlCl$_3$. The resulting solution of polyisobutylene in hexane with suspended aluminum chloride is drawn off, and after it has passed through a heat exchanger an excess of dilute caustic soda solution is added to destroy the catalyst. After thorough mixing in a nozzle mixer, the mixture of polymer solution and dilute caustic soda solution runs into a settling tank, where it forms two layers. The lower layer, consisting of dilute caustic soda solution, aluminum hydroxide, and salts, is discarded. The upper layer, consisting of a solution of polymer in hexane and unreacted starting material, is fed into a degassing vessel, in which the lower boiling components, principally unreacted isobutene, are stripped off at 99 °C and 0.35 bar. n-Butenes if present are removed from the system to avoid poisoning the reaction. In a second degasser, the last traces of volatile constituents are removed. The unreacted isobutene is condensed, dewatered in a water separator, and added as recycle isobutene to the starting mixture.

In its batch process for producing medium molecular mass polyisobutylenes, BASF makes use of the chain-terminating effect of diisobutene [55, pp. 88 – 91], [58].

In a conical reaction vessel with good thermal insulation, pure, dry liquid isobutene and pure, dry liquid ethylene as internal coolant are mixed in the ratio 1 : 2. The quantity of diisobutene added to the isobutene depends on the desired molecular mass of the polyisobutylene. A sufficient quantity of gaseous boron trifluoride for the polymerization is then carefully introduced into the mixture. The mixture starts to boil, and the evaporating ethylene removes the heat of polymerization. The polyisobutylene remains behind in the reaction vessel in the form of a white foam. It is removed from the vessel and freed from all volatile constituents in a degassing machine.

2.2.4. High Molecular Mass Polyisbutylenes

BASF and Exxon have developed continuous processes for producing high molecular mass solid polyisobutylenes [55, pp. 97 – 104].

BASF Belt Process. In the belt process, the conical polymerization vessel of the BASF batch process described above is replaced by a continuous polymerization belt (Fig. 14) [59].

A gently inclined trough-shaped continuous steel belt, 16 – 18 m long and 50 cm wide, runs

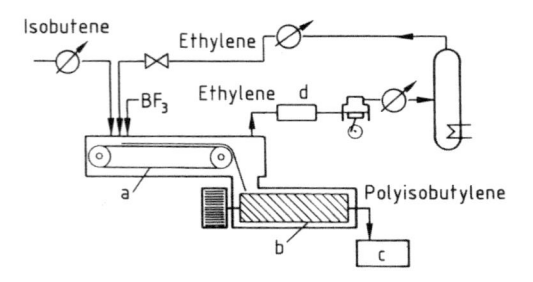

Figure 14. BASF belt process for high molecular mass polyisobutylenes
a) Steel belt; b) Twin-screw machine; c) Cooling and compounding; d) Purification

in a gastight housing over a drive roller and a tension roller. A mixture of pure, dry liquid ethylene and isobutene in the ratio 1:1 flows through a pipe onto one end of the steel belt. From a second pipe a solution of boron trifluoride in the same quantity of ethylene as is mixed with the isobutene is added. The quantities of cocatalyst and chain-length regulator [55, pp. 82 – 91] depend on the desired molecular mass. The polymerization starts instantly. It continues at the boiling point of the liquid ethylene (-103.7 °C) and in a short time is ended. The heat of polymerization is absorbed by the evaporation of ethylene. The vaporized ethylene is recycled via a purification and liquefaction system. The polyisobutylene is stripped from the steel belt with a scraper and degassed and homogenized in a twin-screw machine.

Exxon Process. Polyisobutylene is also produced by the process used for butyl rubber (see → Rubber, 4. Emulsion Rubbers) [60].

2.3. Structure

The polymer obtained by cationic polymerization has the following structure:

$$H_3C-\underset{\underset{CH_3}{|}}{\overset{\overset{CH_3}{|}}{C}}-CH_2-\underset{\underset{CH_3}{|}}{\overset{\overset{CH_3}{|}}{C}}-CH_2-\underset{\underset{CH_3}{|}}{\overset{\overset{CH_3}{|}}{C}}- \cdots\cdots\cdots CH_2-\underset{\underset{CH_3}{|}}{\overset{\overset{CH_2}{||}}{C}}$$

In the stretched state polyisobutylene gives a well-developed X-ray diffraction pattern which shows that 8 monomers with their 16 methyl groups are arranged in an almost regular helix around the principal axis, forming one translation period of the chain after 3 rotations [61].

It has been shown by IR spectroscopy that the principal end group is the methylene group [62].

2.4. Properties

The rubber-like polyisobutylenes, like all other polyisobutylenes, owe their physical and chemical properties to their paraffinic character. This results in a low glass transition temperature, low specific thermal conductivity, very low water vapor permeability, high dielectric strength, high resistivity, low dielectric constant, low dielectric loss factor, high ageing resistance, and rot resistance as well as an unusually wide-ranging chemical resistance. The good solubility in hydrocarbons and chlorohydrocarbons and the insolubility in alcohols, esters, and ketones also result from the paraffinic nature of the polyisobutylenes.

At room temperature, polyisobutylenes are resistant to dilute and concentrated hydrochloric acid, sulfuric acid, phosphoric acid, chlorosulfonic acid, phenolsulfonic acid, naphthalenesulfonic acid, formic acid, acetic acid, ammonia, potassium hydroxide solution, caustic soda solution, as well as aqueous calcium hydroxide, aqueous hydrogensulfite, copper sulfate solution, potassium permanganate solution, hydrogen peroxide, chromic acid, and dichromate solution. Above 80 °C, concentrated sulfuric acid causes carbonization and concentrated nitric acid causes decomposition.

Some properties of polyisobutylene that are independent of the degree of polymerization are as follows:

Density, g/cm^3	0.92
Volume coefficient of thermal expansion at 20 °C, K^{-1}	6.3×10^{-4}
Specific heat, kJ kg^{-1} K^{-1}	2.0
Glass transition temperature (DSC), °C	-60
Thermal conductivity, W K^{-1} m^{-1}	0.19
Refractive index n_{20}^D	1.51
Dielectric constant ϵ_r (50 Hz, 20 °C) DIN 53 483	2.2
Dielectric loss factor tan δ (50 Hz, 20 °C) DIN 53 483	$\leq 5\times10^{-4}$
Water vapor permeability, g m^{-1} h^{-1} mbar^{-1}	2.5×10^{-7}

Further properties are listed in Tables 3 and 4.

Table 3. Properties of polyisobutylenes dependent on degree of polymerization, exemplified by Oppanol B (BASF)

Oppanol	Viscosity, Pa · s		Intrinsic viscosity $[\eta]$, cm³/g	Number-average molecular mass M_n *
	at 20 °C	at 100 °C		
Low molecular mass				
B 3	25	0.2 ± 0.02		820
Medium molecular mass				
B 10	5.0×10^4	2.2×10^2	27.5 – 31.2	24 000
B 12			34.0 – 39.1	30 000
B 15	5.0×10^5	3.0×10^3	45.9 – 51.6	40 000
High molecular mass				
B 50	1.5×10^8	8.0×10^5	113 – 143	120 000
B 80			178 – 236	180 000
B 100	3.6×10^8	6.7×10^7	241 – 294	250 000
B 120			295 – 361	300 000
B 150			416 – 479	425 000
B 200	1.5×10^{11}	1.0×10^9	551 – 661	600 000
Ultra high molecular mass				
B 246			> 770	> 760 000

* $M_n = 0.94 \sqrt{\frac{[\eta]10^3}{2.27}}$

Table 4. Behavior of Oppanol B100 and B200 towards solvents (weight increase by swelling in wt % at room temperature)

Storage time	3 1/2 days		30 days	
Oppanol type	100	200	100	200
Acetone	1.7	1.5	3.9	3.6
Alcohol	0.1	< 0.1	0.35	0.2
Benzene	400	400	500	500
Butyl acetate	34	24	34	37
Carbon tetrachloride			soluble	
Chlorobenzene			soluble	
Chloroform			soluble	
Cyclohexane			soluble	
Cyclohexanone	9	7.4	24	18
Dimethylformamide	0.6	0.6	2.1	1.7
Ether	200	100		
Ethyl acetate	18	12	21	20
Ethylene glycol	0.6	1.4	0.9	2.0
Gasoline			soluble	
Glycerol	0.6	0.6	1.0	1.6
Hydrogen sulfide			soluble	
Isobutanol	1	0.3	2.4	0.7
Isooctane			soluble	
Methanol	< 0.1	< 0.1	0.1	0.1
Methyl acetate	8	5	7.8	10
Methylene chloride	200	200	200	200
Mineral oil			soluble	
Nitrobenzene	3	2	5.4	4.5
Olive oil	4	1.5	4.2	3.1
Paraffin			soluble	
Paraffin oil			soluble	
Tetrahydrofuran			soluble	
Toluene			soluble	
Water	< 0.1	< 0.1	0.1	0.1
Xylene			soluble	

2.5. Uses and Processing

The low molecular mass polyisobutylenes are used as plasticizing components for modifying sealants. In addition, they are used for producing insulating oils.

After further functionalization with polar groups, the low molecular mass polyisobutylenes are used as dispersants in fuel and lubricant additives.

The medium molecular mass polyisobutylenes are chiefly used as raw materials for adhesives and sealants. Due to their very low water-vapor permeability, permanently flexible sealants based on polyisobutylene – carbon black are important in the production of sealed double glazing units. Combinations of polyisobutylene with paraffin or wax are used for impregnating paper and board.

The breaking point of bitumen is lowered and its flexibility improved by addition of medium molecular mass polyisobutylene.

High molecular mass polyisobutylenes are used as blends with medium molecular mass grades for producing self-adhesive compounds, e.g., for adhesive plasters.

Because of their good tolerance by the skin, filled polyisobutylenes are used to produce sealing rings for colostomy bags.

Unstabilized polyisobutylenes are used in chewing gum bases.

Blending polyisobutylenes with paraffin oil and inorganic fillers gives permanently elastic sealants, which are usually marketed in extruded form as strip or round-profile string.

High molecular mass polyisobutylenes with high filler contents are used to produce sheets that are used for sealing buildings against groundwater and seeping water as well as for protection against corrosion and radiation. Conductive and magnetic polyisobutylene sheetings are also available.

The cold impact resistance of polypropylene is improved by blending with polyisobutylene. The sensitivity of polyethylene to environmental stress cracking is reduced by adding polyisobutylene.

Polyisobutylenes are compatible with natural and synthetic rubber and with reclaimed rubber.

Solutions of ultra high molecular mass polyisobutylenes ($\bar{M}_n > 760\,000$) in organic solvents even in low concentrations ($\ll 1\%$), have pronounced viscoelastic properties and distinctly increased extensional viscosity.

Polyisobutylene can be compounded on a roll mill or in an internal mixer. Compounding is carried out at 140 – 170 °C to minimize thermal degradation of the polyisobutylene.

2.6. Commercial Products

BASF, Germany:	Oppanol B 3 (liquid); Oppanol B 10, B 12, B 15 (flexible resin-like, tacky); Oppanol B 50, B 80, B 100, B 120, B 150, B 200, B 246 (solid, rubber-like)
Exxon, United States:	Vistanex LM-MS, LM-MH (flexible resin-like, tacky); Vistanex MM L-80, MM L-100, MM L-120, MM L-140 (solid, rubber-like)

In addition, a range of low molecular mass polyisobutylenes are marketed.

2.7. Copolymers [63]

Isobutene – isoprene – divinylbenzene terpolymers are produced under conditions similar to those used for butyl rubber, with the addition of divinylbenzene. Unlike butyl rubber, they exhibit no cold flow.

Isobutene – styrene copolymers behave similarly to polyisobutylenes or butyl rubber.

The copolymers have no economic significance.

References
Specific References

1. B. Wunderlich: *Macromolecular Physics*, Academic Press, New YorkCrystal Structure, Morphology, Defects, vol. 1, 1973; Crystal Nucleation, Growth, Annealing, vol. 2, 1976.
2. G. Natta, P. Corradini, I. W. Bassi, *Makromol. Chem.* **21** (1956) 240 – 244.
3. A. Turner Jones, *J. Polym. Sci. Part B* **1** (1963) 455 – 456; *Polymer* **7** (1966) 23 – 59.
4. S. W. Cornell, J. L. Koenig, *J. Polym. Sci. Part A-2* **7** (1969) 1965 – 1982.
5. I. T. Luongo, R. Salovey, *J. Polym. Sci. Part A-2* **4** (1966) 997 – 1008.
6. I. D. Rubin in H. Marowetz (ed.): *Polymer Monographs*, vol. 1, Gordon & Break Sci. Publ., New York 1968.
7. H. Tompa: *Polymer Solutions*, Butterworths, London 1956.
8. V. F. Holland, R. L. Miller, *J. Appl. Phys.* **35** (1964) 3241 – 3248.
9. G. Natta, *Angew. Chem.* **68** (1956) 393 – 403.
10. S. W. Cornell, I. L. Koenig, *J. Polym. Sci. Part A-2* **7** (1969) 1965 – 1982.
11. G. Videtto, A. I. Kovacs, *Kolloid Z. Z. Polym.* **220** (1967) 1 – 18.
12. W. Glenz, G. Goldbach, unpublished data from measurements made by Chemische Werke Hüls AG, Marl.
13. C. Geacintow, R. B. Milles, H. I. L. Schuurmans, *J. Polym. Sci. Part A-1* **4** (1966) 431 – 433.
14. G. Goldbach, G. Peistscher, *J. Polym. Sci. Part B* **6** (1968) 783 – 788.
15. F. Danusso, G. Gianotto, *Makromol. Chem.* **61** (1963) 139 – 156.
16. I. Powers, I. D. Hoffman, I. I. Weeks, F. A. Quinn, Jr., *J. Res. Natl. Bur. Stand. Sect. A* **69** (1965) 335 – 345.
17. H. Wilski, T. Grewer, *J. Polym. Sci. Part C* **6** (1964) 33 – 41.
18. Allied Chem. Corp., BE 652 381, 1965.
19. C. Geacintow, R. S. Schotland, R. B. Miles, *J. Polym. Sci. Part B* **1** (1963) 587 – 591; *J. Polym. Sci. Part C* **6** (1964) 197 – 207.
20. I. Boor, Jr., E. A. Youngman, *J. Polym. Sci.* **2** (1964) 903 – 907.
21. J. P. Rakus, C. D. Mason, *J. Polym. Sci. Part B* **4** (1966) 467 – 468.
22. G. Goldbach, *Angew. Makromol. Chem.* **29/30** (1973) 213 – 227.
23. G. Goldbach, *Angew. Makromol. Chem.* **39** (1974) 175 – 188.
24. C. D. Armeniades, E. Baer, *J. Makromol. Sci. Phys. B* **1** (1967) no. 2, 309 – 334.
25. I. D. Rubin, *J. Polym. Sci. Part A* **3** (1965) 3803 –3813.
26. I. Boor, I. C. Mitchel, *J. Polym. Sci.* **62** (1962) 70 –73.
27. J. Haase, R. Hosemann, S. Köhler, *Polymer* **19** (1978) 1358 – 1361.
28. J. Haase, S. Köhler, R. Hosemann, *Z. Naturforsch. A* **33** (1978) 1472 – 1483.
29. M. Goldstein, M. E. Seeley, H. A. Willis, V. J. I. Zichy, *Polymer* **19** (1978) 1118 – 1122.
30. J. Plenikowski, O. Umminger, *Kunststoffe* **55** (1965) 822 – 827.
31. M. L. Kasakevich, *Kunststoffe* **80** (1990) 54 – 57.
32. A. I. Foglia, *Appl. Polym. Symp.* **11** (1969) 1 – 18.

33. J. Plenikowski, *Kunststoffe* **55** (1965) 431 – 437.
34. I. P. Rakus, C. D. Mason, R. J. Schaffhauser, *J. Polym. Sci. Part B* **7** (1969) 591 – 595.
35. G. Goldbach, *Kunststoffe* **64** (1974) 475 – 481; *Angew. Makromol. Chem.* **74** (1978) 123 – 146.
36. W. Bollmann, *NDZ, Neue Deliwa-Z.* **9** (1972) 385 –391.
37. Unpublished data from measurements made on behalf of Shell Research.
38. C. R. Lindgren, *Polym. Eng. Sci.* **10** (1970) 163 –169.
39. J. Park et al., *J. Am. Water Works Assoc.*, Oct. 1991, 71 – 78.
40. J. Park et al., *J. Am. Water Works Assoc.*, Nov. 1991, 85 – 91.
41. M. Kasakevich et al., *Kunststoffe* **80** (1990) 1361 –1362.
42. M. Ifwarson, T. Traenker, *Kunststoffe* **79** (1989) 827 – 830.
43. M. Ifwarson, P. Eriksson, *Kunststoffe* **76** (1986) 245 – 248.
44. M. Ifwarson, *Kunststoffe* **79** (1989) 525 – 529.
45. M. Ifwarson, P. Eriksson, "40A: Experience from 12 Years Evaluation of Cross-Linked Polyethylene," report at 6th International Conference Plastic Pipes VI, York 1985, p. 40A.1.
46. J. Schultze, M. Homann: C_4 *Hydrocarbons and Derivatives*, Springer-Verlag, Berlin 1989.
47. A. Wouters, D. Hopper: "Polybutylene Film for Non-retortable Applications," *Pira 5th International Packaging Conference*, LondonOct. 3 – 4, 1990.
48. A. Wouters: "Polybutylene Film for Non-retortable Applications,"*Packforst Seminar Scanplast '91*, Gothenburg, March 6, 1991. R. Lee: "High Performance Lidding Films from Polybutylene Blends," *Pira International Conference*, Birmingham, Nov. 22, 1990.
49. S. W. Lebedew, E. P. Filonenko, *Ber. Dtsch. Chem. Ges.* **58** (1925) 163 – 168.
50. S. W. Lebedew, G. G. Kobliansky, *Ber. Dtsch. Chem. Ges.* **63** (1930) 103 – 112.
51. De Bataafse, GB 358 068, 1930 (H. D. Elkington).
52. I. G. Farbenindustrie, DT 641–284, 1931 (M. Otto, M. Müller-Cunradi).
53. P. H. Plesch: *The Chemistry of Cationic Polymerization*, McMillan Publ., New York 1963.
54. J. P. Kennedy: *Cationic Polymerization of Olefins*, A Critical Inventory, Wiley-Interscience, New York 1975, pp. 86 – 137.
55. H. Güterbock: *Polyisobutylen and Isobutylen-Mischpolymerisate*, Springer Verlag, Berlin 1959.
56. Cosden Petroleum, US 2 957 930, 1956 (W. K. Jackson); BE 584 683, 1959 (W. K. Jackson); *Chem. Eng. N. Y.* **67** (1960)no. 16, 98 – 101.
57. Standard Oil Development Co., US 2 698 320, 1952 (A. R. Garabrant, H. G. Goering, H. G. Schneider). Esso Research and Engineering, US 2 849 433, 1953 (H. G. Schneider, H. G. Goering, V. F. Mistretta); US 8 42 533, 1955 (R. P. Cahn).
58. Standard Oil Dev., US 2 176 194, 1936 (L. A. Bannon). Jasco, US 2 296 399, 1939 (M. Otto, H. G. Schneider).
59. I. G. Farbenindustrie, DT 496 482, 1937 (M. Otto, H. Güterbock, A. Hellemanns).
60. Exxon Research and Eng. Comp., US 2 893 981, 1954 (J. L. Ernst, A. B. Small, L. W. McLean); US 2 856 394, 1954 (A. B. Small, J. L. Ernst).
61. H.-U. Lenne, *Kolloid Z.* **137** (1954) 65 – 70. A. M. Liquori, *Acta Crystallogr.* **8** (1955) 345 – 347.
62. J. F. Gross, K. W. Nelson, J. M. Slobodin, *Zh. Fiz. Khim.* **25** (1951) 504 – 512. M. S. C. Flett, O. H. Plesch, *J. Chem. Soc.* 1952, 3355 – 3358.
63. J. P. Kennedy, E. Marechal: *Carbocationic Polymerization*, Wiley Interscience, New York 1982.

Further Reading

E. Benham, M. McDaniel: Polyethylene, High Density, *Kirk Othmer Encyclopedia of Chemical Technology*, 5th edition, John Wiley & Sons, Hoboken, NJ, online DOI: 10.1002/0471238961.0809070811091919.a01.pub2.

J. K. Fink: *Handbook of Engineering and Specialty Thermoplastics*, Wiley, Hoboken, NJ 2010.

Y. V. Kissin: Polymers of Higher Olefins, *Kirk Othmer Encyclopedia of Chemical Technology*, 5th edition, John Wiley & Sons, Hoboken, NJ, online DOI: 10.1002/0471238961.1615122511091919.a01.pub2.

R. B. Lieberman et al.: Polypropylene, *Kirk Othmer Encyclopedia of Chemical Technology*, 5th edition, John Wiley & Sons, Hoboken, NJ, online DOI: 10.1002/0471238961.1615122512090502.a01.pub2.

N. Maraschin: Polyethylene, Low Density, *Kirk Othmer Encyclopedia of Chemical Technology*, 5th edition, John Wiley & Sons, Hoboken, NJ, online DOI: 10.1002/0471238961.12152316050219.a01.pub2.

D. Nwabunma: *Polyolefin Blends*, Wiley, Hoboken, NJ 2008.

D. Nwabunma, T. Kyu: *Polyolefin Composites*, Wiley, Hoboken, NJ 2008.

M. Tolinski: *Additives for Polyolefins*, William Andrew, Oxford 2009.

S. C. O. Ugbolue: *Polyolefin Fibres*, Woodhead Publ., Cambridge 2009.

Polycarbonates

GEORG ABTS, Bayer MaterialScience AG, Leverkusen, Germany

THOMAS ECKEL, Bayer MaterialScience AG, Leverkusen, Germany

ROLF WEHRMANN, Bayer MaterialScience AG, Leverkusen, Germany

1.	Introduction	763	3.3. PC–Polyester Blends	772
2.	Properties	764	4. Production	773
2.1.	Bisphenol A Polycarbonate	764	4.1. Interfacial Polycondensation	773
2.2.	Modified Bisphenol A Polycarbonate (Copolymers)	765	4.2. Melt Transesterification	774
			5. Processing	775
2.3.	Bisphenol A Polycarbonate with Additives	767	6. Applications	775
			7. Recycling	778
2.4.	Aliphatic Polycarbonates	768	8. Toxicology and Environmental Aspects	779
3.	Blends	768		
3.1.	General	768	References	779
3.2.	PC–ABS Blends and PC–ASA Blends	769		

1. Introduction

The first aromatic polycarbonates were prepared by EINHORN in 1898, based on resorcinol or hydroquinone, by reaction with phosgene in a pyridine solution. The resorcinol-based polycarbonate decomposed at 190°C, while the hydroquinone-based resin had a melting temperature above 280°C. In 1902, BISCHOFF and VON HEDENSTRÖM obtained the same products, using diphenyl carbonate instead of phosgene.

Aliphatic polycarbonates were prepared by CAROTHERS already in 1930. The crystalline products had a melting temperature below 100°C and resulted from the reaction of diethyl carbonates and glycols. In 1941, the Pittsburgh Plate Glass Company (PPG Industries, Inc.) introduced a product based on the allyl ester of diethylene glycol carbonate cross-linked with peroxides to give a colorless, transparent material (Section 2.4).

Meanwhile, SCHNELL at Farbenfabriken Bayer (now Bayer MaterialScience AG) and independently FOX at General Electric Company were trying to create a polymer with superior properties to poly(ethylene terephthalate) (PET), which had been developed in 1946. SCHNELL turned his attention to polycondensation reactions using 2,2-bis(4-hydroxyphenyl)propane, known as bisphenol A (BPA), which was already being used for the production of epoxy resins. Consequently, in 1953 he discovered 2,2-bis(4-hydroxyphenyl)propane polycarbonate (**1**), also termed bisphenol A polycarbonate [24936-68-3] (BPA-PC), which still is the economically most important polycarbonate. The first trial products had melting temperatures of 225–230°C. Simultaneously, Fox had concentrated his activities on polyester resins. By chance he discovered a transparent material, which was identified as BPA-PC as well [1–4].

Structure 1: Bisphenol A polycarbonate repeating unit

The polymer was first produced industrially and introduced to the market in Europe in 1958 by Farbenfabriken Bayer and in the USA in 1960 by Mobay Chemical Corporation (a former joint venture of Monsanto Company and Bayer AG) and General Electric.

Ullmann's Polymers and Plastics: Products and Processes
© 2016 Wiley-VCH Verlag GmbH & Co. KGaA, Weinheim
ISBN: 978-3-527-33823-8 / DOI: 10.1002/14356007.a21_207.pub2

The great commercial success of BPA-PC is due to its unique combination of key properties: extreme toughness, outstanding transparency and high heat distortion resistance. In addition, polycarbonate has a broad potential as blend partner with other polymers such as polyesters (PET, PBT) and rubber-modified polymers (e.g., ABS, ASA).

In addition to bisphenol A, many other bisphenols have been converted to (co)polycarbonates [5–11]. However, very few of them have been used as (co)monomers in industrial polycarbonate production. Examples include 1,1-bis(4-hydroxyphenyl)cyclohexane [843-55-0] (bisphenol Z, BPZ) (**2**) and 2,2-bis(3,5-dibromo-4-hydroxyphenyl) propane [79-94-7] (tetrabromobisphenol A, TBBPA) (**3**):

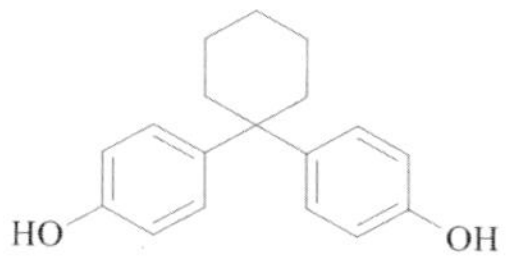

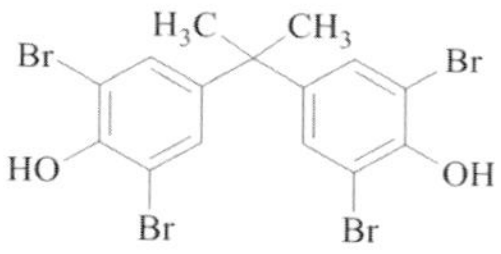

Formerly, BPZ was used as a comonomer for bisphenol A in special polycarbonates for cast film production. It improved the solubility and reduced the crystallization tendency of the polycarbonate during the casting procedure.

TBBPA has been cocondensed with bisphenol A in order to increase the flame resistance of the polycarbonate. Also, its pure oligocarbonate has been used as flame retardant in special polycarbonate compositions (Section 2.3).

In the late 1980s, 1,1-bis(4-hydroxyphenyl)-3,3,5-trimethylcyclohexane [129188-99-4] (BP-TMC) (**4**) has become important for the development of highly heat resistant polycarbonates [12], which have been on the market since 1990.

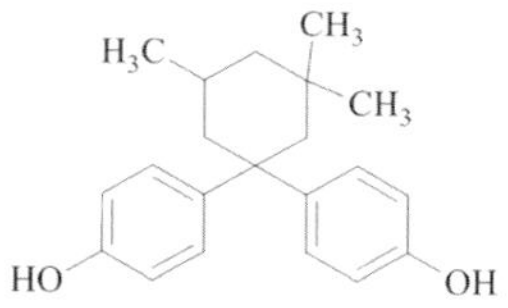

BPA-PC can be modified not only by cocondensation of bisphenol A with other bisphenol monomers, but also by cocondensation with blocks of other polymers.

Polysiloxane blocks [10–15] have become industrially important in this respect (Section 2.2).

2. Properties

2.1. Bisphenol A Polycarbonate

Industrial BPA-PCs for injection molding and extrusion have weight-average molar masses M_w of 18 000–35 000. Compared to other thermoplastics, the degree of polymerization of BPA-PC is relatively low. The number of repeating units in commercial products is only approximately 70–140. Therefore, it is astonishing, that this low degree of polymerization leads to the extraordinary toughness and heat resistance which are typical of polycarbonate. The toughness and impact strength of BPA-PC are considerably higher than those of other amorphous single-phase thermoplastics [e.g., polystyrene, poly(methyl methacrylate)].

The tough fracture of BPA-PC depends on its molecular mass. With increasing molar mass, the tough-to-brittle transition is shifted to lower temperatures. For general purpose grades, brittle fracture is observed below −10°C [16]. Special grades, such as impact-modified polycarbonates and siloxane-modified copolycarbonates, keep their toughness down to temperatures of approximately −40°C or even lower.

The high transparency of BPA-PC and its high heat resistance ($T_g \approx 150°C$, which is 50°C above that of polystyrene) are very important for its uses and applications. Long-term stability of BPA-PC against heat aging in air can be achieved up to 120°C or more, depending on the test method.

Table 1. Typical properties of BPA-PC (Makrolon 2805, Bayer MaterialScience)

Property	Test standard	Value
Tensile modulus, MPa	ISO 527-1, -2	2400
Yield stress, MPa	ISO 527-1, -2	66
Yield strain, %	ISO 527-1, -2	6.2
Stress at break, MPa	ISO 527-1, -2	70
Nominal strain at break, %	ISO 527-1, -2	> 50
Impact strength at 23°C, kJ/m^2	ISO 179-1eU	unbroken
Impact strength at −30°C, kJ/m^2	ISO 179-1eU	unbroken
Notched impact strength at 23°C, kJ/m^2	based on ISO 179-1eA	75
Notched impact strength at 23°C, kJ/m^2	based on ISO 179-1eA	16
Ball indentation hardness, N/mm^2	ISO 2039-1	115
Vicat softening temperature (50 N; 120°C/h), °C	ISO 306	146
Temperature of deflection under load (1.8 MPa), °C	ISO 75-1,-2	125
Coefficient of linear thermal expansion, K^{-1}	ISO 11359-1,-2	6.5×10^{-5}
Thermal conductivity (23°C; 50 % R.H.), W m^{-1} K^{-1}	ISO 8302	0.2
Burning behavior at 0.75 mm	UL 94	Class V-2
Glow-wire flammability test (GWFI) at 0.75 mm, °C	IEC 60695-2-12	850
Oxygen index (method A) in %	ISO 4589-2	28
Relative permittivity (100 Hz/1 MHz)	IEC 60250	3.1/3.0
Dissipation factor (100 Hz/1 MHz)	IEC 60250	$5 \times 10^{-4}/90 \times 10^{-4}$
Volume resistivity, Ω m	IEC 60093	10^{14}
Surface resistivity, Ω	IEC 60093	10^{16}
Electrical strength at 1 mm, kV/mm	IEC 60243-1	34
Moisture absorption (23°C, 50% R.H.), %	analogous to DIN 53 495-1 L	0.15
Refractive index (Procedure A)	ISO 489	1.586
Luminous transmittance (clear transparent materials) at 1 mm, %	ISO 13468-2	89
Density, kg/m^3	ISO 1183-1	1200

BPA-PCs are soluble in chlorinated hydrocarbons (e.g., chloroform, dichloromethane, chlorobenzene) and nonhalogenated solvents (e.g., pyridine, m-cresol). They are insoluble in water, alcohols, and (cyclo)aliphatic hydrocarbons.

BPA-PCs are generally amorphous and thus transparent when processed by injection molding or melt extrusion.

By special means, however, crystallinity can be induced to a certain extent, which can be observed by X-ray diffraction. For this purpose, heating at temperatures between the softening point and the crystalline melting point (ca. 260°C), treatment with solvents (e.g., acetone, toluene), and stretching of fibers or films is applied [6].

Decomposition of polycarbonate depends on temperature, processing conditions and formulation. As a rule of thumb, processing temperatures far beyond 320°C should be avoided for general-purpose grades.

BPA-PC has a low flammability (oxygen index 26). Its flame resistance can be improved considerably by adding small amounts of flame retardants (Section 2.3).

Although the mechanical properties continue to increase above M_w = 35 000, BPA-PC with molar mass greater than 40 000 cannot be processed due to its extremely high melt viscosity. In the past, such high molecular mass polycarbonates, however, have been cast from solution into films with desirable properties. See also → Films.

At molecular masses below 18 000, mechanical and thermal properties (e.g., toughness, elongation at break, heat resistance) deteriorate sharply.

Amongst others, mechanical and thermal properties of BPA-PC are influenced by its end groups. Introducing alkyl phenols instead of phenol as chain terminator leads to improved properties [12, 17].

Some typical properties of a standard BPA-PC are given in Table 1.

2.2. Modified Bisphenol A Polycarbonate (Copolymers)

BPA-PC exhibits tough fracture and outstanding notched impact strength above −10°C.

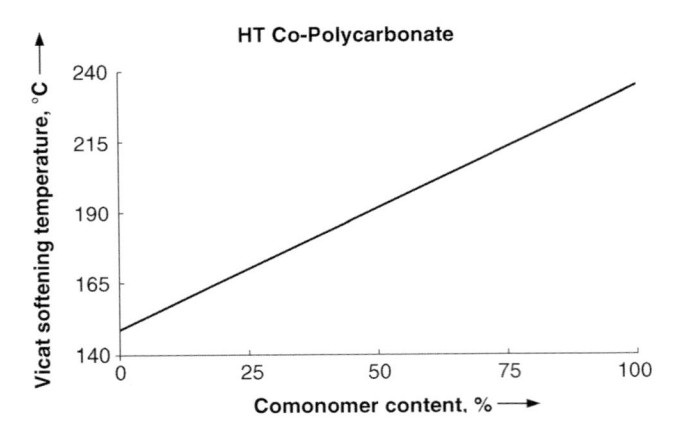

Figure 1. Vicat temperature as a function of comonomer content

Below −10°C however, the tough fracture changes to a brittle fracture and the notched impact strength is considerably reduced.

The notched impact strength at low temperatures can be improved by different means. For example, the copolycarbonate of BPA and 4,4′-dihydroxydiphenyl (DOD) **5** exhibits outstanding toughness at low temperatures, partially improved environmental stress cracking behavior (ESC), and, due to the larger amount of aromatic segments, extended flame retardant properties.

5

Of commercial importance is the incorporation of polysiloxane blocks into the polymer backbone [10, 13]; this shifts the tough–brittle transition to below −40°C. A siloxane content of 2–10 wt% produces a satisfactory effect. Polydimethylsiloxane blocks are coupled to the polycarbonate via the hydroxyphenyl end groups (**6**).

Corresponding products are commercially available, for example Lexan EXL grades from Sabic Innovative Plastics.

Partial replacement of the carbonate groups in BPA-PC by terephthalate or isophthalate groups leads to aromatic polyester carbonates with higher heat distortion resistance[14, 15].

Nowadays, this class of polyester carbonates has been almost completely replaced by high-heat copolycarbonates (PC-HT), e.g., Apec grades from Bayer MaterialScience and Lexan XHT grades from Sabic Innovative Plastics. In these polymers, a second bisphenol is used as comonomer in the polycondensation reaction with BPA. Glass temperatures up to 240°C can be achieved. For example, by using 1,1-bis(4-hydroxyphenyl)-3,3,5-trimethylcyclohexane (BP-TMC) [12], the heat resistance can be adjusted.

Insofar, commercially available products reflect technical requirements, usually expressed by the Vicat temperature, which is directly linked to the amount of comonomer (Fig. 1).

These copolycarbonates are transparent, tough, and colorless, even at high glass transition temperatures [12]. They exhibit better heat and weatherability properties (less yellowing

6

than the polyestercarbonates). Their melt viscosities are comparable with those of polysulfones, polyethersulfones, and polyetherimides having comparable glass transition temperatures.

Several other copolycarbonates have been developed in the past to meet specific requirements such as improved scratch resistance or reduced birefringence. However, none of these products has reached large-volume production so far. Some were already withdrawn from the market.

Besides linear polycarbonates, branched grades have been developed for certain applications. They can be obtained, for example, by cocondensation of multifunctional phenols such as 1,1,1-tris(4-hydroxyphenyl)ethane (THPE) with BPA. The resulting polymers show structural viscosity and provide higher melt stiffness [10, 12, 18–21]. They are particularly suitable for melt extrusion. They have a reduced viscosity under the shear stress in the extruder, and after leaving the extruder nozzle the melt is highly viscous and thus sufficiently stable. Complicated extrudates, such as multiwall sheets, can be produced with high accuracy. The structural viscosity of melts of these polycarbonates is also important for the production of large hollow articles by extrusion blow molding (e.g., 5-gallon water bottles).

2.3. Bisphenol A Polycarbonate with Additives

Most of the additives discussed in this section are described in detail elsewhere (→ Plastics, Additives).

Mold-Release Agents. The addition of small amounts (< 1 wt%) of mold-release agents to the polycarbonate improves the mold release of complicated parts [10]. Most commonly used are fatty acid esters of higher alcohols, especially pentaerythritol tetrastearate (PETS) or glycerol monostearate (GMS).

See also → Release Agents.

Flame Retardants. In the past, flame resistance of BPA-PC has been increased by incorporating small amounts of tetrabromobisphenol A [10]. In rare cases, tetrabromo-BPA oligocarbonates themselves are still used as additives to improve flame retardancy. For ecological reasons, the use of brominated flame retardants is rapidly decreasing.

Today, specific additives are commonly used. For example, flame resistance is significantly improved by the addition of small amounts (< 1 wt%) of special salts, such as potassium perfluoro-n-butane-1-sulfonate or potassium diphenylsulfone-3-sulfonate [22–25]. The tendency to form flaming drips after ignition can be reduced by adding small amounts (< 1 wt%) of polytetrafluoroethylene (PTFE) [26]. To achieve higher surface quality (gloss), PTFE–SAN batches are used. However, these additives lead to nontransparent and opaque compounds.

For transparent applications, only small amounts of the above-mentioned salts can be used. Recently, highly branched, transparent polycarbonates with improved flame-retardant properties have been introduced into the market.

Light Stabilizers. Ultraviolet light, especially in combination with atmospheric oxygen and moisture, causes decomposition of BPA-PC. Decomposition starts with a photo-Fries rearrangement, which leads to the formation of salicylates, followed by benzophenones [10, 11]. These structures absorb visible light in the short-wavelength region and are thus responsible for the yellowing of polycarbonates. Stabilization against degradation by UV light (i.e., against yellowing) is important for exterior uses and applications and can be achieved by adding small amounts (< 1 wt %) of derivatives of benzotriazole, hydroxyphenyltriazine, or benzylidene-bis-manonate [11].

Reinforcement. Glass fibers are generally used to improve rigidity, stiffness, and dimensional stability. In addition, the coefficient of thermal expansion is reduced.

For polycarbonate compounds different kinds of glass fibers are available, such as chopped strands, long fibers, milled fibers, and glass bubbles. Typical ranges cover 10–35 wt%. [10, 11, 27]. As expected, however, glass fibers reduce impact and notched impact strength significantly.

For extreme stiffness, carbon fibers are used.

Recently, polycarbonate composites, e.g., with glass or carbon fiber fabric, are gaining more interest.

Heat and process stabilizers (e.g., organic phosphites or phosphonites) can be added to the polycarbonate to prevent thermal oxidation during manufacturing and processing at high temperature [10]. They react with hydroperoxides formed by autooxidation of the polymer backbone, act as oxygen scavengers, and prevent degradation as they get oxidized. These stabilizers are called secondary antioxidants. They are often used in combination with primary antioxidants like sterically hindered phenols to further improve the performance of the polycarbonate composition.

Colorants and Pigments. A wide range of colored transparent or opaque polycarbonates is obtained by using appropriate amounts of soluble organic dyes or inorganic pigments. For transparent applications, only small amounts (less than 1%) of soluble organic dyes can be used. Special colorants are also approved for medical applications and food contact.

When using inorganic pigments, such as TiO_2, $BaSO_4$ or talc, resulting compositions are translucent or opaque. Concentrations may be up to 15%, for example for highly reflecting white compounds. Dependent on their concentration, these insoluble substances lead to a reduction of mechanical properties, especially impact and notched impact strength as well as elongation at break.

Miscellaneous Additives. Special additives are used for dedicated applications. For example, organic and inorganic infrared absorbers are applied to improve heat management in cars with transparent PC sunroofs. They also fulfill strict absorption requirements of polycarbonate grades for welding goggles or laser marking of injection molded parts. Antistatic agents prevent parts from dust attraction and thermal conductive compounds are gaining importance with respect to heat dissipation of LED lighting assemblies.

Blowing agents (e.g., 5-phenyltetrazole and phenyldihydroxo-oxadiazonone) can be added during thermoplastic processing into moldings to obtain low-weight foamed structures (→ Foamed Plastics, Section 4.15 Polycarbonate Foams) [10].

2.4. Aliphatic Polycarbonates

Besides the thermoplastic aromatic polycarbonates, a group of various aliphatic systems is gaining more and more economical importance. Different molecular architectures are used in various technical applications.

The polymerization of diethylene glycol bis-allylcarbonate (ADC) in presence of a peroxy initiator leads to a highly cross-linked system (CR-39, PPG Industries, Inc.). Due to its outstanding optical properties, it has been used for ophthalmic lenses and sunglasses for decades.

Telechelic aliphatic polycarbonates with a molar mass of approximately 2000 g/mol, obtained in a melt polycondensation process of hexanediol with diphenyl carbonate, are used as raw materials for coatings and thermoplastic polyurethanes (TPU). The carbonate groups provide better stability against hydrolysis than the ester group in polyester diols. The coatings exhibit excellent outdoor weatherability properties. They can be applied from solution or in the form of aqueous dispersions.

A special group of new raw materials are polyether carbonate diols in which carbon dioxide is used as carbonate source (sustainability) [1].

Furthermore, high molecular mass aliphatic PCs for injection molding have been developed. In these materials, isosorbide is employed as biobased comonomer. The resulting products, e. g., Durabio from Mitsubishi Chemical Corporation, combine enhanced optical properties with good mechanical and thermal characteristics. The property profile is intermediate between poly(methyl methacrylate) (PMMA) and BPA-PC.

3. Blends

3.1. General

Initial approaches to using BPA-PC as a blend partner date back to the late 1960s. At that time acrylonitrile–butadiene–styrene (ABS) was the preferred candidate, and the resulting blends saw increasing growth rates in the following decades. Nowadays the most important and commercially established blends are those with ABS graft polymers [28–30] and those with poly(butylene terephthalate) (PBT) or poly(ethylene terephthalate) (PET) in combination with impact-resistance modifiers such as ABS [28, 29, 31]. Of the produced polycarbonate, an estimated more than 20% is used in polycarbonate blends. Both classes of blends add technical benefits to the key properties of BPA-PC. Compared to ABS, BPA-PC–ABS blends have improved heat resistance. They also exhibit good light stability and surface quality. Compared to BPA-PC, processability and high toughness, especially at low temperature (tough fracture in the notched impact strength test) are typical benefits. Blends of BPA-PC with poly(alkylene terephthalates) and impact-strength modifiers have high heat distortion resistance and high toughness, even at low temperature. Their resistance to gasoline and other solvents is better than that of BPA-PC. Both classes of PC blends offer high flexibility in designing properties in wide ranges and thus allow tailor-made products according to customer requests.

3.2. PC–ABS Blends and PC–ASA Blends

BPA-PC is a linear polymer with a low rate of crystallization. Thus, it is regarded as an amorphous polymer with a glass transition temperature T_g of approximately 145°C. Its maximum use temperature is approximately 130°C and, because of its excellent thermal oxidative stability, this almost coincides with its continuous use temperature (125°C). BPA-PC is noted for its excellent impact resistance due to the high segmental mobility of the polymer chain (gamma transition at –100°C [32]).

ABS, on the other hand, is a highly ductile material with a Vicat softening temperature of about 100°C. Typical ABS materials consist of mixtures of a rubber-free SAN copolymer and a SAN-grafted polybutadiene rubber. The low T_g of polybutadiene rubber (below –70°C) is the reason for excellent notched impact strength of this class of polymers, especially at low temperatures (below −20°C). In acrylic–styrene–acrylonitrile polymers (ASA), the SAN-grafted polybutadiene rubber is replaced by SAN-grafted poly(butyl acrylate) rubber with a T_g of ca. −40°C. The benefit of ASA compared to ABS is better UV stability due to the absence of double bonds in the polymer backbone that are sensitive to oxidation.

Polycarbonate blends with ABS have been successful because they combine typical PC properties, e.g., heat distortion, ductility and flame retardancy with those of ABS, e.g., ductility at low temperature, processability, and improved chemical resistance. The mixture of PC and ABS is physically stable because the carbonate group of the PC and the nitrile group of the SAN matrix interact [33] and provide just enough compatibility for a useful blend. Figure 2 shows the typical morphology of a PC–ABS blend with PC as matrix component and SAN copolymer and ABS graft polymers as disperse phase.

Optimum interaction between the polymers is obtained for SAN copolymers containing 25–30 wt% acrylonitrile [33]. Intrinsic compatibility can be enhanced through addition of a third component containing ester groups, e.g., PMMA or poly(butylene terephthalate) (PBT) [34], an effect that can be explained in terms of mutual miscibility.

These PC–ABS blends offer higher heat distortion temperature than ABS, combined with good processing characteristics [35–37]. Depending on PC–ABS ratios, the Vicat softening temperature of the blend can be varied between 100°C (ABS level) and 140°C (PC level). Typical technical data of these blends are summarized in Table 2. All products show ductile impact behavior at –30°C. Polycarbonate blends with ABS resins have been successfully used in automotive interior and exterior applications [35, 38].

PC–ASA blends offer a better resistance towards UV light and weathering compared to PC–ABS. Applications include instrument panels, electrical distribution boxes, and kitchen exhaust hoods [39]. Because this advantage is achieved at the expense of poorer mechanical properties, their share of the total market of PC–styrenics blends is relatively small.

Typical unfilled PC–ABS blends have a tensile modulus in the range of 2000–2300 MPa.

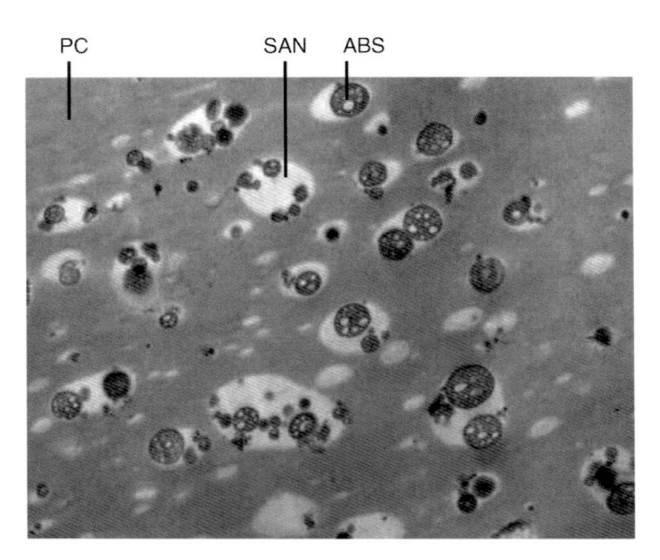

Figure 2. Electron micrograph showing the morphology of a flame-retardant PC–ABS blend (Bayblend, Bayer MaterialScience).

Higher values can be reached by adding mineral fillers (e.g., talc) or glass fibers. With 20% talc a tensile modulus of nearly 5000 MPa and with 20% glass fiber more than 7000 MPa can be obtained.

In general BPA–PC suffers from stress-crack sensitivity coupled with poor chemical resistance. The latter prevents PC from being used in automotive applications where frequent contact with gasoline and oils would ruin its mechanical and thermal properties. Therefore it is one of the most important targets of blending technology to improve this drawback. Chemical resistance of BPA-PC can be improved by blending with ABS or ASA or with semicrystalline polyester resins. PC–ABS blends are less sensitive against aromatics, ketones, esters, and hydrocarbons compared to BPA-PC itself.

PC–ABS blends provide good paintability and better UV stability if PC is the continuous phase [40–43]. The addition of UV absorbers to PC–ABS blends improves the protection of ABS against the damaging effects of UV light. PC–ASA blends provide superior weather resistance, which makes them suitable for outdoor applications.

Table 2. Typical technical properties of PC–ABS blends with different ratios (Bayblend, Bayer MaterialScience)

	Test	Units	PC:ABS 45:55	PC:ABS 60:40	PC:ABS 70:30
Melt viscosity (MVR* 260°C/5 kg)	ISO 1133	cm³/10 min	12	12	12
Injection molding					
melt temperature	ISO 294	°C	260	260	260
molding shrinkage (parallel and transverse)	ISO 294-4	%	0.55–0.75	0.55–0.75	0.55–0.75
Tensile test	ISO 527-1, -2				
modulus		MPa	2100	2200	2300
stress at break		MPa	49	52	55
strain at break		%	>50	>50	>50
Izod notched impact strength	ISO 180/A				
23°C		kJ/m²	40	45	48
−30°C		kJ/m²	36	41	38
Heat distortion					
HDT A	ISO 75-1, -2	°C	95	100	109
Vicat B/120	ISO 306	°C	112	120	131

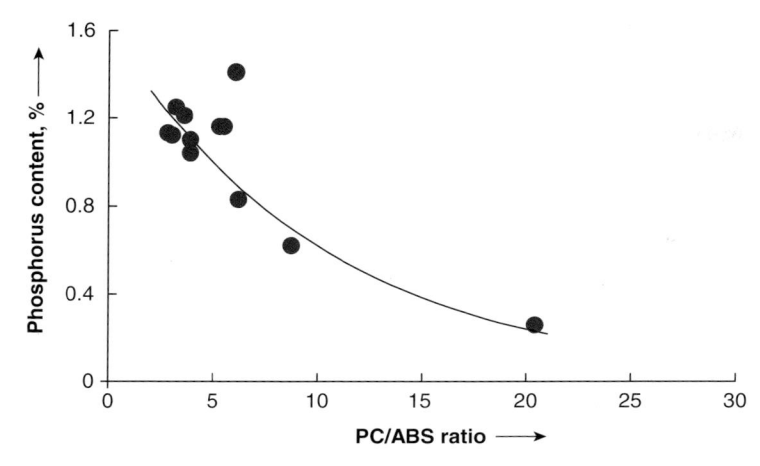

Figure 3. Phosphorus content as function of PC–ABS ratio to achieve a rating of UL 94 V-0 at 1.6 mm (Bayblend, Bayer AG)

Flame-retardant PC–ABS blends were introduced in the 1980s. Due to their balanced properties, they have found widespread applications as housing materials in the electrical and electronics industry, where UL 94 V and UL 94 5V are the most important fire safety specifications. These materials exhibit good electrical insulating capability. Major applications are instrument panels [44] and business equipment housings [45]. Today, flame-retardant PC–ABS blends are firmly established in the market as an independent product category and have reached at least an equal share compared to non-flame-retardant PC–ABS blends.

Until the late 1980s, bromine compounds and antimony oxide as a synergist were the state of the art as flame retardants for PC–ABS. As a result of the ecological discussions in Europe, especially in Germany and the Nordic countries, many global producers of computers and related equipment are seeking to obtain so-called eco-Labels such as the Swedish TCO 99 and German Blue Angel Marks, which require bromine- and chlorine free material for the housings [46]. Aryl phosphates such as triphenyl phosphate (TPP) as well as oligophosphates such as resorcinol bis(diphenyl phosphate) (RDP) and bisphenol A–bis(diphenyl phosphate) (BDP) are utilized as effective flame retardants [47]. Additionally, small amounts of PTFE (< 0.5%) are incorporated as antidripping agents.

The amounts of aryl phosphate necessary to reach a rating of V-0 according to UL 94 at 1.6 mm thickness depends on the PC–ABS ratio. Due to the different inherent flame retardancies of polycarbonate and ABS, the PC–ABS ratio determines the amount of phosphate flame retardant. The higher the PC–ABS ratio, the lower the phosphorus concentration required for a rating of V-0 (Fig. 3).

In the early 1980s, TPP was the most commonly used flame retardant in PC–ABS blends. Due to its high phosphorus content of 9.5% and its volatility, it is highly effective. However, under certain processing conditions TPP tends to migrate to the surface (plate out). To overcome this disadvantage, TPP was replaced by the much less volatile BDP [47]. Mixtures of TPP and oligophosphates such as BDP are also effective and advantageous for certain applications [48].

In order to meet UL 94 V-0 rating at a thickness less than 1.5 mm and to follow the trend to thinner molded articles, certain inorganic materials (i.e., talc, aluminum oxides) are added to improve the effectiveness of the phosphates [49].

Depending on PC/ABS ratios in these flame-retardant blends, the technical properties can be modified over a wide range. Vicat softening temperatures between 90 and 135°C can be achieved. Mechanical properties can be designed between ductile materials with high

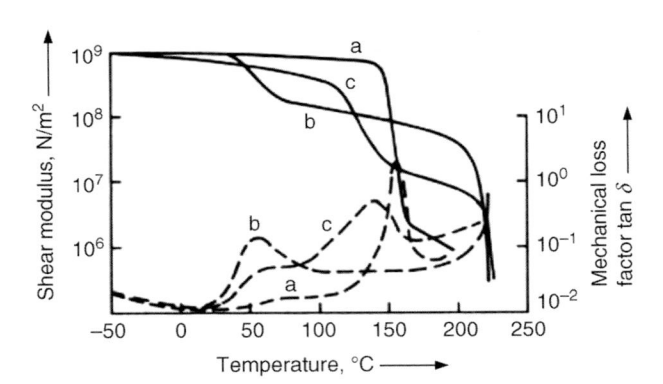

Figure 4. Dynamic mechanical properties of PC (a), PBT (b), and PC-PBT blends (55:45) (c): shear modulus (---) and mechanical loss factor (—) as function of temperature

notched impact strength and materials with tensile modulus up to 7000 MPa (by reinforcement with mineral fillers or glass fibers).

3.3. PC–Polyester Blends

The most important polyesters which are commercially used as blend partners for PC are poly(ethylene terephthalate) [25038-59-9] (PET) and poly(butylene terephthalate) [24968-12-5] (PBT). PBT crystallizes within 1–3 s, whereas PET crystallizes within 15–20 s. PBT's higher melt flow and faster crystallization lead to a shorter cycle time. The advantage of PET over PBT is its higher melt temperature (256 vs. 226°C).

PC–polyester blends combine the benefits of polyesters (chemical resistance due to crystallinity) with those of BPA-PC (impact resistance, also at low temperatures). In the resulting blends the stress-crack sensitivity of PC can be improved significantly. In order to achieve balanced mechanical properties, PC–polyester blends can also contain ABS or methyl methacrylate–butadiene–styrene (MBS) grafted impact modifiers. The most attractive market for PC–polyester blends is automotive applications such as high-class front-end modules and other exterior car body parts. Front-end modules made of PC–polyester blends can withstand an impact with a speed of 5 km/h at high and low temperature.

Blends of PC with polyesters are based on a partial miscibility of the two components in the amorphous phase [50–52]. Thermal and dynamic mechanical analysis shows a sharp T_g transition for PC at about 150°C [10], and a sharp melting point at about 225°C with a broad glass transition for PBT due to the presence of crystalline and amorphous portions of the polyester. The partial miscibility of PC and PBT is indicated by a shift of the PC's T_g to about 140°C in a PC/PBT blend.

Blending the two resins results in a complex co-continuous morphology. Resistance towards chemicals (e.g., oils and lubricants) is significantly enhanced compared to PC. However, the impact properties of these blends are still insufficient for automotive applications. Modification with a rubber compound (frequently core–shell impact modifiers such as MBS or ABS) solves this problem [53 → Polymer Blends].

The key to morphology control and thus mechanical properties is control of transesterification. Care should be taken to prevent reactions (Fig. 5) catalyzed by residual polycondensation catalysts present in the PBT [54–56]. Addition of proprietary stabilizer packages is commonly practiced. In addition, processing temperatures of 270–290°C should not be exceeded.

PC–PET blends have similar properties to their PBT analogs [44, 57, 58]. The lower crystallization rate of PET allows the development of transparent materials for use as housing materials in medical applications by virtue of their good dimensional stability, clarity, and low water absorption. The products have better chemical resistance than PC and are impact-modified to give acceptable low-temperature impact performance.

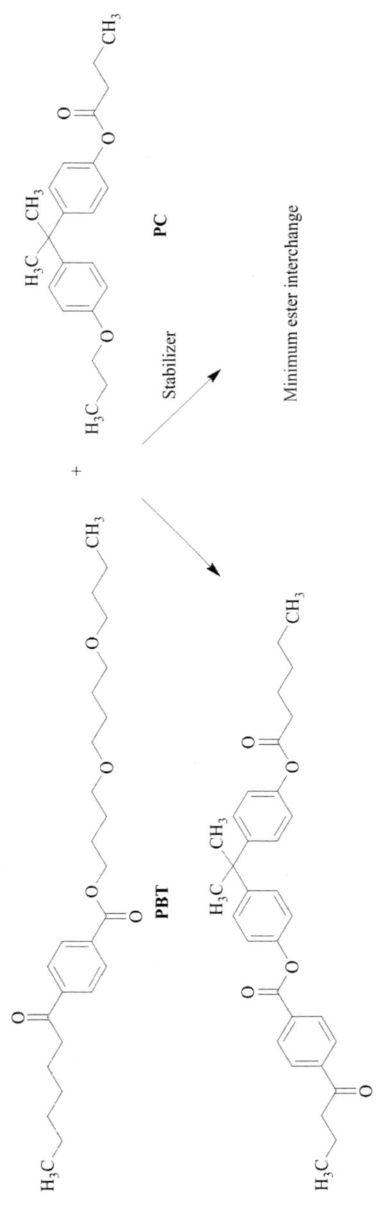

Over the last decade, the portfolio of PC–polyester blends has been extended to mineral- and glass fiber reinforced versions with a tensile modulus of more than 3000 MPa, and to flame-retardant products with V-0 rating according UL 94 V standard (halogen- or phosphorus-based flame retardants).

Highly transparent blends of BPA-PC with amorphous polyesters (e.g., Xylex resins from Sabic Innovative Plastics) exhibit good flowability for thin-wall applications and their impact strength is similar to polycarbonate. In comparison to PC, these products provide better resistance against certain chemicals and solvents.

4. Production

4.1. Interfacial Polycondensation

Interfacial polycondensation is currently used for the industrial production of polycarbonates [6, 9, 10]. Bisphenol A is treated with phosgene at 20–40°C in a two-phase mixture consisting of an aqueous, alkaline phase and an immiscible organic phase:

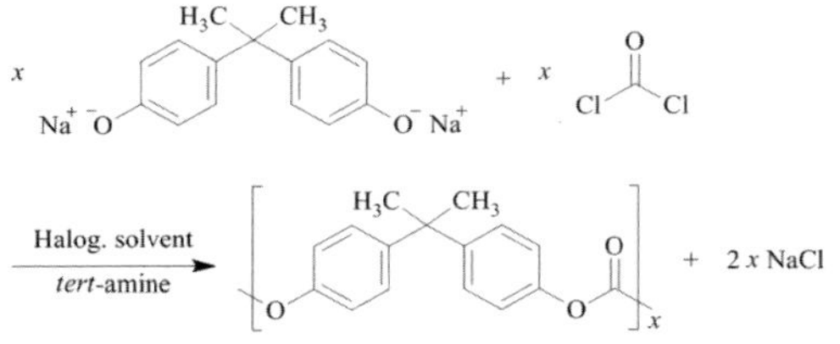

Figure 5. Ester interchange reaction between PC and PBT

BPA is first dissolved in the aqueous phase as sodium bisphenolate and phosgene is dissolved in the organic phase, which generally consists of chlorinated hydrocarbons (e.g., dichloromethane). The reaction occurs at the organic–inorganic interface to produce carbonate oligomers, which then enter the organic phase. A 10–20 mol% excess of phosgene (relative to the amount of bisphenol A) is generally used because a small proportion of the phosgene is hydrolyzed. The hydrolysis products (NaCl and Na_2CO_3) are dissolved in the aqueous phase. The polycondensation is catalyzed with tertiary aliphatic amines (e.g., triethylamine). During the polycondensation reaction the aqueous phase is kept at alkaline pH values (9–14).

Monofunctional phenols, e.g., phenol and 4-*tert*-butylphenol, are added as chain terminators to control the molecular mass. These compounds then form the terminal groups of the polycarbonate bound via carbonate groups.

Polycarbonate production can be carried out batchwise in stirred tanks or continuously in cascades of stirred tanks or in tubular reactors in which the organic and aqueous phases are finely distributed by shear forces. After completion of the polycondensation, the aqueous alkaline phase, which contains sodium chloride and sodium carbonate, is separated from the organic phase (e.g., by centrifugation). The organic phase is washed with dilute acid (e.g., hydrochloric or phosphoric acid) to neutralize the alkali. The organic phase is finally washed free of electrolytes with demineralized water. All wash stages are generally performed in separators.

Industrially, various processes are used to isolate the polycarbonate from solution. The most important are:

1. Precipitation with nonsolvents for polycarbonate (e.g., heptane)
2. Precipitation by introduction of the solution into hot water and subsequent evaporation of the polycarbonate solvent
3. Evaporation of the solvent in extruders or strand evaporators

Both precipitation processes lead to powders, from which granules are obtained by pelletizing extruded strands. If the polycarbonate is isolated by evaporation of the solvent (chlorinated hydrocarbon) in an extruder, the granulate is obtained directly.

Stabilizing or modifying additives are generally fed into the polycarbonate during extrusion at the end of the production process or in a subsequent compounding step.

4.2. Melt Transesterification

Diphenyl carbonate is transesterified in a solvent-free melt with bisphenol A to form polycarbonate [6, 10]:

During this process phenol is split off and removed by distillation. Stirred tanks, helical reactors, or screw reactors (extruders) can be used. The reaction is usually catalyzed with very small amounts of, e.g., alkaline or phosphonium salts at high temperatures and reduced pressure. The temperature is initially low and the pressure is slightly reduced (e.g., 200°C at 20 kPa). During the course of reaction the temperature is raised to ca. 290–310°C and the pressure is further reduced to 0.1 kPa. Towards the end of the reaction the temperature must be high enough to maintain a sufficiently high melt flow and the pressure must be as low as possible to remove all of the eliminated phenol from the viscous melt. The polycarbonate is finally spun in the form of strands and granulated.

The melt transesterification process was the first industrial process for the production of polycarbonate. It is recently again gaining more and more importance due to the development of phosgene-free production processes [12]. Reaction of carbon monoxide with alcohols in the presence of catalysts gives dialkyl carbonates that can, in turn, be treated with phenol to form diphenyl carbonate [12]. The latter can then be treated with bisphenol A, as described above, to produce polycarbonate by melt transesterification. Catalyst systems and the purity of raw materials have been further improved. In contrast to interfacial polycondensation, defect structures are formed during the melt condensation (Fries rearrangements). Nevertheless, the obtained product quality of general purpose grades is comparable.

By applying a dedicated catalyst, even branched grades are accessible without adding a separate branching agent as comonomer, which is not possible in the interfacial process.

5. Processing

Prior to processing, BPA-PC must be dried to a residual moisture content of less than 0.02 wt%. For extrusion and high-quality applications such as transparent goods or optical discs, the residual moisture content should not exceed 0.01 wt %. The use of a dry-air dryer with a dew point below −30°C and a drying time of 3–4 h at 120°C are recommended. For optical applications or glazing products, the drying time may be nearly twice as long. The dried granules should be conveyed with dried air from the dryer to the injection molding machine in order to avoid the absorption of moisture.

For extruded goods with high output volume such as sheets, foregoing drying is not practical. In these cases, moisture is removed during plasticization by a degassing unit in the extruder.

All modern plastics injection molding machines are suitable for polycarbonate processing. Machines with three-section screws with a length/diameter ratio of 20:1 to 23:1 and a screw/channel depth ratio of 2:1 to 2.5:1 are recommended. The heating systems and regulating instruments must be designed for operation up to at least 350°C or up to 400°C (for high-heat copolycarbonates).

Melt temperatures vary from 280 to 320°C, depending on the grade. The mold temperature is usually 80°C.

Due to their different heat resistance, processing temperatures for high-heat copolycarbonates or polycarbonate blends need to be adapted correspondingly (Table 3).

For extrusion, high-viscous or structurally viscous branched polycarbonates are particularly suitable (Section 2.2). Hollow articles, films, sheets, profiled sections, pipes, and strands are examples of extruded products.

Extruder screws should have a length/diameter ratio of 25:1 (three-section screws) or 33:1 (extruders with degassing unit).

Table 3. Typical melt and mold temperatures for the injection molding of polycarbonates

	Melt temperature, °C	Mold temperature, °C
BPA-PC	280–320	80–120
PC-HT	330	100–140
PC–ABS blends	240–280	60–80
PC–polyester blends	240–280	60–80

In general, very high temperatures or long residence times of the polymer melt in the molds should be avoided as this may lead to deterioration due to a decrease in molecular mass.

Semifinished products such as sheets, films, and profiles can also be made from BPA-PC. Slow cooling in air or water is required to obtain low-stress products. Depending on size and output, the cooling temperature can vary from 40°C for thin profiles up to 130°C for massive sheets.

The temperature at the nozzle outlet is ca. 230–300°C, depending on melt viscosity and output volume.

Films and sheets of BPA-PC can be subjected to thermoforming. BPA-PC can also be polished to a high gloss, coated, printed, embossed, or vacuum-metallized and bonded with solvents. Two-pack adhesives (mainly polyurethane adhesives) are suitable for bonding BPA-PC parts to one another and to other materials. Parts made of BPA-PC may also be joined to each other by vibration, ultrasonic, friction-heated tool, and hot-gas welding. The polymer can also be machined, sawed, and drilled.

6. Applications

Electrical/Electronics Sector. The electrical/electronics sector is the classical and a major user of BPA-PC [4, 12, 59] because of the special combination of properties: outstanding electrical insulation, good flame retardance, high heat resistance, excellent toughness, dimensional stability, and transparency. Moreover, these properties are largely independent of temperature and atmospheric humidity.

The trend to minimize the size of electrical devices requires parts with higher heat resistance, because at the same time wall thickness must be reduced. This leads to an ongoing substitution of ABS or acrylate styrene acrylonitrile (ASA) materials by polycarbonate.

Due to higher demands regarding flame resistance, PC–ABS blends are used to a large extent, as the utilized flame retardants meet current directives such as the European Waste Electrical and Electronic Equipment (WEEE), the Restriction of the Use of certain Hazardous Substances (RoHS), and the IEC 60335-1 standard for domestic appliances.

Typical applications include housings for mobile devices such as cell phones, notebooks, and tablets, but also for flat-panel displays, TV sets, and other appliances. Other applications include smart meters, lighting, power supplies, and distributor equipment such as lamp sockets, multiple contact strips, relay parts, power plugs, safety switches, components for electronic calculators, fluorescent tube sockets, current meters, and winding supports.

A relatively new application for BPA-PC is injection-molded complex-shaped lenses and optics for light-emitting diodes (LEDs), which are energy-saving and long-lasting illuminants. Besides the requirement for high transparency, BPA-PC is used due its better thermal and dimensional stability compared to PMMA, as well as its lower water absorption. Special light-management solutions can also be implemented with PC films, such as light decoupling films for organic LEDs (OLEDs) and light-guiding lenticular films having a fine network of lenses that combine the light points of individual LEDs into a lighting band.

Automotive Sector. In automobiles, BPA-PC reduces weight, increases the efficiency and durability of assemblies, and improves the looks of a vehicle due to greater freedom of design compared to glass or metal. Important properties are the high impact strength, heat resistance, and transparency. BPA-PC is used, for example, in headlight covers and reflectors, as well as radiator grills (PC–PET).

In the last years, replacement of metal parts by metallized plastic is gaining more and more attention. This leads to a significantly higher quality ranking of plastics, providing them with a premium surface appearance. The different metallization processes being applied result mainly from the adhesion of the metal to the layer.

Of special importance in this field are PC–ABS blends, which enable electroplating processes through oxidation of the rubber component and subsequent activation of the surface.

BPA-PC and PC-HT can also be metallized by vacuum metallizing and sputter coating. These methods use evaporated metal, which is condensed on the cold surface of the plastic in a vacuum chamber, resulting in a dense film having a high reflectivity (see also → Plastics Processing, 3. Machining, Bonding, Surface Treatment).

Metallized surfaces require high-quality raw materials as well as sophisticated injection molding processes, since surface defects are easily detectable.

While decorative metallic coatings enable thermoplastic parts to function as economical, lightweight alternatives to metals, functional metallic coatings provide EMI shielding, circuit paths, or high reflective surfaces. Major applications are reflectors in automotive lighting systems, e.g., head lamp lenses, radiator grills, and others.

PC–ABS blends have become very important in dashboards, where good heat resistance, high toughness and tough fracture (splinter-free fracture) are required at low temperatures [10, 12, 60]. Blends of polycarbonate and PET or PBT and impact strength modifiers are used in premium front-end modules, where low-temperature impact strength (down to −40°C) and resistance to gasoline are necessary. The latter property is affected favorably by partially crystalline polyesters.

In particular the need to reduce fuel consumption and the corresponding vehicle emissions as well as the trend towards electromobility have given new impetus to lightweight construction with BPA-PC. Hereby, automobile glazing with transparent polycarbonate is particularly attractive, because it permits solutions that are up to 50% lighter than glass–an important fact when considering the size of panorama roofs or complete tailgate modules. Similarly, part of the high battery weight in electric vehicles can be compensated. In addition, the good thermal insulation of PC compared to glass panes reduces the energy required to heat the vehicle interior, as shown by recent developments. This conserves the battery, and extends the range of the electric vehicle. Concepts like electromobility also lead to new application possibilities for PC that are remote from the vehicle. One example is the housing of charging stations, for which a PC-ABS blend is already in series production. For automobile interiors, formable PC hardcoat films are gaining ground for the production of decorated high-gloss components.

Building and Construction. Wide application areas for solid and multiwall sheets of PC continue to be architecture, traffic, safety, and transport. The transparency, extreme toughness, and moderate price of polycarbonates make them suitable for window panes and roofing. Multiple-wall sheets of polycarbonate combine optimum impact strength, good light transmission, light weight, and thermal insulation with minimal material consumption (low wall thickness) [61]. Therefore, polycarbonate is a suitable material that meets the increasing demands for higher energy efficiency in buildings. Sheets made of BPA-PC have become firmly established in industrial buildings such as railway stations and sports stadiums, but are as well used as roofs for greenhouses, safety glazing in ice hockey arenas and as windows for telephone booths and gymnasiums.

Optical Information Storage Systems. Although continuously declining, a very large amount of PC is used for the production of optical data storage devices such as CD, DVD, and Blu-ray discs. These storage devices are also used as supplements for advertising and special offers in magazines with high print runs. Moreover, thanks to their robustness and reliability, they are still popular as media for safe storage of personal data [4, 59].

The use of BPA-PC for optical information storage systems is described in [10–12].

The main requirements fulfilled by polycarbonate are high transparency, low birefringence, good mechanical and dynamic stability, good heat distortion resistance, and excellent flow properties.

Miscellaneous Uses. Further important areas of use of polycarbonates include office equipment (casings, covers, keyboards) and extruded, coextruded or laminated films and sheets (credit cards, cover sheets, capacitor sheets), foams (electrical distributor boxes, housings), medical applications (blood oxygenators, hemodialysis separating membranes in dialyzers, sterilizable equipment), packaging (water tanks, food containers), domestic applications (electric razors, sink units), and photographic or optical applications (camera cases, lenses, optical fibers).

Another application area for PC that is growing in importance is the use in furniture. The transparency lends it an appealing optical weightlessness. Special grades have been developed in order to meet the requirements for flame resistance for use in public buildings, such as fire standards UNI 8456, 8457, 9174, and 9175 [59].

Economic Aspects In 2013, global PC consumption amounted to some 3.62×10^6 t. Global PC consumption for blends grew more than sevenfold between 1990 and 2013, from about 85 000 to 615 000 t/a, corresponding to an annual increase of almost 10%. This exceptional growth was due to automotive applications and the boom in the IT sector.

Estimates predict a worldwide mid-term growth of the PC market in the medium one-digit percentage range. Hereby, the main contribution will come from countries in the Asia-Pacific region (primarily China) and South and East Europe, but also from new applications. The amount of PC used in various sectors is shown in Table 4 [59].

Table 4. Polycarbonate application split in 2013 (source: Bayer MaterialScience estimation, various press announcements)

Application	Market share, %
Electrical, information technology & communications	33
Mobility (mainly automotive)	16
Optical discs	15
Construction (primarily panels)	15
Consumer products	15
Appliances	4
Medical	2

Table 5. Main polycarbonate producers and trade names (source: Bayer MaterialScience estimation, various press announcements)

Producers	Main trade names
Bayer MaterialScience	Makrolon®, Apec®, Bayblend®, Makroblend®
Sabic Innovative Plastics	Lexan®, Cycoloy®
Styron (formerly Dow)	Calibre®, Pulse®
LG Chem (formerly LG-Dow)	Lupoy®
Teijin	Panlite®, Multilon®
Mitsubishi*–Sam Yang	Iupilon®, Novarex®, Trirex®, Triloy®
Idemitsu	Tarflon®, Tariloy®
Chi Mei	Wonderlite®, Wonderloy®
Samsung Cheil	Infino®, (Staren)®, Hopelex®
Kazanorgsintez	Green Tower®
Unigel Plasticos (formerly Pdb)	Durolon®

* Mitsubishi Gas, Mitsubishi Chemical, Mitsubishi Engineering Plastics.

The largest PC consumer in 2013 was the Asia-Pacific region with about 60%. Some 23% were accounted for by countries in the EMEA region (Europe, Middle East, and Africa). The NAFTA area contributed 15% to overall consumption, and Latin America 2%.

The world's largest single PC market is China, which accounted for 36% of global consumption in 2013. Apart from Japan, responsible for 6% of worldwide PC sales, also the other Asian countries have significant PC markets (about 18% of sales).

The number of PC producers is not large, but is growing steadily. Bayer MaterialScience and Sabic are still the two largest producers with each about 27% of global capacities. Other major producers include Mitsubishi (12%), Teijin (10%), and Styron (6%).

The expansion of PC capacities in Asia is proceeding steadily; first Asian producers are starting to export PC to Europe and the USA. Table 5 lists main producers and trade names.

7. Recycling

Even at the end of life phase of a whole product life cycle (including production and use phase), the polycarbonate (PC) in a product is still a valuable resource which can be recovered. Recovery options for PC include mechanical recycling (regaining PC as secondary raw material), feedstock recycling (regaining building blocks for chemical synthesis), and recovery of the embedded energy as a substitute for virgin fossil fuels.

In the past, several recycling technologies have been developed for PC itself as well as for special applications in which PC is used. The majority of published information in this field concentrates on three main areas.

The first area deals with the recovery of PC from optical data storage applications like CDs. Processes and machines for the removal of coatings and metallization are mainly the subject of patents. Examples range from sand blasting [62], crushing and milling [63], to chemical (acidic, basic, oxidative) washing procedures [64]. Material produced in that way can be directly re-used, e.g., for blending with other polymers [65], or further processed to improve its quality, e.g., by an increase in molecular weight [66].

The second area focuses on chemical methods to recover monomers such as BPA or aromatic carbonates as feedstock. A number of methods for the degradation of PC exist, e.g., by hydrolysis, alcoholysis (including glycolysis), or aminolysis [67]. In this field, current research focuses on optimizing the efficiency in recovering valuable building blocks. Especially ways of degrading PC without the need for solvents are subject to current research. For example, degradation with supercritical/critical media or "mild" catalysis has been reported in this field [68–70].

The third area of research is thermal decomposition, i.e., pyrolysis, of PC. Heating waste PC under defined conditions can yield synthesis gas and liquid hydrocarbons for synthesis or fuel applications. Here, increasing the pyrolysis efficiency and limiting the number of byproducts are future challenges which could be solved by newly developed catalysts [67].

However, all recycling operations rely on a certain input amount and quality of waste material which typically depends on the waste management system that is established locally for different waste streams. The quality of PC parts can range from clean and separate (e.g., production waste, so-called post industrial waste) to polluted and comingled collections (e.g., post-consumer mixed-waste bins). The recycling of PC in post-consumer waste remains a challenge due to the partially low PC content and the high demands on sorting technology.

8. Toxicology and Environmental Aspects

Harmful effects of polycarbonates are not known if properly handled [71]. Polycarbonates have not proved toxic when included in rat feed at a concentration of 6% [72].

The toxicity of the gases produced by combustion of flame-resistant polycarbonates (according to DIN E 5346/37) is similar to or lower than that of wood, wool, or cotton [73]. For safety precautions, exhaust gas and wastewater control, and toxicology of phosgene, see → Phosgene.

Several states have regulated the use of polycarbonate for the manufacture of articles which are intended to come into contact with foodstuffs. In the USA, the FDA has published the requirements for polycarbonate resins in its 21 Code of Federal Regulations (CFR), § 177.1580. In the European Union (EU), the Commission Regulation (EU) No. 10/2011 of 14 January 2011 (and its amendments) specifies the requirements for "plastic materials and articles intended to come into contact with food".

The (German) Federal Institute for Risk Assessment (BfR) issues recommendations for the use of certain materials in food-contact applications, e.g., Recommendation XI for "polycarbonates and mixtures of polycarbonates with other polymers or copolymers".

References

1 G. Abts, *Kunststoff-Wissen für Einsteiger*, Hanser-Verlag, München 2014.
2 K. Weirauch, L. Bottenbruch, D. Freitag, C. Wulff: *50 Years of Makrolon*, Gupta-Verlag, Ratingen 2003.
3 D.G. Legrand, J.T. Bendler: *Handbook of Polycarbonate Science and Technology*, Marcel Dekker, New York/Basel 2000.
4 C. Bothur: *Polycarbonates*, Süddeutscher Verlag onpact, München 2008.
5 W.F. Christopher, D.W. Fox: *Polycarbonates*, Reinhold Publishing, New York 1962.
6 H. Schnell: *Chemistry and Physics of Polycarbonates*, Polymer Reviews, vol. 9, Interscience Publishers, New York 1964.
7 G. Peilstöcker, H. Krimm: "Die Polycarbonate" in R. Vieweg, L. Goerden (eds.): *Kunststoff-Handbuch*, vol. 8, Polyester, Carl Hanser Verlag, München 1973, pp. 1–245.
8 V. Serini, D. Freitag, H. Vernaleken, *Angew. Makromol. Chem.* **55** (1976) 175–189.
9 H. Vernaleken: "Polycarbonates" in F. Millich, C.E. Carreher, Jr. (eds.): *Interfacial Synthesis*, vol. II, Marcel Dekker, New York 1977, pp. 65–124.
10 D. Freitag, U. Grigo, P.R. Müller, W. Nouvertné: "Polycarbonates" in H.F. Mark et al. (eds.): *Encyclopedia of Polymer Science and Engineering*, 2nd ed., vol. 11, Wiley, New York 1988, pp. 648–718.
11 R. Pakull, U. Grigo, D. Freitag: "Polycarbonates" in R. Dolbey (ed.): *Rapra Review Reports*, Report 42, vol. 4, no. 6, Pergamon Press, Oxford 1991, pp. 1–30.
12 D. Freitag, G. Fengler, L. Morbitzer, *Angew. Chem. Int. Ed. Engl.* **30** (1991) 1598.
13 Bayer AG, DE-OS3334782, 1983 (W. Paul, W. Nouvertné, H. Steinberger, W. Grape).
14 D. Rathmann, *Kunststoffe* **77** (1987) 10.
15 D. Rathmann, P. Tacke: "Polyarylate" in L. Bottenbruch: *Kunststoff-Handbuch*, 2nd ed., vol. 3/3: Technische Thermoplaste, Carl Hanser Verlag, München 1994.
16 L. Morbitzer, U. Grigo, *Angew. Makromol. Chem.* **162** (1988) 87–107.
17 Bayer AG, DE-OS2842005, 1980 (D. Freitag, W. Nouvertné, K. Burkhardt, F. Kleiner).
18 Farbenfabriken Bayer AG, GB1079821, 1967 (H. Schnell et al.).
19 Farbenfabriken Bayer AG, GB1151304, 1969 (H. Schnell et al.).
20 Bayer AG, DE-OS2113347, 1972 (D. Freitag, U. Haberland, H. Krimm).
21 Bayer AG, DE-OS2500092, 1976 (K. Idel, D. Freitag, W. Nouvertné).
22 V. Mark, *Org. Coat. Plast. Chem.* **43** (1980) 71.
23 J.L. Webb, *Org. Coat. Plast. Chem.* **43** (1980) 79.
24 Bayer AG, US3775367, 1973 (W. Nouvertné).
25 L.S. Thomas, S.A. Ogoe, *Tech. Pap. Reg. Tech. Conf. Soc. Plast. Eng.* **1985**, 100.
26 K. Kircher, *Kunststoffe* **74** (1984) 12.
27 K. Kircher, *Reinf. Plast.* **25** (1981) 382, 420.
28 D. Freitag, U. Grigo, P.R. Müller, W. Nouvertné: "Polycarbonates" in H.F. Mark et al. (eds.): *Encyclopedia of Polymer*

Science and Engineering, 2nd ed., vol. 11, Wiley, New York 1988, pp. 648–718.
29. W. Nouvertné, H. Peters, H. Beicher, *Plastverarbeiter* **33** (1982) 1070.
30. W. Witt, *Kunststoffe* **73** (1983) 498.
31. W. Witt, *Gummi Fasern Kunstst.* **38** (1985) 155.
32. D. Freitag, G. Fengler, L. Morbitzer, *Angew. Chem. Int. Ed. Engl.* **30** (1991) 1598.
33. J. Huang, M. Wang, *Adv. Polym. Technol.* **9** (1989) 293.
34. J.M. Wefer, US4895921, 1985.
35. G. Lindenschmidt, R. Theysohn, *Eng. Plast.* **3** (1990) 1.
36. W.K. Chin, L.J. Hwang, ANTEC '87 (1987) 1379.
37. D.J. Stein et al., *Angew. Makromol. Chem.* **36** (1974) 89.
38. K.W. McLaughlin, *Polym. Eng. Sci.* **22** (1989) 1550.
39. H.F. Giles: *Modern Plastics Encyclopedia*, McGraw-Hill, New York 1987–1988.
40. R. K. Young, C. A. Weber, K. McLaughlin: *Novel Polycarbonate/ABS Blends for Instrument Panel Applications*, SAE Technical paper 860254, SAE International, Washington, D.C. 1986.
41. D. Quintens, G. Groeninckx, M. Quest, L. Aerts, *Polym. Eng. Sci.* **30** (1990) 1474; **22** (1990) 1484.
42. H. Takahashi et al., *J. Appl. Polym. Sci.* **36** (1988) 1821.
43. W. Chiang, D. Hwung, ANTEC '86, 492.
44. Specialty Polymeric Blends and Alloys, Skeist Incorporated, Whippany, N.J. 1991.
45. H. Kress, D. Folajtan, N. Lazear, ANTEC '88, 1834.
46. *Plastics Flammability Handbook*, 3rd ed. Carl Hanser Verlag, Munich 2004, pp. 168–170.
47. S.V. Levchik, E. Weil, *Polym. Int.* **54** (2005) 981–1102.
48. Bayer AG, US5672645, 1997 (T. Eckel, D. Wittmann, M. Öller, H. Alberts).
49. Bayer AG, US5849827, 1998 (M. Bödiger, T. Eckel, D. Wittmann, H. Alberts).
50. G.J. Pratt, M.J.A. Smith, *Polymer* **30** (1989) 1113.
51. W.N. Kim, C.M. Burns, *J. Appl. Polym. Sci.* **41** (1990) 1575.
52. D.C. Wahrmund, D.R. Paul, J.W. Barlow, *J. Appl. Polym. Sci.* **22** (1978) 2155.
53. M.E.J. Dekkers, S.Y. Hobbs, V.H. Watkins, *J. Mater. Sci.* **23** (1988) 1225.
54. S.Y. Hobbs, M.E.J. Dekkers, V.H. Watkins, *J. Mater Sci.* **23** (1988) 1219.
55. A.W. Birley, X.Y. Chen, *Br. Polym. J.* **16** (1984) 77.
56. D. Delimoy et al., *Polym. Eng. Sci.* **28** (1988) 104.
57. W.N. Kim, C.M. Burns, *J. Appl. Polym. Sci.* **41** (1990) 1409.
58. R.W. Avakian, R.B. Allen, *Polym. Eng. Sci.* **25** (1985) 462.
59. K. Horn, F. Schnieders, D. Pophusen, C. Yesildag: "Stable Growth: Polycarbonate (PC)", *Kunstst. int.* (2013) no. 10, 44–50.
60. L. Morbitzer, H.J. Kreß, C. Lindner, K.H. Ott, *Angew. Makromol. Chem.* **132** (1985) 19.
61. K. Kircher, *Kunststoffe* **80** (1990) 1113.
62. H. Aoki, Y. Miyasaka, I. Sato, JP06091652, JP3153882 (1994).
63. A. Defazio, US000005203067 (1993).
64. Y. Gao, CN000102532589 (2012).
65. F. Elmaghor, L. Zhang, R. Fan, R. Run, H. Li, *Polymer* **45** (2004) 6719–6724.
66. Bayer AG, WO2003020805, 2003 (H. Haehnsen, U. Hucks, S. Kratschmer, M. Schraut).
67. E.V. Antonakou, D.S. Achilias, *Waste Biomass Valorization* **4** (2013) 9–21.
68. G. Grause, N. Tsukada, W.J. Hall, T. Kameda, P.T. Williams, T. Yoshioka, *Polym. J.(Tokyo, Jpn.)* **42** (2010) 438–442.
69. V. Margon, U.S. Agarwal, C. Bailly, G. de Wit, J.M.N. van Kasteren, P.J. Lemstra, *J. Supercrit. Fluids* **38** (2006) 44–50.
70. R. Pinero, J. Garcia, M.J. Cocero, *Green Chem.* **7** (2005) 380–387.
71. G. Bornmann, A. Löser, *Arzneim. Forsch.* **9** (1959) 9–13.
72. R.R. Montgomery in *Patty's Industrial Hygiene and Toxicology*, 3rd ed., vol. II c, Wiley-Interscience, New York 1982, p. 4441.
73. G. Kimmerle, F.H. Prager, *J. Combust. Toxicol.* **7** (1980) 42.

Polyester Resins, Unsaturated

HORST KRÄMER, Hüls Aktiengesellschaft, Marl, Federal Republic of Germany

1. Introduction 781
2. Raw Materials.................. 781
3. Production 783
4. Additives...................... 784
5. Curing 785
6. Types of UP Resins 785
7. Testing and Properties........... 786
8. Uses and Economic Aspects 788
9. Occupational Health and Environmental Aspects 789
References..................... 790

1. Introduction

Unsaturated polyester resins (abbreviated as UP resins according to DIN 7728, part 1) undergo cross-linkage to form thermosetting plastics (DIN 16 945). Most UP resins are medium- to low-viscosity, liquid solutions of unsaturated polyesters in copolymerizable monomers (usually styrene) as cross-linking agents.

Unsaturated polyesters are formed by polycondensation of unsaturated and saturated dicarboxylic acids with diols. UP resins are hardened (cured) or cross-linked by free-radical polymerization with initiators (e.g., peroxides) and accelerators at normal temperature (10 –40 °C), or with peroxides alone above 80 °C. The unsaturated groups of the polyester chains react with the double bonds of the copolymerizable solvent monomer. Solid products are formed that are used as molding materials or in paints. They are also reinforced with fibers or filled, and used as putties or mortars.

Vinyl ester resins, which are also termed phenacrylate resins (DIN 16 946, part 2), are polyadducts of liquid epoxy resins and acrylic acid in styrene. Strictly speaking, they are not unsaturated polyester resins, but since they are processed, cured, and used similarly, they are also discussed.

2. Raw Materials

The properties of unsaturated polyesters depend on the nature and quantitative ratios of the raw materials (α-, β-unsaturated dicarboxylic acids, saturated dicarboxylic acids, and diols), as well as on the condensation conditions. Many different materials are quoted in the patent literature and in reference manuals [1–5]. However, only those described here are of practical importance. This limited number of raw materials has hardly changed since the early 1970s and permits wide variation as regards the processing and properties of UP resins.

A larger number of double bonds in the polyester chain increases the reactivity, the heat of reaction released during curing, and the thermal stability of the end product. The incorporation of long-chain dicarboxylic acids or diols increases the flexibility of the cured substance.

Unsaturated Dicarboxylic Acids. Maleic anhydride and fumaric acid are used almost exclusively as unsaturated acid components. They are described in detail elsewhere, see → Maleic and Fumaric Acids.

Maleic anhydride [108-31-6] is not as expensive as fumaric acid and contains one mole of water less than the free dicarboxylic acid, which is favorable for condensation. During condensation maleic acid largely rearranges into the isomeric form, fumaric acid. This rearrangement can also be promoted by catalysts.

Fumaric acid [110-17-8] is only used when the degree of rearrangement of maleic acid to fumaric acid is insufficient for the required properties of the UP resins. Use of fumaric acid instead of maleic anhydride in the condensation improves the thermal stability and mechanical properties of the product. The disadvantages of fumaric acid compared with maleic anhydride are its higher cost and higher water content.

Saturated Dicarboxylic Acids. *Phthalic acid [88-99-3]* is used exclusively in the form of phthalic anhydride [85-44-9] (→ Phthalic Acid and Derivatives). On account of its favorable price and diverse potential uses, it is employed in all standard resins.

Isophthalic Acid [121-91-5] (→ Terephthalic Acid, Dimethyl Terephthalate, and Isophthalic Acid) has a better thermal stability, mechanical strength, and resistance to chemicals than phthalic acid.

Terephthalic Acid [100-21-0] (→ Terephthalic Acid, Dimethyl Terephthalate, and Isophthalic Acid) is used in resins where high impact strength, thermal stability, and weather resistance are required.

Tetrahydrophthalic Anhydride [935-79-5] yields resins having a high impact strength but limited weather resistance. It is used primarily for resins employed in paints and automotive body fillers.

Adipic Acid [124-04-9] (→ Adipic Acid) is used on account of its flexible carbon chain to produce soft resins and mixed resins that are employed to plasticize harder resins.

Hexachloroendomethylenetetrahydrophthalic (HET) acid [115-28-6] yields resins with low flammability and good chemical resistance.

Tetrabromophthalic acid [13810-83-8] also lowers flammability, but tends to produce strong discoloration in sunlight due to the release of bromine.

Diols. *Ethylene glycol* [107-21-1] (→ Ethylene Glycol) is inexpensive, but it sharply reduces the styrene compatibility of UP resins. It is generally used in combination with other diols such as diethylene glycol, dipropylene glycol, or butylene glycol.

1,2-Propanediol [57-55-6], propylene glycol, is by far the most widely used diol for standard resins (→ Propanediols). It has good styrene compatibility, only slight tackiness after curing in air, and good mechanical properties.

Diethylene Glycol [111-46-6] and *dipropylene glycol* [110-98-5] result in plasticization of the resins and increased water absorption. They are used mainly in combination with 1,2-propanediol to adjust the mechanical properties or refractive index of the cured resin.

1,3-Butanediol [107-88-0], butylene glycol, (→ Butanediols, Butenediol, and Butynediol) improves weather resistance, but is not used as often as the aforementioned diols on account of its high price and limited availability.

Neopentyl Glycol [126-30-7], 2,2-dimethyl-1,3-propanediol (→ Alcohols, Polyhydric, Section 2.3.1.) increases the thermal stability, chemical resistance, and weather resistance of UP resins. Particularly when combined with isophthalic acid, it yields high-quality products for use in chemical plant construction.

$$HO-CH_2-\underset{\underset{CH_3}{|}}{\overset{\overset{CH_3}{|}}{C}}-CH_2-OH$$

Ethoxylated or Propoxylated Bisphenol A [80-05-7] is very often used for resins with a high thermal stability and extremely good resistance to chemicals. With resins formulated from these diols it should be borne in mind that traces of phenolic groups can interfere with curing.

Bisphenol A

Although *bis(hydroxymethyl)tricyclodecane* [26896-48-0] is expensive, it imparts high thermal stability and chemical resistance, with tack-free hardening of uncovered surfaces.

HO–CH$_2$–[tricyclodecane]–CH$_2$OH

Dibromoneopentyl Glycol [3296-90-0] is used to reduce combustibility. A disadvantage is that hydrogen bromide readily splits off during polycondensation and attacks the walls of the reaction vessel. Dibromoneopentyl glycol is therefore seldom used.

N,N-Bis(2-hydroxyethyl)-p-toluidine [3077-12-1] is sometimes incorporated into the polyester chain by condensation to obtain resins that cure at normal temperatures with benzoyl peroxide but without an accelerator.

HOH$_4$C$_2$ C$_2$H$_4$OH
\\N/
[phenyl ring]
CH$_3$

Monomers for Cross-Linking UP Resins. Vinyl, allyl, and acrylic compounds are suitable monomers for the cross-linking of UP resins. They are readily miscible with the resins and their homopolymerization rate does not exceed the copolymerization rate with the unsaturated polyester chain. In most cases low- to medium-viscosity UP resins are required. *Styrene [100-42-5]* is by far the most suitable monomer for these products (→ Styrene). The styrene content is normally between 25 and 45 wt %.

The following vinyl compounds are often recommended as alternatives, but are of little practical importance:

α-Methylstyrene [98-83-9] is insufficiently reactive. *4-Methylstyrene [622-97-9]* would be the most suitable alternative to styrene for technical reasons [6], but has an unpleasant odor and is not produced on an industrial scale.

Vinyltoluene, an isomeric mixture mainly of 4- and 3-methylstyrenes, is not very suitable on account of its low reactivity and unpleasant odor.

Divinylbenzene [1321-74-0] is not available in large amounts.

Methyl Methacrylate [80-62-6] copolymerizes with difficulty with unsaturated polyesters. It is used in combination with styrene, however, to improve weather resistance and to lower the refractive index. The ratio of the refractive index of the resin to that of the glass fiber reinforcement determines the transparency of finished parts such as domelights and light panels.

Diallyl Phthalate [131-17-9] is used as a cross-linking agent in the production of pourable, solid UP resin molding compositions. The viscosity of diallyl phthalate is too high for liquid UP resins and copolymerization is too slow for curing at room temperature.

Triallyl Cyanurate [101-37-1] is also a conventional cross-linking agent for solid molding compositions.

3. Production

UP resins are usually produced by batchwise condensation of dicarboxylic acids or anhydrides with diols. Continuous esterification of acid anhydrides and epoxides may, however, also be employed.

Batchwise Condensation. Dicarboxylic acids are esterified with diols at 180 – 230 °C in a nitrogen atmosphere in heatable, coolable reactors while stirring. Esterification is generally performed with a slight excess (up to 10 %) of diols. The water formed in the reaction is distilled off. Passage of nitrogen through the reaction mixture promotes removal of water and increases the esterification rate considerably. The end point of the condensation is determined by measuring the acid number and, possibly, also the hydroxyl number and the viscosity. Commercial UP resins have a mean molecular mass of 2000 – 4000.

The still liquid melt is cooled and then dissolved, while stirring, in styrene in coolable mixers. The styrene contains inhibitors such as hydroquinone or *tert*-butylcatechol (see Chap. 4).

Even if the same raw materials are used, the resin products may differ as regards their processing properties and end properties depending on the amount of excess glycol, the temperature, and reaction conditions.

In the case of raw materials with widely differing esterification rates, two-stage condensation must be used to obtain uniform buildup of the polyester chain and thus favorable end properties. For example, if isophthalic acid and maleic acid are added simultaneously to the reaction vessel, maleic acid is incorporated more quickly into the polyester chain. In the case of isophthalic acid resins, isophthalic acid should therefore first be condensed with glycol in a molar ratio of 1 : 2 to form a bishydroxy ester. Esterification should be continued after adding maleic anhydride.

Azeotropic esterification is important only for certain paint resins. An entraining agent (e.g., toluene or xylene) is added and during condensation the water is distilled off together with the toluene. The toluene is then separated from the water with a dephlegmator and returned to the reactor. At the end of the reaction the toluene must be removed from the resin (e.g., with a thin-film evaporator).

Continuous Esterification of acid anhydrides and epoxides is a very elegant, economic method. However, it is limited to the small number of raw materials that are available as anhydrides or epoxides. It cannot, therefore, meet the demands of the polyester processors for different resin types. Thus, for example, the extremely important isophthalic and terephthalic acid resins cannot be produced by this method because the dicarboxylic acids do not form anhydrides. The method is recommended only for producers with very large sales of standard resins.

The raw materials (e.g., maleic anhydride, phthalic anhydride, and propylene oxide) are converted into polyesters by reacting them for ca. 20 min and 200 °C, generally in the presence of catalysts. In contrast to batchwise condensation, no water of reaction is formed.

4. Additives

See also → Plastics, Additives

Inhibitors are used to prevent premature polymerization and to modify the reaction times. The most important are hydroquinone, 1,4-benzoquinone, *tert*-butylpyrocatechol (*tert*-butylcatechol), toluhydroquinone (methylhydroquinone), and 3,4-di-*tert*-butyl-*p*-cresol (3,4-di-*tert*-butyl-4-hydroxytoluene). All these inhibitors have specific effects and are used for different resin types. In some cases they are added prior to condensation, in other cases they are dissolved in styrene and added after condensation.

Optical Brighteners are added to slightly yellowish resins for certain uses (e.g., transparent products or white buttons). Blue dyes, which become decolorized during curing, are commonly used as curing indicators.

Commercially available *light stabilizers* (e.g., substituted benzophenones and hydroxyphenyltriazoles) prevent premature yellowing of the hardened resins in sunlight.

Thixotropic Agents have to be added to prevent sagging or dripping of the resins when applied to vertical surfaces. Finely dispersed silica is mainly used for this.

Most UP resins have a tacky surface when cured in air. Tackiness can be prevented by adding *skin-forming agents*, generally paraffins. The formation of a paraffin film also reduces styrene emission in pressureless processing. Resins containing such additives are termed "environmentally friendly" or "milieu polyesters".

The additives listed above are generally added to the resin by the resin manufacturer. The following substances are added to the resin by the processor.

Hardened UP resins have a surface resistance of $>10^{13}$ Ω. In order to prevent buildup of electrostatic charge, graphite and metal powders can be added as *antistatic agents* to reduce the surface resistance to ca. 10^4 Ω.

Flame retardants include chlorinated paraffins and antimony trioxide. Aluminum hydroxide (hydrated alumina) is employed on a large scale for parts used in the electrical industry that have to be halogen-free (halogens can attack copper contacts and hydrogen chloride may be released during fires).

Both soluble *dyes* and insoluble *pigments* can be used for coloring. Pigments have to be made into a paste in resins or plasticizers before being added.

Fillers include chalk, quartz flour, and quartz sand. The type and amount of filler depend on the final use and desired properties of the resin. Filled resins serve as putties, compression molding compounds, casting compounds, mortars, and concrete. The fillers must be dry so that they do not to affect curing.

Blowing Agents are used to produce polyester foams, various additives and systems are used for this (e.g., azodicarbonamide, *tert*-butylhydrazinium chloride and iron(III) chloride).

Thickening Agents are required for the production of sheet molding compounds (SMC), and bulk molding compounds (BMC). Magnesium oxide and, in some cases, magnesium and calcium hydroxides are normally used. These agents are mixed with fillers and used to impregnate glass fibers, thereby converting the molding compound into a highly viscous state.

Low-Shrink and Low-Profile Additives are solutions of thermoplastics in styrene and are added to reduce volume shrinkage occurring during curing. Polystyrene, poly(vinyl acetate), poly(methyl methacrylate), saturated polyesters, cellulose acetobutyrate, and polycaprolactams are usually used. Polyethylene with a particularly low melting point is used to improve the properties of sheet molding compounds. On account of its insolubility it has to be added to the resin batch as a powder immediately before processing.

5. Curing

At normal temperatures UP resins can be cured with peroxides and accelerators. The most important systems include hydroperoxides (e.g., methyl ethyl ketone peroxide, cyclohexanone peroxide, and acetylacetone peroxide) with cobalt octanoate or, in exceptional cases, also vanadium octanoates or naphthenates, as well as benzoyl peroxide (dibenzoyl peroxide) with aromatic amines (e.g., *N,N*-dimethylaniline, *N,N*-diethylaniline, and *N,N*-dimethyl-*p*-toluidine).

Moisture slows down or prevents curing with hydroperoxides and cobalt salts, but does not affect reactions with benzoyl peroxide and amines. Complete curing generally requires heating above the glass transition temperature of the cross-linked polyester. Only completely cured products exhibit optimum mechanical properties and resistance to weathering and chemicals.

At elevated temperatures only peroxides are required for curing, accelerators are unnecessary. Examples include dibenzoyl peroxide above ca. 80 °C, *tert*-butyl 2-ethylperhexanoate above ca. 100 °C, and *tert*-butyl perbenzoate above ca. 120 °C.

UP resins can also be completely cured in a few minutes by exposing them to UV light from high-pressure mercury vapor lamps in the presence of photosensitizers, e.g., benzoin ether [24650-42-8] (ω,ω-dimethoxy-ω-phenylacetophenone).

Some acyl phosphine oxide compounds [7] can be used in combination with the aforementioned UV sensitizers to allow curing in normal daylight.

Detailed information about optimum curing systems and reaction conditions for specific uses are given in the data sheets published by resin and reactant manufacturers.

During the transition from the liquid to the solid state, UP resins shrink by 5 – 8 % depending on their double bond content. Compensation of shrinkage by thermoplastics additives is discussed in Chapter 4.

6. Types of UP Resins

Many types of UP resins are available. Selection depends on processing properties and end properties. Processing properties include reactivity, viscosity, and thickening behavior with magnesium oxide. End properties include mechanical behavior, thermal stability, flammability, resistance to chemicals, and color.

Resins may be liquid (low-, medium- or high-viscosity) or solid. Their reactivity may be low, average, or high. A wide range of UP resins can be produced by appropriate choice of the following factors: type and quantity of raw materials, degree of condensation, acid and hydroxyl numbers (and thus the nature of the terminal groups), and the content of cross-linking monomers and other additives. Consequently, comparison or exchange of related resin types from different producers may be difficult.

Standard Resins are formulated from inexpensive raw materials and have a wide variety of uses. They can be supplied with or without additives (light stabilizers, thixotropic agents, accelerators, or UV sensitizers).

Elastic Grades have a high flexibility and elongation at break due to the incorporation of long-chain monomers, and are generally used in admixture with more brittle types. Some highly elastic grades are used for jointless roof coverings on account of their weather resistance and high elongation at break of up to 200 % in tensile tests.

Flame-Resistant Resins generally contain chlorine or bromine in the polyester chain or in added compounds. Flammability can be adjusted to satisfy widely differing regulations and standards by using other additives such as phosphorus compounds, antimony compounds, and aluminum hydroxide.

Resins with a high chemical resistance are mainly synthesized from special raw materials such as isophthalic acid, terephthalic acid, neopentyl glycol, ethoxylated or propoxylated bisphenol A or bis(hydroxymethyl)tricyclodecane. They are used particularly in chemical apparatus (pipes and vessels).

Resins for *sheet molding compounds (SMC) and bulk molding compounds (BMC)* have specially adjusted ratios of acid and hydroxyl numbers and a constant water content. This ensures controlled thickening with magnesium oxide (MgO). Low-shrink and low-profile additives may also be added to the individual types, compatibility depends on the degree of unsaturation [8].

Resins for *fillers and putties* are preaccelerated with aromatic amines. This ensures accurate adjustment of curing with benzoyl peroxide and allows sanding within a short period of time. Rigid and plasticized resins are available for repair putties and light-colored products for marble putties.

Resins for free-flowing *compression molding compounds* are generally supplied monomer-free in powder form to the molding compound producers, who formulate the compounds with cross-linking agents (e.g., diallyl phthalate or triallyl cyanurate) and fillers.

Trade Names. Important trade names of UP resins in Europe and the United States include:

Alpolit	Hoechst
Altek	Alpha Corp.
Aropol	Ashland Chemical Corp.
Beetle	BIP
Civic	Neste Oy
Crystic	Scott Bader
Hetron	Ashland Chemical Corp.
Lamellon	DSM
Leguval	Bayer and DSM
Marco	US-Steel Corp.
Norpol	AS Jotun-Group
Norsodyne	Cray Valley
Palatal	BASF
Polylite	Reichhold Chemicals
Roskydal	Bayer
Selectron	PPG Industries
Sniatron	Alusuisse
Sirester	SIR
Torolithe	Routtand S. A.
Trusurf	OCF Fiberglas Corp.
Uralam	Synthetic Resins
Vestopal	Hüls
Viapal	Vianova Kunstharz

Due to company takeovers or mergers and sales of know-how and trade names, assignment of trade names to manufacturers may change.

7. Testing and Properties

A distinction should be made between the properties of UP resins in the as-supplied state (Table 1) and properties after curing (Table 2).

In order to determine the properties in the as-supplied state, simple, rapid tests are desired by producers and customers for quality assurance. Testing is generally carried out by agreement

Table 1. Properties of styrene-containing UP resins in the as-supplied state

Property	Test method	Test value
Acid number	DIN 53 402	10 – 30 mg KOH/g
Styrene content	DIN 16 945	25 – 45 %
Viscosity at 25 °C	DIN 53 015	0.2 – 4.5 Pa · s
Density at 20 °C	DIN 51 757	1.07 – 1.3 g/cm^3
Flash point	DIN 51 755	35 °C
Refractive index at 25 °C	DIN 53 491	1.524 – 1.559

Table 2. Properties of cured UP resins without fillers or reinforcement fibers (without soft resins)

Property	Test method	Test value
Density at 23 °C	ISO/R 1183	1.20 – 1.40 g/cm^3
Flexural strength	ISO 178	80 – 160 N/mm^2
Outer fiber strain	ISO 178	2 – 7 %
Impact strength	ISO 179 small test bar	6 – 15 kJ/m^2
Tensile strength	ISO/R 527 test specimen 1	40 – 90 N/mm^2
Elongation at break	ISO/R 527 test specimen 1	1.5 – 5 %
Compressive strength	ISO 664	150 – 180 N/mm^2
Ball indentation hardness 10"	ISO 2039	200 – 245 N/mm^2
Barcol hardness (934–1)		35 – 48
Modulus of elasticity (bending)	DIN 53 457 Section 2.3.	3400 – 4300 N/mm^2
Mean specific heat		1.10 – 1.45 kJ kg^{-1} K^{-1}
Thermal stability (heat distortion)	ISO 75 method A	60 – 145 °C
Thermal conductivity	DIN 52 612	0.16 – 0.18 W m^{-1} K^{-1}
Coefficient of linear expansion	VDE 0304 Part 1 § 4	60 – 115 10^{-6} K^{-1}
Volume resistivity	IEC Publ. 93	>10^{15} Ω · cm
Surface resistivity	IEC Publ. 167	>10^{13} Ω
Dielectric dissipation factor		
at 800 Hz	IEC Publ. 250	0.003 – 0.005
at 1 MHz		0.015 – 0.035
Relative permittivity at 800 Hz	IEC Publ. 250	3.0 – 3.5
Comparative tracking indices	IEC Publ. 112	CTI>600 CTI>600M
Dielectric strength	IEC Publ. 245	35 – 45 kV/mm

between suppliers and customers according to national standards, but also according to in-house procedures. The resins are often dispatched with test certificates describing the contents of the delivered batch.

Important properties used to characterize the liquid resins in the as-supplied state include the acid number, solids content, viscosity, refractive index, and reactivity. The acid number, solids content, and viscosity can be used to evaluate the degree of condensation and mean molecular mass. Further important properties of liquid resins are color, flash point, and density.

The acid number is measured by titration with potassium hydroxide solution. The solids content can be determined with automatic thermobalances by evaporating the volatile constituents (basically styrene). Viscosity is checked with rotating-cylinder viscometers, but also by measuring efflux times from temperature-controlled flow cups. The viscosity can be lowered by dilution with styrene or other monomers. The refractive index is measured in refractometers, and is used to monitor the correct composition of the resin.

The reactivity of UP resins is generally determined by measuring the rise in temperature that occurs during curing. This method is more sensitive than differential scanning calorimetry. The initial temperatures chosen often depend on the subsequent processing temperatures (e.g., room temperature or the temperature of hot compression molds). Measurements differ widely as regards heating bath, vessels, thermocouples, amount of resin, and nature and amount of the reactants. Accurate comparisons of product data sheets published by various manufacturers cannot therefore be made. Examples of measurement setups are given in DIN 16 945.

The thickening of resins for SMC and BMC with magnesium oxide is an important property but is difficult to measure quickly. An ISO standard is currently being prepared which involves accelerated thickening at elevated temperature, and allows conclusions to be made about thickening behavior at room temperature. This test is satisfactory for resin processors, but resins with the same results may behave differently when tested under normal thickening conditions at 25 °C.

In most cases cured UP resins are relatively brittle, addition of fillers and reinforcing materials (e.g., glass fibers) is required to obtain technically useful properties. Nevertheless, cured UP resins must be tested in the pure state. Production of test specimens requires care and experience because factors such as internal stress arising during curing can falsify the results. Mechanical, thermal, and electrical properties are measured (Table 2). The tensile strength, elongation at break, and thermal stability are particularly important for comparing the performance of different resins.

8. Uses and Economic Aspects

UP resins are only used without fillers or reinforcing agents in small amounts for embedding decorative objects, in furniture lacquers, adhesives, paints (→ Paints and Coatings, 1. Introduction) and casting resins (impregnating resins) for insulating electrical equipment (→ Insulation, Electric).

The majority of UP resins are reinforced with glass fibers which impart high strength and rigidity to molded parts. Natural fibers (e.g., jute) and synthetic fibers are also used in elastic molded parts.

Casting compounds, putties (→ Paints and Coatings, 1. Introduction), and UP resin concrete are produced using fillers with different particle sizes and compositions.

The most important uses of UP resins follow:

Automobile Construction. Parts for special-order vehicles, caravans, and agricultural vehicles are produced by hand from glass-fiber-reinforced UP resins. Flat sheets and panels for truck superstructures and containers are produced industrially in continuous laminating processes. Body parts for automobiles and commercial vehicles are produced from preimpregnated SMC.

On account of their outstanding resistance and impact strength behavior, pressed parts made from UP resins reinforced with natural or synthetic fibers are employed in car interior fittings, engine compartments, and vehicle underbodies; they have gained a considerable market share. Filled UP resins are used as automobile repair putties.

Building Industry. The use of light panels made of glass-fiber-reinforced UP resins has greatly declined. UP resins with glass fiber reinforcement are used for domelight crowns, facade elements, cellar window frames, and light shafts made from SMC. Beacons, chimneys, and large aerials are wound from UP resins and glass fibers. Highly elastic casting compounds are used for flat roof coverings. Drainage systems with channels, catch basins, and gasoline separators are produced industrially from polyester concrete. Imitation marble and onyx can be produced from UP resins, colorants, and chalk (marble) or aluminum oxyhydrate (onyx); they are used to produce inexpensive, high-quality sanitary ware.

Electrical Industry. The electrical industry is still a main customer for molded parts made from SMC and BMC. Examples include cable distribution cupboards and switching cupboards, skirting boards for lamps, multiway connectors, switch covers, and clamping sleeves. Coils and windings are impregnated with UP resins.

Pipes, Vessels, and Apparatus. Cylindrical molded components reinforced with glass fibers for the chemical industry and oil transportation are fabricated in winding and centrifugal processes from chemically resistant resins whose chemical resistance is appropriate for the intended use. Glass-fiber-reinforced UP resins are also used for fertilizer and fodder silos and cooling towers.

Boat and Shipbuilding. Glass-fiber-reinforced UP resins have long been used for building sports, life-saving, and fishing boats. Larger vessels (e.g., 50-m long minesweepers with 30-mm thick sides) are now also being built from this material.

Miscellaneous Uses. Other molded parts made from glass-fiber-reinforced UP resins include machinery covers and casings, sports equipment, toys, garden furniture, and traffic signs. Important areas of use of nonreinforced UP resins include buttons, buckles, fashion accessories, and furniture finishings.

Economic Aspects. In 1988 ca. 1.3×10^6 t of UP resins were consumed worldwide.

Following a fall in consumption as a result of the oil crises at the end of the 1970s and subsequent stagnation, polyester resins have recently experienced an increase in consumption. This is

Table 3. International distribution of the areas of use for UP resins (10^3 t)

	United States	Western Europe	Japan
Building and construction	241	125	74
Automobile construction	50	78	9
Pipes and vessels	62	61	67
Boats and shipbuilding	155	55	23
Electrical industry	13	28	9
Paints and other nonreinforced products	68	30	30
Miscellaneous uses	31	73	18
Total consumption	620	450	230
of which			
BMC/SMC *	64	58	58
Polymer concrete (in building and construction)	62	36	16
Production capacity	800	540	300

* BMC=bulk molding compounds; SMC=sheet molding compounds.

primarily due to increased use in the automobile industry, especially in compression molding and resin transfer molding.

Consumption according to areas of use is listed in Table 3. The apparent excess capacities in this table are somewhat misleading because they include alkyd resin capacities for the paint sector. The capacity of the UP resin manufacturers is fully utilized and an expansion in capacity can be expected in the next few years.

9. Occupational Health and Environmental Aspects [9]

Toxicological and environmental aspects of unsaturated polyester resins are principally determined by the properties of styrene ($\rightarrow$ Styrene). Styrene is a liquid with a relatively high vapor pressure which means that there may be considerable emissions, particularly when polyesters are processed in open molds.

Irritation of the eyes and mucous membranes of the upper respiratory tract is observed and depends on the exposure period and concentration of styrene in the inspired air. Tiredness, loss of concentration, and similar symptoms have been reported at styrene exposure levels above 100 ppm at the workplace.

Styrene and styrene-containing UP resins were classified in 1988 by the EC as "slightly hazardous" with the danger symbol x_n, R-phrase 10 inflammatory, R-phrase 20 health hazard if inhaled, and R-phrase 36/38 eye and skin irritant. A grading with the R-phrase 45 "Can cause cancer" was specifically excluded.

In 1989 in the United States, styrene was classed by OSHA as not carcinogenic, and was included in the category "Avoidance of narcosis". The TLV-TWA value was reduced from 100 ppm to 50 ppm.

In Germany the MAK value was reduced from 100 ppm to 20 ppm in 1987 on account of reported neurotoxic effects. It was announced that the carcinogenic effects of styrene would be reexamined.

Maximum workplace concentrations for various countries follow (ppm):

Austria	200
Belgium	50
Commonwealth of Independent States	1.2
Czechoslovakia	47
Denmark	25
Finland	20
France	50
Germany	20
Italy	50
Japan	50
Netherlands	100
Norway	50
Poland	12
Sweden	20
Switzerland	50
United Kingdom	100
United States	50

In many countries these concentrations will probably be lowered in the near future.

The observance of low maximum workplace concentrations does not present any particular problem in the production of styrene, polystyrene, and styrene copolymers nor in the processing of polystyrene and styrene copolymers. In UP resin processing, however, it is often almost impossible to satisfy this requirement.

Styrene emissions can be considerably reduced by ventilation devices at the workplace, by using "environmentally friendly resins" that contain a skin-forming agent such as paraffins (Chap. 4), or by completely enclosing automatically operating plants. Statutory regulations (e.g., TA-Luft in Germany) impose limits on the discharge of evaporated styrene and waste-gas treatment plants are necessary if fairly high levels of emissions are reached (e.g., >100 mg/m^3 in TA-Luft).

Prolonged exposure of the skin to liquid UP resins often leads to irritation, and particular care must be taken when working with the peroxides used as curing agents on account of their corrosive effect. Goggles should always be worn to protect the eyes when handling peroxides.

UP resins are flammable and hazard classifications differ in individual countries. The content of volatile monomers (styrene), the flash point, and the viscosity of the resins are usually decisive factors. Details are given in the manufacturers' safety data sheets.

Liquid UP resins containing styrene as crosslinking agent and must not pass into the ground water. Relevant statutory regulations governing the storage and transportation of UP resins must be strictly observed. Storage tanks for UP resins should be installed in catchment basins to trap any leakages or escaping product.

References

1 P. H. Selden: *Glasfaserverstärkte Kunststoffe*, Springer Verlag, Berlin 1967, pp. 8 ff.
2 G. W. Becker, D. Braun (eds.): Duroplaste,*Kunststoff-Handbuch*, vol. 10, Hanser Verlag, München 1988, pp. 94 ff.
3 H. V. Boenig: *Unsaturated Polyesters, Structure and Properties*.Elsevier, Amsterdam 1964.
4 P. F. Bruins: *Unsaturated Polyester Technology*, Gordon and Breach Science Publishers, New York 1976.
5 G. Pritchard: *Developments in Reinforced Plastics-1, Resin Matrix Aspects*, Applied Science Publishers, London 1980, pp. 59 ff.
6 Mobil Oil Corp., EP 0 003 406, 1979 (J. G. Murray).
7 BASF, EP 0 073 413, 1982 (A. Henne *et al.*).
8 Rohm and Haas, US 3 772 241, 1966 (C. H. Kroekel).
9 *AVK-Handbuch*, Arbeitsgemeinschaft Verstärkte Kunststoffe, Frankfurt 1988, chap. 6.

Further Reading

B. L. Deopura, R. Alagirusamy, M. Joshi, B. Gupta (eds.): *Polyesters and polyamides (Woodhead Publishing in Textiles)*, CRC Press, USA 2008.

Polyesters

HORST KÖPNICK, Bayer AG, Dormagen, Federal Republic of Germany

MANFRED SCHMIDT, Bayer AG, Dormagen, Federal Republic of Germany

WILHELM BRÜGGING, Hüls Aktiengesellschaft, Marl, Federal Republic of Germany

JÖRN RÜTER, Hüls Aktiengesellschaft, Marl, Federal Republic of Germany

WALTER KAMINSKY, Universität Hamburg, Institut für technische und makromolekulare Chemie, Hamburg, Federal Republic of Germany

1. Introduction 791
2. Polyesters as Intermediates for Polyurethanes 792
2.1. Raw Materials 792
2.2. Production 792
2.2.1. General Information 792
2.2.2. Industrial Production 793
2.3. Quality Specifications and Testing 794
2.4. Properties 795
2.5. Storage and Transportation 795
2.6. Uses and Economic Aspects 796
3. Thermoplastic Polyesters 796
3.1. Raw Materials 797
3.2. Production 798
3.2.1. General Principles 798
3.2.2. Production of Poly(Butylene Terephthalate) 799
3.2.3. Production of Other Partially Aromatic Polyesters 801
3.2.4. Production of Fully Aromatic, Liquid Crystalline Polyesters 802
3.2.5. Production of Other Partially Aromatic Polyesters 802
3.2.6. Production of Fully Aromatic, Liquid Crystalline Polyesters 802
3.3. Analysis 804
3.4. Co-Components and Additives 805
3.5. Properties 806
3.6. Uses 809
3.7. Toxicology 811
3.8. Economic Aspects 811
3.9. Recycling 812
References 814

1. Introduction

Classification. Polyesters are defined as all polymers which contain repeating units of the ester group

$$-\overset{\underset{\|}{O}}{C}-O-$$

in the polymer chain. Polyesters are generally produced by condensation of polyfunctional carboxylic acids (or their derivatives) with polyfunctional alcohols. Other methods of production include polycondensation of hydroxycarboxylic acids and polymerization of cyclic esters (e.g., lactones).

Polyesters with widely varying properties can be produced depending on the starting materials:

1. High molecular mass linear polyesters ($M_r > 10\,000$) are produced from difunctional alcohols and dicarboxylic acids (or their derivatives, e.g., diesters or dichlorides) or from polymerizable lactones. They can be processed thermoplastically to molding materials (*thermoplastic polyesters*, Chap. 3). The term thermoplastic polyesters also includes polycarbonates, see → Polycarbonates.

2. Low molecular mass polyesters ($M_r \leq 10\,000$) from saturated aliphatic or aromatic dicarboxylic acids and difunctional or mixtures of di- and trifunctional alcohols are linear or slightly branched *intermediates for polyurethanes* (see Chap. 2).

3. Low molecular mass polyesters ($M_r \leq 10\,000$) from di-, tri-, and polyfunctional alcohols and polyfunctional, usually aromatic carboxylic acids or saturated or unsaturated fatty acids are classified as *alkyd resins*, see → Alkyd Resins.

4. *Unsaturated polyesters* are polyesters which can copolymerize with unsaturated

compounds and are formed from polyfunctional alcohols and polyfunctional unsaturated carboxylic acids. After copolymerization with monomers such as styrene they can also be classified as thermosets, see→ Polyester Resins, Unsaturated.

2. Polyesters as Intermediates for Polyurethanes

In 1937 O. BAYER developed an addition polymerization method for the synthesis of polyurethanes that was based on the reaction of diisocyanates with polyols [1]. Hydroxy-functional polyesters were soon introduced as reactants. The only highly branched polyesters known at that time were based on aromatic dicarboxylic acids, and they resulted in hard, highly cross-linked products that were processed in coatings and molding compounds [2]. In 1942 H. PINTEN treated moderately branched aliphatic hydroxypolyesters with a diisocyanate and obtained a rubbery-elastic material ("I-Gummi") [3]. Systematic study of the reactions of linear and moderately branched hydroxypolyesters with excess diisocyanates, and subsequent treatment of these intermediates with water, diols, diamines, and other compounds containing reactive hydrogens, resulted in a wide range of polymers displaying advantageous properties (e.g., thermosets, thermoplastics, cast elastomers, hard and soft foams, molding compounds, hard and flexible coatings). Polyurethanes came to be known as "tailor-made polymers" [4], [5]. This diversity was partly a result of great flexibility with respect to starting materials, mean molecular mass, and mean functionality.

Today, a broad spectrum of hydroxypolyesters with mean molecular masses between 400 and 6000, hydroxyl numbers of 28 – 300 mg KOH/g, and acid values generally under 1 mg KOH/g are employed as intermediates in the manufacture of polyurethanes. Highly branched polyesters, made largely from aromatic or hydroaromatic dicarboxylic acids, find use as binders for polyurethane coatings (DD coatings, Bayer), → Paints and Coatings, 1. Introduction, [6].

The present review is limited to a discussion of linear and moderately branched hydroxypolyesters derived from aliphatic starting materials.

2.1. Raw Materials

Adipic acid is most commonly employed as the acidic component. Other dicarboxylic acids such as oxalic, succinic, and glutaric acid are of lesser importance because the resulting polyesters exhibit lower stability toward hydrolysis. Higher molecular mass aliphatic carboxylic acids (azelaic acid, sebacic acid) are economically unattractive. "Dimeric acids" [7], produced by the dimerization of unsaturated plant fatty acids and hydroxy-functional monocarboxylic acids, find application in special situations. The hydroxy-functional acid dimers afford the corresponding polyesters [8] by self-condensation onto a starter diol. Aromatic dicarboxylic acids such as phthalic acid (in the form of the anhydride), isophthalic acid, and terephthalic acid are sometimes added as modifiers. Typical diol components are ethylene glycol, diethylene glycol, 1,2-propanediol, 1,3-butanediol, 1,4-butanediol, neopentyl glycol, and 1,6-hexanediol. Glycerol, trimethylolethane, trimethylolpropane, and hexanetriol are employed to obtain branched polymers. Developments in the synthesis of hydroxypolyesters include the use of raw materials such as ϵ-caprolactone, which affords polycaprolactone [9] via ring-opening polymerization on a starter polyol. Oligo- or polyethers can also be incorporated into the polymer chain, leading to the corresponding etheresters [10].

2.2. Production

2.2.1. General Information

The only currently economic method for the production of polyesters is direct esterification of dicarboxylic acids with diols [11]. Since esterification is a reversible equilibrium reaction, the process can only be driven to completion if the resulting water of condensation is continuously removed. Polyesterification represents an example of *condensation polymerization*. Not only are monomeric units added to the growing chains, but individual chains also react with one other because they may have both carboxyl and hydroxyl terminal groups. The esterification conditions also permit constant transesterification within the chain itself. Consequently, polyesters exhibit relatively broad molar mass distributions, which can be calculated using the Schulz–Flory method [12], [13]. Transesterification also accounts for the fact that, if more than two starting materials are employed, these become incorporated into the polyesters in a statistical way regardless of their time of introduction.

The esterification of a dicarboxylic acid with a diol represents a higher-order reaction. In the absence of an added catalyst, and if the reverse reaction is prevented by removal of the resulting water, the kinetics of the reaction can be described as third order [14], [15] because the acidic groups themselves exert a catalytic effect on the reaction:

Reaction rate = $k \cdot [OH] \cdot [COOH]^2$

When an esterification catalyst is utilized, its influence becomes dominant and esterification exhibits second-order kinetics. Esterification is catalyzed by acids, bases, and transition-metal compounds. However, the commercial use of catalysts is limited because all catalysts tend to have undesirable effects on subsequent polyurethane reactions. In the case of polyesters that are used in reactions where traces of catalyst do not interfere, synthesis is catalyzed with low concentrations (ppm) of compounds of titanium, tin, or antimony. The appropriate starting ratios for the synthesis of a given hydroxypolyester can, in theory, be calculated from the desired number-average molecular mass (M_n), which can in turn be determined from the hydroxyl number.

Given the simplified formula

$$HO-R-O\left[\begin{matrix}C-R'-C-O-R-O\\ \parallel\quad\quad\parallel\\ O\quad\quad O\end{matrix}\right]_n H$$

a linear polymer would result from n mol of dicarboxylic acid (HOOC–R′–COOH) and $n+1$ mol of diol (HO–R–OH). The degree of polycondensation, n, is calculated as follows:

$$n = \frac{M(\text{polyester}) - M(\text{diol})}{M(\text{dicarboxylic acid}) + M(\text{diol}) - 2M(\text{water})}$$

Losses of diol accompanying the removal of water of condensation and as a consequence of side reactions (formation of ethers and aldehydes) usually require the addition of an empirically determined excess of diol which depends upon the type of diol and the reaction conditions.

The first phase of the esterification is conducted under an inert atmosphere (nitrogen, carbon dioxide) and the water of condensation is removed rapidly by distillation through a column at >130 °C. Once an acid value of about 30 has been reached, the reaction must be accelerated in a second phase by facilitating the removal of water from the esterification equilibrium. This is done by application of a vacuum (vacuum-melt method), introduction of a carrier gas (usually carbon dioxide), or recirculation of an entrainer such as toluene or xylene (azeotropic distillation). The last approach is not applicable when the diol component is ethylene glycol or 1,2-propanediol, which themselves form azeotropic mixtures with the entrainers.

A low acid value (<1 mg KOH/g) is desired. This means that the reaction mixture tends to become highly depleted in one reactant toward the end of the reaction, greatly reducing the rate. This second phase of polyester synthesis therefore requires a large fraction of the overall reaction time (e.g., 20 of 30 h).

2.2.2. Industrial Production

Polyesters are produced industrially either batchwise or continuously according to a variety of methods. The vacuum-melt and the purge-gas-melt methods are preferred, the azeotropic method finds only limited application.

Batchwise Vacuum-Melt Method [16]. The apparatus constitutes a closed system, comprising a stirred-tank reactor equipped with an anchor agitator, a distillation column, a condensor, a receiver, and a vacuum pump. The starting materials are heated in the reactor, with stirring but in the absence of air. Heat is provided by electrical energy, high-pressure steam, or a heat-transfer medium. In the first phase of the reaction the temperature is held between 125 and 220 °C under ambient pressure and ca. 90 % of the resulting water is removed by distillation. A mixture of oligoesters is produced which is immediately condensed to polyester in the second reaction phase at 170 – 230 °C and with stepwise reduction of the pressure to 0.005 bar. Residual water of reaction is distilled from the mixture, together with excess diol and minor byproducts such as low molecular mass aldehydes and ketones (present in ppm amounts relative to the polyester). The process is monitored constantly by measuring the acid value, the hydroxyl number, and the viscosity.

Example of the preparation of a polyester with an average molecular mass of 2000 from ethylene glycol and adipic acid. A mixture of 86.8 kmol of adipic acid and 95.6 kmol of ethylene glycol, including an empirically determined excess of 1.1 kmol ethylene glycol, is heated with stirring under nitrogen to 130 °C. Once distillation begins the temperature of the reaction mixture is raised to 200 °C over the course of 10 h. During this period 156 kmol of water of reaction is distilled into the receiver. When distillation is complete, the acid value of the reaction mixture is ca. 55 mg KOH/g. The pressure in the apparatus is subsequently reduced in stages to 0.01 bar at constant temperature over a period of 15 h. During this period the remainder of the water of reaction is distilled off, along with excess ethylene glycol and byproducts. The reaction mixture has the following properties: hydroxyl number, 55.5; acid value, 0.7; viscosity, 0.515 Pa·s at 75 °C; water content, 0.02 wt %. After pressure equilization with nitrogen and cooling to 110 °C, the polyester product (7.7 kmol) is transferred to a storage tank.

Purge-Gas – Melt Method [17]. This process differs from the vacuum-melt method in that during the second phase of reaction the distillable products are removed by passage of an inert gas (N_2, CO_2) through the mixture rather than by lowering of the pressure. This method increases the loss of diol, a fact that must be taken into account in the raw material calculations. The apparatus must also be suitably adapted: the reactor is equipped with a gas inlet and the distillation column is replaced by a simple distillation tube. A vacuum pump is not required.

Azeotropic Method [18]. This process is conducted at atmospheric pressure in the presence of an inert solvent (e.g., toluene or xylene) which entrains and removes the water of reaction. A vacuum-melt apparatus can be adapted to the azeotropic method by including facilities for removal and continuous recycling of the entrainer solvent. Lower reaction temperatures (125 – 190 °C) are used and permit polyester to be produced under milder conditions. Solvent residues in the final product are removed under vacuum.

Wastewater and Exhaust Air. Both the water of reaction and the process exhaust air contain excess diols and reaction byproducts which are biodegradable in dilute aqueous solution. The processes do not therefore present environmental problems if an exhaust air purifier and a biological water treatment system are employed.

Construction Materials. The materials used for constructing the equipment must satisfy a variety of demands. High resistance to corrosion is essential with respect to the acidic reaction medium and the distillates to prevent contamination of the polyester with traces of metal. Moreover, stress may also arise in the reactor materials due to the high reaction temperatures. These requirements are best met by highly alloyed austenitic chromium – nickel – molybdenum steels, for example V 4 A steel (DIN 1.4571).

The material requirements posed by the raw materials and hydroxypolyester products are less severe. Stainless steel (V 4 A, V 2 A), pure aluminum, $AlMg_3$ alloy, and in special cases even unalloyed steel can therefore be used for pipes and containers for storage and transport.

2.3. Quality Specifications and Testing

Since hydroxypolyesters are subjected to further processing to polyurethanes, they must satisfy stringent quality criteria that can only be obtained through a high degree of product standardization. The most important input data for the polyurethane reaction include the hydroxyl number (DIN 53 240), acid value (DIN 53 402), water content (DIN 53 979), color number (iodine number, DIN 6162; Hazen/APHA, DIN 53 404), and viscosity (DIN 53 015), and are fixed within narrow tolerances for each type of polyester. Individual products often have additional quality criteria that depend on further processing.

The acid value is a measure of the content of terminal carboxyl groups in the polymer and is especially important. A minimum acid value is preferred because terminal carboxyl groups react with isocyanates much more slowly than hydroxyl groups, and can therefore lead to chain termination during polyaddition. Furthermore, the reaction of carboxyl groups with isocyanates produces amide functions and carbon dioxide, which can result in undesirable bubble formation in the polyurethane. Finally, residual acid in a polyester has negative effects on catalysis of the polyurethane reaction and the stability of the polyurethane product toward hydrolysis. Consequently, an acid value limit is mandatory for polyesters (generally <1 mg KOH/g, but often <0.5 mg KOH/g).

The presence of even trace amounts (ppm levels) of metals, acids or bases, fats, silicone oils, and other surface-active agents in polyesters must be avoided. Such impurities can originate from the starting materials or the processing equipment; their removal is extremely difficult, if not impossible. Consequently, both starting materials and apparatus must be subjected to rigid standards. The analytical data cited in Section 3.1 for the thermoplastic polyester raw materials must be regarded as minimum standards.

Manufacturers utilize the test methods already described, and other specially developed tests for product and process quality control. An applications test is often conducted for final quality assessment.

2.4. Properties

Physical Properties . The diverse applications of polyester-polyurethanes require a wide variety of starting materials. Pure polyadipates of ethylene glycol, 1,4-butanediol, and 1,6-hexanediol are waxy crystalline substances with softening points around 40 – 50 °C. Their crystallinity helps to provide the polyurethane products with outstanding physical characteristics, especially with regard to elasticity, tensile strength, and notch resistance. In some application areas, however, a tendency to harden at low temperatures or as a result of stress (crystallization of elongation) is problematic.

Polyesters made from mixtures of the aforementioned diols, alone or in combination with other diols, can be advantageous in such cases, and are either liquid or greasy at room temperature. 1,2-Propanediol, 1,3-butanediol, diethylene glycol, and neopentyl glycol afford liquid polyadipates.

Due to their broad molecular mass distributions (Schulz–Flory distribution) [12], [13], the polyesters have a relatively high viscosity (e.g., ca. 600 mPa · s at 75 °C and a mean molecular mass of 2000). Densities range between 1.1 and 1.5 g/cm^3. The flash points (DIN 51 758) lie above 200 °C, and the ignition points (DIN 51 794) above 400 °C. Polyesters are relatively soluble in common organic solvents, including acetone, ethyl acetate, and toluene.

Chemical Properties. Terminal hydroxyl groups are largely responsible for the chemical properties of the polyesters, which undergo reactions typical of alcohols. The most important reaction from an applications standpoint is the facile, quantitative addition of hydroxyl groups to the isocyanate groups of di- and polyisocyanates, which affords segmented, high molecular mass polyester-polyurethanes (→ Polyurethanes).

The polyesters exhibit a certain degree of sensitivity toward hydrolysis, and newly formed acidic groups have an autocatalytic effect. The rate of hydrolysis is also increased by the same compounds that catalyze esterification. Resistance to hydrolysis is a function of polyester structure and increases with increasing hydrophobic character and with decreasing ester-group content [16]. Polyesters derived from long-chain diols and dicarboxylic acids are more stable than those based on short-chain starting materials. Diethylene glycol polyesters exhibit only marginal resistance to hydrolysis because of the hydrophilic effect of their ether groups. During further processing acid acceptors (e.g., carbodiimide, epoxides) are often added as hydrolysis stabilizers.

2.5. Storage and Transportation

During storage and transportation of polyesters, exposure to moisture and temperatures above 120 °C must be prevented to avoid undesirable effects during polyurethane production. The products are either stored in sealed, air-tight drums at ambient temperature or in heated, insulated containers under nitrogen at 70 – 110 °C. As liquids they are transported in barrels or in heatable, insulated bulk containers; as solids they are transported in drums.

The construction materials mentioned in Section 2.2.2 are generally utilized for storage and transport vessels; barrels are usually constructed from tin-free steel. Turbidity is sometimes observed in polyesters that are liquid at ambient temperature but have been stored for long periods of time at low temperature; this can be alleviated without affecting the quality of the product by heating the polyester briefly to 80 °C.

Recommendations regarding storage and transport conditions for polyesters are provided by manufacturers. Handling of polyesters presents no significant hazards—a consequence of their chemical characteristics. Spilled material is best treated by mixing it with sand and then sending it to an approved waste disposal site, or by burning it in an appropriate incinerator.

2.6. Uses and Economic Aspects

Hydroxypolyesters are sometimes reacted either alone or in combination with diols or diamines to give segmented polyurethanes, which are marketed as aqueous dispersions, solutions, or as powders that are thermoplastically processable or soluble in special solvents (→ Fibers, 6. Polyurethane Fibers; → Polyurethanes). In other cases they are combined on site with polyisocyanates and various additives to form soft to hard foams, elastomers, adhesives, flexible coatings, and joint fillers.

The polyesters described above are marketed by a number of firms; individual product types are further characterized by appended letters and/or numerical indices. Examples of trade names include:

Baycoll, Bayflex, Baygal, Baytec, Baytherm	Bayer
Crestapol	Scott Bader
Daltogel, Daltolac, Daltorez	ICI
Desmophen	Bayer
Diorez	Briggs and Townsend
Estolan	Lancro Chemicals
Fomrez	Witco Chem. Corp.
Lumitol/Elastophen	BASF
Multron/Multrathane	Miles Inc. (formerly Mobay Chem. Corp.)
Nippolan	Nippoly
Polyesteralkohol/Systol	BASF Schwarzheide GmbH
Vulkollan	Bayer

Estimated production capacities in 1983 were (in 10^3 t/a):

Federal Republic of Germany	89
German Democratic Republic and Poland	15
EC	188
United States	44
Japan	21
World	277

Worldwide consumption in 1983 was estimated at ca. 200 000 t.

3. Thermoplastic Polyesters [19–25]

The history of thermoplastic polyesters (see also → Fibers, 5. Polyester Fibers) goes back to 1929 with the pioneering work of CAROTHERS. The first thermoplastically processable polyesters synthesized from adipic acid and ethylene glycol were described by him in 1932. Polyesters only became of industrial interest in 1941, with the synthesis of high melting point products based on terephthalic acid.

The rapid industrial development of polyesters after World War II was initially restricted to polyester fibers based on poly(ethylene terephthalate) [25038-59-9] (PETP), poly(oxy-1,2-ethanediyloxycarbonyl-1,4-phenylenecarbonyl). A polyester based on poly(1,4-dimethylenecyclohexane terephthalate) [25135-20-0], PDCT, occupies a special position (see also → Fibers, 5. Polyester Fibers).

Poly(ethylene terephthalate) was subsequently also used in the production of films (see also → Films, Section 6.8.).

Thermoplastic polyesters were first employed as construction materials in 1966 [26], and were initially based on poly(ethylene terephthalate). In 1970 the more readily processable poly(butylene terephthalate) [24968-12-5] (PBT), poly-(tetramethylene terephthalate), poly(oxy-1,4-butanediyloxycarbonyl-1,4-phenylenecarbonyl) was introduced onto the market.

A short time later these homopolyesters were supplemented by a series of thermoplastic copolyesters which are suitable for specific areas of application due to their special properties [27]. These copolyesters include polyetherester block copolymers as thermoplastic elastomers; copolyesters based on 1,4-cyclohexanedimethanol for hard, glass-clear injection molded articles; and other thermoplastic copolyesters of varying composition for powder coatings, paint binders (→ Paints and Coatings, 1. Introduction), and hotmelt adhesives.

Recently a new class of fully aromatic thermoplastic polyesters has been developed – liquid crystalline polyesters [28], [29]. These polymers have aroused great interest on account of their outstanding mechanical and thermal properties. Two such products are commercially available, one being synthesized from 4-hydroxybenzoic acid and 6-hydroxy-2-naphthoic acid, and the other from 4-hydroxybenzoic acid, terephthalic acid, and 4,4'-dihydroxybiphenyl.

3.1. Raw Materials

Thermoplastic polyesters are generally produced from dicarboxylic acids and dihydric alcohols. The most common raw materials follow:

Dicarboxylic acids (and derivatives)
Dimethyl terephthalate [120-61-6]
Terephthalic acid [100-21-0]
Dimethyl isophthalate [1459-93-4]
Isophthalic acid [121-91-5]
Dimethyl adipate [627-93-0]
Adipic acid [124-04-9]
Azelaic acid [123-99-9]
Sebacic acid [111-20-6]
Dodecanoic diacid [693-23-2]
1,4-Cyclohexane dicarboxylic acid [1076-97-7]
Dimethyl-1,4-cyclohexane dicarboxylate ester [94-60-0]

Hydroxycarboxylic acids and lactones
4-Hydroxybenzoic acid [99-96-7]
6-Hydroxy-2-naphthoic acid [16712-64-4]
Pivalolactone [1955-45-9]
ϵ-Caprolactone [502-44-3]

Dihydric alcohols and bisphenols
Ethylene glycol [107-21-1]
1,2-Propanediol [57-55-6]
1,3-Propanediol [504-63-2]
1,4-Butanediol [110-63-4]
1,4-Cyclohexanedimethanol [105-08-8]
2,2-Dimethyl-1,3-propanediol [126-30-7]
1,6-Hexanediol [629-11-8]
Poly(tetrahydrofurandiol) [9040-43-1]
4,4'-Dihydroxy-1,1'-biphenyl [92-88-6]
1,4-Hydroquinone [123-31-9]

Terephthalic acid is by far the most important dicarboxylic acid. This compound was previously difficult to obtain in sufficiently pure form, dimethyl esters of dicarboxylic acids were therefore mainly used as raw materials. Lactones, hydroxycarboxylic acids, and bisphenols are of lesser importance.

Particularly stringent requirements are placed on the purity of the raw materials because impurities can either interfere with polycondensation via chain termination or branching (e.g., mono- or polyfunctional compounds) or can lead to undesirable secondary reactions and discoloration under the high reaction temperatures. These requirements apply to polyesters used as starting material for fibers, but particularly to the plastics sector where polyesters with substantially higher molecular masses are needed.

Analytical requirements for the four most important raw materials are as follows:

Dimethyl Terephthalate (DMT). For production and properties see → Terephthalic Acid, Dimethyl Terephthalate, and Isophthalic Acid.

Purity (GC analysis)	>99.9 %
Solidification point (min.)	140.63 °C
Saponification value	577.8 mg KOH/g
Acid value (max.)	0.03 mg KOH/g
Acid value after 4 h at 175 °C (max.)	0.06 mg KOH/g
Color of the melt after 24 h at 170 °C (max.)	10 (Hazen number)
Color in H_2SO_4 (max.)	10 APHA
Fe content (X-ray fluorescence) (max.)	1 ppm
Nitro groups and nitroso groups should also be absent	

Terephthalic Acid (TA). For production and properties, see → Terephthalic Acid, Dimethyl Terephthalate, and Isophthalic Acid. The purity of "polymer grade" terephthalic acid must satisfy appropriate requirements for the production of fibers, films, and molding compounds. As a rule terephthalic acid is available with the following specifications:

Acid value	675 ± 2 mg KOH/g
Ash content (max.)	15 ppm
Metals (atomic absorption) (max.)	10 ppm
of which Fe (max.)	2 ppm
Co, Mo, Ni, Ti, Mg (max.)	1 ppm
Ca, Al, Na, K (max.)	2 ppm
4-Carboxybenzaldehyde (polarography) (max.)	25 ppm
Water content (Karl Fischer) (max.)	0.5 %
Color in 5 % dimethylformamide (max.)	10 APHA

Under certain conditions (absence of fluorenone structures, maintenance of a defined color index) the 4-carboxybenzaldehyde content can reach 200 ppm without impairing the polyester quality [30].

Ethylene Glycol. For production and properties, see → Ethylene Glycol.

Relative density d_{20}^{20}	1.1151 – 1.1156
Boiling point range	196 – 200 °C
Solidification point	−11 to −13 °C
Acid content as acetic acid (max.)	0.005 %
Fe content (max.)	0.07 ppm
Chlorine compounds	none
Diethylene glycol content (max.)	0.08 %
Water content (max.)	0.08 %
Ash content (max.)	0.005 g/100 mL
Acetaldehyde content (max.)	30 ppm
Color index after 24 h at 170 °C	<10 (Hazen number)
UV transparency at m	
220 µ	70 %
275 µm	90 %
350 µm	95 %

1,4-Butanediol. For production and properties, see → Butanediols, Butenediol, and Butynediol, Section 1.3.

Purity (GC analysis) (min.)	99.3 %
Water content (max.)	0.05 %
Solidification point (min.)	19.5 °C
Color index (max.)	10 APHA
Ash content	0

3.2. Production

3.2.1. General Principles

Two-Stage Process. Generally, thermoplastic polyesters are produced in two stages. In the first stage a polyester precondensate (M_n 100 – 2000, depending on the molar ratio of the starting compounds) with the formula

HO−Y−O(OC−X−COO−Y−O)$_x$ H, $x \approx 1 - 10$

is produced by transesterification of dicarboxylic diesters or by esterification of the dicarboxylic acid HOOC−X−COOH with excess dihydric alcohol HO−Y−OH. The precondensate then reacts in a second step (i.e., polycondensation) with the elimination of dihydric alcohol to form high molecular mass polyesters (M_n>10 000) of the formula:

HO−Y−O(OC−X−COO−Y−O)$_y$ H, $y \geq 100$

Transesterification and esterification are equilibrium reactions that are accelerated by catalysts. The reactions are continuously displaced towards the polyester precondensate by removing the more volatile alcohol or water.

Polycondensation is also an equilibrium reaction, which is shifted towards the high molecular mass polyester by constant removal of the eliminated diol or water (application of a vacuum or purging with an inert gas), and is accelerated by catalysts. The higher the desired molecular mass of the polyester, the more severe are the conditions required to remove the diol and water. Polycondensation is a step-growth reaction (of the type A − A+B − B=diol + dicarboxylic acid or A − B=hydroxycarboxylic acid), which leads to nonuniform end products. The molecular mass distribution can be calculated from the molar ratio of diol to dicarboxylic acid and/or from the conversion p [31]. With a stoichiometric ratio of the A and B groups, the mass fraction W_x of the degree of polycondensation x can be specified as a function of the conversion p [32]:

$$W_x = \frac{(M_L + M_U x)}{[M_U + M_L(1-p)]} p^{x-1} (1-p)^2$$

M_L and M_U denote the molecular masses of the leaving molecules L and basic units U, respectively. For negligibly small molecular masses of the leaving molecules ($M_L/M_U \to 0$) or as an approximation for high conversion rates ($p \to 1$), the equation transforms to the Schulz–Flory distribution function

$$W_x = xp^{x-1}(1-p)^2$$

Introduction of the mean degree of polycondensation $P_n = 1/(1-p)$ gives

$$W_x = \frac{x}{P_n^2}\left(1 - \frac{1}{P_n}\right)^{x-1}$$

The relationship $P_n = 1/(1-p)$ clearly shows that a minimum degree of polymerization of 100 can only be achieved with conversions of > 99 % (p=0.99).

The kinetics of esterification, transesterification, and polycondensation [33–40] have been investigated in great detail. See also → Polymerization Processes, 1. Fundamentals, Chap. 1., → Polymerization Processes, 1. Fundamentals, Chap. 2.. In general, polycondensation is regarded as a second-order reaction [41], [42], with an activation energy that depends strongly on the catalyst system (PETP: uncatalyzed 96.6 kJ/mol [41], catalysis with manganese acetate 40.8 kJ/mol [42]).

Other Processes. In addition to the conventional two-stage process for producing linear thermoplastic polyesters, many other processes are used with other starting compounds. Addition of ethylene oxide to dicarboxylic acids gives bis(2-hydroxyethyl) esters. In analogy with the previously described esterification and transesterification products, these esters undergo polycondensation with the elimination of ethylene glycol [43].

Fully aromatic polyesters with a high melting point are produced by interfacial polycondensation of dicarboxylic dichlorides with bisphenols in the presence of bases and with elimination of hydrogen chloride [44]. Dicarboxylic acids also react with diesters (preferably acetates) of dihydric alcohols with the elimination of acid [45].

The esterification of an anhydride with a dihydric alcohol or the transesterification of anhydrides with epoxides yields linear polyesters [46], [47].

A special process for producing high molecular mass thermoplastic polyesters involves the polymerization of lactones in the presence of organometallic compounds [48].

Secondary Reactions [49–51]. In all methods for producing linear thermoplastic polyesters, secondary reactions that occur during polycondensation can alter the stoichiometric ratio and thus terminate polycondensation or confer undesirable properties on the end product.

The most important secondary reactions include:

1. Ether formation to produce diethylene glycol units (Y=C_2H_4)

$$2 \sim\!\!\!\sim\!\!X-\underset{\underset{O}{\|}}{C}-O-Y-OH \xrightarrow{-H_2O} \sim\!\!\!\sim\!\!X-\underset{\underset{O}{\|}}{C}-O-Y-O-Y-O-\underset{\underset{O}{\|}}{C}-X\sim\!\!\!\sim$$

2. Dehydration of ethylene glycol to form acetaldehyde, or of 1,4-butanediol to form tetrahydrofuran

3. Ester pyrolysis which in the case of PETP, for example, produces carboxyl groups (poorer resistance to hydrolysis) and olefins (discoloration):

$$\sim\!\!\!\sim\!\!\text{Ar}-\underset{\underset{O}{\|}}{C}-O-CH_2-CH_2-O-\underset{\underset{O}{\|}}{C}-\text{Ar}\sim\!\!\!\sim$$
$$\downarrow$$
$$\sim\!\!\!\sim\!\!\text{Ar}-\underset{\underset{O}{\|}}{C}-OH + H_2C=CH-O-\underset{\underset{O}{\|}}{C}-\text{Ar}\sim\!\!\!\sim$$

4. Ring formation between two adjacent carboxyl groups (particularly pronounced in adipic acid):

$$HOOC-(CH_2)_4-COOH \xrightarrow[-H_2O, -CO_2]{} \bigcirc\!\!=O$$

3.2.2. Production of Poly(Ethylene Terephthalate) from Dimethyl Terephthalate

$$HO(CH_2)_2O\left[\underset{\underset{O}{\|}}{C}-\text{Ar}-\underset{\underset{O}{\|}}{C}-O-(CH_2)_2-O\right]_x\!H$$

Poly(ethylene terephthalate) (PETP) is produced industrially both batchwise and continuously. The apparatus used for these processes is also suitable for the production of other thermoplastic polyesters. For the sake of simplicity, however, the batchwise production of PETP is discussed here. Separate vessels are often used to melt the starting compounds and for the transesterification and polycondensation reactions.

Melting. Dimethyl terephthalate (DMT) is melted at 150 – 160 °C in a stirred tank heated with steam, carrier oil, or electricity in an inert atmosphere (nitrogen).

Transesterification. The molten DMT and ethylene glycol are reacted in heated, stirred transesterification reactors at 150 – 200 °C. At the start of the reaction (at normal pressure in a nitrogen atmosphere) the lower temperature range is preferred to minimize sublimation of

DMT. The methanol released during the transesterification is continuously distilled off from the reaction mixture. The excess of ethylene glycol is normally ca. 0.5 – 1 mol/mol DMT.

The use of catalysts is essential to achieve a reasonable transesterification rate at moderate temperatures. Although the catalysts listed in the patent literature cover practically the whole periodic system, weakly basic compounds such as amines, metal oxides, alkoxides, and acetates are used as transesterification catalysts in practice at a concentration of 0.01 – 0.1 wt % referred to DMT. Systematic investigations showed that the rate constant for the transesterification of DMT with ethylene glycol correlates with the electronegativity of the catalyst metal ion [52]. Metal compounds in which the electronegativity of the metal ion is in the range 1 – 1.7 are most effective (e.g., acetates of Ca, Mg, Zn, Cd, Pb, and Co). An almost identical order is found when the stability constants of the metal dibenzoylmethane complexes are related to the rate constants of the transesterification reaction [53].

The choice of a suitable catalyst is an important factor in the production of PETP as regards the economy of the production process and quality of the end product. Transesterification should proceed as quantitatively as possible because residual methyl ester groups can interfere with subsequent polycondensation [54]. This is particularly true if the activity of the transesterification catalysts is lowered by additives introduced at the polycondensation stage. The course of transesterification is therefore controlled by the amount of methanol released, whereby entrained ethylene glycol has to be taken into account (density measurement or ultrasonic monitoring).

Polycondensation. The transesterification product is added as a melt to the polycondensation reactor, which can be heated to over 300 °C and must be equipped with a very efficient stirrer. Some of the excess ethylene glycol is often first distilled off at normal pressure (protective gas, nitrogen) by gradually increasing the temperature to ca. 250 °C. Polycondensation then follows, with a reduction in pressure and a further increase in temperature. In order to obtain sufficiently high molecular masses, temperatures of ca. 270 – 280 °C at a final vacuum of <1 mbar are required.

Like the transesterification reaction, the polycondensation reaction has to be accelerated by catalysts. All the catalysts used for transesterification are, in principle, also suitable for polycondensation because the two reactions are extremely similar. The catalyst remaining in the reaction mixture after the transesterification can therefore be used further for polycondensation. This approach does not always provide satisfactory products, however, because active transesterification catalysts decrease the thermal stability of PETP [55]. In many cases the transesterification catalyst is therefore masked with suitable additives (mainly phosphorus compounds) and replaced by special polycondensation catalysts, usually antimony, germanium, titanium, or lead compounds at a concentration of 0.005 – 0.05 wt % referred to DMT.

Rapid polycondensation is achieved by quickly removing the eleminated ethylene glycol by means of intensive mixing and a maximum polymer melt surface area. In conventional polycondensation plants this is achieved by intensive stirring. Distribution of the polymer melt in thin-film procedures as in thin-layer evaporators or in annular disk reactors is, however, more effective for obtaining very high molecular masses.

Polycondensation is stopped when a defined melt viscosity (measure of the molecular mass), is reached. The power consumption of the stirrer motor is often an adequate indicator of melt viscosity. In order to terminate polycondensation, the vacuum in the reaction vessel is removed with oxygen-free nitrogen and the melt is expelled from the reactor either under nitrogen pressure or, preferably, with spinning pumps. The melt is directly quenched with water and comminuted into chips or pellets to prevent oxidation by atmospheric oxygen.

Conventional batchwise and continuous polycondensation produces PETP that is processed into threads and fibers (solution viscosity ca. 50 – 80 cm^3/g measured in phenol/tetrachloroethane 60/40 wt % at a concentration of 0.0023 g/cm^3, see Section 3.3).

Before further processing the PETP must be carefully dried at 80 – 130 °C to reduce the water content to below 0.01 wt %. Drying is performed with the same apparatus used for solid-phase postcondensation. Heating during drying causes crystallization of PETP which is necessary to avoid agglomeration of the particles during prolonged storage.

A PETP with a fairly high intrinsic viscosity, such as can be produced with the thin-layer evaporator or annular disk reactor, is required for the plastics sector. Alternatively, however, the comminuted PETP (chips, pellets, etc.) is often subjected to solid-phase postcondensation for a fairly long time (~20 h) at temperatures of up to 250 °C in an inert gas stream or in vacuo, resulting in an increase in molecular mass. The rate-determining step in this reaction is diffusion of the eliminated ethylene glycol from the polymer particles, which depends on the particle size [56]. Postcondensation in the solid phase can be performed either batchwise (in vacuum dryers or in a fluidized bed) or continuously (in countercurrent with hot nitrogen).

Production Example for PETP [57]. 12.8 kg of DMT and 8.1 kg of ethylene glycol are transesterified in the presence of 3.32 g of manganese acetate and 0.83 g of cobalt acetate for 3 h at 195 °C under nitrogen in a 30-L steel reactor equipped with an anchor-type stirrer and a distillation column. The liberated methanol is distilled off. 10.76 g of triphenyl phosphate is then added and stirred in for 30 min at 195 °C; 2.75 g of germanium oxide and 0.5 g of tetraisopropyl titanate are added and stirring is continued for a further 30 min.

The temperature of the reaction mixture is then raised to 280 °C within 6 h and the pressure subsequently reduced to 0.3 mbar in 2 h. After a further 2 h under these conditions the vacuum is removed with nitrogen. The polyester melt is extruded in the form of strands, which are cooled in water and then pelletized. The solution viscosity of the polyester is 62 cm^3/g (in phenol/tetrachloroethane 60/40, concentration 0.0023 g/cm^3).

3.2.3. Production of Poly(Ethylene Terephthalate) from Terephthalic Acid

The direct esterification of terephthalic acid (TA) with ethylene glycol became important when economic processes were developed for producing fiber-grade TA [58]. An important advance was made when the reaction times were reduced by performing esterification under pressure (2.7 – 5.5 bar) at temperatures above the boiling point of ethylene glycol (198 °C) [59], [60].

Table 1. Quality specifications for additive-free PETP used for film production [61]

Specification	Raw material	
	DMT	TA
Intrinsic viscosity, cm^3/g	65 ± 0.015	65 ± 0.015
Carboxyl terminal groups, mmol/kg	20 – 30	20 – 30
Ash content, wt %	<0.04	<0.02
Diethylene glycol content, wt %	0.5 – 1.2	0.7 – 1.2
Filter value (15 µm filter braid), bar cm^2/kg	<30	<30

The TA process now yields products that are qualitatively competitive with the polyesters produced in the DMT process (see Table 1) [61].

In new plants direct esterification is the preferred method of PETP production. The advantages of direct esterification with TA compared with transesterification with DMT are

1. The higher reaction rate
2. The lower weight of TA compared with DMT (storage costs)
3. The use of water instead of methanol as condensation agent
4. No transesterification catalyst is required
5. Higher molecular masses are obtained

Temperatures of around 220 – 260 °C are generally maintained to overcome the sparing solubility of TA. The water formed is continuously removed by distillation. The molar ratio of ethylene glycol to TA should be < 2 (in most patents < 1.5 is specified) to suppress formation of diethylene glycol (see). Direct esterification is possible without catalysts, but can be accelerated by basic catalysts (amines) or transesterification catalysts (usually Sb$_2$O$_3$).

Esterification is followed by gradual pressure release, the temperature is simultaneously increased and excess ethylene glycol is distilled off. The formation of PETP takes place in the polycondensation stage, which proceeds analogously to the DMT – ethylene glycol process as regards temperature, pressure, and catalysts (see Section 3.2.2, a detailed process description is given in [62], [63]).

Direct esterification at normal pressure can be achieved in a PETP precondensate as reaction medium, in which TA is readily soluble [64]. This process is particularly suitable for continuous direct esterification.

The secondary reactions occurring during polycondensation are of decisive importance in determining the quality of the PETP (see). Ether formation results in diethylene glycol units which have adverse effects on dyeing behavior but also lower thermal and UV stability (for the mechanism of diethylene glycol formation, see [65]). Free carboxylic groups formed by ester pyrolysis have a negative influence on the postcondensation and hydrolysis stability of PETP. Ester pyrolysis also produces olefinic decomposition products that lead to discoloration of PETP.

3.2.4. Production of Poly(Butylene Terephthalate)

$$\text{HO(CH}_2)_4\text{O}\left[\overset{\text{O}}{\underset{\|}{\text{C}}}-\underset{}{\bigcirc}-\overset{\text{O}}{\underset{\|}{\text{C}}}-\text{O}-(\text{CH}_2)_4-\text{O}\right]_x\text{H}$$

Poly(butylene terephthalate) (PBT) is produced analogously to PETP. The currently preferred process is still the transesterification of DMT with 1,4-butanediol followed by polycondensation. The main byproduct is tetrahydrofuran, formed from 1,4-butanediol by dehydration (see). Although some alkyl tin compounds lower tetrahydrofuran formation, the resulting polymers have a lower thermal stability [66]. Although tetrahydrofuran formation is catalyzed by acids, TA is increasingly used as a raw material instead of DMT [67], [68]. For the advantages of the TA method, see Section 3.2.3.

Transesterification of DMT is performed with a slight excess of 1,4-butanediol (molar ratio 1,4-butanediol: DMT < 2, preferably 1.3 –1.7) at 160 – 220 °C and normal or excess pressure. Soluble titanium compounds are generally used as transesterification catalysts.

Polycondensation at ≤ 260 °C and < 1 mbar is usually also catalyzed by soluble titanium compounds (generally tetraalkyl titanates).

PBT can generally be subjected to polycondensation in the melt to give products with higher molecular masses than PETP (solution viscosity up to ca. 120 cm^3/g, measured in phenol – tetrachloroethane, weight ratio 3 : 2 at 25 °C, concentration 0.0023 g/cm^3). The product is dried in the same way as PETP, see Section 3.2.2. For some uses in the plastics sector (optical waveguide coverings, coatings for bowden wire, headlight dimmers) a higher molecular mass PBT is required, which can also be obtained by solid-phase condensation.

Production Example for PBT [69]. 8.0 kg of DMT and 5.9 kg of 1,4-butanediol (molar ratio with respect to DMT 1.6) are added to a 30-L reactor equipped with an anchor-type stirrer. After heating the mixture to 150 °C, 160 g of a 7 % solution of titanium tetrabutylate in 1,4-butanediol is added over 100 min while stirring to initiate transesterification. After gradual heating to 200 °C over 1.5 h, transesterification is continued for a further 3.5 h at 200 °C, and the methanol formed is distilled off.

The pressure is reduced from 1 bar to 1.3 mbar over 2 h, and the temperature is subsequently increased from 200 to 245 °C in 1 h. Continued heating at 245 °C and 1.3 mbar for ca. 5 h gives a polymer with an intrinsic viscosity of 124 cm^3/g.

3.2.5. Production of Other Partially Aromatic Polyesters

Other high molecular mass thermoplastic polyesters, such as poly(1,4-cyclohexylenedimethylene terephthalate) [25135-20-0], PDCT, copolyesters or copolyetheresters (see → Fibers, 5. Polyester Fibers) are produced in a similar way to PETP and PBT. The different melting points of the end products and different reactivities of the monomers must, however, be taken into account. Depending on their chemical constitution (diol, diester, and diacid), the cocomponents (see Section 3.4) are added before, during, or after the transesterification phase, or before or during the esterification phase.

Production examples for polyesters based on 1,4-cyclohexanedimethanol [1,4-bis(hydroxymethyl)cyclohexane] and for polyetheresters are given in [70–72].

3.2.6. Production of Fully Aromatic, Liquid Crystalline Polyesters

Fully aromatic, liquid crystalline polyesters can be produced in the melt by three main processes. Direct esterification of phenols with aromatic carboxylic acids in the presence of catalysts such as titanium tetrabutyrate or dibutyl tin diacetate is described in [73]. On account of the high reaction temperatures (>300 °C) there is however the danger of decarboxylation of the carboxylic acids, and the method is therefore restricted to monomers that are stable at the melting point.

A second process involves reacting phenyl esters of aromatic carboxylic acids with relevant phenols. The phenyl esters can be produced by direct esterification or by reaction of diphenyl carbonate with dicarboxylic acids, with the formation of carbon dioxide and phenol.

The commercially available liquid crystalline polyesters Vectra [81843-52-9] and Xydar [31072-56-7] are produced by a third method: acidolysis of phenolic acetates with aromatic carboxylic acids [74], [75].

$$HO-\left[\left(\underset{O}{\underset{\|}{C}}-\bigcirc-O\right)_{0.73}\left(\underset{O}{\underset{\|}{C}}-\bigcirc\bigcirc-O\right)_{0.27}\right]_x C-CH_3$$

Vectra

Vectra is synthesized from 4-hydroxybenzoic acid and 6-hydroxy-2-naphthoic acid. Xydar is prepared from 4-hydroxybenzoic acid, 4,4′-dihydroxy-1,1′-biphenyl, and terephthalic acid. The basic structural formulae shown below can be modified by using other monomers such as isophthalic acid or 4-aminobenzoic acid, or by using a different monomer ratio [75–78]. In order to acetylate the phenolic hydroxyl groups, the monomers are added to a stainless steel stirred tank, followed by an excess of acetic anhydride under an inert gas atmosphere (nitrogen). Too large an excess of acetic anhydride affects the color of the polymer and leads to the formation of mixed anhydride structures. Most patents specify a molar ratio of acetic anhydride to phenolic OH groups of 1.02 – 1.18.

The acetylation reaction is performed at 120 – 150 °C under reflux and is complete after 1 to 5 h. The acetic acid that is formed and any acetic anhydride still present are distilled off. The temperature is then raised in stages to ca. 320 °C and a low molecular mass prepolymer is formed. Polycondensation is then usually carried out by gradual reduction of the pressure (<15 mbar) and distilling off further acetic acid until the desired molecular mass (10 000 – 30 000) is obtained. In contrast to Vectra, the maximum molecular mass of Xydar (*mp* 421 °C) in the melt is limited by the higher melting point and higher melt viscosity. Consequently, the Xydar prepolymer, which is still liquid at 320 °C, is discharged from the reactor under an inert gas, granulated, and postcondensed in the solid phase (inert gas or vacuum) at a maximum temperature of 365 °C.

Vectra is synthesized exclusively from high-purity hyroxycarboxylic acids and can, in theory, undergo polycondensation to yield products of infinitely high molecular mass. The molecular mass can be limited by adding small amounts of 4,4′-diacetoxybiphenyl [79]. Catalysts may, however, be added to accelerate acetylation and polycondensation (ca. 25 – 500 ppm referred to the polymer). Catalysts include alkali-metal acetates (e.g., sodium or potassium acetate) and acids (e.g., *p*-toluenesulfonic acid or 2-naphthalenesulfonic acid). Iron ions have a high catalytic activity in the production of Vectra [74]. The iron content of the polymer should not, however, exceed 50 ppm because this has an adverse effect on its thermal properties.

$$HO-\left[\left(\underset{O}{\underset{\|}{C}}-\bigcirc-O-\underset{O}{\underset{\|}{C}}-\bigcirc-\underset{O}{\underset{\|}{C}}-O-\bigcirc-\bigcirc-O\right)_{2}\right]_x C-CH_3$$

Xydar

Xydar is often synthesized in the presence of potassium sulfate (25 – 50 ppm referred to the polymer) to improve processibility and appearance of the polymer [75].

Production Example for Vectra [74]. 24.05 kg of 6-hydroxy-2-naphthoic acid and 47.85 kg of 4-hydroxybenzoic acid are added to a 151.4-L steel reactor (stainless steel 316) equipped with a helical stirrer, a solids and liquid trap, and a distillation column. The reactor is alternately evacuated (ca. 9 mbar) and flushed with nitrogen three times. 49.90 kg of acetic anhydride (technical quality) is then added. The reaction mixture is heated to 120 – 140 °C and kept at this temperature for 1 h. The temperature is then raised to 200 °C in 1.5 h. Polycondensation takes place at 200 –327 °C over a period of 3 h under nitrogen (2.8 m³/h), followed by 1.25 h (at 327 °C) in vacuo (ca. 13 mbar). The hot polymer melt is then extruded and comminuted. The intrinsic viscosity of the polymer in pentafluorophenol at 60 °C (concentration 0.1 wt %) is 540 cm³/g.

3.3. Analysis

Thermoplastic polyesters have to be analyzed before they are processed. The most important parameter to be determined is the molecular mass because this influences the processability of the molding compounds and the mechanical properties of the article produced from the polyester.

In most cases the molecular mass is determined indirectly via the reduced or specific viscosity (in cm³/g) in a capillary viscometer ($\rightarrow$ Plastics, Analysis). The intrinsic viscosity [η] is thus obtained by extrapolation of the reduced viscosity to zero concentration and from this the molecular mass M [80] according to

$$[\eta] = KM^a$$

For polymers with a range of molecular masses, M becomes M_n (number average), M_w (weight average), or M_v (viscosity average) depending on the method of determination. See also, $\rightarrow$ Plastics, General Survey, Section 2.2.2..

The reduced viscosity depends on the solvent, concentration, and temperature. For quality control it is normally sufficient to specify the solution viscosity under defined conditions.

The constants K and a have been determined for PETP and PBT under various conditions (see Table 2). For PBT a correlation between the reduced viscosity and M_w is also described [86].

The melt viscosity η_m is just as important a quantity as the solution viscosity as regards the processing of thermoplastic polyesters. Measurement under defined conditions (temperature, viscometer, shear rate, etc.) is sufficient for monitoring quality. Like the solution viscosity, the melt viscosity is related to the molecular mass [87]:

$$\eta_{m,0} = KM_w^{3.5}$$

where $\eta_{m,0}$ is the melt viscosity in Pa·s at a shear rate of 0. Values for K in Pa·s are

PETP[88] : $1.13 \times 10^{-14} \exp(-11.9755 + 6082.1/T)$
PBT[85] : 1.02×10^{-14} at 250°C

Investigations on the melt viscosity of polyetheresters and fully aromatic, liquid crystalline polyesters are described in [89].

The molecular mass distribution of thermoplastic polyesters is determined only in special cases (for methods see [90]), for example to establish the influence of process or reactor technology on the polyesters. Most linear polyesters exhibit a Schulz – Flory distribution (see Two Stage Process). Further information on the quality of thermoplastic polyesters is obtained by measuring the melting point, glass transition temperature (both preferably by differential thermal analysis), the hydroxyl terminal groups [91], acid terminal groups [92], and the diglycol content [93].

The water content [94] is also of great importance for the processing of thermoplastic polyesters because water often causes hydrolytic decomposition at high processing temperatures. The equilibrium water content, for a given degree of polymerization P_n must not be exceeded on account of the danger of hydrolytic decomposition. This content can be calculated for high conversions from the approximate formula

$$P_n = \sqrt{\frac{K}{n_{H_2O}}}$$

P where K is the equilibrium constant and n_{H_2O} denotes the number of moles of water per mole of monomer. For example, the maximum water content for PETP ($K \approx 4$) with a molecular mass $M_n \approx 10\,000$ ($P_n = 100$) is 0.01 wt %.

Table 2. Constants K and α for PETP and PBT

Polyester	Solvent	K	α	Method of determination	Reference
PETP	phenol – tetrachloroethane 1 : 1, 20 °C	0.0755	0.685	terminal group determination (M_n)	[81]
PETP	2-chlorophenol, 25 °C	0.030	0.74	terminal group determination (M_n)	[82]
PETP	2-chlorophenol, 25 °C	0.0425	0.69	sedimentation, diffusion (M_w)	[82]
PETP	phenol-1,2-dichlorobenzene 1 : 1, 25 °C	0.0469	0.68	light scattering (M_w)	[83]
PETP	phenol – tetrachloroethane 3 : 2, 25 °C	0.0468	0.68	light scattering (M_w)	[84]
PBT	phenol – tetrachloroethane 3 : 2, 30 °C	0.0116	0.87	terminal group determination (M_n)	[85]
PBT	phenol-1,2-dichlorobenzene 1 : 1, 25 °C	0.0437	0.72	light scattering (M_w)	[83]

A further criterion for the use of thermoplastic polyesters in the fiber or plastics sector is the intrinsic color. This is usually specified as a blue – yellow value (measured with an electric remission photometer) [95].

3.4. Co-Components and Additives [96]

Thermoplastic Copolyesters. The dyeability, elasticity, pilling behavior, shrinkage, hydrophilicity, and flame resistance of polyester fibers can be improved by condensation with appropriate comonomers [97].

The crystallinity, melting point, glass transition temperature, elasticity, flammability, solubility, and other properties of thermoplastic polyesters used in the plastics sector can be modified by selective co-condensation.

Examples of such copolyesters are glass-clear (amorphous) copolyesters based on 1,4-cyclohexanedimethanol, terephthalic acid, and isophthalic acid (i.e., poly(1,4-cyclohexylenedimethylene terephthalate-co-isophthalate) [26124-27-6] [98]); elastic copolyesters based on 1,4-butanediol, poly(tetrahydrofurandiol), and terephthalic acid (i.e., poly(butylene terephthalate-co[multibutylenoxy] terephthalate) [71], [72]; neopentyl-glycol-containing copolyesters with good solution properties in paint solvents [99]; and copolyesters based on 1,4-butanediol, terephthalic acid, and isophthalic acid with relatively low melting points for hotmelt adhesives [100]. Thermoplastic copolyesters with elevated glass transition temperatures (and thus improved heat resistance) can be obtained by incorporating heterocyclic diols during condensation [101]. For the production of copolyesters, see 3.2.5.

Additives. Raw polymers are seldom processed as such into semifinished or finished articles. Additives are usually added to the polymers in a process known as compounding or formulation. Compounding processes (e.g., mixing, kneading, extrusion, pelletization) are of great importance because the properties, processing, and use of thermoplastic polymers are substantially determined by the additives (e.g., nucleating agents, fillers, flame retardants, stabilizers, pigments).

Nucleating Agents [102], [103]. See also → Plastics, Additives, Section 9.1. The processability of PETP in injection molding is mainly determined by its slow crystallization rate and by the associated danger of cold crystallization (tendency to distortion). High molding temperatures (ca. 140 °C) and addition of nucleating agents are therefore necessary to accelerate the nucleation rate. Inorganic substances such as talc, magnesium oxide, or clay are added as nucleating agents in concentrations of 0.001 – 1 % with a maximum particle size of 2 – 3 µm. Organic compounds include salts of aromatic carboxylic acids, fatty acid salts, Montan waxes, and polymers such as polyethylene. The nucleating agents can be added at any stage during PETP production or as a powder to the granulate.

PBT and PBT copolyesters with polytetrahydrofuran generally crystallize so rapidly that nucleation is not necessary.

Glass Fibers [104]. See also → Plastics, Additives, Chap. 7. The heat resistance, rigidity, and hardness of thermoplastic polyesters can be considerably increased by adding glass fibers or beads. For example, the heat distortion temperature (according to Martens, DIN 53 458) of PBT reinforced with glass fiber varies as follows:

Heat distortion temperature, °C	50	75	120	160	190
Glass fiber content, wt %	0	10	20	30	50

The glass reinforcement may be added as beads, ground glass fiber, staple glass fiber, or long glass fiber. Maximum heat resistance, rigidity, and hardness are achieved by adding long glass fibers. These additives are mixed with the thermoplastic polyesters in special extruders in a separate compounding stage; amounts of up to 50 % are commonly employed.

Flame Retardants. See also → Flame Retardants, and → Plastics, Additives, Chap. 6. Thermoplastic polyesters (PETP and PBT) have a relatively low flammability because they escape the action of the flame due to melting and droplet formation (oxygen index of PBT, 23 %; UL 94 rating, HB); see also → Flame Retardants, Chap. 3.). Fillers and, in particular, glass fiber reinforcement prevent droplet formation, resulting in a sharp increase in flammability (oxygen index of PBT reinforced with 10 % glass fiber, 19 %). Legislation is increasingly promoting the use of self-extinguishing, non-drip plastics formulations.

Self-extinguishing thermoplastic polyesters are obtained by incorporating ca. 30 % of flame retardants, which are usually aromatic halogenated compounds combinedwith antimony oxide [105] or with phosphorus compounds [106] (see also → Flame Retardants). Examples are brominated polycarbonates [107] and polystyrenes [108], or ethylene(tetrabromophthalimide) in combination with antimony oxide. A combination of red phosphorus and antimony oxide is also used, particularly with PETP [109].

The flame retardants are not normally added during the production of thermoplastic polyesters, but in a subsequent compounding stage.

The polybrominated diphenyl ethers are suspected of producing polybrominated dibenzodioxins and dibenzofurans in the event of fire [110]; they are therefore no longer used in Germany, and interest in halogen-free flame retardants has greatly increased. Up to now, however, only developments and improvements of conventional systems, such as microencapsulated red phosphorus, ammonium polyphosphates, magnesium hydroxide, and magnesium carbonate have been described [111], [112]. The magnesium compounds have the disadvantage that they have to be added in substantially higher amounts and may therefore seriously affect the polymer properties.

Fully aromatic liquid crystalline polyesters have a low flammability and do not require flame retardants.

Stabilizers. See also → Plastics, Additives, Chap. 2., → Plastics, Additives, Chap. 3.. In contrast to PETP and PBT, thermoplastic polyetheresters are extremely sensitive to photooxidative decomposition [113] and must be stabilized. Stabilizers include sterically hindered phenols or amines, which are added to the starting substances in an amount of 0.1 –0.5 wt % or are incorporated in the polymer melt [114].

The antioxidant 4,4′-bis(α,α'-dimethylbenzyl) diphenylamine has been approved by the FDA for use in phthalate, isophthalate, and terephthalate polyesters with 1,4-butanediol and polytetrahydrofuran.

Pigments [115], [116]. Pigments with an adequate heat resistance (e.g., titanium dioxide, carbon black, phthalocyanines, quinacridones, cadmium yellow) are suitable for coloring thermoplastic polyesters.

Admixture of Other Polymers. The properties of partially aromatic thermoplastic polyesters can be selectively improved by blending with other polymers. See also, → Polymer Blends, Section 4.7.. PBTs whose impact properties have been improved with polybutadiene graft rubbers are used in car bumper systems [117]. Impact strength can also be increased by adding polyethylene [118]. Higher flexibility is obtained by blending with polyether elastomers [119].

New alloying components such as styrene–maleic anhydride copolymers or acrylonitrile–styrene – acrylate copolymers also expand the application range, particularly of PBT [120]. Polymer blends of bisphenol-A-polycarbonate with PBT or PETP have good low-temperature strength and low shrinkage values, they have been commercially successful for several years [121].

Blends with polyurethanes and polyamides [122] or poly-ϵ-caprolactone [123] are also described.

3.5. Properties

The properties of thermoplastic polyesters depend primarily on the starting compounds (dicarboxylic acids, diols), their molecular mass, and added fillers. The properties of PETP, PBT, and PDCT as fiber raw materials are described in → Fibers, 5. Polyester Fibers, Chap. 2..

PETP and PBT. The two most important thermoplastic polyesters for the plastics sector, namely PETP and PBT, can be described as partially crystalline polymers. They have a high hardness and rigidity, good creep strength, high dimensional stability, and very good slip and wear behavior [124], [125]. Whereas PBT occurs almost exclusively in the partially crystalline form, PETP can also be processed into amorphous molded bodies with high transparency; on heating to 70 – 100 °C this transparency is lost due to postcrystallization. Transparency can, however, be retained if the polyester chains are forcibly oriented, for example by biaxial stretching and heat setting (transparent films of PETP, → Films, Section 6.8.).

A special property of PETP is its slow crystallization, which necessitates molding temperatures of ca. 140 °C and the use of a nucleating agent and crystallization accelerator (see).

PBT and PETP are resistant to water, weak acids and bases, alcohols, ketones, ethers, aliphatic hydrocarbons, and chlorinated aliphatic hydrocarbons at room temperature.

Solvents for PBT and PETP include hexafluoro-2-propanol, hexafluoroacetone, and 2-chlorophenol (mixed solvents, see Table 2).

Hydrolysis proceeds rapidly in the melt in the presence of moisture. The dependence of the hydrolysis rate on temperature and moisture content has been investigated [126], [127].

The most important mechanical, thermal, and electrical properties of PETP and PBT are summarized in Table 3.

Polyetheresters [128–130]. Thermoplastic polyetheresters are block copolymers synthesized from mutually incompatible rigid crystalline and soft amorphous segments. The rigid segments generally consist of an aromatic polyester (e.g., PBT), whereas the soft segments consist of an aliphatic polyether [e.g., poly(butylene glycol) or

Table 3. Properties of PETP and PBT [125]

Property	Test method	PETP			PBT		
		Partially crystalline	30 % glass fiber reinforced	Partially crystalline (standard grade)	Partially crystalline (tough grade)	30 % glass fiber reinforced	30 % glass fiber reinforced with flame retardant
Mechanical (23°C)							
Tensile strength and yield stress, N/mm^2	DIN 53 455	73	150 – 170	52	38	130 – 140	110 – 120
Elongation at break, %	DIN 53 455	300	2	>250	>250	2	2
Modulus of elasticity tensile test), N/mm^2	DIN 53 457	2850	10 000 – 12 000	2700	1100	8500 – 10 000	9000 – 11 000
Bending strength, N/mm^2	ASTM D 790	118	230 – 250	85	43	200 – 210	190 – 200
Impact strength, kJ/m^2	DIN 53 453		25 – 40			30 – 40	20 – 25
Notched bar impact strength, kJ/m^2	DIN 53 453	4	6 – 14	5	23	8 – 19 [*]	6 – 14 [*]
Ball indentation hardness, N/mm^2	DIN 53 456	170	200 – 230	139	86	200 – 230	210 – 240
Physical and thermal							
Density, g/cm^3	DIN 53 479	1.37	1.58	1.31	1.29	1.53	1.66
Moisture absorption (saturation 23 °C, 50 % relative humidity), %	ASTM D 570	0.35	0.25	0.20	0.20	0.13	0.10
Deflection temperature under load, °C	ISO/R 75, method A	82	225	50	50	205 – 225 [*]	205 – 225 [*]
Linear coefficient of expansion, 10^{-5} K^{-1}	ASTM D 696	7	3	8	13	3	3
Flammability	UL 94						V-0
Crystallite melting point, °C	differential thermal analysis	255 – 260	255 – 260	220 – 225	200 – 205	220 – 225	220 – 225
Electrical							
Volume resistivity, Ω · cm	VDE 0303, part 3	>10^{14}	>10^{15}	>10^{15}	>10^{15}	>10^{15}	>10^{15}
Surface resistance, Ω	VDE 0303, part 3	>10^{13}	>10^{13}	>10^{13}	>10^{14}	>10^{13}	>10^{13}
Dielectric loss factor tan δ,							
50 Hz	VDE 0303, part 4	0.0015	0.002	0.002	0.007	0.0024	0.0025
10^6 Hz		0.019	0.016	0.020	0.021	0.018	0.016
Dielectricity constant ϵ_r,							
50 Hz	VDE 0303, part 4	4.3	4.3	3.8	3.7	4.4	4.5
10^6 Hz		3.5	4.4	3.2	3.7	3.8	3.8
Dielectric strength, kV/cm	IEC Pub 234, (2 mm plate)	200	170	150	180	170	170

[*] Upper values refer to grades reinforced with long fibers.

poly(ethylene glycol)]. On account of the physical cross-linking of the rigid and soft domains, polyetheresters behave like elastomers at the service temperature ($>T_g$, $<mp$). The cross-linking sites are, however, thermally labile and the polymers can therefore be thermoplastically processed above the melting point.

Important thermal, mechanical, and electrical properties of polyetheresters are summarized in Table 4.

The very low glass transition temperature of the polyether (ca. −50 °C) and the high melting point of the polyester (ca. 200 °C) result in polymers with a very broad service temperature range. Polyetheresters with Shore-D hardnesses of 35 – 72 can be formulated by varying the block length and the ratio of the rigid and soft components.

The special properties depend on the nature of the rigid and soft segments. Copolyesters based on butylene glycol terephthalate crystallize very rapidly from the melt, in contrast to polymers synthesized from ethylene glycol terephthalate blocks. Copolymers with poly-(propylene oxide glycol) and poly(tetramethylene ether glycol) exhibit similar swelling behaviors in water and oil. On the other hand, polymers

Table 4. Properties of polyetheresters [131], [132]; Xydar [133], [134]; and Vectra [134]

Property	Test method	Polyetherester	Vectra		Xydar	
			A 950 (base type)	A 130 (30 % glass fiber reinforced)	SRT 300 (base type, high-melting)	G 330 (30 % glass fiber reinforced)
Mechanical (23 °C)						
Tensile strength, N/mm^2	DIN 53 455	9 – 46	156	188	138	117
Elongation at break, %	DIN 53 455	170 – 700	2.6	2.1	4.9	2.6
Tensile modulus of elasticity, N/mm^2	ASTM D 638		9700	16 600	16 600	12 000
Bending modulus of elasticity,						
−40 °C, N/mm^2	ASTM D 790	160 – 2400				
23 °C N/mm^2	ASTM D 790	50 – 600	9000	15 000	13 800	11 600
100 °C N/mm^2	ASTM D 790	30 – 210				
Bending strengh, N/mm^2	ASTM D 790				131	150
Notched bar impact strength according to Izod, J/m	ASTM D 256					
−40 °C	ASTM D 256	without break				
23 °C	ASTM D 256	without break	520	150	130	176
Hardness, Durometer *D*, points	DIN 53 505	35 – 72				
Physical and thermal						
Density, g/cm^3	ASTM 0792	1.1 – 1.3	1.4	1.6	1.4	1.6
Moisture absorption, %	ASTM D 570	0.2 – 0.6	0.02	0.02	<0.1	<0.1
Heat distortion temperature (HDT/A, 1.82 N/mm^2), °C	ASTM D 648	<50	180	230	355	252
Linear coefficient of expansion, 10^{-6} K^{-1}						
Longitudinal	ASTM D 696	ca. 90 – 200	−5	5		5
Transverse		ca. 90 – 200	75	65		
Flammability	UL 94	HB	V-0	V-0	V-0	V-0
Melting point, °C	differential scanning calorimetry	148 – 219	280	280	421	
Vicat softening point, °C	DIN 53 460 ASTM D 1525	80 – 207	145	177	369	
Electrical						
Volume resistivity, Ω · cm	DIN 53 482	10^{10}–10^{14}	10^{16}	10^{16}		
Surface resistance, Ω	DIN 53 482	>10^{11}	4×10^{13}	8×10^{13}		
Dielectric loss factor tan δ, 10^6 Hz	ASTM D 150	0.036 – 0.06	0.02	0.017	0.041	0.033
Dielectric constant ϵ_r, 10^6 Hz	ASTM D 150		3.3 (kHz)	3.7	3.1	3.6
Dielectric strength, kV/mm	ASTM D 149		39	43	23	35

based on hydrophilic poly(ethylene oxide glycol) swell markedly in hot water, but less in oil. The incorporation of poly(propylene oxide glycol) decreases the thermal stability.

Polyetheresters have a good notched bar impact value at low temperature, high strength, flexibility, and abrasion resistance. On account of their partially crystalline nature they are insoluble in most solvents, exceptions include m-cresol and tetrachloroethane which are used as solvents for viscosity measurements. The swelling resistance in oils, aliphatic and aromatic hydrocarbons, alcohols, ketones, and esters improves with increasing hardness. Polyetheresters are extremely impermeable to fuels. Their good hydrolysis resistance can be improved by adding polycarbodiimides [135].

The resistance of polyetheresters to light, heat, and oxidation is lower than that of PBT or PETP (for stabilization, see).

Liquid Crystalline Polyesters [136–138]. The two partially crystalline polyesters Xydar and Vectra have only been on the market since the mid 1980s. These liquid crystalline polymers (LCPs) represent one of the most interesting developments in the technical plastics sector. See also → Liquid Crystals, Chap. 4.; → Plastics, General Survey, Section 5.3..

On account of their rod-shaped molecular structure, the polymer chains are partially ordered in the melt. The macromolecules adopt a parallel orientation in microscopically small regions, with the result that the melt exhibits properties of both a solid (order) and a liquid (mobility). This intermediate state is termed liquid crystalline. Xydar and Vectra have nematic phases, i.e., the longitudinal axes of the molecules are parallel, but the molecular centers of gravity are irregularly distributed (→ Liquid Crystals, Section 3.1.). Under the influence of a shear force, such as occurs in normal thermoplastic distribution, the microscopically ordered regions can be macroscopically oriented. This orientation is retained in the solid state and is responsible for the outstanding properties of liquid crystalline polyesters. These properties are, however, strongly orientation-dependent due to the anisotropy of the polymers. LCPs cannot therefore be compared with conventional thermoplastic polymers solely on the basis of values measured on uniaxial test bodies. Nevertheless, LCPs differ from "conventional" thermoplastics as regards the following properties (see also Table 4):

1. Their coefficient of thermal expansion is unusually low for polymers and can reach the values of steel; it can be adapted to other materials by varying the processing conditions2) Good processability: on account of their very low melt viscosity, even filigree structural parts can easily be filled
2. Low melt enthalpy (short cycle times)
3. Low flammability: V-0 rating according to UL 94 with wall thickness <0.8 mm
4. High heat resistance
5. Resistance over a wide temperature range to solvents, fuels, and most chemicals, except oxidizing acid and strong alkali
6. Resistant to radiation and transparent to short-wavelength light
7. Low water absorption
8. Very good mechanical properties (strength, impact resistance, modulus of elasticity)

The anisotropy of the mechanical properties and the thermal coefficient of expansion can be reduced by adding fillers (glass fibers, minerals), which simultaneously improve surface quality. LCPs have a poor weld strength, in unfilled LCPs it is about half the strength transverse to the flow direction. To overcome this, special molded parts are constructed in which the weld is located in such a way that it is not subsequently exposed to high stress.

The most important mechanical, thermal, and electrical properties of LCPs are summarized in Table 4.

3.6. Uses

Poly(Ethylene Terephthalate). More than 85 % of PETP is processed into fibers (→ Fibers, 5. Polyester Fibers, Section 3.4.). Poly-(1,4-dimethylenecyclohexane terephthalate) is used for special fibers (see → Fibers, 5. Polyester Fibers, Section 4.2.). The production of high-grade, heat-resistant PETP films is described in → Films, Section 6.8.. A large proportion of PETP is used to produce gastight bottles for carbonated beverages (see Section 3.8). Highly stressed technical molded parts such as bearings, gearteeth, cam wheels, connectors, bolts, screws, and washers are produced from PETP by injection molding [139].

On account of the processing difficulties (see Section 3.4) fewer molded parts are made from PETP than from PBT. If new developments aimed at reducing the high molding temperatures prove to be successful, this proportion may rapidly increase [140]. Injection molding machines are mainly used for processing.

Trade names include Arnite, Akzo; Crastin, CibaGeigy; Grilpet, EMS-Chemie; Impet, Hoechst Celanese; Melinite, ICI; Rynite, Du Pont; Techster, Rhône-Poulenc; Valox, GE Plastics.

Poly(Butylene Terephthalate). PBT too can be spun into fibers (see → Fibers, 5. Polyester Fibers, Chap. 4.), which can be colored, without carriers, with disperse dyes and used in the carpets sector [141]. Trade names are sometimes the same as for PETP fibers. Examples include Trevira, Hoechst; Tergal, Rhône-Poulenc; Fortrell, Celanese.

Most PBT is processed by injection molding and extrusion, a relatively low-viscosity formulation is used for injection molding, and a medium- to high-viscosity formulation for extrusion.

Examples of injection-molded PBT products (unreinforced and reinforced) include housings for heated domestic appliances (e.g., coffee-making machines, egg boilers, toasters, hair-dryers); bearing bushes, trip cams, gear wheels, fan wheels, ropesheaves, and pump impellers for the mechanical engineering sector; coil formers, capacitor covers, plug boards, and lamp sockets (often with flame retardants) for the electrical sector.

Protective covering for optical waveguides, films, sheets, monofilaments, and hydraulic lines are produced by extrusion. Paper and board can also be extrusion coated.

Trade names include B. Arnite, Akzo; Celanex, Hoechst Celanese; Crastin, Ciba-Geigy; Grilpet, EMS-Chemie; Orgater, Atochem; PBT, Toray; Pibiter, Montedison; Pocan, Bayer; Rynite, Du Pont; Techster, Rhône-Poulenc; Tenite, Eastman Kodak; Ultradur, BASF; Valox, GE Plastics; Vestodur, Hüls.

Copolyesters. Reference has already been made to possible uses of *thermoplastic copolyesters* (Section 3.4). Copolyetheresters [synthesized from dimethyl terephthalate, 1,4-butanediol, and poly(tetrahydrofuran)diol] are used on account of their high flexibility in the cold (without plasticizers), particularly for winter sports articles (e.g., ski brakes, soles for cross-country ski shoes, disks for ski poles). A novel field of application is crawlers for snowmobiles, where high abrasion resistance is also required. Copolyetherester elastomers can be processed by rotational casting into tires for vehicles not intended for use on roads (lawnmowers, garden tractors, golf trolleys), and into hydraulic oil tanks. Highly flexible pipes (e.g., for hydraulics lines) can be produced by extrusion.

Trade names include Hytrel, Du Pont; Arnitel, Akzo; Lomod, General Electric Plastics.

Amorphous copolyesters [99] produced from dimethyl terephthalate, terephthalic acid, isophthalic acid, neopentyl glycol, ethylene glycol, and other monomers are soluble in conventional solvents and can be processed into highly elastic, sterilization-resistant coatings (coil coatings). See also → Paints and Coatings, 1. Introduction.

Trade names include Dynapol L, Hüls; Vitel, Goodyear.

Partly crystalline copolymers based on poly(butylene terephthalate-*co*-isophthalate) are valuable hotmelt adhesives [100] for the shoe industry (shoe soles) or furniture industry.

Trade names include Dynapol S, Hüls; Vestamelt, Hüls.

Copolyesters based on similar starting components are used for powder coatings that produce highly flexible, weather-resistant coatings [142].

Trade names include Dynapol P, Hüls.

Liquid Crystalline Polyesters. Xydar and Vectra types cannot be used without reinforcement or fillers on account of their marked anisotropic properties and tendency to fibrillation. They are usually reinforced or filled with glass fibers, carbon fibers, or minerals and are processed by injection molding.

Areas of application exploit the properties described in Section 3.5. Examples include the automotive and aviation industries, where good thermal and mechanical properties, inherent flame resistance, and low fume formation are important for interior fixtures and fittings.

On account of the low thermal expansion, high heat resistance, and favorable electrical properties, LCPs are employed in the microelectronics and optoelectronics sectors, mainly for integrated circuits, connectors, coil formers, lighting fixtures, and sheathing for glass fiber cables.

On account of their good resistance to chemicals and solvents, LCPs are used as packings in distillation columns and for fuel lines. Vectra (Hoechst-Celanese) and Xydar (Amoco) are transparent to microwave radiation, and are therefore used for microwave cookware and structural parts in microwave ovens.

3.7. Toxicology

High molecular mass polycondensates are usually considered to be biologically inert. However, the biological effect of polymers can be influenced by soluble or volatile components of monomers, oligomers, additives, and degradation products.

The toxicology of PETP has been investigated, especially for its use as a packaging film and textile fiber. No toxic effects were observed [143]. Tests on bacteria showed that neither PETP polyester nor other substances contained in PETP bottles extracted with dichloromethane were mutagenic [144].

PETP, BPT, and various copolyesters with 1,4-cyclohexanedimethanol and/or poly(tetrahydrofurandiol) are permitted by the FDA. They can therefore be used in contact with foods under particular conditions (viscosity, extract content, temperature, etc.) in accordance with CFR from 1988: CFR 170.1240; CFR 177.1315; CFR 177.1590; CFR 177.1630; CFR 177.1660.

Polyterephthalate diol esters are approved by the German health authority (Bundesgesundheitsamt) for use according to the food and commodity law (Aug. 15, 1974). The starting materials which can be used are 1,4-butanediol, ethylene glycol, 1,4-cyclohexanedimethanol, poly(ethylene glycol), terephthalic acid, dimethyl terephthalate, isophthalic acid, adipic acid, azelaic acid, dimethyl azelate, sebacic acid, dimethyl sebacate, n-dodecanedioic acid.

3.8. Economic Aspects [140], [145–151]

PETP and PBT. The above-average growth in thermoplastic polyesters that occurred at the start of the 1980s has continued. Double-figure growth rates (ca. 15 %) in the PBT and PETP molding materials sector have been achieved worldwide. Table 5 shows the development of demand for PETP and PBT between 1984 and 1989.

Most thermoplastic molding materials are processed in glass-fiber-reinforced form. PETP molding compounds are difficult to process and are therefore used in substantially smaller amounts than PBT molding compounds. The main areas of use are electrical engineering and electronics (25 – 30 % of overall production) and automobile construction (20 – 25 % without blends).

PBT – PETP blends, which have a good surface quality, as well as PBT – polycarbonate blends with high impact strength have become important. The amount of PBT used for blends in 1988 was ca. 10 000 t/a in Western Europe and ca. 20 000 t/a in the United States. The amount of PETP used for blends in 1988 was ca. 2000 t/a in Western Europe and 10 000 t/a in the United States.

PBT is largely used for construction materials, whereas PETP is mainly processed into fibers, sheets, films, and beverage containers. In 1989 15.6×10^6 t of synthetic fibers were produced worldwide; polyester fibers, primarily PETP, accounted for 8.4×10^6 t (54 %). The development of consumption for PETP sheets, films, and bottles is given in Table 6.

Films and sheets of PETP are chiefly used in the photographic sector, magnetic tapes, and packaging materials.

Table 5. Development of demand for PETP and PBT molding materials (without blends) in 10^3 t

	1984	1986	1988	1989
PETP				
United States	9	13	18	20
Japan	6.5	8	10	12
Western Europe	3.8	5	6	8
PBT				
United States	36	51	60	66
Japan	29	39	49	55
Western Europe	19	23	29	31

Table 6. Development of consumption for PETP sheets/films and bottles in 10^3 t

Region	1984	1986	1988	1989
United States				
Sheets/films	209	200	232	237
Bottles	243	291	334	360
Western Europe				
Sheets/films	130	170	190	195
Bottles	70	110	175	195
Japan				
Sheets/films	93		180	
Bottles	27		65	

Polyetheresters. The worldwide demand for polyetheresters based on dimethyl terephthalate, 1,4-butanediol, and polytetrahydrofuran was estimated to be 38 000 t in 1991. Of this figure, the United States accounts for 22 000 t and Western Europe and Japan 8000 t each. The annual growth rates are 10 – 15 %. The most important areas of use are hoses, cables, and automobile parts.

Liquid Crystalline Polyesters. The plant capacities for LCPs are ca. 9000 t/a, of which Vectra (Hoechst-Celanese) accounts for ca. 2300 –3600 t/a [152] and Xydar (Amoco) for ca. 4500 t/a [153]. The demand in the United States (ca. 90 % of overall demand), Western Europe, and Japan is currently considerably lower, at ca. 2500 t/a. However, high average growth rates, particularly in the electrical engineering and electronics sector, of at least 15 % to more than 20 % per annum are anticipated [154].

The major proportion (ca. 60 %) of LCPs is used for electrical and electronics components as well as for microwave-transparent cook ware. The remainder is used in the industrial, telecommunications, automotive, and aviation sectors.

3.9. Recycling

Polyesters (PETP, PBT) play an important role in the production of beverage bottles, foils, films, and fibers. The local authorities, in particular, protest against the increase in the volume of waste caused by the use of nonreturnable polyester bottles. In countries in which these bottles have already been introduced (e.g., USA, Japan, France, Italy, and the UK) laws have been passed to enforce the return and reutilization of the bottles. Because of this, the recycling rate for PETP bottles in the United States, for example, is 90 % [155]. In Europe the proportion of collected PETP is ca. 4 %. Two methods are used for recycling polyester: (1) direct reuse after comminution and washing, and (2) alcoholysis.

It is more difficult to directly reuse PETP foils and films of which the majority consist of composite materials. Such waste can only be processed by combustion or chemical recycling via pyrolysis.

Reuse. Prerequisites for good recycling of PETP materials are [156]:

1. There must be a sufficient amount of homogeneous recyclable material
2. An economical recycling process must be available for cleaning and separating the materials
3. The recovered plastic must have a sufficiently large market and a sufficiently high price to warrant economical recycling

This means that in designing and manufacturing a product, allowances have to be made for its subsequent recycling product. As a result of its high density (1.32 – 1.40 g/cm^3), PETP can be efficiently separated from most other plastics. Separation from poly(vinyl chloride) (PVC) is possible with the help of flotation. High standards are required during sorting for reuse as material for bottles. Due to the incompatability of PETP with the majority of other plastics the proportion of the latter should not exceed 100 ppm, i.e., only one polyethylene or PVC bottle should be among 10 000 PETP bottles.

In addition to collection, the following costs are estimated:

Sorting	0.3 DM/kg
Grinding	0.2 DM/kg
Washing, dryin	0.4 DM/kg

In the United States a new plant has been built by Du Pont and Waste Management of North America (WMNA) in Chicago (capacity 20 000 t/a); it separates the PETP chips and converts them to clean polymer. Du Pont then uses or sells the final product [155].

In every recycling step some cyclic and linear oligomers are formed. Especially the linear oligomers cause some problems during processing.

Hydrolysis and Alcoholysis. Another way to break down polyester is hydrolysis or alcoholysis which sometimes requires drastic reaction conditions and long reaction times. The hydrolysis of polyesters results in the formation of carboxylic acids and alcohols, from which new polyester can only be produced after separation and purification.

The breakdown of polyesters is more readily achieved by alcoholysis. The diols and dicarboxylates are formed according to the following reaction scheme:

$$\sim\sim\sim R'-CO-O-R-O-CO-R'-CO-O-R\sim\sim\sim \xrightarrow{R'OH}$$

$$\sim\sim\sim R'COOR' + HOROH + R'OOC-R'-COOR' + HOR\sim\sim\sim$$

If methanol is used for the alcoholysis of PETP, the monomers used for its polymerization are reformed. If multifunctional alcohols are employed, polyols can be produced, which can be modified and used for various purposes (e.g., for the production of polyurethane foam). Both processes are carried out on a commercial scale in the United States [157].

Glycolysis to monomer (dihydroxyethyl terephthalate) and low molecular mass oligomers represents a compromise between the regeneration of the starting materials by methanolysis or hydrolysis and direct remelt extrusion. Glycolysis signifies degradation with glycols [158]; it is less expensive than methanolysis or hydrolysis and more versatile than remelt extrusion. The resultant low-viscosity monomer can easily be filtered and is repolymerized to a useful molecular mass; it is used as a comonomer and ingredient for controlling luster, color, etc.

Glycolytic recycling can be a batch or continuous process, the rate depending on temperature, catalyst, state of subdivision of the feedstock and the glycolytic PETP ratio. In addition, the final monomer composition is controlled by the reaction time and holdup time after depolymerization. Low ratios of glycol to ester permit higher temperatures and faster reactions but result in higher molecular mass oligomers. Side reactions occurring during depolymerization are minimized by the addition of a buffer such as sodium acetate and by limiting the storage time at high temperature. PETP scrap suitable for glycolytic recycling includes production waste, fibers, films, flakes, and bottles; however, ingredients leading to side reactions, copolymers, end caps, or color must be minimized for critical products and satisfactory repolymerization rates.

Pyrolysis. If the polyester waste is highly soiled or combined with other materials, then in most cases it can only be chemically recycled by pyrolysis [159]. Pyrolysis denotes thermal decomposition under the exclusion of air. Mixtures of hydrocarbons are obtained, whose value is smaller than that of the glycolysis products but higher than that of the pure calorific value.

In the process developed by the University of Hamburg, polyester and other plastics are broken down in a fluidized bed that is indirectly heated by radial thermal pipes [160]. The fluidized bed consists of fine-grained sand (0.3 – 0.5 mm) and is heated to 650 – 850 °C. The pulverized polyester waste is transferred into the fluidized bed with a screw. After leaving the reactor the pyrolysis products are freed of entrained sand and fillers in a cyclone and cooled, whereby oils are condensed and expelled. The residual gas serves for the fluidization of the bed and as heating gas for the burner and other purposes. The plant has a capacity of 20 – 50 kg/h.

120.5 kg of chips consisting of PETP or exposed photographic films were fed with a screw (capacity 18.9 kg/h) into the fluidized bed and heated to 768 or 700 °C, respectively [161]. Table 7 summarizes the product groups obtained from PETP. The fraction of water, which is low (2.1 wt %), is the result of adhering moisture and formation during the decomposition of PETP. There are more water and solid residues in the pyrolysis products of film materials (Table 8). The

Table 7. Products obtained after pyrolysis of 120.5 kg polyester at 768 °C in a fluidized bed

Products	Content	
	kg	wt %
Gas	61.2	50.8
Benzene- and toluene-boiling fractions	26.7	22.2
High-boiling fraction	12.8	10.6
Tar (*bp* > 340 °C)	8.7	7.2
Soot (organic solids)	8.6	7.1
Water	2.5	2.1

Table 8. Components (wt %) obtained after pyrolysis of PETP at 768 °C and film material (PETP, polyethylene, and gelatin) at 700 °C

Component	Pure PETP	Used film
Hydrogen	0.30	0.37
Methane	3.80	5.28
Ethane	0.24	1.70
Ethylene	1.50	3.44
Propene	0.08	1.73
Butene	0.01	0.41
Butadiene	0.04	0.27
Carbon dioxide	17.20	17.30
Carbon monoxide	24.70	16.39
Benzene	18.30	7.79
Toluene	2.50	1.48
Styrene, ethylbenzene	2.55	6.26
Indene	0.41	0.23
Naphthalene	1.40	2.04
Biphenyl	1.47	0.97
Fluorene	0.18	0.10
Ketones, acetates	0.15	0.12
Acetophenone	1.40	1.81
Other compounds	11.87	11.15
Carbon black, fillers	7.10	13.40
Water	2.10	8.76

main gaseous components are carbon monoxide and dioxide. Of the hydrocarbons, on-ly methane and ethylene have notable percentages. The net calorific value (13.4 MJ/m^3) is ca. 40 % of the value for natural gas. As a result of its hydrogen content of 8.7 vol % the gas burns with a less vigorous flame.

The largest oil fraction is the benzene- and toluene-boiling fraction, in which the benzene, toluene, and xylene (BTX) aromatics account for over 97 %. This means that pyrolysis of 1 t of polyester waste yields 215 kg of the petrochemically valuable BTX aromatics. The concentrations of the aliphatic and heteroatom-containing hydrocarbons are both < 0.2 wt %. These low concentrations make the processing of this pyrolysis oil fraction to BTX aromatics very much easier.

In contrast, the high-boiling components, constitute a multicomponent system. This fraction, including tars, accounts for 17.8 wt %. The main components are naphthalene, biphenyl, indene, and acetophenone. The high content of acetophenone (10 wt %) is characteristic of the pyrolysis of polyester (PETP). This compound is not formed during the pyrolysis of polyolefins. After pyrolysis the remainder of the oxygen bound to the polyester is distributed as follows: 62 % in carbon dioxide, 31.6 % in carbon monoxide, 5.5 % in water, 0.75 % in acetophenone, and 0.15 % in other keto-compounds and carboxylic acids.

The pyrolysis of polyester and composite materials such as films (PETP, polyethylene, gelatin) in the fluidized bed showed that oxygen-containing plastics can be largely defunctionalized and high proportions of mononuclear aromatics can be obtained.

References

1 O. Bayer, *Angew. Chem.* **59** (1947) no. 9, 257 – 272.
2 I. G. Farbenind., DE 756 058, 1940 (W. Bunge, O. Bayer, S. Petersen, G. Spielberger).
3 Dynamit Nobel, DRP-Anm. D 90 260, 1943 (H. Pinten).
4 A. Höchtlen, *Kunststoffe* **40** (1950) 221.
5 E. Müller et al., *Angew. Chem.* **64** (1952) no. 19/20, 523 – 531.
6 H. Wagner, H. F. Sarx: *Lack Kunstharze*, 5th ed., Hanser Verlag, München 1971.
7 R. D. Aylesworth et al., *Mod. Plast.* **35** (1958) no. 9, 145.
8 Bayer, EP 0 180 749, 1988 (G. Weber, K. König, M. Schmidt).
9 Union Carbide, US 2 890 208, 1956 (D. M. Young, F. Hostettler, C. Horn); US 3 284 417, 1966 (F. Hostettler, G. Magnus, H. Vineyard); EP 117 538, 1984 (L. Domeier, E. Hsi).
10 BASF, GB 2 147 595, 1983 (C. J. Reichel). Bayer AG, DE-OS 3 437 915, 1984 (M. Barnes, W. Betz). DE-OS 3 621 039, 1986 (M. Dietrich, M. Kapps, K. König, R. Nast).
11 J. H. Saunders, K. C. Frisch: *Polyurethanes, Chemistry and Technology*, Interscience, New York 1962.
12 P. J. Flory: *Principles of Polymer Chemistry*, Cornell University Press, New York 1953.
13 P. J. Flory, *Angew. Chem.* **86** (1974) 109.
14 W. Griehl, G. Schnock, *Faserforsch. Textiltech.* **8** (1957) no. 10, 408.
15 G. Henrici-Olivé, S. Olivé: *Polymerisation, Katalyse, Kinetik, Mechanismen*, Verlag Chemie, Weinheim, Germany 1969.
16 *Houben-Weyl* **14/2**, 2.
17 BASF, NL 6 505 683, 1965.
18 H. Batzer, *Makromol. Chem.* **7** (1951) 82.

General References

19 V. V. Korshak, S. V. Vinogradova: *Polyesters*, Pergamon Press, London 1965.
20 R. Vieweg, L. Goerden: "Polyester", *Kunststoff-Handbuch,* vol. VIII, Hanser Verlag, München 1973.
21 H. Ludewig: *Polyester-Fasern*, 2nd ed., Akademie-Verlag, Berlin 1975.
22 *Process Economics Program*, Stanford Research Inst.,, Menlo Park, Ca.,vol. **18**, 1966, vol. 18 A, 1972; vol. 96 (1975); vol. 96 A, 1977.
23 B. M. Walker, C. P. Rader: *Handbook of Thermoplastic Elastomers*, Rheinhold Publ., New York 1988.
24 A. Blumstein: *Polymeric Liquid Crystals*, Plenum Press, New York 1985.
25 L. L. Chapoy: *Recent Advances in Liquid Crystalline Polymers*, Elsevier Applied Science Publishers, London 1985.

Specific References

26. D. W. van Krevelen, J. Bussink, F. J., Huntjens, J. L., Voigt: "Processing Polymers to Products," Internat. Congress 1966, Amsterdam Oct. 17 – 19, 1966, N.V.'t Raedthuys, Utrecht, p. 167.
27. H. G. Elias, F. Vohwinkel: *Neue Polymere Werkstoffe für die industrielle Anwendung*, Hanser Verlag, München 1988.
28. J. Economy, *Mol. Cryst. Liq. Cryst.* **169** (1989) 1.
29. W. J. JacksonJr., *Mol. Cryst. Liq. Cryst.* **169** (1989) 23.
30. M. Hizikata, *CEER Chem. Econ. Eng. Rev.* **9** (1977) no. 11, 32.
31. J. Dörffel, J. Rüter, W. Holtrup, R. Feinauer, *Farbe +Lack* **82** (1976) no. 9, 796.
32. P. J. Flory, *J. Amer. Chem. Soc.* **58** (1936) 1877. P. J. Flory: *Principles of Polymer Chemistry*, Cornell University Press, Ithaca N.Y. 1953. H. G. Elias, *Makromoleküle*, vol. **1**, Hüthig & Wepf Verlag, Basel 1990, p. 232.
33. J. Hsu, K. Y. Choi, *J. Appl. Polym. Sci.* **33** (1987) 329.
34. J. Otton, S. Ratton, *J. Polym. Sci. Part A* **27** (1989) 3535.
35. L. Yurramendi, M. J. Barandiaran, J. M. Asúa, *Polymer* **29** (1988) 871.
36. F. Pilati *et al.*, *Polymer* **22** (1981) 799; **22** (1981) 1566.
37. F. Pilati *et al.*, *Polymer* **24** (1983) 1479.
38. T.-Y. Ju *et al.*, *Polymer* **27** (1986) 1111.
39. A. A. Khan, K. Y. Choi, *J. Appl. Polym. Sci.* **37** (1989) 707.
40. A. Fradet, E. Maréchal, *Adv. Polym. Sci.* **43** (1982) 52.
41. G. Challa, *Makromol. Chem.* **38** (1968) 105.
42. G. Reinisch, H. Zimmermann, G. Rafter, *Faserforsch. Textiltech.* **20** (1969) no. 5, 225.
43. T. Tani, K. Enok, *Hydrocarbon Process.* **49** (1970) no. 11, 146.
44. P. W. Morgan: *Condensation Polymers by Interfacial and Solution Methods*, Interscience, New York 1965.
45. M. Levin, S. S. Temin, *J. Polym. Sci.* **28** (1958) 79.
46. R. F. Fischer, *J. Polym. Sci.* **44** (1960) 155.
47. K. Hamann, *Makromol. Chem.* **51** (1962) 53.
48. UCC, US 3 021 313, 1959 (E. F. Cox, F. Hostettler).
49. J. M. Besnoin, K. Y. Choi, *J. Macromol. Sci. Rev. Macromol. Chem. Phys.* **C 29** (1989) no. 1, 55.
50. B. Möller *et al.*, *Acta Polym.* **33** (1982) no. 1, 38.
51. G. Rafler, J. Blaesche, *Acta Polym.* **33** (1982) no. 8, 472.
52. K. Yoda, K. Kimoto, T. Toda, *Nippon Kagaku Kaishi* **67** (1964) 909.
53. K. Tomita, H. Ida, *Polymer* **16** (1975) 185.
54. H. Zimmermann, E. Schaaf, *Faserforsch. Textiltech.* **20** (1969) no. 4, 185.
55. K. Tomita, *Polymer* **17** (1976) 221.
56. L.-C. Hsu, *J. Macromol. Sci. Phys.* **B 1** (1967) no. 4, 801.
57. Hüls, DE-OS 2 811 982, 1978 (G. Horlbeck, K. Burzin, R. Feinauer).
58. *Chem. Eng. News* **49** (1971) no. 1, 31.
59. Goodyear, GB 1 261 529, 1972 (R. V. Rinehart).
60. Goodyear, GB 988 548, 1965 (W. B. Pengilly).
61. U. Tiehle, *Kunststoffe* **79** (1989) 11.
62. Fiber Industries, GB 1 296 242, 1972 (D. J. Lowe).
63. Fiber Industries, CA 962 396, 1975 (P. Schaefer, W. H. Yates, P. A Mason).
64. J. E. Kemkes, *J. Polym. Sci. Part C* **22** (1969) no. 2, 713.
65. S. G. Hovenkamp *et al.*, *J. Polym. Sci. Polym. Chem. Ed.* **8** (1972) no. 3, 679.
66. Standard Oil, US 4 014 858, 1977 (G. R. Chipman).
67. H. D. Schumann, *Chemiefasern/Textilind.* **40** (1990) 1058.
68. Du Pont, EP 0 431 977, A 2, 1989 (D. J. Lowe).
69. Mitsubishi, DE-OS 2 539 249, 1974 (T. Kimura, K. Kohyama, T. Okada, K. Sakata).
70. Eastman Kodak, US 2 901 466, 1959 (C. J. Kibler, A. Bell, J. G. Smith).
71. Du Pont, DE 2 213 128 C 3, 1986 (W. K. Witsiepe).
72. Du Pont, DE 2 265 319 C 3, 1989 (W. K. Witsiepe).
73. Celanese Corporation, EP 0 088 546 B 1, 1983 (A. J. East).
74. Celanese Corporation, EP 0 117 086 B 1, 1984 (L. F. Charbonneau).
75. Dart Industries, DE 3 443 219 A 1, 1985 (A. B. Firestone).
76. Celanese Corporation, EP 0 183 433 B 1, 1989 (G. D. Kiss).
77. Celanese Corporation, DE 2 844 817 C 2, 1989 (G. W. Calundann).
78. P. Edwards, *Mod. Plast. Intern.* **19** (1989) March, no. 3, 32.
79. Celanese Corporation, US 4 567 247, 1984 (H. N. Yoon).
80. H.-G. Elias: *Makromoleküle*, 4th ed., Hüthig & Wepf Verlag, Basel 1981p. 259.
81. H. M. Koepp, H. Werner, *Makromol. Chem.* **32** (1959) 79.
82. G. Meyerhoff, S. Shimotsuma, *Makromol. Chem.* **135** (1970) 195.
83. A. Horbach, R. Binsack, H. Müller, *Angew. Makromol. Chem.* **98** (1981) 35.
84. L. D. Moore, *Polym. Prepr. Am. Chem. Soc. Div. Polym. Chem.* **1** (1960) 234.
85. W. H. F. Borman, M. Kramer, *Org. Coat. Plast. Chem.* **34** (1974) 77.
86. D. Garske, W. Kaufmann, *Kunststoffe* **68** (1978) no. 2, 87.
87. F. Bueche, S. W. Harding, *J. Polym. Sci.* **32** (1958) 177.
88. D. R. Gregory, *J. Appl. Polym. Sci.* **16** (1972) 1479.
89. M. Brown, W. K. Witsiepe, *Rubber Age (N.Y.)* **104** (1972) no. 3, 35. K. F. Wissbrun, G. Kiss, F. N. Gogswell, *Chem. Eng. Commun.* **53** (1987) 149.
90. B. Vollmert: *Grundriß der Makromolekularen Chemie*, vol. **III**, E. Vollmert-Verlag, Karlsruhe 1982, p. 72.
91. A. Conix, *Makromol. Chem.* **26** (1958) 226. H. Zimmermann *et al.*, *Faserforsch. Textiltech.* **18** (1966) 487, 536.
92. H. A. Pohl, *Anal. Chem.* **26** (1954) 1614.
93. L. H. Ponder, *Anal. Chem.* **40** (1968) 229. H.-D. Dinse, E. Tuček, *Faserforsch. Textiltech.* **21** (1970) no. 5, 205.
94. F. Salzer, H. Geldermann, *Kunststoffe* **60** (1970) no. 9, 688.
95. G. W. Becker, D. Braun: Die Kunststoffe, Chemie, Physik, Technologie, *Kunststoffhandbuch*, Hanser Verlag, München 1990, p. 814 ff.
96. J. T. LutzJr., *Thermoplastic Polymer Additives*, Marcel Dekker, New York 1989.
97. P. Ehrler, G. Egbers, *Mitt. Text. Ind.* **79** (1972) no. 10, 361.
98. Eastman Kodak, Kodar Copolyester A 1 50, company brochure TR-50 (1975) and MB-58 B (1976).
99. Dynamit Nobel, DE-AS 1 807 776, 1970. Chem. Werke Hüls, DT 2 521 791, 1975 (F. Blaschke *et al.*).
100. Dynamit Nobel, DE-AS 1 912 117, 1969. K. Brühning, K. G. Sturm, *Adhesion* **3** (1973) 83.
101. Ciba-Geigy, DE-OS 2 436 109, 1973 (L. Buxbaum *et al.*).
102. R. Vieweg, L. Goerden: Polyester, *Kunststoffhandbuch*, vol. **VIII**, Hanser Verlag, München 1973, p. 701.
103. J. M. Dekoninck, R. Legras, J. P. Mercier, *Polymer* **30** (1989) 910 – 913.
104. Celanese, DE-AS 2 042 447, 1970 (D. D. Zimmermann).
105. BASF, DE-OS 2 247 652, 1972 (W. Seydl). Celanese, DE-AS 2 042 450, 1970.
106. P. J. Koch *et al.*, *J. Appl. Polym. Sci.* **19** (1975) 227.
107. Teijin, DE-OS 2 726 110, 1977 (N. Saiki, T. Kaneko).
108. BASF, DE-AS 2 515 473, 1975 (R. Wurmb *et al.*).
109. BASF, DE-OS 2 249 910, 1972 (W. Seydl).

110 D. Bieniek, M. Bahadir, F. Korte, *Heterocycles* **28** (1989) 719.
111 H. Staendeke, D. J. Scharf, *Kunststoffe* **79** (1989) no. 11, 1200.
112 G. Kirschbaum, *Kunststoffe* **79** (1989) no. 11, 1205.
113 A. V. Patsis (ed.): *Advances in the Stabilisation and Controlled Degradation of Polymers,* vol. I, Technomic Publishing Co., Lancaster – Basel 1986.
114 J. J. Zeilstra, *Angew. Makromol. Chem.* **137** (1985) 83–92.
115 S. B. Gordon, *Plast. Eng.* **39** (1983) no. 11, 37.
116 A. S. Wood, *Mod. Plast.* **15** (1985) no. 10, 60.
117 D. Rempel, *Kunststoffe* **76** (1986) 900.
118 Enka, DE 1 273 193, 1965 (W. Rein, E. Siggel).
119 Bayer, DE-OS 2 363 512, 1973 (H. Hespe, D. Rempel, L. Morbitzer).
120 A. S. Wood, *Mod. Plast. Intern.* **19** (1989) Aug., 32.
121 W. Witt, *Kunststoffe* **77** (1987) 1009.
122 BASF, DE-OS 2 350 852, 1973 (R. Wurmb, J. Kinde).
123 General Electric, DE-OS 2 331 826, 1973 (D. W. Fox, A. D. Wambach).
124 K. D. Asmus, *Kunststoffe* **62** (1972) 635.
125 F. Breitenfellner, J. Habermeier, *Kunststoffe* **66** (1976) 610.
126 K. Kishore, S. Sankaralingam, *Polym. Eng. Sci.* **24** (1984) 1043.
127 H. Zimmermann, N. Thalkim, *Polym. Eng. Sci.* **20** (1980) 680.
128 H. Krodel, *Kautsch.+Gummi, Kunstst.* **42** (1989) 304.
129 L. G. P. Dalmolen, *Kautsch.+Gummi, Kunstst.* **42** (1989) 205.
130 H. Dominninghaus, *Plastverarbeiter* **40** (1989) no. 2, 42.
131 Du Pont, Hytrel, technical information, 1989.
132 H. Saechtling: *Kunststoff Taschenbuch,* 24th ed., Hanser Verlag, München 1989.
133 Dartco Mfg., Xydar, technical information, 1985.
134 Hoechst, Vectra, technical information, Sept. 1989.
135 I. Whelan, J. Goff, *Br. Plast. Rubber* **2** (1985) 42.
136 P. D. Frayer, *Polym. Compos.* **8** (1987) no. 6, 379.
137 T.-S. Chung, *Polym. Eng. Sci.* **26** (1986) no. 26, 901.
138 J. R. Dole, *CHEMTECH.* **17** (1987) no. 4, 242.
139 *Gummi, Asbest, Kunstst.* **22** (1969) no. 4, 368. K. D. Asmus, D. Niedernberg, H. Schell, *Kunststoffe* **57** (1969) 266.
140 *Kunststoffe* **79** (1989) no. 10, 925.
141 L. Riehl, *Chemiefasern/Textilind.* **28** (1978) 636.
142 Chem. Werke Hüls, DE-AS 2 611 691, 1976 (H. J. Bax et al.).
143 T. Otaka et al., *Shokuhin Eiseigaku Zasshi* **19** (1978) no. 5, 431.
144 E. Nowak et al., *Tech. Pap. Reg. Tech. Conf. – Soc. Plast. Eng.*1980, no. 5, 543.
145 Euwid Kunststoffdienst (1988) 4.
146 *Textil Prax. Int.* **45** (1990) June, 568.
147 H. Piechota, *Kunststoffe* **79** (1989) no. 10, 893.
148 H. M. Caesar, *Kunststoffe* **77** (1987) no. 10, 1004.
149 H. M. Caesar, F. J. M. M. Schiphorst, *Kunststoffe* **76** (1986) no. 10, 863.
150 *Mod. Plast. Internat.* **16** (1986) no. 1, 24; **17** (1987) no. 1, 20; **18** (1988) no. 1, 20; **19** (1989) no. 1, 26; **20** (1990) no. 1, 31.
151 *Kunststoffe* **79** (1989) no. 10, 896.
152 *Mod. Plast.* **4** (1990) 14.
153 *Chem. Mark. Rep.* (1989) Oct. 3, 35 (from data bank).
154 *High Technol.* (1989) June, 20 (from data bank).
155 R. A. Fleming, *Makromol. Chem. Macromol. Symp.* **57** (1992) 76.
156 J. van der Goorbergh, in *Recycling von Kunststoffen,* G. Menges, W. Michaeli, M. Bittner (eds.), Hanser Verlag, München 1992, p. 457.
157 D. Gintis, *Makromol. Chem. Macromol. Symp.* **57** (1992) 185.
158 G. Bauer in K. J. Thomè-Kozmiensky (ed.): *Recycling of Wastes,* EF-Press, Berlin 1989, p. 293.
159 H. Sinn, W. Kaminsky, J. Janning, *Angew. Makromol. Chem. Int. Ed. Engl.* **15** (1976) 660.
160 W. Kaminsky, *J. Anal. Appl. Pyrolysis* **8** (1985) 439. W. Kaminsky, H. Sinn, C. Stiller, *Chem.-Ing.-Tech.* **57** (1985) 778.
161 W. Kaminsky, *Makromol. Chem. Macromol. Symp.* **48, 149** (1991) 381.

Further Reading

J. M. Asua (ed.): *Polymer Reaction Engineering,* Wiley-Blackwell, Oxford 2007.

M. Chanda, S. K. Roy: *Industrial Polymers, Specialty Polymers, and Their Applications,* CRC Press, Boca Raton 2009.

A. J. East: *Polyesters, Thermoplastic,* Kirk Othmer Encyclopedia of Chemical Technology, 5th edition, John Wiley & Sons, Hoboken, NJ,online DOI: 10.1002/0471238961.1615122505011920.a01.pub2.

C. A. Harper (ed.): *Handbook of Plastics Technologies,* McGraw-Hill, New York, NY 2006.

D. Malpass: *Introduction to Industrial Polyethylene: Properties, Catalysts, and Processes,* Wiley, Hoboken, NJ 2010.

J. M. Margolis: *Engineering Plastics Handbook,* McGraw-Hill, New York 2006.

K. Matyjaszewski, Y. Gnanou, L. Leibler (eds.): *Macromolecular Engineering,* Wiley-VCH, Weinheim 2007.

H. Nava: *Polyesters, Unsaturated,* Kirk Othmer Encyclopedia of Chemical Technology, 5th edition, John Wiley & Sons, Hoboken, NJ, online DOI: 10.1002/0471238961.161512251905 1212.a01.pub2.

C. Vasile, M. Pascu: *Practical Guide to Polyethylene,* Rapra, Shawbury 2005.

M. Xanthos (ed.): *Functional Fillers for Plastics,* 2nd ed., Wiley-VCH, Weinheim 2010.

H. Yamashita, Y. Nakano (eds.): *Polyester,* Nova Science Publ., Hauppauge, NY 2008.

Polyethylene

DUSAN JEREMIC, Borealis Polyolefine GmbH, Linz, Austria

1.	Introduction	817
2.	Properties	819
2.1.	Molecular Structure and Morphology	819
2.2.	General Properties	821
2.3.	Unimodal and Multimodal Polyethylene	825
2.4.	Ultra-High-Molecular-Mass Polyethylene (UHMWPE)	826
2.5.	Ethylene Copolymers	827
3.	Polymerization Chemistry	829
3.1.	Polymerization	829
3.2.	Free-Radical Catalysis	829
3.3.	Coordination Polymerization	833
3.3.1.	Mechanism	833
3.3.2.	Catalysts	834
3.3.2.1.	Phillips Catalysts	835
3.3.2.2.	Ziegler Catalysts	836
3.3.2.3.	Single-Site Catalysts (Metallocenes)	838
4.	Raw Materials	841
4.1.	Ethylene	841
4.2.	Comonomers	841
4.3.	Other Materials	842
5.	Production Processes	842
5.1.	High-Pressure Process	843
5.1.1.	Autoclave Reactor	844
5.1.2.	Tubular Reactor	844
5.1.3.	High-Pressure Copolymers	845
5.1.4.	Linear Low-Density Polyethylene	845
5.2.	Coordination Polymerization—Low-Pressure process	845
5.2.1.	Suspension (Slurry) Process	845
5.2.2.	Gas-Phase Process	847
5.2.3.	Solution Process	849
5.2.4.	Combined or Multimodal Processes	850
6.	Uses	851
6.1.	Films	851
6.1.1.	Extrusion Coating	853
6.1.2.	Blow Molding	853
6.1.3.	Injection Molding	853
6.1.4.	Rotational Molding	854
6.2.	Pipes	854
6.2.1.	PE Pipes	854
6.2.2.	PE used for Steel-Pipe Coating	855
6.3.	Wire and Cable Insulation	856
6.4.	Plastomers	856
7.	Environmental Aspects	856
7.1.	Manufacturing	856
7.2.	Polymer Disposal and Recycling	856
7.3.	Material Replacement and Contribution to the Environment	857
	References	857

1. Introduction

Worldwide polyethylene manufacturing and consumption volume represents approximately 40% of the total demand of all thermoplastic materials and approximately one third of the demand of the overall plastic material (Fig. 1).

With its simple and inexpensive structure, chemical resistance, relatively good physical properties, a density that is lower than that of most other polymers, and relatively low environmental impact, polyethylene is the material of choice for many applications that span from medical implants (ultrahigh-molecular-mass polyethylene, UHMWPE), devices, and packaging, via energy and infrastructure materials (pipe and electrical), to flexible food packaging and many others. The overall capacity (2010) globally installed for manufacturing one of the forms of polyethylene is in the order of 100×10^6 t/a, and its growth rate still approximately equals that of the gross domestic product [1].

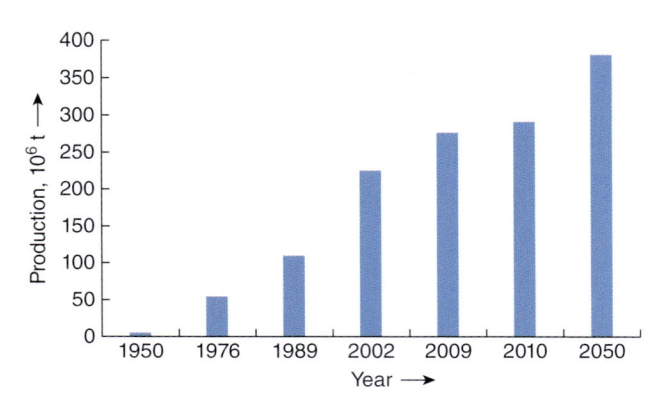

Figure 1. Global manufacturing of polymer materials in 10^6 t/a

Based on their original manufacturing process, polyethylenes can be divided into two large groups of materials: high-pressure and low-pressure `polyethylene. More often, they are devided into classes based on the density of the polymer, typically into three groups: low-density polyethylene (LDPE), which is material made in the high-pressure process, high-density polyethylene (HDPE); and linear low-density polyethylene (LLDPE) The density of polyethylene is in direct correlation with its crystallinity, which further depends on the branching of the polymer chain and the content and nature of the comonomers, which usually are α-olefins or short-chain polar alkyl compounds. Theoretical density of the fully amorphous polyethylene is 880 kg/m^3, and fully crystalline material would have a density of 1000 kg/m^3 [2].

Since the introduction of the single-site catalysts, LLDPE has often been further divided into two other groups: Ziegler–Natta LLDPE (znLLDPE) and single-site LLDPE (mLLDPE, with m standing for metallocene).

Historically, LDPE was the first ethylene polymer invented and manufactured industrially. The invention was made by chance and careful observation by the ICI scientists GIBSON and FAWCETT in the early 1930s [3]. Mixing ethylene and benzaldehyde at 170°C under a pressure of approximately 200 MPa led to the formation of polymeric-like materials that have been identified as polyethylene. The density of the material made at the high temperature and under high pressure was measured to be 915–920 kg/m^3 indicating a relatively low degree of crystallinity owing to a high level of branching. Polymerization of ethylene under these conditions is initiated by free radicals and, as such, the polymerization consists of a few main and side reactions generating radical centers at random molecules and locations in the growing polymer chains. Coupling any two of the radical centers leads to establishing chemical bonds between them that possibly and often result in forming branches or cross-linking of polymer chains.

First initiators or catalysts that enabled polymerization of ethylene at lower pressures and temperatures were discovered in the early 1950s by three research groups working independently at Standard Oil (Indiana, USA), Phillips Petroleum [4], and in the group of KARL ZIEGLER at the Max Planck Institute for Coal Research in Mülheim an der Ruhr, Germany [5]. Phillips and Ziegler catalysts have been commercialized extensively and are still used for manufacturing of almost 90 vol% of all polyolefins. Phillips and Ziegler catalysts are transition-metal-based coordination catalysts that initiate a "one-unit-at-the-time" chain-growing reaction for olefin polymerization that produces essentially linear polyethylene with a density of approximately 960 kg/m^3, HDPE. These catalysts are also called coordination catalysts.

Innovation of another group of coordination catalysts was made possible by the invention of methyl aluminoxane (MAO) by the group of WALTER KAMINSKY in the early 1980s [6]. MAO is a product of partial hydrolysis of trimethylaluminum, a highly reactive compound that explodes in contact with water. Organometallic complexes of transition metals containing one or two cyclopentadienyl rings as ligands, which are described earlier than MAO and found to be largely inactive as catalysts in combination with the "traditional" activators, aluminum alkyls, become very potent polymerization catalysts when mixed with MAO. These are known as single-site catalysts.

LLDPE is the product of a coordinative polymerization initiated by Ziegler or single-site catalysts, by which ethylene is copolymerized with an α-olefin. The product is highly branched containing the alkyl rests of comonomer that are attached to the polymer backbone chain. As a result of the branching, crystallization of the polymer is disturbed resulting in low density of the material. More importantly, many other physical characteristics of the polymer such as toughness are also affected by the incorporation of the comonomer. LLDPE was first introduced in 1960 by DuPont Canada [7] on the small scale, and its commercial exploitation became significant only after Union Carbide introduced their Unipol technology for gas-phase polymerization, a much more economically viable and flexible polymerization process [8]. Gas-phase polymerization was yet another breakthrough in polyethylene polymerization and it was adopted by many other companies. LLDPE is now manufactured worldwide almost exclusively by the low-pressure polymerization processes.

Coordination catalysts are deactivated by most oxygen-containing and/or polar compounds. Therefore, copolymerization of ethylene with polar comonomers is not achievable at low pressure by using coordination catalysts. The free-radical process is used to produce copolymers of vinyl acetate, acrylates, methacrylates, and the corresponding acids, but chain transfer prevents the use of higher olefins because of the drastic reduction in molecular mass of the polymer.

Although some processes and/or catalyst types are more suited for manufacturing certain types of polyethylene described above, many manufacturing plants can produce more than one class of materials. Sometimes, even the same catalyst can be used for manufacturing a wider scope of polyethylene densities as well.

2. Properties

2.1. Molecular Structure and Morphology

For definitions and means of measurements of polymer properties, see → Plastics, General Survey and → Plastics, Analysis.

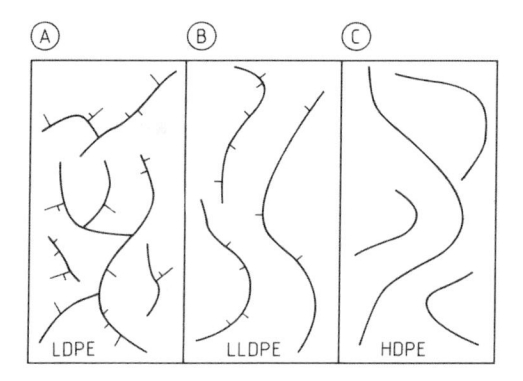

Figure 2. Schematic molecular structure of three types of polyethylene
A) Low-density polyethylene; B) Linear low-density polyethylene; C) High-density polyethylene

Molecular Structure. The molecular structures of the three types of polyethylene classified by their density, LDPE, HDPE, and LLDPE can mainly be distinguished by the branching of the polymer chain that determines the crystallinity of the material. In Figure 2, the structure of three grades is illustrated.

As a product of radical polymerization, the high-pressure product, LDPE, is highly branched and partially cross-linked. The molecular mass distribution (MMD) of polymer chains of different degrees of polymerization and molar masses present in the material is typically broad.

LLDPE is predominantly linear containing short-chain branches equally distributed throughout the chain. The length of the branches depends solely on the comonomer used, which influences the physical properties of the polymer. The homogeneity of the comonomer incorporated is defined as both distribution of the comonomer across the polymer chains of different sizes and uniformity of branching distribution in one chain, which is a significant factor that influences quality and performance of the final material. The higher quality materials, which are tougher, exhibit, for example, more puncture resistance, better processability, and contain branches that are more uniformly distributed across chains of all sizes, whereby the distance between them tends to be equal.

The comonomer distribution depends largely on the catalyst used for manufacturing, but it can be influenced and partially tailored by some advanced multimodal polymerization processes such as the Borstar process invented by Borealis.

Improving comonomer homogeneity is one of the main goals of catalyst research performed by numerous industrial research organizations. One of the main benefits of using single-site catalysts for manufacturing LLDPE is that these catalysts enable improved comonomer utility and incorporation homogeneity. Some single-site catalysts feature even so-called reverse comonomer distribution producing polymers for which the branching frequency and amount increases with the molecular mass of the polymer chains. Those materials are often also superior in their physical performance [9]. The MMD of LLDPE depends on the factors common for all other materials manufactured by coordination polymerization.

HDPE, which is higher in density, contains less comonomer incorporated and fewer branches. Still, for most applications that require higher physical performance such as pipes, distribution of the few branches also plays an important role. Typically, HDPE materials have a much broader MMD than LLDPE. This is needed for keeping these higher density and often higher molecular mass materials processable.

Crystal Structure. Crystallization of polyethylene is a much faster process than of other similar polymers, for example, polypropylene. Crystallization occurs in the form of platelets and lamellae, similar to the crystallite structures of low-molecular-mass paraffin waxes (→ Waxes, 4. Petroleum Waxes). Their thickness is typically in the range of 8–20 nm and the melting point of the polymer depends in part on the thickness of lamellae [10].

Polymer chains typically fold and, thus, tend to be oriented perpendicular to the length of lamella, and not parallel to it as one could expect (Fig. 3).

As the degree of branching influences the folding and crystallization, the thickness of the lamellae also depends on it. Conditions of crystallization are also an important factor that influences their thickness; therefore, slow cooling, also called annealing, leads to thicker lamellae.

Tie molecules are typically long-chain, high-molecular-mass molecules that partially crystallize in more than one lamella tying the two or more units. They are important components for those materials that need to be tough and exhibit higher crystallinity, density, and, at the same time, little brittleness.

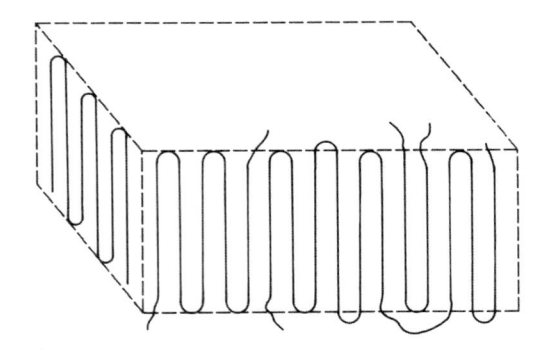

Figure 3. Folded-chain lamellar crystal of polyethylene

Short-chain branches and their specific geometry disturb the crystallization process and are thermodynamically excluded from the crystalline lamellae. As mentioned above, branches cause chain folding leading to thinner lamellae that contain alkyl rests of the branches on the surface of the lamellae. The rate of crystallization influenced by the cooling intensity may be high enough to cause kinetic inclusion of the branches also in the mostly crystalline region creating less than fully defined regions that cannot be simply categorized into amorphous and/or crystalline areas. The significant fraction of such a polymer resides then in the interfacial fraction [11, 12].

Branches can also play a role in "forcing" a part of the long polymer chain to crystallize in more than one lamellae thus creating a tie molecule. Therefore, creating homogeneous comonomer distribution also for HDPE is one of the key features of a polymerization catalyst.

Slower cooling causes a lower number of nucleation sites with crystallization growing from these sites centers outwards until the surfaces meet. The resulting spherulites have then banded-bent structures as seen under a polarizing microscope, which scatter light and produce the effect seen as "milkiness" of polyethylene. The crystallites themselves are much smaller than the wavelength of the light, and, therefore, cannot be the cause for this optical effect [13]. Hence, ethylene copolymers can be at the same time transparent and at least partially crystalline.

With the development of the single-site catalysts based on organometallic Ti complexes containing only one cyclopentadienyl ligand instead of two, which have their active sites

Figure 4. Schematic representation of LCB polyethylene

rather exposed and approachable by the monomer and/or comonomer and are also called constrained geometry catalysts, it was now possible to manufacture polyethylene with so-called long-chain branches, LCBs [14] (Fig. 4).

In the course of polymerization, chain propagation and β-hydrogen rearrangement occurs by leaving a polymer chain with the vinyl terminal group that can become a macromonomer. Incorporation of the vinyl group into another growing polymer chain produces branches with the long polymer chain attached to the backbone structure, similar to what happens in the radical polymerization during manufacturing of LDPE. The presence of long-chain branches has a significant effect on the rheological behavior of the polymers that show pronounced shear thinning (nonlinear decrease of viscosity at higher shear). Consequently, the processability of the material is significantly improved over that of similar materials that contain no LCB.

Block Copolymers. It has been recognized very early that it is difficult to improve all properties of the material at the same time by changing the amount of comonomer, i.e., the branching in the polymeric chain. Therefore, block copolymers have been developed by Dow Chemicals that consist of polyethylene chains containing segments with many short-chain branches attached to the segments of a homopolymer in alteration. Block copolymers result from polymerizations with two single-site catalysts, one very efficient in incorporating the comonomer, and the other not capable of copolymerizing any other olefin than ethylene. These so-called chain shuttling agents are main-group metal–alkyl compounds that function as "transfer vehicles" for growing polymer chains from one transition-metal active site to the other without stopping the growth of the chain. Parts of the chain that grew on one active site then contain more comonomer incorporated than the parts that grew on the other active site, and the result is a block copolymer. The possible advantages of the new materials are higher elasticity, temperature resistance, higher temperature of crystallization, etc. [15] → Plastics, General Survey, Section 5.3.6 Block Copolymers.

2.2. General Properties

Polyethylene in its natural form, free of additives and pigments, is a white material that melts at 120–140°C, does not dissolve in any solvent at room temperature, and typically has no smell or taste. Lower density polyethylenes, LDPE and LLDPE, are softer solids, transparent in bulk. Films made out of the lower density polyethylene are also transparent and have a yielding feel. Higher density material is opaque and more rigid. HDPE film is stiff and crisp in appearance.

Different grades of polyethylene are typically characterized by their density and melt flow rate (MFR) that is also sometimes known as melt flow index (MFI) [16]. Density is most often measured gravimetrically by weighing a polyethylene specimen submerged in water under controlled conditions. Specimens for the density measurement are usually annealed, conditioned at an elevated temperature for a certain time, and then slowly cooled down to room temperature. MFR is described by the amount of polymer heated to 190°C and melted that passes through a die of defined size under the defined load of weights. Polyethylene powder or pellets are placed in heated and tempered cylinders under a piston. After a certain period of time needed to melt the material, the piston is loaded with the prescribed weight and the flow of polyethylene through the die is measured. The amount of material in grams that flows during exactly ten minutes is the MFR of that material at the applied load (Fig. 5).

The ratio between MFR values obtained under different loads of weight is known as the melt flow ratio, usually labeled as flow rate ratio (FRR), and typically the results obtained at a load of 21 kg are divided by the result of the 2.1 kg measurement. This ratio is significant because the dependence of viscosity of the molten polyethylene on the applied shear is not linear but highly depends on the broadness molecular-mass distribution of the material as well as on other characteristics such as content of LCBs. Therefore, LDPE with the high content of LCBs has much higher FRR than typical LLDPE. The FRR is a good indication of the processability of the polymer and the amount of energy needed to extrude the material. MFR and FRR are quick methods to characterize polyethylene and differentiate its grades. They are, however, too crude for making any further conclusions on the viscosity or other rheological properties of the material.

The behavior of polyethylene under shear is shown in Figure 6, in which the dependences of the viscosity on the shear rate of an LDPE and an LLDPE are compared. At sufficiently low shear rates the viscosity of all polyethylenes becomes Newtonian, that is, it is independent of the shear rate. Owing to the narrow MMD, the viscosity of LLDPE is less shear dependent than that of LDPE, and LLDPE has a higher viscosity under the higher shear conditions used in processing equipment. In Table 1 the physical

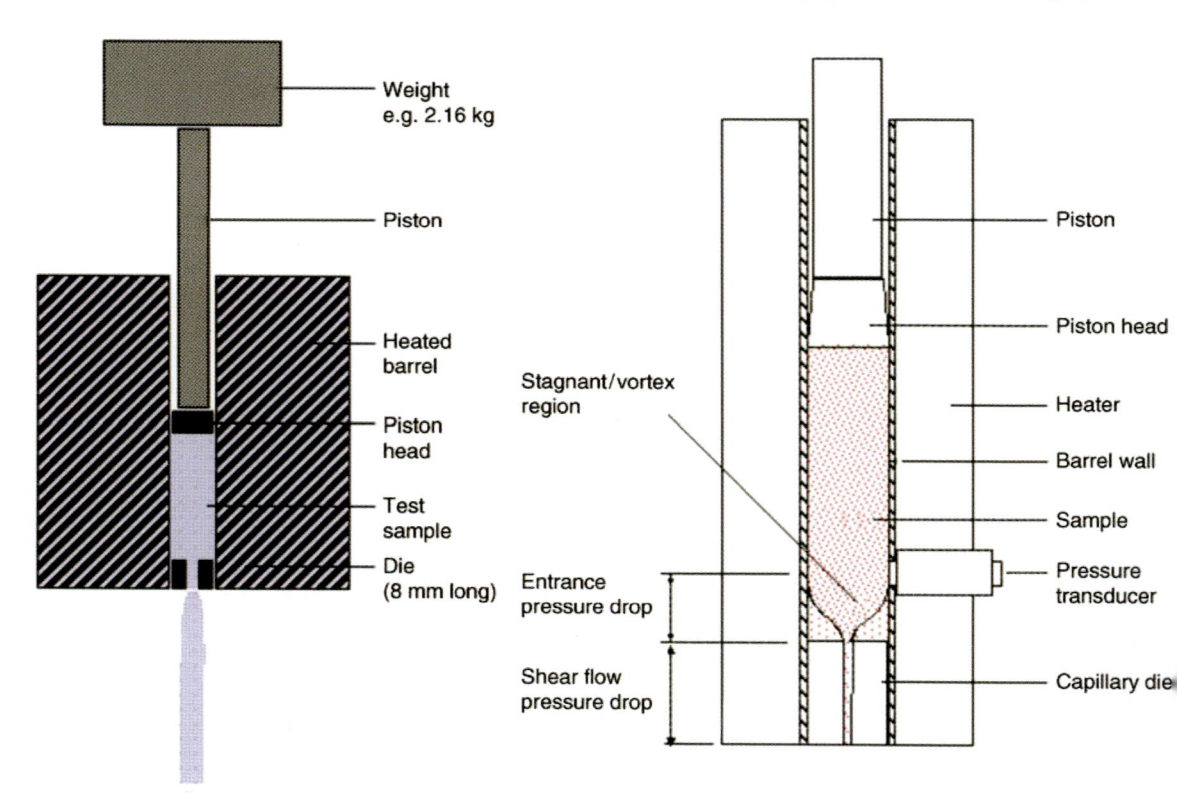

Figure 5. Melt flow indexer and capillary rheometer, which can alternatively be used

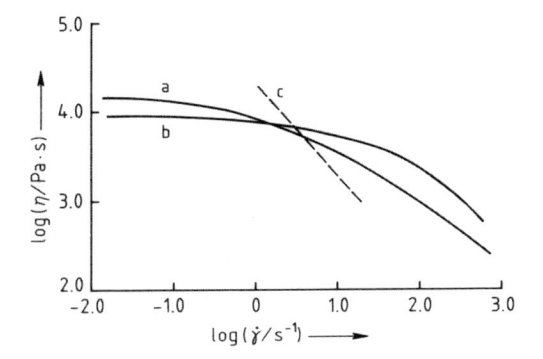

Figure 6. Dependence of viscosity η on shear rate $\dot{\gamma}$ for two polyethylenes
a) Low-density polyethylene; b) Linear low-density polyethylene; c) Apparent shear rate in melt flow index test

properties of the different grades of polyethylene are listed.

By observing the melt behavior during an MFR test, one can obtain an indication on the die swell and die swell ratio under different loads. The observation indicates elastic memory of the material, another important parameter for processing of polyethylene.

While polyethylene is considered as chemically resistant and inert material, it requires stabilization to sustain environmental influences at extended exposure. Free radicals from oxygen can cause uncontrolled cross-linking of polymer chains changing physical characteristic of the material. Also exposure to various other substances can alter the material: surface-active compounds can cause cracks at the stressed parts over time, and certain solvents such as toluene, for example, cause material swelling. Therefore, immediately after polymerization, the polymer is usually compounded with additives that improve long- and short-term stability, and/or modify the properties of the polyethylene. These additives are UV stabilizers, antioxidants, slip agents, antiblock additives, and others.

One of the ways to characterize stability and chemical inertia of polyethylene is by testing its so-called environmental stress crack resistance. The test is performed by placing a polyethylene specimen under a certain stress and immersing it in a solution of a reagent that can cause an impact on its resistance to stress, i.e., detergent,

Table 1. Properties of some typical polyethylenes (data from Repsol Quimica)

Property	LDPE	HDPE	LLDPE	Method	Standard
Polymer grade	Repsol polyethylene 077/A	LyondellBasell Hostalen GD-4755	BP LL 0209		
Melt flow index (MFI), g/600 s	1.1	1.1	0.85	190°C/2.16 kg	ASTM D1238
High load MFI, g/600 s	57.9	50.3	24.8	190°C/21.6 kg	ASTM D1238
Die swell ratio (SR)	1.43	1.46	1.11		
Density, kg/m^3	924.3	961.0	922.0	slow annealed	ASTM D1505
Crystallinity, %	40	67	40	DSC	
Temperature of fusion (max.), °C	110	131	122	DSC	
Vicat softening point, °C	93	127	101	5°C/h	ASTM D1525
Short branches**	23	1.2	26	IR	ASTM D2238
Comonomer		butene	butene	NMR	
Molecular mass*					
M_w	200 000	136 300	158 100	SEC	
M_n	44 200	18 400	35 800	SEC	
Tensile yield strength, MPa	12.4	26.5	10.3	50 mm/min	ASTM D638
Tensile rupture strength, MPa	12.0	21.1	25.3		
Elongation at rupture, %	653	906	811		
Modulus of elasticity, MPa	240	885	199	flexure	ASTM D790
Impact energy,					
unnotched, kJ/m^2	74	187	72		ASTM D256
notched, kJ/m^2	61	5	63		ASTM D256
Permittivity at 1 MHz	2.28				ASTM D1531
Loss tangent at 1 MHz	100×10^{-6}				ASTM D1531
Volume resistivity, Ω · m	10^{16}				
Dielectric strength, kV/mm	20				

*Corrected for effects of long branching by on-line viscometry.
**Number of methyl groups per 1000 carbon atoms.

at a controlled, possibly elevated temperature [17]. The time that passes before the specimen yields to stress is measured and it indicates stability of the material.

As mentioned above, density correlates to crystallinity of polyethylene and is a good indicator of it. Furthermore, crystallinity is also measured by differential scanning calorimetry (DSC) that provides the melting and softening point of the polymer as well as enthalpy of fusion.

FTIR and NMR spectroscopy are the two methods that can accurately measure branching in polyethylene. NMR spectroscopy is also used to measure the amount of long-chain branches [18].

Comonomer distribution is measured by the methods that are based on crystallization and elution, such as temperature-raising elution fractionation (TREF) Polyethylene is completely dissolved at high temperatures and the solution is then introduced to the column packed with an inert support such as glass beads. The polymer solution is then cooled down slowly at a controlled rate, allowing crystallization in an onion-like structure with the branching gradient. The potential for crystal formation is dependent on the branching content and, therefore, the highly branched fractions will form the crystals at lower temperature and will be last to deposit on the inert support. With passing the solvent through the column and by increasing the temperature at controlled rate, fractions are eluted with the increasing crystalline content as a function of time and temperature (Fig. 7) [19].

Typical LLDPE materials prepared by using a Ziegler catalyst will always contain some homopolymer that is eluted at temperatures above 90°C. Branched fractions elute first, and the shape of the peaks depend on their homogeneity.

Finally, so-called hyphenated techniques (such as GPC–FTIR) and cross-fractionation techniques (such as TREF–GPC) have become extremely useful in characterizing and investigating the microstructure of polyethylene. These techniques represent combined methods in which the polymer is fractionated first on the basis of one attribute, i.e., its molecular mass, and then the fractions are further fractionated or tested on the basis of another attribute, i.e. branching. GPC–FTIR technique is extensively used for determining the distribution of SCB across the molecular mass in polyethylene.

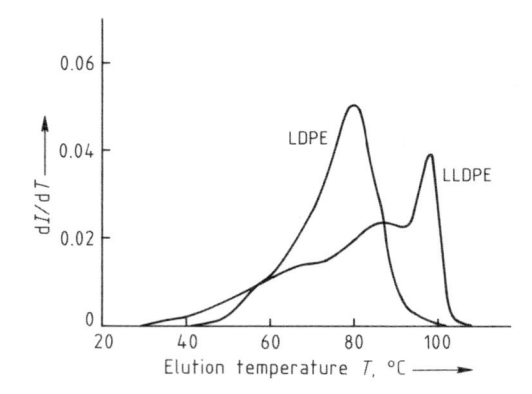

Figure 7. Temperature rise elution fractionation curves (I is the cumulative weight fraction)

An important feature of polyethylene is its relative toughness, i.e., resistance to stress. A standard method to measure toughness of film materials is determination of its dart drop index (DDI). A dart is let to drop from a defined height at a framed film. The weight of the dart needed to puncture the film is measured. DDI gives the weight at which the dart punctures the film in at least 50% of experiments. The DDI depends very much on the density of the material, but also on the comonomer distribution in it. It is known that polyethylene made with single-site catalysts, in which comonomer is more evenly distributed across the molecular mass, has a much higher DDI than materials made with Ziegler catalysts.

Other measurements of polyethylene's toughness are elongation at break and puncture resistance.

The length of the short-chain branches in the polymer chain, and thus the choice of the comonomer, affects the toughness of LLDPE. While the performance of 1-butene copolymers is sufficient for simpler packaging applications, increased DDI values, puncture resistance, and elongation at brake can be achieved with higher α-olefins such as 1-hexene and 1-octene.

Film materials are characterized by their optical performance, gloss, and haze, whereby more gloss and less haze indicate more transparent and better appearing films. Furthermore, sealing temperature (the minimum temperature needed to weld two layers of polyethylene together) and strength are important features of polyethylene films.

Other parameters such as barrier properties of polyethylene films and permeability of oxygen and/or moisture vapors are not at the level of those of other polymers such as ethylene–vinyl acetate copolymer (EVA). However, polyethylene is still an important packaging material used for many applications in the food and other industries.

As a viscoelastic material, polyethylene exhibits creep behavior, which means that strain produced by applying stress on the polyethylene specimen is time dependent and can be multiplied if measured after several days of exposure to stress compared to strain measured after only a few seconds. In Figure 8, the strain-versus-time curve is shown for an HDPE specimen at a constant tensile stress of 6.5 MPa [20].

The creep is a consequence of a relative mobility of amorphous regions of polyethylene at room temperature. If polyethylene is used for long-term applications, creep as well as possible crack formation and failure need to be taken into account. Prolonged high stress may lead to failure under conventionally measured yield stress.

LDPE or high-pressure polyethylene produced with peroxides as polymerization initiators is a very-high-purity polymer that contains low amounts or no traces of metal residuals. Therefore, LDPE is very suitable as insulator for high-frequency cables: With no polar groups in the polymer chain and its high purity, LDPE has a very low loss tangent that is appropriate for the wire and cable applications. Other polyethylenes, LLDPE and HDPE, do contain catalyst residuals, Phillips, Ziegler, or single-site, and are, therefore, less pure and/or suitable for these uses.

Because of its purity and inertia to chemical agents, polyethylene, specifically the ultrahigh-molecular-mass polyethylene, UHMWPE, with polymer chains that have molar masses in excess of $1-2 \times 10^6$, is often used as the material of choice for manufacturing medical prosthesis, artificial joints, knees, and hips. The process of qualifying materials for medical and/or food applications is strictly regulated by regulatory bodies such as Food and Drugs Administration in the USA and other country regulatory bodies.

2.3. Unimodal and Multimodal Polyethylene

As a product of coordination polymerization that is usually terminated by chain transfer or cleavage of the metal–carbon bond, polyethylene is always a mixture of polymer chains of different lengths with different molar masses. Distribution of the polymer chains of different lengths, with different molar masses, is an important descriptor of polyethylene materials.

The most significant parameter is the broadness of the MMD, which originates in the polymerization process or the catalyst used or both. To produce polymers with narrow MMD, the polymerization needs to occur under constant and uniform conditions in one reactor, and the catalyst needs to contain few types of active sites. In contrast, catalysts containing more types of active sites will produce polymer chains of various lengths, depending on the active site nature, and the distribution will be broader.

Typically, different catalyst classes contain more or less types of active sites, and, therefore, produce polymers with broader or narrower MMD. Single-site catalysts are well defined compounds, sometimes also very well characterized, and the polymer that they deliver typically has a very narrow MMD, indicating one only type of active site (a "single-site"). Ziegler catalysts, however, are less well defined chemicals, and tend to contain several types of active sites resulting in broader MMDs in

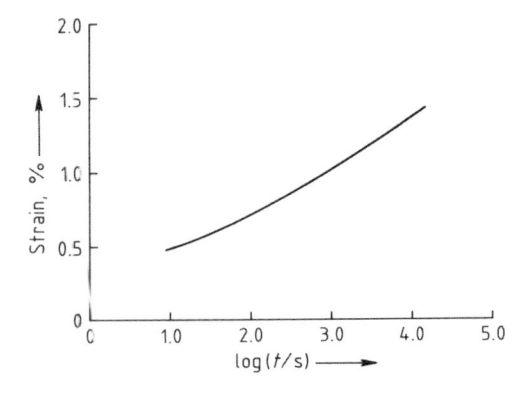

Figure 8. Tensile creep of HDPE (Rigidex 006–60) [20]

the polyethylene for which they are used. Finally, the polyethylene with the broadest MMD is made with Phillips chromium-based catalysts.

Polymer chains of different lengths containing a more or less uniformly distributed comonomer contribute to the performance of the material. For example, higher molecular mass chains are needed to provide toughness and bubble stability in film materials. Lower molecular mass chains improve processing of the material. The broadness of the MMD also determines processing performances of the polymer and, for example, the optical performance.

Polyethylene materials that have a rather narrow MMD are usually called unimodal materials. If the distribution is divided into more than one group of polymer chain lengths (one molecular mass group), it is called multimodal, and if there are two groups, it is called bimodal.

PE manufacturers prefer to use only few catalyst types at one plant with transitions of two different formulations as rarely as possible. Therefore, the MMD, which is defined by the choice of the catalyst, is given and cannot easily be changed by modifying the number of types of active sites. Thus, to increase the number of freedom degrees and improve the ability to design polymer materials as needed, bimodal or multimodal polymerization plants have been developed and implemented. These plants operate with two or more polymerization reactors in series and polymerization occurs in these reactors under different conditions, which are controlled separately. The catalysts used at the beginning of the reaction and growing polymer particles of later stages "travel" through the cascades of reactors and are exposed to different polymerization conditions. By choosing and controlling the conditions, the length of polymer chains and branching content in each polymerization stage can be influenced, and, therefore, molecular mass and comonomer distribution can be actively designed.

The bimodal polymerization process enable manufacturing of advanced polyethylene materials typically used for manufacturing pressure pipes for transportation of gases and/or liquids. These pipes need to sustain pressure, sometimes at elevated temperatures, over an extended period of time. Usual life expectance needed for the pipes is more than 50 years and the pressure is approximately 10 MPa. The material providing the required performance is high-molecular-mass, high-density polyethylene. High mechanical resistance, i.e., resistance to pressure, stems from the highly crystalline part of the polymer that crystallizes as folded-chain lamellae. Given their high crystallinity, these chains do not need to, and should not be very long. To maintain the pressure for long time and avoid rapid crack propagation, by which the crystalline lamellae split from each other and the pipe becomes brittle, the high-density parts need to be connected by tie molecules, long polymer chains that partially cocrystallize in more than one lamellae creating a network between the units. Tie molecules need to be long, have high molecular mass, and are branched to a certain extend to avoid more than partial crystallization [21].

Manufacturing of the described materials with the appropriate combination and design of components to achieve the required performance is performed by polymerization processes with two or more reactors. Lower molecular mass homopolymer is typically produced in the first reactor. The material is then transferred into the second reactor in which the conditions are different and new polymer chains produced there are longer and contain some comonomer. In addition, to achieve the right design and MMD broadness that enables easier processing of the overall high-molecular-mass polymer, two potentially different types of polymer chains are manufactured in a subsequent manner so that every polymer particle eventually contains both types, securing good dispersion in the final material after compounding.

2.4. Ultra-High-Molecular-Mass Polyethylene (UHMWPE)

Ultra-high-molecular-mass polyethylene is a polyethylene with a molecular mass higher than 1×10^6, typically between 3 and 6×10^6. UHMWPE is a product of coordination polymerization at usually lower temperatures with high ethylene concentration in the reactor. Both Ziegler and single-site catalysts can be used for manufacturing UHMWPE.

This type of materials cannot be characterized by the usual size-exclusion technique

(GPC), because it is not completely soluble in the commonly used solvents even at very high temperatures. Its viscosity is extremely high, and thus, it is not possible to measure the MFR of the material either. Viscosity molecular mass can be determined for a certain UHMWPE by a modified method for measuring the relative viscosity of very diluted polymer solutions in decaline at 135°C, provided that the polymer is dissolved completely.

UHMWPE normally contains no comonomer so there is no branching in the linear polymeric chain. Still, the density of the material is only 940 kg/m^3 because the length of the chains and the extremely high viscosity hinder the crystallization process.

The UHMWPE cannot be processed conventionally by melting the polymer in an extruder and shaping it in the molten form. Instead, there are few other methods to convert UHMWPE:

- Sintering into large blocks under pressure and at elevated temperature to obtain a material that can be used for machining objects or cutting tapes out of the solid material
- Ram extrusion into rods and objects that are also used for machining into final products
- Gel spinning for manufacturing fiberlike materials

As sintering is one of the primary ways to process UHMWPE, the size and regularity of the polymer particles that are generated in the polymerization process is of particular importance. Smaller spherical particles with uniform particle size distribution provide for more successful sintering. Therefore, it is advantageous to use a batch polymerization process for manufacturing UHMWPE. Unlike in the continuous process, the residence time in the batch process is identical for all polymer particles and it can be assumed that they grow following the same kinetics, thus taking the same shape and dimensions at the end of the polymerization.

UHMWPE is used for applications for which hardness and durability is needed. Gel-spinning-produced fibers are extensively used for manufacturing bullet proof vests and other impact protection equipment. Films cut from the UHMWPE blocks are used as the lower layer of modern skis, and UHMWPE-made wheels are used for conveyer belts that transport high loads. Because of high purity, hardness, and durability, UHMWPE is used for manufacturing medical prosthesis and artificial joints.

Blending of UHMWPE with other materials, lower molecular mass polyethylene for example, is a rather difficult process. The difference in viscosities of the materials disables good dispersion of the materials. This is particularly due to the fact that the long chains are entangled with each other, which further increasing viscosity of the material. By using diluted catalysts, either in solution or supported on very low loadings, it is possible to obtain so-called disentangled UHMWPE, which contains polymer chains that are separate from each other. Viscosity of such a material is considerably lower than that of the entangled one and, thus, it can be, to a certain extent, blended with other materials [22].

2.5. Ethylene Copolymers

Branching effects crystallinity and further performance of polyethylene. One of the principle reasons for copolymerizing ethylene with α-olefins is to reduce and/or control their crystallinity to gain toughness, softness of the material, and other physical characteristics needed for certain applications. Propene is the only comonomer that, if incorporated, can partake in crystallization.

LLDPE is one of the higher volumes thermoplastic materials used worldwide, mostly for packaging applications. VLDPE or polyethylene-based plastomers and elastomers with densities in the range of 860–900 kg/m^3 are also very important in standalone applications, i.e., films, but also as materials to be blended to other thermoplastic materials as impact modifiers.

Coordination copolymerization of ethylene with polar comonomers is not possible because of the reactivity of the catalysts with the polar groups. The only process that enables incorporation of monomers such as vinyl acetate and other esters and acids is through a high-pressure radical polymerization process. In Table 2 the principal types of ethylene polar copolymers are

Table 2. Principal types of ethylene copolymers

Comonomer	Abbreviation	Feature	Catalysis*
Vinyl acetate	EVA	flexibility	FR
Methyl acrylate	EMA	flexibility, thermal stability	FR
Ethyl acrylate	EEA	flexibility at low temperature, thermal stability	FR
Butyl acrylate	EBA	as for EEA	FR
Methyl methacrylate	EMMA	flexibility, thermal stability	FR
Butene	VLDPE	flexibility at low temperature, thermal stability	Z, SS
Hexene	VLDPE	as for butene copolymer	Z, SS
Octene	VLDPE	as for butene copolymer	Z, SS
Acrylic acid	EAA	adhesion	FR
Methacrylic acid	EMAA	adhesion	FR
Methacrylic acid + Na^+ or Zn^{2+} (ionomer)		adhesion, toughness, stiffness	FR
Acrylic acid + Zn^{2+} (ionomer)		adhesion, toughness, stiffness	FR
Carbon monoxide	PK	polyketone, stiffness, high melting	**
Norbornene	COC	cycloolefin copolymer, transparency	SS

*FR = free radical, Z = Ziegler catalysis, SS = single-site catalyst.
**Novel group VIII catalyst.

listed that are in commercial production. Of these, the vinyl acetate copolymers are produced in the largest quantities.

At room temperature, the amorphous regions of the copolymer (where the comonomer groups are concentrated) are mobile and the effect of introducing a comonomer is to reduce the stiffness progressively (Fig. 9).

In addition to the branches that result from comonomer incorporation, the branches that are formed under the high-pressure synthesis conditions also contribute to the reduction in crystallinity. Below room temperature, there are differences in crystallinity between ester copolymers having different glass transition temperatures T_g of the amorphous material. The T_g of the copolymers follow the T_g of the ester homopolymers and results in EEA and EBA copolymers (see Table 2) that are flexible down to lower temperatures than EVA copolymer, with EMMA copolymer having the highest T_g. The VLDPEs produced by the LLDPE processes should have the best low-temperature properties, particularly in the case of narrow composition distribution polymers. Properties of three typical EVAs and a VLDPE [23] are listed in Table 3.

The methacrylic and acrylic acid copolymers are produced to provide enhanced adhesion, particularly in coextruded films or laminates. Not included in Table 3 are terpolymers, in which acid monomers are used together with ester comonomers to improve adhesive properties.

The ionomers are produced by the partial neutralization of acidic copolymers containing 10 wt % of (meth)acrylic acid by sodium or zinc ions. The ionic salts and the unneutralized acid groups form strong interchain interactions, producing a form of thermally labile cross-linking in both the solid and the molten states. The T_g of the amorphous material in ionomers is slightly above room temperature, resulting in a stiffness similar to that of LDPE at room temperature. However, the stiffness of most ionomers decreases rapidly with increasing temperature above 40°C. The polyketone (PK) and cycloolefin (COC) copolymers are both developments in which ethylene is present as a 50 mol % alternating copolymer [24]. The PK materials are believed to be made with a novel group VIII catalyst [25]. They are targeted for

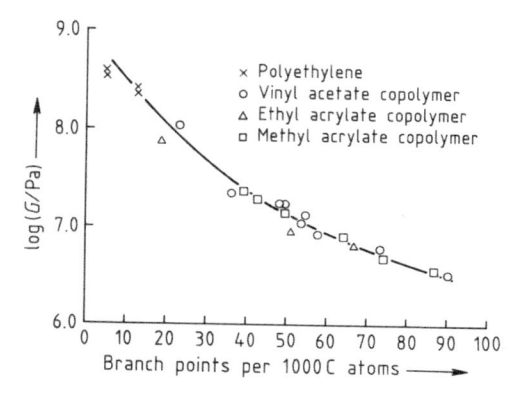

Figure 9. Dynamic modulus G (ca. 2 Hz) at 23°C as a function of total branch points

Table 3. Properties of ethylene copolymers

Property	EVA (Repsol PA-501)	EVA (Repsol PA-538)	EVA (Repsol PA-440)	VLDPE (DSM TMX 1000)[a]	Condition	Standard
Vinyl acetate content, wt %	7.5	18	28			
Melt flow index, g/600 s	2	2	6	3	190°C/2.16 kg	ASTM D1238
Density, kg/m^3	926	937	950	902	rapid annealed	ASTM D1505
Vicat softening point, °C	83	64		66		
Tensile strength, MPa	16	16	11	11.5	50 mm/min	ASTM D638
Elongation at rupture, %	700	700	800		TMX 1000: 0.4 m/s	
Modulus of elasticity, MPa	156[b]	47[b]	24[b]	95[c]		
Permittivity at 1 MHz	2.46	2.70				ASTM D1531
Loss tangent at 1 MHz	0.014	0.035				ASTM D1531
Volume resistivity, Ω · m	2.0×10^{15}	2.5×10^{14}				
Dielectric strength, kV/mm	19	20				

[a] Octene copolymer.
[b] 0.2% strain, 100 s.
[c] ASTM D790.

medium-stiffness applications and also have good barrier properties for hydrocarbons. The carbonyl group, however, makes the polymers susceptible to degradation by sunlight. The COC materials from Topas Advanced Polymers (former Hoechst) make use of the remarkable catalyst activity and copolymerization ability of the single-site catalysts [26] and the high transparency and low birefringence allow its use in compact disks and transparent packaging.

3. Polymerization Chemistry

3.1. Polymerization

(→ Polymerization Processes, 1. Fundamentals). Ethylene polymerization is an exothermic reaction that generates 93.6 kJ/mol (3.34 kJ/g) of energy. Given the specific heat of ethylene of 2.08 JK^{-1} g^{-1}, the temperature in the gas phase rises for approximately 16°C for each 1% conversion to polymer.

Heat management, in particular, removal of heat and controlling the polymerization temperature are among the key challenges in a commercial polymerization process. Depending on the process, there are different strategies to remove heat from polymerization effectively and enable high-throughput polymerization reactions without compromising the operability of the process.

Besides applying cooling agents and removing heat through the walls of the reactor, circulation of gas in the gas-phase process carries the heat away from the polymerization reaction to the heat exchanger, where it is cooled down before it is injected again into the reactor. In addition, higher boiling inert materials are used in the gas-phase process as so-called condensing liquids. Liquids are injected into the reactor in which they evaporate taking considerable amount of heat away. They become a part of the recycle gas that is then passed through the heat exchanger where the liquids are condensed, separated, and then injected into the reactor again.

Solution polymerization process is an adiabatic process in which heat of the reaction is used to maintain the high temperature needed to dissolve polyethylene in the solvent. Eventually, the heat is removed from the polymer by evaporation of the solvent after the polymerization.

3.2. Free-Radical Catalysis

Polymerization Elements. High-pressure polymerization process is initiated by free radicals that are a product of homolytic cleavage of the O–O bonds in added peroxides. Very high pressure of ethylene in the range of 150 and 350 MPa is needed to achieve a reasonable concentration of the monomer. Temperature in excess of 160°C enables the polyethylene product to be dissolved in the unreacted ethylene. In addition, the conditions are optimized for high production and conversion rates to maintain the economical viability of the needed sizeable investment in the polymerization plants. Conversion of 20% of the monomer is typically achieved in approximately 40 s, making the

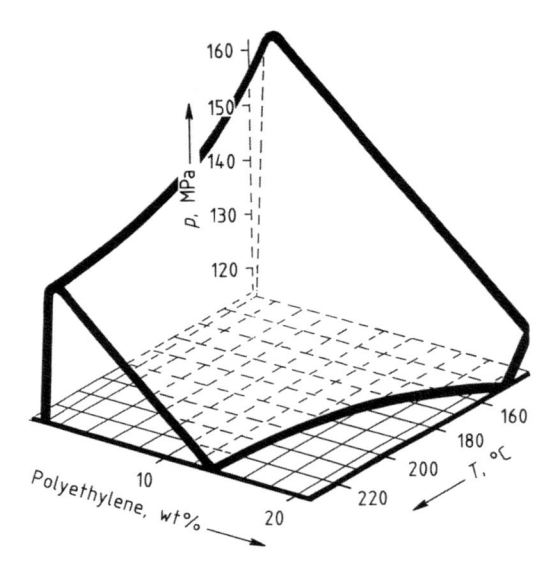

Figure 10. Schematic cloud point surface for ethylene – polyethylene

residence time in the reactor extremely short in comparison with residence times of other processes. A polyethylene schematic phase diagram for an ethylene–polyethylene system is shown in Figure 10.

For a more detailed account of the effect of molecular mass and MMD on the phase equilibrium, see [27].

The single-phase ethylene–polyethylene mixture allows the reaction to take place as a classical free-radical-initiated solution polymerization (→ Polymerization Processes, 2. Modeling of Processes and Reactors). Some aspects, which are particularly important for ethylene systems, are the following:

1. The very high monomer pressure, as stated before, sometimes reaches 350 MPa contributing to the very high reaction rate. In addition to concentrating the monomer that remains in the gaseous phase, it influences the rate constant [28] by enabling the reactant to reach the transition state. Changing the pressure from 0 to 200 MPa is calculated to increase the polymerization rate by a factor of approximately 12 [29].

2. The activation energy for ethylene polymerization is 32 kJ/mol, and the elevated polymerization temperature of ca. 250°C, which, owing to the temperature gradient in the reactor, can in some spots reach over 300°C, certainly contributes to high polymerization rates. As the polymerization is initiated by the radicals that are a product of bond cleavage, there is no risk of deactivating the initiator at the high temperature as it would be in the case of typical Ziegler or single-site catalysts.

3. Chain transfer process (Fig. 11) is the reaction by which a free radical center of an alkyl, for example, a growing polymer

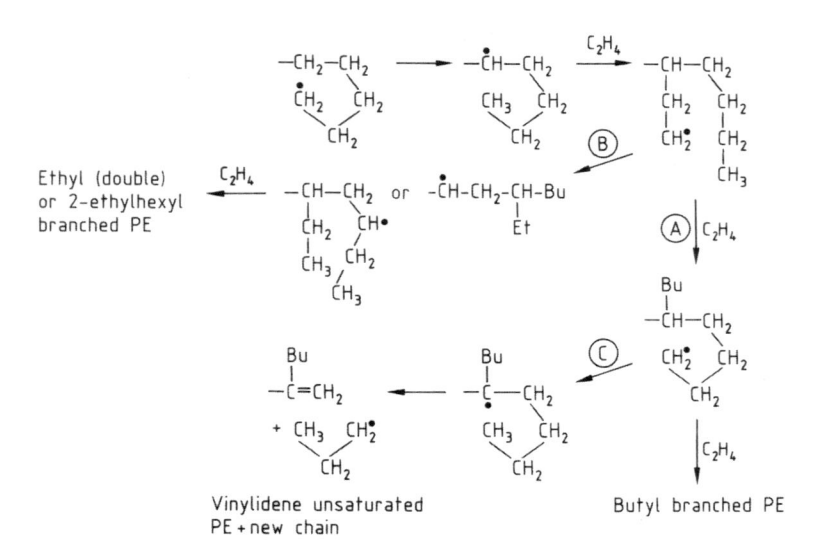

Figure 11. Principal intramolecular chain-transfer reactions

chain, attacks another alkyl group, abstracts a hydrogen atom creating a saturated alkyl chain, and frees another radical. This process has higher activation energy than the polymerization reaction, and it is favored by the increased temperature of the radical polymerization reaction. Consequently, the molecular masses of the polymer chains depend on the reaction temperature and can be limited if the temperature is higher. In addition, purity of the feedstock material, absence of all chemicals that might favor the chain-transfer process is essential for radical polymerization. On the other hand, targeted use of the same agents is an option to control the molecular mass of the polymer. Chain-transfer agents used commercially are hydrogen, propane, propene, acetone, and methyl ethyl ketone. Chain transfer to some compounds with very active hydrogen atoms such as propene and particularly the higher alkenes can lead to radicals that are insufficiently reactive towards ethylene to re-initiate new chains rapidly, and result in reduced reaction rates. The alkyl radical of a growing polymer chain can attack the rest of the chain by forming a ring and abstracting a hydrogen atom from the fourth methylene group down the chain. This is called intramolecular chain transfer. In this so-called back-bite mechanism described by M.J. Roedel of DuPont [30], the chain continues growing from that fourth carbon atom in the chain, and a butyl rest remains as a branch in the backbone polymer chain. If, after the addition of one ethylene molecule to the newly formed secondary radical, a further back-bite occurs (Fig. 11B), a pair of ethyl branches or a 2-ethylhexyl group is formed. A further possibility, shown in Figure 11C, is that back-bite occurs to a branch point, and the tertiary radical then decomposes into a new short radical, leaving a vinylidene group at the end of the polyethylene chain. This process is the principal chain-termination mechanism in LDPE, and concentrations of vinylidene groups approach one per number average molecule for LDPEs produced at high temperature.

4. If the chain transfer takes place between two chains as an intermolecular chain transfer branches are formed that are considerably longer, LCBs. The same process also causes broadening of the MMD of LDPE. Typically, LDPE chains contain more branches if they are longer, which is a consequence of the increased number of potential sites for radical attacks in the longer chains, assuming that every free radical will grow a polymer chain until the polymerization is terminated by the chain-transfer or radical combination reaction. Statistically, the probability for termination is the same irrespective of whether the polymerization and growth of the particular chain has been started by an initiator or by a radical that itself is a product of a chain-transfer reaction. Therefore, statistically all chains, irrespective of whether they are in the polymer backbone or branches, have the same statistical length, but branches are sometimes slightly longer in the absolute length than the backbone chains themselves. In addition to temperature and pressure that accelerate the chain-transfer mechanism, also the concentration of dissolved polymer plays an important role. Polymer that is dissolved in the polymerization media is more accessible to alkyl radicals for attacks, and, therefore, the amount of branching is directly proportional to the concentration. This again is the reason for the significant difference in products between tubular and autoclave high-pressure reactors. Tubular reactors made out of relatively narrow and very long steel pipes behave similarly to plug-flow reactors, and stirred autoclave reactors can be seen as typical continuously stirred tank reactors (CSTRs). Thus, the concentration of dissolved polymer at the site of polymerization is different in the two types of reactors and the LCBs of the products as well. Theoretical analyses for autoclave [31–33], and tubular [34] reactors have been presented.

5. High-pressure polymerization was discovered in 1933 by Gibson and Fawcett of ICI by a mistake that caused ingress of oxygen traces in highly pressurized pipes [3], heated to a high temperature, containing ethylene. Oxygen, or the free radicals resulting from the cleavage of the O−O bond at the high temperature, was the first radical initiators for the radical ethylene polymerization reaction. For some time, oxygen was

used in the early commercial processes. However, the mechanism by which oxygen forms free radicals is rather complicated, and at lower temperatures oxygen can act as an inhibitor [35]. To achieve much better control of the polymerization kinetics, in the modern plants specifically designed organic peroxide molecules or, more often, a mixture of them are used. They have defined temperature of radical formation and lifetime suited for the specific needs of the process. Typical initiators are listed in Table 4. Oxygen is still used in combination with other initiators in the tubular reactors only.

6. Initiators are often called "catalysts" even though they are not catalysts according to the definition of catalysis and catalysts, because the radicals are consumed and destroyed in the process. Still, one molecule that generates radicals initiates polymerization of many ethylene molecules, and the polymerization, although exothermic, is not a spontaneous process. In that sense the role of initiators can be seen as "catalyst-like".

Table 4. List of commercially used initiators for radical polymerization

Chemical description	T_s max.*, °C	T_s min.*, °C	1 h half-life temperature**, °C
2,5-Dimethyl-2,5-di(tert-butylperoxy)hexane	40	10	134
tert-Butylperoxy-2-ethylhexyl carbonate	20		117
tert-Amylperoxy-2-ethylhexanoate	5	−20	91
tert-Amyl peroxypivalate	−10		72
tert-Amylperoxy-2-ethylhexyl carbonate	20		113
2,5-Dimethyl-2,5-di(2-ethylhexanoylperoxy)hexane	15	−20	86
2,5-Dimethyl-2,5-di(tert-butylperoxy)hexyne-3	30	0	141
tert-Butyl peroxy-2-ethylhexanoate	10	−30	91
1,1-Di(tert-butylperoxy)cyclohexane	25		113
tert-Butyl peroxyneodecanoate	−10	−30	64
tert-Butyl peroxyneodecanoate, 75% solution in odorless mineral spirits	−10	−20	64
tert-Butyl peroxypivalate, 25% solution in odorless mineral spirits	−5	−20	75
tert-Butyl peroxypivalate, 40% solution in odorless mineral spirits	−5	−15	75
tert-Butyl peroxypivalate, 75% solution in odorless mineral spirits	−5	−15	75
tert-Butyl peroxyneoheptanoate, 75% solution in odorless mineral spirits	−10		67
tert-Butyl peroxydiethylacetate	15	−30	93
1,1-Di(tert-butylperoxy)-3,3,5-trimethylcyclohexane	25		105
1,1-Di(tert-butylperoxy)-3,3,5-trimethylcyclohexane, 50% solution in odorless mineral spirits	25		105
1,1-Di(tert-butylperoxy)-3,3,5-trimethylcyclohexane, 75% solution in odorless mineral spirits	25		105
1,1-Di(tert-butylperoxy)-3,3,5-trimethylcyclohexane	25		105
3,6,9-Triethyl-3,6,9-trimethyl-1,4,7-triperoxonane, solution in isoparaffinic hydrocarbons	40	10	146
3,6,9-Triethyl-3,6,9-trimethyl-1,4,7-triperoxonane, solution in isoparaffinic hydrocarbons	40	10	
Di(3,5,5-trimethylhexanoyl) peroxide, 37.5% solution in odorless mineral spirits	0	−15	77
Di(3,5,5-trimethylhexanoyl) peroxide, 50% solution in odorless mineral spirits	0	−10	77
Di(3,5,5-trimethylhexanoyl) peroxide, 75% solution in odorless mineral spirits	0	−10	77
tert-Butyl peroxyisobutyrate, 50% solution in odorless mineral spirits	10		98
tert-Butyl peroxy-3,5,5-trimethylhexanoate, 30% solution in odorless mineral spirits	25	−20	114
tert-Butyl peroxy-3,5,5-trimethylhexanoate, 60% solution in odorless mineral spirits	25	−20	114
1,1,3,3-Tetramethylbutyl peroxy-2-ethylhexanoate	5	−20	88
1,1,3,3-Tetramethylbutyl peroxyneodecanoate, 70% solution in odorless mineral spirits	−15		57
1,1,3,3-Tetramethylbutyl peroxypivalate, 75% solution in odorless mineral spirits	−15	−25	66
tert-Butyl peroxy-3,5,5-trimethylhexanoate	25	−20	114
Di-tert-butyl peroxide	40	−30	141
tert-Butylperoxy isopropyl carbonate	25	−20	117
tert-Butyl peroxybenzoate	25	10	122
2,2-Di(tert-butylperoxy)butane	30		116
Di(2-ethylhexyl) peroxydicarbonate,	−20		64
tert-Butyl peroxyacetate	10	−15	119
tert-Butyl cumyl peroxide	40		136
tert-Amyl hydroperoxide, 85% solution in water	30		190
1,1,3,3-Tetramethylbutyl hydroperoxide	25	−5	159

* T_s = storage temperature.
** 1 h half-life temperature = temperature at which concentration of the radical source is reduced by 50% after 1 h of being exposed to it.

7. The mechanism of kinetic chain termination is by combination of radicals. This further widens the MMD in LCB systems if the rate of initiation–combination is high [33].

8. Characteristics of the high-pressure ethylene polymerization manufacturing are also occasional runaway or decomposition reactions, also called "decomps". In these reactions ethylene decomposes into carbon and a mixture of methane and hydrogen. These too are highly exothermic reactions, which, according to the theoretical calculations, could cause increase of the final temperature and pressure in the reactor to 1400°C and 620 MPa (reactors are normally operating at 250°C and 200 MPa). As the decomposition is not otherwise avoidable, the high-pressure plants are designed for this sudden and abrupt increase in temperature and pressure. They are equipped with rapture disks and/or fast pressure release mechanisms. Still, decomps appear as controlled explosion of the plants that are usually followed by discharging the inventory of the reactor into free space emitting a black cloud of carbon. Modern control strategies based on thorough monitoring of conditions in the reactor contribute to less frequent decomps occurrences in high-pressure plants.

$C_2H_4 \rightarrow C + CH_4 \Delta H = 127 \, kJ/mol$
$C_2H_4 \rightarrow 2C + 2H_2 \Delta H = 53 \, kJ/mol$

3.3. Coordination Polymerization

3.3.1. Mechanism

Coordination polymerization is a typical chain polymerization reaction in which the polymer chain grows one monomer unit at the time. The name coordination stems from the principle mechanism that is based on the coordinative bonding of the double-bond containing monomer with the transition metal complex that features a free electron orbital capable of accepting π-electrons of the double bond. After the rearrangements, a σ-bond between the metal and carbon atom is formed and the coordinative site is freed for a new monomer to be bonded with the active site.

The second and all further monomer units are inserted between the first and/or preceding carbon atom, and the chain becomes longer by that one unit. Consequently, polymer chains grow from the side connected to the transition metal [36].

All three types of catalysts that are used for low-pressure polymerization of ethylene, Phillips, Ziegler, and single-site, are based on the same mechanism and they all contain a transition metal as a principle component.

Several reasons exist for termination of the coordination polymerization, for example, β-hydrogen rearrangement, in which the hydrogen atom in the β-position to the active site is abstracted and transferred to the incoming olefin. The chain growth continues with the new monomer unit and a vinyl double bond at the end of the polymer chain on the leaving chain is formed. Alternatively, in the presence of hydrogen in the reactor, hydrogen molecules can coordinate instead of the monomer, which leads to hydrogenolysis of the metal–carbon bond of the growing chain. The products are a saturated polymer chain, an alkane, and metal hydride species [37]. The metal hydride is further active in polymerization, and a new polymer chain can be formed on the active center. This process can occur many times until the active site is deactivated by poisons and/or side reactions.

On the basis of the described chain termination mechanism, hydrogen is often used as the molecular-mass controlling agent in coordination polymerization.

Coordination polymerization mostly produces linear polymer chains and the only way to introduce branching is incorporation of comonomers, typically 1-butene, 1-hexene, or 1-octene. Olefins with internal double bonds, such as 2-octene, cannot be incorporated into the polymer backbone and are catalyst poisons.

They coordinate with the transition metal, but the next step, formation of the σ-bond, is sterically hindered and does not take place. Therefore, the active site becomes blocked for further polymerization.

As the products of the β-hydrogen transfer are polymer chains with a vinyl group at the end of the chain, they could theoretically become re-engaged in the polymerization as so-called macromonomers. However, this is sterically hindered, given the length of the alkyl rest attached to the vinyl group. Single-site catalysts that are based on organometallic complexes containing one cyclopentadienyl ligand only, in which the coordination sphere of the active site is designed to be sterically open and accessible by large molecules, can indeed copolymerize also these macromonomers, and the result of this process are LCBs also in the low-pressure polymerization products.

3.3.2. Catalysts

In the case of coordination polymerization, the use of the term catalyst for the initiators is both somewhat inappropriate and very common. Ziegler catalysts and most Phillips catalysts require activation that occurs in the polymerization reactor. Ziegler catalysts are activated by aluminum alkyls and Phillips catalyst by the monomer. Therefore, more appropriate term would be catalyst precursors, at least for the two classes of catalysts. Single-site catalysts are typically formulated so that they contain the complex precursor and activator, and are ready to polymerize from the beginning. In addition, all coordination polymerization catalysts are destroyed in the process and cannot be recovered. Still, they enable and accelerate polymerization with very large turnover numbers, and, therefore, the term catalysts is commonly accepted and used.

Catalysts for all particle-forming polymerization processes, which include all polymerization processes except solution polymerization, are used in their supported form, impregnated on an inert carrier. This is to achieve good dispersion of the catalyst in the polymerization reactor, improve local heat transfer from the active site, secure active site separation and avoid mutual deactivation by two or more active sites, improve resistance to catalyst poisons and a few other reasons.

Carriers are typically highly porous materials that have a spongelike structure under the optical microscope. Due to the high porosity, carriers have large specific surface areas in the order of 250 m^2/g of material. Metal oxides such as SiO_2 and metal halides such as $MgCl_2$ are often used as typical carriers for olefin polymerization catalysts. Depending on the polymerization process, the carrier particles can have sizes from approximately 5 to approximately 50 μm in diameter. It is always advantageous if the particle size distribution is as narrow as possible.

Polymer particles are formed in the particle forming process following so-called replica mechanism and their basic shape is identical to the shape of the catalyst particle that caused their formation, just several times larger (Fig. 12).

Polymerization occurs on the active sites that are distributed across the carrier particle inside the pores on the surface of the carrier material. After the initial stage of polymerization, the polymer is formed in the pore cavities creating

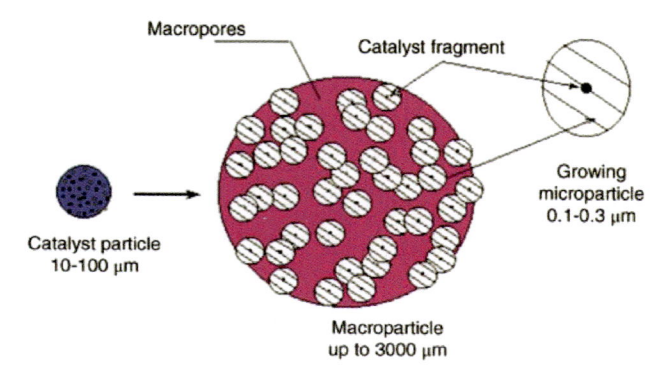

Figure 12. Catalyst particle growth by replication

certain hydrodynamic pressure at the pore walls. Under the pressure, particle structure and pore walls break but the polymer that is bound to all parts of the particle still keeps them together maintaining the form of the primary particle [38].

In the described mechanism, diffusion of monomer, access to the active site is not hindered by the growing polymer mass and the polymerization continues.

3.3.2.1. Phillips Catalysts

Phillips catalysts are extensively used for manufacturing polyethylene worldwide and over one third of the overall polyethylene volume is produced by using one of these catalysts.

They are typically supported catalysts that are manufactured by impregnation of the above-described carriers with an aqueous solution of chromium salts. Owing to its high solubility in water, CrO_3 originally was the main source of Cr for manufacturing Phillips catalysts. Given the high toxicity of Cr^{VI+} compounds, CrO_3 has been exchanged by less poisonous trivalent chromium salts. Loading of chromium in the supported catalyst can vary between 0.2 and 1.0% based on the total catalyst weight. Carrier impregnated with the salt is calcined in an air-driven fluidized bed at the temperature between 500 and 900°C to remove absorbed water and surface hydroxyl groups. By this process, chromium is converted into surface chromate (Fig. 13) by oxidation from the lower oxidation state found in the trivalent salt used as the chromium source.

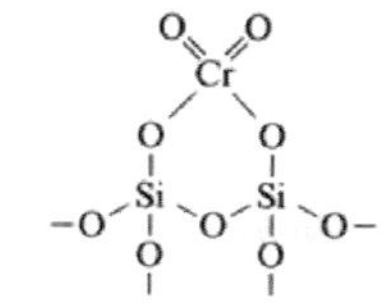

Figure 13. Chromate attached to the silica surface

If it contacts ethylene or some other reducing agent, the catalyst is reduced and activated. The reduction by ethylene is the reason for prolonged initiation time of Phillips catalysts that have not been pretreated with activators such as carbon monoxide. After activation, Phillips catalysts do not need further alkylation or any other co-catalyst to polymerize ethylene. Owing to the large variety of active sites obtained in the process of activation, Phillips catalysts usually produce polyethylene with broad MMD.

Importantly, chromate is only active if attached to the support, which seems to be a part of the active site as well (Fig. 14).

As the characteristics of the support are very important for all coordination polymerization catalysts, choice or treatment of carriers can be the differentiating factor causing diversity in catalyst performance, enabling production of a variety of compounds.

The carriers are commonly supplied already impregnated with chromium, and only the

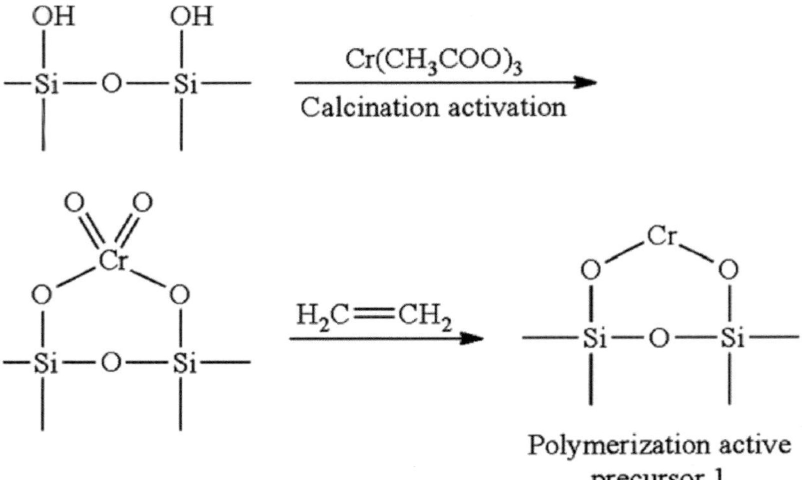

Figure 14. Active sites of a chromium based catalyst

activation step is performed by the polyethylene manufacturer at the plant.

Hydrogen, the common chain-transfer agent used for controlling the molecular mass of the polymer produced with almost all other coordination catalysts, cannot be used with Phillips catalysts because it is oxidized to water by the chromate groups. Very often, supports used for preparing chromium catalysts are titanated, treated with Ti containing compounds to bond chromium atoms to the support through titanate bonds [39]. This treatment enables manufacturing of polymers with a lower molecular mass, higher MFR, and even broader MMD.

Chain termination is the most common end of the polymerization. If chromium catalysts are the initiators, the β-hydrogen transfer process leaves chromium hydride, an active species that can start a new polymer chain, and the polymer chain with the vinyl group at its end (Fig. 15).

β-Hydrogen transfer is favored by high temperatures, and adjusting the polymerization temperature can be used to some extent as the means of molecular mass control.

Productivity of chromium catalysts is in excess of 5 kg/g leaving only traces (ppm) of Cr metal in the polymer.

Chromium catalysts are still very important and widely used in the commercial manufacturing of polyethylene. This, however, is no longer reflected in the intensity of research in the area of catalysis. Most companies and academic organizations are focused on other catalyst classes, mostly single-site, postmetallocene and/or Ziegler catalysts.

3.3.2.2. Ziegler Catalysts

Ziegler catalysts are coordination polymerization catalysts that are based on the transition metals from the fourth group of the periodic system: titanium, zirconium, and hafnium. They were discovered in 1953 at the Max Planck Institute for Coal Research in Mülheim an der Ruhr, Germany, in the research group of KARL ZIEGLER [40]. ZIEGLER was initially interested in polymerization of butadiene with main group metal alkyls as initiators. Attempting the polymerization with aluminum alkyls, he found so-called "Aufbau" reactions, insertion of ethylene molecules into the metal–carbon bond of aluminum alkyls. The reaction occurs at an ethylene pressure higher than 8 MPa and temperature higher than 80°C. Products of the aufbau reaction are oligomers of ethylene that contain up to 100 ethylene molecules. Research of the oligomerisation lead to the discovery of the so-called nickel effect: dimerization of ethylene catalyzed by Ni^0 complexes in the presence of aluminum alkyls. Presence of elemental Ni in the reactor, caused by aggressive cleaning of high-quality steel, turned the aufbau reaction into dimerization resulting in isobutene as the main product. Having discovered and explained the Ni effect, ZIEGLER expanded his interest to other transition metals and their interactions with aluminum alkyls and ethylene, looking for other unexpected effects. When $ZrCl_4$ was exposed to ethylene in the presence of aluminum alkyls, white linear polyethylene was formed, and Ziegler catalysts have been discovered.

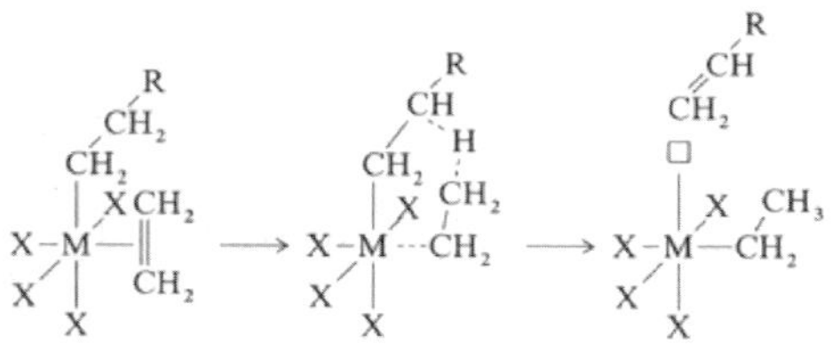

Later on, the catalysts have been used for polymerization of propylene by GIULIO NATTA in Ferrara, Italy. NATTA contributed significantly to the development of polypropylene, as well as to the theoretical explanations of the polymerization mechanism. Therefore, a Nobel Prize for chemistry was awarded to these two researchers in 1963 and the polymerization is known as Ziegler–Natta polymerization.

Early development of Ziegler catalysts was driven by the need to increase catalyst activity and eliminate the need for extraction of catalyst residuals from the polymer. First catalysts used in the commercial ethylene polymerization were based on reduction of $TiCl_4$ with aluminum alkyls resulting in crystalline $β$-$TiCl_3$. This was a self-supported catalyst in which only a very small part of all transition-metal atoms were participating in the polymerization.

Catalyst productivity was improved first by choosing different, longer alkyl aluminum compounds and alkoxytitanium chlorides [41, 42], so that the extraction was not needed any longer. While the activity of titanium was improved, it

Figure 15. β-Hydrogen-transfer-to-olefin mechanism

was still demonstrated that only less than 1% of the transition-metal atoms are active in the catalysis. The explanation for the low-active-site count in the catalysts was in the low accessibility of the titanium atoms due to the low number of them exposed at the surface of the crystallite largely containing titanium chloride.

To increase the accessibility by diluting Ti atoms and making them more effective in polymerization, inert metal oxides like silica and alumina were tried as supports for early supported Ziegler catalysts. Deposition of TiCl$_4$ directly on the surface of the oxides did not produce expected effects regarding activity increase [42], and the first significantly more active catalysts were manufactured by using Mg(OH)Cl as the carrier [43]. Since then, Mg compounds, mostly MgCl$_2$, have been among the important constituents of all Ziegler catalysts. Silica-supported catalysts include Mg-containing complexes of Ti atoms deposited on the carrier. Pure MgCl$_2$, generated either by precipitation or spray drying of Mg alkoxides, is often used as the support for Ziegler catalysts.

Increase in catalyst activity obtained by supporting is believed to be a result of the improved accessibility and larger participation of Ti atoms, which means a larger number of active sites per engaged Ti, and not a result of increased activity of single active sites. Furthermore, the support also affects the number of types of active sites, typically reducing them, which leads to somewhat narrower MMD (Fig. 16).

To further modify MMD, Lewis bases such as THF are often added to Ziegler catalysts. It is believed that the electron donors selectively deactivate some types of active sites effectively narrowing MMD.

Shape and size of supported Ziegler catalysts are, for most processes, very relevant features that determine their usability and range of products that can be manufactured using them. As surface area increases and tortuosity decreases in smaller particles, all resulting in improved accessibility of active sites, catalyst productivity increases by reducing the diameter of the support. At the same time, heat removal from smaller particles is more efficient. On the other hand, a definite size of particles is needed for processes in which entrainment, that is, carry over out of the polymerization reactor, is possible.

Narrow particle size distribution is always advantageous enabling accurate sizing of catalyst particles appropriate to the process needs.

Spherical shape and smoothing of the particle's outer surface play an important role in generating static electricity caused by triboelectric motion in the reactor. Rougher surfaces cause more friction leading to higher charge in the reactor, and static electricity is often the root cause of undesired reactor fouling, that is, deposition of polymer in the reactor that disturbs normal operation of it.

Ziegler catalysts are efficient in incorporating comonomers in the ethylene backbone of the polyethylene chain. They are used for manufacturing the whole line-up of low-pressure polyethylenes from linear, very high density materials to plastomers and elastomers. Distribution of the incorporated comonomer, however, is not ideally homogeneous in polymers made with Ziegler catalysts: shorter chains with lower molecular mass often contain more branches than longer chains. As homogeneity of branching is a key feature providing for improved physical performance of the polymer, the current development of Ziegler catalysts is focused on improving their copolymerization performance.

Ziegler catalysts can usually produce polymers of very high molecular mass that sometimes reaches the UHMWPE range. Molecular mass of the polymer is controlled by use of a chain-transfer agent, typically hydrogen, in the reactor. Relative molar ratio between hydrogen and ethylene determines the reactivity ratio between propagation and chain transfer, and

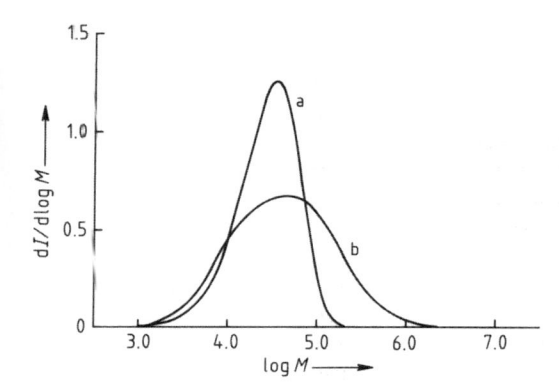

Figure 16. MMD curve of a Ziegler HDPE compared with simple statistical theory
a) Most probable distribution; b) Hostalen GD-4755

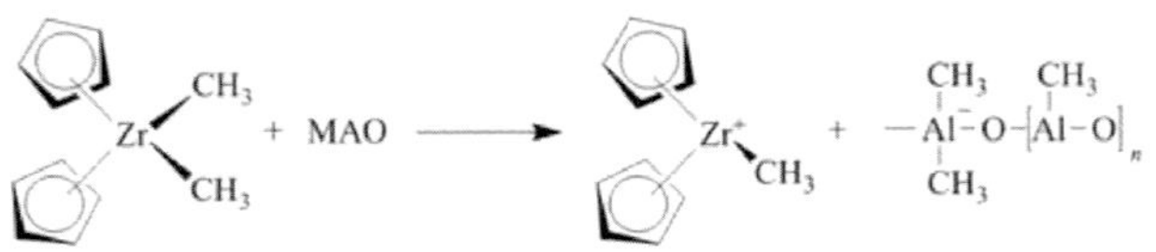

Figure 17. Metallocene activation with MAO

the molecular mass can be fine-tuned by modifying the ratio of concentrations.

Chain termination by β-hydrogen transfer that results in unsaturation occurs also if polymerization is initiated by Ziegler catalysts, but not as often as with other catalysts. Polyethylene prepared with Ziegler catalysts does not usually contain any long-chain branches.

Ziegler catalysts are used in both, supported and unsupported forms. All particle-forming processes need the catalysts to be supported. In the solution process, Ziegler catalysts are formulated in-line by mixing dissolved components of the catalyst formulation immediately before or in the reactor.

3.3.2.3. Single-Site Catalysts (Metallocenes)

For a characterization of metallocenes, see also → Metallocenes. Metallocenes are organometallic compounds, complexes of metals with two cyclopentadienyl ligands creating a sandwichlike structure. Their structure has been accurately described for the first time in the early 1950s by WILKINSON [44] and FISCHER [45] independently. The first characterized metallocene was ferrocene, bis(cyclopentadienyl) iron.

As soluble and well defined compounds, metallocenes of the fourth group of transition metals have been tested early as potential precursors for Ziegler catalysts for ethylene polymerization. With aluminum alkyls as activators, titanocene, zirconocene, or hafnocene exhibit very low if any activity in polymerization.

Invention of aluminoxanes, in particular, methyl aluminoxane (MAO) by KAMINSKY [46, 47], was the key for use of metallocenes in olefin polymerization (for details on MAO, see → Aluminum Compounds, Organic):

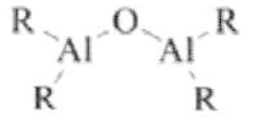

When activated with MAO, metallocenes are highly active polymerization catalysts that produce polymers with features that are advantageous over those of polyethylenes produced by other catalysts (Fig. 17).

The structure of MAO was not well described and understood for a very long time, and it was recognized from the beginning that a rather large excess of MAO is needed to activate metallocene complexes. Still, it was clear that the catalysts originated from the metallocenes containing few, most likely only one, type of active sites: the resulting polyethylene had the theoretically narrowest MMD. This is the reason why this catalyst class is also called single-site catalysts. Furthermore, after the initial invention, many monocyclopentadienyl metallorganic compounds and some other compounds not containing the cyclopentadienyl (Cp) ring at all, none of them formally fitting the name metallocene any longer, were found to be very active in polymerization of ethylene. For that reason the name single-site catalysts is broader and more applicable to the class of these advanced catalysts.

Single-site catalysts have been developed further, and besides MAO activation, activation with Lewis acidic compounds capable of abstracting an alkyl group, typically methyl, of the transition metal and creating an ion pair, became well understood and are now commercially used [48]. Boron compounds such as tris(pentafluorophenyl)borane or trityl tetrakis(pentafluorophenyl) borate are known to react with methyl derivatives of organometallic compounds to create a loosely separated ion pair active in the polymerization of olefins (Fig. 18).

Various other strategies for activation of single-site catalysts have been developed, which include use of mixed activators and or active supports.

There are several features that distinguish single-site catalysts from other coordination polymerization catalysts. The active sites of the single-site catalysts are tetrahedrally coordinated and, therefore, more accessible than the active sites of Ziegler catalysts that are in an

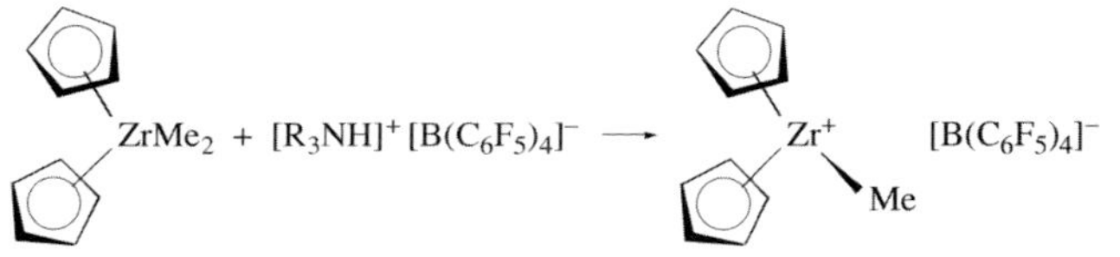

Figure 18. Borate activation of a dimethyl derivative of a metallocene complex (by product aniline not shown)

octahedral coordination. As highly reactive and relatively sterically open compounds, single-site catalysts are very sensitive to impurities. They react with the oxygen-containing compounds, and, of course, moisture, even if they are present in traces, becoming deactivated for polymerization. Unlike Ziegler catalysts, however, containing only one or few types of active sites, poisoning of single-site catalysts does not change the product quality, but only reduces catalyst activity and productivity. MMD and comonomer distribution do not change as a result of poisoning.

Furthermore, so-called hydrogen and comonomer responses of single-site catalysts differ by an order of magnitude from those typical for other catalysts such as Ziegler catalysts. The molar ratio between the chain-transfer agent, hydrogen, and ethylene needed to obtain certain lower molecular mass polyethylenes is in the range of 10^{-7} if a single-site catalyst is used, whereas a Ziegler catalyst would need several percent of chain-transfer agent in the polymerization gas composition. Similar observations are made for the comonomer-to-ethylene molar ratio needed to achieve branching.

Single-site catalysts are very efficient in incorporating comonomers, which means that the reactivity ratios for ethylene and longer α-olefins incorporation into the polymer chain are not only more similar than with other catalysts, but also the incorporation is much more homogeneous and the branches in the resulting polymer chain are spaced evenly across the chains of all lengths.

Some single-site catalysts incorporate comonomers in a reverse mode, that is, the longer chains contain more comonomer than the shorter chains, rendering their physical performance extraordinary well [49].

Single-site catalysts have an excellent comonomer utility, the ratio between the comonomer used in the polymerization reaction and comonomer needed to achieve certain density and/or branching. This advantageous comonomer utility is a consequence of their many advantageous features mentioned above. Firstly, because of the low comonomer concentration needed, losses of comonomer are much smaller than for those catalysts that need an order of magnitude higher concentration of the longer chain comonomers, the residuals of which remain adsorbed in the polymer particles and are often lost. Secondly, because of the improved homogeneity of the single-site-catalysts reaction products, same density is achieved with lower number of branches because the branches, every one of which is functional in disturbing crystallization, are evenly spaced, improving physical performance of the polymeric product. Finally, because of the even distribution of comonomer across the polymer chains of different lengths, the smaller number of branches improves the physical performance more efficiently than if they are concentrated in the shorter chains.

The polymerization reaction mechanism for single-site-catalyzed polymerization is very similar to the mechanism described for coordinative polymerization with Ziegler catalysts, which means that chain propagation and termination reactions follow the same steps. Some single-site catalysts are more prone to the β-hydrogen transfer termination, that is, they are more sensitive to higher temperatures, leading to polymers that contain more vinyl unsaturated end groups. This, in combination with the improved active-site accessibility, enables re-engagement of the macrocomonomers (polymer chains containing a vinyl group) in the polymerization resulting in branches that are as long as the polymer chain previously created (LCBs). As described above, LCBs, even at very low concentration, bring specific rheological and physical properties. The LCB-containing polymers are easier to process and have improved properties.

Since the initial discovery of single-site catalysts, many organometallic compound classes have been developed as precursors for

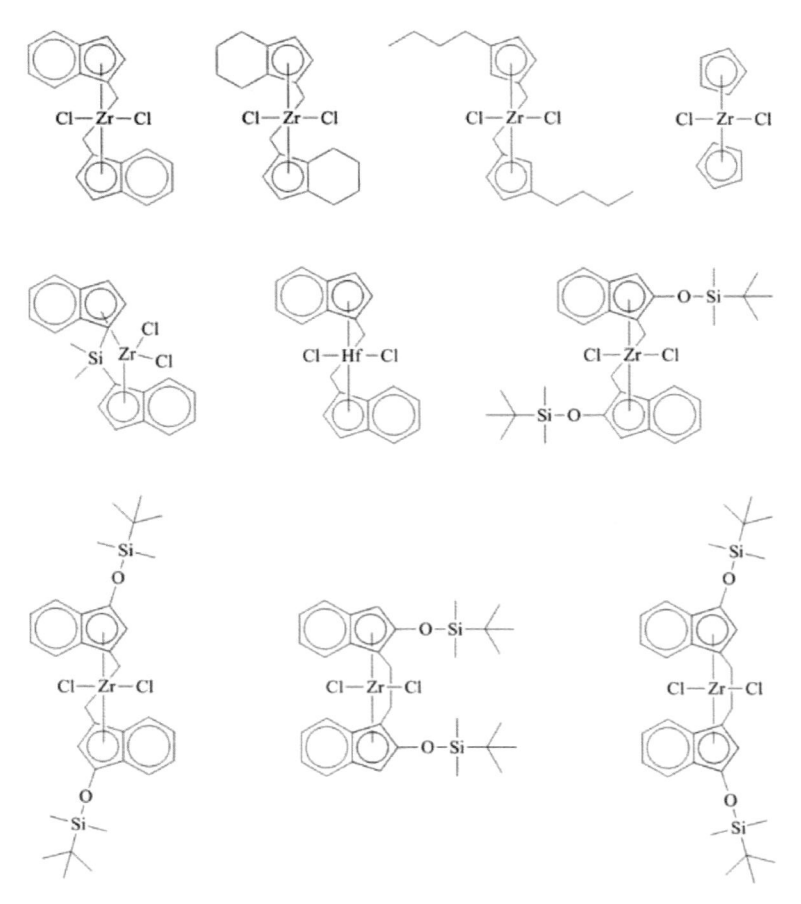

Figure 19. Typical metallocene complexes structures that are used commercially in polyethylene production

polymerization catalysts. As the active sites of the single-site catalysts are chemically well characterized and defined, they can be designed by using advanced molecular modeling techniques to perform in the desired way. Besides, the classic metallocenes and their derivatives are still used commercially. Examples are shown in Figure 19.

Mono-Cp compounds became commercially important with the innovation of the constrained geometry catalysts and phosphinimine ligands (Fig. 20).

Single-site catalysts can also be used in both their supported and unsupported form. If supported, they are usually combined with the activator, and the solid catalyst is formulated so that it is ready to polymerize immediately, and does not require activation and/or pre-contact as other coordination catalysts.

Combining the precursor with the activator and injecting them together can create a challenge, making fast dispersion of the catalyst powder more important after injection into a continuous reactor to avoid agglomerations of polymer particles formed in the first moments of the catalyst's injection in the polymerization media. Supporting single-site catalysts and

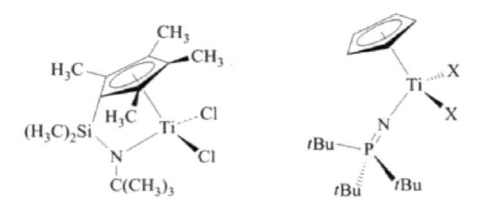

Figure 20. Constrained geometry catalyst and phosphinimine ligand based complex

occasional prepolymerization of the supported material usually stabilizes the catalyst and increases its robustness to impurities. Prepolymerization includes limited growth of the catalyst particles by polymerizing a small amount of monomer, comonomer, or any other olefin, which can be performed on- or offline. The kinetics of the catalyst is also influenced to some extent by its interaction with the support. In some formulations, the support can take an active role in contributing to the catalyst activity increase.

Single-site catalysts are often used in their unsupported form in the solution polymerization process, in which they are dissolved and combined with an alkylating reagent and an activator immediately before adding them to or inside the continuous-polymerization reactor.

4. Raw Materials

All polymerization processes for manufacturing polyethylene are sensitive to impurities. Some initiators already react with oxygen that is present in trace (ppm) amounts or moisture, and many other compounds can decrease or completely stall polymerization. Using high-quality material and purification of the raw material is often the key factor for obtaining good-quality material in an economically viable way. Purity specifications for polymerization-grade ethylene are listed in Table 5.

Most processes for polymerization of ethylene used commercially are continuous with the constant flow of raw materials through the polymerization reactors. Conversion of ethylene, depending on the process, is usually rather low, which is similar for other ingredients. Therefore, the raw materials leaving the reactor are purified, sometimes separated, and the polymerization composition is rebuilt before it is injected back into the reactor.

4.1. Ethylene

For further information on ethylene and ethylene production see → Ethylene.

Worldwide yearly production of ethylene was 109×10^6 t in 2006. Most of ethylene used for manufacturing polyethylene is produced by steam cracking. In the last decade, shale-gas exploitation and use for ethylene production became a significant factor in the polyethylene industry. The abundance and relatively inexpensive exploitation of it is influencing the polymer price and availability fostering new developments, particularly in North America [27].

In an effort to reduce environmental impact of manufacturing polyethylene, some polyethylene producers have developed so-called "green polyethylene" manufactured by using ethylene produced from renewable sources. Sugarcane is used as the base raw material for producing ethanol that is further dehydrated into ethylene. Ethylene is then polymerized to polyethylene in the conventional way.

Purity of ethylene is an important factor for manufacturing polyethylene. For most processes and catalysts the maximum amount of oxygen and/or moisture in ethylene cannot exceed several ppm, which is similar for all other oxygen-containing impurities. Most industrial crackers deliver ethylene that is rather pure enough, but methods to purify the feed stream online are often used. These are purification columns in which the gas is passed through a material that selectively binds impurities. Zeolites with defined pore size are used to physically remove moisture from ethylene, and various de-oxo beds that are based on metal ions in their lower oxidation state that oxidize in contact with oxygen are used to remove traces of oxygen [2].

4.2. Comonomers

The most important comonomers used for manufacturing various grades of polyethylene are

Table 5. Specifications for polymerization-grade ethylene*

Ingredient	Concentration
C_2H_4	>99.9 vol %
CH_4, C_2H_6, N_2	<1000 vol ppm
Olefins + diolefins	<10 vol ppm
Acetylene	<2 vol ppm
H_2	<5 vol ppm
CO	<1 vol ppm
CO_2	<1 vol ppm
O_2	<5 vol ppm
Alcohols (as MeOH)	<1 vol ppm
H_2O	<2.5 vol ppm
Sulfur	<1 vol ppm
Carbonyl sulfide	<1 vol ppm

*Data obtained from Repsol.

linear α-olefins, butene, hexene, and octene. There are several methods used for industrial manufacturing of the linear α-olefins: oligomerization of ethylene, Fischer–Tropsch synthesis, and dehydratation of higher alcohols. Most of the processes are not selective, producing a mixture of olefins, and require careful separation and purification. Methods for purification are similar to those used for purification of ethylene with the added care needed to avoid isomerization of the α-olefins. Internal olefins effectively deactivate coordination catalysts by their ability to establish a coordinative bond between the double bond and the metal active site, blocking sterically and electronically further rearrangements that should generate a σ-bond between the metal and carbon atom. De-oxo beds can, in a longer contact with a-olefins, accelerate isomerization and, therefore, it is important to use them in a dynamic mode.

Polar comonomers can be polymerized in radical polymerization only. The vinyl acetate and acrylate esters used as comonomers in the free-radical process are normally commercial-quality materials typically already stabilized to prevent homopolymerization while kept in storage or during pumping, but still active for the copolymerization reaction. In addition, they must be freed of dissolved oxygen by nitrogen sparging.

4.3. Other Materials

Initiators, coordinative catalysts and radical sources, are manufactured mostly by very specialized producers, very often also as custom synthesis.

Catalysts for solution process polymerization are formulated online by mixing solutions of the catalyst ingredients immediately before adding to or in the polymerization reactor.

Hydrogen is commonly used as the chain-transfer agent needed for molecular-mass control.

Besides the raw materials described above, there are other materials that are used either as polymerization process aids, post-reactor additives needed for product stabilization, pigmenting, or providing other final product properties.

Important components of all commercial polymerization processes are the inert materials used as diluents in slurry and solution processes, and as heat removing agents in the gas-phase process. These are typically hydrocarbons that are volatile and easily removable from the final product, having as high as possible heat capacity, and are not interacting with the polymerization process and/or the polymer.

Antifouling agents are often needed to avoid polymer deposition on the walls of polymerization reactors. These often are as antistatic agents that prevent static electricity charge of polymer.

Stabilizers are antioxidants needed to quench possible radical reactions in the formulated product leading to cross linking of the polymer chains.

One of the most commonly used pigments is carbon black that also serves as the UV-light stabilizer of polyethylene.

5. Production Processes

Commercial polyethylene manufacturing plants are among the largest industrial manufacturing processes. Current world's scale plants are capable of manufacturing 650 000 t of polyethylene per year, corresponding to the hourly production rate of more than 75 t.

Most processes are operated in the continuous mode, 24 hours per day and without any unnecessary interruptions. They are automated to the extent that they can be operated in the normal production mode with very few operators, with shifts seldom exceeding ten people.

Many manufacturing plants are integrated with the production of raw materials, located in the immediate vicinity of ethylene crackers that supply the monomer directly to the polymerization unit. This is advantageous for reasons such as transportation savings but also quality assurance coming from the direct supply.

Given the manufacturing volume, it is obviously beneficial that the plants are located close to the customers and markets where the material is sold and consumed.

The modern plants are rather versatile and can produce a variety of grades, sometimes also different types of polyethylene. Use of more than one type of catalysts is as well possible, further increasing the flexibility. Transitions between the products and/or catalysts is, however, very costly and a risky operation that requires high operational skills and careful planning to avoid further costs and operability issues with the manufacturing.

With the ever-increasing demand for product quality and improved physical performance, the need for additional degrees of freedom needed for tailoring the product, that is, designing its molecular characteristics that provide for the improvements, is growing as well. The plants are, therefore, becoming increasingly complex in terms of the number of reactors and possibilities to control the polymerization conditions.

Polyethylene plants are divided into two principal groups: high-pressure and low-pressure polymerization processes.

5.1. High-Pressure Process

High-pressure processes are used mostly for free-radical-initiated polymerization that requires high monomer concentration and temperature, but can also be used for polymerization initiated with coordination catalysts.

A typical high-pressure process is shown schematically in Figure 21.

There are two types of high-pressure processes that are distinguished on the basis of the form of the polymerization reactor: autoclave and tubular. An autoclave reactor is a typical CSTR reactor, and a tubular reactor behaves at least partially as a plug–flow reactor. To achieve the desired conversion and reaction rate, both types of reactors need to operate under a very high pressure that is in the range 150–200 MPa for autoclave reactors and 200–350 MPa for tubular ones. This high pressure presents a particular challenge for the technology that needs to operate continuously in a safe and reliable way.

Besides the size of the reactor walls, special construction expertise is needed to provide for increased fatigue resistance of the materials. Joints and connections in a high-pressure plant are designed in a way that uses the pressure to increase sealing forces [50].

The intellectual property obtained from the development of the high-pressure polymerization technology is very often kept by the owners as trade secret, deliberately choosing to not seek for a patent protection that would mean eventual exposure of the knowledge to the competition.

After the invention and commercial implementation of LLDPE in the early 1960s, the overall assumption in the industry was that the time of the high-pressure process has passed

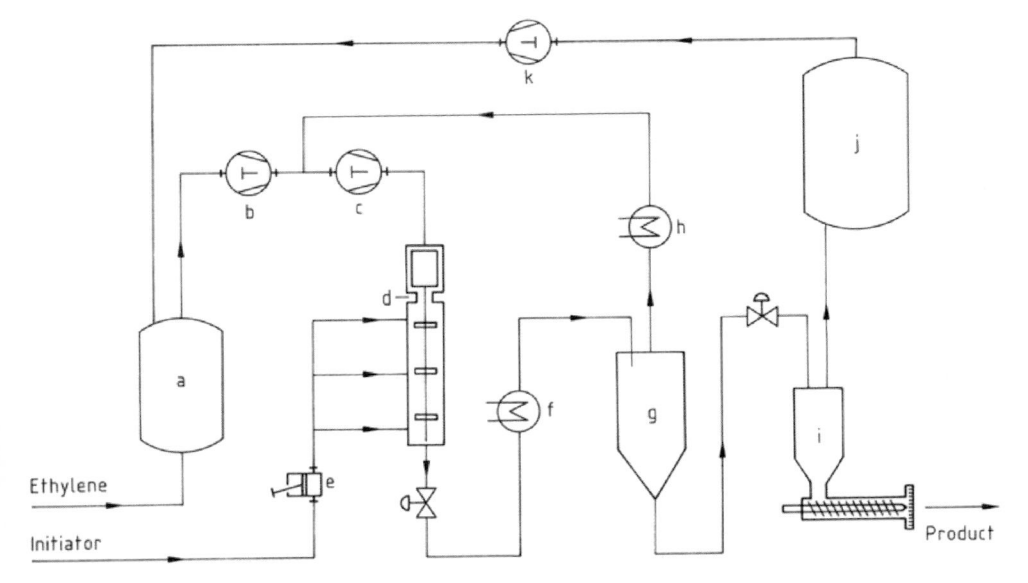

Figure 21. High-pressure process
a) Ethylene stock tank (5 MPa); b) Primary compressor; c) Secondary compressor (200 MPa); d) Autoclave reactor; e) Initiator pumps; f) Product cooler; g) Separator (25 MPa); h) Recycle cooler; i) Low-pressure separator and melt extruder; j) Low-pressure stock tank (0.2 MPa); k) Booster compressor

with LLDPE expected to fully take over the market and applications served by the LDPE material. On the contrary, owing to the specific characteristics of LDPE, that is, physical performance coming particularly from the long-chain branches and highly branched structure of the polymer, LDPE remained irreplaceable. It is still routinely blended to almost all LLDPE film materials to improve and ease processability of a final product or in multilayer application. Opposite to the 1960s' expectations, a few world-scale high-pressure plants have been built since then and a few other companies are considering expanding the capacities even further.

5.1.1. Autoclave Reactor

An autoclave high-pressure reactor as shown in Figure 22 is a typical adiabatic CSTR reactor, stirred mechanically with the feed stream inlet and product outlet to secure continuous operations of the reactor.

Heat of the reaction is not actually removed but rather adsorbed by the fresh monomer entering the reactor. Therefore, temperature control is achieved by maintaining the balance between the feed of the fresh monomer and conversion that again is achieved by dosing the initiator or mixture of them to different zones of the reactor. For a range of commercial initiators used in both, autoclave and tubular processes, see Table 4.

Most autoclave reactors are multizone reactors with two or more segments operating at different, increasing temperatures from 180°C to 290°C. Residence time in an autoclave reactor is typically between 30 and 60 s, and the volume of it is up to 1 m^3. Per-pass conversion of ethylene in an autoclave reactor is at the level of 20%.

Autoclave reactors are particularly prone to the above described runaway decomp events and, therefore, have to be equipped with rupture disks and possibilities for fast discharge of the content.

5.1.2. Tubular Reactor

Tubular reactors are typical plug-flow reactors, sometimes also called piston-flow reactors. The key assumption for the plug-flow reactors is that the fluid in them is perfectly mixed in the radial direction and not mixed in the axial direction.

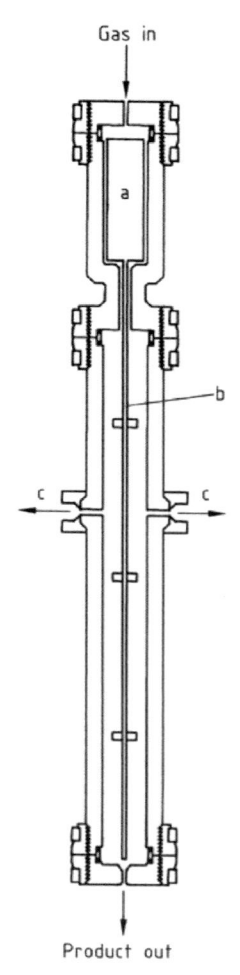

Figure 22. High-pressure autoclave reactor
a) Stirrer motor; b) Stirrer shaft; c) Bursting disk ports

Consequently, the composition of the fluid can be better described by the position of it in the reactor than by the time that the fluid has spent in it.

To assure mixing in radial but not in axial direction, tubular reactors need to have a very high l over d ratio, so most tubular reactors used for high-pressure polymerization are very long with narrow tubes coiled by combining straight segments that are connected by 180° bends. The inner diameter of the tubes is approximately 60 mm and they are as long as up to 1000 m, with thick tube walls needed to sustain high pressure of the reaction.

To achieve the required temperature control, the tubes are jacketed and equipped with numerous thermocouples and inlets for initiator, gas, or other ingredients that are located mostly in the junctions.

Automatic temperature control is even more critical for the tubular reactors than for the autoclave ones. Due to the length of the reactor and the plug-flow nature of the polymerization, it is not always and everywhere possible to achieve absolutely efficient heat removal through the reactor wall, and temperature spikes can occur at almost any position of the tube when the reaction temperature is reached.

An important element of this rather sophisticated temperature control is the combination of different initiators, liquid peroxides, and oxygen, and their dosing to the reactor at the chosen spots.

Tubular reactors have two significant advantages over the autoclave ones: due to the advanced temperature control it is possible to recognize temperature increases early enough to react timely and reduce the reaction rate so that the decomp occurs less frequently, leading to overall higher conversion of ethylene. Most of the newly built high-pressure polymerization plants are based on the tubular technology.

5.1.3. High-Pressure Copolymers

Ethylene copolymers and terpolymers with polar comonomers are becoming increasingly important owing to their very specific performance that cannot be achieved by any other materials not even by blending of grafting different polymers in the post-reactor process.

These copolymers show particularly interesting performance when combined with cross-linking agents in the post-reactor treatment. Due to the improved heat deformation resistance and thermal oxidative stability, combined with the very low impurities level in the polymer, these materials are very well suited for manufacturing the insulation for medium- and high-voltage cables.

Polar monomers, often oxygen containing, deactivate all coordinative catalysts and can be copolymerized with ethylene in the radical polymerization high-pressure process only.

Some obvious modifications to the polymerization plant are needed to enable copolymerization: supply of the comonomer in its purified form, precise dosing in terms of amounts and reactor position, separation and recycling of the unreacted comonomer, removal and recycling of the unreacted comonomer from the product, etc.

5.1.4. Linear Low-Density Polyethylene

A typical high-pressure plant can be converted and/or used for coordination polymerization, manufacturing high- or linear low-density polyethylene. Several companies have at least explored the possibility, particularly as a means to extend the lifetime of a mature high pressure plant or as a response to the anticipated diminishing demand for LDPE.

There are no obvious advantages in terms of catalyst performance or product quality of utilizing a high-pressure technology for coordination polymerization.

5.2. Coordination Polymerization—Low-Pressure process

Polymerization of ethylene at low pressure became possible in the 1950s, after the coordinative catalysts have been invented. Besides the substantially lower complexity of the hardware needed for the polymerization, the fundamentally different polymerization mechanism also delivers fundamentally different products, linear polyethylenes, unlike highly branched and cross-linked LDPE.

Introduction of the crystalline polyethylene and further developments of copolymers that have a controllable and tunable degree of crystallinity significantly widened the use of polyethylenes for those applications that could not be serviced by LDPE.

5.2.1. Suspension (Slurry) Process

Polymerization of ethylene in a slurry, in which supported or solid catalyst and growing polymer particles are suspended in an inert hydrocarbon, has some important advantages from the process point of view: mixing is relatively readily achievable; heat removal from the polymer particles is very efficient; in case a low-boiling solvent is used, removal of the hydrocarbon at the end of the polymerization is relatively simple; it is inexpensive; etc. Therefore, the first process developed for commercial polymerization with Ziegler catalysts was a slurry process.

At the beginning of its commercial exploitation, the polymerization with chromium-based (Phillips) catalysts required removal of the catalyst residuals from the final polymer. After development of higher activity and well dispersed catalysts that could be left in the product in the 1960s, the use of the slurry polymerization technology also for Phillips catalysts became economically viable (Fig. 23).

The slurry process used today can be operated with coordinative catalysts from all classes. Owing to the partial solubility of polyethylene comonomers with low crystallinity, the slurry process is somewhat limited and not totally suited for manufacturing LLDPE.

Batch Slurry Polymerization Process. Polymerization in the slurry can be performed in the batch or in the continuous mode. Batch mode polymerization occurs in a stirred autoclave reactor in which temperature and pressure are maintained constant throughout the polymerization run. The main reason for operating a commercial manufacturing plant in the batch mode is the need for uniform product particles of UHMWPE products. UHMWPE is often processed by compression molding by which the material is not melted but sintered. Shape and size of the polymer particles play a key role in this kind of product processing, justifying the less economically attractive batch polymerization process.

Continuous Slurry Polymerization Process. The most common form of continuous polymerization processes that are operated in the uninterrupted mode is based on the loop reactors. Loop reactors are made out of the relatively large diameter jacketed pipes that, connected together, form a loop. The polymer suspension is pumped continuously through the loop with the fluid velocity between 5 and 10 m/s.

The diameter of the typical slurry polymerization reactor is 50 cm and they could be up to 300 m long. Concentration of the solids in the slurry is up to 60%. Depending on the polymer manufactured and the diluents used, the operating temperature and pressure can be between 70 and 105°C and 5 and 10 MPa, respectively. Residence time in a slurry loop reactor is between 40 min and 3 h, depending on the process.

There are several ways how the catalyst can be supplied to a continuous or any other slurry reactor. The simplest way is by suspending of supported or solid catalyst in a mineral oil and then pumping this suspension into the circulating fluid in the polymerization reactor. Alternatively, catalyst powder can be mixed with the liquid diluents to create viscous suspension, mud. The catalyst mud is then transferred through a rotary valve into the diluents stream that is pumped into the reactor.

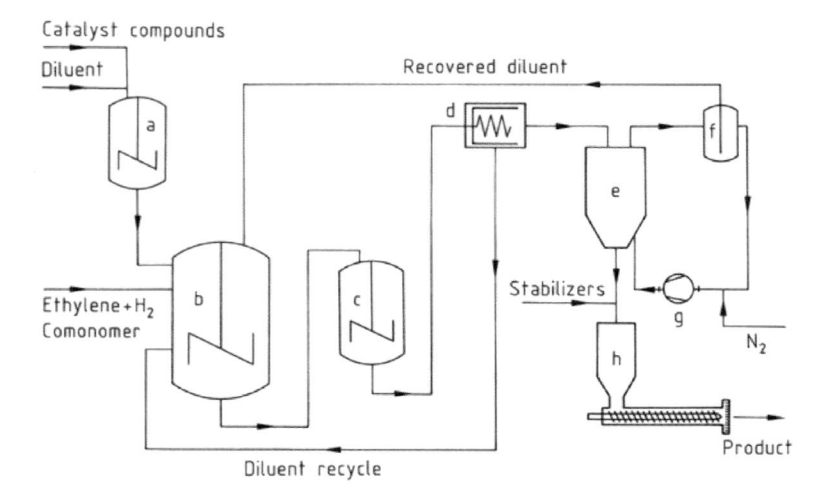

Figure 23. Hoechst suspension polymerization process
a) Catalyst preparation vessel; b) Polymerization reactor; c) Run-down reactor; d) Centrifuge; e) Fluidized-bed drier; f) Diluent condenser; g) Nitrogen circulator; h) Powder-fed extruder

Very often, slurry loop processes contain at least two loop reactors: prepolymerization reactor, which is much smaller in size, and the main reactor. The catalyst is exposed to the limited amount of monomer, usually at somewhat lower temperature, to grow 5–20 times in weight before it is transferred into the main reactor. Both of these reactors are operated in the continuous mode. The purpose of the prepolymerization step is in initiating the polymerization under mild conditions under which the particles start forming rather slowly and the risk of creating fines is smaller. At the same time, some catalysts become even more productive because of the controlled particle growth achieved by the prepolymerization.

Most slurry loop reactors are equipped with the settling legs, extension traps for sedimentation of the solid polymer material resulting in more concentrated slurry. Removal of the product then occurs from the settling legs, so that a limited amount of the diluents leaves the reactor and needs to be flushed out.

Slurry loop reactors are operating in the CSTR mode and the mixing should be effective in both radial and axial direction. This, however, holds true only to a certain extent. It is known that under the conditions for the production of low-molecular-mass polymer, that is, when concentration of hydrogen and/or of some more active catalysts in the reactor is relatively high, a concentration gradient is established and monomer- or chain-transfer concentration decreases in the parts of the reactor that are further away from the catalyst and or feed stream inlet. The gradient can cause certain broadening of the MMD of the product.

Disadvantages of the slurry polymerization reactors are operability problems typically caused by fouling of the reactor walls as a result of polymer agglomeration and stickiness to the wall. More often, the reactor starts operating in an unstable mode, in which the polymerization rate is oscillating and becomes difficult to control.

Still, slurry loop reactors are among the most used forms of the ethylene polymerization processes in the world.

5.2.2. Gas-Phase Process

Another commonly used process for the industrial ethylene polymerization is the gas-phase or fluidized-bed process. Besides the economic advantages, which are relatively low investment cost, absence of a liquid diluent, and large-scale plants, the abundance of the process is at least partially a consequence of the very successful licensing strategy and implementation of Union Carbide. The company developed the Unipol process and distributed the licence to it more than 120 times worldwide. The Unipol process was introduced in the late 1970s, originally for manufacturing LLDPE, but it is currently often used for manufacturing both HDPE and LLDPE materials. Following Union Carbide, other companies developed their original gas-phase processes. Naftachemie (later BP and INEOS) developed the Innovene G process, and Montell (later LyondellBasell), developed the Spherilene process. Both processes have been successfully licensed as well.

Schematics of the gas-phase polymerization process is presented in Figure 24.

The gas-phase process is the most versatile process in terms of the products that can be manufactured by it. As the polymer particles are in their granular form in the reactor, no solubility and viscosity constraints exist such as in the solution process, and the potential for agglomeration, risk of operating soluble materials, is reduced because of the absence of diluents. Therefore, the density range of the products is broad, from 890 to 970 kg/m^3 with an MFR that varies between nonmeasurable to over 100. The process is also versatile in terms of the comonomers that can be used. 1-Butene and 1-hexene are the most often used comonomers, and BP/INEOS used 4-methylpentene as well. If single-site catalysts are used for initiation of the polymerization, owing to their high comonomer response, high reactivity, and need for very low concentration of comonomers in the reactor, also a high-boiling comonomer such 1-octene can be utilized.

Finally, the gas phase process can be used with all kinds of coordination catalysts: Phillips, Ziegler, and single-site. The catalysts need to be supported on silica or $MgCl_2$, and the particle size needs to be sufficiently large to avoid carryover of the small particles into the process. They are usually larger than 40 μm in diameter. The shape of the particles must be as regular as possible. Sphericity and smoothness of the surface is required to avoid static electricity build up

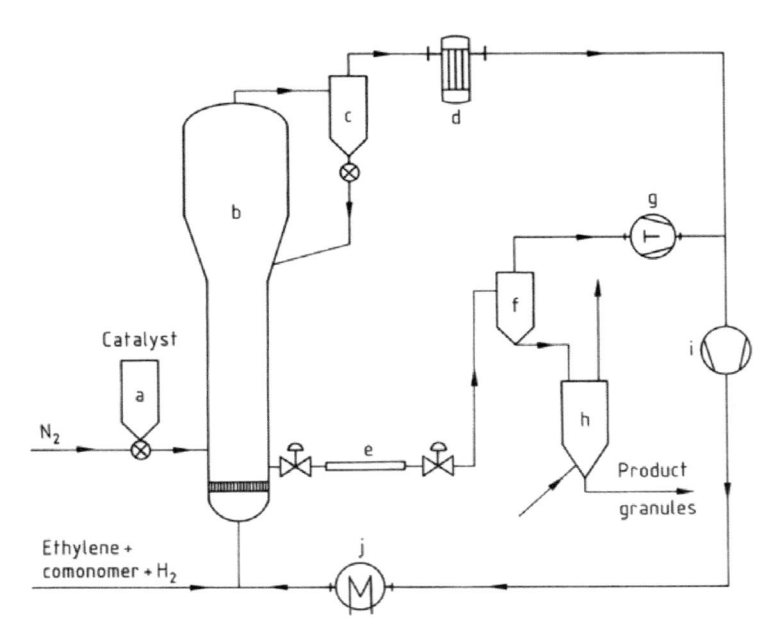

Figure 24. Fluidized-bed process
a) Catalyst hopper and feed valve; b) Fluidized-bed reactor; c) Cyclone; d) Filter; e) Polymer take-off system; f) Product recovery cyclone; g) Monomer recovery compressor; h) Purge hopper; i) Recycle compressor; j) Recycle gas cooler

through the triboelectric effects, and to provide for higher powder bulk density in the reactor.

Most often, the catalyst is injected into the fluidized-bed reactor in its dry form, as powder. This is needed to achieve fast and efficient dispersion of the catalyst into the polymerization bed in the shortest time possible.

The fluidized-bed reactor, which is an essential part of the gas-phase process, is a vertically mounted, rather simple large steel cylinder with the typical widening of the upper area, so-called disengagement zone. Polymerization occurs between a perforated steel plate mounted in the lower part of the cylinder, distributor plate, and the disengagement zone. The polymerization bed of polymer particles is fluidized by the circulation gas traveling from the bottom of the reactor up with the velocity of 0.3–0.8 m/s. When the gas–polymer mixture reaches the disengagement zone, the velocity is reduced because of the volume enlargement and the gas is not capable to carry the particles any longer and they drop back into the main zone of the reactor. The polymer-free cycle gas travels then further to the heat exchanger and compressor, which sends it back into the reactor. Before the circulation gas reaches the distributor plate, the gas composition is rebuilt, the converted monomer, comonomer and chain transfer agent are refurbished to the needed concentrations.

Gas-phase fluidized-bed reactors can have different sizes, and the largest ones are >6 m in diameter and ≥16 m. They can hold more than 100 t of polymer in the polymerization bed, and the production rate in the process can be as high as 80 t per hour.

The fluidized-bed reactors normally operate with the total pressure of about 2 MPa and in the temperature range between 70 and 105°C, depending on the polymer that is produced. C_2 partial pressure in the reactor can be up to 1.5 MPa, and the per-pass conversion of the monomer is between 2 and 10%.

Circulation gas serves a few purposes in the gas phase process and one of them is the heat removal. The gas enters the reactor at the temperature that is up to 20 or more degrees lower than the operating temperature and it absorbs the heat of the reaction almost instantaneously. The gas is cooled down in the heat exchanger.

To further increase efficiency of the heat removal in the gas-phase process, Union Carbide later developed the so-called condensing mode of operating the gas-phase process. In this process, an inert hydrocarbon, pentane or hexane, is injected into the circulation gas, prior to entering the reactor. The liquid hydrocarbon evaporates in the reactor and absorbs the heat of the reaction. Hydrocarbon vapors are carried over as a part of the circulation gas to the heat exchanger in which they are condensed into the liquid. The liquid is then reinjected in the reactor. By applying the condensing mode of operations, the capacity of the same reactor can be sometimes doubled, which increases the production rate and reduces the residence time.

Mixing in a fluidized-bed reactor is provided by the fluidization and the typical fluidization motion of the polymer powder. The circulation gas travels from the bottom to the top carrying only a part of the polymer powder. The other part of it, separated from the gas in the disengagement zone, falls down along the wall of the reactor, and travels in the other direction, from the top to the bottom. By this, the polymer particles create a certain shear that scrubs the polymer from the wall on which it tends to stick, potentially causing fouling. In addition, the circulation gas tends to create gas bubbles in the reactor. These bubbles coalesce into larger bubbles that burst at the very top of the reactor, placing polymer powder into turbulent motion.

Due to the rather efficient mixing, it is believed that there is little to no gradient in temperature or gas composition throughout a fluidized-bed reactor.

5.2.3. Solution Process

In the solution process, the polymer remains dissolved in an inert diluent throughout the polymerization. As polyethylene is not really soluble in any solvent, the process must be operated at a very high temperature so that melted polymer mixes with the diluent and creates a solution. Because of the high temperature, speed of the polymerization reaction is significantly increased so the residence time in a solution process reactor is rather short, from 90 to 1000 s. This is also very well in line with the decreased stability of most coordination catalysts at the temperatures higher than 150°C sometimes reaching 200°C or more. The solution process, same as the high-pressure process, operates in the adiabatic regime. Per-pass conversion of ethylene is high and it reaches 95%. Schematics of the solution process is presented in Figure 25.

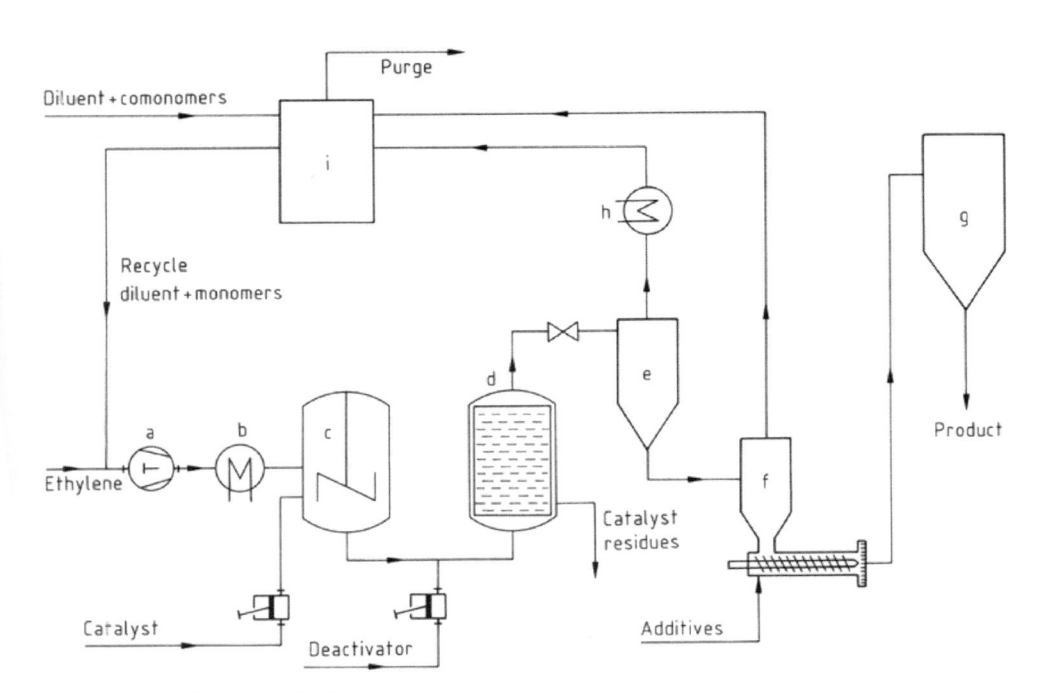

Figure 25. Solution polymerization process
a) Reactor feed pump; b) Temperature control; c) Reactor; d) Catalyst adsorber; e) First separator; f) Low-pressure separator and melt-fed extruder; g) Purged product hopper; h) Recycle cooler; i) Diluent and monomer purification unit

The range of products that can be manufactured in solution is constrained by the viscosity of the melt solution that needs to flow through the process. Depending on the branching of the product that can decrease the viscosity to some extent, the MFR of the material manufactured in solution must be higher than 0.2.

Products of the widest density range can be produced in the solution process, form the lowest at 860 kg/m^3 to homopolymers. The process is also very versatile in terms of comonomer choice. It can produce butene, hexene, or octene copolymers, and it is also used for manufacturing EPDM rubber.

Catalysts used for polymerization in solution are Ziegler and single-site and they are typically formulated by mixing solutions of the individual components of the catalyst system immediately before or after they are injected into the reactor. If operated with the VOCl$_3$ or Ziegler catalysts, the catalyst must be deactivated after the melt solution leaves the reactor, so that the polymer remains well defined and no uncontrolled side reactions in the downstream piping takes place. If the vanadium catalyst is utilized, residuals of the heavy metal need to be removed from the polymer by passing the melt solution through an alumina column. Finally, the melt solution is depressurized and the unreacted monomer and comonomer are collected and sent for purification and recycling in the process. Residual product is then extruded into pellets.

One of the advantages of the solution process is its short residence time and the short time needed to transition from manufacturing one product grade, or using one catalyst, to the other. The size of the reactor and low material inventory in it increase the flexibility of the process so that the transition time becomes on the order of couple of hours, whereby a gas-phase process, for example, might need up to 12 or more hours for the same type of change. This and the versatility of the materials that can be produced in the solution process make the technology interesting for manufacturing speciality products, needed in relatively smaller quantities, in a still economically viable manner.

5.2.4. Combined or Multimodal Processes

MMD of polyethylene chains and distribution of branching in them are the two most important factors that influence the physical performance of the material. The manufacturing processes described so far, in which polymerization occurs in one single reactor, are by themselves providing only very limited possibilities to influence the distribution and the product quality depends mostly on the catalyst used. Different catalysts, depending on the number and nature of the types of active centers, produce different products in terms of molecular mass and comonomer distribution, but that cannot be controlled very easily, especially because plants are typically limited to using only few catalysts and it is not reasonable to switch the catalyst whenever a different grade is needed to be manufactured.

To increase the number of freedom degrees and enable effective tailoring of the polymer so that some enhanced performance is obtained, several companies have developed processes that include two or more polymerization reactors, typically in series, that can, but do not have to, be different in their nature. These processes and/or plants are often called bimodal (or multimodal) processes and/or plants because they can produce bimodal (or multimodal) polymers.

By combining the reactors, polymerization conditions can be chosen so that different materials are produced in each of them so that the combination, the final product, obtains the desired molecular mass broadness and the comonomer is placed in the polymer chains of the chosen length. This is typically achieved by making two or more distinct polymer components that are substantially different in average molecular mass and/or in the amount, sometimes even type, of comonomer incorporated. If, for example, the material produced in the first reactor contains little or no comonomer and the chains produced are rather short and in the second reactor the conditions are adjusted to make a polymer that has higher molecular mass and contains more branching, the final product will have a bimodal, broad overall MMD providing for a good processability, and will contain a high-molecular-mass component that will provide for melt stability and physical performance. The distribution of comonomer will be skewed toward longer chains, further contributing to the good physical performance.

Although the same or similar effects could also be achieved by blending two materials in a post-reactor step, the bimodal process have substantial

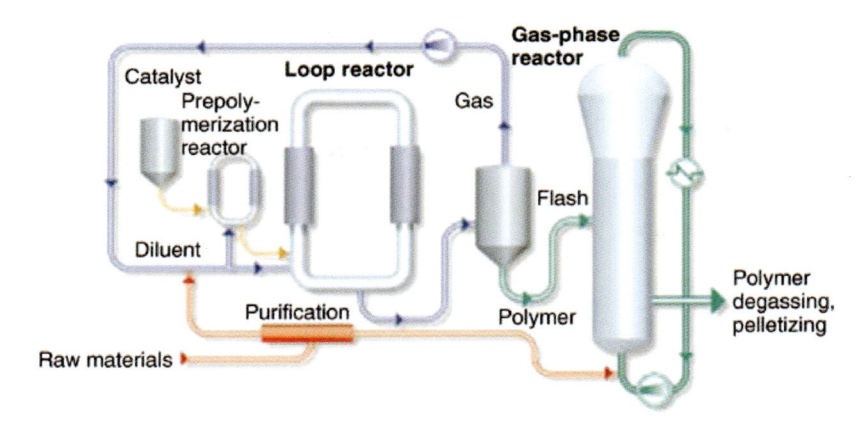

Figure 26. Borstar process

advantages. The "onion" structure that is obtained in a particle-forming process by passing the same catalyst-growing polymer particle through the reactors in series contributes to a substantially better miscibility of the two or more relatively different polymer components, and one avoids the complexity of keeping and managing the inventory needed for the blending step.

The first bimodal plants were developed by combining two slurry-loop reactors in series, and later processes were developed in which different types of reactors are connected. The Borstar process developed by Borealis is a combination of a slurry-loop reactor and a gas-phase reactor. The Borstar process also contains a third step, a smaller slurry-loop prepolymerizing reactor preceding the main slurry loop (Fig. 26).

Most modern solution plants are built with two reactors that can, but do not have to, be operated in series. As the blending occurs in solution, dispersion and mixing works well and one can operate the two reactors also in a parallel configuration.

6. Uses

Polyethylene is one of the most abundant man-made materials overall. It can be found in a very broad range of applications that span from medical prosthesis, artificial joints made of UHMWPE and surgically implanted as replacement for the natural ones, via structural applications that include pipes and cable insulation, to waste bags.

The specific properties of polyethylene including low density, mechanical strength, sealing properties, flexibility, easy processing, low cost, and low environmental impact make the material still increasingly attractive even 80 years after it was industrially manufactured for the first time.

The variety of now well established polymerization and post-reactor treatment processes that are continuously further developed enable the industry to develop specialized products for targeted applications with the properties that are constantly enhanced.

6.1. Films

A large volume of polyethylene production is used in applications in which the material is extruded into thin, flexible foils or films. Film is the preferred material for many forms of packaging, but is also used for other purposes for which separation and/or containment are needed. Typical examples of the nonpackaging polyethylene film applications are agricultural film used for the construction of greenhouses and geomembrane film, thick material that provides containment of the landfill sites and long-term separation of the disposed solids and liquids from the soil.

The two principal manufacturing methods for polyethylene films are blown-film extrusion and cast-film extrusion.

Blown-film extrusion is the method by which molten polyethylene is continuously extruded

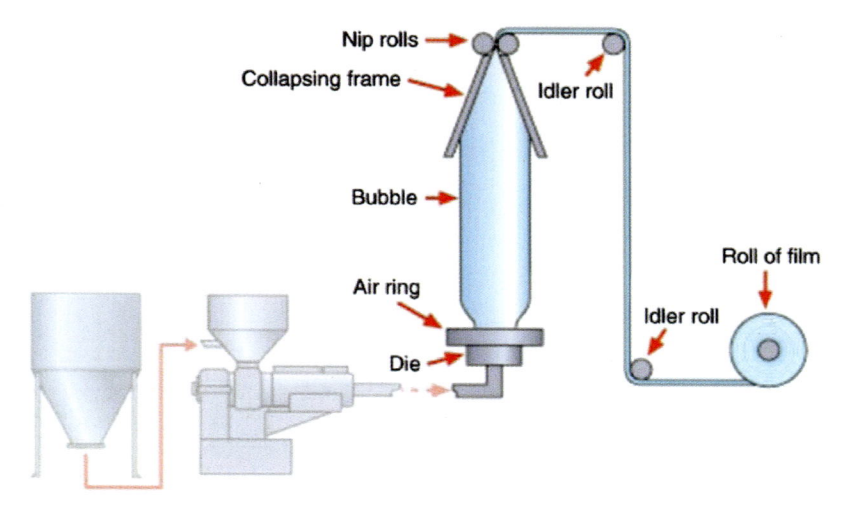

Figure 27. Blown-film line

through a circular die to form a bubble filled with air or nitrogen. The bubble is collapsed at the top by rollers into a two-dimensional tube that is then rolled on the role. The tube is usually further processed into bags by sealing and cutting, or simply cut to create a flat foil (Fig. 27).

The diameter of the bubble can be up to several meters and it is larger than the diameter of the die; the ratio between the two is called "blow-up" ratio. Thickness of the blown film can be between 10 μm and several hundreds of micrometers. Depending on the blow-up ratio, film thickness, and the cooling of the bubble, which is the time allowed for the material to crystallize, the polyethylene chains could be more or less oriented in the final product. The orientation substantially influences the physical performance of the film increasing for example machine direction and decreasing the transverse direction tear resistance.

Most modern blown-film extrusion processes are capable of extruding multilayers from which different materials are coextruded into one film product. This is performed by using multiple extruders and complicated dies that have overlapping openings so that the molten polymers meet and form the bubble all together without mixing with each other. Some of the machines can handle up to eleven layers and the materials that are combined can be very different as long as the adhesive layers are separating two otherwise less compatible layers. The advanced film blowing enables creation of packaging materials that have a combination of properties originating from the polymers specifically designed to provide only some of them. This is particularly important because the properties very often cancel each other and it is not possible to have a certain combination in one polymer only. Consequently, it is difficult to achieve somewhat higher stiffness and improved oxygen and moisture barrier performance in the material that needs relatively low temperature for sealing two layers into one, which is an important requirement for manufacturing bags. Higher density, more crystalline material is stiffer and presents a better barrier, and more mobile polymer chains that are less organized into crystallite, higher branched (lower density) materials, provide easier sealing of the layers.

Cast film is a product of extruding a thin layer of polymer onto chilled rotating rollers (Fig. 28). The material solidifies very fast in the process so that the polymer chains remain largely oriented in the machine direction only. As a consequence, cast films have improved clarity and gloss, can be stretched easily, cling to the surface, and have reasonably high puncture resistance. This makes them an ideal packaging material with which single or more objects need to be protected or held together by wrapping the film material around them. Cast film, also cold-stretch film, is used for industrial packaging of very large objects like vehicles and helicopters,

Figure 28. Cast-film line

but also in common households as food packaging.

All polyethylene grades can be used for manufacturing film materials. LDPE, HDPE, and LLDPE are used for blown-film materials, and LLDPE of somewhat higher density, 930–935 kg/m^3, is used for manufacturing cast films. Manufacturers chose different polyethylene materials depending on the application of the film and the specific properties needed for it.

6.1.1. Extrusion Coating

Very often, a combination of materials is needed to achieve the right combination of properties required for flexible packaging. As in the multilayer film extrusion, in which different polymers are combined, in the extrusion-coating process, polyethylene can be combined with other nonpolymeric materials that are needed for esthetical and traditional reasons, or because polymers cannot provide for the performance that is required.

Mostly, LDPE, owing to its high branching content that leads to a high die swell, is extruded through a thin, wide slit die over a moving layer of paper or other material. The polymer clings to the material and the products are polyethylene-coated cardboard, paper, or metal foils used for packaging of liquids or extremely oxygen-sensitive products.

6.1.2. Blow Molding

Blow molding is a conversion process for forming hollow objects, for example, bottles, in a two-stage process that involves formation of a so-called parison in an extrusion step. A parison is a piece of melted plastics with the hole on the end that is then placed into a mold. In the next step, a gas (usually air) under pressure is injected through the hole into the parison, so that a molten polymer takes the form of the mold and cools down (Fig. 29).

Blow molding is used for manufacturing bottles and containers for packaging of aggressive liquids, household cleaning materials, solvents, or gasoline. HDPE is often used in the process providing for a good environmental-stress crack resistance, a property that is reinforced by higher crystallinity of the polymer.

6.1.3. Injection Molding

Injection molding is one of the more versatile processing methods in terms of the products that are manufactured. Molten polymer is injected under pressure into a mold at one or more injection points at which it solidifies by cooling. The process is used for manufacturing caps and closures of bottles, toys, houseware, containers, and many other objects. The current European beverage market consumes 222×10^9 of bottle

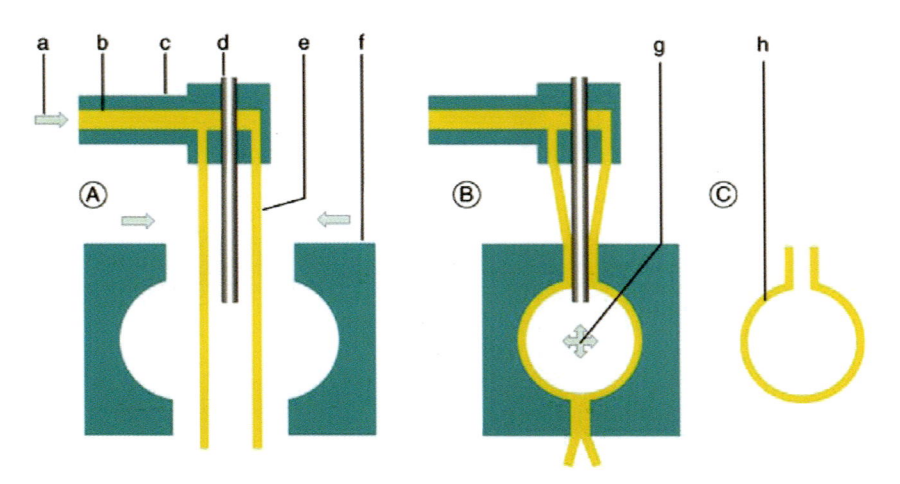

Figure 29. Blow molding
A) Parison formation; B) Blow molding step; C) Final product removal
a) Injection point; b) Molten material; c) Die for parison extrusion; d) Air inlet; e) Parison; f) Mold; g) Expanded gas; h) Final product

caps and closures per year, and most of them are made out of a polymer, very often polyethylene.

Owing to the nature of the process, relatively low molecular mass material that can flow fast through the mold is needed for injection molding. Mostly HPDE and LDPE with the MFR that can be up to 100 or more are used.

6.1.4. Rotational Molding

Rotational molding is the least investment-intensive method for processing of polyethylene, and it is, therefore, used for manufacturing of large hollow parts that cannot be manufactured by using injection or blow molding in an economically viable manner. Molds used for rotational molding are produced out of the inexpensive sheet metal and can be easily modified or reproduced if needed.

The manufacturing process is discontinuous, and it starts with placing a certain amount of polymer powder into the mold that is then closed. The mold is mounted on an arm that rotates and tumbles it in all directions making sure that all inner surfaces of the mold get in contact with the fine polymer powder. The mold is either placed in an oven or is heated by gas flames to the temperature that exceeds the melting point of the polymer. By this, polymer melts still covering the surface. Finally, the mold and the polymer are cooled down so that the object solidifies, and the mold is opened (Fig. 30).

The process is slow and not very efficient in terms of the numbers of objects that can be produced. Still, as molds for injection or blow molding are very costly and even much more so if they exceed certain size, rotational molding is used for small series of objects. Most water-sport crafts, such as kayaks, are manufactured from polyethylene in a rotational molding process.

Single-site polyethylene having narrow MMD is particularly suited for rotational molding processing. The melting point of the polymer is exceptionally sharp so that all material melts at once and the surface of the mold is covered evenly.

6.2. Pipes

Polyethylene is used for manufacturing plastic pipes for water and gas carrying, and as a protection of the steel pipes used for construction of oil and gas long-distance pipelines.

6.2.1. PE Pipes

Polyethylene pipes are increasingly used in the construction industry for hot and cold water and for gas installations. Light weight, flexibility, simplicity of installation, low cost, and good performance of the polyethylene pipes make the material a good replacement for metal or pipes produced from other polymers.

Figure 30. Rotational molding with four molds mounted on the rotating arm

The key feature that is needed for pipes is their long-term pressure resistance. To avoid very costly damages and possible serious incidents, pipes used for gas and water supply need to be able to sustain elevated and constant pressure of up to 10 MPa over a period that is not shorter than 50 years. Materials used for building such pipes are carefully designed, selected, and certified, and the pipes made out of it are tested under stress. To enable development of the materials that, of course, need feedback on the performance, accelerated test methods have been developed that provide sufficient confidence in the data extrapolated from the short-term stress exposure to the 50 years period. Still, developing pipe polymer is a very time-consuming process and changes on the market happen less frequently than in some other applications of polyethylene.

Pipes are produced from the very-high-molecular-mass and high-density polyethylene. To sustain pressure at all, highly crystalline structure organized in thick lamellae is needed not to allow water or gas molecules to pass through the material. Failures that occur short time after pressurizing the pipe because of the insufficient thickness of the crystallites are called ductile failures.

To avoid cracking of the pipes under pressure, longitudinal splits of the material attributable to slippage of the crystallites, long, "free" polymer chains are needed that partially crystallize in more than one lamella, connecting them, and keeping them from slipping. These tie molecules should be branched, containing comonomers, so that the probability of them crystallizing in more than one lamellae is increased.

Pipe materials are normally classified and certified according to the pressure they can resist. PE80 material can resist 8 MPa, and PE100 is good up to 10 MPa. PE80 can still be manufactured by using chromium catalysts in a unimodal polymerization process. To achieve the PE100 performance, bimodality has to be introduced. For bimodal polyethylenes, two kinds of polymer chains can be produced on purpose: relatively long with no branching in one reactor, and very long chains with some comonomer content in the other. By broadening the MMD using two polymerization conditions, processability of the otherwise tough material became possible, and because of the onion structure of the polymer particles, the two components can be homogenized effectively through the extrusion process.

In addition to the described pipe material, postreactor-treated polyethylene can also be used for manufacturing of pipes. In this processing, the material still containing a certain amount of vinyl groups is treated so as to build chemical bonds between the chains that contain the double bonds. This process is also called cross-linking, and it can be achieved by co-extruding initiators, free radical sources, or by irradiating manufactured pipes.

6.2.2. PE used for Steel-Pipe Coating

Steel pipes used for constructing oil and gas pipelines are exposed to the environmental conditions that can cause severe and rapid corrosion of the metal. Polyethylene is used to coat the steel pipes protecting them from moisture and oxygen.

Steel pipes are typically coated by using three layers of polymers. Already manufactured pipes are first coated with a thin layer of epoxy resin. The pipe is then heated by using magnetic induction, and the epoxy resin polymerizes on the surface of the pipe. The next layer, polyethylene adhesive, is extruded as a tape to the surface of the rotating pipe and it is followed by the layer of top coat, polyethylene material that is the actual protection. The adhesive resin is needed to provide peeling resistance and tight connection between the epoxy layer and the polyethylene top coat.

Alternatively, the two polyethylene layers can be extruded axially to cover the pipe, through a circular die mounted around it.

6.3. Wire and Cable Insulation

With its high purity and outstanding dielectric properties, LDPE has been used for insulation of high-frequency and medium-voltage cables since the beginning of the industrial manufacturing of polyethylene. In the structure of modern electrical and information transmitting cables, for which several metal layers need to be separated by several polymer layers and the entire cluster has to be protected from the environmental influences, polyethylene is still used very extensively as an insulation material and as the top coat for jacketing the cables.

Also in this application, cross-linking is an important method for postreactor modification of the material to secure higher molecular mass of polymer and its higher resistance properties.

6.4. Plastomers

Polyethylene that is highly branched, i.e., it contains large amount of comonomer, and has, therefore, low density, lower than 910 kg/m^3, is distinguished from other typical polyethylene grades through its thermoplastic properties.

Plastomers are usually produced in a solution process by using Ziegler or single-site catalysts using octane, hexane, or butene as a comonomer. They are used for blending with other polymers, polyamide or polypropylene, to enhance the impact resistance of the otherwise brittle matrix. These blends are essential in the automotive industry, for example, for manufacturing bumpers.

Another significant application of plastomers is film. Because of its nonmigrational adhesion properties, so-called clinging, plastomer film is used as a temporary protection of the sensitive surfaces in the electronic industry.

7. Environmental Aspects

7.1. Manufacturing

Polyethylene manufacturing processes are highly sensitive to all kinds of impurities, they operate at high or very high pressure, and the processing conditions need to be very strictly controlled. Most chemicals used in the polymerization process, initiators and catalysts, are temperature, moisture, and oxygen sensitive and need to be stored, transferred, and used with no exposure to the open environment. This also means that the polyethylene production environment is rather well protected from the process itself, because losses of any containment cannot be tolerated at all.

Some limited, controlled emission to the environment could occur during the flaring process, which is controlled burning of excess flammable gases that need to be eliminated from the process. This is, however, avoided whenever possible, for environmental and regulatory reasons, but also because flaring unnecessarily increases the manufacturing costs.

7.2. Polymer Disposal and Recycling

Incineration of polyethylene to recover thermal energy is the most efficient way of its disposal because the products of burning polyethylene are almost exclusively CO_2 and water. In some European countries, as much as up to 76% of heat needed for household heating is generated by alternative fuel, which is collected waste material containing polyolefins as a large part.

Other ways of recycling polyethylene are also practiced and developed, although not yet to a very large extent. Postconsumer waste is contaminated, and it is difficult to separate polymers effectively, which are not always well miscible. Contamination also renders the recycled material for any packaging use, especially for food packaging, so the number of applications that can tolerate recycled polyethylene is significantly reduced. Examples in which recycled polyethylene can be applied are manufacturing of outdoor urban furniture or composites of polyethylene with wood used for construction of platforms.

Industrial waste is a somewhat better source of recyclable polyethylene. Cast film is used for wrapping shipments of pallets carrying raw materials for many industries, and if collected, the films are relatively clean and narrow in comonomer distribution composition.

Finally, landfills still remain a viable option for polymer disposal for many countries. The environmental impact of polyethylene in landfills is, however, significant, because the material remains solid for a very long time, but it is also limited to the solid polymer, because there are no dangerous decomposition products of polyethylene.

As described earlier, much effort has been made to improve the environmental impact of polyethylene on the monomer level, that is, ethylene generated from renewable sources. The primary source, sugarcane, consumes CO_2 in the photosynthesis, shifting the balance of overall CO_2 emission from using and burning polyethylene. Bio- or green polyethylene, manufactured from such ethylene, has become a significant factor on the polyethylene market with the consumption of approximately 500 000 t/a (2013).

7.3. Material Replacement and Contribution to the Environment

To assess the environmental impact properly, the net result of replacing other materials by polyethylene, which have much higher impact or present other risks to the environment, have to be taken into account. Examples are polyethylene pipes that need no plasticizing and can be burned with no aggressive residuals. In the food industry, polyethylene packaging is essential in the food preservation and for reduction of losses.

References

1 palstermat.com , Strong growth in global polyethylene industry expected to continue at a CAGR of 3.5% over next 5 years, http://www.plastemart.com/Plastic-Technical-Article.asp?LiteratureID=1971&Paper=strong-growth-global-polyethylene-industry-3.5-percent-over-5-years (accessed February 10, 2014).
2 Kerry, F.G: "*Industrial Gas Handbook: Gas Separation and Purification*", CRC Press, Boca Raton 2010.
3 TCE Today, http://www.tcetoday.com/~/media/Documents/TCE/Articles/2011/845/845cewctw.pdf (accessed February 10, 2014).
4 H.A. Wittcoff, B.G. Reuben, J.S. Plotkin: *Industrial Organic Chemicals*, J. Wiley & Sons, New York 2012.
5 G. Wilke: 50 Jahre Ziegler-Katalysatoren, *Angew. Chem.* **115** (2003) 5150–5159.
6 R. Hoff, R.T. Mathers: *Handbook of Transition Metal Polymerization Catalysts*, J. Wiley & Sons, New York 2010.
7 D.B. Malpass: *Introduction to Industrial Polyethylene*, J. Wiley & Sons, New York 2010.
8 V.K. Jolly: *Commercializing New Technologies: Getting from Mind to Market*, Harvard Business Press, Harvard 1997.
9 L.D. Cady, C.J. Frye, P. Howard, T.P. Karjala, P.J. Maddox, I. M. Munro, D.R. Parikh, R.S. Partington, K.P. Peil, L. Spencer, W.R. Van Volkenburgh, P.S. Williams, D.R. Wilson, M.J. Winter, US6462161 B1, 1998.
10 D.C. Basset: "Lamellae and Their Organization in Melt-Crystallized Polymers", *Self-order and Form in Polymeric Materials*, Springer, Heidelberg 1995, p. 181.
11 K. Begman, K. Nawotki, *Kolloid-Zeitschrift und Zeitschrift für Polymere* (1967) 132.
12 S.-D. Clas, D.C. McFaddin, K.E. Russell, *J. Polym. Sci., Part B Polym. Phys.* **25** (1987) 1057.
13 R.S. Stein, R. Prud'homme, *J. Polym. Sci., Part B Polym. Phys.* **9** (1971) 595.
14 Y. Kissin: *Alkene Polymerization Reactions with Transition Metal Catalysts*, Elsevier, Amsterdam 2008.
15 T.T. Wenzel, D.J. Arriola, E.M. Carnahan, P.D. Hustad, R.L. Kuhlman: "Chain Shuttling Catalysis and Olefin Block Copolymers (OBCs)", *Topics in Organometallic Chemistry*, Springer, Heidelberg 2008.
16 T. Whelan: *Polymer Technology Dictionary*, Springer, Heidelberg 1994. p. 159.
17 Prospector ESCR measurement ASTM Method, http://www.ides.com/property_descriptions/ASTMD1693.asp (accessed June 6, 2014).
18 J.P. Blitz, D.C. McFaddin: *The characterization of short chain branching in polyethylene using Fourier transform infrared spectroscopy*, J. Wiley & Sons, New York 2003.
19 P. Starck: "Studies of the comonomer distributions in low density polyethylenes using temperature rising elution fractionation and stepwise crystallization by DSC", *Polym. Int.* **40** (1990) 111–122.
20 M. Philip, I.M. Ward, B. Pearson, *J. Material Sci.* **21** (1986) 879.
21 Plastic Pipe Institute, Handbook of polyethylene pipe, 2014, http://plasticpipe.org/publications/pe_handbook.html (accessed February 10, 2014).
22 S.M. Kurtz: *The UHMWPE Handbook: Ultra-High Molecular Mass Polyethylene in Total Joint Replacement*, Academic Press, New York 2004.
23 DSM, *Teamex Octene Based VLDPE*, company brochure, Geleen 1990.

24. P. Maplestone: *Modern Plastics International*, Canon Communications LLC, Santa Monica, CA 1996, pp. 5–61.
25. J.A.M. Van Broekhoven, M.J. Doyle, US4808699, 1987.
26. W. Hartke, F. Osan, *Cycloolefin polymers*, US5610253, 1995.
27. Economics & Statistics Department, American Chemistry Council, *Shale Gas, Competitiveness, and New US Chemical Industry Investment: An Analysis Based on Announced Projects*, American Chemistry Council, Washington, DC 2013.
28. K.E. Weale: *Chemical reactions at high pressures*, Spon, London 1967.
29. P.-C. Lim, G. Luft, *Macromol. Chem.* **184** (1983) 849.
30. M.J. Roedel, *J. Am. Chem. Soc.* **75** (1953) 6110.
31. Nicolas, L., *J. Chim. Phys.* **55** (1958) 185.
32. Tobita, H., *J. Polym. Sci. Polym. Phys.* **33** (1995) 841–853.
33. R.A. Jackson, P.A. Small, K.S. Whiteley, *J. Polym. Sci. Polym. Chem. Ed.* **11** (1973) 1781.
34. P. Ehrlich, et al., *A.I.Ch.E. Journal* **22** (1976) 463.
35. Y. Tatsukami, T. Takahashi, H. Yoshioka, *Macromol. Chem.* **181** (1980) 1107.
36. Cossee, et al., *J. Catal.* **3** (1964) 80, 99.
37. T. Ziegler, et al., *J. Am. Chem. Soc.* **121** (1999) 154.
38. T.F.L. McKenna, A. DiMartino, G. Weickert, J.B.P. Soares: "Particle Growth During the Polymerisation of Olefins on Supported Catalysts, 1 – Nascent Polymer Structures", *Macromol. React. Eng.* **4** (2010) 40–64.
39. C.E. Marsden, R.J. Parker, US8039564 B2, 2011.
40. Martin, H.: *Polymers, Patents, Profits: A Classic Case Study for Patent Infighting*, John Wiley & Sons, New-York 2007.
41. B. Diedrich, *Appl. Polym. Symp.* **26** (1975) 1.
42. B. Diedrich, K.D. Keil, DE1595666 A1, 1966.
43. Solvay, BE650679, 1963.
44. G. Wilkinson, M. Rosenblum, M.C. Whiting, R.B. Woodward: "The Structure of Iron Bis-Cyclopentadienyl", *J. Am. Chem. Soc.* **74** (1952) 2125–2126.
45. E.O. Fischer: "Zur Kristallstruktur der Di-Cyclopentadienyl-Verbindungen des zweiwertigen Eisens", Kobalts und Nickels. *Z. Naturforsch.* **B7** (1952) 377–379.
46. W. Kaminsky: "The Discovery of Metallocene Catalysts and their Present State of the Art", *J. Polym. Sci., A: Polym. Chem.* **42** (2004) 3911.
47. W. Kaminsky: "Olefin Polymerisation Catalysed by Metallocenes", *Advances in Catalysis* **46** (2001) 89.
48. G.G. Hlatky: "Heterogeneous Single-Site Catalysts for Olefin Polymerization", *Chem. Rev.* **100** (2000) no. 4, 1347–1376.
49. P. Hoang, G. Baxter, US20060189769 A1, 2005.
50. J.P. Marano Jr, J.M. Jenkins III: "Polymerization at High Pressure", in *High Pressure Technology*, vol. 2, CRC Press, Boca Raton 1977.

Polyimides

ROBERT G. BRYANT, NASA Langley Research Center Hampton, VA 23681, United States, USA

1.	Introduction	859	3.2.1.	Polymerization Processes	870
2.	Physical and Chemical Properties.	862	3.2.2.	Thermosetting	872
2.1.	Electronic Interactions	862	3.3.	Polymer Conversion	873
2.2.	Structural Features	862	4.	Uses	873
2.3.	Thermal Stability	862	4.1.	Films	873
2.4.	Environmental Resistance	862	4.2.	Coated Wires	874
2.5.	Electrical Properties	863	4.3.	Membranes, Tubes, Fibers, and Foams	874
2.6.	Mechanical Properties	863	4.4.	Melt-Processable Polyimides	875
3.	Synthesis	863	4.5.	Direct-Formed Polyimides	876
3.1.	Amic Acid Route	863	4.6.	Matrix Composites and Adhesives	876
3.1.1.	Poly(Amic Acid) Formation	864	5.	Photoimageable Polyimides	877
3.1.2.	Polyimide Formation	867	5.1.	Negative Photoresists	878
3.1.3.	One-Step Formation of Polyimides	870	5.2.	Positive Photoresists	879
3.2.	Synthesis Using Imide-Containing Monomers	870		References	879

1. Introduction

Polyimides are a class of high-performance polymers that contain an imide group, defined as an sp^3 nitrogen bonded to two adjacent carbonyls. This imide structure is most commonly found as part of a five- or six-membered ring. Aromatic polyimides (PIs) polyetherimides (PEIs) and polyimide-amides (PIAs) are noted for their mechanical properties, chemical resistance and dielectric strength. They exist as thermoplastics and thermosets and are available in many formats. Polyimides are used as hot-melt adhesives, matrix resins for composites, dielectric films, photoimageable coatings, flex circuits, foams, wire insulation, thin-walled tubing, molding resins, high-voltage connectors, and high-performance bushings and seals. Although polyimides command a premium price, their versatility, reliability, and demonstrated performance makes them suitable use in high-performance applications where failure and replacement costs are high. As the requirements for lighter weight, broader temperature performance, increased strength, durability, and inertness continue to drive technology, polyimides will play an increasing role in the innovations of the future (Table 1).

The first reported synthesis of an aromatic polyimide was in 1908 [1]. However, most of the initial development and commercialization of polyimide products was performed by researchers at DuPont, who benchmarked this endeavor in the 1960s with the release of Kapton H film, Vespel molded parts, and Pyre-ML wire varnish [2]. This effort inspired other researchers in academia, industry, and government laboratories to pursue the chemistry, fabrication, and applications of polyimides not envisioned several decades ago. There are several excellent books and review articles that address the extensive topic of polyimides [3–8]. Herein, an overview of the field of polyimides is given from synthesis and basic kinetic behavior to fabrication and articles of manufacture.

Table 1. Feedstock commodity polyimide thermoplastic resins and powders (2014)

Polyimide resin manufacturer	Trade name(s)	Polyimide structure
DuPont (USA) http://www.dupont.com/	Kapton Vespel Pyre-ML	
Imitec Inc. (USA) http://www.imitec.com/	I-XXX	PMDA=2XX, BPDA=9XX, ODPA=5XX, and 6FDA polyimide thermoplastics, thermosets, and bismaleimides (BMI)
Kaneka Corp. (Japan) http://www.kaneka.co.jp/kaneka-e/	Apical-AH	
	Apical-NPI	
Ube Industries (Japan) http://www.ube-ind.co.jp/english/index.htm	Upilex-S	
	Upilex-R	

AURUM — Mitsui Chemicals (Japan) http://www.mitsuichem.com/index.htm

Ultem — Saudi Basic Industries Corp. (SABIC) (Saudi Arabia) http://www.sabic-ip.com/gep/en/Home/Home/home.html

Extem

YS-20 — Shanghai Research Institute of Synthetic Resins (China) http://www.chem-syn.com/en/default.asp

YS-30

2. Physical and Chemical Properties

2.1. Electronic Interactions

One of the features of polyimides is that the polyimide macromolecules interact with themselves and each other by either an electronic polarization or charge transfer mechanisms [9–11]. The degree to which these mechanisms play a role depends on the electron affinity of the imide ring (electron deficiency) and the electron donating ability of the N-substituent (ionization potential) [12]. The extent to which these electronic interactions effect neighboring polymer molecules influences the polyimide's color, glass transition temperature (T_g), crystallinity, mechanical properties, and chemical resistance [13, 14]. Clearly, the intrachain electronic polarization affords the yellow to dark orange color that arises in polyimides [13, 14]. The amount of chromophores present can be controlled to some extent by the selection of the cyclodehydration environment, choice of monomers, polymerization conditions, and especially processing methods, and is not related to the presence of unconverted isoimide [13].

2.2. Structural Features

Generally, the bulk structure–property relationships that apply to most polymers apply to polyimides [15]. The thermal transitions are influenced by the selection of monomers and processing techniques [16–19]. Monomers that have kinks, offsets, swivel units, pendent substituents, aliphatic moieties, and bulky units tend to lower the T_g, melt viscosity, crystallinity, and increase solubility, as these units take up more free volume by introducing a more random structure than their symmetric counterparts [20–31]. The proper selection of fluorinated or phenylphosphine oxide containing monomers has produced nearly colorless soluble polyimides [14, 17, 32]. One interesting polyimide developed by NASA, LaRC-SI, does not use the standard approaches to develop solubility. This polyimide starts from a mixture of two dianhydrides, biphenyltetracarboxylic dianhydride (BPDA) and oxydiphthalic dianhydride (ODPA), and a diamine o-dianisidine (3,4′-ODA) to afford an amorphous high-molecular-mass polyimide that is soluble in common high-boiling amide solvents [33–35]. Remarkably, the two homopolymers, ODPA/3,4-ODA (LaRC-IA) and BPDA/3,4′-ODA, and the copolymer using ODPA/BPDA/4,4′-ODA are all insoluble and semicrystalline [33, 34]. There have been successful attempts of creating thermotropic polyimides. These liquid crystalline polyimides rely on extended arylene ether-type diamines to create polyimide thermoplastics and blends, or thermosetting polyimides [36, 37]. The advantages of such systems are that they can be more easily melt-processed, the reduced thermotropic melt viscosity, and can undergo shear-induced molecular orientation to gain additional property enhancements.

2.3. Thermal Stability

Aromatic polyimides are highly thermally stable because of their degree of aromaticity, ring structure, and packing density, which are also a function of cure cycles [38–40]. Typical decomposition points for polyimides, not subject to thermally initiated retro reactions, are above 450°C in air and above 500°C in nitrogen for short time intervals [5]. Endcapping a polyimide with a stable monofunctional unit enhances thermal stability by protecting chemically active chain ends that are subject to thermal and chemical instability [41, 42]. Treating polyimides at temperatures in excess of 500°C, under nonoxidizing conditions, results in high yields of approximately 60 wt% of carbonaceous material. The main volatile evolved from pyrolysis is the CO from the imide ring [43]. These polyimide-based carbon films and preforms have been derived for use as membranes and carbon/carbon structural components [44–49].

2.4. Environmental Resistance

Polyimides are resistant to chemical attack, radiation and moisture. An exception is the PMDA-containing polyimides that are hydrolyzed under basic conditions [50–52]. The effect of degradation of one imide ring, in PMDA, changes the characteristics of the other ring by resonance through the common aromatic ring. This is also the case with monoaromatic diamines. However, most polyimides are highly resistant to caustics, acids, organic solvents, specialty fluids, and salts, even at elevated temperatures [26, 53–55].

Polyimides are typically bioinert, that is, they are not toxic or bioactive. However, Material Safety Data Sheets of every single article of manufacture should be consulted for detailed specification.

2.5. Electrical Properties

The electrical properties and temperature range of polyimides have led to their widespread use as dielectrics, flexible substrates, electronic packaging, and wire insulation for the electronics industry [56–64]. An excellent review on some of the issues for reliability in electronic assemblies, from a materials standpoint, was published indicating moisture as one of the main mechanisms of electrical failure [65, 66]. Although the imide ring is polar and electron-deficient, the two dipoles of the carbonyl groups oppose each other and the imide rings along the polyimide backbone also directionally oppose each other. This geometry simulates the bulk effect of a nonpolar polymer making polyimides good dielectric materials, when combining these properties with their high temperature and chemical stability [60, 62–65, 67–69]. Typical dielectric constants range from below 3 to above 6, depending on structure, frequency, temperature, and moisture absorption. As expected, fluorinated polyimides display dielectric constants typically 0.5 lower than the nonfluorinated polyimides, as the fluorine atom imparts low polarizability, water uptake, and surface energy [70–75]. Similar comparisons can be made from the fluorinated olefinic systems (PTFE and PFE) with their corresponding hydrocarbon analogues.

2.6. Mechanical Properties

The bulk mechanical properties of aromatic polyimides are among the highest for commercial plastics [76]. The various processed forms of solid polyimides have excellent mechanical properties with elongations from 3 to 120%, tensile strengths and moduli 170–270 MPa and >3 GPa, respectively, over a wide range of temperatures from cryogenic to well above ambient. This is a result of the interchain interactions of polyimides and their structural integrity imparted by their aromatic character, morphology, and sub-T_g transitions, that may impart additional toughness [77–82]. Those polyimides that can undergo melt processing tend to make excellent hot-melt adhesives.

3. Synthesis

The synthetic approach to the formation of polyimides is divided into three groups. The first group involves the synthesis of the polyimide by increasing the molecular mass of the monomer(s) through the formation of the imide ring through the amic acid/ester route. The second approach involves increasing the molecular mass of a precursor already containing the imide ring. The last method involves the conversion of a polymer into a polyimide. The requirement is that these approaches must be based on high-yield organic reactions.

3.1. Amic Acid Route

The two-step approach to polyimides is the most general approach used in the synthesis of linear polyimides [83–86]. The first step involves the treatment of a diamine either with a dianhydride (the most popular), a tetracarboxylic acid, a diester diacid, a diester diacid chloride, or a trimellitic anhydride chloride (TAC) as in the case of polyamide imides (Amoco's Torlon) [87, 88] (Fig. 1, step 1). The reaction schemes are generally performed in a high-boiling polar protic (phenolic) or aprotic (N-methyl-2-pyrrolidinone, NMP; N,N-dimethyl acetamide, DMAc; N,N-dimethyl formamide, DMF; etc.) solvent at low to moderate temperatures, resulting in the formation of the poly(amic acid) or poly(amide acid), amic ester, amic acid ester, etc. through nucleophilic attack by the amine on the carbonyl carbon [89]. Next, the imide ring is formed by thermal treatment, chemical dehydration, or both by a second nucleophilic attack occurring at the adjacent carbonyl carbon by the amide nitrogen. This is followed by the elimination of a condensate (typically water or alcohol). Removal of this condensate forces the equilibrium toward the right, driving the polymerization to completion (Le Chatelier's principle) thus forming the imide ring (step 2) [84].

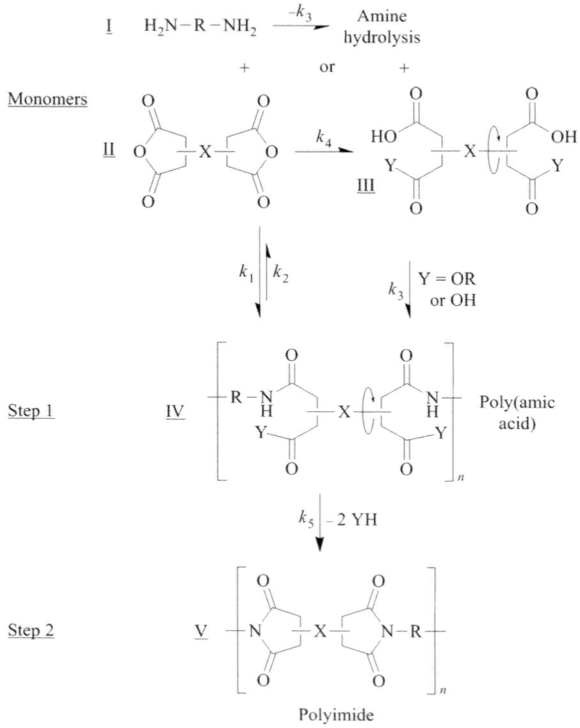

Figure 1. Reaction diagram for thermal imidization via poly(amic acid/ester)

Alternatively, a "one-step" method or direct conversion is used where a diamine and a dianhydride are placed in a solvent and heated to a temperature at which the formation of the amic acid moiety rapidly dehydrates to the resulting imide. Thus, the polyimide is formed in "one step" [90].

3.1.1. Poly(Amic Acid) Formation

To successfully carry out this synthetic approach, the following requirements must be met: chemical purity, the proper type and percent solids of reaction medium (facilitating solubility of the resulting poly(amic acid)), reaction temperature, and a method allowing for the removal of any byproducts of polymerization (heat, condensates, etc.) [91, 92]. Most commonly, linear polyimides involve treating a diamine with a dianhydride or tetracarboxylic acid/ester to afford the amic acid or amic ester, followed by imide ring formation during the final polymerization step. The routes to achieving a high-molecular-mass polyimide, starting from the monomers via the poly(amic acid), are shown in the reaction diagram in Figure 1 [93].

When a diamine (I) is treated with either a dianhydride (II), tetraacid or diacid ester (III), alternative routes may occur depending on the reaction kinetics, in which k_1 and k_3 are desirable and k_2 and k_4 are not. Table 2 summarizes the approximate kinetic k_x values.

The pathway to k_4 depends on: the amount of water or alcohol in the reaction medium; the reactivity (electrophilicity) of the dianhydride or tetraacid/diacid ester; and the nucleophilicity of the diamine (Table 3). Steric hindrance and isomerization; polarity and autocatalytic effects

Table 2. Typical rate constants for poly(amic acid) synthesis [94–96]

Reaction	Rate constants
Propagation (k_1)	0.5–6.01 mol^{-1} s^{-1}
Depolymerization (k_2)	10^{-5}–10^{-6} s^{-1}
Imidization (k_5)	10^{-8}–10^{-9} s^{-1}
Anhydride hydrolysis (k_4)	0.1–0.41 mol^{-1} s^{-1}
Amine hydrolysis ($-k_3$)	0–10^{-6} s^{-1}

Table 3. Electron affinity and basicity values of several common dianhydrides and diamines used in the synthesis of polyimides [13]

Dianhydride	Abbr. name	E_a [eV]	Diamine	Abbr. name	pK_a (H_2O)
	PMDA	1.9		pDA	6.08
	BPDA	1.38		mDA	4.08
	BTDA	1.55		DAB	4.60
	DSDA	1.57		DABP	3.10
	ODPA	1.30		DDSO2	2.15
	6FDA	ND*		ODA or DAPE	5.20
	BPADA or LDA	1.12		Bis-DA	ND*

*ND = Not determined.

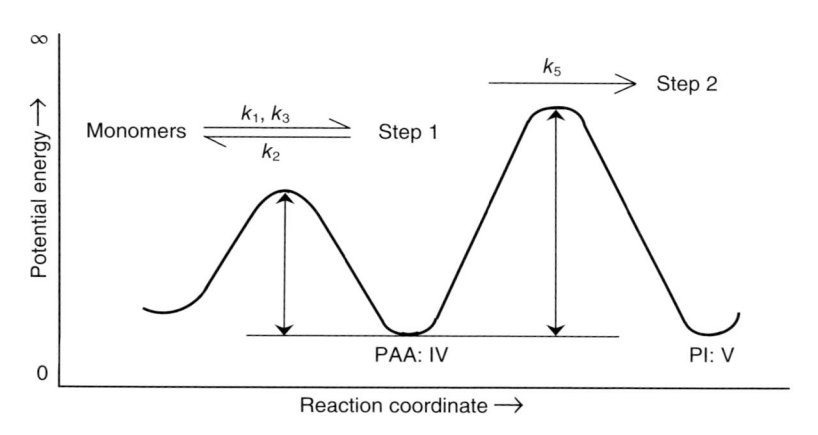

Figure 2. Qualitative reaction coordinate diagram for polyimide synthesis via poly(amic acid) (PAA) [122]

of the solvent; and the time/temperature profile during the poly(amic acid) formation also affect the pathway to k_4 [97, 94, 95]. Conversion of the anhydride moiety, k_4, has the effect of introducing another chemical functionality, with a different reactivity that can affect both k_1 and k_3 through resonance. As shown on the reaction coordinate (Fig. 2), the effect of k_2 competes with k_1 and k_3 when the reaction temperature is raised and the thermal dehydration k_5 conversion to imide is occurring. Once the imide ring is formed, depolymerization is severely limited through removal of the condensate byproduct. The choice of the diamine and dianhydride affects the rate of poly(amic acid) formation both sterically, and through the electronic interactions of the monomers [98–100].

As seen in Table 2, pyromellitic acid dianhydride (PMDA) is the dianhydride with the highest E_a value. Any diamine that is treated with PMDA should polymerize faster than if it is treated with a dianhydride having a lower affinity. However, once one of the anhydride units of PMDA has reacted, the second anhydride will display a lower reactivity because the former anhydride group, now an amic acid, has less electron withdrawing ability. This effect on other dianhydrides is less pronounced as the anhydride groups are separated by additional moieties that decrease the influence of the reacted anhydrides [97]. A similar trend is seen with diamines. The more basic the amine, the faster it can react with a moiety that is susceptible to nucleophilic attack. Additionally, any chemical groups mitigating the conjugative resonance effects of the amines on each other, the less of an effect the conversion of one amine will have on the reactivity of the other [101].

Solvent effects depend on several factors, including, the polarity of the medium: the percent solids; and the solubility of the reactants. As the polarity of the medium changes with conversion to poly(amic acid), the viscosity and potential autocatalytic effects (kinetic conversion to poly(amic acid)) are observed [102]. These effects are caused by the increasing molecular mass of the poly(amic acid) that results in the buildup of carboxylic acid. The hydroxyl portion of the amic acid can form additional hydrogen bonds to adjacent poly(amic acid) groups, thereby artificially increasing the observed viscosity [86, 91]. These additional carboxyl acid protons can assist nucleophilic attack (on the carbonyl carbon) by coordination with the anhydride carbonyl oxygen increasing the step growth kinetics to poly(amic acid) formation. An accurate kinetic profile has been difficult to obtain as a result of the solvent effects, the changing reaction medium, and the changing pH. Some investigators claim that irreversible second-order kinetics is followed, while others have observed reversible autocatalytic kinetics when tetrahydrofuran (THF) is the solvent. Typically, polymerizations carried out in amide solvents (polar aprotic) do not display autocatalytic behavior [99, 100]. These solvent molecules closely associate with the hydroxyl protons, effectively isolating them from interacting with other anhydrides and amic acids. The reaction rate generally increases as the solvents become more polar and more basic [103, 104]. A model compound study showed that the acylation rate increased from THF < acetonitrile < DMAc < m-cresol. The polar protic solvent, m-cresol was claimed to increase the rate because it functions as an acidic catalyst [102, 105]. Similar observations were noted when acids (benzoic acid) were introduced into a poly(amic acid) medium employing an amide solvent. THF, not being very basic, does not associate with the poly(amic acid) and some reversal occurs. This "retro" reaction (k_2) occurs when a proton is transferred from a carboxylic acid to an amide nitrogen (or any other basic species), followed by the subsequent attack of the carboxylate oxygen on the adjacent carbonyl carbon to reform the anhydride or the rare isoimide [103, 104]. This reverse reaction can be suppressed by replacing the hydroxyl proton of the poly(amic acid) with amine salts or esters causing a marked decrease in conversion to monomeric species [86, 106–108].

The effects of adventitious water on both the starting anhydride and poly(amic acid)s has been shown to decrease their molecular mass over time through conversion to the diacid acid and hydrolytic cleavage, respectively [86, 93, 109–112]. This drop in molecular mass appears to be more dramatic the more dilute the solution and may be caused by the apparent increase in water content as the percent poly(amic acid) decreases. Additionally, the conversion from poly(amic acid) to monomeric species is unimolecular, whereas the reverse reaction is bimolecular, shifting the equilibrium to the left towards the reactants [84]. Evidence has indicated that further imidization of the poly (amic acid) occurs upon standing over time, at various temperatures, releasing water into the solution, resulting in further hydrolysis of the poly(amic acid) [86, 93, 113].

Lastly, monomer addition (sequence and rate) has shown to affect both thermodynamic equilibrium, and the resulting molecular mass of the poly(amic acid). This condition is a result of the difference in solubility of the monomers (dianhydride and diamine) in the polymerization solvent [84, 93]. As the dianhydride slowly dissolves in a medium containing the faster dissolving diamine, it maintains an intrinsic stoichiometric balance with the diamine, setting the condition for an infinite degree of polymerization. This "pseudo-interfacial" polymerization generates high-molecular mass poly(amic acid) until one of the monomers is consumed. Thus, near the completion of the reaction, the medium contains a wide bimodal distribution, high polymer and unreacted monomer contents. As the unreacted or excess monomer has time to come in contact with the high-molecular mass poly(amic acid), thermodynamic equilibrium (k_1/k_2) occurs and the molecular masses redistribute towards a unimodal distribution [114–116].

3.1.2. Polyimide Formation

The three methods to form the polyimide from the poly(amic acid) are thermal, chemical, and a combination of both.

Thermal Imidization. The thermal conversion of poly(amic acid) to polyimide requires the removal of the condensate produced during cyclodehydration (Fig. 1). This is achieved by elevated temperatures and, whenever possible, the addition of a solvent that facilitates azeotropic removal of the condensate. A typical solution thermal conversion reaction setup contains approximately 30% solids poly(amic acid) in NMP, 10% solids xylenes (azeotrope to remove water), and the reaction temperature is maintained at about 165°C for several hours with water distilling out rapidly early in the reaction [84, 53, 33]. Thermal conversion of poly(amic acid) in solution is fairly straightforward and only becomes complex when the resulting polyimide precipitates out of solution prior to complete imidization. The kinetics is more complex when the poly(amic acid) is isolated as a powder, spun as fiber, or cast as a film prior to imidization [117]. Generally, the azeotropic cosolvent is not present in these "solid-state" preforms, and the effects of the advancing T_g, solvent evaporation rate, the percent solids, steric effects, side reactions, cure temperature, and surface-to-volume ratio of the polymeric preform effect the overall kinetics [85, 118, 119, 16]. Thus, attempts to follow the imidization rates using spectroscopy have been confusing as these factors are difficult to account for. One reaction that does not appear to occur during thermal imidization is cross-linking between diamide groups [120–122]. This is not surprising because there are soluble polyimides cited in the literature, and no spectral evidence exists that supports this mechanism. However, thermal aging above 350°C for a period of time may result in cross-linking, but this is most likely the result of a free radical mechanism (thermal degradation) and is influenced by film thickness, atmosphere, and the casting surface [123–125].

Two pathways of thermally induced imidization are (1) the loss of the proton on the carboxylic acid group after cyclization or (2) removal of the proton on the carboxylic acid prior to or during ring closure. As the conjugate base of the amide is more nucleophilic than the amide itself, ring closure should be faster in the latter case [85].

When thermal imidization occurs in the presence of an amide solvent, the conversion to polyimide proceeds more readily, as the loss of the amic acid proton is facilitated through solvent hydrogen bonding. In some cases, the rate of thermal imidization occurs in two steps [85, 118]. The first step is the rapid conversion to imide above 150°C, and the second step is a slower process during the latter stages of imidization. The following scenario can explain this. During the early stages of imidization, more solvent is present to assist ring closure and the T_g of the polymer is low [85]. As the thermal conversion continues, the T_g is increased and the solvent content decreases [118]. Chain mobility and solubility is decreased with increasing conversion, especially if the growing polyimide is semicrystalline [96]. Further depletion of solvent decreases the amount of solvent–polymer complexation that normally facilitates the rate of ring closure [85]. This two-step rate effect is seen for fibers (high surface area) and cast films for which solvent is lost during thermal processing. It has been shown that thicker films form imides at faster initial rates. These thicker films retain solvent longer, because of the extended migration/evaporation time, allowing the chains more mobility and increased solvent–polymer complexation allowing for additional conversion at temperature [126]. However, the effects of the surface imidizing faster (skinning) than the bulk must be taken into account for thermal convection curing [93]. The resulting molecular mass of the polyimide depends on the molecular mass of the poly(amic acid) [93]. Although the kinetics follows that in Figure 2, the resulting physical properties of the polyimide films are different as displayed in Figure 3.

Three molecular-mass divisions are plotted as a function of temperature versus qualitative physical properties, demonstrating the effect of the poly(amic acid) molecular mass on the polyimide during conversion. This effect can be rationalized knowing that the amount of chain scission is independent of the molecular mass, resulting in a more dramatic effect on lower molecular mass poly(amic acid)s. The lower molecular mass poly(amic acid) can be reduced to monomers, where these units may

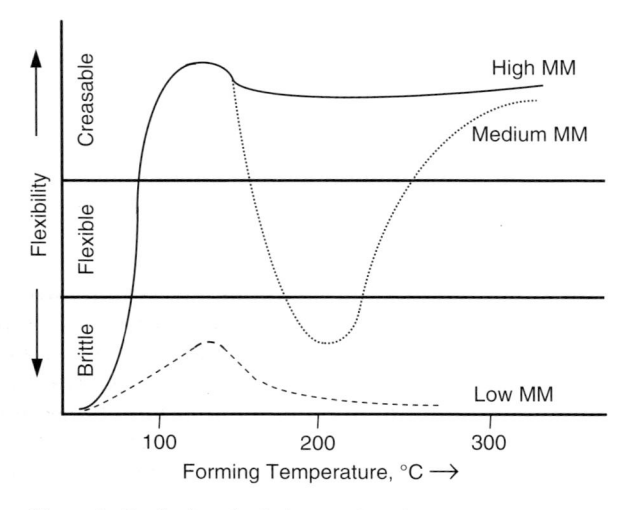

Figure 3. Qualitative physical properties of thermally formed polyimide films of different molecular masses (MM) [93]

be thermally unstable and susceptible to oxidation. When this occurs, the adulterated monomers upset the stoichiometric balance resulting in low-molecular-mass polyimide. Higher molecular mass poly(amic acid)s are less likely to be converted down to monomeric levels, and the oligomers will repolymerize to form medium- or high-molecular-mass polyimide thereby maintaining their desired physical properties.

An interesting reaction sequence, parallel to the two-step synthesis of polyamides through the amide-salt process, has been outlined [7, 20, 127]. Here, the diamine and tetracarboxylic acid are first mixed in water to form the salt precipitate that is collected and dried. The resulting amine salt is placed in water and heated under pressure to 130°C for 1 h, then to 180°C for 2 h and cooled to ambient temperature. The resulting polyimides have molecular masses and physical properties equivalent to those made by the conventional two-step methods.

Chemical Imidization. Chemical imidization is not widely used, except in gel processing, as it employs additional reagents. However, it has the advantage of low-temperature imidization and can be used to directly form fine polyimide molding powder or uniquely structured materials [93, 128, 129]. A typical chemical imidization reaction employs a 20–30% solids poly(amic acid) in an amide solvent with a slight molar excess of acetic anhydride and a molar equivalent of a triamine (triethyl amine, pyridine, or β-picoline) [130–133]. The percent conversion for chemical imidization is a function of polyimide solubility. If the polymer crystallizes and/or precipitates from the reaction medium, imidization will be incomplete [93, 128]. Those systems that remain soluble must undergo thermal treatment to convert any isoimide, and remove residual solvent. The mechanistic routes of chemical imidization are shown in Figure 4, and involve using a triamine to form

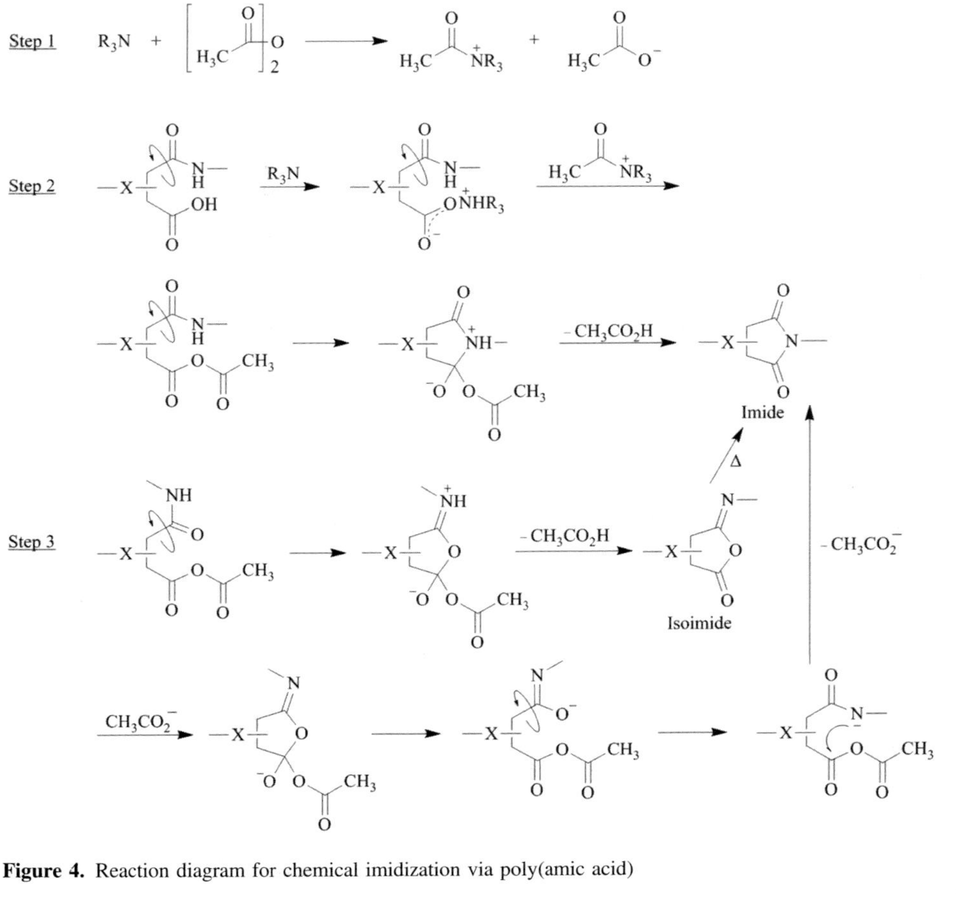

Figure 4. Reaction diagram for chemical imidization via poly(amic acid)

the amine salt from acetic anhydride, step 1, and the amine salt of the poly(amic acid), step 2. Next, the salt is replaced with the acetyl group (from the dianhydride salt) to form a mixed anhydride. The acetate group from the mixed anhydride group leaves as the amide nitrogen attacks the carboxylic acid carbonyl. The acetate group picks up the amide proton to form acetic acid resulting in the formation of the imide ring. If the poly(amic acid) is not in the proper conformation (i.e., the amide group may be rotated out of position) the carbonyl oxygen may attack the mixed anhydride to form the corresponding isoimide, step 3. The isoimide is converted to the imide thermally or by the presence of acetate ions that attack the ester carbonyl carbon, closely following the later stages of Step 2. The important step is the conversion of the poly(amic acid)'s hydroxyl group into a better leaving group, the acetate [134, 135].

Combinational Process. One system that takes advantage of both the chemical and thermal imidization is the gel casting of non-melt-processable polyimide film. These systems start as poly(amic acid) in NMP or DMAc to which is added β-picoline and acetic anhydride. This mixture starts to imidize and forms a gel as it is coated onto a rotating heated drum. The swollen gel film has the mechanical integrity allowing it to be stripped off the drum and gripped by a tenter frame. Final conversion and solvent removal is achieved thermally (infrared and convection) while the film is undergoing mechanical orientation [136].

3.1.3. One-Step Formation of Polyimides

Polyimides that are soluble in high-boiling organic solvents can be directly converted from their corresponding monomers. This is done by mixing a diamine and a dianhydride into a solvent above 180°C. Typically, solvents such as *m*-cresol with catalytic isoquinoline, benzoic acid, nitrobenzene, or NMP are used depending on the polyimide formed [137–139]. This system is useful for diamines and dianhydrides having low reactivity and those that are sterically hindered, such as perylenic dianhydrides [140, 141]. The conversion of monomer to polyimide proceeds rapidly with the intermediate poly(amic acid) being short-lived to the point that it is virtually undetectable. The degree of imidization is nearly 100% [142–145].

3.2. Synthesis Using Imide-Containing Monomers

Two methods may be employed to form a polyimide when the imide ring is intact. The first method uses a high-yield organic reaction by which imide-containing monomers polymerize by a different mechanism (aromatic nucleophilic displacement, addition, and free-radical reactions). The second approach involves thermosetting imide oligomers.

3.2.1. Polymerization Processes

Aromatic Nucleophilic Displacement. Aromatic nucleophilic displacement is used in the synthesis of polyarylene ethers. Halogenated bisimides are polymerized with bisphenolates and activated A–B phthalimides to form polyetherimides [146–150]. The reaction proceeds through the Meisenheimer-type transition by lowering the activation energy of the displacement reaction. Table 4 shows the relative rates of displacement on substituted phthalimides with phenoxide. Remarkably, the displaced nitro group may affect the growing polyimide chain if not handled properly [146, 151].

A related synthetic method has been employed specifically for both bisphenols and halogenated

Table 4. Relative reactivities of substituted phthalimides with phenoxide in DMSO at room temperature [152]

Position–X	R = –CH$_3$	R = –C$_6$H$_5$
3-Cl	1	–
3-F	4	20
3-NO$_2$	37	130
4-F	–	65
4-NO$_2$	170	520

diimides or the halogen-containing A–B-phenolated phthalimide monomers. Bisphenol or the N-phenol unit of the fluorinated phthalimide is converted to its corresponding trimethylsilated analogue. Polymerization of these silated monomers into polyimides are achieved in the melt at high temperature by using cesium fluoride as a catalyst [152].

Addition Reactions. This approach involves using maleimide-terminated monomers (bismaleimides, BMIs) in equal molar ratios with a dinucleophile such as a diamine, or a diene such as dibenzocyclobutane (BCB) and substituted dicyclopentadienes to form the corresponding linear polyaspartimides through Michael-type addition or the polyimides by a Diels–Alder [4+2] cycloaddition (Table 5) [153–162].

Linear polyimides can be synthesized directly from the A–B maleimidobenzocyclobutane monomers [175]. These reactions proceed without volatile evolution, but undergo thermally initiated retrograde reactions that lead to change in physical properties, chemical cross-linking and aromatization of the cyclohexene ring (BCBs) [176].

Reactions involving BMIs with dicyclopentadieneones also proceed by a cycloaddition mechanism [177, 178]. This has the same advantage inherent to all Diels–Alder reactions in that no volatiles are generated. However, with these pentadieneone dienophiles, CO is eliminated forming a cyclohexadiene moiety. The cyclohexadiene can undergo further addition reactions or aromatize with the abstraction of two hydrogen atoms depending on the reaction

Table 5. Chemical groups used for functionalizing or reacting with polyimides [163–174]

Reactive group		Onset cure temp., °C	Comments on long-term stability at 177°C
cyanate		200–230	poor thermal stability
benzocyclobutene		200–250	co-reacts with maleimide, poor long-term stability
trifluorovinyl ether		250	small processing window, unknown thermal stability
ethynyl		200	small processing window, unknown thermal stability
phenylethynyl		320–370	excellent thermal stability
nadimide		300–350	cyclopentadiene evolution, low toughness
maleimide		200–230	poor thermal stability
chloromaleimide		200–230	poor thermal stability
methylmalimide		220–250	poor thermal stability
phenylmaleimide		340–370	unknown thermal stability

conditions, and the added stability of any pendent substituents [177, 178]. This type of chemistry can be used to create photosensitive polyimides for imaging.

Free-Radical Polymerization. The last approach utilizes free-radical polymerization of substituted maleimides to afford ethylenic-type imide polymers [154, 179]. Maleimide groups are copolymerized with unsaturated monomers such as styrene and other maleimides to afford copolyimides [180]. Polymerization of these systems is typically carried out with azobisisobutyronitrile (AIBN) in polar solvents such as THF [181]. These types of polyimides can be used for a variety of applications from optical and photosensitive resins to structural materials, depending on the selection of the N-substituent [175, 180].

3.2.2. Thermosetting

Thermosetting systems involve the placement of reactive moieties with the polyimide, terminal, pendent, or blended as copolymers with the reactive imide monomer (Table 4). These systems contain maleimide, norbornene (nadic), vinyl and allylic, acetylenic, phenylethynyl, or other types of unsaturated groups that react to cross-link the polyimide [182, 163–172]. When bismaleimides are used alone or in excess with a difunctional (at least) nucleophilic monomer, the resulting system is thermosetting with chain extension (through Michael-type addition) occurring prior to the thermal cross-linking of the maleimide functionality (via Michael addition) [117, 183–186]. Although the cross-linking of the maleimide group is thermally initiated, it is catalyzed with both free radicals and bases [180, 164, 165]. Typical synthesis involves 30% (or more) total solids diamine and two equivalents of maleic anhydride in acetone at ambient temperature. After the bismaleamic acid precipitates (several hours), triethylamine is slowly added to the mixture until dissolution occurs (approximately 0.5 equivalents). Acetic anhydride (slightly over 1 equivalent) is added and the mixture is heated at reflux for several hours and the crude product is isolated by water precipitation. The intermediate maleamic acid is susceptible to isomerization during cyclodehydration resulting in the production of the less desirable acetanilide [178, 164]. Metal acetate catalysts have been known to increase the yields of BMI [187]. Blends of BMIs with aromatic allyls and phenylpropenyl oligomers, such as a bisallylphenol, have been used as thermosetting systems that undergo complex reaction involving Ene addition followed by Diels–Alder addition, respectively [117, 183–185, 188–194].

The nadic group has been extensively explored for creating polyimide thermosetting resins including the PMR-type composite materials. These materials include the nadic-endcapped LaRC 160, PMR-15, PMR-II, RP-46, Superimide 800 and others, and the nadic/amine terminated systems which includes AFR-700B. The nadic functionality is similar to the maleimide functionality in that it can undergo similar addition-type reactions and can form the maleimide upon rearrangement and release of cyclopentadiene. A study comparing the thermochemistry of these two moieties is described wherein time, temperature, and free-radical initiation were detailed [191]. These nadic-terminated oligomers are formed from nadic anhydride (NA), a diamine and a dianhydride [195–203]. These systems (like their aromatic counterparts) do not exhibit the same behavior as maleamic acid during imidization as structural conformation remains in place. Unfortunately, the nadimides generate cyclopentadiene at the elevated temperatures required to cure them. Hence, these systems must be cured under pressure to avoid volatilization of the cyclopentadiene that copolymerizes with itself, other nadic moieties and the resulting maleimide [204–206].

In an attempt to create a high-temperature thermosetting imide system that can be handled as a room-temperature liquid, the polymerization of monomeric reactants (PMR) approach was developed. Here, the dianhydride and nadic anhydride are treated with molar equivalents of an aliphatic alcohol and allowed to form the liquid diacid ester isomers. The diamine and the diacid esters are then dissolved in methanol, and the liquid system is solution-cast or coated onto a preform to create a prepreg that is dried and cured under pressure to form a composite part. The advantage of this system is that the monomers remain intact at or below room temperature [207–210]. The molar ratios of the diacid-methylester of BTDA, 4,4′-methylenedianiline

(MDA) and the acid methylester of NA of 2.087:3.087:2 are used to make a 1500 g/mol imidized reactive oligomer referred to as PMR-15. There are several studies that characterize the reactive monomers, the cure chemistry and kinetics, cure cycles, and the resulting thermomechanical stability of the various nadic systems made by the PMR approach [205, 206, 208–216]. Acetylenic groups on polyimides are used to induce thermal cross-linking [169, 217]. The actual chemical structures developed during the acetylenic cross-linking can vary, because the reactivity of these groups is high enough to allow attack on chemical moieties other than acetylenic units [218]. National Starch's Thermid IP-600 is an acetylene-terminated thermoset that has enhanced solubility, as it is an isoimide that thermally converts to imide when cured [219, 77]. The highest temperature thermosetting matrix resin/structural adhesives were those based on phenylacetylene (phenylethynyl) [220–222]. NASA's phenylethynyl-terminated imide (PETI) series are systems that cure at temperatures above 320°C [223–230].

3.3. Polymer Conversion

One method has already been introduced through the chemistry of isoimide synthesis and conversion (Section 3.1.2 polyimide formation). If the isoimide is desired directly, a poly (amic acid) is dissolved in a solvent system that facilitates removal of the byproduct and a molar equivalent of dicyclohexylcarbodiimide or similar cyclodehydrating agent. These isoimides are then converted by thermal conversion to the corresponding imide at elevated temperatures [231]. Other methods involve transimidation and treatment of dianhydrides with diisocyanates. Transimidiation involves the exchange of one imide nitrogen functionality (as the amine) for another. This type of reaction is favored when the attacking amine is more basic than the amine that comprises the imide ring. Because of the lower reactivity of aromatic amines, in comparison to aliphatic amines, this type of reaction offers certain advantages. When the less basic 2-aminopyridine terminates imide oligomers, this amine can be readily exchanged for other aromatic diamines to form a high-molecular-mass polyimide through proper stoichiometric control [232–235]. This scenario allows for the formation of otherwise infusible polyimides, by melt processing techniques, and also permits the incorporation of chemical moieties that are susceptible to hydrolysis via the poly(amic acid) route. Another conversion route involves the reaction of diisocyanates with dianhydrides to form the polyimide followed by elimination of CO_2 [236–243]. There is no general consensus on the mechanism of this reaction. Some researchers contend that the isocyanate reacts with water to form the amine or alcohol to form the urethanes that react with the dianhydride to form the polyimide [238]. Others postulate a seven-membered ring formation that liberates CO_2 forming the imide ring [239]. Metal salts have catalyzed these imide-forming reactions, and some have been polymerized in the melt [244, 245].

4. Uses

Polyimides are manufactured in a variety of forms, finding utility ranging from electronics and medical to structural and adhesive applications. The major forms of polyimide are films, molded and extruded stock, fiber-reinforced adhesive tape, varnish, foam, and composite prepreg.

4.1. Films

One of the more prevalent forms of polyimides is thin films. The most popular films are Kapton by Dupont, Upilex by Ube Industries, Ultem by SABIC, and Apical by Kanekafuji, to name a few. In film form, some of these materials do not display a T_g, have good radiation resistance, and are used as bagging materials for composites, belts, insulative wire wrapping, and flexible circuits. They are coated with acrylics and silicones to make pressure-sensitive tape [246]. Polyimide film is also adhesively bonded to metal foil (copper) to one or both sides, and sold to PCB manufactures for flexible cabling and circuits (examples include DuPont's Pyralux and clad derived from Kaneka's Apical series of PI films). Metal is also vapor-deposited onto polyimide film for applications requiring thin copper or other metals (less than 0.5 oz/12 µm, i.e., 14 g/12 µm) for which the additional adhesive is undesirable. As these films and clads serve as dielectric materials, they are tested under IPC TM-650 tests and display breakdown voltages above 200 V/μm (5 kV/mil, where 1 mil is 0.001 inch or 25.4 μm) and 1 MHz dissipation factors of 3.2 on average [13]. Several methods to create metallic coatings in situ involves the introduction of silver organic salts (complexed silver(I) acetates) or other metallic complexes into the poly(amic acid). Through thermal imidization, which may be photoassisted, the silver metal is reduced and migrates to the surface to form a reflective coating [247–252]. Another method involves chemically etching the polyimide surface forming the carboxylic acid groups and exchanging these protons with silver ions followed by UV radiation [253]. This has the advantage of not depositing metal ions throughout the film thickness. The latest trend in polyimide dielectrics is the introduction of nanotechnology in order to improve the insulative properties or electrical conductively. A nanoporous polyimide film was developed by blending the poly(amic acid) with PEO and POSS, followed by thermal imidization and thermal oxidation, resulting in a nanoporous silicon oxide polyimide film with a reduction in the dielectric constant [254]. A 3D flex circuit was created by embedding nanowires into polyimide film by heavy-ion implantation followed by surface etching and gold deposition to create conductive patches. The goal was to develop Z-axis conduction for flex circuitry and magnetic sensing applications [255]. Polyimide films and coatings have been used for space applications involving thermal and electrical insulation, reflectors and proposed for clear overcoats structures exposed to the space environment. Their inherent aromatic structure affords radiation and UV resistance. However, as organic materials, they suffer from the effects of atomic oxygen (AO), which is more prevalent in low earth orbit. In order to combat the effects of radiation and UV in the vacuum of space, studies of the effect of imide group location, using fluorinated monomers or phenylphosphine oxide (PPO) and silicon oxide (SiO) containing monomers, for optical clarity and AO protection, and nanoparticles have been studied. The results being that polyimides that have the imide units in direct aromatic coupling (e.g., PMDA) are more susceptible, and those polyimides with inorganic filler particles tend to form an in situ barrier when exposed to AO [32, 256–258].

4.2. Coated Wires

For years, the electrical and aerospace industry has used polyimides as high-performance wire insulation. These coatings are used in combination with other materials and may either be biaxially wrapped onto the wire, solution coated and dried in multiple passes to form a varnish, or directly extruded over the wire. The polyimide insulation must survive tight radial bends, chaffing, and electrically induced oxidation and carbon tracking [259–267]. Since polyimides possess high dielectric strength, thermal stability, radiation resistance, and are self-extinguishing, they are employed in both air and spacecraft [263–267]. The low-earth-orbit conditions are especially devastating for organic polymers because of the presence of atomic oxygen and UV radiation [268–270]. This has led to the incorporation of silicone and phosphine oxide units in the polyimide that resists the atomic oxygen degradation [256]. The NASA LaRC-TOR series polyimides, which contain phosphine oxide units in the backbone, have been one of the leading polyimides for low-earth-orbit satellite and vehicle applications [271, 272].

4.3. Membranes, Tubes, Fibers, and Foams

The fabrication of microporous membranes, thin-walled tubes, fibers, and films is performed by solution-casting or dry-jet spinning followed by heat treatment. These asymmetric flat or hollow fiber membranes are used for the selective separation of gas and liquid mixtures. The main parameters for polymeric-based membrane material selection are the free volume, the molecular distance, which relates to porosity, the shape of the molecules, the burst pressure, temperature range, and, of course, the mixtures that require separation. Here, polyimides, with bulky substituents, display the highest fluxes and selectivity when compared to other glassy polymeric materials including polyphenylene oxides [273]. The asymmetry is introduced by selective solution blending of solvent/nonsolvent mixtures to generate a porous gradient, during polymer precipitation, across the membrane wall. This is followed by thermal evaporation techniques to set the final microstructure [274–282]. In some cases, the surface wettability is improved by treating the polyimide surface with acids to increase the hydrophilicity without sacrificing thermal stability, which is especially important for fuel cells [283, 284]. Lastly, one surface may consist of porous ceramic coating to facilitate ion transport [285]. Thin-walled polyimide tubes created from PMDA and BPDA–ODA polyimides cannot be made by melt processing and are created by successive layering and drying to build up thickness on or inside a fused silica tube [286, 287]. These polyimide-coated tubes find use as GC columns for chemical separation and as catheter medical tubing [288, 289]. Continuous fibers, solution-spun and dried from poly(amic acid) or as the polyimide, present a challenge as the solvent, cyclodehydration and shrinkage problems associated with curing the fiber result in a technical challenge. Although the properties of these fibers are excellent when the proper draw ratio is used, they have not achieved a high-performance niche market able to economically compete with polyesters or polyaramides [289–292].

The extensive processing methodology and physical characteristics make polyimides excellent resins for the design of foamed preforms. The thermal range, density variance and char characteristics make polyimide foams choice candidates for structural insulation, flexible gaskets, acoustical insulation and thermal protection for aerospace and naval vessels [293]. These foams can be created from momomer solutions, amic acid gel or powder with blowing agents and solvent complexation. These components may undergo volatile release on thermal conversion to help develop and control the resulting cellular structure [294–299]. These methods are used when creating open-cell polyimide foams of varying density. The final cell size and type depends on both the thermal cycle and the diffusion rate of the agent out of the particle cell [300]. One of the leading candidate polyimide foams for future cryo-tankage insulation is NASA's TEEK series that have applications for the next generation of aerospace vehicles [301–303]. A more recent development is the synthesis of polyimide aerogels made by using chemical imidization followed by solvent extraction-replacement prior to the final step of CO_2 critical fluid extraction [129, 304, 305]. The physical properties of these aerogels can be controlled by monomer selection and processing conditions.

4.4. Melt-Processable Polyimides

Polyimides that display a T_g and T_m well below their decomposition or thermosetting temperatures are candidates for melt-processing techniques that include extrusion, injection, and compression molding. Several other factors play into the selection of melt-processable polyimides and include melt viscosity, work time, or pot life at temperature and the ability to carry fillers or blends. Melt viscosity is crucial as it directly relates to molecular mass, affecting the processing temperature [241, 306–308]. The higher the molecular mass, the higher the temperature and pressure needed to create the part. As the polymer's molecular mass increases, the mechanical properties increase until a critical molecular mass is obtained, at which the melt viscosity increases while the mechanical properties start to level out. This "knee" or bend in the mechanical and physical properties versus molecular mass curve is where any further increase in molecular mass increases the complexity of the melt forming without a significant improvement in the overall properties of the neat resin [306–308]. The influence of molecular mass and its relationship to melt viscosity are affected by the structure of the polyimide in a fashion similar to the basic structure–property relationships discussed above. Several resin systems developed at GE, NASA Langley, Furon, Mitsui, and others have found utility as thermoplastic molding resins. The T_g for these thermoplastic polyimides are in the range of 210–270°C and the polymers are processed 50–150°C above their glass-transition temperature [306, 307]. The molar masses of these resins range from several thousand to above 30 000 g/mol. These systems display excellent fracture toughness and high modulus and have been extruded, compression molded, machined and polished neat, or filled with graphite, PTFE, metal, and ceramic powders. These thermoplastic polyimides are sold as rod, sheet and film stock or as final stock parts and composites [49, 309–317].

Thermosetting polyimides solve the molecular mass effect on melt viscosity, but add the additional problems associated with curing exotherms and time–temperature-limited processing windows. Thus, a wide processing window from the melt flow of the monomer(s) to the onset cure temperature should be selected. The melt-processable thermosetting systems are typically BMIs, PMRs, or BCB-maleimides, and these can be blended with chopped fibers or other fillers, ground or extruded, and pelletized for moldings [318–324]. These imide oligomers are then cured into high-performance PCB substrates and composite forms. NASA Langley has developed and commercialized several PETI-based low viscosity (LV) thermosetting resins, PETI-298 and 330, that can be injection-molded or resin-infused into carbon fiber preforms and then cross-linked at elevated temperatures [324, 325].

4.5. Direct-Formed Polyimides

There are several polyimides that do not undergo melt flow, but are directly formed into parts requiring performance under extremely harsh conditions [326–328]. These direct-formed polyimides include DuPont's Vespel (includes the SP product line), Ube's Upimol-S and Imitec's I-903 and I-772. These polyimides undergo a forming process similar to that of ceramic sintering. The starting fine-particle polyimide powder is either hot isostatically pressed (HIPed) or ram-extruded and hot-formed to generate a polyimide stock shape that is machined to size, with a density in the range of 1.4–1.5 g/cm^3 for the neat system [329]. These polyimide powders or films have also been blended/impregnated with graphite, PTFE, and boron nitride for dry lubrication for which tribology and abrasion issues are important [330–332]. These parts have a very low coefficient of thermal expansion and do not smear under load at elevated temperatures, making them excellent choices for high-temperature pressure–velocity requirements [333–337]. These molded parts have found utility in rotating equipment, silicon wafer and chip processing, machine tooling, and chemical and material processing.

4.6. Matrix Composites and Adhesives

(→ Composite Materials; → Adhesives, 1. General) Polyimides display properties that make the resulting advanced composites excellent choices as primary structural members for aerospace applications [338–340]. The fibers in the composite are the load-carrying members and the resin serves to transfer the load to the fibers and hold them in place. Thus, the composite is only as good as the resin [341]. Therefore, the selected matrix resin must have excellent thermal, chemical and mechanical toughness, high compressive strength and adhesive properties. One of the key elements in creating a strong composite is the wetting out of the fibers with the proper amount of resin. This is achieved by two methods, dry and wet (or solution) prepregging [342]. The dry-coating method may consist of coating the fibers with fine polyimide powder and melt shearing the powder to form a film about the fibers [343, 344]. Another dry process involves melt infusing a polyimide film into a fiber matrix affording prepreg tape. Solution prepregging involves the coating of the fibers with a solution containing the polymer or reactive monomers and removing the solvent at a later stage to form the prepreg tape. The resulting prepreg should display tack and drape, i.e., tack to allow it to remain in place prior to and during consolidation, and drape to allow it to contour to the tooling surface. These prepregs are then cut into the desired shapes, stacked, and laminated together on a tool in a hot press or autoclave to form the composite part. In order to achieve excellent consolidation, these polyimide laminates must undergo melt flow and adhere [332]. Hence, the polyimides must display high adhesive strength under a variety of conditions [345–348]. Initial criteria for screening the hot-melt aerospace polyimide adhesives is the Titanium–Titanium or Aluminum–Aluminum Lap Shear test (ASTM D-1002), and for the electronic industry the T-peel and 180° peel tests are used (ASTM D-1976 & D-3330). Well prepared titanium lap shear strengths for polyimide adhesives range from 2000 psi (~138 bar) to more than 8000 psi (550 bar) under ambient conditions. It is worth noting that surface preparation plays a key role in determining the adhesive strength of the resin system. The polyimide resin systems that have been used for aerospace composite programs include BMIs, PMR-15, DuPont's Avimide series (based on the PMR approach) and NASA's PETI-5 (developed under the HSR program) that exceeds the performance criteria for the High Speed Civil

5. Photoimageable Polyimides

Photoimageable polyimides are mainly used in microelectronic processing as solder masks, protective coatings, and negative and positive photoresists for the selective etching and layering of complex circuitry on multilayered printed circuit boards (PCBs) and silicon wafers [352]. For such applications, these resins must have good substrate adhesion and wettability, a long storage life and high quantum yields upon exposure, undergo changes in solubility without swelling, and serve as a thermal protective coating, which is an inherent characteristic of polyimides. There are several ways (Fig. 5) in which polyimides (PIs) have been used in

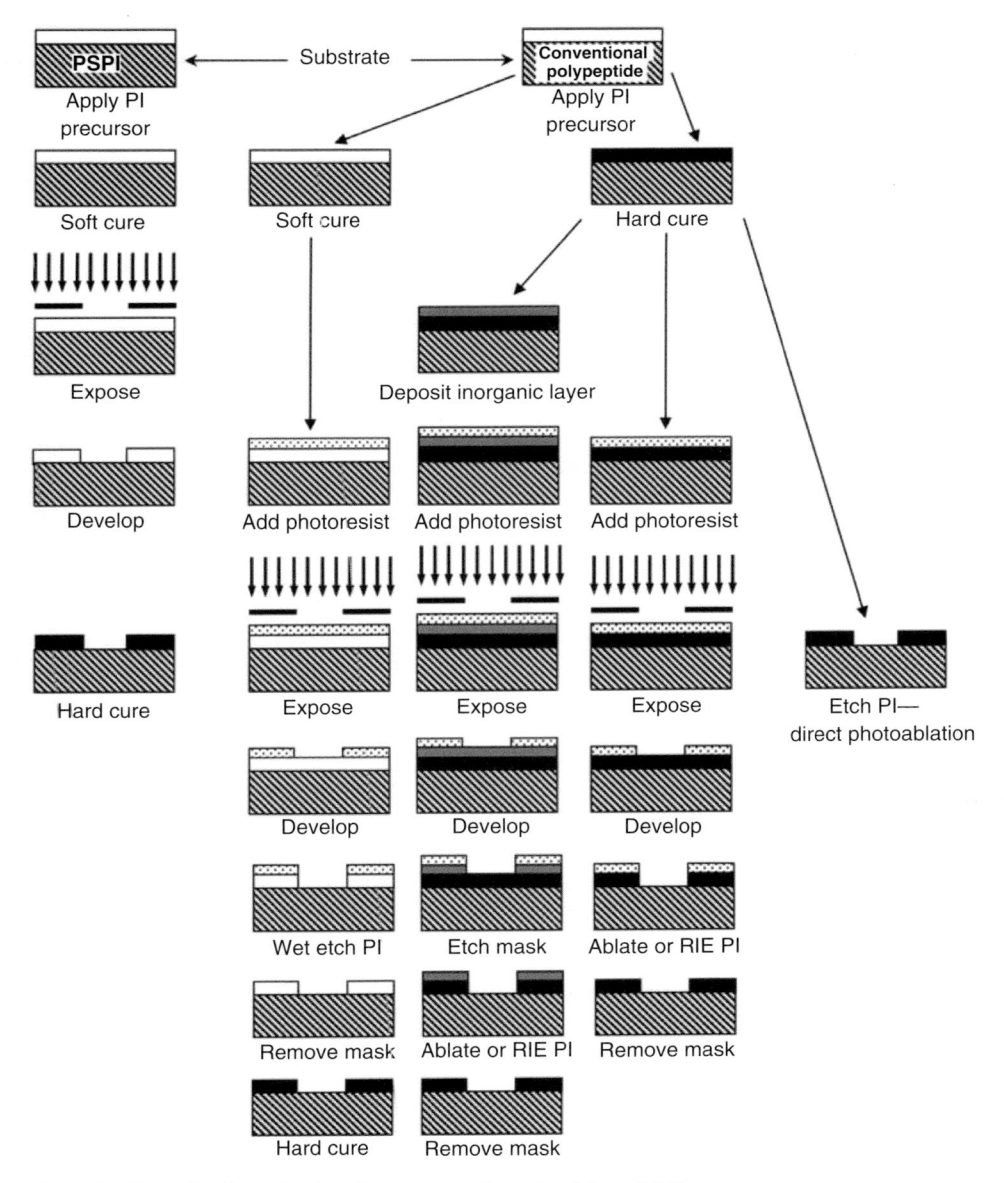

Figure 5. Chart of various photoimaging processes (reproduced from [352])

applications where "photoimaging" is required to produce a pattern on a substrate (typically silicon, or other materials including polyimide itself [352]. The first method is reactive ion etching (RIE, includes plasma etching) or photoablation. This method does not require a photosensitive material, only a mask that can withstand the reactive ion gas or photons and the material to be patterned, that is protected by the mask. In RIE, the reactive ions, generated by a radiofrequency field, attack the exposed polyimide to etch out the pattern of the mask. This process can be used with conventional polyimide film and requires few steps, but is not highly accurate and requires a hard (metal, silicon, etc.) mask. The other methods involving the conventional polyimide, as a dielectric, require an additional photosensitive material as the overcoat. The method also uses a mask, but higher degrees of accuracy are obtained as the photosensitive material's exposure is performed with coherent short-wavelength light. The combination of step reduction, increased accuracy, and less complexity while maintaining good dielectric, mechanical, and thermal properties have spurred the development of photosensitive polyimides, (PSPIs) used with these photoimaging methods [352–354].

5.1. Negative Photoresists

The two types of photoimageable resists are positive and negative. The negative resist undergoes a reaction when exposed that typically decreases solubility, in comparison to the unexposed region, that remains in place during solvent development (a negative image is produced). Negative PSPIs are often used in their soluble poly(amic acid) form with photosensitive units either attached to the polymer backbone, in conjugation with the carboxyl unit of the poly (amic acid), or incorporated into the polyimide structure. Specific areas of the cast photosensitive film are exposed to light, thereby becoming insoluble through a cross-linking mechanism, and the masked portions (unexposed areas) are washed away with solvent. The remaining cross-linked (exposed area) is then baked to thermally cure the poly(amic acid), releasing the cross-linked portion. Subsequent washing removes these cross-linked photocured regions. There are several approaches used to create negative PSPIs, with the more popular being those of the covalent, ionic, and intrinsic types [355].

The covalent type was one of the first and still widely used commercial approaches. A photosensitive unit is attached to the carboxyl group of the starting dianhydride prior to polymerization with a diamine (Fig. 1, k_4 followed by k_3). The resulting sensitized poly(amic acid) is then exposed and the resulting sensitized ester unit photo-cross-links. The exposed film is then developed and baked (Fig. 1, k_5). Such photosensitive units are of the acrylate or cinamyl variety [356, 357]. A system that uses a soluble fluorinated polyimide involves the chloromethylation of the aromatic ring on the arylether portion, followed by treatment with cinnamic acid to form the pendent photosensitive cinnamate group. This resulting PSPI was photo-cross-linked, and the unexposed portion was washed away with solvent [358]. The ionic type is similar to the covalent type with the pre-made poly(amic acid) sensitized by using a tertiary amine having a photoreactive group and a photoinitiator. The interaction of the tertiary amine and the poly(amic acid) hydroxyls form a photosensitized salt that is ionically bound to the poly(amic acid) (similar to Fig. 4, Step 2). The ionic PSPI is then exposed, washed, and baked to develop the image. The last type is the intrinsic PSPIs, meaning that they are photosensitive as made. These PSPIs are generally based on BTDA and *ortho*-alkylated diamines. When these intrinsic systems are exposed to 365 nm wavelength, the hydrogen atoms of these pendent alkyls are abstracted through the excitation (C=O diradical formation) of the benzophenone radicals inducing cross-linking of the pendant methylene radicals. Recently, a hyperbranched PSPI was synthesized terminated by 3-aminophenylacetylene. This system takes advantage of the charge-transfer complex developed between the electron-acceptive imide group and the electron-donating phenylene terminal units. When exposed to 365 nm UV light, a clear negative image is resolved [359]. Overall, resolution of negative PSPIs is controlled by several factors including film thickness, exposure wavelength, and the selectivity of the developer, but the typical resolving limit that can be obtained is a 3:1 width to resist thickness ratio [82]. These drawbacks of the negative PSPIs manifest themselves through pattern distortion, shrinkage, pinhole formation, and thermal stability effects during wafer processing.

5.2. Positive Photoresists

The positive PSPIs undergo a chemical change in solubility on exposure, versus a molecular-mass change. Typically, the masked positive PSPI is exposed to light and a pendent photosensitive unit is converted into a carboxylic acid or hydroxyl unit that allows the exposed portion polyimide to be removed under aqueous conditions (producing a positive image). The advantage at this point is that since the exposed and unexposed PSPI are close in structure, there is reduced physical swelling on exposure, and the unexposed areas can be re-exposed with a different mask. Hence, these positive PSPIs lead to more accurate patterning and tighter line widths. Lastly, the positive PSPIs are baked to volatilize the pendent photosensitive units to afford the polyimide dielectric. Most of the chemistry of these positive resists use variations of the Wolff rearrangement of diazo systems [82]. As an example, a polyimide with pendent hydroxyl groups was synthesized and blended with diazonaphthoquinone (DNQ). Here, the DNQ complexes with the pendent hydroxyl portion of the polyimide, decreasing the solubility of the polyimide towards aqueous base. Upon exposure, the DNQ undergoes the Wolff rearrangement, forming the indenecarboxylic acid and decomplexing from the polyimide. This results in increasing the solubility of the exposed polyimide mixture towards aqueous base solution [360, 361]. Other methods involve PSPIs made with acetal-containing diamines. These are photocatalytically depolymerized in the presence of p-toluenesulfonic acid, resulting in derivatives of benzaldehyde and hydroxyphthalimides that are readily removable [362]. PSPIs with pendent nitrobenzylester groups can be photodegraded to the corresponding aryl–acid functionality to enhance the resulting converted polyimides solubility in caustic solutions, with the additional advantage of subsequent thermal removal of the nitrobenzylester group form the unexposed portion [363]. Most of these post-thermally developed PSPIs, negative and positive, have mechanical properties close to that of standard polyimide film, from which they are derived,

thus demonstrating excellent planarity (surface flatness), high-voltage breakdown resistance, and low dielectric constants.

References

1. T.M. Bogert, R.R. Renshaw, *J. Amer. Chem. Soc.* **30** (1908) 1130.
2. E.I. DuPont de Nemours & Co., US3179614, US3179634, 1965 (W.M. Edwards).
3. M.W. Ranney, *Polyimide Manufacture*, Noyes Data Corp, Park Ridge 1971.
4. M.K. Ghosh, K.L. Mittal (eds.): *Polyimides, Fundamentals and Applications*, Marcel Dekker, New York 1996.
5. M.I. Bessonov, M.M. Koton, V.V. Kudryavtsev, L.A. Laius: *Polyimides: Thermally Stable Polymers*, Plenum Publ. Corp, New York 1987.
6. D. Wilson, H.D. Stenzenberger, P.M. Hergenrother: *Polyimides*, Blackie, London and Chapman & Hall, New York 1990.
7. C.E. Sroog, *Prog. Polym. Sci.* **16** (1991) no. 4, 561.
8. P.M. Hergenrother, *High Perform. Polym.* **15** (2003) no. 1, 3.
9. General Electric Co., US3833544, 1974 (T. Takekoshi, J.S. Manello).
10. T.L. St. Clair in D. Wilson, H.D. Stenzenberger, P.M. Hergenrother: *Polyimides*, Blackie, London and Chapman & Hall, New York 1990, Chap. 3.
11. T.A. Gordina, B.V. Kotov, O.V. Kolninov, A.N. Pravednikov, *Vysok. Soyed., B*, **237** (1977) no. 3, 612.
12. C. Cui, S.J. Cho, K.S. Kim, C. Baehr, J.C. Jung, *J. Chem. Phys.* **107** (1997) no. 23, 10201.
13. M.I. Bessonov, M.M. Koton, V.V. Kudryavtsev, L.A. Laius: *Polyimides, Thermally Stable Polymers*, Plenum Publ., New York 1987, Chap. 1, Tables 1.2 and 1.5.
14. A.K. St. Clair, W.S. Slemp, *SAMPE J.* **21** (1985) no. 4, 28.
15. D.W. Van Krevelen: *Properties of Polymers, Their Estimation and Correlation with Chemical Structure*, Elsevier, New York 1976.
16. K.-M. Chen, T.-H. Wang, J.-S. King, A. Hung, *J. Appl. Polym. Sci.* **48** (1993) 291.
17. NASA, US4595548, US4603061, 1985 (A.K. St. Clair, T.L. St. Clair).
18. T. Hou, N.T. Wakelyn, T.L. St. Clair, *J. Appl. Polym. Sci.* **25** (1988) 1731.
19. J.B. Friler, P. Cebe, *Polym. Eng. Sci.* **33** (1993) no. 10, 587.
20. R. Ginsburg, J.R. Susko in K. Mittal (ed.): *Polyimides: Synthesis and Characterization*, Vol. 1, Plenum Publ., New York 1984, pp. 237–247.
21. C.-P. Yang, Y.-Y. Su, F.-Z. Hsiao, *Polymer* **45** (2004) 7529.
22. V.L. Bell, B.L. Slump, H. Gager, *J. Polym. Sci., Polym. Chem. Ed.* **14** (1976) 2275.
23. H.H. Gibbs, C.V. Breder, *Polym. Prep.* **15** (1974) no. 1, 775.
24. F.W. Harris, W.A. Feld, L.H. Lanier, *Polym. Prep.* **16** (1975) no. 1, 520.
25. T.L. St. Clair, D.J. Progar, *Polym. Prep.* **16** (1975) no. 1, 538.
26. P.M. Hergenrother, T. Wakelyn, S.J. Havens, *J. Polym. Sci.* **25** (1987) 1093.
27. M.K. Gerber, J.R. Pratt, T.L. St Clair in C. Feger, M.M. Khojasteh, J.E. McGrath (eds.): *Polyimides: Materials, Chemistry and Characterization*, Elsevier, Amsterdam 1989, pp. 487–496.
28. J.R. Pratt, T.L. St. Clair, M.K. Gerber, C.R. Gautreaux in C. Feger (ed.): *3rd Int. Conf. Polyimides*, Ellenville 1988, pp. 223.
29. C. Koning, L. Teuwen, E.W. Meijer, J. Moonen, *Polymer* **35** (1994) no. 22, 4889.

30. C.E. Lee, *J. Macromol. Sci. – Rev. Macromol. Chem. Phys. C* **29** (1989) no. 4, 431.
31. Y. Yin, S. Chen, X. Guo, J. Fang, K. Tanaka, H. Kita, K-I. Okamoto, *High Perform. Polym.* **17** (2005) no. 2, 317.
32. C.M. Thompson, J.G. Smith, Jr., J.W. Connell, *High Perform. Polym.* **15** (2003) no. 2, 181.
33. R.G. Bryant, *High Perform. Polym.* **8** (1996) 607.
34. R.G. Bryant, *Polym. Prep.* **35** (1994) no. 1, 517.
35. NASA, US5639850, 1997 (R.G. Bryant).
36. P. Alder, J.G. Dolden, P. Smith, *High Perform. Polym.* **7** (1993) 421.
37. M. Tanaka, M. Konda, M. Miyamota, Y. Kimura, A. Yamaguchi, *High Perform. Polym., Comm.* **10** (1998) 93.
38. K.-M. Chen, T.-H. Wang, J.-S. King, *J. of Appl. Polym. Sci.* **46** (1993) 291.
39. J.-H. Chang, K.M. Park, *Euro. Polym. J.* **36** (2000) 2185.
40. J.A. Hinkley, B.J. Jensen, *High Perform. Polym.* **7** (1993) 1.
41. E.I. DuPont de Nemours & Co., US3234181, 1966 (K.L. Olivier).
42. D.A. Scola, R.A. Pike, J.H. Vontell, J.P. Pinto, C.M. Brunette, *High Perform. Polym.* **1** (1989) no. 1, 17.
43. S.F. Dinetz, E.J. Bird, R.L. Wagner, A.W. Fountain III, *J. Anal. Appl. Pyrolysis* **63** (2002) 241.
44. S.M. Saufi, A.F. Ismail, *Carbon* **42** (2003) 241.
45. Y. Xiao, T.-S. Chung, M.L. Chng. S. Tamai, A. Yamaguchi, *J. Phys. Chem. B.* **109** (2005) 18741.
46. Y. Isono, A. Yoshida, Y. Hishiyama, Y. Kaburagi, *Carbon* **42** (2003) 241.
47. T. Takeichi, M. Zuo, M. Hasegawa, *J. Polym. Sci. Part B: Polym. Phys.* **39** (2001) 3011.
48. V.E. Yudin, M.Y. Goykhman, K. Balik, P. Glogar, P. Polivka, G.N. Gubanova, V.V. Kudryavtsev, *Carbon* **40** (2002) 1427.
49. M.M. Angelovici, R.G. Bryant, G.B. Northam, A.S. Roberts Jr., *Materials Letters*, **36** (1998) 254.
50. R. Delasi, *J. Appl.1972 Polym. Sci.* **16** (1972) no. 11, 583.
51. J. Seo, A. Lee, J. Oh, H. Han, *Polym. J.* **32** (2000) no. 7, 583.
52. C.I. Croall, T.L. St. Clair, *J. Plast. Film Sheeting* **8** (1992) 172.
53. W. Volksem, P.M. Cotts: *Polyimides: Synthesis, Characterization and Properties*, Vol. 1, Plenum Press, New York 1984, pp. 163–179.
54. S.J. Havens, R.G. Bryant, B.J. Jensen, P.M. Hergenrother, *Polym. Prep.* **35** (1994) no. 1, 553.
55. R.G. Bryant, B.J. Jensen, P.M. Hergenrother, *J. Appl. Polym. Sci.* **59** (1996) 1249.
56. C.J. Wolf, R.S. Soloman, *SAMPE J.* **20** (1984) no. 1, 16.
57. J. Munson in K. Gilleo (ed.): *Handbook of Flexible Circuits*, Van Nostrand Reinhold, New York 1992, Chap. 2.2.1.
58. D.D. Denton, D.R. Day, D.F. Priore, S.D. Senturia, E.S. Anolick, D.J. Scheider, *Elec. Mat.* **14** (1985) 119.
59. M.C. Tucker, *Hybrid Circuit Technol.* **8** (1991) no. 10, 23.
60. J. Fjelstad: *An Engineer's Guide to Flexible Circuit Technology*, Electrochemical Publ., Isle of Man 1993, p. 23.
61. UpJohn Co., US4001186, 1977 (B.K. Onder).
62. G. Raju, A. Katebain, S.Z. Jahi, *IEEE Trans. Dielectr. Elctr. Insul.* **10** (2003) no. 1, 117.
63. S. Muruganand, S.K. Narayandass, D. Mangalana, T.M. Vijayan, *Polymer Int.* **50** (2001) 1089.
64. H. Czarczyńska, A. Dziedzic, B.W. Licznerski, M. Łukaszewicz, A. Seweryn, *Microelectronics J.* **24** (1993) 689.
65. D.E. Herr, N.A. Nikolic, R.A. Schultz, *High Perform. Polym.* **13** (2001) 79.
66. F.W. Mercer, T.D. Goodman, *High Perform. Polym.* **3** (1991) no. 4, 297.
67. M. Hasegawa, *High Perform. Polym.* **13** (2001) no. 2, S93.
68. M.K.M. Hasegawa, *High Perform. Polym.* **15** (2003) 47.
69. H. Wang, X. Tao, E. Newton, *High Perform. Polym.* **15** (2003) 91.
70. K. Goto, T. Akiike, Y. Inoue, M. Matsubara, *Macromol. Symp.* **199** (2003) 321.
71. F.W. Mercer, M.T. McKenzie, *High Perform. Polym.* **5** (1993) 97.
72. A.K. St. Clair, T.L. St. Clair, W.P. Winfree, *Polym. Mater. Sci. Eng.* **59** (1988) 28.
73. T. Chino, S. Saski, T. Matsura and S. Nishi, *J. Polym. Sci., Polym. Chem. Ed.* **28** (1990) 323.
74. M.K. Gerber, J.R. Pratt, A.K. St. Clair, T.L. St. Clair, *Polym. Prep.* **31** (1990) no. 1, 340.
75. L.V. Ng, H.J. Fick, M. Bui, *Adv. Packaging.* **29** (2000) 29.
76. C.E. Sroog in D. Wilson, H.D. Stenzenberger, P.M. Hergenrother (eds.): *Polyimides*, Blackie, London, and Chapman & Hall, New York 1990, Chap. 9.
77. N. Bilow, A.L. Landis, R.H. Boschan, J.G. Fasold, *SAMPE J.* **18** (1982) no. 1, 8.
78. V.M. Svetlichnyi, K.K. Kalninish, V.V. Kudryavtsev, M.M. Koton, *Dokl. Akad. Nauk.* **237** (1977) no. 3, 612.
79. G.A. Bernier, D.E. Kline, *J. Appl. Polym. Sci.* **12** (1968) 593.
80. K. Nakamae, T. Nishino, N. Miki, *High Perform. Polym.* **7** (1995) 371.
81. J. Hansen, T. Kanaya, K. Nishida, K. Kaji, K. Tanaka, A. Yamaguchi, *J. Chem. Phys.* **108** (1998) no. 15, 6492.
82. H. Steppan, G. Burh, H. Vollmann, *Angew. Chem. Int. Ed. Engl.* **21** (1982) no. 7, 455.
83. DuPont de Nemours, US 2710853, 1955; W.M. Edwards, I.M. Robinson 2867609, 1959.
84. C.E. Sroog, A.L. Endrey, S.V. Abramo, C.E. Berr, W.M. Edwards, K.L. Oliver, *J. Polym. Sci.: Part A* **3** (1965) 137.
85. J.A. Kreuz, A.L. Endrey, F.P. Gay, C.E. Sroog, *J. Polym. Sci.: Part A-1* **4** (1966) 2607.
86. Y. Tong, L. Lui, S. Veeramani, T.-S. Chung, *Ind. Eng. Chem. Res.* **41** (2002) 4266.
87. Y. Imai in M.K. Ghosh, K.L. Mittal (eds.): *Polyimides, Fundamentals and Applications*, Mercel Dekker, New York 1996, Chap. 3.
88. J.F. Dezern, *SAMPE J.* **24** (1988) no. 2, 27.
89. F.W. Harris in D. Wilson, H.D. Stenzenberger, P.M. Hergenrother (eds.): *Polyimides*, Blackie, London and Chapman & Hall, New York 1990, Chap. 1, 2.
90. General Electric Co., US4324884, US4324885, US4330666, 1982 (D.M. White).
91. G.M. Bower, W.L. Frost, *J. Polym. Sci., Part A* **1** (1963) 3135.
92. M.L. Wallach, *Polym. Prep.* **6** (1965) no. 1, 3135.
93. R.A. Dine-Hart, W.W. Wright, *J. Appl. Polym. Sci.* **5** (1967) 609.
94. V.I. Kolegov, *Polym. Sci. USSR* **18** (1976) no. 8, 1929.
95. V.I. Kolegov, S. Ya. Frenkel, *Polym. Sci. USSR* **18** (1976) no. 8, 1919.
96. M.I. Bessonov, in Y. Takahata (ed.): *Int. Symp. Polym Microelectron.*, Kodansha, Tokyo 1990, pp. 721–734.
97. V.W. Svetlichnyi, K.K. Kalnin'sh, V.V. Kudryavtsev, M.M. Koton, *Dokl. Akad. Nauk. SSSR (Engl. Transl.)* **237** (1977) no. 3, 693.
98. J.H. Hodgkin, *J. Polym. Sc., Polym. Chem. Ed.* **14** (1977) 409.
99. D.G. Hawthorne, J.H. Hodgkin, M.B. Jackson, J.W. Loder, T. C. Morton, *High Perform. Polym.* **6** (1994) 287.
100. D.G. Hawthorne, J.H. Hodgkin, *High Perform. Polym.* **11** (1999) 315.
101. V.A. Zubkov, M.M. Koton, V.V. Kudryavtsev and V.M. Svetlichnyi, *Zh. Org. Khim. (Engl. Trans.)* **17** (1982) no. 8, 1501.

102. V.A. Solomin, I.E. Kardish, Y.S. Snagovskii, P.E. Messerle, B.A. Zhubanov, A.N. Pravendnikov, *Dokl. Akad. Nauk SSSR (Engl. Trans.)* **236** (1977) no. 1, 510.
103. A.Ya. Ardashnikov, I.Ye Kardash and A.N. Pravednikov, *Polym. Sci. USSR* **13** (1971) no. 8, 2092.
104. Ya.S. Vygodskii, T.N. Spirina, P.P. Nechayev, L.I. Chudina, G. Ye. Zaikov, V.V. Korshak, S.V. Vinogradova, *Polym. Sci. USSR* **19** (1977) no. 7, 1738.
105. R.L. Kaas, *J. Polym. Sci., Polym. Chem. Ed.* **19** (1981) 2255.
106. A.N. Pravednikov, I.Ye. Lordash, N.P. Glukhoyedov and A.Ya. Ardashnikov, *Polym. Sci. USSR* **18** (1973) no. 8, 1903.
107. P.P. Nechayev, Ya.S. Vygodskii, G.Ye. Zaikov, S.V. Vinogradova, *Polym. Sci. USSR* **18** (1976) no. 8, 1903.
108. R.J.W. Renyolds, J.D. Sheldon, *J. Polym. Sci. C* **23** (1968) 45.
109. J.B. Haung, B.M. Gong, *J. Vac. Sci. Technol. B* **3** (1985) no. 1, 253.
110. M.L. Bender, *J. Am Chem. Soc.* **79** (1957) 1258.
111. M.L. Blender, *J. Amer. Chem. Soc.* **80** (1958) 5380.
112. E.J. Soichi, S.J. Havens, P.R. Young, P.M. Hergenrother, *High Perform. Polym.* **7** (1995) 55.
113. E.I. Siochi, NASA/CR-1998-207633, NASA, Washington, D.C. May 1988.
114. L.W. Frost, I. Kesse, *J. Appl. Polym. Sci.* **8** (1964) no. 3, 1039.
115. R.A. Orwoll, T.L. St. Clair, K.D. Dobbs, *J. Polym. Sci., Polym. Phys. Ed.* **19** (1981) 1385.
116. C.C. Walker, *J. Polym. Sc., Polym. Chem. Ed.* **26** (1988) 1649.
117. H. Furutani, J. Ida, K. Tanaka, H. Nagano, *High Perform. Polym.* **12** (2000) no. 4, 461.
118. M.I. Bessonov, M.M. Koton, V.V. Kudryavtsev, L.A. Laius: *Polyimides, Thermally Stable Polymers*, Plenum Publ., New York 1987, p. 75, Table 1.2.
119. L.A. Laius, M.I. Bessonov, Ye.V. Kallistova, N.A. Adrova, F.S. Florinskii, *Polym. Sci. USSR* **9** (1967) no. 10, 2470.
120. J.W. Verbicky, Jr., L. Williams, *J. Org. Chem.* **48** (1981) 175.
121. S.V. Vinogradova, V.V. Korshak, Ya. Vygodskii, *Polym. Sci. USSR* **8** (1966) 888.
122. F.W. Harris, L.H. Lanier in F.W. Harris, R.B. Seymour (eds.): *Structure-Solubility Relationships in Polymers*, Academic Press, New York 1977, pp. 183–198.
123. C.A. Pryde, *J. Polym. Sci. A* **27** (1989) 711.
124. S. Kuroda, K. Terauchi, K. Nogami, I. Mita, *Eur. Polym. J.* **25** (1989) 1.
125. J.A. Cella, *Polymer Degrad. Stabil.* **36** (1992) 99.
126. D.K. Brandon, G.L. Wilkes, *Polymer* **36** (1995) no. 21, 4083.
127. B. Dao, J. Hodgkin, T.C. Morton, *High Perform. Polym.* **11** (1999) 205.
128. J. Chiefari, B. Dao, A.M. Groth, J.H. Hodgkin, *High Perform. Polym.* **15** (2003) 269.
129. H. Guo, M.B. Meador, L. McCorkle, D.J. Quade, J. Guo, B. Hamilton, M. Cakmak, *ACS Appl. Mater. Interfaces* **4** (2012) 5422–5429.
130. D.P. Heberer, S.Z.D. Cheng, J.S. Barley, S.H.-S. Lien, R.G. Bryant, F.W. Harris, *Macromolecules* **24** (1991) 1890.
131. E.I. DuPont de Nemours, US1379630, US1379631, US1379633, 1965 (A.L. Endry).
132. E.I. DuPont de Nemours, US3179632, 1965 (W.R. Hendrix).
133. E.I. DuPont de Nemours, US3282898, 1966 (R.J. Angelo).
134. S.V. Vinogradova, Ya.S. Vygodskii, V.D. Vorob'ev, N.A. Churochkima, L.I. Cludina, T.N. Spirina and V.V. Korshak, *Polym. Sci. USSR* **16** (1974) no. 3, 584.
135. M.M. Koton, T.K. Meleshko, V.V. Kudryavtsev, P.P. Nechayev, Ye.V. Kamzolkina, N.N. Bogorad, *Polym. Sci. USSR* **24** (1982) no. 4, 791.
136. R.J. Cotter, C.K. Sauers, J.M. Whelan, *J. Org. Chem.* **26** (1961) 10.
137. J.G. Stephanie, P.G. Rickerl in M.K. Ghosh, K.L. Mittal (eds.): *Polyimides, Fundamentals and Applications*, Marcel Dekker, New York 1996, p. 9.
138. S.V. Vinogradova, G.L. Slonimskii, Ya.S. Vygodskii, A.A. Askadskii, A.I. Mzhef'skii, N.A. Churochkina, V.V. Korshak, *Polym. Sci. USSR* **11** (1969) no. 12, 3098.
139. S.V. Vinogradova, Ya.S. Vygodskii, V.V. Korshak, *Polym. Sci. USSR* **12** (1970) no. 9, 2254.
140. A.A. Kuznetsov, *High Perform. Polym.* **12** (2000) no. 3, 445.
141. F.W. Harris, S.L.-C. Hsu, *High Perform. Polym.* **1** (1989) no. 1, 3.
142. D. Dotcheva, M. Klapper, K. Muller, *Macromol. Chem. Phys.* **195** (1994) 1905.
143. S.V. Vinogradova, Z.V. Gersashehenko, Ya.S. Vygodskii, F.B. Sherman, V.V. Korshak, *Dokl. Akad. Nauk SSSR (Engl. Trans.)* **203** (1972) no. 4, 285.
144. Z.V. Gersashehenko, Ya.S. Vygodskii, G.L. Slonimskii, A.A. Askadskii, V.S. Papkov, S.V. Vinogradova, V.G. Dashevskii, V.A. Klimova, F.B. Sherman and V.V. Korshak, *Polym. Sci. USSR* **15** (1973) no. 8, 1927.
145. D.S.E. Schab-Bakcerak, *High Perform. Polym.* **13** (2001) 45.
146. F.W. Harris, Y. Sakagauchi, M. Shibata, S.Z.D. Chang, *High Perform. Polym.* **9** (1997) 251.
147. General Electric Co., US3787364, US3838097, 1974 (J.G. Wirth, D.R. Heath).
148. General Electric Co., US3847869, 1974 (F.J. Williams).
149. T. Takekoshi, J.G. Wirth, D.R. Heath, J.E. Kochanowski, J.S. Manello, M.J. Webber, *J. Polym. Sci., Polym. Chem. Ed.* **18** (1980) 3069.
150. D.M. White, T. Takekoshi, F.J. Williams, H.M. Relles, P.E. Donahue, H.J. Klopher, G.R. Louks, J.S. Manello, R.O. Mathews, R.W. Schluent, *J. Polym. Sci., Polym. Chem. Ed.* **19** (1981) 1635.
151. J.A. Orlicki, L.J. Thompson, L.J. Markoski, K.N. Sell, J.S. Moore, *J. Polym. Sci., Part A: Polym Chem.* **40** (2003) 936.
152. F.J. Williams, P.E. Donahue, *J. Org. Chem.* **42** (1977) 3414.
153. R.G. Bryant, T.L. St. Clair in C. Feger, M.M. Khojasteh, M.S. Htoo (eds.): *4th Int. Conf. Polyimides*, Technonic Publishing Co., Ellenville 1991, pp. II–69.
154. B.C. Trivedi, B.M. Culbertson: *Maleic Anhydride*, Plenum Publ., New York 1982, Chap. 12.
155. V.L. Bell, P.R. Young, *J. Polym. Sci., Part A* **24** (1986) 2647.
156. C. Digiulio, M. Gauther, B. Jasse, *J. Appl. Polym. Sci.* **29** (1984) 1771.
157. V.S. Volkov, S.A. Dolmatov, L.V. Yudina, V.S. Levshanov, Marinyuk and L.A. Sholokhova, *Polym. Sci. USSR* **25** (1983) no. 2, 404.
158. J.V. Crivello, *J. Polym. Sci., Polym. Chem. Ed.* **11** (1973) 1185.
159. J.E. White, M.D. Scala, D.A. Snider, *J. Appl. Polym. Sci.* **29** (1984) 891.
160. W. Oppolzer, *Synthesis* **11** (1978) 793–802.
161. L.-S. Tan, E.J. Soloski, F.E. Arnold, *Polym. Prep.* **27** (1986) no. 1, 453.
162. L.-S. Tan, F.E. Arnold, E.J. Soloski, *J. Polym. Sci., Polym. Chem. Ed.* **26** (1988) 3103.
163. E.I. DuPont de Nemours & Co., US2444536, 1948 (N.E. Searle).
164. E.I. DuPont de Nemours & Co., US3929713, 1975 (G.F.D. Alelio).
165. C.E. Sauers, *J. Org. Chem.* **21** (1969) no. 2, 189.
166. D. Kumar, *Chem. Ind.* **21** (1981) no. 2.

167. C.P. Reghumadhan Nair, T.V. Sebastian, S.K. Nema, K.V.C. Rao, *J. Polym. Sci., Part A, Polym. Chem. Ed.* **24** (1986) 1109.
168. L.C. Cessna, H.H. Jablomer, *31st SPE. Technol. Conf.* 1973, p. 57.
169. Hughes Aircraft Co., US3845018, 1974 (N. Bilow, A.L. Landis, L.J. Miller).
170. A.L. Landis, N. Bilow, R.H. Boschan, R.E. Lawrence, T.J. Aponyl, *Polym. Prep.* **15** (1974) no. 2, 533.
171. N. Bilow, A.L. Landis, *Polym. Prep.* **19** (1978) no. 2, 23.
172. N. Bilow, L.B. Keller, A.L. Landis, R.H. Boschen, A.A. Castillo, *Sci. Adv. Mater. Proc. Eng. Ser.* **23** (1978) 791.
173. A.O. Hardy, T.L. St. Clair, *Sci. Adv. Mater. Proc. Eng. Ser.* **28** (1983) 711.
174. F.W. Harris, A. Pamidimukka, R. Gupta, S. Das, T. Wu, G. Mock, *Polym. Prep.* **24** (1983) no. 2, 324.
175. J.K. Stille, F.W. Harris, H. Mukamal, R.O. Rakutis, C.L. Schilling, G.K. Norem, I.A. Reeds, *Am. Chem. Soc., Adv. Chem. Ser.* **91** (1969) 628.
176. R.A. Kirchhoff, K.J. Bruza, *Progress Polym. Sci.* **18** (1993) 85.
177. Union Carbide Corp., US2890206, US2890207, 1959 (E.A. Kraiman).
178. F.W. Harris, S.O. Norris, *J. Polym. Sc., Polym. Chem. Ed.* **11** (1973) 2143.
179. K.J. Bruza, K.A. Bell, M.T. Bishop, E.P. Woo, *Polym. Prep.* **35** (1994) no. 1, 373.
180. P.O. Towney, R.H. Snyder, R.P. Conger, K.A. Leibbrand, C.H. Stitteler, A.B. Williams, *J. Org. Chem.* **26** (1961) 15.
181. Z.M.O. Razev, *Prog. Polym. Sci.* **25** (2000) 163.
182. C. Hulubei, S. Morario, *High Perform. Polym.* **12** (2000) no. 4, 525.
183. M. Bergen, A. Combet, P. Grosjean, BP1190718, 1970.
184. C.M. Tung, C.L. Long, T.T. Liar, *Polym. Mater. Sci. Eng.* **52** (1983) 138.
185. Toray Industries Inc., US3669930, 1972 (T. Asahara, N. Yoda, N. Minami).
186. T.M. Donnellan, D. Roylance, *Polym. Eng. Sci.* **32** (1996) no. 6, 409.
187. A.V. Tungare, G.C. Martin, *Polym. Eng. Sci.* **33** (1993) no. 10, 614.
188. T.T. Serafini, P. Delvigs, G.R. Lightsey, *J. Appl. Polym. Sci.* **16** (1972) 905.
189. W. Vaucracynest, J.K. Stille, *Macromolecules* **13** (1980) 1361.
190. T. Takekoshi, J.M. Terry, *Polymer* **35** (1994) no. 22, 4875.
191. C. Gouri, C.P. Reghumadhan Nair, R. Ramaswarmy, *High Perform. Polym.* **12** (2000) 497.
192. H. Furutami, J. Ida, K. Tanaka, H. Nagano, *High Perform. Polym.* **12** (2000) no. 4, 461.
193. E.I. DuPont de Nemours & Co., US4154737, 1979 (G.G. Orphanides).
194. H.D. Stenzenberger, K.U. Heinen, D.O. Hummel, *J. Polym. Sci., Polym. Chem. Ed.* **14** (1976) 2911.
195. M.A. Chandhari, *32nd Int. SAMPE Symp. Exhib.* **32**, 597 (1987).
196. E.A. Burns, H.R. Lubowitz, J.F. Johns, NASA CR72460, NASA, Washington, D.C. 1968.
197. M.J. Marks, D.C. Scott, B.R. Guilbeaux, S.E. Bales, *J. Polym. Sci., Part A: Polym. Chem.* **35** (1997) 385.
198. H.R. Lubowitz, *Org. Coat. Plast. Chem.* **31** (1971) no. 1, 561.
199. R.W. Vaughan, J.F. Jones, H.R. Lubowitz, *15th Natl. SAMPE Technol. Conf.* **15** (1969) 59.
200. R.H. Pater, *Polym. Eng. Sci.* **31** (1991) no. 1, 20.
201. K.C. Chuang, J.E. Waters, D. Hardy-Green, *42nd Int. SAMPE Symp. Expo. Exhib.* **42** (1997) no. 2, 1283.
202. M.A. Meador, M.A.B. Meador, J. Petkovsek, T Oliver, D.A. Scheiman, C. Gariepy, B. Nguyen, R.K. Eby, *46nd Int. SAMPE Symp. Expo. Exhib.* **46** (2001) no. 1, 497.
203. P. Deligs, D L. Kloptek, P.J. Cavano, *High Perform. Polym.* **6** (1994) 209.
204. P. Deligs, D.L. Kloptek, P.J. Cavano, *High Perform. Polym.* **9** (1997) 161.
205. A.C. Wang, M.M. Ritchey, *Macromolecules* **14** (1981) 825.
206. D. Wilson, *Brit. Polym. J.* **20** (1988) no. 5, 405.
207. TRW Inc., US3528950, 1970 (H.R. Lubowitz).
208. P.R. Young, NASA TM-83192, NASA, Washington, D.C. 1981.
209. R.D. Vanucci, W.B. Alston, NASA TMX-71816, NASA, Washington, D.C. 1976.
210. T.L. St. Clair, R.A. Jewell, *23rd Natl. SAMPE Symp. Exhib.* **23** (1978) 320.
211. R.H. Pater: "Rheological, processing, and 371 deg C mechanical properties of Celion 6000/N-phenylnadimide modified PMR composites", NASA Lewis Res. Center, NASA, Washington, D.C. 1983, pp. 23.
212. G.D. Roberts, R.W. Lauver, *J. Appl. Polym. Sci.* **33** (1987) no. 8, 2893.
213. A. Lee, *High Perform. Polym.* **8** (1996) 475.
214. D. Dean, M.A. Abdalla, U. Vaidya, R. Ganguli, C.J. Battle, M. Abdalla, A. Hauqe, S. Campbell, *High Perform. Polym.* **17** (2005) no. 4, 497.
215. J.Y. Hao, A.J. Hu, S.Y. Yang, *High Perform. Polym.* **14** (2002) 325.
216. C.D. Simone, Y. Xiao, X.D. Sun, D.A. Scola, *High Perform. Polym.* **17** (2005) 51.
217. W. Xie, W.-P. Pan, K.C. Chuang, *Thermochimica Acta* **143** (2001) 367–368.
218. S. Alan, L.D. Kandpul, I.K. Varma, *J. Macromol. Sci. – Rev. Macromol. Chem. Phys. C* **33** (1993) no. 3, 291.
219. M.D. Sefcik, E.O. Stejskal, R.A. McKay, J. Shaefer, *Macromolecules* **12** (1979) 423.
220. H.D. Stenzenberger, P. Konig, M. Herzog, W. Römer, S. Pierce, M. Camming, *18th Int. SAMPE Technol. Conf.* **18** (1986) 500.
221. A.L. Landis, A.B. Naselow, *Natl. SAMPE Technol. Conf. Ser.* **14** (1982) 236.
222. J.A. Johnston, F.M. Li, F.W. Harris, T. Takekoshi, *Polymer* **35** (1994) no. 22, 4865.
223. R.G. Bryant, P.M. Hergenrother, *High Perform. Polym., Comm.* **7** (1995) 121.
224. R.G. Bryant, B.J. Jensen, P.M. Hergenrother, *Polym. Prep.* **34** (1993) no. 1, 566.
225. P.M. Hergenrother, R.G. Bryant, B.J. Jensen, S.J. Havens, *J. Polym. Sci., Part A* **32** (1994) 3061.
226. P.M. Hergenrother, J.G. Smith, Jr., *Polymer* **35** (1994) no. 22, 4857.
227. P.M. Hergenrother, R.G. Bryant, B.J. Jensen, J.G. Smith, Jr., S. P. Wilkinson, *39th Int. SAMPE Symp. Exhib.* **39** (1994) no. 1, 961.
228. NASA, US5412066, 1995 (P.M. Hergenrother, R.G. Bryant, B. J. Jensen, S.J. Havens).
229. J.W. Connell, J.G. Smith, Jr., P.M. Hergenrother, J.M. Criss, *High Perform. Polym.* **15** (2003) 375.
230. R. Yokota, S. Yamamoto, S. Yano, T. Sawaguchi, M. Hasegawa, H. Yamaguchi, H. Ozawa, R. Sato, *High Perform. Polym.* **13** (2001) 861.
231. A. Mochizuki, T. Tadashi, M. Ueda, *Polymer* **35** (1994) no. 18, 4022.
232. W.X. Huang, S.L. Wunder, *J. Appl. Polym. Sci.* **59** (1996) 511.
233. E.I. DuPont de Nemours & Co., NL6413552, 1965.

234. Y. Imai, *J. Polym. Sci. B* **8** (1970) 555.
235. M.E. Rogers, D. Rodrigues, G.L. Wilkes, J.E. McGrath, *Polym. Prep.* **32** (1991) no. 1, 176.
236. Z.Y. Wang, T.P. Bender, H.B. Zheng, L.Z. Chen, *Polym. Adv. Technol.* **11** (2000) 652.
237. W.J. Farrissey, J.S. Rose, P.S. Carleton, *Polym. Prep.* **9** (1968) 1581.
238. R.A. Meyers, *J. Polym. Sci., Part A-1* **7** (1969) 2757.
239. P.S. Carleton, W.J. Farrissey, J.S. Rose, *J. Appl. Polym. Sci.* **16** (1972) 2983.
240. N.D. Ghatge, U.P. Mulik, *J. Polym. Sci., Polym. Chem. Ed.* **18** (1980) 1905.
241. G.D. Khune, *J Macromol Sci., Chem. A* **14** (1980) no. 5, 687.
242. UpJohn Co., US3708458, 1973 (L.M. Alberino, W.J. Farrissey, J.S. Rose).
243. UpJohn Co., US3562189, 1971 (W.J. Farrissey, A. McLaughlin, J.S. Rose).
244. Bayer Corp., US3445477, 1969 (G. Mueller, R. Merten).
245. UpJohn Co., US4001186, 1977 (B.K. Onder).
246. 3M Corp.: Technical Data Sheets, Chap. 5413, 5433, 1994 and 5419, St. Paul, MN 1995.
247. R.E. Southward, D.S. Thompson, D.W. Thompson, A.K. St. Clair, *Chem. Mater.* **9** (1997) 1691.
248. R.E. Southward, D.W. Thompson, A.K. St. Clair, *Polym. Prep.* **39** (1998) no. 1, 423.
249. R.E. Southward and D.M. Stoakley, *Prog. Org. Coat.* **41** (2000) 99.
250. R.E. Southward, J. Dean, J.L. Scott, S.T. Broadwater, D.W. Thompson, *Polym. Mater.: Sci. Eng.* **84** (2001) 339.
251. D.S. Thompson, L.M. Davis, D.W. Thompson, R.E. Southward, *ACS Appl. Mater. Interfaces* **1** (2009) no. 7, 1457.
252. G.A. Gaddy, E.P. Locke, M.E. Miller, R. Broughton, T.E. Altbrecht-Schmitt, G. Mills, *J. Phys. Chem. B.* **108** (2004) no. 45, 17378.
253. K. Akamatsu, S. Ikeda, H. Nawafune, *Langmuir* **19** (2003) 10366.
254. Y.-J. Lee, J.-M. Huang, S.-W. Kuo, F.-C. Chang, *Polymer* **46** (2005) 10056.
255. K.H.M. Lindeberg, *Sensors and Actuators A Physical* **105** (2003) 150.
256. T.K. Minton, M.E. Wright, S.J. Tomczak, S.A. Marquez, L. Shen, A.L. Brunsvold, R. Cooper, J. Zhang, V. Vij, A.J. Guenthner, B.J. Petteys, *ACS Appl. Mater. Interfaces* **4** (2012) 492–502.
257. S. Devasahayam, D.J.T. Hill, J.W. Connell, *High Perform. Polym.* **17** (2005) no. 4, 547.
258. K. Yokota, S. Abe, M. Tagawa, M. Iwata, E. Miyazaki, J-I. Ishizawa, Y. Kimoto, R. Yokota, *High Perform. Polym.* **22** (2010) no. 2, 237.
259. V.K. Agarwal, *IEEE Trans. Dielectr. Electr. Insul.* **24** (1989) no. 5, 741.
260. F. Li, Z.-Q. Wu, E.P. Savitski, X. Jing, Q. Fu, F.W. Harris, S.Z. D. Cheng, R.E. Lyon, *42nd Intl. SAMPE Symp. Exhib.* **42** (1997) no. 2, 1306.
261. T.J. Stueber and C. Mundson, *38th Int. SAMPE Symp. Exhib.* **38** (1993) no. 1, 641.
262. P.R. Young, A.K. St. Clair, W. Slemp, *38th Intl. SAMPE Symp. Exhib.* **38** (1993) no. 1, 664.
263. F. Hoertz, D. Koenig, J. Hanson, M.V. Eesbeek, *High Perform. Polym.* **13** (2001) S517.
264. A.R. Frederickson, C.E. Benson, J.F. Bockman, *Nucl. Instr. Meth. Phys. Res. B.* **208** (2003) 454.
265. M.-H. Kim, S.A. Thibeault, J.W. Wilson, L.C. Simonsen, L. Heilbrom, K. Chang, R.L. Kiefer, J.A. Weakley, H.G. Maahs, *High Perform. Polym.* **12** (2000) 13.
266. C.C. Fay, D.M. Stoakley, A.K. St Clair, *High Perform. Polym.* **11** (1999) 145.
267. C.O.A. Semprimoschnig, S. Heltzel, A. Polsak, M.V. Eesbeek, *High Perform. Polym.* **16** (2004) 207.
268. K. Yokota, N. Ohmae, M. Tagawa, *High Perform. Polym.* **16** (2004) 221.
269. J.A. Dever, R. Messer, C. Powers, J. Townsend, E. Woodridge, *High Perform. Polym.* **13** (2001) S391.
270. S. Devasahayam, D.J.T. Hill, J.W. Connell, *High Perform. Polym.* **17** (2005) 547.
271. J.W. Connell, K.A. Watson, *High Perform. Polym.* **13** (2001) 23.
272. M.A. Alexander, D.J.T. Hill, J.W. Connell, K.A. Watson, *High Perform. Polym.* **13** (2001) 55.
273. S.I. Semenova, *J. Membrane Sci.* **231** (2004) 189.
274. Ube Industries Ltd., US4512893, 1985 (H. Makino).
275. W.J. Koros et al., *Prog. Polym. Sci.* **13** (1988) 339.
276. W.J. Koros, D.R.B. Walker, *Polym. J.* **23** (1991) no. 5, 481.
277. K. Matsumoto, K. Ishii, T. Kuroda, K. Inoue, A. Iwama, *Polym. J.* **23** (1991) no. 5, 491.
278. Y. Li, C. Cao, T.-S. Chung, K.P. Pramoda, *J. Membrane Sci.* **245** (2004) 53.
279. K.C. Khulbe, C.Y. Feng, F. Hamad, T. Matsura, M. Khayer, *J. Membrane Sci.* **245** (2004) 53.
280. D. Punsalan, W.J. Koros, *J. Appl. Polym. Sci.* **96** (2005) 1115.
281. M. Niva, H. Kawakami, M. Kanno, S. Nagaoka, T. Kanamori, T. Shinbo, S. Kubota, *J. Biomater. Sci., Polym. Ed.* **12** (2001) no. 5, 533.
282. J. Fang, X. Gou, S. Harada, T. Watari, K. Tanaka, H. Kita, K. Okamoto, *Macromolecules* **35** (2002) 9022.
283. N. Asano, K. Miyatake, M. Watanabe, *Chem. Mater.* **16** (2004) no. 15, 2841.
284. D.M. Sullivan, M.L. Bruening, *J. Am. Chem. Soc.* **123** (2001) 11805.
285. T.W.F. Baeuml, *J. Chromatogr. A.* **961** (2002) 35.
286. M.B.J. Balla, *J. Chromatogr. A.* **299** (1984)139.
287. J. Blomberg, P.J. Schoenmakers, U.A.T. Brinkman, *J. Chromatogr. A.* **972** (2002) 137.
288. R. Roth, *Medical Dev. Diag. Ind.* **23** (2001) no. 1, 102.
289. M. Eashoo, Z. Wu, A. Zhang, D. Shen, C. Tse, F.W. Harris, S.Z. D. Cheng, *Macromol. Chem. Phys.* **199** (1994) 2207.
290. S.K. Park, R.J. Farris, *Polymer* **42** (2001) no. 26, 10087.
291. Q.-H. Zhang, M. Dai, M.-X. Ding, D.-J. Chen, L.-X. Gao, *Euro. Polym. J.* **40** (2004) 2487.
292. J.-H. Zhuang, K. Kimura, C. Xia, Y. Yamashita, *High Perform. Polym.* **17** (2005) 35.
293. F.-Y. Hshieh, D.B. Hirsch, H.D. Beneson, *Fire Mater.* **27** (2003) 119.
294. Imi-tech Corp., US4535101, 1985 (R. Lee, U.A.K. Sorathia, D. A. Okey).
295. Imi-tech Corp., US4539342, US4535099, US4562112, 1985 (R. Lee, G.A. Ferro, D.A. Okey).
296. Sorrento Eng. Co., US4980102, 1990 (F.U. Hill).
297. Imi-tech Corp., US4900761, 1990 (R. Lee, M.D. O'Donnell).
298. W.J. Farrissey, Jr., J.S. Rose, P.S. Carleton, *J. Appl. Polym. Sci.* **14**, 1093 (1970).
299. T.-H. Hou, E.S. Weiser, E.J. Siochi, T.L. St Clair, *High Perform. Polym.* **16** (2004) 487.
300. C.J. Cano, E.S. Weiser, R.B. Pipes, *Cell. Polym.* **23** (2004) no. 5, 299.
301. E.W. Weiser, T.F. Johnson, T.L. St. Clair, Y. Echigo, H. Kaneshiro, B.W. Grimsley, *High Perform. Polym.* **12** (2000) 1.
302. M.K. Williams, G.L. Nelson, J.R. Brenner, E.S. Weiser, T.L. St Clair in G.L. Nelson and C.A. Wilkie (eds.): "Fire and

303. A. Kuwabara, M. Ozasa, T. Shimokawa, N. Watanabe, K. Nomoto, *Adv. Compos. Mater.* **14** (2005) no. 4, 343.
304. M.B. Meador, S. Wright, A. Sandberg, B.N. Nguyen, F.W. Van Keuls, C.H. Mueller, R. Rodríguez-Solís, F.A. Miranda, *ACS Appl. Mater. Interfaces* **4** (2012) 6346–6353.
305. M.B. Meador, E.J. Malow, R. Silva, S. Wright, D. Quade, S.L. Vivod, H. Guo, J. Guo, M. Cakmak, *ACS Appl. Mater. Interfaces* **4** (2012) 536–544.
306. L.M. Nicholson, K.S. Whitley, T.S. Gates, J.A. Hinkley, *J. Mater. Sci.* **35** (2000) 6111.
307. L.M. Nicholson, K.S. Whitley, T.S. Gates, *Mech. Time-Depend. Mater.* **5** (2001) 199.
308. E.J. Siochi, P.R. Young, R.G. Bryant, *40th Int. SAMPE Symp. Exhib.* **40** (1995) no. 1, 11.
309. T.H. Hou, E.J. Soichi, N.J. Johnston, T.L. St. Clair, *Polymer* **35** (1994) 4956.
310. T.H. Hou, N.J. Johnston, T.L. St. Clair, *High Perform. Polym.* **7** (1995) 104.
311. B.J. Jensen, T.H. Hou, S.P. Wilkinson, *High Perform. Polym.* **7** (1995) 11.
312. T.H. Hou, R.G. Bryant, *High Perform. Polym.* **8** (1996) 169.
313. T.H. Hou, R.G. Bryant, *High Perform. Polym.* **9** (1997) 437.
314. A.P. Deshpande, J.C. Seferis, *J. Thermoplast. Compos. Mater.* **12** (1999) 498.
315. M. Namkung, R.G. Bryant, B. Wincheski, A. Buchman, *J. Appl. Phys.* **81** (1997) no. 8, 4112.
316. M. Namkung, R.G. Bryant, B. Wincheski, A. Buchman, *J. Appl. Phys.* **83** (1998) no. 11, 6474.
317. A. Buchman, R.G. Bryant, *Appl. Comp. Mater.* **6** (1999) 309.
318. L.J. Laursen, M.T. Bishop, D.R. Calhoun, S.A. Laman, W.W. Lee, S.R. Wood, P.C. Yang, *39th Int. SAMPE Symp. Exhib.* **39** (1994) no. 1, 951.
319. D. Wilson in D. Wilson, H.D. Stenzenberger, P.M. Hergenrother (eds.): *Polyimides*, Blackie, London and Chapman & Hall Inc., New York 1990, Chap. 2.
320. B.J. Jensen, S.E. Lowther, A.C. Chang, *Polym. Prep.* **38** (1999) no. 2, 305.
321. D.G. Hawthorne, J.H. Hodgkin, T.C. Morton, *High Perform. Polym.* **11** (1999) 15.
322. L.M. Nicholson, T.S. Gates, *J. Thermoplast. Compos. Mater.* **14** (2001) 477.
323. T.A. Bullions, J.M. Jungk, A.C. Loos, J.E. McGrath, *J. Thermoplast. Compos. Mater.* **13** (2000) 460.
324. J.W. Connell, J.G. Smith. Jr., P.M. Hergenrother, J.M. Criss, *High Perform. Polym.* **15** (2003) 375.
325. R.J. Cano, S. Ghose, K.A. Watson, P.B. Chunchu, B.J. Jensen, J.W. Connell in S. Van Hoa, P. Hubert (eds.): *Proceedings of the International Conference on Composite Materials (ICCM-19)*, Curran Associates, Red Hook 2013, pp. 6795.
326. E.I. DuPont de Nemours & Co., US3413394, 1968 (T.F. Jordan).
327. http://www.dupont.com/products-and-services/plastics-polymers-resins/parts-shapes/brands/vespel-polyimide.html (accessed March 22nd, 2014).
328. http://www.ube.es/specialty/producto.asp?id=159&marcarr=159 (accessed March 20th, 2014).
329. J.W. Carr, C. Feger, *Prec. Eng.* **15** (1993) no. 4, 221.
330. T. Kobayashi, A. Nakao, M. Iwaki, *Nucl. Instr. Meth. Phys. Res. B.* **306** (2003) 1110.
331. K.G. Budinski, *Wear* **203–204**, (1977) 302.
332. M.J. Devine, A.E. Kroll, *Lubr. Eng.* **20** (1964) no. 6, 223.
333. Y.H.M. Kitoh, *Thin Solid Films* **271** (1995) 92.
334. P. Samyn, P. De Baets, G. Schoukens, B. Hendrickx, *Polym. Eng. Sci.* **43** (2003) no. 9, 1477.
335. P. Samyn, J. Quintelier, P. De Baets, G. Schoukens, *Materials Letters* **59** (2005) 2850.
336. A. Tanaka, K. Umeda, S. Takatsu, *Wear* 257, 1096 (2004).
337. Z.H. Tan, Q. Guo, Z.P. Zhao, H.B. Liu, L.X. Wang, *Wear* **271** (2011) nos. 9 and 10, 2269.
338. N.L. Hancox, *Materials and Design* **12** (1991) no. 6, 317.
339. S.P. Wilkinson, J.M. Marchello, D. Dixon, N.J. Johnston, *38th Intl. SAMPE Symp. and Exhib.* **38** (1993) no. 1, 1.
340. D.R. Tenney, J.G. Davis, Jr., N.J. Johnston, R B. Pipes, J.F. McGuire, NASA/CR–2011-217076, NASA Langley Research Center, Hampton 2011.
341. M. Gentz, B. Benedikt, J.K. Sutter, M. Kumosa, *Compos. Sci. Technol.* **64**, 1671.
342. A. Ramasamy, Y. Wang, J. Muzzy, *38th Intl. SAMPE Symp. Exhib.* **38** (2), 1993, 1882–1891.
343. B.J. Jensen, R.G. Bryant, J.G. Smith, Jr., P.M. Herrgenrother, *J. Adhesion* **54** (1995) 57.
344. J.P. Nunes, J.F. Silva, A.T. Marques, N. Crainic, S. Cabral-Fonseca, *J. Thermoplast. Compos. Mater.* **16** (2003) 231.
345. E.C. Millard, *Int. J. Adhes. Adhes.* **4** (1994) no. 4, 171.
346. S.J. Shaw, *Polymer International* **41** (1996) 193.
347. N. Furukawa, Y. Yamda, Y. Kimura, *High Perform. Polym.* **9** (1997) 17.
348. B.J. Jensen, T.H. Hou, S.P. Wilkinson, *High Perform. Polym.* **7** (1995) 11.
349. T.H. Hou, B.J. Jensen, P.M. Hergenrother, *J. Comp. Mater.* **30** (1996) no. 1, 109.
350. T.H. Hou, S.P. Wilkinson, N.J. Johnston, R.H. Pater, T.L. Schneider, *High Perform. Polym.* **8** (1996) no. 4, 491.
351. J.G. Smith, Jr., H.L. Belvin, E.J. Soichi, R.J. Cano, N.J. Johnston, *High Perform. Polym.* **14** (2002) no. 2, 209.
352. J.-M. Bureau, J.-P Droguet in M.K. Ghosh, K.L. Mittal (eds.): *Polyimides, Fundamentals and Applications*, Marcel Dekker, New York 1996, pp. 23–744.
353. S. Fujimura, K. Komo, K. Hikazutani, Y. Yano, *Jpn. J. Appl. Phys.* **28**(10), 2310.
354. T.O. Herndon, R.L. Burke, J.A. Yasaitis, *Solid State Technol.* **27** (1989) no. 11, 179 (1984).
355. N. Yoda, H. Hiramoto, *J. Macromol. Sci.-Chem. A* **21** (1984) nos. 13 and 14, 1641.
356. J.-S. Li, Z.-B. Li, P.-K. Zhu, *J. Appl. Polym Sci.* **77** (2005) 943.
357. S.L.-C. Hsu, M.H. Fan, *Polymer* **45** (2004) 1101.
358. A. Zhang, X. Li, C. Nah, K. Hwang, M.-H. Lee, *J. Polym. Sci., Part A, Polym. Chem. Ed.* **41** (2003) 22.
359. C. Liu, X. Zhao, Y. Li, D. Yang, L. Wang, L. Jin, C. Chen, H. Zhou, *High Perform. Polym.* **25** (2013) no. 3, 301.
360. S.L.-C. Hsu, P.-I. Lee, J.-S. King, J.-L. Jeng, *J. Appl. Polym. Sci.* **98** (2005) 15.
361. X.Z. Jin, H. Ishii, *J. Appl. Polym. Sci.* **15** (2005) 67.
362. L. Lu, A. Sun, M. Jikei, M.-A. Kakimoto, *High Perform. Polym.* **15** (2003) 67.
363. K.H. Choi, J.C. Jung, K.S. Kim, J.B. Kim, *Polym. Adv. Technol.* **16** (2005) 367.

Polymethacrylates

Klaus Albrecht, Evonik Industries AG, Darmstadt, Germany

Manfred Stickler, Röhm GmbH, Darmstadt, Germany

Thomas Rhein, Röhm GmbH, Darmstadt, Germany

1.	Introduction	885	5.1.1.	Standard Molding Compounds	892
2.	Properties	885	5.1.2.	Heat-Resistant Molding Compounds	893
3.	Raw Materials	888	5.1.3.	Impact-Resistant Molding Compounds	894
4.	Production Processes	888			
4.1.	Bulk Polymerization	889	5.1.4.	Molding Compounds for Special Optical Uses	894
4.1.1.	Casting of Acrylic Glass	889			
4.1.2.	Continuous Production of Cast Acrylic Glass	890	5.2.	Semifinished Products	895
			5.2.1.	Extruded Acrylic Glass	895
4.1.3.	Tube Polymerization	890	5.2.2.	Cast Acrylic Glass	895
4.2.	Suspension Polymerization	891	5.3.	Binders and Paints	896
4.3.	Continuous Polymerization of Molding Compounds	891	5.4.	Additives for the Petroleum Industry	896
4.4.	Batchwise Solution Polymerization	892	5.5.	Miscellaneous Uses	896
4.5.	Emulsion Polymerization	892	6.	Toxicology	897
5.	Uses	892	7.	Environmental Aspects	897
5.1.	Molding Compounds	892		References	897

1. Introduction

Poly(methyl methacrylate) [9011-14-7] (PMMA) has been produced industrially since the early 1930s. The history and development of this polymer are very closely associated with the chemist Otto Röhm [1, 2]. In 1932, Röhm and his collaborator W. Bauer were the first to polymerize methyl methacrylate [36426-74-1] (MMA) into transparent sheets [3]. Almost parallel to these developments at Röhm & Haas in Germany, corresponding activities were in progress at Rohm & Haas and Du Pont in the United States and at ICI in England. A few years later, the first processable thermoplastic molding compounds based on PMMA were available.

PMMA occupies an intermediate position between "commodities" and "engineering plastics"; its market has in the meantime developed greatly. In 2010, the PMMA production worldwide exceeded 1.6×10^6 t/a with roughly 60% of the market in Asia and 20% each in NAFTA and Europe [4]. Acrylic glass and molding compounds amount to a consumption of about 50% of produced MMA; the other half is used for the synthesis of other monomers, dispersions, paints, binders, adhesives, additives, etc.

PMMA is the main representative of this polymer class; polymers and copolymers of higher alkyl methacrylates are of minor economic importance.

2. Properties

The properties that make PMMA an attractive polymeric material are its good mechanical strength, its outstanding optical properties (clarity, brilliance, transparency), and its extremely good weather resistance [5, 6]. This favorable profile can be supplemented by the surface modification of semifinished products (e.g., scratch-resistant or antistatic coating), coloring or pigmenting, and treatment with flame retardants to open up a wide variety of potential uses for this polymer (see Chap. 5).

Table 1. Physical properties of PMMA (standard values at 23 °C and 50% R.H. for Plexiglas GS 233 and Plexiglas XT 7H)

Property	Cast (extruded)[*]	Test procedure
Density, g/cm^3	1.19	DIN 53 479
Charpy impact strength, kJ/m^2	15	ISO 179/1 D
Izod-test value, kJ/m^2	1.6	ISO 180/1 A
Tensile strength (1/1 test bar 3, $v = 5$ mm/min), MPa	80 (72)	DIN 53 455
Elongation at break (1/1 test bar 3, $v = 5$ mm/min), %	5.5 (4.5)	DIN 53 455
Flexural strength (standard bar 80 × 10 × 4 mm), MPa	115 (105)	DIN 53 452
Long-term tensile strength (1/1 test bar 3, $t = 10\,000$ h), MPa	38 (30)	DIN 53 444
Modulus of elasticity (short-term value), MPa	3300	DIN 53 457
Shear modulus (ca. 10 Hz), MPa	1700	DIN 53 445
Ball indentation hardness $H_{961/30}$, MPa	200 (190)	DIN-ISO 2039–1
Degree of transmission (3 mm thickness, illumination D 65, visible region, $\lambda = 380–780$ nm), %	92	DIN 5036
Refractive index n_D^{20}	1.491	DIN 53 491
Linear thermal expansion coefficient (0–50 °C), K^{-1}	70×10^{-6}	DIN 53 752-A
Shrinkage onset temperature, °C	> 80	
Vicat softening point (method B), °C	115 (102)	DIN-ISO 306
Heat distortion temperature (at flexural stress of 1.8 MPa), °C	105 (90)	ISO 75
Water absorption (compared with dry state after 24 h, specimen 50 × 50 × 4 mm), mg	30	DIN 53 495

[*] Extruded from molding compound.

Pure PMMA is an amorphous polymer, which can exhibit a high surface gloss, high brilliance, and crystal-clear transparency. It is classified as a hard, rigid, but brittle material with a maximum use temperature of just below 100 °C. The tensile, compressive, and flexural strengths are considered satisfactory; scratch resistance is also good and can be improved further by the use of special coatings. The most important physical properties of PMMA are summarized in Table 1; a distinction is made between cast PMMA ($M_r > 10^6$) and PMMA extruded from molding compounds ($M_r < 5 \times 10^5$). The properties listed in Table 1 were determined according to the so-called basic value table [7]; further data can be found in the CAMPUS database [8, 9]. This data base also includes information about commercial copolymers of MMA with small proportions of acrylate esters (e.g., < 10% methyl acrylate). Incorporation of acrylate allows the rheological properties of the thermoplastics to be varied within a wide range to satisfy the processing requirements during injection molding and extrusion.

PMMA is resistant to aliphatic hydrocarbons, nonpolar solvents, aqueous alkalis, and aqueous acids. Aromatic hydrocarbons, chlorinated hydrocarbons, esters, ketones, and other polar solvents attack PMMA, causing it to dissolve or swell. Whereas alcohols and water alone are regarded as nonsolvents, their mixtures may have a solubilizing effect (cosolvency). Detailed information on solvents and precipitating agents for PMMA can be found in [10]. Stress cracking is an important failure mode when molded parts are being exposed to poor solvents, such as ethyl acetate or nonsolvents, such as methanol.

Most physical properties E of polymers depend on the molecular mass. For example, the viscosity of a polymer melt increases continuously with the molecular mass. Some other properties (e.g., glass transition temperature, density, refractive index) show an asymptotic behavior, which can be described by the following equation:

$$E = E_\infty + \frac{k}{M_n^-}$$

For sufficiently high number-average molecular masses M_n^-, a polymer-specific limiting value E_∞ is accordingly reached (typically at ca. $M_n^- > 50000$); the constant k is influenced by the terminal groups of the polymer.

Table 2 summarizes some physical properties of the most important polymethacrylates. The values were obtained at sufficiently high molecular mass (i.e., E_∞), so that the influence of the alcohol residue of the ester group is apparent.

The glass transition temperatures, T_g, decrease sharply with the length of the linear, aliphatic side chain. Whereas PMMA, the first member of this homologous series, is a hard, rather rigid material, the polymers with residues larger than n-hexyl are rubberlike. The glass transition temperature decreases until poly(n-dodecyl methacrylate). With longer linear ester residues, side-chain crystallization occurs,

Table 2. Physical properties of the most important polymethacrylates

Polymer	CAS registry no.	ρ (20 °C), g/cm^3	n_D^{20}	T_g, °C	δ, (J/cm^3)$^{1/2}$	References
Poly(methyl methacrylate)	[9011-14-7]	1.190	1.490	125	18.6	[11, 12, 15]
Poly(ethyl methacrylate)	[9003-42-3]	1.119	1.485	66	18.3	[11]
Poly(n-propyl methacrylate)	[25609-74-9]	1.085	1.484	35	18.0	[11]
Poly(isopropyl methacrylate)	[26655-94-7]	1.033	1.4728	81		[11, 12]
Poly(n-butyl methacrylate)	[9003-63-8]	1.058	1.483	27	18.0	[11]
Poly(isobutyl methacrylate)	[9011-15-8]	1.045	1.477	53	17.7	[11]
Poly(sec-butyl methacrylate)	[29356-88-5]	1.052	1.480	62		[11, 12]
Poly(tert-butyl methacrylate)	[25189-00-8]	1.022	1.4638	122	17.0	[11, 12]
Poly(n-hexyl methacrylate)	[25087-17-6]	1.007*	1.481	−5	17.6	[11]
Poly(cyclohexyl methacrylate)	[25768-50-7]	1.100	1.5065	92		[11, 12]
Poly(n-octyl methacrylate)	[25087-18-7]	0.971*		−20	17.2	[11]
Poly(2-ethylhexyl methacrylate)	[25719-51-1]			−10		[11]
Poly(n-decyl methacrylate)	[29320-53-4]			−70		[13]
Poly(dodecyl methacrylate)	[25719-52-2]	0.929*	1.4740**	−55	16.8	[11]
Poly(hexadecyl methacrylate)	[25986-80-5]		1.4750**			[11]
Poly(octadecyl methacrylate)	[25639-21-8]			−100	16.0	[11, 15]
Poly(bornyl methacrylate)			1.5059			[11]
Poly(isobornyl methacrylate)	[28854-39-9]	1.060	1.500	110	16.6	[11, 15]
Poly(menthyl methacrylate)	[30847-50-8]	1.010	1.4890	90		[11]
Poly(phenyl methacrylate)	[25189-01-9]	1.210	1.5715	120		[11]
Poly(1-naphthyl methacrylate)	[31547-85-0]		1.6410	75		[11]
Poly(benzyl methacrylate)	[25085-83-0]	1.179	1.5680	54	20.3	[11]
Poly(2-hydroxyethyl methacrylate)	[25249-16-5]		1.5119	55		[11]
Poly(ethylthioethyl methacrylate)	[27273-87-6]		1.5300	−20		[15]
Poly(furfuryl methacrylate)	[29320-19-2]	1.2603	1.5381	60		[11, 12]
Poly(tetrahydrofurfuryl methacrylate)	[25035-85-2]		1.5096	60		[11]
Poly(3,3,5-trimethylcyclohexyl methacrylate)	[75673-26-6]		1.485	79		[11]
Poly(tricyclodecyl methacrylate)				150		[14]
Poly(2,2,3,3-tetrafluoropropyl methacrylate)	[29991-77-3]	1.4934	1.4215	64		[12]

*At 25 °C.
**n_D^{30}.

which masks the glass transition. Polymers with linear alkyl ester residues always have a lower glass transition temperature than their branched and alicyclic isomers. Large bulky residues considerably restrict the mobility of the polymer main chain, particularly if they are joined directly to the ester group.

When discussing the data in Table 2 one has to keep in mind that the properties of a polymer, in particular the glass transition temperature, also depend on its tacticity. If a tactic polymer is regarded as a terpolymer of isotactic, syndiotactic, and heterotactic triads, then its glass transition temperature can be described by the following linear combination:

$$T_g = x_{ii}T_{g,i} + x_{is}T_{g,h} + x_{ss}T_{g,s}$$

where $T_{g,i}$, $T_{g,h}$, and $T_{g,s}$ denote the glass transition temperatures of the (imaginary) polymers constructed purely from isotactic, heterotactic, or syndiotactic triads, respectively, and x_{ii}, x_{is}, and x_{ss} denote the associated mole fractions of these triads in the relevant real polymer. In the case of PMMA, the values $T_{g,i} = 42$ °C, $T_{g,h} = 139$ °C, and $T_{g,s} = 124$ °C were determined by fitting experimental T_g values to x_{ii}, x_{is}, and x_{ss} mole fractions as obtained from NMR analysis [16]; data for other polymethacrylates can be found in the literature [15]. The mole fractions x_{ii}, x_{is}, and x_{ss} depend on the polymerization temperature and the polymerization mechanism (and are possibly influenced by any solvent used in the polymerization). These relationships have been investigated in detail in the case of PMMA [17].

Not all the influences on the glass transition temperature described above have been taken into account in Table 2. Most of the listed T_g values were measured on atactic polymers synthesized by radical polymerization. This explains to some extent the considerable differences researchers find in one and the same polymethacrylate. In addition, variations also arise due to the use of different methods to measure T_g (dilatometry, differential scanning calorimetry) and evaluation procedures (e.g., "midpoint" and "on-set" evaluation in differential scanning calorimetry).

The presence of low molecular mass molecules, such as residual monomer in the polymers represents a large source of errors in the determination of T_g; even relatively small proportions can lead to considerable reduction of the glass transition temperature. Bulk synthesis of polymethacrylates with bulky side groups (e.g., isopropyl, *tert*-butyl, or cyclohexyl methacrylate) is especially critical because even in the case of a very sharp end polymerization they yield products that still contain a few percent of residual monomers [18]. The values listed in Table 2 were obtained on reprecipitated, carefully dried material and are considerably higher than previous literature values.

Polymers used as materials for optical application require good transmission behavior; the refractive index and Abbe number are other important parameters. Polymethacrylates cover a considerable range of refractive indices; plastics with a particularly low refractive index have been in demand for polymer optic fibers (POF). Fluorinated polymethacrylates are particularly suitable in this connection. A summary of relevant data (T_g, n_D^{20}) is given in [19].

The influence of tacticity on the refractive index and density should also be taken into account, but up to now, little information has been available.

3. Raw Materials

The starting material for PMMA is the monomer methyl methacrylate (MMA). One important route to this monomer (as well as to methacrylic acid) is methanolysis (or hydrolysis) of methacrylamide sulfate. The latter is produced from acetone and hydrogen cyanide via the intermediate acetone cyanohydrin. Alternative synthesis pathways have also been adopted: the C-4 route involves two-stage oxidation of isobutene (or *tert*-butanol) to methacrylic acid and MMA. In the C-3 route, propene is reacted with carbon monoxide to give isobutyric acid, from which methacrylic acid is then obtained by oxidative dehydrogenation. The C-2 pathway involves condensation of propanal (or propionic acid) with formaldehyde to methacrolein (or methacrylic acid). Propanal can be obtained from ethylene by hydroformylation. New routes include the fermentation of sugar by modified *E. Coli* bacteria where 3-hydroxyisobutyric acid is an important intermediate. For further details of the synthesis of methacrylic acid and MMA, see → Methacrylic Acid and Derivatives.

MMA for polymerization is usually available with a purity of >99.9%; impurities are mainly water, methanol, methyl formate, and traces of acid.

MMA is stabilized for storage or transportation with phenol derivatives (e.g., hydroquinone or hydroquinone monomethyl ether) in amounts of 10–100 ppm; the stabilizing action of these compounds requires the presence of oxygen. Nitrogen oxides or aromatic nitro compounds can be used in the absence of oxygen. MMA should be stored under exclusion of light and at a temperature as low as possible (<10°C).

To obtain higher methacrylate esters, methacrylic acid is esterified with the relevant alcohols. Longer-chain methacrylate esters are also synthesized by transesterification of MMA.

4. Production Processes

Methyl methacrylate can be polymerized by a free radical mechanism [20, 21] or an anionic mechanism [22–25]. Anionic polymerization is of little industrial importance because the monomer has to be subjected to extensive purification and polymerization has to be performed at low temperatures. The group transfer polymerization technique (GTP) is mechanistically very similar to anionic polymerization, and is less complicated [26, 27], but has only found little applications. Atom transfer radical polymerization (ATRP) as a form of living radical polymerization is gaining industrial importance [28].

Radical polymerization of MMA may be carried out homogeneously (bulk or solution polymerization) or heterogeneously (suspension or emulsion polymerization). It can be initiated with radiation (light, γ-radiation), heat, or radical-forming chemical agents. Spontaneous thermal initiation of MMA polymerization is extremely slow [29] and is therefore of no industrial relevance. Photoinitiation of MMA polymerization can be performed without sensitizers, but photochemically labile compounds are generally added. On exposure to light these compounds form radicals either directly (e.g., azo compounds, disulfides, derivatives of benzoin or acetophenone) or are converted into an excited state and then form radicals after abstracting a hydrogen atom from the monomer or solvent (e.g., derivatives of benzophenone or anthraquinone) [30–34]. Initiation of polymerization by γ-rays (e.g., ^{60}Co source) may be of interest if the addition of a thermal initiator is undesirable and a conventional light source cannot be used because the polymerization batch absorbs too strongly (e.g., colored solutions, porous materials, such as wood or stone impregnated with monomer [35]).

The polymerization of MMA is most commonly initiated by using thermal radical-forming agents (i.e., initiators). Azo compounds [36, 37] and organic peroxy compounds [37] (diacyl peroxides, dialkyl peroxides, alkyl peresters, peroxy dicarbonates, perketals, alkylhydroperoxides) are often employed. A wide range of initiators is available so that an initiator can be selected with an appropriate half-life for the relevant polymerization temperature. Redox systems also play an important role in the formation of initiating radicals, particularly in polymerizations in aqueous solution [38].

The polymerization of MMA up to high conversion (particularly in bulk polymerization) is governed by the gel (Trommsdorff) effect [39, 40]. The extremely large increase in the viscosity of the reaction medium during polymerization hinders diffusion of the polymer radicals considerably. Thus, bimolecular chain termination becomes increasingly difficult and the concentration of the growing radicals increases by several orders of magnitude. As a result, the polymerization rate and mean molecular mass increase sharply [41], leading to a further increase in the viscosity of the reaction batch. Under these circumstances, removal of the heat of polymerization (57.7 kJ/mol for MMA [42]) may become a serious problem or even a potential hazard in a batch bulk reaction as it is done for the production of cast sheet.

The considerable decrease in volume associated with polymerization of the monomer may create additional problems. For the bulk polymerization of MMA shrinkage is 25.6% at 60 °C and 100% conversion. Shrinkage decreases with an increase in size of the alcohol residue of methacrylate.

The method selected for the industrial polymerization of MMA (e.g., bulk, suspension, emulsion, or solution polymerization) is generally governed by the form in which the polymer is finally used.

4.1. Bulk Polymerization

Bulk polymerization can be carried out to produce sheets, rods, and tubes of PMMA or can be used to produce molding compounds by extrusion or injection molding processes.

4.1.1. Casting of Acrylic Glass

The conventional casting process is still the most important method for producing high-quality acrylic glass with a mirror-smooth surface (e.g., Plexiglas GS, Perspex, Lucite, Altuglas, Shinkolite, Sumipex). Casting molds consist of two plane-parallel silicate glass plates (e.g., float glass with a typical length, width, and thickness of 3 000, 2 000, and 8–12 mm, respectively) that are held apart by an elastic spacer (gasket) and clamped together with metal clamps. The elastic gasket runs around the edge of the plates and is usually made of poly(vinyl chloride) or ethylene–vinyl acetate copolymer. It must have sufficient compressibility to allow for the shrinkage occurring during polymerization so that the acrylic sheet does not become detached from the glass plates. The thickness of the resulting sheets is between a few millimeters and up to 30 cm.

The reaction mixture consisting of monomer, initiator, and other additives (e.g., comonomers, cross-linking agents, stabilizers, flame retardants, dyes, or pigments) is introduced into the chamber through a small hole in the gasket. Care must be taken to avoid formation of air bubbles; if necessary, the solution should first be evacuated. A monomer–polymer solution (syrup) may be used instead of the monomer solution. This syrup contains 5–20% PMMA and can be obtained by polymerization of MMA in a stirred-tank reactor. The benefit is the reduced shrinkage and heat of polymerization in the chamber and the reduction of polymerization time. Furthermore, the viscous syrup helps to avoid pigment sedimentation and leakage from the chamber.

The contents of the sealed cells are polymerized in horizontal support racks mounted in hot-air ovens with a high air circulation rate, in autoclaves using spray water (90–100 °C), or in water tanks (20–60 °C) with circulating pumps. The residence time may range from a few hours to several days depending on the sheet thickness. However, conversion only reaches 80–90%, because the glasslike solidification of the reaction mixture that occurs during polymerization below the glass transition temperature of the pure polymer prevents further reaction (glass effect). Subsequent polymerization at 110–120 °C is therefore necessary to obtain maximum conversion [18, 44]. After cooling to room temperature, the polymer sheet is separated from the glass plates and can be removed manually or by handling robots. The glass plates can be reused.

Because the casting process is labor-intensive and time-consuming, alternative methods have been developed for more efficient production of cast PMMA glass. One approach is to stack several plane-parallel glass plates and spacers together. The simplest case is a double cell (three plates with two spacers). Constructions that are more complex have been known for a long time and include a process patented in 1933 in which several glass plates with intermediate seals are stacked in vertical rows. The plates are enclosed on both sides by heatable and coolable metal chambers through which a heat-transfer agent flows and are held together by a screw press [45, 46]. The heat-transfer agent flowing through the metal chambers allows accurate temperature control. Shrinkage is compensated for by compression of the packet, resulting in tighter thickness tolerances of the cast sheets.

In a special variant of this process, the entire apparatus is tightly encapsulated to facilitate degassing of the reaction batch by evacuation. The system is then pressurized to suppress the formation of boiling bubbles caused by local overheating during polymerization. The advantages of this method are the smaller space requirement and reduced workforce. A disadvantage is that the use of metal chambers limits the sheet sizes. The process is still used on a small scale in Europe (Rostero) and in the United States (Polycast Corp).

The production of molding compounds in a casting operation similar to the above described process but in cells made from metal an polyester film pouches has been used in the past but has been abandoned for cost reasons.

4.1.2. Continuous Production of Cast Acrylic Glass

Not all improvements in the conventional chamber process described in Section 4.1.1 can overcome the fact that it is a batch production with variation and cost limitations. The attempts to develop continuous processes have led to a double-belt process that allows efficient production of large series of sheets up to ca. 6 mm thick [47–50]. The MMA–PMMA syrup is continuously fed between two endless, highly polished steel belts (width up to 3 m, length ca. 250 m). The belts are sealed at the edge with flexible plastic hoses (e.g., of PVC) and guided at a rate of about 1–3 m/min over roller systems through several accurately controlled temperature zones. The continuous product sheet can either be wound on rollers or cut into rectangles. This process was first used in the United States by Swedlow and subsequently by Rohm & Haas, Du Pont, and Mitsubishi. The surface quality of the acrylic glass product depends solely on the quality of the polishing of the steel belt; the brilliance of acrylic glass produced by the cell polymerization process is rarely obtained. Polymerization in the double-belt process is faster than in the conventional casting process, resulting in somewhat lower molecular masses and therefore some minor deterioration of mechanical properties. Besides the production of high quality optical sheets the biggest application is in so-called solid surface materials, such as Corian (PMMA–aluminum hydroxide composite).

4.1.3. Tube Polymerization

Tubes of PMMA can be produced by extrusion from a molding compound, but also by a special type of bulk polymerization, i.e., centrifugal polymerization [51]. Monomer or syrup is fed into tubular molds (e.g., of stainless steel) that are rotated during polymerization. A film is produced on the internal wall of the pipe by centrifugal forces and polymerizes to form pipes of specific diameter (13–650 mm) and wall thickness (1–10 mm). This technique is extremely labor-intensive and requires high-precision apparatus. Centrifugal polymerization is therefore seldom used on an industrial scale.

4.2. Suspension Polymerization

PMMA can be produced in the form of small beads by suspension polymerization [52, 53]. The monomer (with or without comonomer) is stirred with about twice its volume of water and dispersants (distributors). Droplets of monomer are dispersed in the water and are stabilized by dispersing aids preventing the droplets from coalescing. The droplets become increasingly viscous during the course of polymerization. Gelatin, cellulose derivatives, synthetic, water-soluble polymers [e.g., poly(vinyl alcohol) or alkali salts of (co)polymers of methacrylic acid] [54], and finely dispersed inorganic substances (kaolin, magnesium carbonate, precipitated aluminum hydroxide) have proved suitable as distributors [55, 56]. The particle size can be controlled within wide limits (50–1000 µm) by choosing appropriate stirring conditions and distributor systems [57].

Suspension polymerization is preferably initiated by adding radical-forming agents, whose distribution equilibrium between the aqueous and the monomer phases should be in favor of the monomer. This also applies to other polymerization auxiliaries, such as chain-transfer agents. The mechanism of suspension polymerization corresponds to that of bulk polymerization. One of the main advantages of suspension polymerization is the ease of heat removal because of an extremely high surface to volume ratio of the polymerizing phase and the high specific heat of water compared to that of the monomer. This permits higher polymerization rates and thus higher space–time yields than in typical batch bulk polymerizations. Another advantage is the substantially lower viscosity of the suspension and the simple isolation of the polymer by filtration.

After the completion of polymerization, the polymer beads are separated from the aqueous phase, washed with water to remove adhering distributor and salts, and dried.

4.3. Continuous Polymerization of Molding Compounds

While in the production of cast acrylic glass continuous processes play a minor role compared with the batch casting process (see Section 4.1.2). the polymerization processes for the production of PMMA used as injection molding and extrusion material are, however, mostly continuous processes. These molding compounds often contain a small proportion of acrylate ester as comonomer (up to ca. 20%) and have a relatively low mass average molecular mass (100 000–200 000) when compared to cast acrylic glass.

The oldest continuous process for producing molding compounds was similar to the double-belt process described in Section 4.1.2 [58, 59]: syrup is metered into an endless hose of sheet material and polymerized by passage through a water bath (ca. 75 °C) and an annealing oven (ca. 110 °C). In a plant with a total length of ca. 50 m the residence time was about 1 h.

Today, there are two main types of processes for the production of thermoplastic PMMA. One type uses the monomer as solvent for the polymerization process in a continuous stirred-tank reactor. Depending on the reaction control scheme, reaction times vary between 45 min and several hours. The typical temperatures are always significantly above T_g and in the range of 130–170 °C. The conversions reach 40–80%. The highly viscous polymer is fed into flash chambers with gear pumps or extruders for degassing of unconverted monomer, which is recovered and reused after refining. In older processes, the aim was to achieve maximum polymerization in the extruder [60, 61]. These techniques are, however, relatively complex and the high polymerization temperatures cause problems with the ceiling equilibrium [42] and thus result in a lower efficiency. More economic processes [62–69] dispense with high conversion in the extruder. The second type of process uses an inert solvent, such as toluene or butyl acetate to reduce the viscosity of the reaction mixture at higher conversions. The polymer separation is carried out in the same way as described above.

The polymer melt can be extruded and pelletized. In order to produce colored pellets, dyes or pigments are mixed with the polymer melt before it is fed into the degassing extruder. Further additives, such as UV-absorbers, lubricants, and stabilizers are added either to the melt or to the stirred-tank reactor.

To avoid fouling of the reactors, it is necessary to operate them hydraulically filled because MMA has a strong tendency to form deposits on reactor walls [70].

4.4. Batchwise Solution Polymerization

The continuous two-stage polymerization of MMA described in Section 4.3 can be carried out both in bulk and in solution. Batchwise solution polymerization of MMA and in particular of longer-chain methacrylates is also of interest, especially if the resultant polymer solutions constitute the form in which the polymer is used. Typical examples include adhesives and paint resins (see also → Adhesives, 1. General; → Paints and Coatings, 2. Types, Chap. 5; → Polyacrylates) or additives for the mineral oil sector. Criteria for the choice of solvents include price, solvent quality, toxicity, flammability, boiling point, and the chain-transfer action during polymerization. Solution polymerization is performed batchwise or with a continuous feed. In order to remove the heat of polymerization, it may be advantageous to operate under evaporative cooling conditions.

4.5. Emulsion Polymerization

Methacrylates play an important role as comonomers for acrylates (see → Polyacrylates) and other vinyl compounds (e.g., vinyl acetate) in the production of aqueous plastics dispersions, which are used as paint resins, paper coating agents, textile binders, paper processing agents, and auxiliaries for leather treatment (→ Leather, Section 8.5). Colloidal dispersion and film-forming properties can be widely varied by adding small amounts of methacrylic acid, methacrylamide, or its methylol or methylol ether derivatives. However, dispersions of pure polymethacrylates do not form films.

Emulsion polymers with a core–shell structure that contain methacrylates as monomers have become important. Such emulsion polymers are used as impact-resistant modifiers [71] for PMMA [72–74] or other thermoplastics [75, 76]. The polymers can be isolated as solids from dispersion by spray drying, chemical or freeze coagulation, or direct coagulation in an extruder [77].

5. Uses

The price of PMMA polymers depends significantly on the cost of the monomer and therefore is somewhat higher than that of other plastic materials. Thus, the materials are used where their specific advantages including low density, resistance to fracture, and, above all, excellent optical properties (transparency, gloss) and weather resistance play a role. PMMA also has favorable processing properties (especially good thermoforming) and can be modified in many ways (e.g., coloring, pigmenting, or surface treatment with scratch-resistant coatings).

PMMA has a wide range of applications, particularly in glazing, roofing, vehicles, machinery, and instruments. Pipelines and vessels made of acrylic glass are also of interest for the storage and transportation of food due to their transparency and lack of smell and taste. Optical components, such as lenses, reflectors, prisms, and spectacle lenses are also made from polymethacrylates. The advent of LCD flat panel displays has become a dominant application with demands for PMMA light guides illuminated with LEDs.

5.1. Molding Compounds

In comparison to the cast sheet products, which can only be mechanically formed, the application and uses of PMMA molding compounds are much more versatile. The thermoplastically deformable polymers can be processed by injection molding and extrusion. The molecular masses (M_r) of PMMA molding compounds are in the range $1-2 \times 10^5$. The recommended processing temperatures are generally between 200 and 280 °C.

5.1.1. Standard Molding Compounds

Standard molding compounds consist of MMA homopolymers and copolymers containing up to 20% acrylate esters. The comonomer permits variation of the softening temperature and thus the melt flow rate of the polymers.

Standard molding compounds are processed by injection molding. As a fully amorphous polymer, PMMA shows very little shrinkage during the molding process. Typical uses include precision parts, such as dials, rulers, and other drawing instruments; optical components, such as lenses or spectacles; plastic optical fibers (→ Fiber Optics, Section 3.2.3); and domestic and household articles. Colored injection-molded parts (e.g., covers for automobile rear lights) represent a very important market. The development of LED edge-lit light guides to illuminate large flat panel displays has increased the demand. The absence of chromophores together with an unsurpassed light transmission is extremely important. These sheets are being extruded and injection molded in dimensions of up to 53.3 cm diagonal. During injection molding a surface pattern can be embossed for control of the light outcoupling, which is achieved on extruded sheet mainly by screen printing.

Standard PMMA molding compounds are also extruded on a large scale into flat sheets (width up to 3 m, thickness up to 25 mm), sinusoidally profiled sheets and rods, as well as hollow shapes, multiwall-sheets, and tubes (up to 30 cm diameter). If necessary the extruded sheets are also subjected to surface treatment by coating or coextrusion (e.g., scratch-resistant, antistatic, or nondrip water spreading coating or antialgae growth). After cutting to size and covering with a protective masking film (e.g., polyethylene) they are marketed as semifinished products.

Trade names of standard PMMA molding materials include Acrylic/Acrycal (Continental Polymers); Acrylite (Evonik CYRO); Acrypet, Shinkolite (Mitsubishi); ; Delpet (Asahi); Diakon (Lucite); Oroglas (Altuglas, in the US: Plexiglas); Plexiglas (Evonik); Lucite (Lucite).

The worldwide market for PMMA standard molding compounds is ca. 500 000 t/a, of which Western Europe and the United States each account for ca. 90 000 t/a.

5.1.2. Heat-Resistant Molding Compounds

A limiting factor in use is that the long-term service temperatures of articles made of standard molding compounds are considerably below 100 °C. Heat-resistant molding compounds can have service temperatures of up to 150 °C. They consist of copolymers of MMA with monomers that stiffen the polymer chains. Examples of such comonomers include maleic anhydride [78] (trade names of copolymers: Acrypet ST, Mitsubishi; Oroglas HT 121 Altuglas; Delpet 980 N, Asahi; Plexiglas hw 55, Evonik) or also methacrylates with bulky ester groups, such as tricyclodecyl methacrylate [14] (Optorez OZ 1000, Hitachi Chemical).

Chain stiffening can also be achieved by polymer-analogous reaction of the functional groups in the PMMA-(co)polymer. For example, adjacent acid groups of a copolymer of methyl methacrylate and methacrylic acid are converted into polymethacrylic anhydride structures under the influence of a strong base [79]. The resulting polymers have Vicat softening temperatures of up to 150 °C, depending on the degree of cyclization. These products have only reached a niche status due to their relatively high cost.

Poly(N-methyl methacrylimide) [119499-71-7] (PMMI), also known as poly(N-methylglutarimide), a thermoplastic molding compound, is obtained by reacting PMMA with methylamine, → Polyimides, Chap. 5, [80–82].

PMMI, marketed under the trade names Kamax (Rohm & Haas), or PMI resin (Mitsubishi Rayon) is now only produced by Pleximid (Evonik). The product has a favorable combination of optical properties, light resistance, and stiffness (modulus of elasticity, ca. 4.8×10^3 N/mm^2). The Vicat softening temperature of PMMI can be selectively varied via the degree of imidation and exceeds 170 °C in highly imidized products.

The imidation reaction can be performed in the melt or in solution. Other primary amines besides methylamine can be used. In addition to the glutarimide units resulting from the cyclization and unreacted methyl methacrylate groups, PMMI may also contain varying proportions (depending on the reaction conditions) of structural elements derived from methacrylic acid and methacrylic anhydride.

The market volume for heat-resistant methacrylate molding compounds is estimated at ca. 2 000 t/a worldwide and remains limited. The most important customers for heat-resistant molding compounds are the automobile and lighting industries. In combination with high intensity LED lighting, there is an increasing interest in such polymers.

5.1.3. Impact-Resistant Molding Compounds

Impact-resistant PMMA molding compounds are multiphase polymer blends based on a soft elastomer phase dispersed in the rigid PMMA matrix. The elastomer phase may consist of EPDM rubbers, poly(butyl acrylate), polybutadiene, and, in rare cases, polysiloxanes. The choice of rubber is especially important for the notched impact resistance at low temperature and weathering performance. Most widely used are acrylic rubbers [poly(butyl acrylate], which have a good weather resistance.

The bonding of the disperse elastomer phase to the rigid matrix and its morphology play an important role in determining the mechanical properties of impact-resistant thermoplastics [71]. If the impact-resistant thermoplastics are to be transparent, the refractive indices of the rigid and soft phases must be identical (as long as the domain size of the elastomer phase does not remain below a critical value, e.g., significantly below the wavelength of visible light).

Various methods can be used to produce impact-resistant PMMA: in older processes, a suitable type of rubber was grafted with MMA (and comonomers to adjust the refractive index) by bulk or solution polymerization. Increasing conversion leads to phase inversion, whereby the elastomer phase, which frequently includes PMMA (or, in the presence of comonomers, the corresponding copolymer), becomes the disperse phase. This desired morphology and the size of the elastomer particles can be controlled by the viscosity of the reaction medium and the shear forces resulting from stirring. They are then fixed by cross-linking when conversion is nearly complete. Graft polymerization by two-stage suspension polymerization is carried out on an industrial scale, too [52].

Emulsion polymerization is particularly effective if particle size and particle morphology have to be selectively modified (see Section 4.5). Impact-resistance modifiers with a core–shell structure have therefore been developed for PMMA [72–74].

Block copolymers of acrylates and methacrylates have been introduced into the market by Altuglas (Nanostrength) and Kurary (Kurarity). Their claimed advantage is that the transparency does not vary with the temperature due to the domain size of dispersed rubber [83, 84].

Worldwide demand for impact-resistant PMMA is estimated to be ca. 65 000 t/a. Products in which butadiene constitutes the tough phase include methacrylate–butadiene–styrene (MBS) [25053-09-2] (e.g., Plexalloy-F-PAB, Evonik; Cyrolite G 20, Evonik-CYRO) and methacrylate–acrylonitrile–butadiene–styrene (MABS) [9010-94-0] (e.g., Terluran, BASF; Cyrolite XT Polymer, CYRO). PMMA modified with butyl acrylate is commercially available under the trade names Delpet SR (Asahi), Diakon TD (Lucite), Oroglas DR (Altuglas), Plexiglas zk (Röhm), and Shinkolite (Mitsubishi Rayon Corp.).

5.1.4. Molding Compounds for Special Optical Uses

PMMA is mainly used because of its very good weather resistance, outstanding optical quality, hardness and good processability (e.g., by injection molding). Optical components (e.g., lenses, prisms, mirrors) represent an important area of use for PMMA [85, 86]. The relatively high water absorption of PMMA (ca. 2% when immersed in water) causes dimensional changes and change in refractive index. This disadvantage can be overcome by using cycloaliphatic methacrylates, such as cyclohexylmethyl methacrylate [87] or tricyclodecyl methacrylate [14] as comonomers. Hydrophilic methacrylates such as poly(2-hydroxyethyl methacrylate) are used in soft contact lenses [88, 89].

The excellent optical properties of PMMA (especially its low orientation birefringence) seem to have favored its use as a substrate for optical data storage media. However, in asymmetric, single-sided audio compact discs the relatively high water absorption of the polymer and the associated warpage may detract from these advantages [90, 91]. This problem did not arise with the two-sided video disc; PMMA molding materials were therefore used for these products [92]. See also → Information Storage Materials.

Extremely pure PMMA is used as plastic optic fibers (POF) [93, 94] → Fiber Optics, Section 3.2.3. The theoretical limiting values of transmission loss of PMMA in the 650 and 567 nm minima of the damping spectrum are 100 and 27 dB/km, respectively [95]. Limits of 5–6 dB/km at 680 and 650 nm can be achieved with special deuterated or fluorinated methacrylates [96]. Fiber to the home (FTTH) projects in Asia and United States drive the demand for POFs, which are only made by three Japanese companies in significant quantities. There are a number of Chinese manufacturers trying to enter this market as well. They also produce POF for decorative purposes.

PMMA is also employed in lithographic materials in semiconductor technology. Use as a high-resolution positive resist is based on the depolymerization of this polymer by radiation, see → Imaging Technology. The irradiated regions are accurately dissolved out with suitable solvents. Sensitivity to electron beams and UV light can be increased further by copolymerization with suitable monomers [97, 98].

5.2. Semifinished Products

5.2.1. Extruded Acrylic Glass

A substantial proportion of standard PMMA molding compounds (and in the United States particularly impact-resistant PMMA) is extruded into semifinished products (sheets, multiwall-sheets, corrugated profile sheets, tubes) [99]. Reference has already been made to possible surface modifications of these extrudates (see Section 5.1.1). Cast acrylic glass with a matt surface requires matt glass for production and is difficult to produce; extruded semi-finished products with matt surface structures are therefore an interesting alternative. It would be desirable if such products could also be thermoplastically formed without losing their matt structure. This can be accomplished by adding highly cross-linked bead polymer to the extrusion compound in total or only to a coextrusion layer. These materials show interesting light diffusing properties and are being marketed for application in furniture and lighting. One advantage is their reduced sensitivity to scratching.

Besides extruded semifinished products, illuminated advertising signs and glazing (including lighting domes) dominate the European market with a volume is ca. 70 000 t/a. Trade names include Acrylite FF (Evonik CYRO), Altuglas (Altuglas), Perspex TX (Lucite), Plexiglas XT and Plexiglas SDP (Evonik). The backlight unit (BLU) has become the single biggest application for extruded sheet in Asia. With tens of millions of TV sets being manufactured, the capacities for sheet grew dramatically in Asia making Asia the dominating region for PMMA consumption.

5.2.2. Cast Acrylic Glass

The substantially more economic production of extruded acrylic glass has led to a considerable decline in the production of cast semifinished products. Nevertheless, cast material has retained its position in applications with stringent requirements with regard to surface quality and optical properties. For example, cast acrylic glass has kept a significant position in the sanitary ware sector (e.g., baths, shower units, and washbasins). Due to certain deficiencies revealed by the hot water alternating test (DIN EN 198), (co)extruded products have not found broad acceptance in this sector.

Another application of cast acrylic glass (optionally with flame retardants) is in noise protection walls on motorways. This market sector grows due to the increasing motorization all around the globe. Large sheets ($\geq 2 \times 4$ m) are needed to prevent drivers from being disturbed by the frames used for mounting. Very large size thick-wall (up to 120 mm) aquarium glazing is another domain of cast acrylic.

Semispherical lenses for touristic submarine is another interesting application.

Aviation glazing, one of the oldest applications of cast acrylic glass, has become increasingly important because of the enormous increase in air traffic. Acrylic glass has the advantages of low mass and outstanding resistance to extreme weather conditions (high UV radiation intensities at high altitudes). It receives the required strength by biaxially stretching [100]. This step also increases dramatically the resistance to crack growth. The extremely stringent optical requirements [e.g., freedom from optical distortion, low deflection, lack of internal defects (e.g., bubbles, inclusions), lack of surface defects (e.g., pimples)] predestine cast acrylic glass for aviation glazing. Aviation standards prohibit the use of extruded material.

The total market volume for cast acrylic glass in Europe is ca. 53 000 t/a. Important trade names and manufacturers include Acrylite (CYRO); Altuglas, Casocryl, Paralinx (Kyowa Gas); Perspex (Lucite); Plexiglas GS (Röhm); Plexiglas (Altuglas); Shinkolite (Mitsubishi); Sumipex (Sumitomo Chemical).

5.3. Binders and Paints

Methacrylates are used in large amounts as comonomers in the production of acrylate–methacrylate copolymers for binders and paint resins (→ Paints and Coatings, 2. Types, Chap. 5). Methacrylate is used to obtain the desired glass transition temperature.

The use of such copolymers in the form of solution polymers, aqueous dispersions, or solid products extends over a wide field (e.g., weather-resistant paints, textile binders, paper coating materials, and adhesives); further details are given elsewhere, see → Polyacrylates.

5.4. Additives for the Petroleum Industry

Long-chain poly(alkyl methacrylates), PAMA, are used as additives in the petroleum sector for many purposes [101–104]. The best-known application is probably their use as viscosity index improvers in engine and gear oils (→ Lubricants, 2. Components, Section 3.2). At low temperatures viscosity index improvers increase the viscosity of the mixture only slightly. At high temperatures, however, they compensate for the sharp decrease in viscosity of the base oil by virtue of their intrinsically high contribution to the solution viscosity of the whole system [105, 106].

This is because the mineral oil becomes an increasingly better solvent for PAMA at higher temperatures and the polymer coils expand greatly. Viscosity index improvers thus prevent destruction of the lubricating oil film and resultant damage to the engine. Viscosity index improvers are generally constituents of an extremely complex additive formulation and interact with other components not only in improving flow behavior, but also in cleaning the engine, dispersing dirt and contaminants, and preventing oxidation of the oil. This dispersing effect can be achieved by incorporating polar comonomers (e.g., N-vinylpyrrolidone) into the PAMA chain.

PAMA-based additives also act as pourpoint depressants (→ Lubricants, 2. Components, Section 3.3). The pour point denotes the temperature at which oil solidifies as a result of precipitation of paraffin. At concentrations of less than 1%, PAMAs hinder the growth of paraffin crystals and thus prevent the formation of aggregates of crystal needles. The oil remains free-flowing and the oil pump is able to function during cold starting.

5.5. Miscellaneous Uses

Polymethacrylates are also used in many other applications.

Most poly(vinyl chloride) processing auxiliaries consist of high molecular mass polymers based on MMA. They are generally marketed in powder form as spray-dried emulsion polymers [107]. Important examples of such products include Diakon APA (Lucite), Kane ACE (Kanegafuchi Chemical), Metablen (Mitsubishi Rayon), and Plastistrength (Altuglas) Paraloid K (Dow Chemical).

Methacrylate reaction resins (monomer–polymer systems) are used in the building industry in mortars and concretes, and in synthetic floor screeds [108, 109].

Alkali-soluble suspension polymers, e.g., Rohagit (Evonik), are used as thickeners for aqueous dispersions and cosmetics, as sizings and finishes in the textile industry, and as dispersion and stabilizing auxiliaries in suspension polymerization. Cation-exchange resins can be obtained by hydrolysis of PMMA, thus producing weakly acidic carboxylic functions (→ Ion Exchangers).

Polymethacrylates are used as dental materials for fillings, false teeth, and prostheses [110], and in surgery as bone cement for implants. Examples of such products include Kallodoc (ICI) and Plexidon (Evonik).

Coatings of polymethacrylate copolymers, such as Eudragit E-30 D [9010-88-2] or

Eudragit RS [39316-06-8] (Evonik) are used as galenic auxiliaries for controlled-release tablets [111, 112].

6. Toxicology

No toxic effects of PMMA have been described, and the polymer is regarded as inert and nontoxic. In individual cases, however, contact of the mucous membranes with polymer powders or dusts may cause allergic and irritating reactions. Skin reactions (presumably due to residual monomer) and asthma have also been noted [113, 114].

Because PMMA is used as bone cement, its potential carcinogenic effect has been monitored in animal implant tests. Malignant tumors were found in some cases at the implant site, but also occurred with inert materials of the same shape. Such tumors are therefore most likely to be foreign-body carcinoma resulting from mechanical irritation. Tumors have not yet been observed in humans after many years of use [113].

7. Environmental Aspects

In contrast to most other polymers, pure PMMA can be almost quantitatively depolymerized to the monomer at high temperatures (ca. 450 °C). Conversion of production waste and waste from semifinished product blanks to MMA by cracking had already been investigated at the start of large-scale industrial production in the 1930s [115, 116]. In these earlier processes, the polymer was mixed with sand and depolymerized batchwise in retorts or with superheated steam (to avoid local overheating and thus uncontrolled pyrolysis). Continuous processes now operate with an indirectly heated fluidized bed and are substantially more economical [117–119].

In 2010, the first fully transparent blends of PMMA with PLA were commercialized following a trend to build chemical products on renewable resources [120].

References

1 H. von Pechmann, O. Röhm, *Ber. Dtsch. Chem. Ges.* **34** (1901) 427.
2 E. Trommsdorff: *Dr. Otto Röhm, Chemiker und Unternehmer*, Econ, Düsseldorf 1976.
3 Röhm & Haas, DE 724 229, 1932 (O. Röhm, W. Bauer).
4 *Kunststoffe* **10** (2011) 70.
5 G. Schreyer, *Angew. Makromol. Chem.* **11** (1970) 159.
6 G.W. Ehrenstein: *Beständigkeit von Kunststoffen*, Carl Hanser Verlag, München 2007.
7 K. Oberbach, *Kunststoffe* **79**, (1989) 713.
8 J. Schmitz, E. Bornschlegl, G. Dupp, G. Erhard, *Plastverarbeiter* **39/4** (1988) 50.
9 F. Johannaber, W. Michaeli: *Handbuch Spritzgießen*, Carl Hanser Verlag, München 2004.
10 O. Fuchs: *Gnamm-Fuchs: Lösungs- und Weichmachungsmittel*, 3 vols., Wissenschaftl. Verlagsgesellschaft, Stuttgart 1980.
11 O.G. Lewis: *Physical Constants of Linear Homopolymers*, Springer Verlag, Berlin 1968.
12 J. Brandrup, E. H. Immergut, E. A. Grulke (eds): *Polymer Handbook*, 4th ed., John Wiley & Sons, Hoboken, NJ 2003.
13 *Kirk-Othmer*, 3rd ed., 15, 377.
14 Perspex Acrylic Sheet, ICI Bulletin.
15 H. Kawai, F. Kanega, H. Kohkame, *Proc. SPIE*, **896** (1988) 68.
16 M. Stickler, unpublished results.
17 M.A. Müller, M. Stickler, *Makromol. Chem., Rapid Commun.* **7** (1986) 575.
18 W. Wunderlich, M. Stickler in B.M. Culbertson, J.E. McGrath (eds.): *Advances in Polymer Synthesis*, Plenum Publishing, New York 1985, p. 505.
19 B. Boutevin, Y. Pietrasanta: *Les Acrylates et Polyacrylates Fluorés Dérivés et Applications*, EREC, Puteaux 1988, p. 235.
20 G. Odian: *Principles of Polymerization*, John Wiley Sons, Inc, Hoboken, NJ 2004.
21 G.C. Eastmond in C.H. Bamford, C.F.H. Tipper (eds.): *Comprehensive Chemical Kinetics*, vol. 14 A, Elsevier, Amsterdam 1976, p. 196.
22 H. Yuki, K. Hatada, *Adv. Polym. Sci.* **31** (1979) 1.
23 A. H. E. Müller in J.E. McGrath (ed.): Anionic Polymerization, Kinetics, Mechanisms, and Synthesis, *ACS Symp. Ser.* **166** (1981) 441.
24 M. Morton: *Anionic Polymerization: Principles and Practice*, Academic Press, New York 1983.
25 M.G. Dhara, S. Sivaram: *Living Anionic Polymerization of Methyl Methacrylate*, VDM Verlag Dr. Müller, Saarbrücken 2010.
26 O.W. Webster et al., *J. Am. Chem. Soc.* **105** (1983) 5706.
27 W.R. Hertler, D.Y. Sogah, O.W. Webster, G.M. Cohen, *Macromolecules* **20** (1987) 1473.
28 L. Zhang, Q. Xu, J. Lu, X. Xia, L. Wang, *Europ. Polym. J.* **43** (2007) no 6, 2718.
29 M. Stickler, G. Meyerhoff, *Makromol. Chem.* **179** (1978) 2729.
30 G. Oster, N.-L. Yang, *Chem. Rev.* **68** (1968) 125.
31 S.S. Labana, *J. Macromol., Sci.-Revs. Macromol. Chem.* **C11** (1974) 299.
32 A. Ledwith, *J. Oil Colour Chem. Assoc.* **59** (1976) 157.
33 A. Ledwith, *Pure Appl. Chem.* **49** (1977) 431.
34 J.P. Fouassier, P. Jacques, D.J. Lougnot, T. Pilot, *Polym. Photochem.* **5** (1984) 57.
35 F. Gütlbauer, E. Proksch, H. Bildstein, *Österr. Chem. Z.* **67** (1966) 349.
36 P.S. Engel, *Chem. Rev.* **80** (1980) 99.
37 G. Moad, D.H. Solomon in G. Allen, J.C. Bevington (eds.): *Comprehensive Polymer Science* vol. 3, part I, Pergamon Press, Oxford 1989, p. 97.
38 C.H. Bamford in G. Allen, J.C. Bevington (eds.): *Comprehensive Polymer Science*, vol. 3, part I, Pergamon Press, Oxford 1989, p. 123.

39. G.P. Gladyshev, K.M. Gibov: *Polymerization at Advanced Degrees of Conversion*, Israel Program for Scientific Translations, Jerusalem 1970.
40. N.M. Bityurin, *Polym. Sci. U.S.S.R* **23** (1981) no 8, 1873.
41. K. F. O'Driscoll in G. Allen, J.C. Bevington (eds.): *Comprehensive Polymer Science*, vol. 3, part I, Pergamon Press, Oxford 1989, p. 161.
42. H. Sawada: *Thermodynamics of Polymerization*, Marcel Dekker, New York 1976.
43. D. Panke, M. Stickler, W. Wunderlich, *Makromol. Chem.* **184** (1983) 175.
44. M. Stickler, *Makromol. Chem.* **184** (1983) 2563.
45. Röhm & Haas, DE 659 469, 1933 (O. Röhm).
46. Röhm & Haas, DE 639 095, 1934 (W. Bauer).
47. ICI, US 2 500 728, 1945 (E.G. Williams).
48. Swedlow Incorp., BE 687 405, 1966 (C.J. Opel, P.H. Bottoms).
49. Swedlow Incorp., BE 687 406, 1966 (O.H. Heilsund).
50. Swedlow Incorp., US 3 376 371, 1968 (C.J. Opel, P.H. Bottoms).
51. Röhm & Haas, DE 673 394, 1936 (W. Bauer).
52. M. Munzer, E. Trommsdorff in C.E. Schildknecht (ed.): High Polymers, *Polymerization Processes*, vol. 29, J. Wiley & Sons, New York 1977.
53. *Winnacker-Küchler*, 4th ed., 6, 414.
54. ICI, GB 427 494, 1933 (J.W.C. Crawford, J. McGrath).
55. Röhm & Haas, DE 735 284, 1935 (O. Röhm, E. Trommsdorff).
56. Röhm & Haas, DE 747 596, 1935 (O. Röhm, E. Trommsdorff).
57. H. Hopf, H. Lüssi, P. Gerspacher, *Makromol. Chem.* **78** (1964) 37.
58. Degussa, DE 1 212 301, 1963 (E. Bäder, W. Unseld, G. Morlock).
59. Degussa, DE 1 923 393, 1969 (E. Bäder, H. Amann, W. Unseld).
60. ICI, DE-OS 1 253 459, 1967 (F. Bild, A. Jukes).
61. W.A. Mack, *Chem. Eng. (N.Y.)* **79** (1972) 99.
62. Dow Chemical, GB 715 666, 1952 (A. Handsen).
63. Monsanto, CA 493 550, 1953 (H.F. Park, R.M. Dickey).
64. ICI, GB 875 853, 1960 (F. Bild, A. William).
65. Amer. Cyanamid, US 3 252 950, 1966 (J. Schmitt, J. Terenzil).
66. Mitsubishi Rayon, DE-OS 2 341 318, 1973 (K. Shimada et al.).
67. Mitsubishi Rayon, DE-OS 2 504 417, 1975 (K. Shimada et al.).
68. Bayer AG, DE-OS 2 724 360, 1977 (F. Wingler et al.).
69. Mitsubishi Rayon, DE-OS 2 438 927, 1974 (T. Maeda, T. Nishizawa, M. Toda, Y. Iwaoka).
70. Sumitomo Chemical, EP 1642603A1, 2007 (H. Nakazawa, H. Takimoto).
71. C.B. Bucknall: *Toughened Plastics*, Applied Science Publ., London 1977.
72. Rohm & Haas, US 3 793 402, 1971 (F. Owens).
73. Röhm, EP 0 113 924, 1983 (N. Sütterlin et al.).
74. P.A. Lovell et al., *Proc. ACS Div. Polym. Mat. Sci. Eng.* **63** (1990) 583.
75. Röhm, EP 0 326 938, 1989 (J. Fischer, W. Siol, M. Munzer, T. Rhein).
76. Röhm, DE 3 743 199, 1987 (J. Fischer, W. Siol, M. Munzer, T. Rhein).
77. Röhm GmbH, WO1998/050212, 1998 (H. Vetter).
78. B.C. Trivedi, B.M. Culbertson: *Maleic Anhydride*, Plenum Press, New York 1982.
79. Sumitomo, EP 0 264 508, 1986 (Y. Kato et al.).
80. Röhm & Haas, DE 1 113 308, 1960 (G. Schröder, K. Tressmar, U. Baumann).
81. Röhm & Haas, DE 2 652 118, 1976 (R.M. Kopchik).
82. Mitsubishi Rayon, EP 0 200 530, 1986 (H. Anzai et al.).
83. Atofina, WO/2004/014926A2,(J.L. Couturier)
84. Kuraray Co Ltd., WO/2010/055798, 2010 (H. Oshima).
85. R.H. Waxler, D. Horowitz, A. Feldmann, *Appl. Opt.* **18** (1979) 101.
86. C. Hofmann, *Plaste Kautsch.* **36** (1989) 338.
87. G. Klepek, *F & M Feinwerktech. Meßtechn.* **96** (1988) 43.
88. H.K. Singer, E.C. Bellantoni, A.R. LeBoef in J.I. Kroschwitz (ed.): *Encyclopedia of Polymer Science and Engineering*, vol. 4, Wiley, New York 1985, p. 164.
89. T.B. Harvey, W.E. Meyers, L.M. Bowmann, *Polym. Mater. Sci. Eng.* **59** (1988) 73.
90. J. Hennig, *Jpn. J. Appl. Phys.* **26** (1987) 26 (Suppl.).
91. H. Ryn, K. Akagane, H. Kori, *CEER Chem. Econ. Eng. Rev.* **18** (1986) 26.
92. G. Kämpf, *Polym. Eng. Sci.* **27** (1987) 1421.
93. R.M. Glen, *Chemtronics* **1** (1986) 98.
94. C. Emslie, *J. Mater. Sci.* **23** (1988) 2281.
95. T. Kaino, M. Fujiki, K. Jinguji, *Rev. Electr. Commun. Lab.* **32** (1984) 478.
96. T. Kaino, Y. Katayama, *Polym. Eng. Sci.* **29** (1989) 1209.
97. R. Pethrick, *Progr. Rubber Plast. Technol.* **2** (1986) 1.
98. E. Reichmanis, L.F. Thompson, *Chem. Revs.* **89** (1989) 1273.
99. M. Buck, *Kunststoffe* **80** (1990) 1132.
100. Röhm GmbH, EP 0716100B1, 1999 (M. Krieg, C. Weber).
101. C.W. Georgi: *Motor Oils and Engine Lubrication*, Reinhold Publ. Co., New York 1950.
102. C.V. Smalheer, K. Smith: *Lubricant Additives*, The Lezius-Hiles, Cleveland 1967.
103. A. Schilling: *Motor Oils and Engine Lubrication*, 2nd ed., Scientific Publication, London 1968.
104. M.W. Ranney: *Lubricant Additives*, Noyes Data Corp., Park Ridge, N.Y. 1973.
105. Rohmax Additives, WO2012/013432, 2012 (B. Eisenberg).
106. Rohmax Additives, WO2009/07147, 2009 (T. Stoehr).
107. D.L. Dunkelberger, *J. Vinyl Technol.* **9** (1987) 173.
108. H.J. Peschke, *Zentralbl. Industriebau* **33** (1987) 10.
109. H. Schuhmann, *BmK (Bauen mit Kunststoffen)* **5** (1988) 5.
110. I.E. Ruyter, H. Oeysaed, *CRC Crit. Rev. Biocompat.* **4** (1988) 247.
111. R.K. Chang, C.H. Hsiao, J.R. Robinson, *Pharm. Technol.* **11** (1987) 56.
112. K.O.R. Lehmann, *Drugs Pharm. Sci.* **36** (1989) 153.
113. D. Henschler, *Gesundheitsschädliche Arbeitsstoffe, Toxikologisch-arbeitsmedizinische Begründungen von MAK-Werten*, VCH Verlagsgesellschaft, Weinheim, 1988.
114. B. Kennes, P. Garcia-Herreros, P. Dierckx, *Clin.Allerg.* **11** (1981) 49.
115. Röhm & Haas, DE 642 289, 1935 (P. Weisert).
116. Röhm & Haas, DE 729 730, 1940 (K.-Th. Kautter, W. Leitenberger).
117. H. Sinn, W. Kaminsky, J. Janning, *Angew. Chem.* **88** (1976) 737.
118. W. Kaminsky, J. Menzel, H. Sinn, *Conserv. Recycling* **1** (1977) 91.
119. W. Kaminsky, H. Sinn, J. Janning, *Chem. Ing. Tech.* **51** (1979) 419.
120. Arkema, WO2010/040955, 2010 (P. Gerard).

Further Reading

R.V. Slone: *Methacrylic Polymers*, Kirk Othmer Encyclopedia of Chemical Technology, 5th edition, John Wiley & Sons, Hoboken, NJ, online DOI: .

Polyoxyalkylenes

FREDERICK E. BAILEY†, Union Carbide Technical Center, Charleston, WV, United States

JOSEPH V. KOLESKE, Consultant, 1513 Brentwood Road, Charleston, WV, United States

1.	Introduction	899	5. Handling.	905
2.	Properties	899	6. Uses	906
3.	Production	903	7. Economic Aspects	907
4.	Toxicity	905	References.	908

1. Introduction

This article deals with polyoxyethylene, polyoxypropylene, and polytetrahydrofuran or poly(1,4-epoxybutane). Other names include poly(ethylene oxide), PEO; poly(propylene oxide), PPO; and poly(tetramethylene oxide), PTMO. Strictly speaking, the latter polymer is not derived from an alkene and thus is not a poly(alkylene oxide); however, it is included here, and the term poly(alkylene oxides) is used as a collective term for these three polymers. The poly(alkylene oxides) have an ether linkage in their main chain backbone and thus are polyethers. They are described by the repeat unit

$-[(CH_2)_x - CHR - O]-$

Poly(ethylene oxide):	R=H, x=1
Poly(propylene oxide):	R=CH_3, x=1
Poly(tetramethylene oxide):	R=H, x=3

Many details about the alkylene oxides and their polymers are given in [1].

2. Properties

Physical Properties. Typical physical properties of PEO, PPO, and PTMO are presented in Table 1. Except for PTMO, these polyethers can be linear or branched and are commercially available in a broad range of molecular masses from less than a few hundred to several million.

Low molecular mass polymers of ethylene oxide are clear, essentially colorless liquids and become partially crystalline solids at a molecular mass of ca. 800. As the molecular mass increases, they change from soft waxy solids to hard waxes. At high molecular mass, they are tough thermoplastics. Polyoxyethylenes of less than ca. 20 000 number-average molecular mass are usually referred to as poly(ethylene glycols). The crystalline polymer is spherulitic, has a unit cell (see Table 2) with four molecular chains, and has a fiber identity period of seven repeat units in two helical turns. The molecules are in an array of dihedral symmetry with twofold axes. One axis bisects the carbon – carbon bond and the other passes through the ether oxygens.

The glass transition temperature of PEO is listed as $-52\,°C$ in Table 1; however, this is the value for a high molecular mass, partially crystalline polymer. This parameter is complicated by crystallinity in the molecular mass range 10^3-10^5 [22], and abnormal values as high as $-17\,°C$ are found for the polymer with $M = 6000$, which is known to have the highest crystallinity (ca. 95 %). Macrospherulites or spherulites that can be seen without magnification have been prepared [23].

PEO is soluble in water, chlorinated hydrocarbons, aromatic hydrocarbons, methyl ethyl ketone, 2-ethoxyethyl acetate, butyl acetate, cyclohexanone, esters, dimethylformamide, and

Table 1. Selected physical properties of the poly(alkylene oxides)

	Poly(ethylene oxide)	Poly(propylene oxide)	Poly(tetramethylene oxide)
Density, g/cm^3			
Amorphous	1.127–1.128^a [2]	ca. 1.01–1.025^a [3]	0.975 (25 °C) [4]
Crystalline	1.21 (20 °C) [5]	1.096 (25 °C) [6]	1.07–1.08 [7]
Melting point, °C			
High molecular mass	65 [5]	75 [8]	58–60 [9]
Oligomeric^b	30–50 [10]	^{c}	10–45
Glass transition			
temperature, °C	−52 [11]	−73 to −78 [12]	−86 [4], [12]

^{a} d_{20}^{20}.
^{b} M_n ca. 400–3000.
^{c} Liquid under ambient conditions.

other solvents [1]. It has a solubility parameter of 10.3 [24]. It is insoluble in ethers and aliphatic hydrocarbons. Mark – Houwink constants for the intrinsic viscosity/molecular mass relationship $[\eta] = KM^a$ are given in Table 3. In water, PEO exhibits an inverse solubility – temperature behavior, becoming less soluble as the temperature is increased.

Intermolecular or association complexes between PEO and other compounds are readily formed due to the strong hydrogen bonding character of the ether oxygens. These complexes, which involve PEO and polymeric acids such as poly(acrylic acid) [35], urea, and thiourea [36], as well as various other inorganic and organic compounds [1], [37], have potential applications in mechanochemical energy conversion systems [38], chemical valves for altering the porosity of membranes [39], and biosensors and carriers for controlled drug delivery [40]. Alkali-metal salt complexes have electrical properties that make them of interest as battery electrodes [41].

Polymers of propylene oxide remain clear, almost colorless liquids or amorphous solids over broad ranges of molecular mass. They can be prepared as amorphous or stereoregular polymers, with stereoregularity resulting in crystallinity. Polyoxypropylenes with $M <$ ca. 20 000 are usually referred to as poly(propylene oxide) polyols, and these polymers are by far the most important products from a commercial standpoint. They are usually marketed as di- and trihydroxyl polyols that are available in a broad range of molecular mass. (The third hydroxyl group comes from the initiator; for example if glycerine, which is the most common initiator for PPO polyols, is used, the PPO grows off all three hydroxyl groups and a trifunctional polyol results.) High molecular mass PPO polymers are used as specialty rubber and represent only a small worldwide demand of about 1000 t/a. The crystalline polymer has a unit cell (see Table 2) that contains two chains in a bowed zigzag conformation, with the methyl group of one molecule located between the ether oxygen and methylene group of the adjacent molecule.

PPO is soluble in most organic solvents. Very low molecular mass polyols (<1000) are insoluble in aliphatic hydrocarbons such as hexane due to the high hydroxyl content. However, polyols with such high hydroxyl contents do have some solubility in water. Mark – Houwink constants for the intrinsic viscosity/molecular mass relationship $[\eta] = K M^a$ are given in Table 3.

Although not so thoroughly studied as those of poly(ethylene oxide), association complexes

Table 2. Crystal data for poly(alkylene oxides)

	Poly(ethylene oxide) [13–15]	Poly(propylene oxide) [16], [17]	Poly(tetramethylene oxide) [18–21]
Unit cell	Monoclinic	Orthorhombic	Monoclinic
a, nm	0.796	1.04	0.548–0.561
b, nm	1.311	0.467	0.873–0.892
c, nm	1.939	0.716	1.207–1.225
β	124.8°		134.2–134.5°

Table 3. Mark – Houwink constants for poly(alkylene oxides)

Solvent	Temperature, °C	$K \times 10^5$	a	Approximate molecular mass range	Average molecular mass	Reference
Poly(ethylene oxide)						
Water	30	12.5	0.78	10^4-10^7	Weight	[25]
Water	35	6.4	0.82	10^4-10^7	Weight	[26]
Aq. 0.45 M K_2SO_4	35	130	0.50	10^4-10^7	Weight	[26]
Aq. 0.39 M $MgSO_4$	45	100	0.50	10^4-10^7	Weight	[26]
Benzene	25	39.7	0.69	$8 \times 10^4-5 \times 10^6$	Weight	[27]
Carbon tetrachloride	20	69	0.61	200–10 000	Number	[28]
Toluene	35	145	0.70	400–4000	Number	[29]
Poly(propylene oxide)						
Acetone	25	75.5	0.56	1000–4000	Number	[30]
Benzene	20	11.1	0.77	700–3300	Number	[31]
Benzene	25	11.2	0.77	$3 \times 10^4-7 \times 10^5$	Weight	[32]
Toluene	20	20.8	0.72	700–3300	Number	[31]
Toluene	25	12.9	0.75	$3 \times 10^4-7 \times 10^5$	Weight	[32]
Poly(tetramethylene oxide)						
Benzene	30	131	0.60	$3 \times 10^4-10^6$	Weight	[33]
Cyclohexane	30	176	0.54	$3 \times 10^4-10^6$	Weight	[33]
Ethyl acetate	30	42.2	0.65	$3 \times 10^4-10^6$	Weight	[33]
Toluene	28	25.1	0.78	$3 \times 10^4-10^5$	Number	[34]

of poly(propylene oxide) are also known. These complexes are formed with a variety of salts such as lithium perchlorate [42] and metal halides, particularly chlorides [43]. Complexes of propylene oxide – ethylene oxide block copolymers with sodium and lithium salts exhibit marked increases in electrical conductivity [44] as compared to the uncomplexed copolymers.

Tetrahydrofuran can be polymerized to high molecular mass, but the commercially available poly(tetramethylene oxide) diols have molecular masses of ca. 650–3000 and a narrow molecular mass distribution. Although these polyols are usually crystalline in nature, their narrow molecular mass distribution [45] makes them easy to use, since even partial melting yields a product with effectively the same molecular mass distribution as that of a completely melted product. The crystalline polymer has a monoclinic unit cell whose characteristic dimensions are given in Table 2.

Solvents for PTMO, which has a solubility parameter of 17.3–17.6 [46], include aromatic hydrocarbons such as benzene and toluene, chlorobenzene, dichloromethane, chloroform, carbon tetrachloride, esters such as ethyl acetate, and nitrobenzene. Low molecular mass products are soluble in acetone, alcohols, and similar polar solvents. The polymers are insoluble in aliphatic hydrocarbons such as heptane, hexane, pentane, and petroleum ether. Only cursory studies of association complexes of PTMO have been conducted [43].

Chemical Properties. Poly(alkylene oxides) are chemically characterized as usually having primary (PEO and PTMO) and secondary (PPO) hydroxyl end groups [47]. PEO and PPO have been prepared with a variety of end groups including allyl or propenyl ethers, which arise during polymerization of ethylene oxide and propylene oxide through the termination reaction [48]; carboxyl [49], [50]; amine [51]; cyano [52]; and vinyl ether [53]. Alkyl or aryl ester terminated PEOs are used as nonionic surfactants [54], [55], and alkoxy-terminated PEOs prepared from alcohols or other monohydroxyl compounds are used for similar purposes as well as in cosmetic and pharmaceutical preparations. In addition to primary hydroxyl termination, PTMOs have been prepared with a majority of primary amino terminal groups [56]. Isocyanate-terminated poly(propylene oxides) and poly(tetramethylene oxides) are known as isocyanate-terminated prepolymers for preparation of polyurethanes.

Copolymers of various types can be prepared from the alkylene oxides. The best known of these are the polyurethanes, though they are not normally prepared from the poly(ethylene oxides). Polyurethanes are formed in a rearrangement

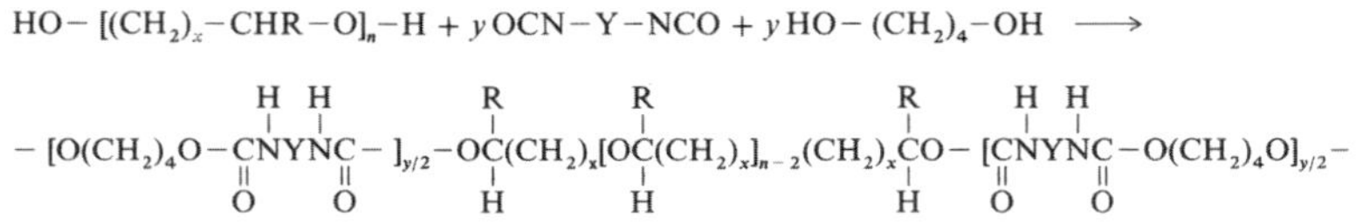

$$HO-[(CH_2)_z-CHR-O]_n-H + y\,OCN-Y-NCO + y\,HO-(CH_2)_4-OH \longrightarrow$$

$$-[O(CH_2)_4O-\underset{\underset{O}{\|}}{\overset{\overset{H}{|}}{C}}N\underset{}{Y}\underset{\underset{O}{\|}}{\overset{\overset{H}{|}}{N}C}-]_{y/2}-\underset{\underset{H}{|}}{\overset{\overset{R}{|}}{O}C}(CH_2)_x[\underset{\underset{H}{|}}{\overset{\overset{R}{|}}{O}C}(CH_2)_x]_{n-2}(CH_2)_x\underset{\underset{H}{|}}{\overset{\overset{R}{|}}{C}O}-[\underset{\underset{O}{\|}}{\overset{\overset{H}{|}}{C}}N\underset{}{Y}\underset{\underset{O}{\|}}{\overset{\overset{H}{|}}{N}C}-O(CH_2)_4O]_{y/2}-$$

Figure 1. Formation of polyurethanes

reaction from poly(alkylene oxide) polyols, short chain diols such as 1,4-butanediol and ethylene glycol, and multifunctional isocyanates. The reaction takes place at relatively low temperatures, including room temperature, as depicted in the general, idealized manner (Fig. 1) where Y represents the residue of a polyfunctional isocyanate without its isocyanato groups. (In this generalization, it should be noted that the starter molecule for the poly(alkylene oxide) is omitted. If the alkylene oxide involved were propylene oxide, on the average both end groups would have been secondary hydroxyl groups.)

This unit would be the average repeating unit in the final polyurethane. A wide variety of isocyanates are used, including toluene-2,4-diisocyanate, 4,4'-diphenylmethane diisocyanate and its polymeric forms, 4,4'-dicyclohexylmethane diisocyanate, and isophorone diisocyanate. The latter two isocyanates are used when light-stable polyurethanes are required.

Urethane acrylates are formed by preparing an isocyanate-terminated prepolymer and then capping it with a hydroxyalkyl acrylate. These acrylate-terminated oligomers are widely used in radiation curable coatings.

Isocyanate-terminated PTMO polyols have been marketed as Adiprene (Du Pont) and Vibrathane (Uniroyal) precursors. Block copolymers of poly(tetramethylene oxide) and poly (butylene terephthalate) are marketed as Hytrel (Du Pont) for a variety of end uses.

Other copolymers include the very important ethylene oxide – propylene oxide copolymers [57], which occur in random, block, and graft form. Random copolymers are usually prepared from mixtures of monomers, and block copolymers by alternately feeding the two monomers. Block copolymers are often used as surfactants, metal lubricants, and functional fluids. Most of the block copolymers have poly(ethylene oxide) as the hydrophilic block and poly(propylene oxide) as the hydrophobic block. Other hydrophobic blocks include poly(styrene oxide), poly (tetramethylene oxide), and polystyrene [58].

ABA block copolymers of poly(1-proline) and amine-terminated PEO [59], as well as those of poly(ϵ- caprolactone) [60] and polyesters [61] grown onto hydroxyl-terminated PEO, have been prepared. A well-known graft copolymer, hydroxyethyl cellulose ($\rightarrow$ Cellulose Ethers) [62], [63] is formed by swelling cellulose in a caustic solution and then adding ethylene oxide to the hydroxyl groups of the anhydroglucose units. Silica [64], gelatin and collagen [65], and chitin [66] are other polymeric substrates to which ethylene oxide has been grafted. Low molecular mass PEOs terminated with 1,2,4-triazoline-3,5-dione have been grafted onto polybutadiene [66].

A particularly useful class of grafted poly (propylene oxide) polyols was discovered in 1966 [67], [68]. These so-called polymer polyols are composed of vinyl polymers (e.g., acrylonitrile, styrene) and copolymers which are grafted to PPO or to PEO- capped PPO. They are produced by adding the vinyl monomers to the PPO polyol and dispersing them by agitation. A catalyst is then added to polymerize the vinyl monomers, which are believed to graft or partially graft to the polyol. The polymer polyol retains the end-group hydroxyl functionality, which is thus available for reaction with, for example, isocyanates. The vinyl polymer, which is incompatible with the polyol, is present in the form of a stable dispersion that acts as a reinforcing filler. These polymer polyols are used to manufacture polyurethane foams that have enhanced moduli while maintaining other desirable characteristics. Commercial products contain either grafted polyacrylonitrile or acrylonitrile – styrene copolymers, and they are used in high-resiliency automotive seating [69], microcellular elastomers [70], and other high-performance applications.

A deficiency of polyethers is their susceptibility to oxidation. PEO degrades by an autoxidation mechanism that is associated with formation and cleavage of intrachain hydroperoxides [71], [72]. Mechanical actions such as high shear can also cause degradation [73].

Hydroperoxide formation is facilitated by heavy metal ions, strong acids, and ultraviolet light. PPO and PTMO degrade by both chain scission and cross-linking [74], [75]. A number of investigations indicate that aromatic compounds such as substituted phenols, cresols, and quinones [76–78], mixtures of aliphatic amines and dihydric phenols (e.g., hydroquinone, *tert*-butylhydroquinone) [79], carbamates [80], and similar compounds function as stabilizers. A prime stabilizer for the poly(alkylene oxides) is 2,6-di-*tert*-butyl-4-methylphenol. Stabilization during storage is usually effected with 200–300 ppm of the stabilizer. However, in end uses involving solutions or slab stock polyurethane foam produced in bun form [1, p. 237] from poly(propylene oxide) polyols, 1000–5000 ppm are used to compensate for the long times at high temperature that are encountered due to retention of the heat of polymerization caused by the excellent insulation properties of the foam.

3. Production

Ethylene oxide was first polymerized in 1863 by WURTZ [81], who heated the oxide with water in a sealed tube. Earlier, in 1859, LOURENÇO [82] had obtained the polymer from ethylene glycol and ethylene bromide. Seven decades later STAUDINGER and LOHMANN [83] prepared a number of polymers with a variety of catalysts, including alkali and alkaline earth metals.

There are three main ring-opening reactions for the polymerization of ethylene oxide and propylene oxide: acid, base, and coordinate initiation. Acid or cationic initiation [84–86] with an acid, HZ, involves addition of an active-hydrogen compound such as an alcohol or glycol, $(HO)_mX$ ($m = 1 - 4$), to the 1,2-epoxide ring. In the initiation step, an oxonium complex forms and ring opening takes place by cleavage of the oxygen – carbon bond to form an initiated species that is available for propagation.

In the propagation step, n molecules of oxide are added to give a polymer having the formula $HO-[(CHRCH_2O)_n]_m-X$. Termination can occur by proton transfer, which yields an unsaturated product, or by anion transfer, which generates the starting acid.

Brønsted or Lewis acids such as $AlCl_3$, $SnCl_4$, PF_6, and BF_3 are used as initiators. The reaction is very fast, molecular mass is relatively low, and molecular-mass distribution is broad.

Base or anionic initiation [87–89] takes place in the following manner, where R' is usually an alkyl group.

$$m\ \underset{O}{CH_2\!-\!\overset{R}{CH}} + R'(ONa)_m \longrightarrow R'(O-CH_2\overset{R}{CH}-)_mO^-Na^+$$

Propagation takes place in a manner similar to that described above. Termination can be by hydrogen abstraction from the backbone, which results in terminal unsaturation and formation of a new initiation site, or by rearrangement of the active species.

Polymerization rates for the two mechanisms when perchloric acid and sodium methoxide are used are given in Table 4 for a few selected alkylene oxides.

Anionic initiators are usually metal alkoxides such as sodium or potassium methoxide or ethoxide. The reaction is usually slower than that obtained with acid initiation, except for epichlorohydrin and glycidol [90], [91]; relatively high molecular masses can be obtained and the molecular mass distribution is relatively narrow.

Ionic coordinative polymerization [92–95] involves a complex of a metal, M, such as Al, Sn, Fe, Mg, Li, etc., with ligands, L (e.g., OH, OR, Cl) that coordinates in an ionic manner with the oxygen atom of the epoxide. Complexation is followed by nucleophilic attack by the

$$\underset{Oxide}{\underset{O}{CH_2\!-\!\overset{R}{CH}}} \overset{HZ}{\rightleftharpoons} \underset{Complex}{\underset{O\cdots H^+Z^-}{CH_2\!-\!\overset{R}{CH}}} \xrightarrow{(HO)_mX} HO-(\overset{R}{CH}CH_2O)_m-X$$

Table 4. Polymerization rate for selected alkylene oxides relative to ethylene oxide [90], [91]

Oxide	Initiator	
	Perchloric acid	Sodium ethoxide
Epichlorohydrin	0.065	4.8
Ethylene oxide	1.0	1.0
Propylene oxide	54	0.5
Isobutylene oxide	5100	0.06

ligand, which results in opening of the oxirane ring to give an initiated species that can undergo propagation to form the polymer chain.

$$CH_2\text{-}\overset{R}{CH} + M^+L^- \longrightarrow CH_2\text{-}\overset{R}{CH} \longrightarrow L\text{-}CH_2\text{-}\overset{R}{CH}\text{-}OM$$
$$\phantom{CH_2\text{-}\overset{R}{CH}}\ \ \ \ O\cdots M^+L^-$$
Oxide   Initiator   Complex

There are two categories of ionic coordination initiators [94], [96]. Initiators in the first category involve the formation of metal – oxygen coordination bonds, and those in the second category are alkaline earth compounds that have an anionic surface capable of coordinating with the oxide monomer.

In the case of propylene oxide, the monomer has an asymmetric carbon atom (C*)

$$CH_2\text{-}\overset{H}{\underset{O}{C^*}}\overset{CH_3}{}$$

and when the ring is opened this asymmetric center is incorporated in the polymer chain. PPO can contain both head-to-tail and head-to-head sequences.

$$-CH_2\overset{CH_3}{CH}-O-CH_2\overset{CH_3}{CH}-O-\quad -CH_2\overset{CH_3}{CH}-O-\overset{CH_3}{CH}CH_2-O-$$
Head-to-tail   Head-to-head

The head-to-tail sequences are formed by the expected carbon – oxygen cleavage between the methylene group and the oxygen atom of the epoxide, whereas the head-to-head sequences occur when unexpected cleavage takes place between the methyl-substituted carbon and the oxirane oxygen. Up to 30 % head-to-head structure has been found in certain polyoxypropylenes [97], [98].

Tetrahydrofuran was first polymerized by MEERWEIN and KRONING in 1937 [99], [100]. An I.G. Farbenindustrie patent [101] indicates that polymers of this monomer were the subject of numerous investigations from 1939 to 1943. Ring strain, a necessity for favorable polymerization, is small in this monomer compared to other cyclic ethers [102], and thus THF has less tendency to polymerize than ethylene oxide and propylene oxide. Although commercial polymers are dihydroxyl functional and low in molecular mass, a high molecular mass polymer with a melting point above 100 °C has been reported [103].

THF is a nucleophilic monomer that forms a tertiary oxonium ion during initiation and only polymerizes by a cationic ring-opening mechanism [104–108]. A variety of mechanisms for formation of the tertiary oxonium ion are known. These include direct alkylation or acylation, in situ formation from reactive organic halides and metal salts or Lewis acids, and direct addition of strong protic acids to the oxygen atom. For example [109], direct alkylation of the monomer by exchange with trialkyloxonium salts illustrates the initiation reaction.

Propagation then takes place by collisions between the initiated species and THF, whereby

$$R_3O^+PF_6^- + O\underset{H_2C\text{-}CH_2}{\overset{H_2C\text{-}CH_2}{\diagup\diagdown}} \longrightarrow$$

$$\underset{H_2C\text{-}CH_2}{\overset{H_2C\text{-}CH_2}{\diagup\diagdown}}O^+\text{-}R\ PF_6^- + (R)_2O$$
Initiated Species

only collisions that occur at the α- carbon atom of the THF molecule result in propagation.

$$\underset{H_2C\text{-}CH_2}{\overset{H_2C\text{-}CH_2}{\diagup\diagdown}}O^+\text{-}R\ PF_6^- + n\,O\underset{H_2C\text{-}CH_2}{\overset{H_2C\text{-}CH_2}{\diagup\diagdown}} \longrightarrow$$

$$R[O(CH_2)_4]_n^-\text{+}O\underset{H_2C\text{-}CH_2}{\overset{H_2C\text{-}CH_2}{\diagup\diagdown}}\ PF_6^-$$

Water, oxygen, and other nucleophilic reagents can be used to terminate the polymerization.

Industrial poly(tetramethylene oxides) are hydroxyl terminated, and control of the termination reaction is important to the process. Hydroxyl groups, which have a profound effect on molecular mass, can be formed if strong protic acids are used to form the initiated species [110], [111]. The proton gives rise to an hydroxyl group on the end of the propagating species, and when water is used to terminate the polymerization, a second hydroxyl group is formed and the initiator acid is regenerated.

$$CF_3SO_3H + x\ O\underset{H_2C-CH_2}{\overset{H_2C-CH_2}{\Big\langle}}$$

$$\downarrow$$

$$\underset{H_2C-CH_2}{\overset{H_2C-CH_2}{\Big\langle}}O^+(CH_2)_4[O(CH_2)_4]_{x-2}-OH \quad CF_3SO_3^-$$

$$\downarrow + H_2O$$

$$H-[O(CH_2)_4]_x-OH + CF_3SO_3H$$

4. Toxicity [112]

Poly(ethylene oxides) have a very low single-dose oral toxicity that decreases with increasing molecular mass. Single-dose oral LD_{50} values in g/kg, for polymers of 200, 1000, 2000, and 6000 are, respectively, 34, 45, 45, and > 50 for male rats; 28, 32, > 50, and > 50 for female rats; 14, (not available), > 50, and > 50 for male rabbits; and 34, > 50, > 50, and > 50 for mice. PEOs of 400, 1540, and 4000 molecular mass were fed to dogs at a 2 % level in the diet with no adverse effect. When fed to rats at 2 %, PEO of 400 molecular mass had no effect, and at 4 % PEOs of 1540 and 4000 molecular mass had no effect. Doubling the dietary amounts for rats resulted in weight loss, an increase in liver weight, and slight histologic changes in the liver with the 400 compound, an increase in liver weight with the 1540 compound, and weight loss with the 4000 compound. Feeding of a 4 000 000 molecular mass compound at 5 % (rat) and 2 % (dog) diet had no detectable effect after 24 months. Testing with a radioactively labeled sample of the high molecular mass PEO indicated no significant absorption of the polymer from the gastrointestinal tract of the rat or dog.

PEOs do not appear to cause appreciable irritation to the eyes of rabbits. Skin irritation is apparently absent, which is attested to by the wide use of these compounds in cosmetic preparations. Overall, there are no reported adverse effects or human injuries from handling of PEOs [111].

Poly(propylene oxides) of molecular mass ca. 200 – 2000 have a single-dose oral toxicity that is noticeable, and they have mild eye irritability. They are not irritating to the skin, and even though there is some absorption through the skin, they do not appear to present a serious industrial health hazard. Inhalation of mists, particularly of the low molecular mass compounds (ca. 400 to 1200), could be hazardous. The low molecular mass compounds are rapidly adsorbed from the gastrointestinal tract and are potent CNS stimulants that readily cause cardiac arrhythmia. Excitement and convulsions are apparent within minutes after oral administration. Higher molecular mass compounds (>2000) are very low in toxicity by all administration routes and do not stimulate the CNS. Oral LD_{50} values (rat) in g/kg for polymers of 400, 1200, and 4000 molecular mass are 1.6, 1.4, and >15 for males and 2.1, 1.1, and >15 for females, respectively.

Poly(tetramethylene oxides) are commercially available in four number-average molecular masses (650, 1000, 2000, and 2900) as dihydroxyl functional compounds [113], [114]. Acute oral toxicity (LD_{50} in g/kg) for albino rats was 11.3, 18.8, and >34 for the 650, 1000, and 2000 molecular mass polyols, respectively. Acute dermal toxicity in albino rabbits was 8.37, 10.25, and >10.25 g/kg for the same polyols. The poly-ols are rated as being minimally irritating to the eyes and skin of albino rabbits [112]; however, a Material Safety Data Sheet indicates that the products may irritate the skin and eyes [115].

5. Handling

Most manufacturers and suppliers of poly(alkylene oxides) have detailed, current handling methods available for bulk, drum, and laboratory handling and storage, which should be consulted when working with the chemicals.

The poly(ethylene oxides) range from liquids at low molecular mass to hard waxes at moderate molecular mass, and to tough thermoplastics at high molecular mass (> ca. 100 000). They are not corrosive to most common construction materials. Although not strongly corrosive to copper and its alloys, some discoloration may occur if the compounds are kept in contact with copper-containing materials for prolonged times. Aqueous solutions of PEOs are corrosive to regular carbon steel; hence, stainless steel should be used for their storage. Tanks for bulk storage are usually made of plain carbon steel. For smaller quantities, lined steel drums are used to minimize the pickup of traces of iron [116]. When short storage times of a month or less are involved, storage under air is acceptable. For prolonged storage, particularly in hot climates or under heated conditions, the products should be kept under an inert (nitrogen) atmosphere. Since PEOs are hygroscopic, dry conditions are important for most uses. PEOs have low vapor pressures and are not fire hazards unless high temperatures are involved. For example, polymers of molecular mass 200, 1000, 4000, and 9000 have Cleveland Open Cup flash points of 171–182 °C, 254–265 °C, 271 °C, and 271 °C, and Cleveland Closed Cup flash points of >148 °C, >177 °C, >177 °C, and >177 °C, respectively.

The poly(propylene oxides) are usually liquid, stable, noncorrosive compounds that have a very low vapor pressure. They are preferably stored at temperatures below 50 °C, usually in carbon steel tanks. However, if there is concern about the pickup of iron, tanks protected with a baked phenolic coating can be used [117]. PPOs should be blanketed with dry air or nitrogen to prevent absorption of moisture during storage. They are usually dried before use in polyurethane manufacture. Nitrogen is preferred for prolonged storage or storage at elevated temperature. They represent fire hazards only at high temperatures. For example, 400, 2000, and 4000 molecular mass polymers have Cleveland Open Cup flash points of 166 °C, 229 °C, and 229 °C, respectively.

The poly(tetramethylene oxides) are liquids or low-melting solids under ambient conditions. Mild steel and phenolic-lined steel tanks are suitable for storage. These products are not considered fire hazards unless high temperatures are encountered, and they have a Tag Open Cup flash point of >163 °C. The compounds are unstable and may decompose at temperatures above 100 to 200 °C or at lower temperatures if in contact with high surface area substrates such as fibrous insulation [115]. When they decompose, very flammable tetrahydrofuran and carbon monoxide can be released.

6. Uses

Low molecular mass poly(ethylene oxides) ($M<800$) are usually used as intermediates for the manufacture of fatty acid ester surfactants that are used in a variety of formulated products. Illustrative of these products are soap or detergent suds control agents, foam stabilizers, fabric softeners, and optical brighteners. Typical detergents contain up to 20 % of a nonionic or anionic surfactant. Polymers with molecular mass of ca. 1000–2000 are low-melting waxes used in pharmaceutical ointments, cosmetic creams, and lotions. Their low toxicity, nonirritating nature, and smooth, creamy texture along with liquefaction at body temperatures make them particularly useful in cosmetic and pharmaceutical applications such as formulations for salves, suppositories, and ointments. Esters of these compounds are used as thickeners in cosmetic formulations.

The medium molecular mass PEO compounds, up to about 20 000, are hard, brittle waxes. They are used as adhesives, binders, plasticizers, molding compounds, wood impregnants, stiffening agents, and metal-working lubricants. Various PEOs are used in artifact preservation, particularly for items recovered after prolonged water immersion [118], [119]. In particular, copolymers of ethylene oxide and propylene oxide are often used in metal-working fluids and as quenchants, lubricants, and hydraulic fluids.

High molecular mass PEOs are used as textile sizes, as hydrodynamic friction reduction agents, in cosmetics and toiletries, as water-soluble films for seeds, in packaging, in adhesives, and in dentistry. The ability of the high molecular mass polymer to reduce hydrodynamic friction has led to applications in fire fighting and municipal sewage discharge control. The polymer is so effective in reducing the friction of water that it has been banned from use in competitive skiing and boating, where it markedly improved speed when incorporated into waxes and polishes. The association complexes of poly(ethylene oxide) have found use in controlled release of drugs, as adhesives, in artificial kidneys, and microencapsulation.

Poly(propylene oxides) are used as plasticizers, lubricants, solvents, softening agents, antifoaming agents, surface active agents, mold release agents, hydraulic fluid compositions, quenchants, and as intermediates in the production of resins (polyurethanes and alkyds). The largest use of the poly(propylene oxides) and ethylene-oxide-capped PPO polyols is in the production of polyurethane foams and elastomers. The main polyurethanes are those based on toluene-2,4-diisocyanate and 4,4′-diphenylmethane diisocyanate, although light-stable versions based on cycloaliphatic polyisocyanates are made for specialty end uses. The polyurethanes are used as automobile seating, automotive crash pads, furniture cushioning, insulation, bedding, packaging, carpet underlay, molded parts, and many other applications.

The low molecular mass poly(tetramethylene oxide) polyols are used in coatings, to make block copolymers that are used for fabricated parts, and polyurethanes used as elastomeric fibers (spandex fibers). PTMOs have FDA approval for use in polyurethanes and adhesives that come into direct contact with dry foods [120]. The fibers have low hysteresis loss, very good tensile strength, and good tear and abrasion resistance. They are particularly useful for clothing, support garments, knitwear, swimwear, and other similar end uses. No commercial uses are known for high molecular mass PTMO.

7. Economic Aspects [121]

Estimates of the worldwide production of poly(alkylene oxide) polyols for polyurethanes is given in Table 5. Poly(propylene oxide) polyols account for about 95 percent of the polyols produced for polyurethanes, and poly(tetramethylene oxide) polyols account for about 2 percent of the total.

It is difficult to estimate amounts produced for nonpolyurethane end uses since major companies use the same facilities to produce for both markets. The production for nonpolyurethane end uses is given in Table 5. The majority of the polyol for nonpolyurethane use is for surface active agents (40 %) and for functional fluids and lubricants (31 %). The remainder is used in a wide variety of end uses (see Chap. 6). In the United States, production is about 18.6 % poly(propylene oxides), 31.4 % poly(ethylene oxides), and 50 % ethylene oxide – propylene oxide copolymers. In Western Europe production is about 38 % PPO and PEO, and the remainder copolymers. In Japan production is about evenly divided between PPO, PEO, and copolymers.

The largest manufacturers are:

United States	ARCO, BASF, Dow, E. R. Carpenter, Mobay, Olin, and Union Carbide
Western Europe	ARCO, BASF, Bayer, Dow, Montedipe, and Shell
Japan	Asahi Denka, Asahi Olin, Dow, and Mitsui Toatsu
Republic of Korea	Korea Polyol and Miwon
Taiwan	Chiunglong Petrochemical

Although production of poly(tetramethylene oxide) polyols is included in the production values for polyols as given in Table 5, production of this specialty polyol is delineated by company in Table 6. Polyurethane fibers and elastomers produced from these polyols have tensile properties as good as or better than those of polyesters, together with especially good hydrolytic stability and excellent low hysteresis loss [122].

Table 5. Major worldwide production or consumption of poly(alkylene oxide) polyols [121]

Country	Production, 10^3 t
Production for polyurethane end uses	
Western Europe	931
United States	835
Japan	260
Korea and Taiwan	54
Total	2080
Production for nonpolyurethane end uses	
United States	159
Western Europe	92
Japan	57
Total	308

Table 6. Worldwide production of poly(tetramethylene oxides) [121]

Company	Production, 10^3 t	Trade name
United States		
Du Pont	30	Terethane
QO Chemicals	11	Polymeg
Wesern Europe		
BASF	4	PolyTHF
Du Pont (Netherlands)	30	Terethane
Japan		
Hodogaya Chemical	3	
Mitsubishi Kasei	4	
Sanyo Chemical	1	

References

1. F. E. Bailey, Jr., J. V. Koleske: *Alkylene Oxides and Their Polymers*, Marcel Dekker, New York 1991.
2. G. M. Powell in R. L. Davidson (ed.): *Handbook of Water-Soluble Gums and Resins*, McGraw-Hill, New York 1980.
3. Union Carbide: Specification Polyols, United States 1987.
4. G. S. Trick, J. M. Ryan, *J. Polym. Sci C* **18** (1967) 93.
5. D. R. Braun in R. L. Davidson (ed.): *Handbook of Water-Soluble Gums and Resins*, McGraw-Hill, New York 1980.
6. G. Natta, L. Porri, P. Corradini, D. Morero, Atti Accad. Naz. Lincei, *Cl. Sci. Fis. Mat. Nat. Rend.* **20** (1956) 408.
7. P. Dreyfuss, M. P. Dreyfuss, *Adv. Polym. Sci.* **4** (1967) 528.
8. C. C. Price, M. Osgau, R. E. Hughes, C. Shambelan, *J. Am. Chem. Soc.* **78** (1956) 690.
9. A. C. Farthing in N. G. Gaylor (ed.): *High Polymers*, vol. 13, part I, Interscience, New York 1963.
10. B. Boener, H. Heitz, W. Kern, *J. Chromatogr.* **53** (1970) 51.
11. J. A. Faucher et al., *J. Appl. Phys.* **37** (1966) 3962.
12. J. Lehmann, P. Dreyfuss, *Adv. Chem. Ser.* **176** (1979) 587.
13. A. H. Kovacs, B. Lotz, A. Keller, *J. Macromol. Sci.-Phys.* **B 3** (1969) no. 3, 385.
14. J. L. Koenig, A. C. Angood, *J. Polym. Sci. Polym. Phys. Ed.* **8** (1970) 1787.
15. J. F. Rabot, K. W. Johnson, R. N. Zitter, *J. Chem. Phys.* **61** (1974) 504.
16. E. Stanley, M. Litt, *J. Polym. Sci.* **43** (1960) 453.
17. M. Cesari, G. Perego, A. Marconi, *Makromol. Chem.* **99** (1966) 194.
18. M. Cesari, G. Perego, A. Mazzei, *Makromol. Chem.* **83** (1965) 106.
19. K. Imada et al., *Makromol. Chem.* **83** (1965) 113.
20. H. Tadokoro, *J. Polym. Sci. Part C* **15** (1966) 1.
21. S. Kobayashi, K. Murakami, Y. Chatani, H. Tadokoro, *J. Polym. Sci. Polym. Lett. Ed.* **14** (1976) 591.
22. F. E. Bailey, Jr., J. V. Koleske: *Poly(ethylene oxide)*, Academic Press, New York 1976.
23. F. E. Bailey, Jr., J. V. Koleske, *J. Chem. Ed.* **55** (1973) 761.
24. P. M. Galin, *Polymer* **24** (1983) 323.
25. F. E. Bailey, Jr., J. L. Kucera, L. G. Imhof, *J. Polym. Sci.* **32** (1958) 517.
26. F. E. Bailey, Jr., R. W. Callard, *J. Appl. Polym. Sci.* **1** (1959) 56, 73.
27. G. Allen et al., *Polymer* **8** (1967) 391.
28. C. Sadron, R. Rempp, *J. Polym. Sci.* **29** (1958) 127.
29. D. K. Thomas, A. Charlesby, *J. Polym. Sci.* **42** (1960) 195.
30. G. Meyerhoff, U. Maritz, *Makromol. Chem.* **107** (1967) 143.
31. W. Scholtan, S. Y. Lie, *Makromol. Chem.* **108** (1967) 104.
32. G. Allen, C. Booth, M. N. Jones, *Polymer* **5** (1964) 195.
33. M. Kurata, U. Utiyama, K. Kamada, *Makromol. Chem.* **88** (1965) 281.
34. S. U. Ali, M. B. Huglin, *Makromol. Chem.* **84** (1965) 117.
35. K. L. Smith, A. E. Winslow, D. E. Peterson, *Ind. Eng. Chem.* **51** (1959) 1361.
36. F. E. Bailey, Jr., H. G. France, *J. Polym. Sci.* **49** (1961) 397.
37. E. Tsuchida, K. Abe, *Adv. Polym. Sci. Ser.* **45** (1982) 1.
38. Y. Osada, *J. Polym. Sci. Polym. Chem. Ed.* **15** (1977) 255; **17** (1979) 3485.
39. S. Nishi, T. Kotaka, *Macromolecules* **19** (1986) 978; *Polym. J. (Tokyo)* **21** (1989) no. 5, 393.
40. R. K. Drummond, J Klier, J. A. Alameda, N. A. Peppas, *Macromolecules* **22** (1989) 3818.
41. R. L. Peck, US 4 797 190, 1989.
42. J. Moacanin, E. F. Cuddihy, *J. Polym. Sci. Part C* **14** (1966) 313.
43. D. B. James, R. E. Wetton, D. S. Brown, *Polymer* **20** (1979) 187.
44. R. Xue, C. A. Angell:Report TR-10, (1987) Order no. AD-A 183502, *Chem. Abstr.* **108** (1988) 205 458.
45. T. G. Croucher, R. E. Wetton, *Polymer* **17** (1976) 205.
46. M. B. Huglin, D. J. Pass, *J. Appl. Polym. Sci.* **12** (1968) 473.
47. J. D. Ingram et al., *J. Macromol. Chem.* **1** (1961) 75.
48. D. M. Simmons, J. J. Verbanc, *J. Polym. Sci.* **44** (1960) 303.
49. K. B. Wagener, C. Thompson, S. Wanigatunga, *Macromolecules* **21** (1988) 2668.
50. S. Matsumura, N. Yoda, S. Yoshikawa, *Makromol. Chem. Rapid Commun.* **10** (1989) 63.
51. Dow Chemical, US 3 236 895, 1966 (J. M. Lee, J. C. Winfrey).
52. Japan Kokai Tokkyo Koho, JP 63 210 128, 1988 (A. Suekane, T. Masuda, H. Kawakami); *Chem.Abstr.* **110** (1989) 76 274.
53. K. S. Kazanskii, N. V. Ptitsyna, *Makromol. Chem.* **190** (1989) 255.
54. W. B. Satkowski, S. K. Huang, R. L. Liss in M. J. Schick (ed.): *Nonionic Surfactants*, Marcel Dekker, New York 1967.
55. A. N. Wrigley, F. D. Smith, A. J. Stirton, *J. Am. Oil Chem. Soc.* **34** (1957) 39.
56. S. Smith, A. Hubin, *J. Macromol. Sci.* **A 7** (1973) 1399.
57. F. T. Wall, *J. Am. Chem. Soc.* **63** (1941) 803.
58. R. A. Quirk, J.-J. Ma, *J. Polym. Sci. Polym. Chem. Ed.* **26** (1988) 2031.
59. S. H. Jeon, S. M. Park, T. Ree, *J. Polym. Sci. Polym. Chem. Ed.* **27** (1989) 1721.
60. R. Perret, A. Skoulois, *C. R. Heled. Seances Acad. Sci.* **268** (1969) 230.
61. Union Carbide, US 3 670 045, 1972 (J. V. Koleske, R. M.-J. Roberts, F. D. DelGuidice).
62. R. T. K. Cornwell in R. L. Whistler, J. N. BeMiller (eds.): *Industrial Gums*, Academic Press, New York 1959.
63. H. M. Spurlin in E. Ott, H. M. Spurlin, M. W. Graffin (eds): *Cellulose and Cellulose Derivatives,* part II, Interscience, New York 1954.
64. T. Tajouri et al., *Chim. Phys. Phys.-Chem. Biol.* **84** (1987) no. 2, 243. A. Gantar et al., *Polymer* **28** (1987) 1403.
65. T. Okujama, *Tanpakushitsu Kakusan Koso* **15** (1970) 47.
66. M. Weber, R. Stadler, *Polym. Prepr. Am. Chem. Soc. Div. Polym. Chem.* **30** (1989) no. 1, 109.
67. W. C. Kuryla, F. E. Critchfield, L. W. Platt, P. Stamberger, *J. Cell. Plast.* **2** (1966) 84.
68. Union Carbide, US 3 304 273, 1967 (P. Stamberger).
69. W. Patten, D. C. Priest, *J. Cell. Plast.* **8** (1972) 134.
70. R. A. Dunleavy, *J. Elastoplast.* **2** (1970) 1.
71. C. W. McGary, Jr., *J. Polym. Sci.* **46** (1960) 51.
72. L. Reich, S. S. Stivain, *J. Appl. Poly. Sci.* **13** (1969) 977.
73. A. Nakano, Y. Minoura, *J. Appl. Poly. Sci.* **15** (1971) 927.
74. L. E. St. Pierre, C. C. Price, *J. Am. Chem. Soc.* **78** (1956) 3432.
75. A. Davis, J. H. Golden, *Makromol. Chem.* **81** (1965) 38.
76. P. A. Okunev, *Nauchni Tr. Ivanov. Energ. Inst.* **14** (1972) 130.
77. Union Carbide, US 3 388 169, 1968 (J. E. Tyre, F. G. Willeboordse).
78. K. Konishi, Y. Sano, E. Ajisaka, JP 7 0 03 224, 1970.
79. California Research, US 2 687 378, 1954 (A. Goldschmidt, W. T. Stewart, E. Cerrito, O. L. Harle).
80. Celanese, US 3 144 431, 1964 (T. J. Dolce, F. M. Berardinelli, D. E. Hudgin).
81. A. Wurtz, *Ann. Chim. Phys.* **69** (1863) 330, 334.
82. A. Lourenço, *Compt. rend.* **49** (1859) 619.
83. H. Staudinger, H. Lehmann, *Ann. Chim.* **505** (1933) 41.
84. P. H. Plesch: *The Chemistry of Cationic Polymerization*, Pergamon Press, New York 1963.

85 J. P. Kennedy, E. Marachal: *Carbocationic Polymerization*, J. Wiley & Sons, New York 1968.
86 S. Sakai, T. Sugiyama, Y. Ishii, *Kogyo Kagaku Zasshi* **69** (1966) 699.
87 G. Gee, W. C. E. Higginson, K. Taylor, M. W. Trenholme, *J. Chem. Soc.* 1961, 1345.
88 B. Wojtech, *Makromol. Chem.* **66** (1966) 180.
89 S. Shigematsu, Y. Miura, Y. Ishii, *Kogyo Kagaku Zasshi* **64** (1961) 153; **65** (1962) 360.
90 R. E. Parker, N. S. Isaacs, *Chem. Rev.* **59** (1959) 737 (1959).
91 G. Gee, W. C. E. Higginson, P. Levesley, K. J. Taylor, *J. Chem. Soc.* 1959, 1338.
92 Union Carbide, US 2 706 181, 1955 (M. E. Pruitt, J. M. Baggett).
93 Union Carbide, US 2 914 419, 1959 (F. E. Bailey).
94 J. Furukawa et al., *Kogyo Kagaku Zasshi* **62** (1959) 1269.
95 Union Carbide, BE 557 883, 1957; US 2 987 988, 1961 (F. N. Hill, F. E. Bailey, Jr., J. T. Fitzpatrick).
96 J. Furukawa, T. Saegusa: *Polymerization of Aldehydes and Oxides*, Wiley-Interscience, New York 1963.
97 E. J. Vandenberg, *J. Polym. Sci. Part B* **2** (1964) 1085.
98 C. C. Price, R. Spector, *J. Am. Chem. Soc.* **87** (1965) 2069.
99 H. Meerwein, J. Kroning, *J. Prakt. Chem.* **2** (1937) no. 147, 257.
100 I.G. Farbenindustrie, DE 741 478, 1939 (H. Meerwein).
101 I.G. Farbenindustrie, FR 898 269, 1945.
102 H. de V. Robles, *Rec. Trav. Chim. Pays-Bas* **59** (1940) 184.
103 Du Pont, US 2 856 370, 1958 (E. L. Muetterties).
104 P. Dreyfuss, M. P. Dreyfuss, *Adv. Polym. Sci.* **4** (1967) 528.
105 G. Pruckmayr in K. C. Frisch (ed.): *Cyclic Monomers*, Wiley-Interscience, New York 1972.
106 B. A. Rosenberg, W. I. Ioshak, N. S. Enikolopian: *Interchain Reactions in Polymeric Systems*, Chemistry, Moscow 1975.
107 Y. Eckstein, P. Dreyfuss, *Anal. Chem.* **52** (1980) 537.
108 K. Yasuda, Y. Yokoyama, S. Matsumoto, K. Harada, in E. J. Goethals (ed.): *Cationic Polymerization and Related Processes*, Academic Press, New York 1984.
109 P. Dreyfuss, M. P. Dreyfuss, *Adv. Chem. Ser.* **91** (1969) 335.
110 G. Pruckmayr, T. K. Wu, *Macromolecules* **11** (1978) 265, 662.
111 K. Matsuda, Y. Tanaka, T. Sakai, *J. Appl. Polym. Sci.* **20** (1976) 2821.
112 *Patty* 3rd ed., **2 C**.
113 Quaker Chemical, POLYMEG Polyols, Applications and Properties Data, United States, 1976.
114 Du Pont, TERETHANE Polyether Glycol, Properties, Uses, Storage and Handling of Du Pont Glycols, United States, 1985.
115 Du Pont, Material Safety Data Sheet, TERATHANE Polyether Glycols, Wilmington 1985.
116 Union Carbide, CARBOWAX Polyethylene Glycols, United States, 1986.
117 Union Carbide, Specification Polyols, United States, 1987.
118 A. M. Rosenqvist, *Stud. Conserv.* **4** (1959) 62.
119 K. Morris, B. L. Siefert, *J. Am. Inst. Conserv.* **18** (1978) 33.
120 United States Food and Drug Administration Regulations 121.2550 and 121.2562.
121 SRI International: *Chemical Economic Handbook*, Menlo Park 1991.
122 W. J. Pentz, R. G. Krawiec, *Rubber Age* 1975, Dec., 31.

Polyoxymethylenes

MICHAEL HAUBS, Ticona GmbH, Frankfurt am Main, Germany

KLAUS KURZ, Ticona GmbH, Frankfurt am Main, Germany

GUENTER SEXTRO, Hoechst AG, Frankfurt, Germany

1.	Introduction	911
2.	Physical and Chemical Properties	912
2.1.	Morphology	912
2.2.	Mechanical Properties	914
2.3.	Thermal Properties	914
2.4.	Chemical Stability	915
2.5.	UV and Weather Resistance	915
2.6.	Permeability	915
2.7.	Fire Behavior	916
2.8.	Optical Properties	916
2.9.	Electrical Properties	916
3.	Production	916
3.1.	Polymerization of Formaldehyde and Trioxane	916
3.2.	Homopolymers	916
3.3.	Copolymers	917
3.3.1.	Polymerization of Trioxane	917
3.3.1.1.	Initiation	917
3.3.1.2.	Chain Propagation	918
3.3.1.3.	Chain Termination by Chain Transfer	918
3.4.	Additives	919
3.4.1.	Stabilizers	919
3.4.2.	Nucleation Agents	920
3.4.3.	Release Agents	920
3.5.	Impact-Resistant Types	920
3.6.	Low Volatile Organic Carbon (VOC) Grades	921
3.7.	High-Strength POM Copolymer	921
4.	Economic Aspects	921
5.	Quality Specifications and Analysis	921
6.	Product Range and Processing	922
7.	Uses	923
8.	Recycling	923
9.	Toxicology and Occupational Health	923
	References	924

1. Introduction

Polyoxymethylenes (POM) are thermoplastically processable plastics, whose structure is predominantly composed of unbranched chains of oxymethylene units:

Polyoxymethylene

The term polyacetals is also commonly used for such products. The first systematic investigations of the production and properties of polyoxymethylenes were carried out by STAUDINGER and KERN [1] in the 1920s. Basic properties such as the relatively high softening temperature of ca. 175°C, high crystallinity, and the ability to form films and fibers were already recognized at that time. However, the insufficient thermal stability of polyoxymethylenes at processing temperatures prevented any practical use of the products.

Following intensive work begun at the end of the 1940s, DuPont made a breakthrough in 1956 with the production of a high-molecular, thermally stable polyoxymethylene that could be processed thermoplastically [2]. The decisive step consisted of blocking the unstable hemiacetal terminal groups by esterification with acetic anhydride. A production plant with ca. 7000 t/a capacity came on stream in the United States in 1959. The product was introduced on the market under the name Delrin.

Polyoxymethylene (homopolymer)

A few years later, Celanese in the United States and Hoechst in Europe succeeded in producing a thermally stable acetal copolymer [3, 4] by copolymerizing trioxane with small amounts of cyclic ethers or cyclic acetals. Compared with homopolymers with blocked terminal groups, the copolymers have the advantage that they are resistant to alkalies, though they have a somewhat lower crystallinity.

$$R-O-[-(-O-)_n-O-]_m-R^*$$

R, R*: $-CH_3$, $-CH_2CH_2OH$ Polyoxymethylene (copolymer)

Production of the acetal copolymer Celcon came on-stream in 1962 in a 7000 t/a plant at the Celanese plant in Bishop, Texas. The year before Celanese and Hoechst in Germany had already established the joint venture Ticona Polymerwerke, which commenced production of the acetal copolymer Hostaform in 1964. In 1962 Celanese and Daicel formed the joint company Polyplastics in Japan and commenced production of the acetal copolymer Duracon. This was followed in the 1970s by Ultraform (established jointly by BASF and Degussa; product name: Ultraform), the Asahi Chemical Industry Co. (product name: Tenac), and the Mitsubishi Gas Chemical Co. (product name: Iupital). Further plants were built in Poland (product name: Tarnoform) and the Soviet Union. In the 1980s and early 1990s new plants were constructed mostly in Asia (Korea Engineering Plastics: Kepital; Lucky-Goldstar: Lucel; Taiwan Engineering Plastics).

New capacities are coming on-stream in the new decade in China, e.g., PTM Engineering Plastics (joint venture of Polyplastics, Ticona and Mitsubishi Eng. Plastics), Shanghai Bluestar New Chem. Materials and from the Yuntianhua Group. It was also announced [5] that Ibn Sina, a joint venture between Celanese and Sabic, is planning to build a production plant in Jubail, Saudi Arabia.

Worldwide demand in the recession year 2009 [5] was at ca. 650×10^3 t, with ca. 25% in Europe, 20% in USA and 55% in Japan and China.

Worldwide ca. 27% of the material are used in automotive applications (biggest single market segment for POM).

Polyoxymethylenes are now widely used in areas that were for a long time the preserve of metals. Their low specific weight (compared with metals) and economic processing together with an advantageous combination of mechanical, thermal, electrical, and chemical properties have led to a wide range of applications. In Asia, especially China, main applications are in consumer articles as well as in electronic areas.

Further details of production, properties, processing, and uses of polyoxymethylenes can be found in [6–8].

2. Physical and Chemical Properties

Important physical properties of homopolymers are summarized in [9], and those of copolymers in [10, 11]. Some physical data for polyoxymethylene are given in Table 1 and [12, 22]. Materials properties can be obtained from the Western European producers (Material Data Bank CAMPUS: Computer Aided Material Preselection by Uniform Standards [23]).

2.1. Morphology

Polyoxymethylene crystallizes in a hexagonal structure with the lattice constants $a=474$ pm and $c=1739$ pm. The polymer chains have a helical conformation with an identity period of 9/5 [24, 25].

The crystalline density is 1.49 g/cm^3 [24], and that of the amorphous state 1.324 g/cm^3 (homopolymer) and 1.320 g/cm^3 (copolymer) [15]. The density of the commercially available partially crystalline polyacetals is 1.42 g/cm^3 and 1.41 g/cm^3 for homopolymers and copolymers, respectively [9, 10, 26].

The crystallinity of polyoxymethylenes varies between 50% and 80%, depending on the comonomer content, molecular mass, and sample pretreatment. Crystallinities of 64–69% for homopolymers and 56–59% for copolymers were found for identical thermal histories [15]. The degree of crystallinity can be increased by thermal posttreatment. Considerably higher values for the crystallite melting point and degree of crystallinity have been reported for the as-polymerized virgin polymer [27].

During crystallization from the melt spherulitic superstructures form. The size of the spherulites depends, among others, on the

Table 1. Physical properties of polyoxymethylene

Property	Value	Reference
Activation energy of polymerization, kJ/mol		
of cationic polymerization of trioxane in dichloroethane	79.5	[13]
in the solid state	36	[13]
Ceiling temperature, °C		
polymerization of formaldehyde (gas) → polyoxymethylene (solid) at 1 bar	116–137	[12]
Coefficient of linear thermal expansion, K^{-1} (ASTM D 696)		
copolymer	0.85×10^{-4}	[14]
homopolymer	0.81×10^{-4}	[14]
Crystallinity, %		
Delrin 550, 20°C, from density data	68.7	[15]
Hostaform C 9020, 20°C, from density data	58.7	[15]
Density, g/cm^3		
homopolymer, Delrin 550, 20°C	1.435	[15]
copolymer, Hostaform C 9020, 20°C	1.416	[15]
Dielectric constant at 10^3 Hz (DIN 53 483, ASTM D 150–59 T)	3.6–4.0	[12]
Dielectric loss tan δ at 10^3 Hz	$(15-50) \times 10^{-4}$	[12]
Dielectric strength, kV/mm (DIN 53 481, 50 Hz, 0.5 kV/s, 0.2 mm film)	50–70	[12]
Enthalpy, kJ/kg		
homopolymer, 300 K	222.5	[16]
copolymer, 300 K	211.4	[16]
Entropy, kJ kg^{-1} K^{-1}		
homopolymer, 300 K	1.487	[16]
copolymer, 300 K	1.439	[16]
Heat capacity, J mol^{-1} K^{-1}		
homopolymer, 300 K	42.79	[16]
copolymer, 300 K	41.11	[16]
molten polyoxymethylene, 300 K	57.67	[17]
Heat distortion temperature under load, °C (ASTM D 648)		
homopolymer	124	[12]
copolymer	110	[12]
Heat of combustion of polyoxymethylene, kJ/mol	-511.6 ± 1.7	[18]
Heat of fusion, kJ/kg		
homopolymer	236	[19]
copolymer	181–192	[20]
Heat of polymerization, kJ/mol		
formaldehyde (gas, 1 bar, 298 K)	-55 to -72	[12]
trioxane (liquid → solid)	-22.6	
Melting point, °C		
Delrin 500	175	[9]
Hostaform C 9021	164–167	[10]
Specific volume, cm^3/g		
Delrin 550, at 20°C	0.6968	[15]
Hostaform C 9020	0.7064	[15]
Surface resistivity, Ω		
homopolymer (ASTM D 257)	2×10^{13}	[14]
Thermal conductivity at 20°C (VDE 0304), W m^{-1} K^{-1}	0.292	[21]
Volume resistivity (film 0.2 mm, DIN 53 482/ASTM D 257–61), Ω·cm	$> 10^{15}$	[10]
Water absorption (ASTM D 570), %		
after 24 h immersion		
homopolymer	0.25	[14]
copolymer	0.22	[14]
equilibrium, 50% R.H.		
homopolymer	0.2	[14]
copolymer	0.16	[14]

crystallization temperature and number of nucleation centers. The overall crystallization rate and number of spherulites formed can be increased by adding nucleation agents (see Section 3.4).

2.2. Mechanical Properties

Polyoxymethylenes exhibit high hardness and rigidity, high toughness down to −40°C, high thermal dimensional stability, low water absorption, good electrical and dielectric properties, good elasticity, favorable slip and wear behavior, and are easy to process. This combination of favorable properties means that polyoxymethylenes are widely used construction materials.

The structural difference between homopolymers and copolymers results in a lower degree of crystallinity for copolymers. As a consequence homopolymers have about 10–12% higher hardness, rigidity, and strength [6]. However, copolymers possess advantages with regard to thermal and chemical stability.

Standardized test methods are used to determine the physical properties. Since polyoxymethylenes are generally processed by injection molding, the mechanical properties are usually measured on injection-molded test specimens. Slight differences in test specimen production and mechanical testing can result in changes in the measured values. For this reason it is only possible to a limited extent to make an absolute comparison of the property values given in the manufacturers' product data sheet [9–11]. Further details are given in [22].

The mechanical properties are measured by the conventional methods used for polymer materials (→ Plastics, Properties and Testing). Test methods and values can be obtained from the brochures of the raw material manufacturers [9–11]. See also Table 1 as well as [22] and Chapter 5.

The suitability of a structural component for a given purpose depends not only on the mechanical properties of the selected raw material, but to a large extent on its structural form and shape [28, 29]. In addition there are influences resulting from the production and environmental conditions, as well as time, temperature, and stress level.

2.3. Thermal Properties

The thermal properties of polyoxymethylenes essentially depend on the crystallinity and chemical structure, which are also responsible for differences in the thermal behavior of homopolymers and copolymers.

Homopolymers have higher crystallite melting points (175°C) than copolymers (164–167°C). Further information on structural changes with temperature are obtained from the torsion vibration test (DIN 53 445), from the temperature function of the mechanical shear modulus G', and the mechanical loss factor d. The transition region (relaxation region) at −60°C represents an important embrittlement limit under impact conditions. A second relaxation region exists at 0°C. A third, subsidiary maximum above 100°C is attributed to molecular motion in the crystalline regions [22] and is important for the thermal dimensional stability.

The continuous use temperatures are, depending on the type of stress, about 90–110°C [22] and 80–100°C [30]. According to Underwriters Laboratory, Hostaform is permitted for use up to 80–105°C [10]. Experience shows that short-term thermal loading (several hours) up to 140°C is permissible [30].

The continuous use temperature is largely determined by thermal aging processes, which generally involve thermal-oxidative damage at high temperature. Thermal-oxidative decomposition as well as decomposition caused by acidic degradation products can be suppressed by adding oxidation stabilizers and costabilizers (see Section 3.4 and [31]).

Thermal stability and continuous use temperatures are, however, not accurately definable material-specific properties and must always be considered in the context of the relevant application.

Temperatures of 190–210°C are generally recommended for processing polyoxymethylenes. Higher temperatures can be used with shorter residence times (e.g., up to 230°C for copolymers) [6, 10].

The specific heat of copolymers is 1.47 kJ kg^{-1} K^{-1}, and the thermal conductivity is 0.31 W m^{-1} K^{-1} (at 20°C) [10]. Further thermal data are given in Table 1 and [12, 22].

2.4. Chemical Stability

Owing to their high degree of crystallinity, polyoxymethylenes exhibit good resistance to numerous chemicals. Copolymers are also resistant to strong alkalies. Polyoxymethylenes are not resistant to strongly acidic and oxidizing substances.

Only hexafluoroisopropanol and hexafluoroacetone sesquihydrate are suitable as solvents at room temperature. Most other solvents dissolve polyoxymethylenes only when the temperature is approaching its melting point; solvents include polar substances such as benzyl alcohol, dimethylformamide, γ-butyrolactone, and phenol [32].

Various investigations on the stability towards numerous chemicals have been published, for example [9–11].

Conventional organic solvents such as alcohols, esters, ethers, ketones, aliphatic and aromatic hydrocarbons, as well as perhalohydrocarbons do not attack polyoxymethylenes at room temperature even under prolonged contact. The mechanical properties are only slightly affected, and the degree of swelling is small.

The good resistance to all common fuels (regular, premium, also fuels containing 15–20% methanol, diesel), as well as to oils and greases, coolants, and brake fluids means that polyoxymethylenes are widely used as construction materials in the automotive sector [10, 33].

Polyoxymethylenes swell only very slightly in contact with water; the water absorption is 0.6% at 20°C, and 1.6% at 100°C [10]. Copolymers in particular are extremely stable at high temperatures. For example, the tensile strength decreases by only 25% after two years in boiling water. The copolymers are therefore widely used in washing machines and dishwashers, as well as in water boilers. Alkaline detergents and rinsing agents do not affect the stability of the copolymers. Polyoxymethylenes are generally not subject to stress-cracking.

As regards use in the sanitary sector, varying chlorine contents in tap water, as well as temperature and pressure fluctuations mean that the suitability must be individually assessed in each case.

2.5. UV and Weather Resistance

Polyoxymethylenes can be protected to a certain extent against UV radiation by the use of additives. Suitable UV absorbers include those based on benzophenone or benzotriazole derivatives. In practice, they are combined with HALS stabilizers, which give a synergistic effect. A further improvement can be achieved by adding carbon black or pigments such as titanium dioxide (see Section 3.4).

Weathering results in a loss of surface gloss, followed by so-called chalking, caused by a layer of degraded, low molecular mass polyoxymethylene. This layer protects deeper-lying layers, so that the reduction in the mechanical properties is greater at the start of weathering, but then decreases. The damage to thin-walled parts is thus greater than to thick-walled parts. The behavior of non-light-stabilized copolymers under natural and artificial weathering is described in [34]. Investigations of the mechanism are given in [35–38].

Various tests are used to evaluate the UV and weathering resistance. Artificial irradiation in the Xenotest and in the Fade-O-Meter produces similar results to natural weathering [37]. Weathering in the Florida test can, depending on the intended use, be performed in the open air or behind window glass. It is not possible to interconvert the various results.

Polyoxymethylenes are only moderately stable to γ-radiation. Medical equipment can be sterilized by using conventional radiation doses of 2.5×10^4 J/kg [10]. The deterioration in the mechanical properties increases with increasing radiation dose [39].

2.6. Permeability

Polyoxymethylenes are only very slightly permeable to many gases and solvents. In particular, the very low permeability to propane; butane; aliphatic, aromatic, and halogenated hydrocarbons; alcohols; and esters should be emphasized. The permeability to hydrogen, CO_2, water, and methanol [21, 33, 40] is higher. A comparison with other plastics is given in [22].

Numerous uses have accordingly been found for polyoxymethylenes; for example, cigarette lighters, aerosol containers, and automotive parts in contact with fuels.

2.7. Fire Behavior

Polyoxymethylenes burn slowly with a pale blue flame, forming droplets. The external ignition temperature is 350°C, and the self-ignition temperature is 360°C.

According to Underwriters Laboratories polyoxymethylenes are classified in Class HB (UL Safety Standard 94). According to the U.S. Federal Motor Vehicle Safety Standard FMVSS 302 and DIN 75 200, the combustion rate does not reach the maximum permitted combustion rate of polyoxymethylenes at wall thicknesses of ≥1 mm, and accordingly use in automobiles is permitted. Further information on flammability can be found in [11].

A formulation with the classification V–0 is not known.

2.8. Optical Properties

Polyoxymethylene moldings are translucent to opaque-white, depending on the thickness. The light transmission of 2 mm thick injection-molded plates is 50%. The refractive index n_D^{20} is 1.48. The gloss depends mainly on the surface quality of the mold.

2.9. Electrical Properties

Polyoxymethylenes have good electrical insulation properties and a favorable dielectric behavior. This, together with the good mechanical properties, means that polyoxymethylenes are useful materials in electrical engineering.

The specific resistivity is $>10^{15}\,\Omega\cdot\text{cm}$, and the surface resistance is $>10^{13}\,\Omega$. The relative dielectric constant ϵ_r is 3.6–4.0. Since the loss factor tan δ is low, high-frequency heating and high-frequency welding are not possible [10].

The dielectric strength describes the behavior under short-acting stress produced by high voltages. Values of 600–700 kV/cm have been measured for polyoxymethylene sheets [10].

Polyoxymethylenes generally do not accumulate static charge. Antistatic and electrically conducting grades have been developed for specialty uses (e.g., Hostaform C 27021 AS and Hostaform EC 140 XF or EC 270 TX [10]).

3. Production

3.1. Polymerization of Formaldehyde and Trioxane

The polymerization of formaldehyde and trioxane proceeds according to an ionic mechanism. Whereas formaldehyde can be polymerized anionically as well as cationically, trioxane, being a cyclic acetal, can be polymerized only under cationic conditions. Polymerization can be performed in the gas phase, liquid phase, in dispersion, and in solution. Trioxane also polymerizes in the solid state.

In practice homopolymers are produced by anionic polymerization of formaldehyde, while copolymers are mainly produced by cationic copolymerization of trioxane with cyclic ethers or cyclic acetals. Processes for producing copolymers that use formaldehyde as starting material have also been described [41].

A review of the historical development of polyoxymethylenes is given in [42–44]. Comprehensive summaries of the homopolymerization and copolymerization of trioxane can be found in [45–47].

For information on the most important process steps in the production of homopolymers and copolymers, together with numerous literature and patent references, see [48].

3.2. Homopolymers

Terminal-group-stabilized homopolymers of formaldehyde with adequate thermal stability were first developed by DuPont [2, 48–50].

The production process can be divided into the following steps:

1. Oxidation of methanol to formaldehyde
2. Purification of formaldehyde
3. Polymerization
4. Terminal group capping
5. Homogenization of the polymer in the melt; addition of auxiliaries

The production of formaldehyde from methanol is described in → Formaldehyde. The formox or silver catalyst process is usually used; see also [51]. The Asahi process, in which formaldehyde is prepared by oxidation of methylal rather than methanol [52] leads to higher formaldehyde concentrations [53].

High-purity formaldehyde is required for the polymerization step. Purification on an industrial scale can be performed, for example, by absorbing the reaction gases from the formox or silver catalyst process in cyclohexanol [54, 55]. Cyclohexanol hemiformal is formed from formaldehyde and cyclohexanol; water and other impurities are then removed by vacuum distillation. Vacuum distillation is followed by thermal cracking. The resultant formaldehyde gas can then be passed after further purification (e.g., removal of reactive impurities by partial polymerization [56, 57]) to the actual polymerization process [49]. Amines [57]; alkoxides [58]; ammonium and sulfonium salts [59]; metal carbonyls [60]; and phosphines and arsines are suitable initiators for anionic polymerization.

The high heat of polymerization (ca. 60–70 kJ/mol [12, 61]) must be removed. Suspension polymerization has proved convenient in practice. Aliphatic or cycloaliphatic hydrocarbons are suitable suspension agents, for which the same purity requirements apply as for the monomer.

The purified formaldehyde is added to a reaction vessel containing the suspension agent (e.g., cyclohexane) and initiator (e.g., tri-n-butylamine). The polymerization can be carried out at room temperature and under normal or reduced pressure. The polymer is formed as a fine suspension.

In the next step the thermally unstable hemiacetal terminal groups of the polymer are converted into alkyl or acyl groups. In particular esterification with acetic anhydride has become industrially important [2]. The reaction can, for example, be carried out in the presence of small amounts of sodium acetate as catalyst in boiling acetic anhydride.

This conversion stage is followed by several purification steps. Finally, the polymer powder is melted in extruders with the addition of antioxidants, costabilizers and other additives, and converted into granules.

Asahi has proposed a modified process in which acetic anhydride is used as a chain-transfer agent. Acetate-terminated polymers are obtained directly in the polymerization process [62].

3.3. Copolymers

The copolymer production process is divided into the following steps:

1. Oxidation of methanol to formaldehyde
2. Trimerization of formaldehyde to trioxane
3. Purification of trioxane
4. Copolymerization of trioxane with cyclic acetals
5. Removal of unstable terminal groups
6. Homogenization of the polymers in the melt; addition of auxiliaries

The production of formaldehyde is described in → Formaldehyde, Chap. 4.. The trioxane must satisfy strict purity requirements for the subsequent cationic polymerization. The concentration of compounds that are chain-transfer agents such as water, methanol and formic acid should be below 80 ppm. The final purification to polymerization grade trioxane is generally carried out by crystallization [63, 64] or distillation.

3.3.1. Polymerization of Trioxane

The production of POM copolymers can best be understood by a brief look at the mechanism of trioxane polymerization. A detailed discussion of the mechanism is given in [47].

Polymerization of trioxane takes place through a series of chemical reactions which can be categorized into initiation, chain propagation and chain termination.

3.3.1.1. Initiation

A large number of compounds may be used as initiators for the cationic polymerization. Important catalysts include

1. Lewis acids, such as BF_3, and their complexes (e.g., $BF_3 \cdot Bu_2O$) [65]

2. Protic acids such as perchloric acid, perfluoroalkanesulfonic acids, as well as their alkyl esters or anhydrides [66]
3. Heteropoly acids such as tungstophosphoric acid [67, 68]

Carbenium salts such as triphenylmethyl hexafluoroarsenate [69], or oxonium compounds, and metal complexes [70].

The following describes the initiation process using protic acids:

Initiation starts with the formation of an acetal-oxonium ion:

Acetal-oxonium ions have the unique feature of isomerizing to oxocarbenium ions in a ring-opening reaction:

This reaction is made possible by the resonance stabilization of the oxocarbenium ions. Nonetheless the equilibrium of reaction (2) is shifted far to the left [71]

Oxocarbenium ions are the decisive species of all chemical reactions taking place during trioxane polymerization. They first react with the counterion and form an acetal ester [72]:

The equilibrium of this strongly exothermic reaction is far on the right side. At very low concentrations however a part of the ester dissociates. The order of magnitude of the oxocarbenium ion concentration can be estimated at 10^{-10} mol/L; the formation of the oxocarbenium ion is an entropy-driven process.

Reaction (3) can explain the strong influence of the counterion (X^-) of the acid as the equilibrium constant of reaction (3) strongly depends on the nature of X^-. Due to the high enthalpy of reaction (3) the equilibrium concentration of the oxocarbenium ion increases strongly with temperature.

Reaction (3) can be used to influence the activity of the initiator. Addition of X^- as salt of the corresponding acid reduces the concentration of the oxocarbenium ion and the speed of polymerization slows down. The activity of the initiator can be varied over a wide range in two ways: by changing the acid concentration and the salt-to-acid ratio. This "two-dimensional" initiator has proven valuable in the industrial production of POM [73].

3.3.1.2. Chain Propagation

Again, the reactive species in the polymerization process is the resonance-stabilized oxocarbenium ion which reacts with trioxane to form an oxonium ion. The polymerization process continues via isomerization to an oxocarbenium ion.

3.3.1.3. Chain Termination by Chain Transfer

The molecular mass of POM is controlled by chain-transfer reactions. This is accomplished by the addition of acetals like methylal or butylal to the polymerization mixture [74]. Nearly all chains (>99.9%) are terminated by chain-transfer reactions. The oxocarbenium ion of the growing polymer chain reacts with methylal to form a polymer with a stable methoxy end group and another oxocarbenium ion which starts a new chain:

If all chains were terminated by chain transfer with methylal then all polymer chains would have stable methoxy end groups and the resulting polymer would be perfectly stable. But due to impurities in the monomer (water, methanol and formic acid) and because of a certain side reaction ("hydride shift") some chains have unstable hemiacetal and formate end groups that are removed in a separate step by heat treatment (in a process called "hydrolysis").

A special case of chain transfer involves the polymer itself as chain-transfer agent. This transacetalization results in a Schulz–Flory distribution for polymer chain lengths and a statistical distribution of comonomer units in the copolymers.

POM copolymers are produced by adding comonomers to the polymerization mixture prior to polymerization. Suitable comonomers include cyclic ethers and cyclic acetals. In particular 1,3-dioxolane, diethylene glycol formal [75], and 1,4-butanediol formal are used. The comonomers are generally used in amounts of 2–5 wt%.

The polymerization of trioxane can be performed in bulk, solution [76], or suspension [77]. Bulk polymerization is the preferred method in practice. It is generally carried out in self-cleaning kneaders or double-screw reactors [78, 79] at 70–115°C. Polymerization is complete within a few minutes, and a very hard, crystalline polymer is formed. The heat of polymerization, ca. 23 kJ/mol [80], leads to evaporation of trioxane and formaldehyde.

Other polymerization processes have been reported, including solid-state copolymerization of trioxane/1,3-dioxolane [81], homogeneous-phase polymerization [82], polymerization in an open vessel [83], and a drum process [84]. After polymerization, the acid catalyst is deactivated and the unstable hemiacetal and formate terminal groups are removed by heat treatment. In practice, both solution [85, 86] and melt [87, 88] processes have proved suitable. Solution processes are generally carried out at 150–200°C, and melt processes at 170–250°C. Amines or phosphines, for example, are added to deactivate the acid catalysts. They also catalyze the removal of unstable end groups. The objective of all the processes is to completely remove unstable terminal groups and residual monomers from the polymer.

The solution processes are very efficient but have the disadvantage of high energy consumption and higher capital cost. After thermal treatment, the product is precipitated in powder form by lowering the temperature. The adhering residual solvent is removed mechanically and thermally. The dry powder, together with stabilizers and other optional additives, is melted and homogenized in extruders and processed into granules. The melt processes do not achieve the efficiency of the solution processes, but require less energy. After addition of stabilizers, the melt can be processed directly into granules.

A large number of patent applications have been filed by various companies for the polymerization as well as the work-up process, e.g., [89–97].

3.4. Additives

3.4.1. Stabilizers

Polyoxymethylenes can be decomposed by the action of high temperatures, oxygen, shear forces, and/or acidic substances. These influences are especially important in the processing of the products in the melt. Further influences may also occur when using the finished parts.

In the case of terminal-group-stabilized polyoxymethylenes, decomposition begins by chain scission. Above the melting point, chain scission is followed by depolymerization with elimination of formaldehyde. For copolymers, this involves elimination of formaldehyde units until the next stabilizing comonomer unit is reached. In the case of homopolymers complete decomposition of the polymer chain occurs. Formic acid, which can initiate an acidolytic chain scission, is formed as byproduct of the decomposition by oxidation of formaldehyde or by disproportionation. Acidic catalyst residues can also produce chain scission [98, 99]; for further details, see [100].

Only multicomponent systems are used to stabilize polyoxymethylenes. Sterically hindered phenolic compounds in particular have proved suitable against thermal-oxidative decomposition (antioxidants). Alkaline-earth metal salts and nitrogen-containing compounds, among others, are used to prevent acidolytic decomposition (acid trapping agents). Formaldehyde trapping agents (scavengers) can be used to further improve the stability.

The following compounds are used as antioxidants:

1. 4,4'-Butylidenebis(6-*tert*-butyl-3-methylphenol) [101]
2. Triethylene glycol bis[3-(3,5-di-*tert*-butyl-4-hydroxyphenyl)propionate] [102]

3. 1,6-Hexamethylene bis[3-(3,5-di-*tert*-butyl-4-hydroxyphenyl)propionate] [103]
4. 2,2′-Methylenebis(4-methyl-6-*tert*-butylphenol) [104]

The following compounds, among others, are used as acid trapping agents:

1. Dicyanodiamide, melamine [105]
2. Calcium carbonate, stearate, or montanate [106]
3. Polycondensates of formaldehyde, isophthalamide and 2-imidazolidinone [107]
4. Polyamides [108]
5. Triazine derivatives [109]

The nitrogen-containing compounds also act as formaldehyde scavengers. Other formaldehyde scavengers include polyhydroxy compounds such as poly(vinyl alcohol) [110], hydantoins [111], glyoxyldiureide compounds [112], hydrazides of carboxylic acids [113] or guanamine compounds [114].

The aforementioned classes of stabilizers can be combined in numerous ways. Optimum utilization of the possible synergistic effects can be achieved only by carefully matching all the components.

Damage caused to polyoxymethylenes by UV radiation and weathering is described in Section 2.5. The addition of organic light stabilizers and special pigments has proved useful in improving the resistance to UV radiation and weathering.

Compounds from the following classes are used as light stabilizers [115–117]: benzotriazoles, benzophenones, benzoates, oxanilides, and sterically hindered amines (HALS stabilizers). Synergistic effects can also be achieved here by suitable combinations of stabilizers of different classes of compound, in particular combination with HALS stabilizers. Titanium dioxide and, in particular, carbon black are suitable pigments.

All the above-mentioned stabilizers and pigments are added during compounding.

3.4.2. Nucleation Agents

The following are suitable as nucleation agents for polyoxymethylenes:

1. Branched or cross-linked polyoxymethylenes [118]
2. Trioxane block copolymers [119]
3. Special melamine–formaldehyde resins [109, 120]
4. Boron nitride [121]
5. Talc [122]
6. Antimony trioxide [123]

Even in amounts of less than 1 wt% these compounds accelerate crystallization of the melt. Nucleated polyoxymethylene molding compounds often allow earlier removal from the mold, and thus a reduction of the cycle time, in injection molding processing. The resulting spherulite structure is finer and more uniform, which results in increased strength and toughness [22, 124].

3.4.3. Release Agents

The addition of release agents may be advantageous for the production of complicated molded parts. Suitable additives include, for example, polyethylene waxes, montan acid esters and salts, N,N'(bis-stearoyl)ethylenediamine, fatty alcohols, and fatty acid salts.

3.5. Impact-Resistant Types

Polyoxymethylenes are materials that already have a high basic toughness, without the need for further modification. However, the toughness can be improved further by impact resistance modification with elastomers.

The first commercial impact-resistance-modified polyoxymethylenes were introduced in 1982, and in the meantime several manufacturers have developed products of this type. The various products differ according to their flow behavior and degree of impact resistance modification. Compared with the unmodified products the impact-resistant types have slightly reduced rigidity, hardness, and thermal dimensional stability, though this is compensated by a substantially higher energy absorption under impact and tensile stress.

The conventional types are based on blends of polyoxymethylene and thermoplastic polyurethane (POM-TPU) elastomers [125–129].

Products in which cross-linked rubbers having a particularly low glass transition temperature and a defined particle size are used for the impact strength modification have been developed as well [130].

Further development of POM-TPU blends with significantly improved impact resistance and weld-line strength, without sacrificing other key properties, has been commercialized by Ticona. This grade, manufactured by a special backbone-modified POM, exhibits increased impact-performance, excellent friction and wear properties. It also makes possible overall optimized part designs as well as reduced part costs. Speed-up of cooling cycles in injection molding of up to 30% has been reported [131]. In most recent developments, these grades are penetrating areas that were once preserved of impact-modified PA66.

Impact-resistant polyoxymethylenes are produced by mixing the components in the melt, using single-screw or double-screw extruders. Processing into moldings is performed by the same methods as are used for unmodified polyoxymethylenes. Whereas polyurethane-modified products have, depending on the extrusion and processing conditions, a variable morphology of the impact-resistant component (size, shape, and distribution of the particles), the rubber-modified products have a uniform particle morphology. In addition to the particularly high low-temperature impact resistance, these products have further advantages, including the extremely good weld line strength and the absolute absence of delamination during processing [132].

3.6. Low Volatile Organic Carbon (VOC) Grades

Due to higher requirements in the automotive interior, low-emission and high-performance materials are in high demand (especially in Europe and Asia). All major POM-manufacturers have now added special low-emission grades to their range of high-quality products. This helps the car industry to implement its emission minimization strategy.

Most recent developments have shown, that special grades are at least 50% lower than the self-imposed limits specified in the guidelines (VDA 275 and VDA 270) of the German Association of the Automotive Industry [133]. These improvements were achieved by optimized manufacturing technologies as well as the use of new additives.

3.7. High-Strength POM Copolymer

As discussed in Section 2.2, POM homopolymer grades have advantages in mechanical properties, while the copolymer grades are superior in thermal stability and chemical resistance. This is mainly due to differences in polymer structure, molecular-mass distribution and end groups between the two families. Because of more ether-type end groups of copolymers, they typically exhibit greater thermal and chemical (hot water and alkaline) stability. The development of a new POM copolymer grade with a unique combination of the best in mechanical properties and the best in thermal and chemical stability has recently been accomplished. The enhancement was achieved by improvements in the cationic polymerization reaction used to make the POM copolymers. A more detailed comparison between the different materials in regards to mechanical properties, creep resistance and flex fatigue of the materials was presented [134].

4. Economic Aspects

Table 2 lists producers of polyoxymethylenes, together with their capacities and trade names [135]. A demand for polyoxymethylenes of about 700×10^3 t/a can be compared with a nominal capacity of ca. 1000×10^3 t/a for 2008. Homopolymers accounted for about 25% of the total market.

5. Quality Specifications and Analysis

Residual monomer content, molecular mass, thermal stability, and mechanical properties are important quality criteria for polyoxymethylenes. For the copolymers there is in addition the comonomer content.

Table 2. Major world producers of polyacetal resins in 2008 [135]

Company and plant location (Trade name)	Annual capacity as of August 2008, 10^3 t	Share of world capacity, %
North America		
DuPont, DuPont Engineering Polymers Parkersburg, WV, United States (Delrin)	73	6.8
Ticona LLC Bishop, TX, United States (Celcon)	102	9.5
Western Europe		
DuPont Performance Elastomers BV Dordrecht, Netherlands (Delrin)	107	9.9
Ticona GmbH Frankfurt am Main, Germany (Hostaform)	100	9.3
Asia		
Korea Engineering Plastics., Ltd Ulsan, Kyongsangnam-do, Republic of Korea (Kepital)	80	7.4
Polyplastics Co., Ltd. Fuji, Shizuoka Prefecture, Japan (Duracon)	100	9.3
PTM Engineering Plastics (Nantong) Co., Ltd. Nantong, Jiangsu, China	60	5.6
Thai Polyacetal Co., Ltd. Rayong, Thailand (Iupital)	60	5.6
Total	682	63.3

The so-called lutidine or acetylacetone method [136], which is specified by the BfR (Bundesanstalt für Risikoforschung, Berlin, Germany) to evaluate the formaldehyde content of plastic molded parts that come into contact with foods [137, 138], is generally used to measure the content of extractable formaldehyde in granules or in molded parts. Formaldehyde can also be determined with chromotropic acid or by titration with sodium sulfite solution. The tests are normally performed on aqueous extracts of the materials.

The molecular mass is monitored in practice via the melt viscosity (melt flow index, MFI). Various η–M relationships are given in the literature for molecular mass determination by means of the solution viscosity. Usual solvents are dimethylformamide and hexafluoroacetone–water mixtures [139, 140].

The weight loss that occurs in the melt at a specific temperature over a certain time and in a defined atmosphere (air, nitrogen) can be used to evaluate the thermal stability [141, 142].

NMR spectroscopy is suitable for monitoring the comonomer content in the end product [143]. When the comonomer is ethylene oxide, the frequency of occurrence of mono-, di-, and triethylene glycol units can also be determined with this method.

IR-spectroscopic methods have been described for the determination of terminal groups in acetylated homopolymers [144, 145].

Conventional analytical techniques such as UV, IR, NMR, and ashing, optionally in combination with suitable separation methods, can be used for the qualitative and quantitative determination of additives, fillers, and reinforcement agents in the end product.

6. Product Range and Processing

Polyoxymethylenes are produced in various degrees of polymerization. The individual types differ as regards their flow behavior (melt index) and in the choice and concentration of additives and colored pigments. Reinforced types, types with improved slide and wear behavior and impact resistance, and products with anti-UV additives are also available [9–11]. The usual supply form is granules of about 3 mm grain size.

Polyoxymethylenes can be processed with all conventional machinery for thermoplastics, such as injection molding machines, extruders, injection and extrusion blow units, as well as by compression → Plastics, Processing, 1. Processing of Thermoplastics. Temperatures between 190°C and 210°C are generally recommended for processing. The temperature in the melt should not exceed ca. 230°C [10, 11].

Posttreatment of injection moldings is generally not necessary. Machining, in which parts are fashioned from the bulk plastic (semi-finished product), has, however, become important. Machining is economically practicable up to medium-size production runs, particularly if computer numerical control techniques are used.

The "outsert" technique has found widespread use in the past. With this technique a wide range of polyoxymethylene functional components can be bonded to a metal blank in a single step in an injection molding process [146]. Such components include guiding and sliding elements, bearings, axles, columns, snap connections, and spring components. They are used, for example, in compact disc drives and in cassette and video recorders.

Further information on processing is given in [6, 7, 22].

7. Uses

Polyoxymethylenes have a balanced combination of favorable properties for technical use, such as hardness, rigidity, toughness, spring elasticity, and resistance to fuels. The range of uses has been broadened further by a large number of modified special grades. Polyoxymethylenes are used in almost all sectors of industry and technology. The main areas of polyoxymethylene consumption in Western Europe are the automotive sector (36%), consumer goods, industrial applications, fluid handling and household appliances (36%), and in electrical engineering and the electronics industry (17%) [5]. Further uses include mechanical engineering, medical technology, toys, and cosmetics.

Utility articles and parts fabricated from polyacetals vary widely as regards their range of application and are accordingly so numerous that it is only possible to list the most important uses. Some examples of applications are as follows [6]:

Automotive Sector: parts for gasoline pumps, gasoline gauges, carburetor floats, brake boosters, chain wheels, clips, tank closures, water separators, lock housings, fan blades, loudspeaker grills, fuel tank module and safety belt parts.

Electrical Industry: gearwheels, bearings, springs, radial cams, lifters, tappets, axles, pinions, gears, transmissions, worm gears, computer keyboards, water boilers, and parts for leakage current protection circuits.

Mechanical Engineering: valves, pistons, sliding bearings, and ball-type nipple cages.

Sanitary Sector: dispenser, fittings, control flaps for WC flush units, and functional parts in water and gas meters.

8. Recycling

On account of the high commercial value of polyoxymethylenes a functioning material reuse recycling system for polyoxymethylenes already exists. Many plastics processors use material residues directly themselves and contribute substantially by means of this internal recycling process to reduce waste. Polyoxymethylene waste is also processed by compounders into regranulate. This regranulate is, however, of poorer quality than the original material.

Hoechst has developed a closed-loop recycling process, in which polyoxymethylene waste is decomposed into the monomer starting materials. The recovered monomers are purified and polymerized [147]. The cycle can be repeated as often as desired without affecting the quality.

9. Toxicology and Occupational Health

Polyoxymethylenes are odorless and tasteless. Feeding tests have not produced any positive clinical results. Skin compatibility investigations showed that polyoxymethylene does not irritate intact skin [40, 148].

There are no objections to the use of polyoxymethylene moldings that come into contact with foods, as long as the regulations of the licensing authorities (BGA [137], FDA [149, 150]) are satisfied and the parts are suitable for the intended use. The regulations cover the nature and amount of monomers, comonomers, catalysts, deactivators, regulators, stabilizers, lubricants, and fillers, and specified contents of fluorine, boron, zinc, and amines must not be exceeded. A maximum permissible upper limit is specified for formaldehyde traces that can be extracted from finished parts.

The maximum permitted workplace concentration for formaldehyde should also be observed when processing polyoxymethylenes.

In Germany the MAK is 0.3 ppm (0.37 mg/m^3) [151], and in the United States the PEL-TWA is 0.75 ppm (0.9 mg/m^3) [152]. These limits may be exceeded for a short period only if certain preconditions are satisfied. For further details see → Formaldehyde.

References

1. W. Kern in H. Staudinger (ed.): *Die hochmolekularen organischen Verbindungen Kautschuk und Cellulose*, Springer Verlag, Berlin 1932.
2. C.E. Schweitzer, R.N. MacDonald, J.O. Punderson, *J. Appl. Polym. Sci.* **1** (1959) no. 2, 158–163.
3. Celanese, US 3027352, 1960 (C.T. Walling, F. Brown, K.W. Bartz).
4. Hoechst, DE 1124703, 1959 (K. Küllmar, E. Fischer, K. Weissermel).
5. T. Vaahs, D. Smeets, *Kunststoffe* **10** (2010) 124.
6. H.D. Sabel et al.: *Kunststoff-Handbuch*, vol. 3/1, Hanser Verlag, München 1992, p. 300–395.
7. H. Schmidt, U. van Spankeren: Polyacetale, in: *Kunststoff-Handbuch*, vol. XI, Hanser Verlag, München 1971.
8. K. Kurz, K.-U. Tönnes: *Polyacetal (POM) Kunststoffe, Eigenschaften und Anwendungen*, 7th ed., VDI-Verlag, Düsseldorf 2008, p 546–574.
9. DuPont: Delrin(r) acetal resin,http://www2.dupont.com/Plastics/en_US/Products/Delrin/Delrin.html (accessed Sept. 16, 2011).
10. Ticona: Hostaform(r) POM, http://www.ticona.com/products/hostaform (accessed Sept. 16, 2011).
11. BASF: Characteristic properties of Ultraform(r) (POM),http://www.plasticsportal.net/wa/plasticsEU~en_GB/portal/general/content/products/engineering_plastics/ultraform (accessed Sept. 16, 2011).
12. G. Sextro: Physical Constants of Poly(oxymethylene), inJ. Brandrup,E.H. Immergut (eds.): *Polymer Handbook*, 3rd ed., vol. 1, J. Wiley & Sons, New York 1989, V/p. 87–V/p. 99.
13. S. Okamura et al., *J. Polym. Sci. Part C* **4** (1963) 827–838.
14. J.C. Bevington, H. May: "Aldehyde Polymers", *Encycl. Polym. Sci. Technol.* **1** (1964) 609–628.
15. H. Wilski, *Makromol. Chem.* **150** (1971) 209.
16. F.S. Dainton, D.M. Evans, F.E. Hoare, T.P. Melia, *Polymer* **3** (1962) 263.
17. H. Suzuki, B. Wunderlich, *J. Polym. Sci. Polym. Phys. Ed.* **23** (1985) 1671.
18. F.S. Dainton, K.J. Ivin, D.A.G. Walmsley, *Trans. Faraday Soc.* **55** (1959) 61.
19. F. Salaris, A. Turturro, U. Bianchi, E. Martuscelli, *Polymer* **19** (1978) 1163.
20. H. Wilski, *Kolloid-Z.Z. Polym.* **248** (1971) 867.
21. VDI/VDE: *Werkstoffe der Feinwerktechnik-Polyacetal-Formstoffe*, Richtlinie 2477, Beuth Verlag, Berlin 1974.
22. H. Dominghaus: *Die Kunststoffe und ihre Eigenschaften*, 3rd ed., VDI-Verlag, Düsseldorf 1988, pp. 307–336.
23. H. Breuer, G. Dupp, J. Schmitz, R. Tüllmann, *Kunststoffe* **80** (1990) 1289–1294.
24. T. Uchida, H. Todokoro, *J. Polym. Sci. Polym. Phys. Ed.* **5** (1967) 63.
25. U. Gaur, B. Wunderlich, *J. Phys. Chem. Ref. Data* **4** (1981) no. 10, 1001.
26. B.E. Read, G. Williams, *Polymer* **2** (1961) 239.
27. M. Dröscher, K. Hertwig, H. Reimann, G. Wegner, *Makromol. Chem.* **177** (1976) 1696–1706.
28. Hoechst, company information, no. HBKB 0395 C.3.3, *Gestalten von Formteilen aus technischen Kunststoffen* Frankfurt, 1991.
29. Hoechst, company information, no. HBKP 0395/C.3.4 D, *Richtlinien für das Gestalten von Formteilen aus technischen Kunststoffen*.(English version HBKP 0395/C.3.4 E) Frankfurt, 1997.
30. E. Wolters, H. Racké, *Kunststoffe* **63** (1973) 608–612.
31. F. Gügümus in R. Gächter,R. Müller (eds): *Taschenbuch der Kunststoffe–Additive*, Hanser Verlag, München 1989.
32. J.F. Walker: *Formaldehyde*, 3rd ed., reprint of orig. ed. 1964, Krieger Publ. Comp., Malabar, Florida,p. 179.
33. K. Kohlhepp, R. Orth, *Kunststoffe* **77** (1987) 686.
34. R. Vesely, M. Kalenda, *Kunststoffe* **59** (1969) 107–110.
35. J.-L. Gardette, H.-D. Sabel, J. Lemaire, *Angew. Makromol. Chem.* **188** (1991) 113–128.
36. N. Grassie, R.S. Roche, *Makromol. Chem.* **112** (1968) 34–39.
37. H. Schmidt, E. Wolters, *Kunststoffe* **61** (1971) no. 4, 261–265.
38. W. Kern, *Chem. Ztg.* **91** (1967) 255–262.
39. E. Wolters, S. Rösinger, *Kunststoffe* **63** (1973) 605–608.
40. Du Pont, company information, *Delrin-Polyacetal*, no. A 84 029–2.
41. Du Pont, US 3183211, 1961 (K.C. Brinker, C.E. Schweitzer).
42. W. Kern, H. Cherdron, V. Jaacks, *Angew. Chem.* **73** (1961)no. 6, 177–224.
43. W. Kern, V. Jaacks, *Kolloid-Z.Z. Polym.* **216–217** (1967) 286–298.
44. W. Kern, H. Deibig, A. Giefer, V. Jaacks, *Pure Appl. Chem.* **12** (1966) 371–386.
45. K. Weissermel et al., *Angew. Chem.* **79** (1967) no. 11, 512–520; *Angew. Chem. Int. Ed. Engl.* **6** (1967) no. 6, 526–533.
46. K. Weissermel, E. Fischer, K. Gutweiler, H.D. Hermann, *Kunststoffe* **54** (1964)no. 7, 410–415.
47. J. Masamoto, *Progr. Polm. Sci.* **18** (1993) 1–84.
48. G.E. Haddeland in: *Process Economics Program*, no. 69: Acetal Resins, Stanford Res. Inst., Menlo Park, California, 1971.http://chemical.ihs.com/PEP/Public/Reports/Phase VI/RP069 RP069 toc.pdf
49. Du Pont, US 2768994, 1954 (R.N. MacDonald).
50. W.H. Linton, H.H. Goodman, *J. Appl. Polym. Sci.* **1** (1959) no. 2, 179–184.
51. J.F. Walker: *Formaldehyde*, 3rd ed., reprint of orig. ed. 1964, Krieger Publ. Comp. Malabar, Florida.
52. J. Masamoto et al., *Macromol. Chem. Macromol. Symp.* **42/43** (1991) 409–423.
53. W.S. Fong in:Process Economics Program, no. 89–2-3: "Acetal Copolymers by an Asahi Chemical Process", SRI Int., Menlo Park, California, 1990.http://chemical.ihs.com/PEP/Public/Reports/Phase 89/RW89-2-3/
54. Du Pont, US 4539387, 1984 (L.M. Blair).
55. Du Pont, US 2943701, 1957 (D.L. Funck).
56. Du Pont, DE 1000152, 1953 (R.N. MacDonald).
57. Du Pont, US 2841570, 1953 (R.N. MacDonald).
58. Du Pont, US 2848437, 1954 (W.P. Langsdorf, G.S. Stamatoff).
59. Du Pont, GB 793673, 1956 (H.H. Goodman, L.T. Shelwood).
60. Du Pont, US 2734689, 1954 (F.C. Starr).
61. J.F. Walker: *Formaldehyde*, 3rd ed., reprint of orig. ed., 1964, Krieger Publ. Comp., Malabar, Florida,p. 159.
62. J. Masamoto, *Progr. Polm. Sci.* **18** (1993) 18.
63. Mitsubishi Gas Chem. Ind., DE-OS 2855710, 1978 (A. Sugio et al.).

64. Hoechst, EP 162252, 1985 (S. Rittner, K.-H. Burg, H. Schlaf).
65. Celanese, US 2989507, 1957 (D.E. Hudgin, F.M. Berardinelli).
66. Degussa, DE 1152818, 1960 (W. Kern, A. Wildenau, V. Jaacks).
67. Polyplastics, EP 325052, 1988 (K. Yamamoto, H. Sano, S. Chino).
68. S. Penczeck et al,*Makromol. Chem.* **190** (1989) no. 5, 929–938.
69. British Ind. Plastics, GB 1054205, 1964 (H. May, B.J. Kendall-Smith, J.A. Dodd).
70. BASF, US 7112651, 2006 (H.-H. Goertz et. al.).
71. S. Penczek et al.: Cationic Ring Opening Polymerization: Acetals, in *Comprehensive Polymer Science*, vol. 3, Pergamon Press, London 1989, p. 796.
72. S. Penczek et al.: Cationic Ring Opening Polymerization: Acetals, in *Comprehensive Polymer Science*, vol. 3, Pergamon Press, London 1989, p. 792.
73. Ticona, US 7902324, 2011 (M. Haubs, M. Hoffmockel, H. Roeschert).
74. Degussa, DE 1194145, 1959 (W. Kern, V. Jaacks).
75. Hoechst, US 3194788, 1959 (K. Küllmar, E. Fischer, K. Weissermel).
76. Celanese, US 2989511, 1958 (A.W. Schnizer).
77. St. Penczek, J. Fejgin, W. Sadowska, M. Tomaszewicz, *Makromol. Chem.* **116** (1968) 203.
78. Hoechst, DE 1161421, 1961 (K. Küllmar, H. Schmidt, G. Roos, E. Fischer).
79. Celanese Corp., US 4 10563156, 1974 (M. Semanchik, D.M. Braunstein).
80. H. Sawada: *Thermodynamics of Polymerization*, Marcel Dekker, New York 1976.
81. Hoechst, DE 2263300, 1972 (G. Sextro, K.-H. Burg, K. Gutweiler).
82. BASF, EP 80656, 1982 (J. Sadlowski, M. Walter, K. Hinselmann).
83. BASF, DE 2003270, 1970 (E. Bäder, H. Amann).
84. BASF, DE 1720300, 1967 (H. Heel, H. Klingenspohr, W. Klee, G. Göschel).
85. Celanese, DE 1445294, 1961 (F.M. Berardinelli, J.E. Wall, E.T. Smith, G.J. Fisher).
86. Hoechst, DE 2452736, 1974 (G. Sextro, K.-H. Burg, H.J. Leugering, H. Schlaf, A. Heller).
87. Celanese, DE 2319972, 1973 (M.D. Golder).
88. BASF, DE 1963402, 1969 (B. Sander, H. Sperber).
89. Asahi, JP 59159812, 1983. *Chem. Abstr.* **102** (1985) 204 475 t.
90. Asahi, JP 57139113, 1981. *Chem. Abstr.* **98** (1983) 35 186 r.
91. Asahi, JP 63162686, 1986 (H. Hata, J. Masamoto). *Chem. Abstr.* **109** (1988) 231 712 u.
92. Mitsubishi Gas, DE 3617754, 1986 (T. Unemura, I. Masumoto, Y. Iha).
93. Mitsubishi Gas, DE 3738632, 1987 (T. Suzumori, I. Masumoto, M. Nakagawa, Y. Iha).
94. UBE, GB 2132213, 1983 (K. Tsunemie, T. Suzuki, Y. Yamamoto).
95. UBE, JP 01108211, 1987 (K. Suzuki, K. Hisatomi, K. Kamimura). *Chem. Abstr.* **111** (1989) 215 164 c.
96. UBE, JP 01121313, 1987 (K. Suzuki, K. Tokunaga, N. Seki). *Chem. Abstr.* **111** (1989) 215 147 z.
97. UBE, JP 61151218, 1984 (K. Hisatomi, H. Ninomiya, Y. Yamamoto). *Chem. Abstr.* **105** (1986) 227 877 b.
98. W. Kern, H. Cherdron, *Makromol. Chem.* **40** (1961) 101.
99. W. Kern, H. Cherdron, *Makromol. Chem.* **40** (1961) 177.
100. G. Opitz, *Plaste Kautsch.* **28** (1981) no. 1, 15–19.
101. Du Pont, DE 1082404, 1957 (R.D. Kralovec, P.N. Richardson).
102. Ciba Geigy, DE 2133374, 1971 (E.K. Kleiner, J.D. Spivack).
103. Ciba Geigy, DE 2203836, 1971 (H. Müller).
104. Du Pont, GB 748856, 1953 (M.A. Kubico, R.N. MacDonald, R.L. Stearns, F.A. Wolff).
105. Celanese, US 3313767, 1967 (F.M. Berardinelli, J. Kray, T.J. Dolce).
106. Hoechst, US 3743614, 1973 (E. Wolters, G. Roos).
107. BASF, DE 1669692, 1967 (E. Schwartz, F. Schmidt, B. Sander, H. Sperber, H. Wilhelm, E. Ricker).
108. Du Pont, DE 1066739, 1957 (R.G. Alsup, P.E. Lindvig).
109. Degussa, EP 0202530, 1986 (H. Amann, G. Morlock, E. Schöla).
110. Du Pont, EP 242176, 1987 (N.E. West).
111. DuPont, US 586 671, 1999 (K. Shinohara et al.).
112. Polyplastics, US 7041718, 2006 (H. Harashina, H. Kurita, T. Yamada).
113. Asahi, US 7816433, 2010 (S. Kumatsu, Y. Sasaki).
114. Polyplastics, US 7183340, 2007 (H. Harashina, H. Kurita).
115. Du Pont, US 3043709, 1958 (L.E. Amborski).
116. Du Pont, GB 921808, 1961 (D.B. Carlson).
117. F.R. Stohler, K. Berger, *Angew. Makromol. Chem.* **176/177** (1990) 323–332.
118. Hoechst, DE 2101817, 1971 (R. Kern, H. Schmidt, K.-H. Burg, E. Wolters).
119. Hoechst, DE 2142091, 1973 (G. Sextro, K.-H. Burg).
120. G. Morlock: *6. Internationales Makromolekulares Symposium*, Interlaken, Switzerland, June 8-9., 1978.
121. Hoechst, US 3767610, 1971 (K.-H. Burg, L. Brinkmann, E. Wolters).
122. Bayer, DE 1247645, 1964 (W. Hechelhammer).
123. Hoechst, US 3783154, 1971 (H.-H. Grossmann, K.-H. Burg, E. Wolters).
124. H.J. Leugering, K.-H. Burg, H. Cherdron, *Makromol. Chem. Suppl.* **1** (1975) 621–636.
125. F. Kloos, E. Wolters, *Kunststoffe* **75** (1985) 735–739.
126. Hoechst, EP 115847, 1984 (R. Reske, E. Wolters).
127. Celanese, DE 2051028, 1970 (F. McAndrew).
128. Bayer, DE 1193240, 1962 (K. Wagner, H. Röhr).
129. Du Pont, EP 0116456, 1984 (E.A. Flexman, Jr.).
130. Hoechst, EP 0156285, 1985 (K.-H. Burg, H. Cherdron, F. Kloos, H. Schlaf).
131. L. Larson, J. Lipke: *ANTEC*, Orlando, FL 2010, p. 1344.
132. H.-D. Sabel, U. Struth, *Kunststoffe* **80** (1990) no. 10, 1118–1122.
133. U. Ziegler: *13. Workshop "Geruch und Emission bei Kunststoffen"*, Kassel, Germany, March 28-29, 2011.
134. R.M. Gronner: "A New High Strength POM Copolymer", *ANTEC*, Boston, MA, May 4, 2011.
135. SRI Consulting, *Polyacetal resins*, October 2008, http://www.sriconsulting.com/CEH/Public/Reports/580.0950/ (accessed Sept. 16, 2011). http://chemical.ihs.com/CEH/Public/Reports/580.0950/
136. T. Nash, *Biochem. J.* **55** (1953) 416–421.
137. R. Franck, H. Wieczorek: *Kunststoffe im Lebensmittelverkehr*, Teil A, XXXIII: Acetalharze, Stand: 1. Sept. 1987, 38. Lieferung, 1989, Carl Heymanns Verlag, Köln.
138. R. Franck: *Kunststoffe im Lebensmittelverkehr*, Teil B II, XVIII: Bestimmung von Formaldehyd in Kunststoffgefäßen aus Melaminharz, Stand 1. Sept. 1961, 33. Lieferung, August 1984, Carl Heymanns Verlag, Köln.
139. L. Hoehr et al., *Makromol. Chem.* **103** (1967) 279.
140. I.M. Bel'gorskii, N.S. Enikolopyan, L.S. Sakhonenko, *Vysokomol. Soedin.* **4** (1962) 1179; *Polym. Sci. (USSR) (Engl. Transl.)* **4** (1963) 367.

141. Du Pont, US 2993025, 1958 (R.G. Alsup, P.E. Lindvig).
142. Celanese, FR 1221148, 1959 (T. Walling, F. Brown, K.W. Bartz, G.W. Polly).
143. D. Fleischer, R.C. Schulz, *Makromol. Chem.* **176** (1975) 677–689.
144. H. Frank, V. Jaacks, W. Kern, *Makromol. Chem.* **114** (1968) 92–112.
145. T.A. Koch, P.E. Linvig, *J. Appl. Polym. Sci.* **1** (1959) 164–168.
146. Hoechst, company information, no. C.3.5, *Outsert-Technik mit Hostaform*.
147. Hoechst, DE 4035495, 1990 (K.-F. Mück, G. Reuschel, D. Fleischer).
148. R.B. Akin: *Acetal Resins*, Reinhold, New York 1962.
149. Federal Register, Code of Federal Regulations, Title 21, Part 177, Section 2470, U.S. Government Printing Office, Washington 1989.
150. Federal Register, Code of Federal Regulations, Title 21, Part 177, Section 2480, U.S. Government Printing Office, Washington 1989.
151. http://gestis.itrust.de/nxt/gateway.dll/?f=templates&fn=default.htm (Jan. 18, 2012).
152. Formaldehyde, Occupational Safety and Health Standards, http://www.osha.gov/&vid=gestisdeu:sdbdeu/SLTC/formaldehyde/index.html (accessed Sept. 16, 2011).

Poly(Phenylene Oxides)

JAN BUSSINK, GE Plastics BV, Bergen op Zoom, The Netherlands

HENDRIK T. VAN DE GRAMPEL, GE Plastics BV, Bergen op Zoom, The Netherlands

1.	Introduction	927
2.	Production	927
3.	PPE Blends	929
3.1.	PPE – PS Blends	929
3.2.	PA – PPE Blends	931
3.3.	PPE – Polyolefin Blends	933
4.	Processing	933
5.	Uses	934
6.	Acknowledgement	936
	References	936

1. Introduction

Poly(phenylene oxides) [9041-80-9], also known as poly(phenylene ethers) or polyoxyphenylenes, refer to a class of polymers made by the oxidative polymerization of substituted phenols. A large variety of polymers have been investigated based on differently substituted phenols (Table 1). However, only poly(2,6-dimethyl-1,4-phenylene ether) [25134-01-4] (PPE) has significant utility, albeit in blends with other polymers such as polystyrene or polyamide:

History. The discovery of poly(2,6-dimethyl-1,4-phenylene ether) in 1956 by HAY (General Electric) [1] marked the introduction of the catalyzed oxidative coupling of phenols as a means of obtaining engineering resins. This poly-(phenylene oxide) made from 2,6-dimethylphenol can be regarded as amorphous for all practical purposes [2], and possesses a high glass transition temperature (T_g 211 °C) in addition to an extremely high melt viscosity, which requires high processing temperatures that lead to side reactions [3],[4]. Only through the exceptional miscibility of PPE and polystyrene (PS) [5] could acceptable processing temperatures and melt viscosities be achieved. This development led to the marketing of the first commercial poly(phenylene oxide) in 1964—Noryl, a blend of PPE and impact modified PS (HIPS). These PPE – HIPS blends are unique because their ultimate use temperature can be easily adapted to application requirements by adjustment of the PPE – HIPS ratio. GE Plastics was the sole supplier of modified PPE until 1983 when its original patents expired; other companies (e.g., BASF with Ultranyl and Hüls with Vestoran) have now introduced their own PPE grades. A blend of HIPS and a PPE copolymer with a similar property profile to poly(2,6-dimethyl-1,4-phenylene ether) was introduced in the early 1970s under the trade name Prevex (Borg – Warner and Mitsubishi Chemicals):

Today, over 200 000 t/a of modified PPE are sold worldwide, mainly for applications in the automotive, electronics, and electrical industries.

2. Production

The synthesis of PPE from 2,6-dimethylphenol, using an amine complex of a basic cupric salt as a homogeneous oxidation catalyst, follows a

Table 1. 2,6-Substituted aromatic polyethers formed by oxidative coupling of di-*ortho*-substituted phenols

R_1	R_2	Polymer	Diphenoquinone compounds	CAS registry no.	T_g, °C	T_m, °C
H	H	branched	branched	[25667-40-7]	82	298
H	CH_3	branched	branched			
CH_3	CH_3	X		[25134-01-4]	211	268
CH_3	C_2H_5	X		[26635-16-5]		
CH_3	$CH(CH_3)_2$	X		[31985-12-3]	144	
CH_3	$C(CH_3)_3$		X	[35916-68-8]		
CH_3	$CH_2C_6H_5$	X		[26545-37-9]	99	
CH_3	C_6H_5			[25805-39-4]	169	
CH_3	OCH_3	X				
CH_3	Cl	X		[26498-58-8]		
C_2H_5	C_2H_5	X			120	
$CH(CH_3)_2$	$CH(CH_3)_2$		X			
$C(CH_3)_3$	$C(CH_3)_3$	X				
Cl	Cl	X		[26023-26-7]	228	269
OCH_3	OCH_3		X	[25667-13-4]	167	
C_6H_5	C_6H_5	X		[24938-68-9]	225	460 (decomp.)
C_6H_5	$3\text{-}CH_3C_6H_4$	X		[79569-12-3]	219	
C_6H_5	$4\text{-}CH_3C_6H_4$	X		[79569-09-8]	218	
C_6H_5	$4\text{-}(CH_3)_3C_6H_4$	X		[79569-10-1]	240	
C_6H_5	2-naphthyl	X		[79569-11-2]	234	

complicated polymerization mechanism which has not been completely resolved [6],[7].

$$n \underset{CH_3}{\underset{|}{\overset{CH_3}{\overset{|}{\bigcirc}}}}\!\!-\!\!OH + \tfrac{1}{2}nO_2 \xrightarrow[\text{Toluene, 30--50 °C}]{\text{Cu/OH/amine}} \left[\!\!\begin{array}{c} CH_3 \\ \bigcirc\!\!-\!\!O \\ CH_3 \end{array}\!\!\right]_y + nH_2O$$

The main byproduct of the polymerization is 3,3′,5,5′-tetramethyldiphenoquinone (TMDPQ):

$$HO\!\!-\!\!\underset{CH_3}{\underset{|}{\overset{CH_3}{\overset{|}{\bigcirc}}}}\!\!-\!\!\underset{CH_3}{\underset{|}{\overset{CH_3}{\overset{|}{\bigcirc}}}}\!\!-\!\!OH$$

TMDPQ

The oxidative coupling chemistry is very versatile because different di-*ortho*-substituted phenols can be used for the formation of 2,6-sub-stituted aromatic (co)polyethers (Table 1). However, only PPE is of commercial importance.

Commercial PPE production can be performed batchwise or continuously. Different reactor designs are possible depending upon the choice of the solvent and introduction of oxygen (pure oxygen or air – oxygen mixtures). A gas – liquid reactor is used to avoid low reaction rates associated with diffusion of air into the liquid. Reaction temperatures are typically 20–65 °C. At higher temperatures, formation of TMDPQ increases and should be avoided. The solubility and viscosity of the final polymer solution depend on the solvent, temperature, polymer concentration, and its molecular mass. Commercial PPE is produced with a weight-average molecular mass of 32 000–40 000.

The following stages can be discerned (Fig. 1):

1. Preparation of the catalyst solution
2. Preparation of reaction mixture
3. Polymerization using oxygen or oxygen – air mixtures
4. Termination of polymerization at a certain degree of polymerization

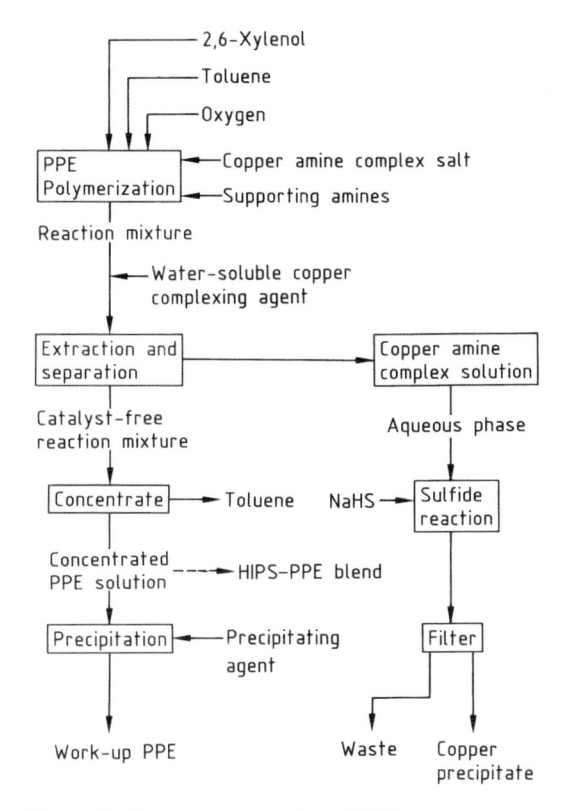

Figure 1. Commercial production of PPE

5. Removal of copper catalysts from the reaction mixture using a copper complex as complexing agent and liquid – liquid extraction
6. Concentration of PPE solution
7. Precipitation of PPE in a nonsolvent
8. Repeated washing and drying (purification of resin)

3. PPE Blends

3.1. PPE – PS Blends

A processable, commercially attractive material is obtained by blending PPE with PS (Fig. 2). The unique, complete miscibility of this blend presumably results from interaction between the π-electron donor PS and the electron-deficient methyl groups of PPE [8]. The T_g of the blend can be varied almost linearly between 100 and 215 °C over the complete composition range of PS and PPE (Fig. 3). Table 2 shows some important properties of PPE – HIPS blends.

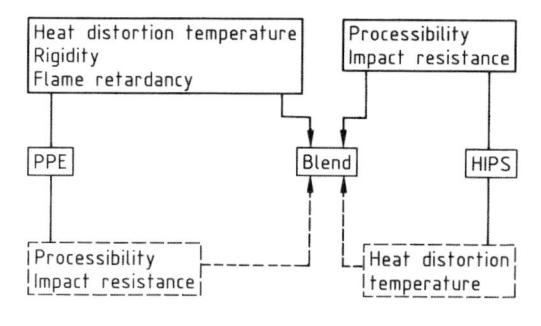

Figure 2. Blending criteria for PPE – PS

Commercial blends such as Noryl (GE Plastics) or Luranyl (BASF) make use of PS impact-modified with polybutadiene (high-impact polystyrene, HIPS). Impact modification of these blends comprising a ductile resin (PPE) and a brittle one (PS) poses an interesting problem because the two components require different rubber particle sizes for optimal impact modification (Fig. 4). Diluting HIPS with PPE yields a more ductile matrix in which smaller rubber particles (diameter 0.5–2 μm) are more effective in impact modification than the larger particles (2–12 μm) [9],[10]. The larger particles are more effective at higher HIPS levels [11]. Processing temperatures of PPE-rich compositions above 80 °C cause additional cross-linking of the polybutadiene rubber phase, rendering it less effective.

PPE – HIPS grades offer relatively constant mechanical properties over a wide temperature

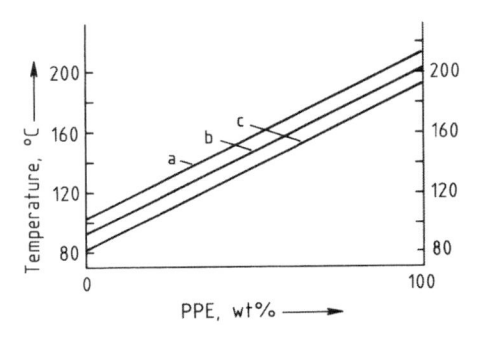

Figure 3. Thermal properties of PPE – HIPS blends
a) Glass transition temperature, T_g; b) Vicat B softening temperature; c) Heat distortion temperature

Table 2. Selected properties of PPE – HIPS blends (Noryl, GE Plastics)

Property	Standards ASTM[f]	Standards DIN[f]	Unreinforced blend[a]	Unreinforced flame retardant[b]	Glass-reinforced blend[c]	Glass-reinforced flame retardant[d]	Structural foam[e]
Mechanical							
Tensile strength (break), MPa	D 638	53 455	50	45	90	77	22
Tensile elongation (break), %	D 638	53 455	50	50	3	2	8
Tensile modulus, MPa	D 638	53 457	2500	2400	6500	7300	1600
Flexural strength (yield), MPa	D 790	53 452	98	90			
Flexural strength (break), MPa	D 790	53 457			140	120	46
Flexural modulus, MPa	D 790	53 457	2500	2500	6000	5000	1800
Hardness (H 358/30), MPa		53 456	100	87	122	147	
Hardness, Rockwell	D 785		R 115	R 115	L 106		
Flexural creep (300 h, 6.9 MPa), %	D 674						0.60
Impact strength							
Charpy impact (notched), kJ/m^2		53 453	15	15	9	9	5
Izod impact (notched), J/m	D 256		200	250	80	60	
Izod impact (notched, −40 °C), J/m	D 256		140	120	70	50	
Thermal							
Vicat softening point (rate B), °C	D 1525		148	105	145	148	
Vicat VST/B/120, °C		53 460	135	100	140	135	
Deflection temperature under load (0.45 MPa), °C	D 648	53 461				133	96
Deflection temperature under load (1.82 MPa), °C	D 648	53 461	130	90	132	126	83
Thermal conductivity, W m^{-1}K^{-1}	C 177		0.22	0.16	0.24		0.12
Linear thermal expansion coefficient (flow), K^{-1} (7–8)×10^{-5}	D 696	VDE 0304/1		6×10^{-5}	7×10^{-5}	4×10^{-5}	2.4×10^{-5}
Physical							
Relative density	D 792	53 479	1.06	1.10	1.21	1.45	0.88
Water absorption (24 h), %	D 570		0.07	0.08	0.06	0.05	
Water absorption (equilibrium), %	D 570		0.14	0.37	0.14		
Water absorption (equilibrium, 100 °C), %	D 570		0.30	0.55	0.32		
Mold shrinkage (flow), %	D 955		0.5–0.7	0.5–0.7	0.2–0.4	0.1–0.3	0.6–0.9
Electrical							
Volume resistivity, Ω · m	D 257		>10^{15}	>10^{15}	>10^{15}	>10^{15}	
Dielectric strength (3.2 mm), kV/mm	D 149		22.0	16.0	17.0		7.5 (6 mm)
Dielectric constant (50 Hz)	D 150		2.70	2.70	2.90		2.30
Dielectric constant (1 MHz)	D 150		2.60	2.60	2.90		2.20
Dissipation factor (50 Hz)	D 150		0.0004	0.0004	0.00008		0.0047
Dissipation factor (1 MHz)	D 150		0.0009	0.0009	0.0014		0.0039
Arc resistance (tungsten), s	D 495						
Flame characteristics							
Oxygen index, %	D 2863		24	33	26		28
Flame retardancy (mm thickness)	UL-94		HB (1.65)	V-0 (1.52) 5 V (3.12)	HB (1.47)	V-0 (1.60) 5 V (3.20)	V-0 (6.29) 5 V (6.29)

[a] Noryl 731.
[b] Noryl N 190, V – 0 rating according to UL 94; V-1 and V-2 grades also available.
[c] 20 wt% glass-filled grade Noryl GFN 2; 10, 20, and 30 wt% glass-filled grades also available.
[d] V-0 rating according to UL 94; 35 wt% glass-filled grade Noryl VO 3525.
[e] Noryl FN 150 D.
[f] Unless otherwise stated.

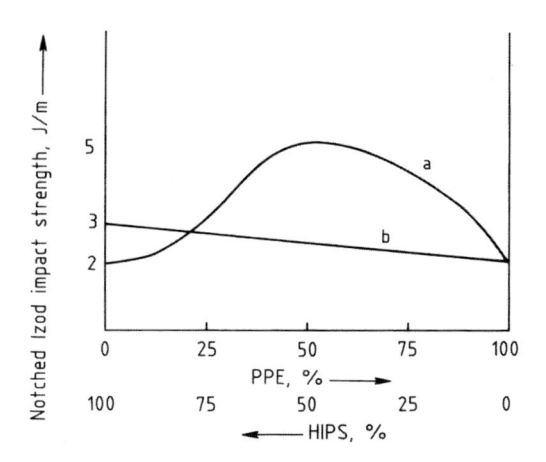

Figure 4. Notched impact strength of PPE – HIPS blends impact-modified with polybutadiene rubber (10 wt %, content in HIPS)
a) 0.5–2 µm polybutadiene rubber; b) 2–12 µm polybutadiene rubber

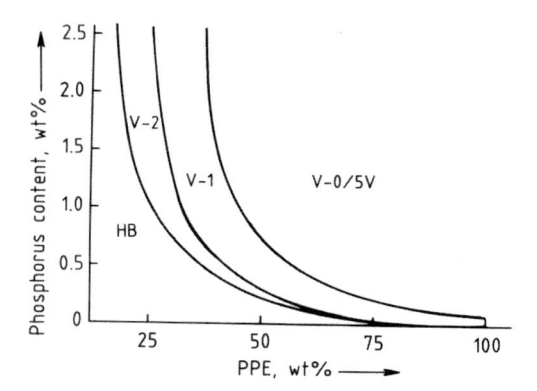

Figure 5. UL-94 flammability ratings of PPE – HIPS blends with different triaryl phosphate contents

range. Ultimate use temperatures of PPE – HIPS blends, as estimated from UL temperature ratings, are between 90 and 100 °C (see Table 2).

The very low water absorption of PPE –HIPS blends and their favorable mechanical properties contribute to a high dimensional stability under load, the most important property of engineering thermoplastics.

Polymer miscibility is not restricted to the PPE – HIPS system, but also holds for combinations of other substituted poly(1,4-phenylene oxides) (see Table 1) with styrene homopolymers such as poly(α-methylstyrene), poly(4-methylstyrene), and styrene copolymers in which the comonomer content is limited to maintain full miscibility with PPE.

Adjustment of the maximum use temperature offered by blend technology also applies to the UL-94 flammability ratings of the PPE – HIPS grades (Fig. 5). For a description of UL-94 ratings, see → Flame Retardants. PPE is intrinsically flame extinguishing as exemplified by its high limiting oxygen index (LOI) value of 29. HIPS, on the other hand, has a LOI of 17 and burns easily and completely [12]. Blending these two miscible polymers provides an intermediate flammability behavior which can be further improved by adding triaryl phosphates to produce flame-retardant PPE – HIPS grades that meet required industrial specifications without utilizing halogen-containing compounds [13]. Since the organic phosphate also functions as a plasticizer, a higher percentage of PPE is required to maintain a given heat distortion temperature (HDT) (Fig. 3). This leads to a well-balanced relation between flame retardancy behavior, composition, and HDT. The flame-retardant PPE – HIPS grades also have excellent melt processability.

Enhanced flame retardancy, however, decreases the comparative tracking index (CTI), a criterion used for applications in the electrical and electronics industry (→ Insulation, Electric, Section 2.1.). This problem can be solved by the addition of inorganic fillers (e.g., glass fibers) and has led to the development of special grades.

PPE – HIPS blends are, like most polymers, excellent electrical insulators. The dielectric loss factor, a measure for the loss of electrical energy through heat buildup, is <0.002 in the range 1 – 16 MHz over a wide temperature span [14]. For other electrical properties of poly(phenylene oxides), see → Insulation, Electric.

Chemical resistance of PPE – HIPS blends is summarized in Table 3. Temperature, strain level, contact time, and the chemical nature of the contact liquid strongly affect the related environmental stress cracking phenomena. PPE –HIPS blends have an intrinsically high resistance against hydrolysis under acidic and basic conditions.

3.2. PA – PPE Blends

PPE can also be used as an organic filler in semicrystalline matrices such as polyamides (PA 66 or PA 6). PA – PPE blends have

Table 3. Chemical resistance of PPE – HIPS (Noryl) and PA – PPE (Noryl GTX) blends

Chemical	PPE – HIPS blend	PA – PEE blend
Water and aqueous solutions		
Cold water	excellent	very good
Hot water	excellent	good
Cleaning agent	very good	satisfactory/good
Salts	very good	good
Acidic solutions	good	good
Basic solutions	good	satisfactory
Concentrated bases	good	satisfactory
Concentrated acids	satisfactory	bad
Organic solvents		
Alcohols	very good	very good
Ketones	very good	satisfactory
Aliphatic hydrocarbons	swelling	very good
Aromatic hydrocarbons	bad	good
Chlorinated hydrocarbons	bad	bad
Esters	bad	satisfactory
Miscellaneous		
Gasoline	bad	(very) bad
Fats	type-dependent	(very) good
Oils	type-dependent	(very) good
Freon	type-depentent	(very) good

been specifically developed for on-line painted exterior automotive parts with excellent ("class A") surface quality.

The favorable processability properties, chemical resistance, and impact resistance of PA are combined with the high-temperature performance of PPE. PA – PPE blends were the first commercial blends of two completely incompatible polymers and are different from the fully miscible PPE – HIPS system. Blends of immiscible polymers are thermodynamically unstable and lack morphological stability, which leads to phase separation and delamination. This can be overcome by the use of compatibilizers (concen-tration usually $\geq$ 10 wt %), which are frequently A–B block copolymers. The compatibilizers reduce the interfacial tension between the constituent polymers, resulting in greatly improved dispersion of PPE in PA. They also provide strong bonding between the dispersed and the continuous phases (important for mechanical properties), and increase morphological stability by preventing agglomeration of dispersed particles. The latter is important as it allows processing of the blends under practical shear conditions such as exist in injection-molding equipment.

Compatibilization is achieved either by adding PA – PPE block copolymers or by in situ formation of these block copolymers. The latter is achieved by addition of PPE end-capped with functionalized maleic anhydride (Fig. 6A) or by addition of PS grafted with maleic anhydride (Fig. 6B). The anhydride groups react with the amine end groups of the PA polymers and form a compatibilizing linear PA – PPE block copolymer (Fig. 6A) or a grafted PS – PA block copolymer (Fig. 6B). Figure 7 shows the effect of compatibilization as determined by scanning electron microscopy[15]. Higher concentrations of PA – PPE copolymers reduce the dimensions of the dispersed PPE phase.

PPE is a ductile material which can be impact-modified to form extremely tough blends. The high glass transition temperature of PPE also increases the strength, rigidity, and dimensional stability of the PA – PPE blends up to the temperature required for automotive lacquering. The blends are usually impact-modified phase selectively with various rubbers located in the PPE or PPE – PS phase dispersed in the PA phase. Two-phase impact modification of the PPE (or PPE–HIPS) and the PA phases is also possible by appropriate selection

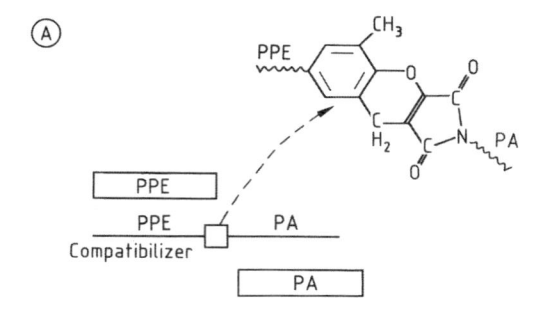

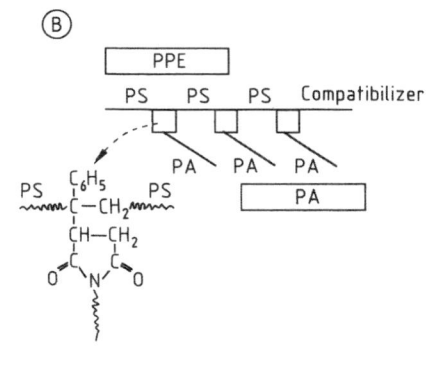

Figure 6. Schematic representation of compatibilization of PA – PPE using PPE end-capped with functionalized maleic anhydride (A) or PS grafted with maleic anhydride (B)

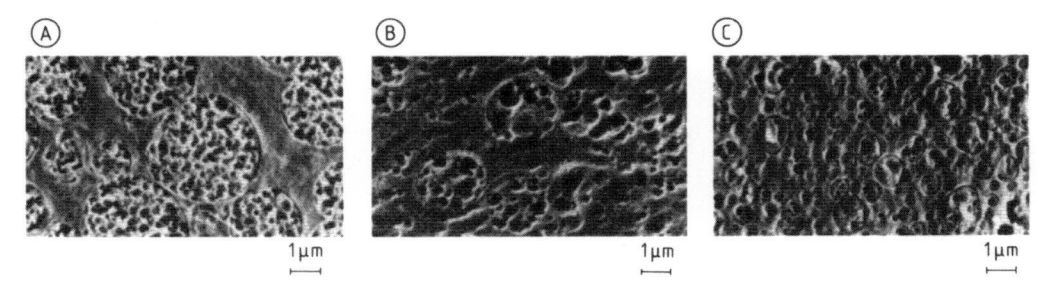

Figure 7. Effect of compatibilization on PA – PPE morphology as determined by scanning electron microscopy with PA as the continuous phase (courtesy of S. Y. Hobbs, General Electric)
A) No compatibilizer; B) 8 wt % compatibilizer; C) 15 wt % compatibilizer

of the rubbers. The PA – PPE blends can thus easily be adapted to give the ultimate use temperatures and impact levels required for specific lacquering and painting applications of exterior automotive parts (see Fig. 8). Ranges of temperature requirements for various coating techniques are indicated.

Table 4 shows important properties of PA –PPE blends. Polyamides are known for their high moisture absorption which reduces their dimensional stability. Addition of PPE to PA lowers the water absorption but, remarkably, PA – PPE blends absorb less moisture than that calculated from their PA volume fraction (Fig. 9). The resulting blend thus has a greater dimensional stability than PA, providing a more constant mechanical property profile with temperature and relative humidity. This dimensional stability can be further enhanced by reinforcing the material with glass fibers. Special grades exist for bumper beams that combine high-temperature paintability with low-temperature impact performance.

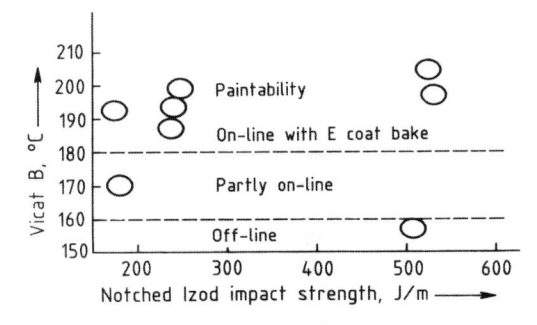

Figure 8. Ultimate use temperature plotted against impact strength of commercially available PA – PPE blends (Noryl GTX, GE Plastics), ranges of possible coating techniques are indicated

Advantages of PA – PPE blends over PPE – HIPS blends are their better chemical resistance (see Table 3), improved environmental stress cracking resistance, and a higher ultimate use temperature range.

3.3. PPE – Polyolefin Blends

In 1989 GE Plastics introduced a blend series (Noryl Xtra) containing $\geq 90\%$ PPE. The high melt viscosities of PPE (see Chap. 1) were reduced using plasticizers and olefinic components of proprietary composition as flow promoters. The resins combine good processability with high dimensional and temperature stability up to 195 °C as well as an UL-94 V-0 flammability rating. The materials have a high surface gloss and can be used as light reflectors.

In addition to excellent intrinsic flame retardancy properties, the special-grade Noryl Xtra LS generates an extremely low amount of smoke during burning because it contains PPE-specific char promoters. This feature is extremely important for its use in electrical insulation applications in enclosed or inaccessible areas (airplanes, ships, public buildings).

4. Processing

Melt viscosities of all PPE – HIPS and PA – PPE grades conform to the processing capabilities of conventional melt-processing equipment (injection molding, extruding etc.). However, PA – PPE blends require a more stringent drying procedure, and care should be taken with PPE – HIPS blends not to exceed the recommended

Table 4. Properties of PA – PPE blends (Noryl GTX, GE Plastics)

Property	Standards		Unreinforced[b]	30 wt % glassfiber-reinforced[c]
	ASTM[a]	DIN[a]		
Mechanical				
Tensile strength (yield), MPa	D 638	53 455	55	
Tensile strength (break), MPa	D 638	53 455	50	160
Tensile elongation (yield), %	D 638	53 455	5	
Tensile elongation (break), %	D 638	53 455	100	3
Tensile modulus, MPa	D 638	53 457	2000	9000
Flexural stength (yield), MPa	D 790	53 452	85	
Flexural strength (break), MPa	D 790	53 452		220
Flexural modulus, MPa	D 790	53 457	2100	7000
Hardness (H 358/30), MPa		53 456	115	141
Impact strength				
Charpy impact (notched), kJ/m^2		53 453	15	7
Izod impact (notched), J/m	D 256		240	85
Izod impact (notched, −20 °C), J/m	D 256		190	75
Thermal				
Vicat softening point (rate B), °C	D 1525		220	240
Vicat VST/B/120, °C		53 460	190	240
Deflection temperature under load (1.82 MPa), °C	D 648	53 461		225
Thermal conductivity, Wm^{-1}K^{-1}	C 177		0.23	0.26
Linear thermal expansion coefficient (flow), K^{-1}	D 696	VDE 0304/1	9×10^{-5}	$(2-3) \times 10^{-5}$
Physical				
Relative density	D 792	53 479	1.10	1.31
Water absorption (24 h), %	D 570		0.40	0.50
Water absorption (equilibrium), %	D 570		3.50	3.60
Mold shrinkage (flow), %	D 955		1.2 – 1.6	0.3 – 0.5
Flame characteristics				
Oxygen index, %	D 2863		20	29
Flame retardancy (mm thickness)	UL-94		HB (1.60)	HB (1.60)

[a] Except where otherwise stated.
[b] Noryl GTX 900.
[c] Noryl GTX 830.

drying time of 4 h at 110 °C, since this leads to loss of impact resistance. Processing temperatures of standard grades are typically 220–270 °C for PPE – HIPS, 280–300 °C for PA – PPE, and 320–340 °C for PPE – olefin blends. Injection molding and glass-filled grades require slightly higher temperatures.

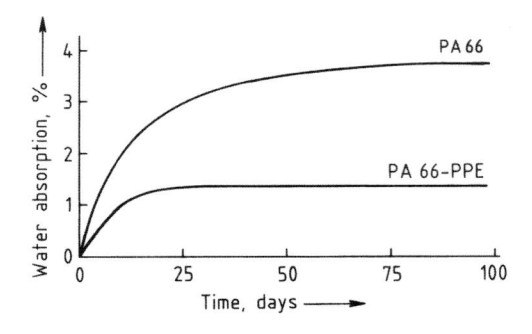

Figure 9. Water absorption of PA and PA – PPE at 23 °C and 50 % R.H.

5. Uses

The most important application areas for PPE blends are in the automotive industry (35 %), in electric and electronic appliances (25 %), office equipment (15 %), and (hot) water distribution and environmental applications (8 %) [13].

Interior Automotive Applications. The development of PPE – HIPS blends marked the break-through of plastics for interior automotive applications. The most significant applications are dashboards of PPE – HIPS (Fig. 10), a development that started in Europe and was followed worldwide. Other applications include radio housings, loudspeakers, and heating and ventilation systems. A recent development are the so called "antisqueak" materials used for handgrips and ventilation grills.

Figure 10. PPE – HIPS dashboard (Noryl, GE Plastics)

Exterior Automotive Applications. The on-line and off-line paintability of PA – PPE blends provided by their high-temperature performance, their "class A" surface with excellent dimensional stability, and good mechanical properties in combination with low-temperature impact resistance makes them suitable for vertical body panels, bumpers, petrol filling caps, wheel trims, grilles, spoilers, electromechanical systems (EMS) and, most recently, fenders for the 1991 Renault Clio 16 S (Fig. 11). These applications allow for easy styling changes, weight savings, and, most importantly, cost savings. The latter is particularly important for "small" production series of 50 000 vehicles per year.

Electrical and Electronics Industry. Flame-retardant PPE – HIPS grades are used mostly on account of their good processability, good mechanical properties, and flame retardancy behavior without using halogens. Typical applications are housings for electrical equipment (Fig. 12), connectors, switches, and interior parts in laundromats. PA – PPE blends can be used in paintable extrusion and blow molding parts.

Figure 11. Fenders of 1991 Renault Clio 165 (Noryl GTX, GE Plastics)

Figure 12. PPE – HIPS housing (Noryl, GE Plastics)

Data Processing and Office Equipment Industry. Flame-retardant PPE – HIPS blends or, alternatively, painted structural foams are used for in-house appliances such as computer housings. These materials are provided with a UV-stabilizing system to comply with the most stringent demands for color stability.

The hydrolytic and dimensional stability of PPE – HIPS blends and the reduced stress – crack sensitivity provided by glass fiber reinforcement makes them suitable for applications in water and air purification systems. All blends in applications involving food contact or drinking water are approved by the FDA.

Building and Construction. This industry segment constitutes a new market for PPE blends. Current low sales percentages result from the fact that traditional materials are still used in potentially large-volume applications such as insulation systems or solar collectors.

New Developments. A few promising new developments can be mentioned. In situ PPE – PS blends used to make expandable beads give strong foams with a low density (to 0.02 g/cm^3) and a heat distortion temperature above 100 °C, a sterilizable and reusable material (Fig. 13).

The use of PPE as an organic, highly heat-resistant, flame-retardant filler that does not affect mechanical properties at high filling ratios, as demonstrated for PA – PPE blends, will be applied to other blend systems (Section 3.2). These new blends include a blend of PPE and poly(butylene terephthalate) which has an even lower moisture absorption than PA – PPE blends and thus higher dimensional stability. Blends of chemically modified PPEs will

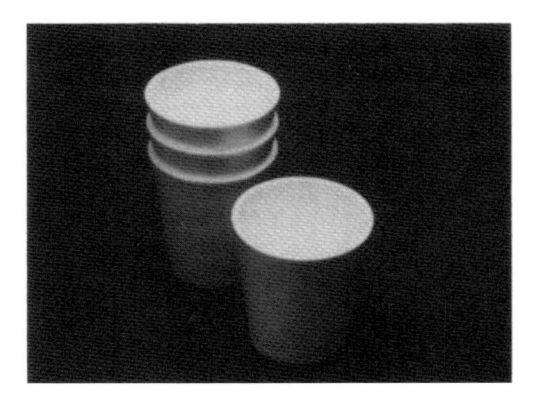

Figure 13. Beakers made from PPE – PS expandable beads (Caril, Shell)
Bead size 300 – 500 μm, density 60 g/L, T_g 110 °C

also find increased use, aimed at applications with specific requirements.

6. Acknowledgement

The authors thank Veronique Hopmans for her contributions.

References

1. General Electric, US 3 306 874, 1967 (A. S. Hay).
2. D. Aycock, W. Abolius, D. M. White: Poly(Phenylene ether), *Encyclopedia of Polymer Science and Engineering*, 2nd ed., vol. **13**, John Wiley & Sons, New York 1988, p. 1.
3. Z. S. Slama, *Acta Polym.* **31** (1980) 746.
4. M. Kryszewski, J. Jachowicz in N. Grassie (ed.): *Developments in Polymer Degradation*, vol. **4**, Applied Science Publishers, London 1982p. 1.
5. General Electric, US 3 063 872, 1962 (E. M. Boldenbruch).
6. H. L. Finkbeiner, A. S. Hay, D. M. White: "Polymerization by Oxidative Coupling," in C. E. Schildknecht, I. S. Skeist (eds.): *Polymerization Processes, High Polymers*, vol. 29, John Wiley & Sons, New York 1977, p. 537.
7. A. S. Hay, *Polym. Eng. Sci.* **16** (1976) 1. A. J. Schouten, G. Challa, J. Reedijk, *J. Mol. Catal.* **9** (1980) 41. C. E. Koning, R. Brinkhuis, R. Wevers, G. Challa, *Polymer* **28** (1987) 2310. F. J. Viersen, G. Challa, J. Reedijk, *Polymer* **31** (1990) 1369.
8. J. R. A. Pearson, *Polym. Eng. Sci.* **18** (1978) 222.
9. C. B. Bucknall, D. Clayton, W. E. Keast, *J. Mater. Sci.* **7** (1972) 1443.
10. General Electric, US 4 128 602, 1978 (G. Lee, A. Katchman).
11. C. B. Bucknall: *Toughened Plastics*, Applied Science Publishers, London 1977.
12. D. W. van Krevelen: *Properties of Polymers*, 3rd ed., Elsevier, Amsterdam 1990.
13. H. Frank, *Kunststoffe* **76** (1986) 859.
14. R. I. Warren, *Polym. Eng. Sci.* **25** (1985) 477.
15. J. R. Campbell, S. Y. Hobbs, T. J. Shea, V. H. Watkins, *Polym. Eng. Sci.* **30** (1990) 1056.

Further Reading

H. F. Mark (ed.): *Encyclopedia of Polymer Science and Technology*, 3rd ed., Wiley, Hoboken, NJ 2005.

D. M. White: *Polyethers, Aromatic*, Kirk Othmer Encyclopedia of Chemical Technology, 5th edition, John Wiley & Sons, Hoboken, NJ, online DOI: 10.1002/0471238961.0118151323080920.a01.

Polypropylene

Markus Gahleitner, Borealis Polyolefine GmbH, Linz, Austria

Christian Paulik, Institute of Chemical Technology of Organic Materials, Johannes Kepler University Linz, Linz, Austria

1.	Historical Overview	937
2.	Polymer Structure	939
2.1.	Molecular and Chain Structure	939
2.2.	Crystallization and Morphology	941
2.3.	Multiphase Copolymers	942
3.	Raw Materials	942
3.1.	Propene and Comonomers	942
3.2.	Polymerization Diluents	943
3.3.	Catalysts	944
3.3.1.	$TiCl_3$-Based Catalysts	944
3.3.2.	$MgCl_2$-Supported $TiCl_4$ Catalysts	945
3.3.3.	Single-Site Catalysts	946
3.3.4.	Aluminum Alkyl Cocatalysts and Donors	946
3.4.	Hydrogen	947
4.	Polymerization Mechanism	947
5.	Industrial Polymerization Processes	949
5.1.	Liquid-Phase Processes	949
5.2.	Gas-Phase Processes	952
5.3.	Hybrid Processes	954
6.	Product Finishing	956
6.1.	Additives	957
6.2.	Compounding and Blending	959
6.3.	Reactive Modification	959
7.	Properties	960
7.1.	Homopolymers	960
7.2.	Random Copolymers	961
7.3.	Impact Copolymers	961
7.4.	Composites and Blends	964
8.	Processing and Applications	965
8.1.	Injection Molding	965
8.2.	Blow Molding	967
8.3.	Fibers and Flat Yarns	968
8.4.	Films and Sheets	970
8.5.	Foaming and Coating	972
8.6.	Technical Applications	973
8.7.	Pipes	974
9.	Environmental Aspects	974
9.1.	Life-Cycle Analysis	975
9.2.	Mechanical Recycling	976
9.3.	Chemical and Energetic Recycling	977
	References	978

1. Historical Overview

The history of polyolefins effectively started with the discovery of high-pressure ethylene polymerization at ICI in the UK in 1933, which initiated the development and production of low-density polyethylene (LDPE). In the search for a more energy efficient route to polyethylene and the development of fully linear polymers of this monomer, Karl Ziegler discovered in 1953 new catalysts based on the combination of transition metal halides (e.g., titanium chloride) with aluminum alkyls (e.g., triethylaluminum) [1]. With these he demonstrated that ethylene could be polymerized to a high molecular mass, crystalline polymer at moderate pressure and temperature and laid the foundation of a revolution for the plastics industry. Before that, propene could not be polymerized to high molecular mass products with the catalysts then available, namely, free-radical, anionic, and cationic systems. Even the most favorable of these yielded only liquids consisting of many isomers and greaselike materials unsuitable for making hard plastic products.

Ziegler's group at the Max-Planck-Institut für Kohlenforschung in Germany had for some years been using aluminum alkyls to convert ethylene to a range of oligomers having an even number of carbon atoms. Unexpectedly, almost total conversion to butenes occurred in one experiment. The cause was eventually traced to nickel-ion contamination in the autoclave. Tests with many other metal compounds revealed that titanium ions were even more active in producing polyethylene (PE) with a high molecular mass, resulting in a first patent [2]. Such combinations of transition metal compounds with an aluminum alkyl later became known as Ziegler–Natta (ZN) Catalysts, thereby acknowledging the immense contribution of GIULIO NATTA to discovering and characterizing polypropylene (PP) and other (stereoregular) α-olefin polymers in 1954.

NATTA, Professor and Director of the Milan Institute of Industrial Chemistry, was a consultant to Montecatini, to whom he assigned his patents. The close collaboration between ZIEGLER and NATTA was acknowledged in 1963 by the joint award of a Nobel Prize in Chemistry for their outstanding contributions to polymer science (a fascinating account of these discoveries is provided by MCMILLAN [3]). NATTA's discovery of the stereospecific polymerization of propene to produce isotactic (and hence crystalline) polypropylene (iPP) [4] resulted first in products with an isotacticity of only around 40%. This regularity was increased to 80% within one year through modification of the catalyst, allowing the first Montecatini production facility for iPP with a production volume of 6.10^3 t/a to go into operation in Ferrara, Italy, in 1957.

In retrospect, it is clearly recognizable that with the chain branching of PE and the stereostructure of PP, two crucial general control factors for the crystallinity and mechanics of polymers had been discovered [5]. Finally, the crystallization structures of the polymers were also elucidated around the same time; ANDREW KELLER presented his concept for chain folding in PE crystals in Bristol in 1957 [6].

The patent dispute over crystalline polypropylene in the USA between NATTA and Montecatini on the one side and Phillips Petroleum Company on the other was clearly a result of different legislation between the USA and Europe. The dispute started in 1958 and became one of the longest in industrial history, only ending 20 years later [7]. Phillips' polymerization catalyst, consisting of chromium-ion-promoted silica–alumina, has not been used in any commercial plant to manufacture iPP.

New applications for the polymer with the impressive combination of a melting point of 165°C and a density of 900 kg/m^3 were constantly being found, in 1959 with the production of the PP fiber Meraklon, and in 1962 with first attempts at PP suitcases made by Hoechst. The ZN catalysts also proved to be suitable for the production of elastomers based on ethylene and propene [8]. For a polymer having its glass transition around 0°C this turned out to be a decisive factor for evolving into technical applications requiring mechanical stability over a wide temperature range. Together with the availability of an initially cheap monomer, this resulted in a rapid growth of the production volume from less than 100×10^3 t/a in 1960 to 6.1×10^6 t/a in 1980 to more than 53×10^6 t/a in 2012, making it the most produced single thermoplastic material (see Fig. 1) [9]. In the same period, the production volume of new plants

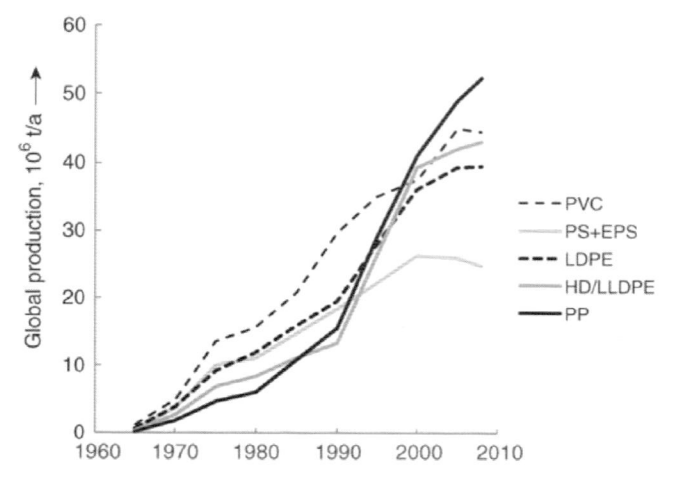

Figure 1. Production-volume development of standard thermoplastics (redrawn after [9])

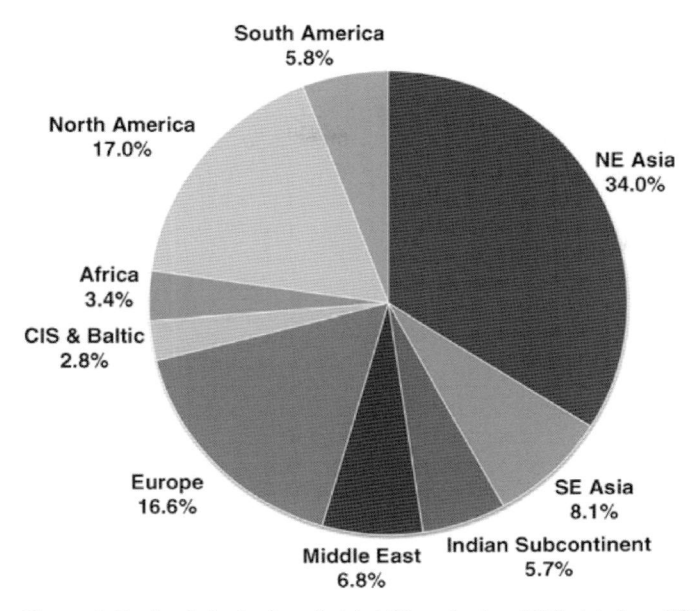

Figure 2. Regional distribution of global PP production (2012, data from [10])

increased to 200–400 × 10³ t/a. A massive regional shift to Asia has occurred in recent years, with Northeast Asia and especially China now having the biggest share in production (see Fig. 2) and fewer than ten companies producing more than 60% [10].

Technical factors for this massive volume increase are numerous, but mostly related to a clever combination of catalyst chemistry and polymerization technology. A key step in terms of application range was certainly the development of products long referred to as "block copolymers". These combine a crystalline PP matrix with embedded particles of ethylene–propene rubber (EPR) and PE, which determine impact resistance and low-temperature resistance [11], and should more correctly be referred to as heterophasic copolymers (HECOs) or PP impact copolymers. Important for their development were improved catalysts and the availability of multireactor polymerization plants. With the second generation of ZN catalysts it became possible from the mid-1970s to produce "technical" HECOs, but with only a moderate EPR content. The introduction of the third generation of ZN catalysts at the end of the 1970s together with the breakthrough of bulk processes for polymerization without solvents enabled reliable synthesis of HECOs with high impact strength. Modern ZN catalysts allow the production of reactor-based thermoplastic elastomers with high impact toughness in a temperature range from +80 to −30°C.

In the field of catalysts, NATTA had already discovered the fundamental suitability of metallocenes for the homogeneous polymerization of olefins. In 1975 a crucial step forward in the practical use was taken with the discovery by KAMINSKY in Hamburg of methylaluminoxane (MAO) as a massively reaction accelerating cocatalyst [12]. This enabled MC-based PE, and BRINTZINGER in Konstanz extended the concept with the use of bridged MCs for iPP in 1984. While the market penetration of metallocenes and other single-site catalysts has been significantly slower than anticipated, due in part to the continuous improvement of the traditional—and cheaper—ZN systems, the former have enabled unprecedented fine-tuning of chain microstructure by ligand design.

2. Polymer Structure

2.1. Molecular and Chain Structure

Since PP is the largest single thermoplastic material [9] the structure, morphology, and properties of isotactic polypropylene (iPP) have been studied extensively. As the propene

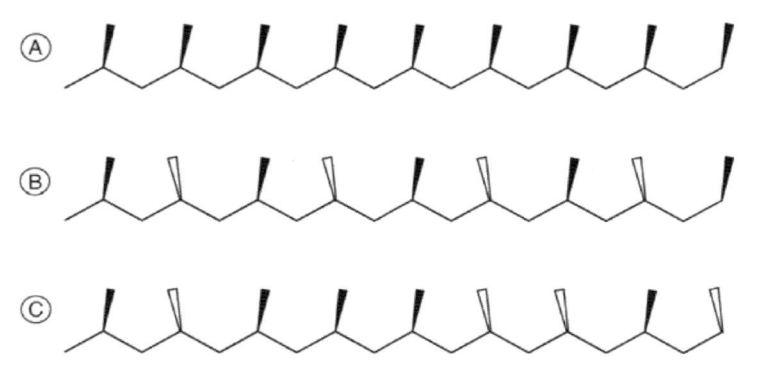

Figure 3. Main polypropylene types
A) Isotactic; B) Syndiotactic; C) Atactic

monomer is asymmetric, polypropylene can be produced with different stereochemical configurations [13]. The different stereoisomers of PP were isolated and characterized by NATTA et al. in 1954–1955. The most common types of PP, shown in Figure 3, are isotactic (iPP), syndiotactic (sPP), and atactic (aPP). In iPP all methyl groups are located on the same side of the backbone, in sPP on alternating sides, and in aPP the methyl groups are arranged randomly along the polymer chain.

Atactic PP is an amorphous material with much less practical importance than its regular counterparts, but it has some niche applications like hot-melt adhesives. The isotactic and syndiotactic stereoisomers of PP are both semicrystalline polymers. The melting points are in the range of 165°C for iPP and 130°C for sPP. Isotactic PP dominates the polypropylene market because it is easily produced with heterogeneous ZN and metallocene catalysts. In addition, it shows easier processing behavior and better mechanical properties. Syndiotactic PP shows a slower crystallization rate and can only be produced with some metallocene catalysts; therefore, it is commercially almost nonexistent.

State-of-the-art ZN catalysts make isotactic PP with only a minor atactic fraction. The industrially used ZN systems favor the 1,2-insertion and head-to-tail enchainment (Fig. 4). Defects such as a 2,1-insertion following a 1,2-insertion create irregularities along the PP chain resulting in a lower melting point and lower crystallinity of the material [14]. With metallocene catalysts it is possible to produce polypropylene with high isotacticity but lower regioregularity, responsible for their lower melting point compared to PP

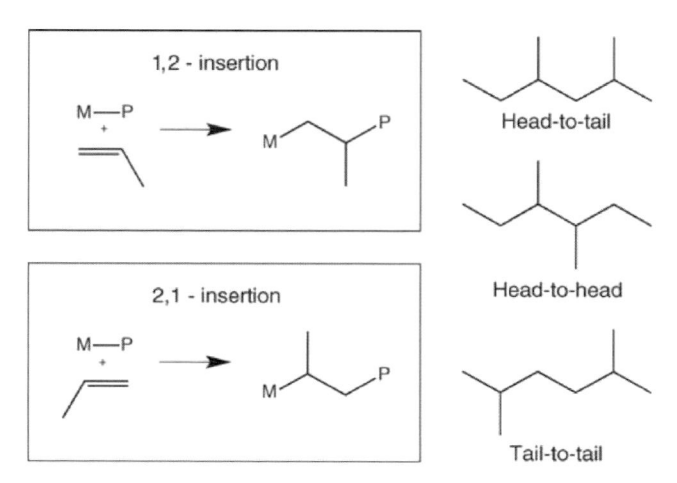

Figure 4. Regioirregularity in polypropylene polymerization

made with a ZN catalyst. With the help of metallocene catalysts also other stereostructures such as block polymers of isotactic and atactic sequences can be obtained, but these materials are not yet in commercial use.

2.2. Crystallization and Morphology

Polypropylene molecules associate to form supramolecular structures when cooled to temperatures below the melting point (crystallization temperature). At the crystallization temperature, the macromolecules begin to arrange themselves into crystals, and ordered crystalline regions and disordered amorphous regions are formed. The growth of the crystals may happen spontaneously (when a highly ordered crystal structure is favorable) or may be induced by the presence of a foreign particle (e.g., a nucleating agent or the metal mold surface). Typically, the onset of crystallization occurs at ca. 110–120°C in a differential scanning calorimeter (DSC) at a cooling rate of 10 K/min. The observed crystallization rate is dependent on the nucleation rate and the rate of crystal growth. Crystallization is generally favored by slower cooling from the melt, and the degree of crystallization can be controlled by the rate of melt quenching and subsequent annealing. Very rapid cooling can even suppress crystallization. The degree of crystallinity can vary and different types of crystal structures depending on the stereochemical structure, processing conditions, and additives exist [15]. The observed morphology of PP exhibits a hierarchy of characteristic scales, as shown in Figure 5. In the cross section of a molded part one can see skin–core structures, which on a finer scale reveal a spherulitic structure on the order of 1–50 μm. The size of these structures can be determined by means of optical microscopy, small-angle light scattering (SALS), and scanning (atomic) force microscopy (SFM). The individual spherulite is composed of lamellar-shaped crystals. The lamellar periodicity is given by the long spacing and is in the range of 10–30 nm, depending on processing and thermal history. Typically used protocols are small-angle X-ray scattering (SAXS), higher-resolution electron microscopy, and SFM. The lamellae are composed of crystallographically ordered regions [16]. The individual macromolecule chains form a helical arrangement packed together in unit cells with specific dimensions. The unit-cell dimensions (0.6–2 nm) are shown in Figure 5 for the α form of iPP. Commonly wide-angle X-ray scattering (WAXS),

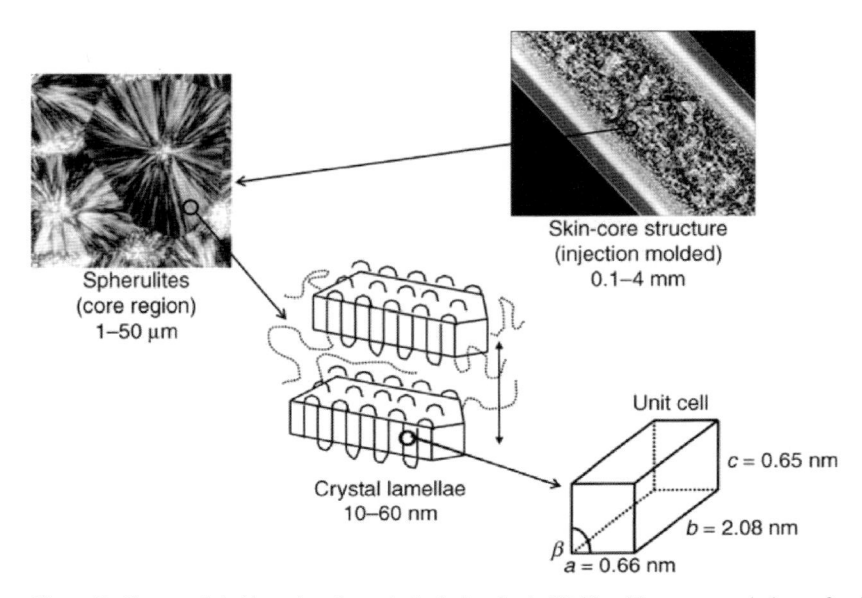

Figure 5. Characteristic hierarchy of morphological scales in PP. The skin–core morphology of an injection-molded specimen is shown to illustrate the morphology on the visual scale (redrawn after [16])

Table 1. Crystal characteristics of PP [17–19]

Crystal form	System	ρ (20°C), g/cm^3	Chains per unit cell	T_m^*, °C
i-α	monoclinic	0.932–0.943	4	171
i-β	pseudohexagonal	0.922	9	150
i-γ	triclinic	0.939	1	131
Smectic		0.916		
Amorphous		0.85		
sPP	orthorhombic	0.93	2	138

*The enthalpy of melting for 100% crystalline iPP is variously reported, due to different methods, e.g., 165±18 J/g [17] and 209 J/g [18].

electron diffraction techniques, and SFM are used to evaluate the structure on this scale. The α form of isotactic polypropylene is the primary form of PP obtained under normal processing conditions. Besides the α form, β, γ, and δ crystal modifications are known for PP. The basic features of the crystal forms of PP are given in Table 1 [17–19]. The β form of polypropylene exhibits a lower tensile strength at a given strain rate and higher impact strength than the α form. The β form can specifically be obtained by the use of nucleating agents [20]. It converts to the α form only on melting and recrystallization.

2.3. Multiphase Copolymers

Commercially PP is available in different forms, dependent on the desired properties. Polypropylene homopolymers only consist of propene monomer units and provide stiffness, but their glass transition temperature T_g is about 0°C and thus are not usable in subzero environments. By addition of an elastomeric phase the impact resistance of the material is increased [21]. The addition of an elastomeric component leads to a multiphase structure often referred to as impact or block copolymer. The achieved impact resistance is dependent on the type, amount, and morphology of the elastomeric phase, while the polypropylene matrix determines the stiffness.

Generally multiphase copolymers are produced in at least two reactors in series. In the first reactor the isotactic polypropylene matrix is made, and the second produces an ethylene–propene (EP) elastomeric copolymer. The dispersed copolymer dissipates energy during impact, increasing the toughness of the material.

The dispersion of an elastomeric phase in a PP matrix can also be achieved through melt mixing (compounding) of rubber materials with polypropylene. An example of the achieved morphology is given in Figure 6.

3. Raw Materials

3.1. Propene and Comonomers

Due to the high sensitivity of ZN catalysts to impurities, which lead to a drastic reduction of the reaction rate and also negatively affect the product quality (isotacticity) of the polymer, the used propene and comonomers require high purity [22]. Compounds that will bind strongly to the active centers (Ti) of the catalyst or react with the alumina-based cocatalyst, such as polar and highly unsaturated compounds (e.g., acetylenes, dienes, carbon monoxide and dioxide, carbonyl sulfide, water, alcohols, ammonia, and arsine) can only be tolerated at parts-per-billion levels [23]. In addition, compounds that do not influence the chemistry but negatively affect the process by increasing the reactor pressure also must be avoided. These inert substances, such as methane, ethane, propane, butane, and nitrogen, also accumulate in the process and complicate the monomer recovery section and recycle streams.

For the production of polypropylene materials polymer-grade monomers are used as feedstock, which can be further refined by passing the monomers through special purification columns. A typical design consists of four purification steps to scavenge and absorb impurities [22].

Table 2 lists the typical specification for propene quality. Certain catalyst systems may require the lower levels listed for carbonyl

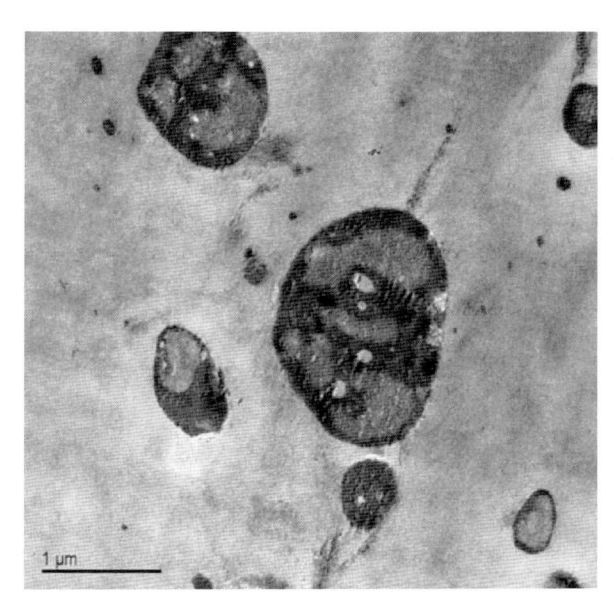

Figure 6. TEM image of the morphology of a commercial high-impact EP copolymer with 16.5 wt% elastomer (Borealis BD310MO). Dark regions correspond to the elastomeric phase with crystalline PE inclusions; contrasted with RuO_4 (scale bar = 1 μm)

sulfide, carbon monoxide, and oxygen (this quality is said to be suitable for metallocene catalysts) [24]. Propene properties relevant to PP production are listed in Table 3.

To broaden the property window of PP certain α-olefins are used in copolymerization with propene. The commercially most important comonomer is ethylene, but 1-butene and higher olefins are reported as well [25, 26].

Propene is normally produced outside the PP plant by steam cracking of naphtha or gas oil at 700–950°C (→ Propene, Section 3.1.) yielding an ethylene-to-propene weight ratio of approximately 2:1. The increasing demand for propene derivatives, especially PP, made other routes such as the dehydrogenation of propane from natural LPG fields commercially interesting (→ Propene, Section 3.3.1).

3.2. Polymerization Diluents

The polymerization of propene can be performed under slurry and gas-phase conditions and combinations thereof. The first commercial processes used inert hydrocarbons ranging from butane to dodecane and refined petroleum fractions as diluents. Typically there is twice as much diluent in the reactor as polymer, and consequently the demands on the purity of the diluent are as stringent as for the monomers. The inert diluent helps to transfer the propene to the solid catalyst and to convey the heat of

Table 2. Propene quality requirements

Type of component	Component	Recommended concentration
Monomer	Propene	99.5 wt% min.
Inert	Ethane	500 vol ppm
	Propane	0.5 wt%
	Butane	500 vol ppm
	N_2 + CH_4	300 vol ppm
	Ethylene	50 vol ppm max.
	1-Butene	50 vol ppm max.
	Butadienes	10 vol ppm max.
Copolymerizing monomer	Isobutene	50 vol ppm max (estimated)
Poison	Propyne	5 vol ppm max.
	Propadiene	5 vol ppm max.
	Oxygen	2–5 vol ppm max.
	Carbon monoxide	0.3–3 vol ppm max.
	Carbon dioxide	5 vol ppm max.
	Carbonyl sulfide	0.03–0.5 vol ppm max.
	Total sulfur	1 wt ppm max.
	Water	5 wt ppm max.
	Methanol	5 vol ppm max.
	Ammonia	1 vol ppm max.
	Hydrogen	10 vol ppm max.

Table 3. Propene properties relevant to PP production

Property	Value
Heat of polymerization	2514 kJ/kg
Boiling point	−47.7°C
Critical temperature	92°C
Critical pressure	4.6 MPa
Vapor pressure at 20°C	0.98 MPa

polymerization to the water-cooled reactor jacket. Polar impurities, such as alcohols, carbonyl compounds, water, and sulfur-containing compounds, must be kept below 1–5 ppm. The quality of the diluent is monitored regularly by means of standard analytical methods, preferably on-line analytics, to prevent accumulation of oxidized species and catalyst fragments.

At the end of the polymerization process the used diluent has to be removed completely to guarantee compliance of the PP material with food and drug regulations. Since state-of-the-art polymerization plants are based on gas-phase and bulk (polymerization in liquid propene) technologies the used diluents can be removed more easily.

3.3. Catalysts [27]

Heterogeneous ZN catalyst systems still largely dominate commercial production of highly isotactic PP. The catalyst system consists of a solid transition metal halide, usually $TiCl_4$, a Lewis base (referred to as internal electron donor, e.g., 1,3-diethers) on a support (mostly $MgCl_2$), and an organoaluminum alkylating agent such as triethylaluminum (TEA), and a second Lewis base (usually called external donor, e.g., alkoxysilanes) added separately to the polymerization mixture. An overview on the development phases of propene polymerization catalysts is given in Table 4.

3.3.1. $TiCl_3$-Based Catalysts

$TiCl_3$ can form four different crystalline modifications (i.e., α, β, γ, and δ) dependent on the preparation method used. The α, β, and δ forms are purple, while the fourth is brown (β) and not used for PP production due to its poor stereospecificity. All purple forms have a layer lattice consisting of chlorine atoms with hexagonal (α), cubic (γ), or random succession of hexagonal and cubic close packing (δ) (Fig. 7).

The most important process of industrial $TiCl_3$ production has been the reduction of $TiCl_4$ by means of Al metal or Al alkyls. A major refinement in Al alkyl-reduced $TiCl_3$ systems has been made by Solvay [28, 29] in their three-stage process, which gives a four- to fivefold increase in activity:

1. Preparation of a reduced solid by reducing $TiCl_4$ with diethylaluminum chloride (DEAC) at 0–2°C, and then heating the mixture to 65°C. The obtained brown solid is a β-$TiCl_3$ ($TiCl_3 \cdot x$ $AlCl_3 \cdot y$ $AlEtCl_2$, where $x \approx 0.15$ and $y \approx 0.20$).
2. The brown β-$TiCl_3$ solid is treated with diisoamyl ether to dissolve out most of the Al compounds.
3. The above solid is treated with excess $TiCl_4$ diluted in hydrocarbons at 60–70°C to give a

Table 4. Performance development of ZN catalysts for PP [a]

Year	Catalyst system	Productivity, $kg_{PP}/g_{Cat.}$	I.I.[b], %	mmmm[c], %	M_w/M_n	H_2 response[d]
1954	δ-$TiCl_3 \cdot 0.33$ $AlCl_3$ + $AlEt_2Cl$	2–4	90–94			low
1970	δ-$TiCl_3$ + $AlEt_2Cl$	10–15	94–97			low
1968	$MgCl_2/TiCl_4$ + AlR_3	15	40	50–60		
1971	$MgCl_2/TiCl_4$/benzoate + AlR_3/benzoate	15–30	95–97	90–94	8–10	low
1980	$MgCl_2/TiCl_4$/phthalate + AlR_3/Silane	40–70	95–99	94–99	6.5–8	medium
1988	$MgCl_2/TiCl_4$/diether + AlR_3	100–130	95–98	95–97	5–5.5	very high
1988	$MgCl_2/TiCl_4$/diether + AlR_3/silane	70–100	98–99	97–99	4.5–5	high
1999	$MgCl_2/TiCl_4$/succinate + AlR_3/silane	40–70	95–99	95–99	10–15	medium

[a]Polymerization conditions: liquid propene, 70°C, 2 h.
[b]Isotactic index (% insoluble in boiling heptane).
[c]Content of isotactic pentads.
[d]Melt flow rate change as function of H_2 concentration.

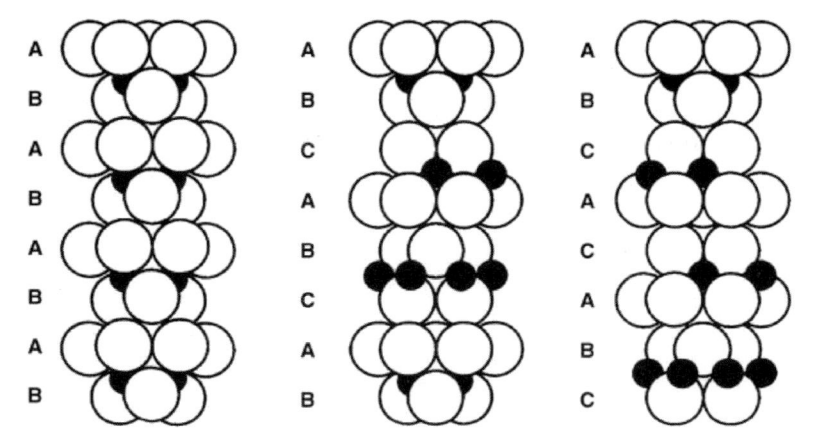

Figure 7. Models of chlorine packing in the different TiCl$_3$ modifications: α (left), γ (center), δ (right) (adapted from [27])

violet δ-TiCl$_3$·x (AlEt$_n$Cl$_{3-n}$)·y (i-Am$_2$O) composition ($n = 0$–2, $x < 0.2$, and $0.01 < y < 0.11$). After washing with hydrocarbons to remove adsorbed TiCl$_4$ and other byproducts, a catalyst with high surface area (150–200 m^2/g), high porosity (> 0.2 cm^3/g), and very high activity is obtained.

Similar catalysts have also been described by other companies, e.g., the Chisso Corporation, Japan. Adding a Lewis base (benzoic esters, alkoxysilanes) is claimed to afford high isotacticity and activity.

3.3.2. MgCl$_2$-Supported TiCl$_4$ Catalysts

A decisive step in the research on ZN PP catalysts was the development of supported titanium catalysts by Montedison (now LyondellBasell) in Italy and Mitsui Petrochemical Industries (now Mitsui Chemicals) in Japan [30]. Collaboration and cross-licensing between these two companies eased the commercial success and avoided incipient interference claims. Starting with finely milled MgCl$_2$, which has a similar structure to violet TiCl$_3$, in 1968, routines were identified to create highly active and stereoselective catalysts. Though other supports with layered structures (e.g., MgBr$_2$, MnCl$_2$, and others) in principle can also be used to make stereospecific catalysts for PP, MgCl$_2$ has been used almost exclusively because it gives catalysts the highest activity and stereospecificity.

For the preparation of particle-form supported catalyst three main procedures have been developed:

1. Mechanical routes: the different catalyst components (usually MgCl$_2$, TiCl$_4$, and a Lewis base) are milled together in suitable ratios.
2. Combined mechanical and chemical routes: MgCl$_2$ or precursors of MgCl$_2$ are co-milled with the Lewis base, followed by one or more treatments with excess TiCl$_4$ at temperatures > 80°C and washing with hydrocarbons to remove unconverted TiCl$_4$.
3. Chemical routes: the active MgCl$_2$ is formed simultaneously with the incorporation of the Ti compound and the Lewis base.
   a. MgCl$_2$ and the Lewis base (e.g., an alcohol) react to form a complex, which is subsequently treated with the internal donor (D$_i$) and excess TiCl$_4$ at temperatures > 80°C, followed by washing with hydrocarbons.
   b. Solid Mg(OR)$_2$ or Mg(OR)Cl is treated with D$_i$ and excess TiCl$_4$. The MgCl$_2$ is formed by the reaction of the Mg compound with TiCl$_4$, and the byproducts (Ti alkoxides) are eliminated by subsequent washing.
   c. MgR$_2$ or MgRCl (optionally dispersed on SiO$_2$, Al$_2$O$_3$, or other carriers) react with a chlorinating agent to form active MgCl$_2$, followed by hot treatment with D$_i$ and excess TiCl$_4$ (as above).

d. First a solution of $MgCl_2$ or other Mg compounds (e.g., $Mg(OR)_2$, $Mg(OCOR)_2$, MgR_2, or Mg silylamide) in suitable solvents such as ROH, trialkyl phosphate, $Ti(OR)_4$, epoxychloropropane, and others is prepared. The solution is then either treated with a chlorinating agent to precipitate $MgCl_2$, which is then loaded with Ti and D_i as described above, or directly treated with D_i and excess $TiCl_4$.

e. In a recent development, Borealis described a method [31] for in situ formation of the catalyst carrier particle and the actual active species by solidifying the catalyst from solution. A solution of a complex of a group 2 metal (Mg compound) and an electron donor in an organic liquid is mixed with a transition metal compound (Ti) to produce an emulsion. By changing temperature and stirring conditions solid catalyst particles with regular spherical shape are obtained.

3.3.3. Single-Site Catalysts

Metallocenes have been used for 30 years as model compounds for ZN reactions. They were quite unsuitable for commercial reactors because of their extremely low activity and poor stereocontrol. During the mid-1970s several groups discovered accidentally that traces of water improve the catalyst activity of metallocene-based homogenous catalysts in the presence of trimethylaluminum ($AlMe_3$). SINN and KAMINSKY first identified the potential of $AlMe_3/H_2O$ activators for metallocene-catalyzed ethylene polymerization [32]. Since the 1980s the performance of metallocene-based catalyst systems been improved substantially to produce isotactic, syndiotactic, and stereoblock PP on an industrial scale [33]. Metallocenes suitable for isotactic PP manufacturing generally seem to be based on zirconocenes supported on inert solids to preserve particle size and shape. This makes them more compatible with the advanced process technologies of the major operators, who refer to them as "drop-in catalysts". The concept of single-site catalysts has been expanded to bridged half-sandwich Ti amide complexes, which became known as constrained-geometry catalysts [34]. For many years the focus of metallocene catalyst development was on group 4 transition metals such as Ti, Zr, and Hf. Today the potential of late transition metal complexes of Ni, Pd, Co, and Fe is recognized as well, with heavy patenting in the field. With the development of single-site catalysts it became possible to specifically design polyolefins with narrow molecular mass distribution (polydispersity $M_w/M_n = 2$), well defined regio- and stereoregularity, and molar mass independent random or sequenced comonomer incorporation (see Fig. 8).

Although isotactic PP remains the polypropylene with the highest commercial interest, the opportunities of controlling the microstructure by catalyst design have helped to achieve a better understanding of basic structure–property correlations of PP and to expand its property range.

3.3.4. Aluminum Alkyl Cocatalysts and Donors

The cocatalysts used with $MgCl_2$-supported catalyst systems are preferably Al trialkyls, triethylaluminum (TEA), and triisobutylaluminum (TIBA). Al alkyl chlorides afford a much poorer performance and can only be used in combination with trialkyls. TEA is made commercially from aluminum, ethylene, and hydrogen. Trimethylaluminum is available as methyl aluminoxane (MAO), made through controlled addition of water to the trialkyl.

The external donor (D_e) that can be used is dependent on the type of internal donor (D_i). In the early 1980s alkyl phthalates were introduced as D_i together with alkoxysilanes as D_e, affording improved productivity and isotacticity (see Table 4). These catalysts are still in use in many industrial PP manufacturing processes. Later 1,3-diethers were introduced as internal donors providing even better activities and isotacticities without the need for any external donor. The most recent developments use succinates as D_i; as in the case of phthalates, also here alkoxysilanes are used as D_e. The performance of these catalyst systems is comparable to that of the phthalate systems, but they lead to much broader molecular weight distributions.

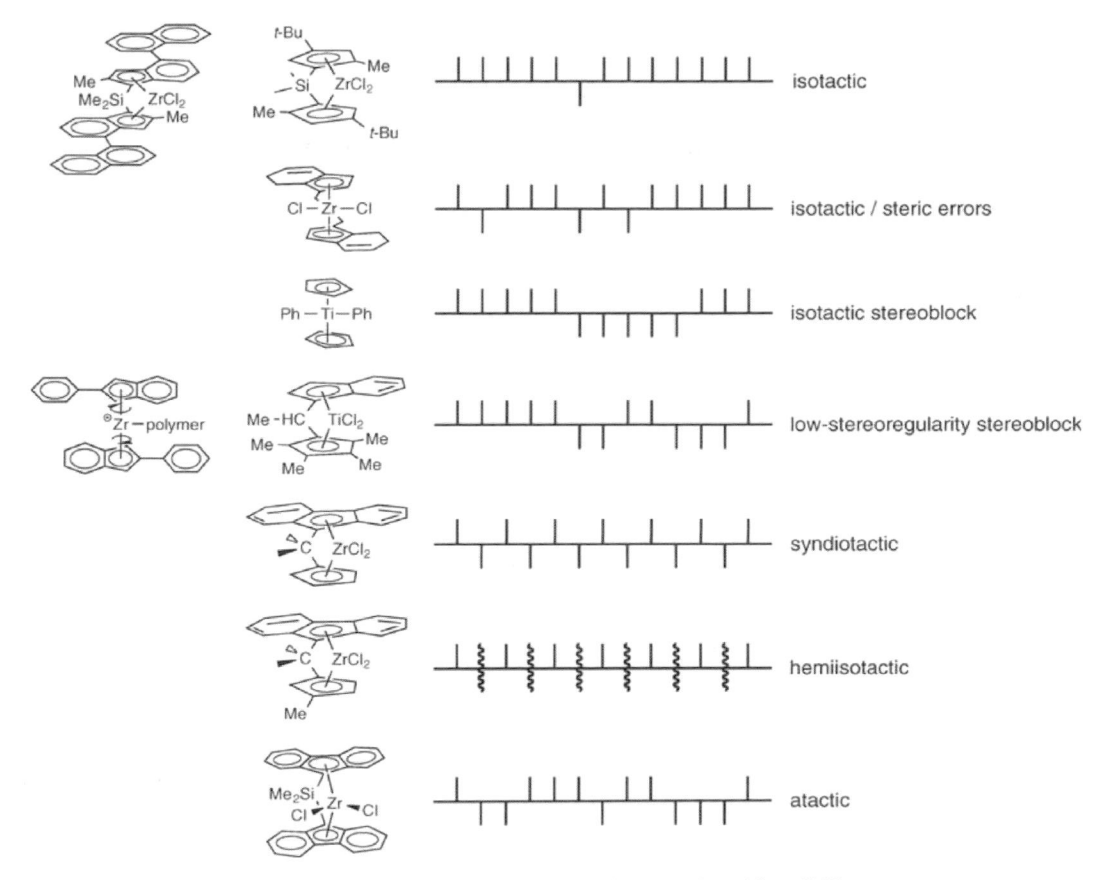

Figure 8. Correlations between metallocene structures and PP architecture (adapted from [32])

3.4. Hydrogen

Hydrogen is used as molar mass regulator in the reactor and its concentration must be accurately controlled.

4. Polymerization Mechanism [35]

For the catalytic polymerization of α-olefins different models and reaction mechanisms were developed. COSSEE [36–38] developed a model to explain the catalytic polymerization process of ethylene or 1-alkenes at a titanium center. The catalytic cycle is shown in Figure 9. It starts with the side-on complexation of the α-olefin, which activates the C–C double bond. This is followed by insertion into the Ti–C bond through a cyclic transition state, by which the polymer chain grows by one monomer unit. Simultaneously, the vacant site becomes free again for complexation of a monomer molecule.

Together with ARLMAN, COSSEE explained how the active sites are formed at the surface of $TiCl_3$. The chloride anions form a closely packed array with gaps where the much smaller Ti cations are located. Each Ti cation is then surrounded octahedrally by six chloride anions. At the surface, the titanium centers and the chloride anions are exposed (Fig. 10), and thus accessible to other components. In the proposed mechanism one chloride ion reacts with the alkyl aluminum compound and is replaced by an alkyl group. In this way, a Ti–C bond is generated at the surface of the $TiCl_3$ particle. The propene (olefin) can insert into this Ti<C->C bond provided there is a vacant site for olefin complexation. This incoming propene molecule approaches the active site, whose immediate crystal geometry controls

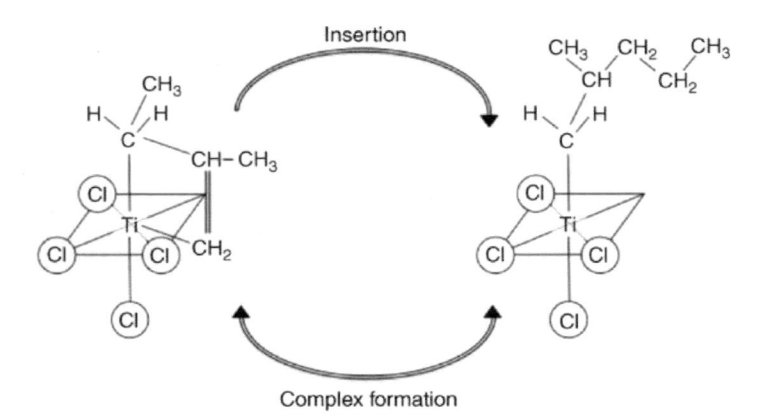

Figure 9. Catalytic cycle for propene polymerization at a heterogeneous TiCl$_3$ catalyst according to Cossee and Arlman (adapted from [35])

both the initial coordination and the configuration as the monomer inserts into the titanium–alkyl σ bond. Accordingly, this is a template-type polymerization controlled by the surface shape, and not by the previously inserted monomer unit. Atactic polymer is formed at more open surface sites having two vacancies or two weakly bonded chlorine atoms. On complexation of one propene molecule to the Ti center, the complex shown in Figure 10 (the figure shows ethene complexation) is formed. In Figure 10, the three chlorine ions at the surface of the TiCl$_3$ particle are shown together with the CH$_2$ group of the growing chain; the nearest-neighbor chlorine ion inside the crystal cannot be seen. A propene molecule occupies the sixth Ti coordination site. Starting from this transition state, propene is inserted into the Ti–C bond. One catalytic cycle, as shown in Figure 9, is thus completed.

In addition to the main reaction, chain propagation (with the monomer as the reaction partner), there are further reactions at the active site, such as β-hydride elimination, which leads to removal of the polymer chain from the active center along with the formation of a vinyl group at the chain end, or chain-transfer reactions with hydrogen and alkyl aluminum compounds.

It is very unlikely that the active sites on the TiCl$_3$ crystal are all uniform, so a distribution of imperfections results. This is why ZN catalysts are referred to as multisite catalysts to emphasize the observed wide distribution of polymer chain lengths and stereoregularities. In MgCl$_2$-supported titanium chloride systems, it is assumed that about 10–20% of the Ti is in the form of active sites. Surface geometry still controls stereoregularity, but Lewis bases are much involved in this regulation [39] (see also Chap. 3).

Metallocene and other single-site catalysts have well-defined structures, which in turn lead to narrow molecular mass distributions and better control of chain irregularities, including new types of chain defects obtained with selected combinations of metallocenes and aluminoxanes. Metallocenes can be used in solution, but for better control of particle size and shape, heterogenized systems are preferred. Supported single-site catalysts can be used as

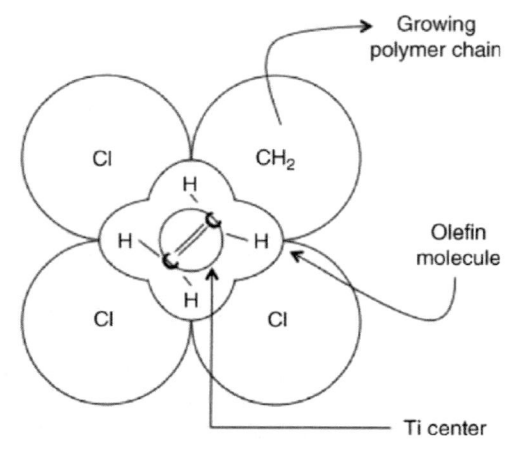

Figure 10. Structure of the active center with a complexed ethylene molecule

so-called drop-in catalysts to emphasize the ease of using these systems in existing plants without spoiling catalyst or plant performance.

5. Industrial Polymerization Processes [40]

Since the start of industrial PP production in 1957 the advances in catalyst technology and the demand for improved product performance have been the main drivers for the development of propylene polymerization processes [41]. The processes can be grouped into three main categories: gas-phase, bulk, and improved slurry. All state-of-the-art process technologies employ a gas-phase or bulk reactor system for the production of homopolymers and random copolymers. For the polymerization of impact copolymers (heterophasic copolymers) an additional gas-phase reactor is connected in series. The typical current PP processes are listed in Table 5.

Polyolefin production lines have been developed to extremely efficient and large machines, starting with annual capacities of about 5×10^3 t in 1963 to more than 400×10^3 t for the newest plants being built now.

Historically the suspension processes are the oldest ones; three facts are decisive for the design:

1. The limited polymerization yield of the catalyst, which on the one hand resulted in the necessity to remove the catalyst residues. On the other hand, it implied the use of serial reactors for improving productivity.
2. The necessity to remove amorphous fractions (atactic PP, which is found at ZN catalysts in low molecular masses and waxy form) because of the limited stereocontrol in the polymerization.
3. The fact that complex copolymers with multiphase structure and various solubility were soon considered as important products to expand the application range of PP.

In Figure 11 the block design for an early PP process is compared with a fourth-generation ZN catalyst setup. Especially the advances in catalyst technology led to the design of more efficient processes, and thus the elimination of some process steps.

For slurry- and gas-phase reactors the polymer is formed around heterogeneous catalyst particles. Slurry technologies can be divided into diluent and bulk processes. In diluent technologies, an inert diluent (typically a C_4–C_6 alkane) is used to disperse the growing polymer particles, while the monomer (propene and ethylene as comonomer) is introduced as a gas and other comonomers (higher α-olefins) are fed as liquids. In bulk technology the polymerization takes place in the liquid propylene monomer. Slurry technology can be used for the production of PE and PP, while bulk technology is restricted to polypropylene and its copolymers. In slurry polymerizations autoclaves and loop reactors are used. Ethylene, propene, and higher α-olefins can also be polymerized in gas-phase reactors. They can be classified into fluidized-bed reactors (FBRs) and stirred-bed reactors (SBRs). In FBRs a gaseous stream of monomer and an inert gas (nitrogen) fluidizes the polymer particles, while mechanical stirring is responsible for suspending the polymer particles in gas-phase SBRs. This reactor type can be subdivided into horizontal and vertical reactors.

5.1. Liquid-Phase Processes

The first commercial PP production processes were based on slurry technology. The main advantage of using slurries rather than the gas phase is the far better heat-transfer capacity when the polymer particles are suspended in a liquid. However, slurry processes require a workup section for the diluent, leading to extra costs in the investment and operation of the plant.

Table 5. Some typical PP processes (homopolymer reactor)

Regime	Process (technology supplier)
Gas phase	Unipol (Dow)
	Innovene PP (Ineos)
	Horizone (JPP)
	Spherizone (LyondellBasell)
	Novolen (Novolen)
Bulk	Spheripol (LyondellBasell)
	Borstar (Borealis)
	Hypol II (Mitsui)
	ExxonMobil
Slurry	Several

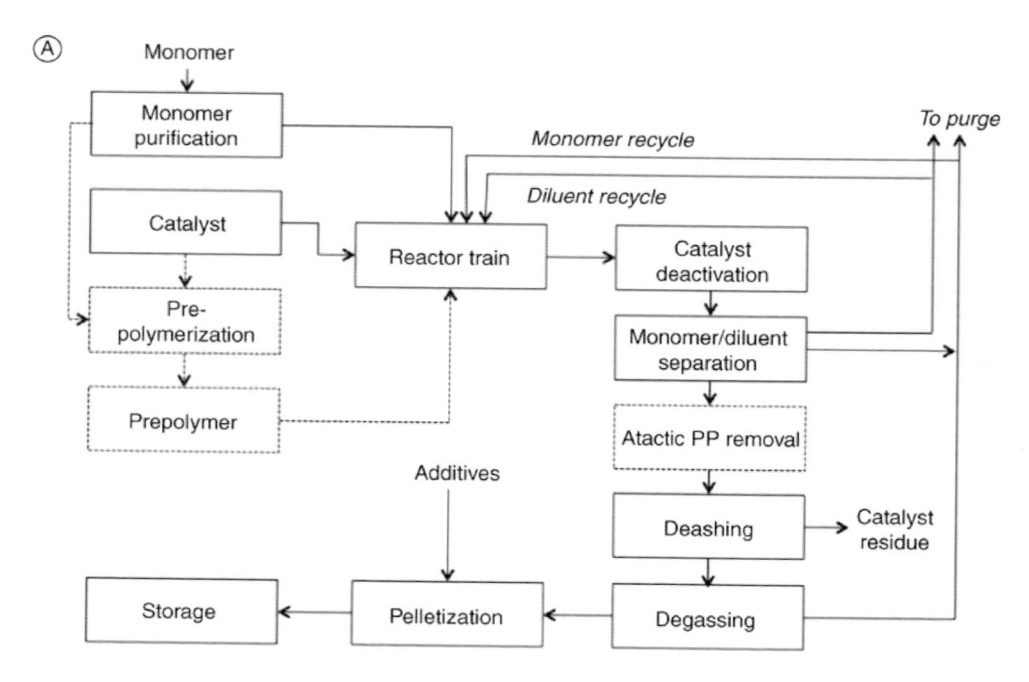

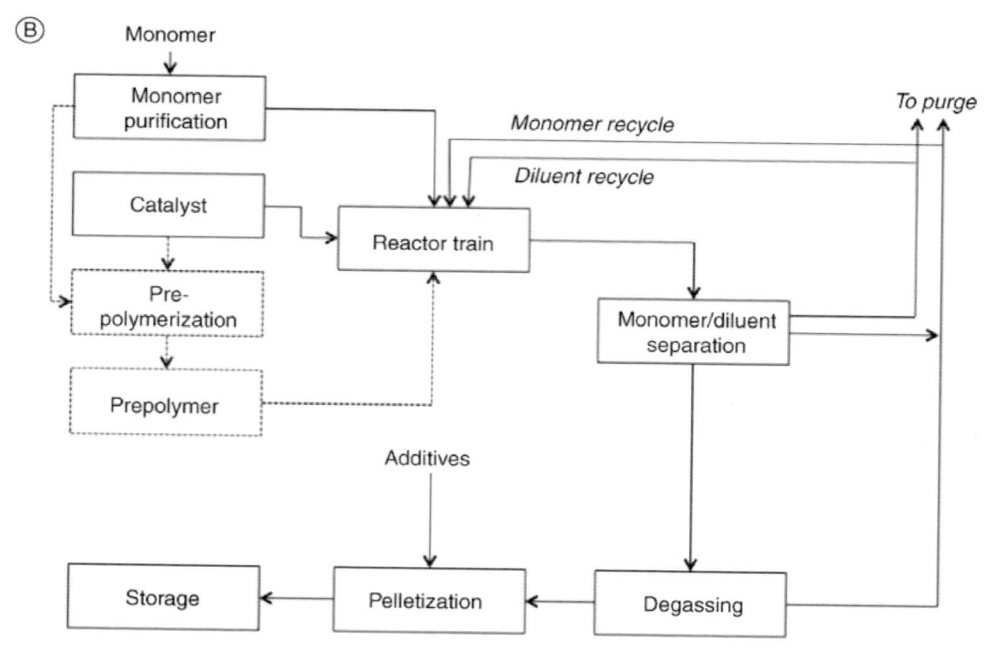

Figure 11. Polyolefin plant designs
A) Early PP process; B) State-of-the-art PP process

Most first-generation PP processes used slurry conditions (Hercules and Montecatini) with hexane as diluent. Two to four stirred tank reactors (autoclaves operated at 60–80 °C and 5–15 bar) not only allowed good monomer conversion, but also the production of copolymers with multiphase structure. The limits are defined here by the copolymer solubility but also the available heat-transfer surface per unit volume compared with loop reactors.

The workup of the reaction mixture in these early processes was rather tedious. Both the atactic fractions and the catalyst, which is still used today for producing PP grades with very high purity, which are, for example, required for producing capacitor films, had to be removed. The catalyst is decomposed with a mixture of alcohol and $NaOH/H_2O$, followed by separation of the organic and aqueous phases. The aqueous stream is distilled, leaving the catalyst residue in the wastewater. The organic stream is then sent to a centrifuge for removing the iPP part. From the diluent stream the aPP is recovered in a thin-film evaporator, while the iPP is purified with steam and dried.

Polymerization in liquid propene as polymerization medium (bulk process) can be performed in continuous stirred-tank reactors or loop reactors. In both cases the use of liquid monomer as polymerization medium maximizes the polymerization rate due to the high monomer concentration. Bulk processes are used to produce homopolymers and random copolymers with less than 5 wt% of ethylene (or other comonomers). Higher concentrations of the comonomer or the production of rubber phases (ethylene–propylene rubber) in heterophasic copolymers are not feasible due to the solubility of the rubber in the liquid monomer. For the production of these materials hybrid processes (see Section 5.3 or multistage gas-phase reactors) are used.

Loop reactors account for production of about 50% of all commercial polyolefins (PE and PP) and many different reactor configurations are used in industrial processes. The loop reactor can be placed in the vertical (*Phillips*, *Spheripol*, *Borstar* processes) or horizontal position (*USI process*). The loops can have up to 12 legs (the loop shown in Fig. 12 has four legs), and each leg can be up to 60 m tall. The volume of a typical loop reactor is on the order of 100 m^3, which corresponds to a capacity of roughly 250×10^3 t/a. The solid content is in the range of 40–50 vol%, mainly limited by the viscosity of the slurry. A pump provides for circulation of the slurry at the bottom of one of the legs. High fluid velocities (10–30 m/s) create a turbulent flow, which helps to prevent particle settling and also improves the heat-transfer coefficient.

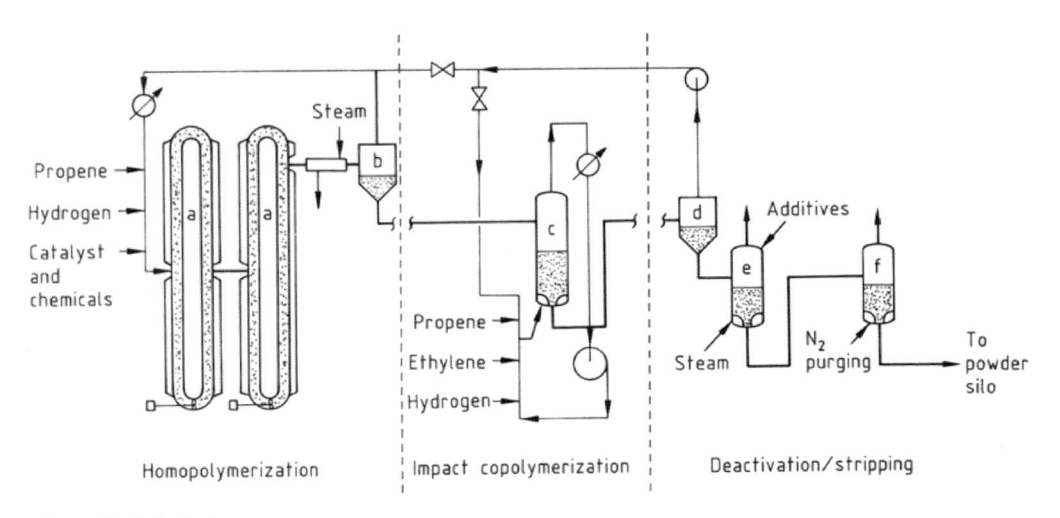

Figure 12. Spheripol process
a) Loop reactors; b) Primary cyclone; c) Copolymer fluidized bed; d) Secondary and copolymer cyclone; e) Deactivation; f) Purging

5.2. Gas-Phase Processes

Gas-phase processes are an economical and energy-efficient alternative to liquid-phase polymerization. Separation of polymer from the unconverted monomer is easy, since the monomer is in the gas phase. There is no need to flash off large amounts of liquids, which means a significant cost reduction compared to slurry processes. Another advantage of gas-phase processes is their potentially broader product window, as there are no solubility limits for hydrogen and monomers in the reaction medium. Polypropylenes with very high melt flows (low viscosities) and high comonomer contents can be produced. However, the actual flexibility of the process depends on the reactor size and the residence-time distribution (RTD). A drawback of gas-phase reactors is their limited heat-transfer capability. To enhance the thermal characteristics of these reactor types special actions must be taken. One possibility is to inject small amounts of liquid components below their dew points or the use of inert gas-phase compounds with a higher heat capacity than nitrogen.

From a process design perspective gas-phase reactors are available as FBRs and SBRs in vertical and horizontal configuration, as well as the multizone-circulating reactor (MZCR) introduced by LyondellBasell.

The *Unipol process* by Dow was initially developed for PE production in the late 1970s and later adapted for the polymerization of PP by introducing Shell high-activity catalysts (SHAC) [23]. The process is based on a large FBR for the production of homopolymers and random copolymers and can be extended by a second, smaller, gas-phase reactor for the production of heterophasic copolymers. The basic design of the process is shown in Figure 13. The expanded upper section in the reactor is designed to reduce gas velocity and powder entrainment. Reaction conditions are reported as 60–70°C for the homopolymer reactor and a pressure in the range of 25–30 bar. The residence time per reactor is typically 1 h. Thus, the grade transition times are comparable with those of loop reactors. The Unipol process has a unique product discharge system, using a cyclone separator.

The *Innovene* and the Japan Polypropylene (JPP) *Horizone process* technology are based on a horizontal stirred gas-phase reactor. The reaction conditions are 65–85°C for the homopolymer reactor and a pressure in the range of

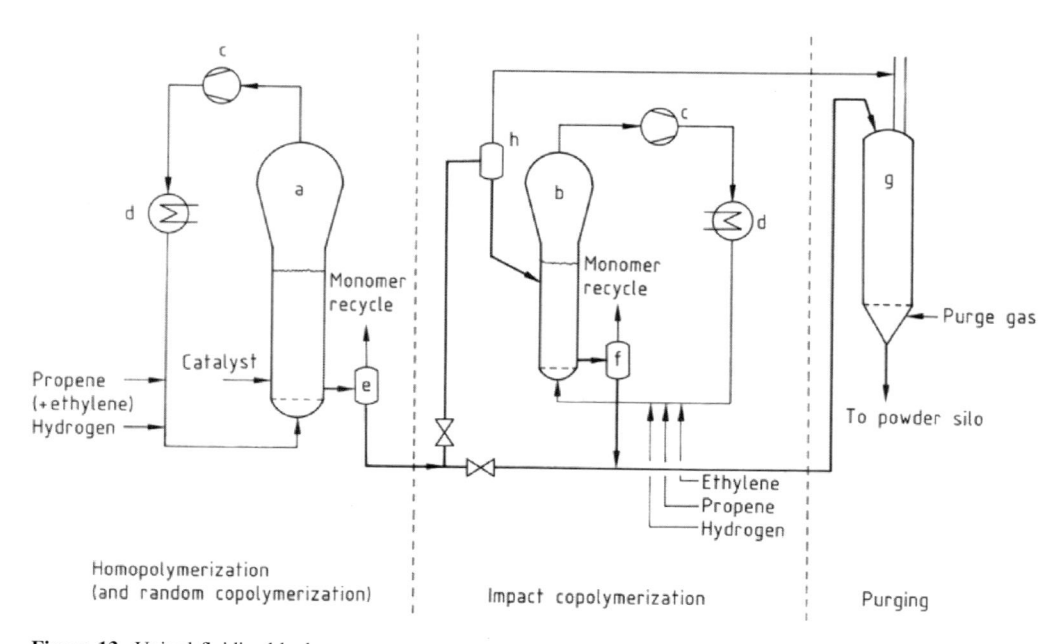

Figure 13. Unipol fluidized-bed process
a) Primary fluidized bed; b) Copolymer fluidized bed; c) Compressors; d) Coolers; e), f) Discharge cyclones; g) Purge

25–30 bar. The residence time per reactor is typically 1 h. The spacing of the injection points for the catalyst and cocatalyst appear to be critical for this type of reactor. If they are to close one encounters lump formation in the feed zone; if they are too far apart the catalyst is not properly activated. Adding the ethylene–propene rubber in a second gas-phase reactor can make heterophasic copolymers.

Vertical stirred gas-phase reactors were developed first by BASF, *Novolen technology*, and put into operation in 1967. Originally the design was that of a stirred autoclave with a bottom-mounted helical stirrer. This stirrer is designed to convey the particles up the reactor wall and let them fall down through the center of the bed. The special design of the agitation system is considered to be responsible for forming subsegments in the reactor. Thus, it is claimed that the reactor behaves like a cascade of a large number of CSTRs. The heat is removed by circulating monomer gas through an external heat exchanger. For impact-modified PP, a rubber phase is polymerized in the second gas-phase reactor connected in series. Figure 14 shows the design of the Novolen process.

The *Spherizone* multizone circulating reactor (MZCR), presented in Figure 15, was developed by Basell and first commercialized in the early 2000s. This reactor concept builds on the riser-tube design known from fluid catalytic cracking. The Spherizone technology consists of two interconnected reaction zones—a riser and a downer—separated by a cyclone. In the riser (the first reaction zone) the gas velocity is high and thus this zone acts like a fast-fluidized bed. In the connected cyclone the polymer particles are separated from the gas and then enter the second reaction zone (downer). The downer can be characterized as a moving packed bed [42]. The two different sections of the reactor can be operated at different concentrations of hydrogen and comonomer, leading to bimodal polymer qualities (with respect to molecular mass and/or comonomer concentration). This is achieved by separating the composition of two reaction zones with the help of a barrier fluid introduced

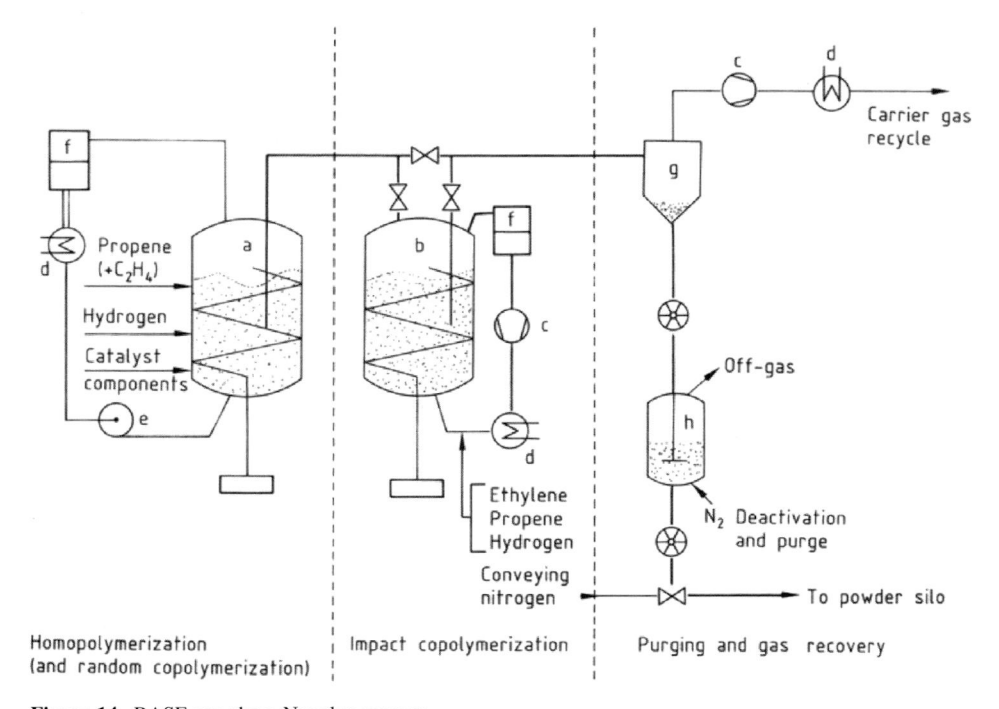

Figure 14. BASF gas-phase Novolen process
a) Primary reactor; b) Copolymerizer; c) Compressors; d) Condensers; e) Liquid pump; f) Filters; g) Primary cyclone; h) Deactivation/purge

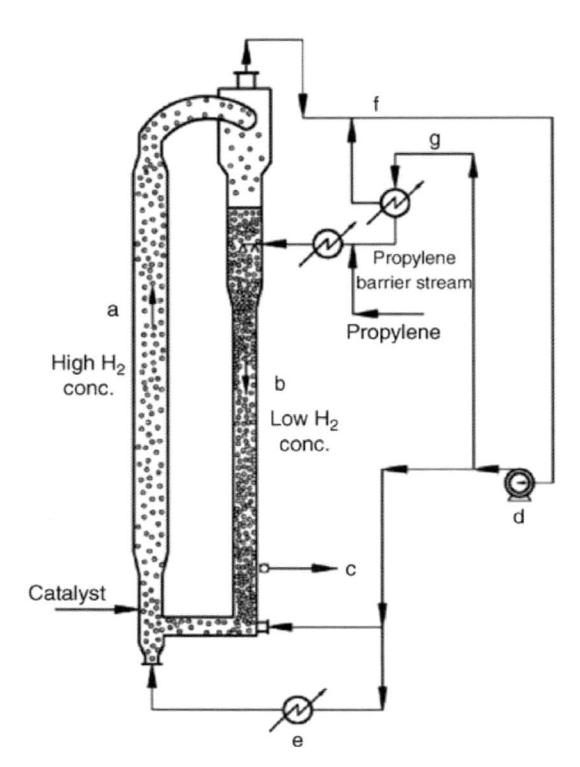

Figure 15. LyondellBasell Spherizone multizone circulating gas-phase process (adapted from [42])
a) Riser: upward pneumatic transport; b) Downer: packed bed moving downward; c) Product discharge; d) Gas fan; e) Heat exchanger; f) Internal gas/solid separator; g) Condenser

at the top of the downer. The fast circulation of the growing polymer particles between the different reaction zones makes the residence time per pass in each zone one order of magnitude smaller than the overall residence time. This leads to a very good homogeneity of the final product even for bimodal compositions.

The multizone circulating reactor can be connected with an additional gas-phase reactor (FBR) for producing heterophasic copolymers.

5.3. Hybrid Processes

Hybrid processes combine the production of a homopolymer or random copolymer with low ethylene content in liquid propene with the production of a rubber phase (ethylene-propene rubber) in a sequential gas-phase reactor.

The dominant hybrid process is the *Spheripol process* by LyondellBasell. Figure 12 shows a two-loop and single-FBR Spheripol design. Roughly one-third of today's world PP production is based on this technology. In this process monomer and catalyst components are fed into the first loop reactor for homo- and random copolymerization. Due to the large surface-to-volume ratio of loop reactors, the heat-removal capacity is high and enables high specific outputs exceeding 400 kg h^{-1} m^{-3} of PP. The shape of the spherical catalyst particles, consisting of many microscopic MgCl$_2$ single crystals with active surface, is replicated to give spherical polymer particles with a narrow particle size distribution. Coupled with the high liquid velocities in the loop reactor, reactor fouling can be avoided. Operating conditions are in the range of 60–80°C and 35–50 bar.

For impact (heterophasic) copolymers, a gas-phase reactor is required. Upon exiting the loop reactors, the propene–PP slurry is depressurized and flashed. The pressure must be adjusted to allow recycling of the vaporized monomer by condensation and to be sufficient for gas-phase copolymerization. The ethylene–propene rubber needed for impact copolymer is produced in an FBR at temperatures up to 100°C and a pressure of 15 bar. A second (or even third) FBR can be added to the process for tailoring the rubber phase with respect to molecular weight and comonomer concentration.

The Mitsui *Hypol* bulk polypropylene process was originally based on two stirred autoclave reactors in place of the loop reactors shown for the Spheripol process in Figure 12, followed by two stirred FBRs. In the Hypol II process loop reactors replace the stirred autoclaves. The gas-phase reactors are also used to produce homopolymers, enabling a broader polymer-property window.

The design and operation of the ExxonMobil PP process technology was originally based on the Hypol design but has been modified with respect to gas-phase reactor operation.

Borealis has developed the *Borstar* polyolefin process technology for the production of PE and PP. A basic design of the Borstar PP technology is given in Figure 16. The platform for making homopolymers and random copolymers is based on a loop reactor followed by an FBR. The loop reactor allows polymerization under sub- or supercritical conditions. For PE this is done in propane and PP is

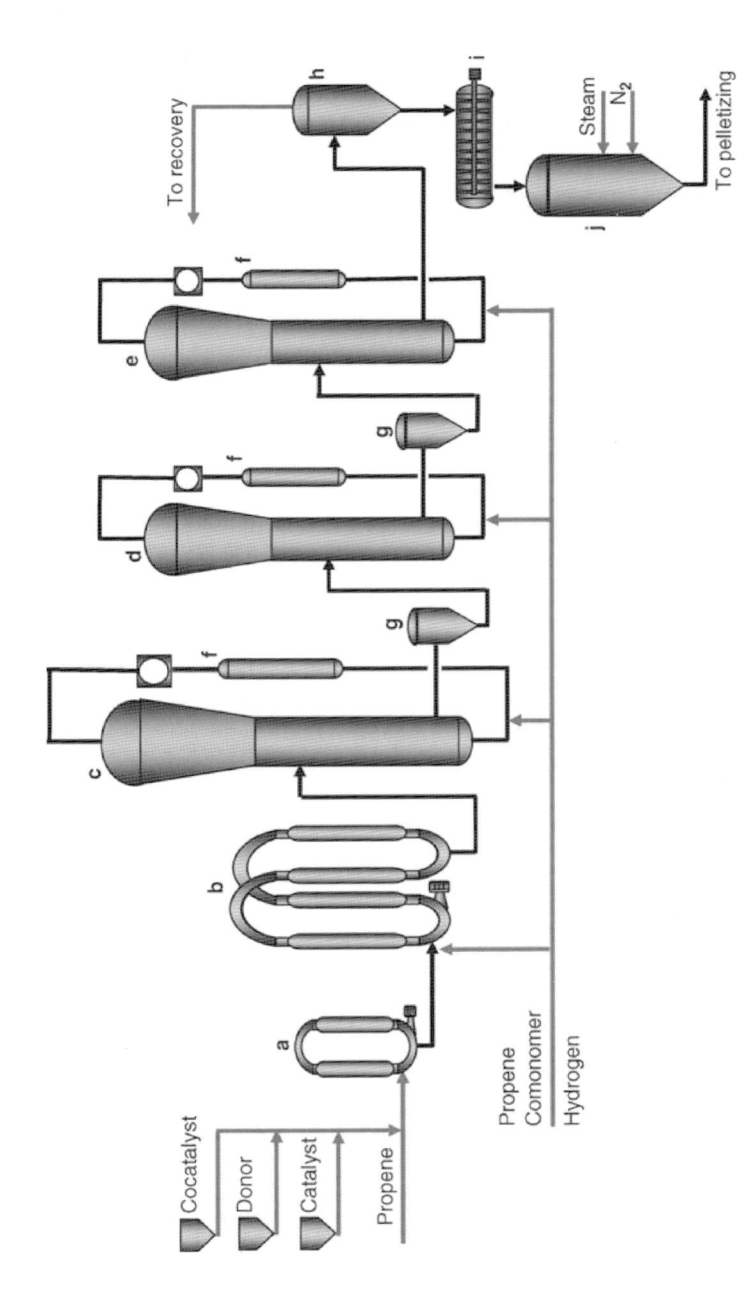

Figure 16. Borstar PP process scheme (four-reactor setup)
a) Prepolymerizer; b) Loop reactor; c) First gas-phase reactor; d) Second gas-phase reactor; e) Third gas-phase reactor; f) Coolers; g) Separators; h) Low-pressure degasser; i) Dryer; j) Purge bin

polymerized in liquid propene. The advantage of running the loop reactor under supercritical conditions is the possibility of increased hydrogen and comonomer concentration without the formation of gas bubbles. In the next section an ethylene–propene rubber is made in an FBR for producing impact copolymers. Powder transfer from the homopolymer FBR to the rubber gas-phase reactor allows complete gas separation for better design of the rubber. An additional rubber FBR can be added to the reactor train if high elastomer contents are required.

The hybrid processes are the workhorses of the polypropylene industry. Loop reactors dominate the bulk polymerization processes due to the better heat-transfer capacity and thus higher production rates. Hybrid processes contain up to six reactors split between a sector dedicated to homopolymers and random copolymers and a second sector used for the production of a rubbery phase (ethylene–propylene rubber) to increase the toughness of the polymers. This second block contains only FBRs. The two blocks are almost always separated by a kind of gas lock/flash in order to better control the properties of the different phases. With the described process design PP manufacturers can produce a wide range of product qualities, but this must be balanced with the capital investments needed for full-fledged reactor train.

6. Product Finishing

The so-called dry end of a polymerization plant starts at the discharge point from the last polymerization reactor and involves deactivation, powder purification, and in most cases melt compounding. Even if the polymer is planned to be sold as powder, it must be stabilized, as PP is far more susceptible to radical attack and degradation than PE. In any case, the transport and sale of PP can be handled both in bulk form (silo trucks or railway cars) and in packages of various sizes, ranging from 25 kg bags to big-bags and octabins (corrugated cardboard containers with PE-film inlays) containing up to 1000 kg.

Melt homogenization can provide several functions for the final product, ranging from homogenization (especially relevant for polymers with a bimodal or generally broad molecular mass distribution (MWD) and heterophasic copolymers [43]) through the dispersion of stabilizers and other functional additives to generating a product with homogeneous particle (pellet) size. With increasing plant size, the respective equipment has also increased in throughput from 15–20 t/h in 1990 to 50–75 t/h in 2010. In addition to cost, extruders are selected according to the following factors: powder morphology, product melt flow rate (MFR) range, melt filtration needs, devolatalization options, throughput flexibility, temperature control, mixing needs, and additive feeds. Single-screw machines are hardly used any more, while twin-screw extruders with co-rotating and segmented screws dominate the market because of their high flexibility and good mixing efficiency. A balance between mixing quality and polymer degradation must always be kept, with temperature (220–260°C) and specific energy input (SEI) as decisive factors. A gear pump between the die plate and the discharge end of the screw can be used both to reduce the SEI and to increase the throughput.

Pelletization is possible in strand form or in underwater systems. Modern polymerization plants are, however, too big for the former in which several melt strands from the die plate are quenched in water for solidification and then chopped by high-speed cutters into 2–5 mm long cylinders. In underwater systems, high-speed knives rotate against the extruder die plate to cut off short lengths of the molten extrudate. The insulated die plate is immersed in water, which solidifies the lens-shaped pellets and also transports them to a sieving and drying device. While nitrogen must be used for powder conveying up to the extruder, air can be used for pneumatic pellet transport afterwards.

The direct sale of nonstabilized or (mostly) spray-stabilized PP powder is very limited. While in the 1990s Himont (presently LyondellBasell) still assumed that the spherical powders resulting from fourth-generation catalysts and the Spheripol process would be suitable in nonpelletized form for large parts of the market, applications are largely limited to rotomolding, melt-blown fiber spinning, and use as supports for liquid additives like peroxides. While energy-intensive extrusion and pelletization operations could be saved, the limited choice of additives and the homogenization problems more than compensate these advantages in most areas.

6.1. Additives

Nonpolymeric additives generally allow further modification of the properties of polyolefins. An approximate distinction is drawn between additives (up to 1 wt% addition, typically ca. 0.05–0.3 wt%) and fillers or reinforcements (5–40 wt% addition). The most important additives [44] are antioxidants and UV stabilizers.

More than PE, PP must be protected against oxidation, particularly above 100°C. The dominant reaction is chain scission by free-radical attack at the tertiary carbon atoms of the backbone. The different types of *antioxidants* can be divided into:

- H donors/primary antioxidants (e.g., sterically hindered phenols; active mostly at application temperatures)
- Hydroperoxide decomposers/secondary antioxidants (e.g., phosphites; active mostly at processing temperatures)
- Radical scavengers [hindered amine light stabilizers (HALS); specific against UV-induced radicals]

In Figure 17 the autooxidation process and the principle of antioxidants in polymers is illustrated. The reaction starts on the left-hand side of the scheme, where scission of the chain (R) initiated by, e.g., heat and shear to give the alkyl radical (R$^•$) is shown. The alkyl radical can be directly deactivated with a radical scavenger or forms the peroxy radical (ROO$^•$) in the presence of oxygen. Both the H donor (primary antioxidant) and the radical scavenger can work as antioxidant here. Otherwise, ROO$^•$ will react with other chains by hydrogen abstraction (cleavage of the weakest C–H bond). This leads to the formation of hydroperoxide (ROOH), which is relatively unstable and decomposes into the reactive alkoxy radical (RO$^•$) and the hydroxy radical ($^•$OH). ROO$^•$ reacts with the H donor to form stable ROOH. Hydroperoxide decomposers (secondary antioxidants) are used to induce (from ROOH) the formation of inert alkoxy (RO$^•$) and hydroxy radicals ($^•$OH) to avoid chain fracture, which otherwise results in discoloration and brittleness.

A further contribution to long-term stability is the use of acid scavengers for neutralizing decomposition products of ZN catalysts, especially HCl. The most common is Ca stearate (0.05–0.1 wt%), which also prevents equipment corrosion. It behaves as a mild slip agent and suppresses attack of HCl on certain antioxidants. Stearates or oxides of zinc and magnesium are also in use, and synthetic hydrotalcite (a hydrated magnesium aluminum hydroxycarbonate) is the latest addition to this class.

Besides the thermomechanical loading to be expected for a specific application, additive packages are also defined by critical environmental interactions, e.g., NO$_x$ with fibers or detergent solutions with appliance components. The choice of stabilizer is further influenced by regulatory constraints such as global or specific migration limits.

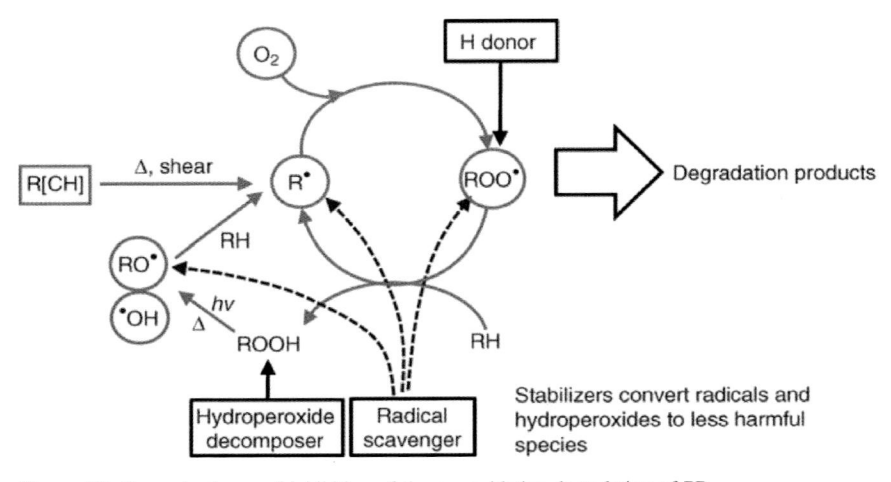

Figure 17. General scheme of inhibition of thermo-oxidative degradation of PP

Continuous outdoor exposure also requires protection from the damaging effects of UV radiation, for which UV absorbers (in the simplest case carbon black or TiO_2) or radical scavengers are used. Polymer photooxidation results from the combined action of light and oxygen. First, the polymer chain is excited by the absorption of light ($h\nu$), and UV absorbers act in this stage of degradation. In the presence of oxygen and further light, an alkyl radical (R$^\bullet$) is formed, and radical scavengers (HALS) are used to scavenge radicals. In contrast to UV light, the effect of high-energy radiation like electrons or γ rays cannot be prevented by additives alone; polymer design measures must be considered here.

Antistatic agents act against static charges and resulting dust deposits on many plastics; 0.2–1.0 wt% of polar additives such as polyether fatty amide and fatty amine condensates, but also simple glyceryl monostearate, reduce the surface resistivity from $>10^{13}$ to ca. $10^7 \, \Omega$. The mechanism involves slow diffusion of the additive to the surface, where it picks up atmospheric moisture. Time and humidity are important, and while at 50% RH most of the recommended additives are satisfactory, dryer environments require the use of fatty amine derivatives.

Slip and antiblocking agents are mainly used in film production where high-speed handling requires a controlled reduction in frictional forces; 0.1–0.5 wt% of oleamide, erucamide, or mixtures thereof is usually sufficient. Fine, spherical particles of silica or other minerals prevent layers of film adhering to each other during storage.

Nucleating agents for PP crystallization [45] are applied both for improving processability and final performance. The combination of moderately slow crystal growth at large undercoolings together with the practical absence of sporadic nucleation makes iPP, which melts at 160–170°C but crystallizes at 100–130°C, an ideal material for controlled nucleation. In addition, the different crystal modifications of iPP (α, β, and γ) can be controlled in their respective expression, and thus affect the mechanical and optical profile of the material. A wide range of chemically very different substances (see Fig. 18) can be used; decisive is the crystalline and not the molecular structure, because nucleation happens by epitaxy and "lattice matching". These additives must be distributed very well. Different routes can be applied here, and the Borealis nucleation technology (BNT) has a special position: Because the dispersion occurs already during the polymerization step, a very low concentration of a polymeric nucleating agent is sufficient for optimum performance [46]. Improvements of stiffness (modulus) and the related heat deflection temperature by selectively nucleating the α modification is one of the key targets. Another target of α nucleation is improved transparency and clarity; here nucleation will even help to prevent haze increase during steam sterilization (normally caused by postcrystallization at the elevated temperature

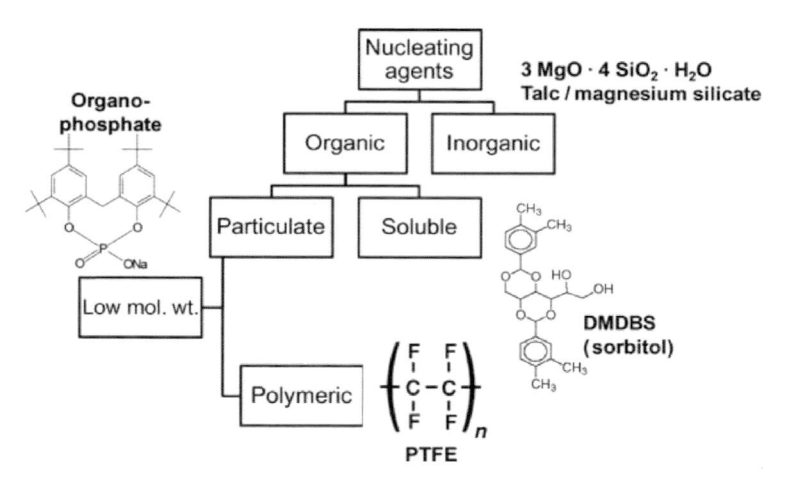

Figure 18. Classification of nucleating agents for PP

used). As crystallization is also accelerated, nucleation also helps to reduce cycle times in injection molding and to allow higher speed in extrusion processes. Specific β nucleation is a useful tool for improving impact strength and reducing crack growth in pipe materials, where it has already been used for many years [20]. Careful balancing of the overall additives recipe is essential in any case, because both adverse and beneficial interactions occur between some of the components.

6.2. Compounding and Blending

For PP, modification after the polymerization process plays a major role in expanding the property range. Besides the already mentioned co-rotating twin-screw extruders with special mixing elements for improving the quality of dispersion and resulting phase structure, other kneaders and mixers are used for special requirements. The Farrel kneader, for example, has a large blending chamber for adding EPR, EPDM, and other elastomers in form of a bale, while the Buss co-kneader allows more gentle dispersion of glass fibers by combining rotating and pulsating melt movement.

Modifications of mechanical properties as well as shrinkage and thermal expansion are possible through the addition of fillers and reinforcements of mostly mineral nature. The respective effects are determined by the properties of the base polymer as well as by the quantity and nature of the filler [47]. Talc with an average particle size of 0.6–10 μm is the most important filler, while calcium carbonate, kaolin, wollastonite, and mica are used to a lesser extent.

When using glass fibers (GF), which give the highest strength, further improvements are possible by modifying the fiber surface (sizing) and using compatibilizers (adhesives). Both short (S) GF having an initial fiber length of 3–5 mm and long (L) GF are applied. While SGF compounds are produced in normal melt mixing, LGF–PP is produced by impregnating continuous GF rovings with a high-MFR matrix and cutting the resulting strand after solidification to 10–15 mm length. Special processing operations are required in the latter case to avoid excessive fiber breakage [48], but the resulting material is well suited for metal replacement, especially in the automotive area.

Organic reinforcing fibers are obtained from natural (regenerative) sources such as hemp, flax, sisal, and wood, including high-crystallinity man-made cellulose fibers [49]. The strength of glass fiber reinforced materials is not achieved with such additives and the lot-to-lot variations of natural fibers are a problem; the full combustibility and improved sustainability of such composites are, however, seen as an advantage.

Modification or blending with elastomeric or polymeric components for further expanding the property range of PP is discussed in Section 7.4.

6.3. Reactive Modification

An even broader variety of structural modification is possible by reactive post-reactor processes. The earliest version was targeted at modifying the molecular mass distribution, whereby the preferential splitting of PP chains in radical reactions allows tailored production of narrow molecular mass distributions [50]. This process, known as visbreaking or (controlled rheology) CR process, is best applied to PP homopolymers or random copolymers, leading to products with higher MFR and broader Newtonian region of the viscosity curve. Often liquid peroxides are in use, which are sprayed under nitrogen onto the warm reactor powder before it is fed to the pelletization extruder. In large-scale plants, visbreaking also allows the production of a multiplicity of grades from one reactor setting. While for reactor grades of PP (with broad MWD), stiffness increases with increasing MFR, it remains constant or is reduced for visbroken (narrow-MWD) grades. At the same time, transparency and surface gloss are rather improved, and these grades have significant advantages in fiber spinning and cast-film processes.

Other reactive modification steps for PP are also mostly based on radical-initiated grafting reactions [51]. The potential applications of this technology range from the production of long-chain branched PP with high melt strength (strain hardening) offering advantages in foaming [52] and other processes with strong extensional flowlike extrusion coating, via the stabilization or partial crosslinking of phase structures (to the point of thermoplastic vulcanizates, TPVs) up to polar modifiers (e.g., by grafting with maleic anhydride).

7. Properties

7.1. Homopolymers

For PP homopolymers (PP-H), the mechanical and optical performance is determined by:

- The average molecular mass and molecular mass distribution
- The chain structure, i.e., the presence of stereo- and regiodefects
- Nucleating additives (see Section 6.1) [53]

As for all semicrystalline polymers, processing history and component geometry will play a further role in affecting the final properties, and the evolution of performance over time will depend on a proper stabilization.

While a proper parameter for the average chain length is the weight average of the molecular mass distribution M_w in industrial practice polymers are normally defined by their melt flow rate (MFR), which in the case of PP is determined at 230°C and 2.16 kg load. The two parameters are related via the melt viscosity, where the zero-shear viscosity η_0 is proportional to the 3.4-th power of M_w [43]. At higher shear rates normally encountered in processing and more characteristic for MFR determination, the dependence is still greater than a power of two because much chain entanglement still exists. The exact correlation, however, also depends on the broadness (polydispersity, defined as ratio between M_w and the number average of the molecular mass distribution M_n) and shape of the molecular mass distribution. PP-H grades are commercially available in an MFR range of 0.3–1000, corresponding to an M_w of 500–20 kg/mol and covering applications from sheet and pipe extrusion to high-speed fiber spinning.

With modern ZN catalysts a single-step reaction gives a polydispersity M_w/M_n of 4–6, while with older catalyst types (TiCl$_3$, used in diluent-based slurry processes) M_w/M_n can be up to 8. Broad or bimodal molecular mass distributions are achieved by multistage polymerization with varying hydrogen feed and/or temperature, while narrow molecular mass distributions with M_w/M_n down to ca. 2 are mostly achieved by visbreaking (see Section 6.3). Both versions have specific advantages: High polydispersities deliver higher crystal orientation through flow-induced crystallization [54, 55], especially in injection molding, which results in higher stiffness and heat resistance and is consequently of interest for technical applications. Narrow-MWD grades crystallize more slowly [53] and have advantages in film casting and fiber spinning, but also for special molding applications like film hinges.

The chain structure of PP homopolymers is affected by stereo- and regiodefects, the latter only being relevant for single-site catalysts (SSC). An absolute quantification of stereoregularity is only possible by ^{13}C NMR spectroscopy in solution, with isotactic triads (mm) or pentads (mmmm) being used for quantification. Commercial PP has a pentad regularity of >92%, and very high stereoregularity of >97% is achievable with modern ZN catalysts and suitable external donors, even without removing any soluble (atactic) byproduct. Besides a broad MWD, high isotacticity is a further tool for reaching highest stiffness [54]. Alternative measures for isotacticity are FTIR spectroscopy (in the solid state, where sample preparation must be considered) and the content of heptanes-soluble fraction. For some applications like nonwovens with thermobonded fibers or biaxially oriented PP (BOPP) film, processing can be facilitated by reduced isotacticity or chain regularity in general. This is readily achieved by lowering the donor feed in polymerization, or by introducing small amounts of copolymerized ethylene.

When isotactic PP is produced by SSC or, more specifically, metallocene catalysts (MC), three major product-related differences over ZN catalysts are evident [56, 57]:

1. The polymers have a much lower polydispersity, with M_w/M_n in the range of 1.8–3. Also, the amount of solubles, measured both in cold xylene (XCS) and in boiling hexane (C6) is reduced significantly, as is the level of hydrocarbon emissions. All of these advantages can be explained by the reduced amount of oligomers generated in SSC-based polymerization.

2. Besides the stereoregularity, the presence of regiodefects (misinsertions of monomers)

must be considered. Most relevant are the 2,1 regiodefects commonly defined as the sum of 2,1-*erythro* and 2,1-*threo* regiodefects, which can be present up to 2 mol% depending on the MC type. While MC-based PP-H types can have a high isotacticity, these defects cause a significant reduction of the melting point and limit both heat resistance and stiffness [57].

3. All chain defects, but also the insertions of comonomers—and specially ethylene—are much more randomly distributed along the chain. At the same time, higher α-olefins like 1-butene and 1-hexene can be copolymerized more easily.

A very special case of PP-H is syndiotactic polypropylene (sPP), which is characterized by significantly lower crystallinity and stiffness [58]. It can be polymerized with specific SSCs, both MC and constrained-geometry types, and varied over a wide range by controlling the relative content of isotactic, syndiotactic, and atactic sequences. Its commercial relevance is, however, so far rather limited, largely as a consequence of the high solubility in various organic solvents.

An overview of the most relevant properties of different PP product families can be found in Table 6.

7.2. Random Copolymers

Similar to the variation of density and mechanical properties when moving from high density polyethylene (HDPE) to linear low density polyethylene (LLDPE) by copolymerization with higher α-olefins like 1-butene (C4) and 1-hexene (C6), the properties of PP can be modified already in one-step polymerizations. Incorporating ca. 2–5 wt% of ethylene (C2) into the chain lowers the overall crystallinity, gives a broader softening range with reduced melting points (see Figure 19), lowers the glass transition temperature, increases the fraction of soluble polymer, and improves transparency and surface gloss [59]. Ethylene–propene random copolymers (EP-RACOs) are somewhat tougher than homopolymers, and do not exhibit the familiar stress-whitening behavior of the tougher copolymer impact grades. These changes for EP-RACOs are determined by the amount of comonomer and its distribution along the chain, with significant differences resulting from catalyst and donor type. The randomness or rather the probability for isolated C2 units in the chain decreases with increasing amount, leading to phase-separation phenomena in case of ZN catalysts at contents >5 wt%; higher values can be achieved with metallocene catalysts.

All changes in mechanical properties and optics are related to a reduced crystallization speed and a lower lamellar thickness caused by the comonomer-related chain defects [60]. This also results in a better quenchability of EP-RACOs to the mesomorphic state, making them ideal materials for highly transparent cast films, where also the broader melting or sealing range is an advantage.

The increased amount of XCS- and C6-soluble material in EP-RACOs is, however, limiting their applicability in this and other areas, especially when packaging of food or pharmaceutical products is the intended use. A better balance between softness and solubles content is possible with C4 copolymers, perhaps because the solubles are produced at semi-exposed catalyst sites more easily accessible to ethylene [59]. The higher T_g and lower toughness of C3/C4 copolymers can be compensated by producing terpolymers with C3, C4, and C2, for which also other advantages like improved processability are found [61]. Typical applications for these specialty grades are sealing layers in coextruded films.

Common to all random copolymers is a shift in crystalline structure from the predominant monoclinic α modification to the orthorhombic γ phase. A new crystal modification was identified in MC-based copolymers with higher α-olefins [62]. The trigonal δ modification allows better incorporation of side chains resulting from these comonomers in the crystal lattice. While for the γ phase some advantages regarding transparency and ductility can be found and promotion by α-nucleating agents is possible [45], no specific technical benefits of the δ phase have been identified so far.

7.3. Impact Copolymers

Sections 5.2 and 5.3 describe multistage copolymerization systems for producing

Table 6. Overview of most relevant PP properties for various grade families

		Unmodified PP grades*					Compounds		
		Soft (1)	Standard (2)	Random (3)	Impact (4)	HCPP (5)	Talc	Glass fiber	Filler & elastomer
Filler content, wt%		0	0	0	0	0	20–40	20–50	10–25
MFR (230°C/2.16 kg), g/10 min	ISO 1183	0.5–15	0.3–1000	0.3–100	0.5–100	2–50	0.5–15	1–10	0.5–50
Density, g/cm³		0.900	0.905	0.900	0.900	0.905	1.05–1.25	1.04–1.22	0.95–1.15
Melting point, °C	ISO 3146	135–150	160–167	135–150	160–167	165–169	164–168	164–168	164–168
Glass transition, °C	ISO 6721-7	–10 to 0	0–4	–10 to 0	0–4	0–4	0–4	0–4	0–4
		–60 to 35			–60 to 35	–60 to 35(5)			–65 to 45
Flexural modulus, MPa	ISO 178	300–700	700–1500		1200–1900	>1900	2500–5000	5000–9000	1000–2000
Yield stress, MPa	ISO 527	15–25	25–40	20–30	25–40	40–50	25–35	70–140	15–30
Extension at yield, %	ISO 527	15–25	5–15	15–25	5–15	4–10	2–4	2–4	5–15
Extension at break, %	ISO 527	>300	20–800	30–1000	20–200	5–40	15–30	3–5	6–10
Charpy IS**, kJ/m²	ISO 179 1eU								
+23°C		n.b.	10–n.b.	15–n.b.	50–n.b.	30–n.b.	10–n.b.	25–50	n.b.
–20°C		30–n.b.	3–80	5–n.b.	10–n.b.	3–40	2–20	20–40	20–n.b.
Charpy NIS**, kJ/m²	ISO 179 1eA								
+23°C		10–n.b.	2–25	5–50	10–n.b.	2–4	2–12	7–20	15–35
–20°C		5–50	1–10	2–15	5–60	1–3	1–5	4–15	5–25
Vicat softening temperature, °C	ISO 306								
Vicat A		80–120	100–140	85–110	140–150	150–160	150–160	140–160	100–130
Vicat B		30–60	50–90	35–65	70–100	100–120	80–100	100–130	40–60
Heat deflection temperature, °C	ISO 75								
HDT-A		35–45	40–60	40–55	50–70	60–80	60–90	130–150	40–60
HDT-B		50–70	70–100	55–75	100–120	110–130	120–140	150–160	70–100
Gloss 20°, %	ISO 2813	30–60	60–80	60–80	70–80	70–80	5–30	20–40	20–40
Ball indentation hardness, N/mm²	ISO 2039-1	20–40	30–80	30–50	80–90	>90	80–90	100–120	40–60

*1: High-EPR heterophasic and random-heterophasic copolymers; 2: Homopolymers; 3: Random co- and terpolymers; 4: Heterophasic copolymers; 5: Homopolymers and low-EPR HECOs.
**IS: impact strength; NIS: notched impact strength; n.b.: not broken.

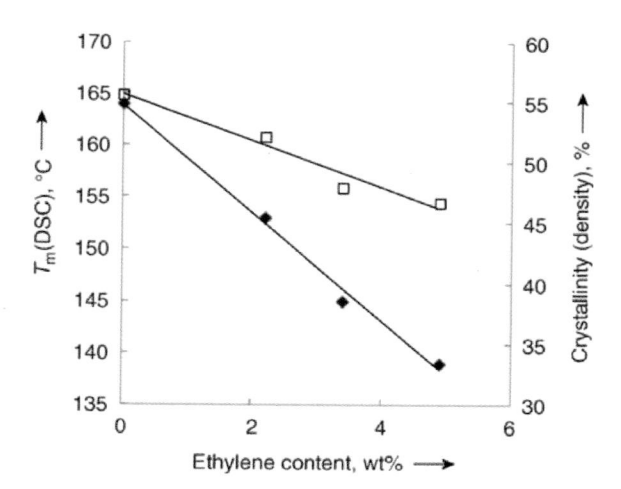

Figure 19. Effect of comonomer effect in EP random copolymers on melting point (♦) and crystallinity (□) (data from [60])

high-impact copolymers of PP. These combine a crystalline PP matrix (produced in the first 1–2 reactors) with embedded particles of EPR and PE (produced in one or more following reactors) giving impact and low-temperature resistance [63]. While often still referred to as block copolymers, the correct term for such products is heterophasic copolymers (EP-HECOs) or PP impact copolymers.

In-reactor production of EP-HECOs has been found to be more effective than compounding (see Section 6.2) for producing impact-modified PP. On the one hand, cost and energy are saved by eliminating the external elastomer components and the compounding step, and on the other hand morphology and performance advantages are achieved. The structure of these reactor blends has been shown to be rather complex: Besides the PP matrix and the amorphous ethylene–propene rubber (EPR), they can also comprise crystalline copolymers having both PP and PE crystallizable segments, and even neat high-density PE (HDPE). The elucidation of this complex structure has been progressing by combining cross-fractionation with various thermal, spectroscopic, and chromatographic techniques correlating the molecular structure to the performance of these materials [64, 65].

Since their development in the 1970s [11], these materials have conquered a wide range of applications from seemingly trivial packaging uses to technically complex parts. The matrix consisting of a PP homopolymer or, in the case of random heterophasic copolymers, an EP-RACO is responsible for retaining good high-temperature performance and adequate stiffness, while the particulate EPR/PE phase contributes good toughness. Depending on the composition of the EPR component the respective T_g can be as low as −60°C, resulting especially in improved impact performance at subzero temperatures. While such low glass transitions can be achieved by a high ethylene content in the EPR phase, high propene content in the same will enhance the compatibility to the matrix, allowing better transparency [21]. Overall, the size, shape, internal structure, and spatial packing of the dispersed EPR domains (which generally range from 0.2 to 4 μm in average diameter under equilibrium conditions) are critical parameters affecting not only the mechanical performance but also properties like surface appearance and migration. The morphology is a complex result of the rheological parameters of the single components, the matrix–dispersed phase compatibility, and the processing conditions.

One critical factor is already the catalyst morphology. In the production of EP-HECOs, the morphology of the produced polymer particles replicates the morphology of the employed catalyst particles. Spherical aggregates of microcrystalline $MgCl_2$ are conventionally used as support material for the $TiCl_4$ in ZN catalysts [63], offering a porous structure with large surface area. The maximum amount of

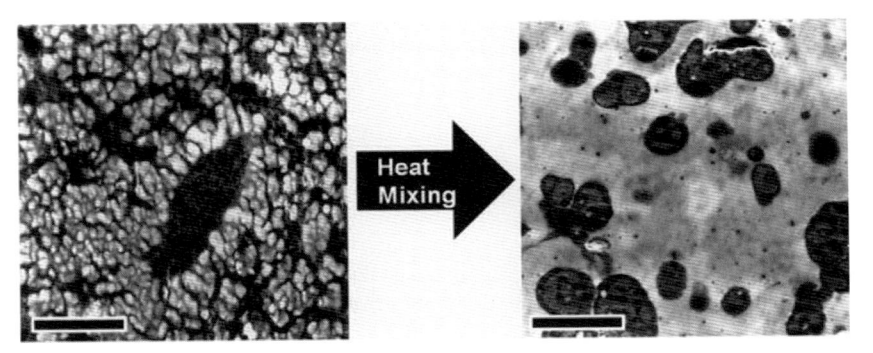

Figure 20. Relation between powder morphology (left) and granule morphology (right), revealed by RuO_4-contrasted transmission electron microscopy (TEM); high-impact EP copolymer type with 26 wt% ethylene and 22 wt% elastomer (EPR; xylene-soluble fraction); EPR is dark because of lower crystallinity; PP matrix and PE inclusions in particles are light; scale bars 2 μm.

EPR/PE which can be accommodated in the polymer granule without causing stickiness appears to primarily depend on the porosity of the catalyst particle. Figure 20 shows the difference in distribution of the EPR phase between the original reactor "powder" (defined by the catalyst structure) and the pelletized or processed polymer [66].

While details of the EPR composition and its viscosity ratio relative to the matrix are highly relevant for product design in detail, a key factor for increasing the energy absorption capacity is to increase the EPR weight fraction in EP-HECOs. Here, the toughness is improved by reducing the interparticle distance, resulting in a stepwise brittle-to-tough transition as a function of EPR concentration, combined with linear reduction of tensile modulus (see Fig. 21). The blend concentration at which this transition occurs is determined by the chemical composition of matrix and dispersed phase as well as their viscosity ratio, but also on the geometry and test conditions. For PP/EPR blends the phase transition and the respective development of the mechanical performance has been comprehensively discussed [67].

7.4. Composites and Blends

An extension or even an alternative to multistage polymerization with the production of EPR can be the addition of external elastomers. This was historically the first approach to impact modification of PP, for example, blending of PP-H with 10–20 wt% of polyisobutylene (PIB) elastomer. Nowadays this route is instead used for fine-tuning the property profile. High impact strengths can be achieved with ethylene-co-octene or -butene plastomers and elastomers based on polymerization with single-site catalysts in solution [68]. The comonomer content and the molecular mass of these homogeneous copolymers can be varied over a wide range, allowing applicability in a wide range of stiffness and MFR. For special purposes, styrene-based elastomers of the di- or triblock type have found uses despite their higher price. Mostly hydrogenated styrene–butadiene triblock systems (SEBS) are used, which can be designed in a wide range of compatibility and viscosity by varying the molecular mass and individual block lengths.

For the still-used external EPRs or ethylene–propylene–diene elastomers (EPDM) based on vanadium catalysts, a two-stage process in which the elastomer is first compounded with PP to give a 50–70 wt% masterbatch of the rubber as free-flowing granules is commonly used. Only rather C2-rich EPR types are sufficiently crystalline to be pelletized in pure form [69]. Also different PE types can be added in minor amounts for property design. In the presence of an EPR phase, preferably reactor-made, the PE will be located inside and towards the center of these rubber droplets [70], and mostly enhances the room-temperature impact strength and limits shrinkage. Addition of HDPE is especially suitable for the reduction of stress whitening or scratch sensitivity.

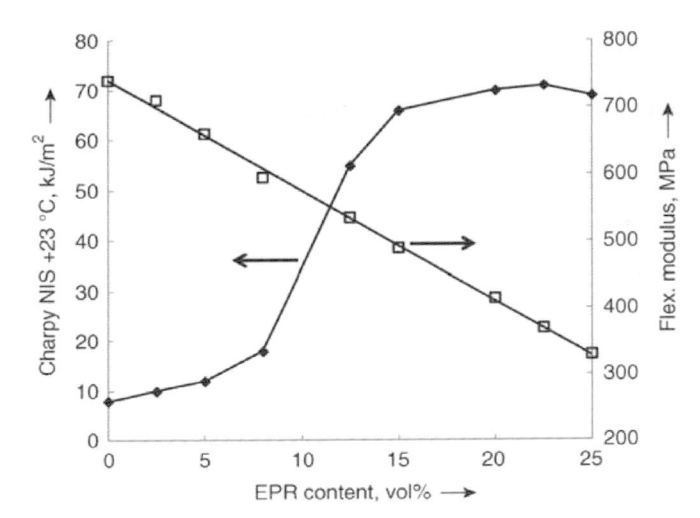

Figure 21. Effect of EPR concentration on stiffness (flexural modulus, ISO 178) and toughness (Charpy notched impact strength, ISO 179 1eA) at 23°C for a series of random-heterophasic copolymers with constant matrix and EPR composition (data from [21])

Blends of PP with other polymers were studied intensely in the 1990s and were even commercialized in some cases. Three blend partners were most studied (in order of increasing polarity): polystyrene (PS), poly(methyl methacrylate) (PMMA), and polyamide-6 (PA-6). In all cases, besides viscosity matching as with other multiphase PP systems, also the use of an appropriate block or graft copolymer as compatibilizer is required for delivering good property combinations [71, 72]. While PS can add stiffness, surface gloss, and print- as well as paintability, PA-6 was shown to deliver a good combination of modulus and temperature resistance, especially when combined with glass-fiber reinforcement. The most relevant example of blend commercialization was the Hivalloy product line of Himont (now LyondellBasell), based on a reactive modification concept with PS and PMMA as blend partners. Introduced on the market in 1994 and produced as pure blends but also reinforced with glass fibers, the product line was stopped in 2001 and finally considered a commercial failure.

8. Processing and Applications

In the global application distribution for PP (see Fig. 22) the extrusion segments (films and fibers, but also coating and pipe extrusion) used slightly more than 50%, while injection and blow molding made up less than 50% of the consumption. In 2010, the global PP market was about 50×10^6 t/a, with more than 50% of this being used in the Asian-Pacific region. Further perspectives for volume growth can be illustrated best with the annual per-capita consumption: The global average in 2010 was 7 kg, with a wide span between developing countries, the BRIC-countries (e.g., India at 2.1 kg), and highly industrialized countries (Western Europe 18.4 kg) [73].

The wide range of PP applications [74] clearly results from the possibility to adapt the polymer structure, its processability, and final performance to many conversion routes and requirement profiles. A general trend towards material substitution has allowed PP to grow by replacing more costly polymers like polycarbonate (PC) and acrylonitrile–butadiene–styrene terpolymers (ABS), but also "traditional" materials like glass and metal. This process, often involving a redesign of the respective application (e.g., going from metal cans to stand-up pouches for food packaging) and can result in a significant overall reduction of raw material and energy consumption (see Chap. 9).

8.1. Injection Molding

Injection molding is used for a wide range of packaging material and general purpose items

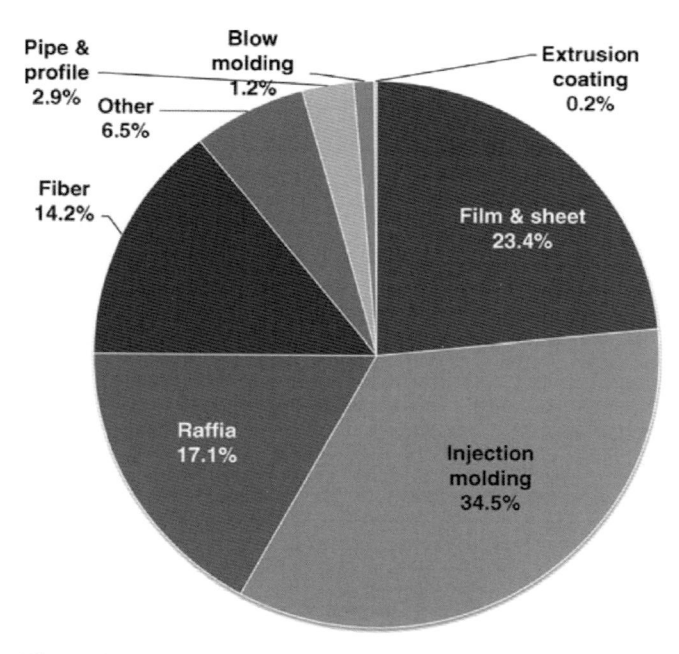

Figure 22. Global application distribution for PP in 2010 (CMAI data, after [73])

with limited requirements, as well as for most of the technical uses of PP. As these two segments require quite different materials and processing considerations, the simpler segment is discussed here, and the technical segment in Section 8.6.

Two main groups of factors define the selection of PP grades for injection molding applications: part geometry, especially the flow ratio (i.e., the ratio of longest path to the section thickness) and the complexity of the part, and the thermomechanical requirements in the respective application. A simple increase of the MFR, for which a range between 1 and 100 g/10 min (230°C/2.16 kg) is presently available for injection molding, will allow thin and narrow cavities to be filled even over longer distances, but the impact strength of such materials will be greatly reduced. Compensating this by moving from PP homopolymers to random or heterophasic copolymers will in turn compromise stiffness, heat resistance, and transparency.

In addition, the formation of crystalline superstructures like shear-induced skin layers [75] is decisive both for the solidification speed and the mechanical strength of the molded parts. The presence of a high molecular mass fraction is decisive here, which makes highly polydisperse grades more prone to skin-layer formation. Further polymer-related factors are anisotropic fillers and nucleating agents [76], but melt temperature and flow speed are also relevant. Thus, like in all semicrystalline polymers, the processing parameters become a part of the final property design. Standard processing conditions used for PP are [77]:

- Melt temperature T_m: 200–265°C
- Injection pressure p_{inj}: 50–150 MPa
- Injection speed v_{inj}: 100–400 mm/s
- Mold (wall) temperature T_w: 30–60°C

As a rule of thumb, lower T_m values are used for higher MFRs, and lower T_w values for thinner parts. This may even lead to partial quenching, a phenomenon which is discussed in Section 8.4. Besides mechanics and optics, the resulting crystalline structure also determines the extent and anisotropy of in- and post-molding shrinkage. Linear post-molding shrinkage of PP is 1.0–1.5%, of which 85% occurs within the first 24 h. Distortion or warpage in molded products results from anisotropic shrinkage and reflects both internal stresses and crystal orientation. PP grades with high MFR and/or narrow MWD (low polydispersity) can help to reduce this problem [60].

An important practical aspect of PP concerns its ability to form strong integral hinges, such as in a lidded box, where the lid is permanently attached to the base by a thin web of polymer along the whole of one edge (Fig. 23). This web, about 0.25–0.6 mm thick, can be produced directly in the molding process. An initial flexing of the hinge while slightly warm induces the correct molecular orientation to permit repeated opening, even at subzero temperatures. In the laboratory, such hinges have withstood 23×10^6 flexes without failure [74].

Apart from this extreme case, injection-molded parts of PP range in wall thickness from 0.3 mm (e.g., for thin-wall dairy cups) to 10 mm (e.g. for pressurized ion-exchanger cartridges in coffee vending machines). Mostly, however, wall thicknesses above 4 mm are avoided by partial foaming (see Section 8.6) or by gas- or water-assisted injection molding. These seemingly complex processes have been developed to a degree of practical simplicity, making them useful even for mass production of articles like beer crates and toys.

General purpose items injection molded from PP can largely be categorized into three groups [74]:

1. Household articles and general consumer products, covering such diverse uses as kitchenware, garden furniture, toys, and luggage. Kitchen bowls and storage containers were among the first articles produced from PP, and outdoor furniture such as stadium seats also already has a long application history. Nevertheless, the development of highly transparent molding grades and improved UV stabilization concepts have improved the possibilities significantly.

2. Packaging systems, ranging from thin-walled cups and beakers through buckets and pails to folding boxes and crates for transport packaging. The requirements vary broadly, but weight reduction and materials savings have been the main factor in recent years. High-crystallinity PP grades, very often as heterophasic copolymers with a good subzero impact performance, dominate the transport packaging segment, while nucleated random copolymers give the best product visibility.

3. Medical articles and pharmaceutical packaging systems have special requirements in terms of allowed additives and required purity. Disposable syringes were the first PP application here, and already required sterilizability as a further parameter. The entrance barriers for polymer producers are higher here, mostly due to the long approval processes [78].

8.2. Blow Molding

For producing bottles, canisters, and other containers, blow molding processes have been employed mostly for PE for a long time already. While PP has been slower in the initial market development, advantages of this polymer, such as better heat resistance and transparency, have enabled the gradual development of a significant application volume. Two different processes with different degrees of complexity and different requirements for the material used are distinguished here:

In *extrusion blow molding* (EBM) an extruded molten parison (tube) is transferred to a mold in which the end is clamped and it is inflated with low-pressure air at 0.4–1.0 MPa [79]. High molecular mass polymers having an MFR of ca. 1.0–2.5 g/10 min, combined with low melt temperatures at the die of 200–210 °C,

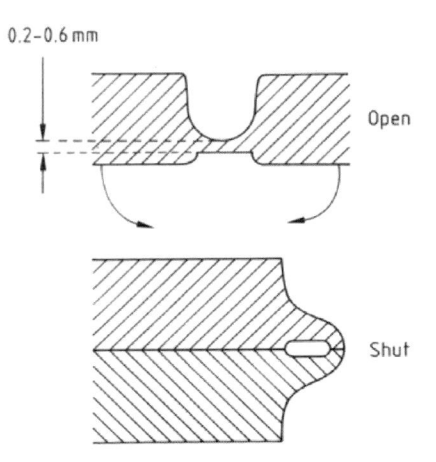

Figure 23. Integral hinge (Courtesy of the "Propathene" Business, ICI Chemicals & Polymers)

provide a suitably stiff melt. If the melt stress in a heavy PP parison exceeds ca. 20 kPa, then the resulting tension thinning will be troublesome as regards parison stability and uniform wall thickness. EBM is cheaper and simpler to operate, but the produced articles will always show a certain degree of surface roughness limiting the transparency and gloss. In modern processing equipment, control of the final container wall thickness by modifying the parison thickness along its extrusion direction can significantly improve appearance and performance [80]. A variation of the EBM process is the blow, fill, and seal (BFS) process, mostly used in the medical area for ampoules and infusion bottles. While LDPE is mostly used in this process, special PP grades are also gradually becoming available.

Injection blow molding (IBM) and the more common *injection stretch blow molding* (ISBM) are two-stage processes in which first a preform is produced by injection molding. Blowing (in IBM) or a sequence of stretching and blowing (in ISBM) takes place in a second mold which maintains the high-definition neck and thread of the preform. In both cases, the produced bottles or wide-mouth containers are characterized by better optical and mechanical performance, but also by improved dimensional accuracy. Significantly different polymers are required here [81] which have an MFR of 10–25 g/10 min and are preferably nucleated random copolymer types in order to facilitate processing. While PP is used in much smaller volumes than poly(ethylene terephthalate) (PET) for ISBM bottles, it has a number of clear advantages in water-vapor permeability and heat resistance, which enable hot filling and/or steam sterilization.

Products from these processes are mostly used for packaging in the food area (dairy products, fresh juices, still water), the nonfood area (household cleaners, dishwashing liquids, cosmetics, body-care products), and also the medical/pharmaceutical area. EBM processes can also be used for technical applications like fluid containers for cars, dishwashers, and other household equipment. A special use for PP-based ISBM bottles are sterilizable baby bottles, as well as reusable sports bottles.

Like in films or injection molding, multilayer constructions resulting from coextruded parisons (EBM) or two-component preforms can be used as precursors for the bottles when special barrier properties are required. Most commonly, a central layer of ethylene–vinyl alcohol copolymer (EVOH) to reduce oxygen diffusion is combined with moisture-protective outer layers of propene–ethylene random copolymer to give containers suitable for storing oxygen-sensitive food products.

8.3. Fibers and Flat Yarns

One of the biggest application areas of PP homopolymers are fibers and tapes as well as their derivatives. The respective applications cover a wide range of physical forms, including increasing amounts of versatile nonwoven fabrics [82, 83]. Monoaxial orientation in the solid or semisolid state can be applied to conventional spinneret yarns, as with polyamides and polyesters, and to flat tapes made from extruded film. These application forms differ in various respects listed in Table 7.

The three conventional PP fiber operations are continuous filament (CF), bulked continuous filament (BCF), and staple fiber (SF). In any case the formation of the primary molten fiber by extrusion through a die plate perforated by

Table 7. Polypropylene fiber processes

Process	Filament count, tex*	Product
Long spin	0.2–3.0	high-tenacity monofilament; drawing integral or separate; high output
BCF yarn	0.2–2.0	special case of long spin making only bulked continuous fiber
Spunbonded	0.2–2.0	venturi haul off; no 2nd stage draw; bonded mat output
Shortspin	0.2–40	compact unit; tow stretched and cut in line for staple
Melt blown	0.002–0.02	low orientation, very fine fiber; only bonded mat output
Fibrillated yarn	110–500	oriented slit film; fibrillated for baler twine, rope, etc.
Weaving tape	ca. 110	nonfibrillated slit film for carpet backing, sacks, etc.
Strapping tape	500–1000	thick, oriented tape as a steel alternative

*tex = weight in grams of 1000 m of yarn; equivalent cylindrical fiber diameter in $\mu m = 37\sqrt{\text{tex}}$.

many small holes is crucial. High melt temperatures (230–280 °C) and melt flow rates (MFR 10–30 g/10 min; preferably combined with narrow molecular mass distribution for high extensibility [84]) are applied on short extruders, mostly combined with manifolds and melt pumps before the 8–20 spinnerets having 50–250 holes each. For the CF process, a long spinline is preferred which can combine spinning and finishing stages as well as out-of-line drawing options. An air-cooling gap of 2–5 m is needed between the die plate and the wind-up roll. Line speeds up to 1000 m/min at the spinning stage, increasing to 3000 m/min during solid drawing over hot rolls, require complex and expensive haul-off, drawing, and windup sections. The product is preferably sold as continuous yarn or tow.

The BCF process is mostly used for carpet face and upholstery yarns and characterized by the addition of a "bulking" step after spinning or spinning and orientation. The bulking or texturizing at 140–160 °C in a jet with compressed air has a similar effect to the crimping process used for other textile fibers, causing fibers to have a wavy to curly structure for higher volume. While the MFR range is the same as for CF, broader molecular mass distribution and higher crystallinity can give advantages here.

In spunbonding, filament spinning and web formation are integrated into one process, yielding a continuous bonded mat of partially oriented yarn as a consequence of some draw in the venturi-type haul-off operating at up to 5000 m/min. There is no further drawing as the fiber is collected as a mat on the take-off belt for conversion into nonwoven fabrics. These are used, for example, in geotextiles, furnishings, and carpet backing [85]. The balance between drawdown rate and orientation in this process is influenced by the polymer molecular mass distribution.

Figure 24 depicts a compact shortspin process for staple fiber production using a die containing ca. 40 000 holes. Cooling air jets freeze the melt within about 20 mm so that godet haul-off rolls can be placed only 1 m away from the die. Orientation is achieved by the in-line stretching of this tow in an air oven, followed by chopping in line to produce staple [82]. Broad-molecular mass distribution polymers again give the necessary melt strength, while MFR requirements increase from 7 to 20 g/10 min as the filaments become finer. These find use in carpet face fiber, and diaper cover stock applications. A useful feature of PP fabrics made from fine fibers is the ability of underwear garments to reduce clamminess by transferring moisture away from the skin to an outer absorbent layer. Polypropylene-based fabrics are also used for sports clothing and socks, but the low melting point of PP calls for special attention to the heat setting of the average domestic iron.

For the most recent production process (Fig. 25), melt-blown (MB) fibers, air is blown through a very fluid polymer melt maintained at high temperature to assist chain scission in the spinning process itself [74]. Polymer MFR at the nozzle can range from 200 to 800 g/10 min or higher, achieved by visbreaking with some residual peroxide in the polymer before extrusion. The specially designed die, having a row

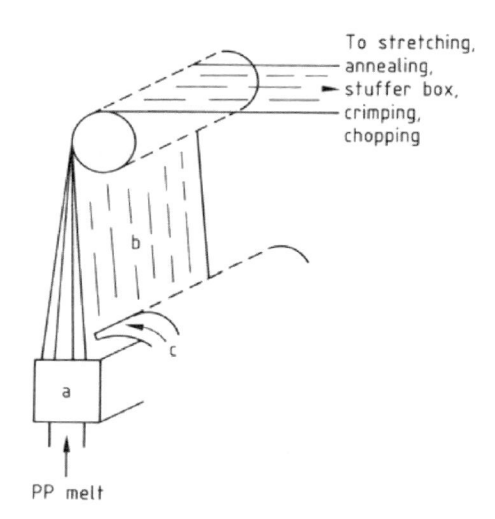

Figure 24. Shortspin process
a) Spinneret; b) Quenched fibers; c) Air-jet cooling; Courtesy of the "Propathene" Business, ICI Chemicals & Polymers

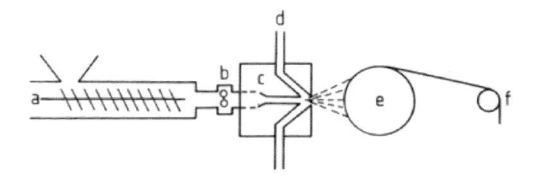

Figure 25. Melt blowing process
a) Extruder; b) Gear pump; c) Heated die; d) Hot air manifold; e) Collector; f) Take-off roll

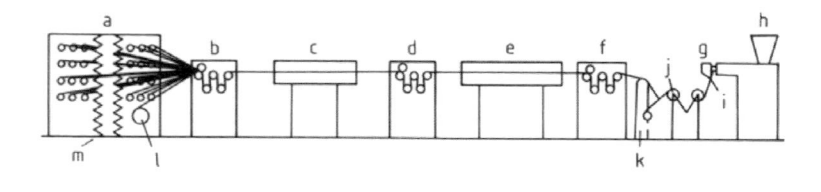

Figure 26. Flat tape/yarn production unit
a) Windup frame; b) Third godet stand; c) Hot-air annealing oven; d) Second godet stand; e) Hot-air orientation oven; f) First godet stand; g) Simple film costing unit; h) Extruder; i) Film die; j) Water-cooled chromium-plated rolls or water bath; k) Slitting unit; l) Starting drum; m) Individual precision flangeless bobbin wind up; Courtesy of the "Propathene" Business, ICI Chemicals & Polymers

of nozzles along its width, sprays a stream of short, low-orientation fibers of very low diameter onto a moving steel-web belt, where they are bonded together and wound onto a roll. This process produces nonwoven fabrics which are of increasing importance for industrial wipes and surgical wrap, and for a variety of filtration applications, such as face masks made from bonded mats of very fine fibers. MB fiber webs have only limited mechanical strength and can be combined with spunbonded webs in a so-called SMS structure.

For producing flat yarns, extruded film is used as the basis. Melt from a conventional slot die is rapidly quenched with cold water or water-cooled steel rolls to give low-crystallinity film, 50–250 μm thick (see Fig. 26). Razor-blade-type knives then slit this broad film into strips 5–20 mm wide, which are oriented in a hot air oven at 120–180°C by drawing between godet rolls at ratios in the range 5 : 1 to 10 : 1. Both film thickness and width diminish in this operation by approximately the square root of the draw ratio. These yarns are preferably woven into cheap textile structures used for bigbags or food transportation like rice or grain bags.

Fibrillated tape is another variety in the "low-performance" area. While flat yarns are quite strong in the direction of draw, they are very weak in the transverse direction. With normal PP homopolymer, simply twisting high-draw-ratio tape induces spontaneous fibrillation. A web of irregular fine fibers is formed as the film splits down its length. The resulting product resembles a coarse sisal twine, for which it is a good substitute. Such fibrillated yarns can be tailored to specific applications by controlling draw ratios, thickness, and polymer. Nowadays, pin fibrillation, in which a co-rotating spiked roll contacts the film just after slitting, offers more precise control over this step. The products are familiar as baler twine, string, and some ropes.

Highly oriented fibers or tapes are also the starting point for the production of self-reinforced PP parts [86, 87]. The ability to reach very high specific strength in fiber spinning processes by post-drawing in the solid or semi-solid state is used in combination with weaving and sintering of such high-strength PP fibers to plates for later thermoforming. These products are marketed under the trade name Curv and can reach modulus values of up to 5000 MPa. The geometrical design is however limited to thermoformed parts.

Thick, 0.3–0.9 mm, oriented tapes are established alternatives to steel bands in many strapping applications. They are generally made by the water-quench process operating at draw ratios of 9–10. Thicker tapes are made from foil or individually extruded film tapes. High molecular mass, MFR 0.4–2.0 g/10 min, PP homopolymers and EP impact copolymers with added chalk and polyethylene largely suppress fibrillation. A further precaution here is to emboss the finished tape with a diamond pattern. A useful aspect of PP strapping tapes is their higher elongation and elastic recovery which helps them to remain tight on packages prone to shrink or settle.

8.4. Films and Sheets

Depending on the processing technology employed, PP can be processed into films and sheets ranging in thickness from 5 to 2000 μm and converted further to a wide variety of products [88].

The more common processing step is film casting with extrusion from a slit die onto a (relatively) cold roll. The so-called chill roll is kept at temperatures between 15 and 80°C by liquid cooling or heating, depending on the polymer processed and its final destination. If the film is applied directly after solidification in this step, in which only a limited degree of orientation in machine direction is achieved, it is known as cast film, which covers a thickness range of ca. 30–300 µm. The rapid cooling leads to films of low crystallinity and good optics, which can be influenced in performance both by a choice of polymer and additives and by variation of the processing conditions [89].

If the film runs from the chill/take-up roll further through a calendar-like arrangement of rolls for solidifying and annealing, it is known as roll-stack film with a primary thickness range of 300–2000 µm (see Fig. 27). These films or sheets are primarily used for subsequent thermoforming applications. For this process a variety of methods can be used depending on the desired shape and dimension. Both in-line and off-line processes are applied with two essential steps, heating and shaping, the former mostly by infrared irradiation and the latter by mechanical force, air pressure, or vacuum. Thermoforming is the only film processing segment in which nucleation is applied to a significant extent [90].

In the thinner part of this thickness range, special high-efficiency cooling setups can be used to produce the lowest crystallinity and highest transparency. With water-bath or steel-belt technology, the ideal materials for pharmaceutical tablet blisters or transparent folding boxes are produced. The application of non-nucleated PP homopolymers is advisable in these special cases.

The thinnest films (5–30 µm) are produced by postdrawing (postorientation) of a thicker cast film, whereby monoaxial orientation (MOPP) and biaxial orientation (BOPP) is possible [91]. The BOPP films are characterized by a much higher orientation, stiffness, gloss, and transparency in comparison to cast film. While in the packaging area, the low water-vapor permeability and the excellent optical appearance are especially appreciated, the high mechanical and thermal resistance make BOPP also suitable for electrical capacitors.

A similar thickness range to cast films can also be reached by film blowing, which for PP is still of much lower importance than PE. Air-cooled blown films are only possible with specific grades combining a certain amount of melt strength (at a general MFR range of 1.5–3 g/10 min) with fast crystallization, and on modern processing lines. Water-quenched blown film, originally the only processing method suitable for PP, has been making a comeback recently, especially for applications with increased optical performance requirements [92, 93]. Air-cooled blown film is also the first step of an alternative BOPP process, the so-called double-bubble technology with sequential longitudinal and transverse stretching of the film tube (see Fig. 28).

In general, materials with MFR values of 1.5–4 are used for BOPP, thermoforming, and blown film applications, while for cast films polymers with an MFR of 7–20 are preferred (always 230°C/2.16 kg). While the low-MFR grades should have a rather broad molecular mass distribution to have sufficient melt strength, the higher MFR cast film grades preferably have a narrow molecular mass distribution, for example, visbroken or SSC-PP types. The film area is largely dominated by PP homopolymers and random copolymers with ethylene, but for sealing layers also terpolymers with ethylene and butene are used. Impact copolymers are limited mostly to nontransparent applications with low-temperature toughness requirements, unless special polymer designs are applied which allow a combination of transparency and toughness (e.g., "interpolymer" concept with high propene content in the EPR phase [74]).

In any case multilayer constructions are possible and even necessary in case of packaging applications requiring the combination of different requirements, such as sealability, sterilizability, and integration of barrier layers (for this,

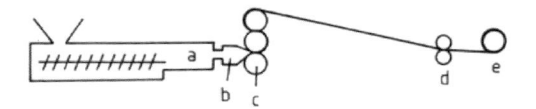

Figure 27. Three-roll stack extrusion
a) Extruder; b) Slot die; c) Polishing stack; d) Nip rolls; e) Wind up

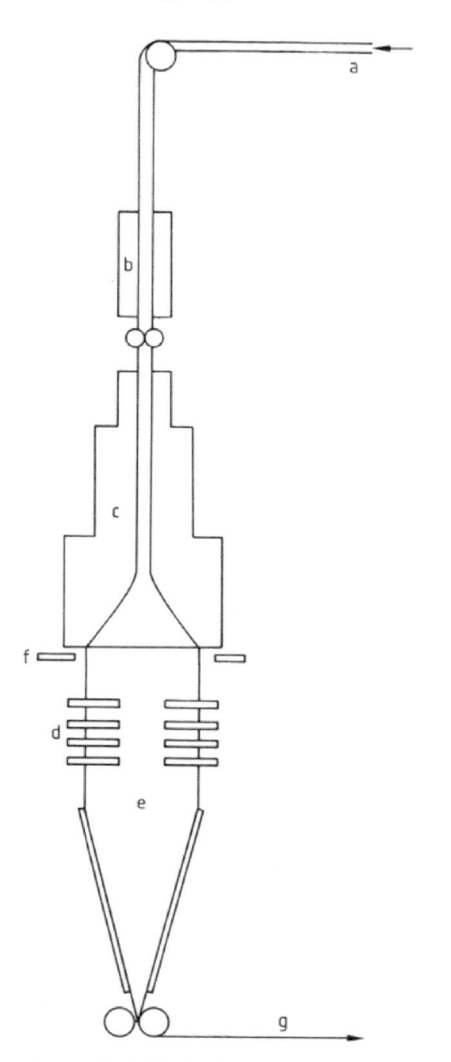

Figure 28. BOPP bubble process
a) Cast tube; b), c) Stretching heaters; d) Bubble guides; f) Air ring; g) Wind up, slitting, etc

polar polymers like EVOH or oriented PA are used) or peel-layers for a more easy opening of a packaging (here multiphase systems like PE–PP blends are applied). Through metallization (e.g., by chemical vapor deposition, CVD) and lamination even more complex structures can be designed. Preferably multilayer co-extrusion with specially design film extrusion heads ahead of the die is used for producing such constructions. Lamination with hot-melt or reactive adhesives is applied whenever nonpolymeric (e.g., paper or metal), specially oriented (e.g., oriented PA) or already printed (e.g., PET) layers are required. In this case, and also whenever printing or gluing of the film is intended, the film surface is activated by corona, plasma, or flame treatment. Corona treatment is the most common process here, but the fact that surface polarity generated with this method will decay over time must be considered when planning production processes [94].

In summary PP films cover an extremely broad area that reaches from textile packaging through medical applications up to electronic components like capacitors. Special requirements can be achieved partly by specific processing or material combinations, but in other areas like sterilizability [95], low-temperature toughness, or resistance to aggressive media, special polymer design is required.

8.5. Foaming and Coating

The foamability of normal PP with linear chain structure is limited and even in specially designed foam injection technologies like the MuCell process only rather high foam densities can be achieved [96]. Good foam cell nucleation, which is always given in fiber or mineral reinforced materials, can improve the situation sufficiently here, in combination with a broad MWD and especially the presence of a high molecular weight component.

A low MFR (0.5–1.5 g/10 min) is also sufficient for particle foaming processes in which, as in the case of polystyrene, massive PP particles are impregnated with a solvent (hydrocarbon or halogenated hydrocarbon). These particles are then foamed and fused under elevated temperatures into the desired shape. Normal pipe or even thermoforming grades can be used here, but an addition of some long-chain branched PP (high melt strength PP, HMS-PP [51, 52] allows for a lower final foam density.

In contrast, HMS-PP is absolutely required for extrusion foaming processes and especially solvent-free processes applying carbon dioxide [97]. Extruded PP foam has excellent mechanical properties including high temperature resistance and can be tailored in its toughness by blending HMS-PP homopolymers with EP impact copolymers. The final application ranges from extruded sheets thermoformed into food trays or packaging components to pipe insulation layers.

Other processes facilitated or even enabled by HMS-PP are thermoforming at high draw ratios (deep shapes) and especially extrusion coating [98], which is otherwise a clear domain of long-chain branched LDPE. The Recart process of Tetra-Pak, in which sterilizable coated cardboard packages are produced for food packaging, is based on such materials which have a higher MFR than for foaming (8–15 g/10 min as compared to 2–4 g/10 min) and must be free of crosslinked fractions which would form gels in the coating process.

8.6. Technical Applications

As a rule, technical parts from PP and its composites are produced by injection molding or its process variations like inlay, gas-injection, and two-component molding. Elaborate part and mold design are required both to fulfill the requirements to surface quality and to process special materials like short and long glass-fiber-reinforced PP (SGF- and LGF-PP) [99]. LGF-PP can be further diluted to vary the final fiber concentration and the related mechanics in the actual processing step, in which special mold and machine design is required.

Especially in the development of modern cars PP plays an important role due to its combination of density (part weight) and mechanics (strength) [100]; out of the ca. 17% mass fraction of thermoplastics in passenger cars, 70% is polypropylene (see Fig. 29). An overall important contribution of the polymer fraction is the weight reduction compared to metals, which is essential for reducing the fuel consumption. As about 90% of lifespan energy demand of a car occurs in the usage phase; replacing 350–500 kg of other materials in a car with about 200 kg of polymers results in about 750 L of fuel saved in a car's lifespan.

SGF-PP and LGF-PP, targeted at metal replacement and often also component integration, find numerous uses in modern car design, generally supported by modeling [101]. From front-end modules, integrating headlight and bumper support as well as the cooling system, through dashboard carriers and underbody protection elements, the range extends to complete

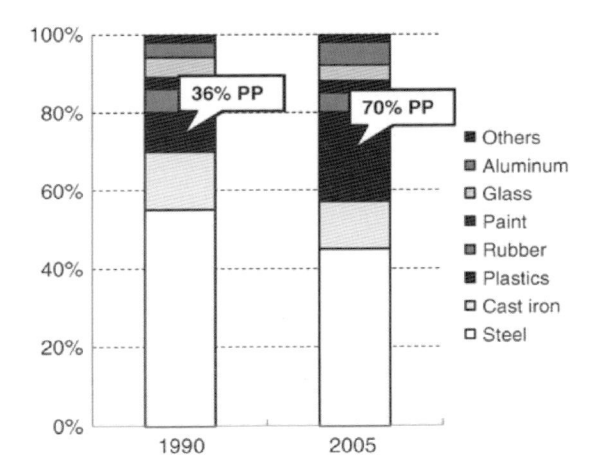

Figure 29. Composition change of a European passenger car from 1990 to 2005 (internal data of Borealis AG)

door and hatch modules, in which LGF-PP frames can be combined with injection-molded cover elements to achieve largely metal free constructions. For the latter, as generally for automotive exterior and interior elements, composites based on high-impact EP copolymers are applied.

Automotive exterior applications started at side trims and door steps, progressed into bumpers, and now include fenders and other parts of the car body. Basically the only part of a passenger car's body which has not yet been designed and produced from a PP-based material is the roof. Base materials for the composites used here are EP impact copolymers with an EPR content of 20–30 wt% and frequently a high-crystallinity matrix, and the mechanical performance is further balanced by external elastomers and mineral reinforcements. Surface quality is important both for painted and the minority of nonpainted applications and it is clearly related to phase structure [102]. Another critical aspect for gap-free combination with metal parts is the thermal expansion of the material, which is normally closely related to shrinkage [103].

For automotive interior applications, slightly lower impact resistance and especially no toughness below −20°C is acceptable, leading to modified recipes for composite design. Here, however, scratch resistance [104] and low levels of emissions and odor are important factors. A variety of techniques ranging from back injection molding onto foam/textile combinations over metal or modified wood inlays to the application of soft varnishes are used for design; the contribution of these components to overall stability must be considered.

One of the oldest areas of PP application in car design is the under-the-hood segment, where air ducts, battery cases, and fluid containers were already produced from PP homopolymers in the 1960s. The most recent innovation in this segment are air-intake manifolds made of GF-reinforced PP, replacing GF reinforced PA-6 with advantages in water adsorption and sound damping.

Due to its good chemical and electrical resistance PP is inherently suitable also for applications in the electrical and electronics sector. In practice, uses range from housing and internal components in household equipment via connector systems for low voltage up to BOPP capacitor films. For the last-named PP homopolymers with extremely high purity are applied. A common key requirement for housings is the surface quality. As the "reference material" is mostly ABS, a combination of gloss and surface hardness or scratch resistance is required here, which can only be achieved with high-crystallinity PP grades.

The white-goods industry with washing machines and dishwashers as main products is an especially important application area of PP-based composites with mineral fillers and glass fibers [105]. SGF-PP and LGF-PP are used in areas with high dynamic loads, like the tubs of washing machines, and special lye-resistant stabilization systems must be used.

8.7. Pipes

The highest molecular masses of PP are used in the pipe area where, compared to PE, the application volume is rather small [106]. A normal MFR range (230°C/2.16 kg) for PP used in pipe extrusion is 0.3–0.8 g/10 min, while still higher levels of processability are used for injection-molded fittings complementing pipe installation systems. The applied processing temperatures range from 220 to 280°C, the latter mostly occurring in multilayer extrusion heads, and must be controlled rather well both to avoid degradation and to ensure proper crystallization, whereby PP is generally far more sensitive than PE.

For pressure pipes, two polymer modifications play an important role: Random copolymers with ethylene are used for hot-water and underfloor heating installation, while β-nucleated PP homopolymers find use in industrial piping applications and are especially suitable for contact with aggressive chemicals like concentrated acids. Pressure, creep, and crack-propagation resistance of these materials is specified in relation to pipe dimensions in ASTM and ISO standards [107]. Here, also the new material class for hot- and cold-water pipes, known as PP-RCT (RCT stands for random, crystalline, temperature-resistant), is specified. The combination of EP random copolymers with β nucleation allows pipes to be used at higher pressures or wall thickness to be reduced.

Another application area which has experienced significant volume growth in recent years is nonpressurized sewage pipes for domestic and infrastructure areas [108]. Especially in the case of large diameters the external load on a pipe in the ground is bigger than the internal load, a problem that can be solved, for example, by multilayer, rib, or corrugated pipe constructions. High-modulus impact copolymers are nearly exclusively used here, as the toughness requirements are rather high. The development presently continues in the direction of even larger parts like shafts and manholes to achieve full mono-material solutions for sewage systems. For large cross-sectional dimensions fiber-glass-reinforced types are in use here.

Pipe jointing uses all the common techniques, with the exception of solvent welding. Ring seals with rubber inserts allow some movement in domestic waste pipes. Also used are flanged joints, hot-plate butt welding, and socket-fusion welding, the last-named being optionally supplemented with hot-gas bead welding. Compression joints involving copper alloys should not be used at temperatures exceeding 60°C because of reduced life expectancy resulting from Cu ions; the use of metal deactivators in stabilization can only partly remedy this hazard.

9. Environmental Aspects

After some decades of success, polymers have been going through a hard time recently. For the first time since the oil crisis of the early 1970s, global production decreased in 2008 as a result of the economic crisis. At the same time, public perception of "plastics" deteriorated, with critical views on issues like the release of endocrine disruptors, littering, and waste. While some of these arguments may be justified, many remain superficial like the quarrel about single-use plastic shopping bags. Being less visible as a packaging material, PP has not been affected as much as PE by this discussion, but the widespread idea that all petro-based polymers can and will be quickly replaced by "bio-based" or even biodegradable materials of course affects customers' choices when looking for materials for new applications.

In reality, bio-based polymers account for less than 0.5% of the global polymer production, and besides the widely discussed question of resources for their production and the related competition with nutritional usage there are also technical issues [109]. While bio-based PE (or even PP which may become available as well) will of course not exhibit different properties than its petro-based counterpart, most biodegradable polymers suffer from significant shortcomings in terms of processing stability, mechanical and optical performance but mostly long-term stability. Life-cycle assessments [110] need to include these factors, which frequently lead to heavier and more complex solutions being taken into consideration in order to be realistic.

Even apart from the discussion of bio- versus petro-based materials, the question of long-term monomer supply is clearly an issue for the polymer industry as a whole [111], even though presently only ca. 6% of total crude oil consumption is employed for polymer production. Natural-gas-based production, including olefin conversion steps such as olefin metathesis, will therefore become more important. Feedstock recycling can also contribute, as can improved polymer and component design to reduce the amount of material necessary for fulfilling a certain function. Overall, however, the numerous advantages of polymers over traditional materials must be kept in mind [112].

9.1. Life-Cycle Analysis

When comparing the environmental impact of different materials, numerous factors must be considered. Life-cycle analysis (LCA) is a common and widely used approach for quantifying energy and resource usage as well as gaseous, liquid, and solid waste streams associated with producing a specific material or article, assuming identical boundary conditions [113]. In contrast to a complete LCA, other measures like carbon footprint (essentially specific CO_2 number per unit weight or application; see Fig. 30 for a comparison of standard thermoplastics) or water footprint can only cover some aspects of the situation.

For most standard thermoplastics, PlasticsEurope (the Association of Plastics Manufactures in Europe) publishes LCAs or "environmental product declarations" at regular intervals. For PP, this was last updated in 2008

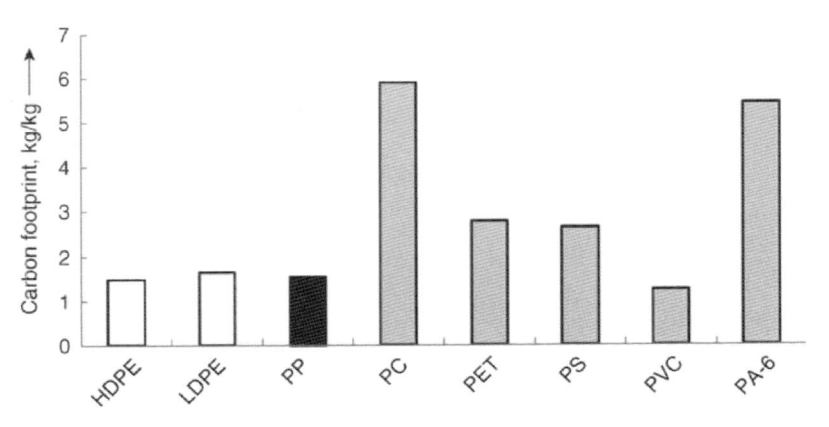

Figure 30. Comparison of the carbon footprint (CO_2 equivalents) of different standard and advanced thermoplastics (data collected from www.plasticseurope.org in July 2013)

[114] and provides an overview of all relevant input and output parameters in PP production, a key figure of which is the average amount of 1.565 kg fossil fuel equivalents needed for producing 1 kg of PP. From an alternative angle the energy demand can be broken down into 52.6 MJ/kg for feedstock and 20.4 MJ/kg for process energy.

Because LCAs should rather be done on an application level, both processing and usage aspects must be considered there. Conversion energy demand is generally lower for polymers than for metals or glass due to their lower melting point and processing temperature, and emissions from conversion are generally not critical for PP, like for other polyolefins [115]. The possibility for saving energy by using PP in automotive applications is discussed in Section 8.6, but also in the packaging area there are numerous positive examples. Realistic LCAs in this area [113, 116] actually conclude that, depending on the boundary conditions, even biopolymers may only offer marginal (if any) advantages over polyolefins.

While the PP base polymer will normally define the major part of energy demand, fillers or reinforcements as well as additives also contribute. For standard additives like the necessary antioxidants this is considered in the PlasticsEurope evaluation [114]. For reinforcing fibers, although glass fibers still dominate the market, the use of organic fibers from renewable sources is increasing in recent years [49]. Whether this improves the LCA and the sustainability in general depends, however, on the achieved thermomechanical performance of the thus-produced composites [117].

Normal LCAs will find their limits of comparability whenever toxic waste streams are involved in the production process. Propene and ethylene monomers themselves are non-toxic asphyxiant gases whose flammable nature calls for the usual extensive fire precautions, as ignition by static electricity is possible with high-pressure leaks and accidental discharges. Monomer recycling is a standard element in industrial production processes, but some waste gas will still end up in traditional flare stacks or modern ground flares. In the latter, the visual impact of flares as well as the noise level is reduced significantly. Since emission certificates for CO_2 are a relevant cost factor in many countries worldwide, producers are working continuously on reducing flaring volumes.

Although catalyst components constitute only a very small proportion of plant materials, both their manufacture and disposal involve noxious substances. Titanium tetrachloride is used for all catalysts, sometimes in considerable excess for high-activity systems. Whereas expensive recycle and recovery stages may be installed as capacities increase, some remaining complex residues will require disposal. Often, this involves hydrolysis with water to generate corrosive HCl and titanium dioxide suspensions. After neutralization with alkali, these slurries are either discharged as aqueous effluent or settled out and used in landfill. Lower aluminum alkyls are pyrophoric, react violently with water, and are intensely aggressive towards exposed skin. Safe handling usually calls for full protective clothing. Disposal can be by burning in special facilities to generate alumina, or quite often, by controlled hydrolysis with water after deactivation with alcohols. Some alkyl manufacturing plants will accept returned waste alkyls for disposal in their own dedicated facilities.

In contrast to those toxic ingredients the catalyst residues in the final polymer are not critical as:

- The Mg, Ti, and Al residues are mostly oxides or hydroxides from the deactivation step with water vapor (free HCl or other residual chlorides are handled by acid scavengers, see Section 6.1)
- The amount is very low (<400 ppm) due to the high productivity of modern catalysts.

9.2. Mechanical Recycling

The chemical structure and inherent properties of PP make it well suited to recycling operations [118]. There is no fear about cross-linking, nor of complications with plasticizers or chlorine-rich species. Re-extruding PP usually lowers its molecular mass and narrows the molecular mass distribution, particularly in aged feedstock. This can be helpful in some applications requiring enhanced flowability, but it negatively affects the mechanical performance and especially the impact strength. Other factors to be considered in reprocessing and aging are loss of oxidation stability, discoloration, contamination, and odor development.

As a rule, increasing heterogeneity (even within the same polymer class), contamination, and age of the starting material for mechanical recycling will reduce the possibilities for producing high-value secondary grades. Even elaborate cleaning and sorting operations have been generally found to be unable to really purify and upgrade highly mixed waste streams [119].

In PP production, off-specification materials will result from grade transitions, production faults, technical flaws, and test runs for the introduction of new grades. Resulting deviations in MFR, copolymer characteristics, and the like will determine whether the respective material can be added to prime products in minor amounts in a process normally called trimming. Otherwise, these are pelletized separately and re-classified as low-requirement products to be sold to processors or independent compounders.

Recycling in the conversion industry is based on the fact that turning PP into moldings, pipes, film, etc. invariably involves some scrap production. Apart from substandard articles, there are sprues from moldings, edge trim from film and sheet, parison waste, and stamped sheet from thermoforming. These are recovered for recycling by the converter himself if the application allows, or sometimes by an external compounder. Problems may arise here due to stabilizer consumption in multiple melting processes as well as by the presence of foreign polymers like in multilayer co-extrusion.

A quite different situation arises when trying to recycle PP after its service life, as in recycling of post-consumer plastic waste [118, 119]. Unless collected separately, thermoplastic materials make up 8–12 wt% of the municipal solid waste (MSW) stream in industrialized countries and will only contribute positively to the calorific value of MSW if incinerated (see below). Recovery of a useable plastic fraction from MSW after undifferentiated disposal is a hopeless task, but even the recovery and reuse of mixed plastics is quite difficult. The development of specific low-profile applications and fabrication processes for such comingled polymers, which are largely incompatible and contaminated, has natural limits. Using mixed thermoplastic waste for massive products, such as posts, construction boards, and other bulky structures for which resistance to water and decay is more important than high inherent strength is an established process but certainly not the perfect solution for efficient recycling.

Separation of specifically collected plastic waste is possible by a combination of sorting, grinding, density separation, and compounding processes, and specialized recycling companies are able to produce a number of secondary plastic materials. Both mixed polyolefins with a PP content of 40–70 wt% and nearly pure secondary PP materials are available on the European market, but always limited in terms of applicability in a number of respects: The color is necessarily dark (from gray to black), properties can vary significantly from lot to lot, and the smell is frequently problematic [120]. All of these factors limit applicability.

A number of measures appear possible to improve the potential of post-consumer plastic waste recycling. In the packaging area, the development of mono-material solutions or at least a reduction of the nonthermoplastic content will facilitate recycling [121]. Closed-loop recycling systems are another option for large-volume applications and especially helpful in recycling high-performance materials in technical applications.

9.3. Chemical and Energetic Recycling

PP can be cracked to liquid hydrocarbons by thermal pyrolysis at 400–550°C or even lower temperatures in the presence of suitable catalysts. Other thermoplastics, such as PE and PS, can also be pyrolyzed to liquid products, allowing mixed feedstock recycling for MSW fractions not suitable for mechanical recycling processes [122].

Modern MSW incineration technology, with energy recovery in the form of electricity generation, can take advantage of the high calorific value of thermoplastics. Municipal waste calorific values average 10 kJ/g, compared with 30–35 kJ/g for coal. At 44 kJ/g, PP and other polyolefins have the same calorific value as fuel oil. Moreover, PP has an exceptionally low sulfur content, which makes it a very clean fuel with no apprehensions about toxic flue-gas emissions.

References

1. R. Mülhaupt, *Angew. Chem.* **116** (2004) 1072–1080.
2. DE973626, 1953 (K. Ziegler, H. Breil, E. Holzkamp, H. Martin).
3. F.M. McMillan: *The Chain Straighteners*, McMillan Press, London 1979.
4. N. Pasquini, E.P. Moore, Jr.: "Introduction" in N. Pasqini (ed.): *Polypropylene Handbook*, Hanser Publishers, Munich 2005, pp. 3–13.
5. M. Gahleitner, J.R. Severn: "Designing Polymer Properties" in J.R. Severn, J.C. Chadwick (eds.): *Tailor-Made Polymers Via Immobilization of Alpha-Olefin Polymerization Catalysts*, Wiley-VCH, Weinheim 2008.
6. A. Keller, *Phil. Mag.* **2** (1957) 1171–1175.
7. P. Pino, G. Moretti, *Polymer* **28** (1987) 683–692.
8. Montecatini SGpI, DE1293453, 1956 (G. Natta, G. Mazzanti, G. Boschi).
9. M. Gahleitner, C. Paulik, W. Neissl, *Kunstst. Int.* **100** (2010) 18–26.
10. IHS Chemical (formerly CMAI) database; Supply/demand balances issued February 2013 (accessed July 2013).
11. H. Schwager, *Kunststoffe* **82** (1992) 499–501.
12. P.S. Chum, K.W. Swogger, *Prog. Polym. Sci.* **33** (2008) 797–819.
13. J.B.P. Soares, T.F.L. McKenna: *Polyolefin Reaction Engineering*, Wiley-VCH, Weinheim 2012, p. 11 ff.
14. R. Silvestri, L. Resconi, A. Pelliconi: "Metallocenes '95", *1st Int. Congr. Metallocene Polym.*, Brussels, April 26–27, 1995, p. 209.
15. C. Maier, T. Calafut: *Polypropylene: The Definitive Users Guide and Databook*, Plastics Design Library, Norwich 1998, p. 11.
16. R.A. Phillips, M.D. Wolkowicz: "Polypropylene Morphology" in N. Pasquini (ed.): *Polypropylene Handbook*, Hanser Publishers, Munich 2005, pp. 147–264.
17. B. Wunderlich: *Macromolecular Physics*, vol. 3, Academic Press, New York 1980, p. 63.
18. J. Brandrup, E.H. Immergut: *Polymer Handbook*, 3rd ed., Wiley, New York 1989, V27.
19. A. Turner-Jones, *Makromol. Chem.* **75** (1964) 134–158.
20. C. Grein, *Adv. Polym. Sci.* **188** (2005) 43–104.
21. M. Gahleitner, P. Doshev, C. Tranninger, *J. Appl. Polym. Sci.* **130** (2013) 3028–3037.
22. M. Ruff, C. Paulik, *Macromol. React. Eng.* **6** (2012) 302–317.
23. D. Bigiavi, M. Covezzi, M. Dorini, R.T. LeNoir, R.B. Lieberman, D. Malucelli, G. Mei, G. Penzo: "Manufacturing" in N. Pasquini (ed.): *Polypropylene Handbook*, Hanser Publishers, Munich 2005, pp. 361–380.
24. H. Mark et al. (eds.): *Encyclopaedia of Polymer Science and Engineering*, vol. 13, Wiley, New York 1987, pp. 472–473.
25. A. Garcia-Penas, J.M. Gomez-Elvira, E. Perez, M.L. Cerrada, *J. Polym. Sci. A Polym. Chem.* **51** (2013) 3251–3259.
26. Sasol, US6111047, 2000 (D.J. Joubert, A.H. Potgieter, I.H. Potgieter, I. Tincul).
27. E. Albizzati, G. Cecchin, J.C. Chadwick, G. Collina, U. Giannini, G. Morini, L. Noristi: "Ziegler-Natta Catalysts and Polymerizations" in N. Pasquini (ed.): *Polypropylene Handbook*, Hanser Publishers, Munich 2005, pp. 15–106.
28. Solvay et Cie, US4210738, 1980 (J.P. Hermans, P. Henriqulie); US3769233, 1973 (J.P. Hermans, P. Henriqulie).
29. R.P. Nielson in R.P. Quirke (ed.): *Transition Metal Catalysed Polymerizations*, vol. 4, part A, Harwood Academic Publishers, New York 1983, p. 61.
30. P. Fiasse, International Conference, Polypropylene–the Way Ahead, Madrid, Nov. 9–10, 1989, paper 4/1.
31. Borealis, WO/2003/000757, 2002 (P. Denifl, T. Leinonen).
32. R. Mühlhaupt, *Macromol. Chem. Phys.* **204** (2003) 289–327.
33. W. Kaminsky, *Catal. Today* **62** (2000) 23–34.
34. P.S. Chum, W.J. Kruper, M.J. Guest, *Adv. Mater.* **12** (2000) 1759–1767.
35. L.L. Böhm, *Angew. Chem. Int. Ed.* **42** (2003) 5010–5030.
36. P. Cossee, *J. Catal.* **3** (1964) 80–88.
37. E.J. Arlman, *J. Catal.* **3** (1964) 89–98.
38. E.J. Arlman, P. Cossee, *J. Catal.* **3** (1964) 99–104.
39. B. Liu, T. Nitta, H. Nakatani, M. Terano, *Macromol. Chem. Phys.* **204** (2003) 395–402.
40. J.B.P. Soares, T.F.L. McKenna: *Polyolefin Reaction Engineering*, Wiley-VCH, Weinheim 2012, Chap. 4, pp. 87–129.
41. J. Qiao, M. Guo, L. Wang, D. Liu, X. Zhang, L. Yu, W. Songab, Y. Liuab, *Polym. Chem.* **2** (2011) 1611–1623.
42. M. Covezzi, G. Mei, *Chem. Eng. Sci.* **56** (2001) 4059–4067.
43. M. Gahleitner, *Prog. Polym. Sci.* **26** (2001) 895–944.
44. H. Zweifel (ed.): *Plastics Additives Handbook*, Hanser Verlag, Munich 2001.
45. M. Gahleitner, C. Grein, S. Kheirandish, J. Wolfschwenger, *Int. Polym. Proc.* **26** (2011) 2–20.
46. A. Menyhárd, M. Gahleitner, J. Varga, K. Bernreitner, P. Jääskeläinen, H. Øysæd, B. Pukánszky, *Eur. Polym. J.* **45** (2009) 3138–3148.
47. Y.W. Leong, M.B. Abu Bakar, Z.A. Mohd-Ishak, A. Ariffin, B. Pukánszky, *J. Appl. Polym. Sci.* **91** (2004) 3315–3326.
48. J.H. Phelps, A.I.A. El-Rahman, V. Kunc, C.L. Tucker III, *Compos. A* **51** (2013) 11–21.
49. H. Ku, H. Wang, N. Pattarchaiyakoop, M. Trada, *Compos. B* **42** (2011) 856–873.
50. C. Tzoganakis, J. Vlachopoulos, A.E. Hamielec, *Polym. Eng. Sci.* **28** (1988) 170–180.
51. M. Rätzsch, M. Arnold, E. Borsig, H. Bucka, N. Reichelt, *Prog. Polym. Sci.* **27** (2002) 1195–1282.
52. N. Reichelt, M. Stadlbauer, R. Folland, C.B. Park, J. Wang, *Cellular Polym.* **22** (2003) 315–327.
53. M. Gahleitner, J. Wolfschwenger, K. Bernreitner, W. Neißl, C. Bachner, *J. Appl. Polym. Sci.* **61** (1996) 649–657.
54. R. Phillips, G. Herbert, J. News, M. Wolkowicz, *Polym. Eng. Sci.* **34** (1994) 1731–1743.
55. R. Pantani R., L. Balzano, G.W.M. Peters, *Macromol. Mater. Eng.* **297** (2012) 60–67.
56. H.H. Brintzinger, D. Fischer, *Adv. Polym. Sci.* (2013) DOI: 10.1007/12_2013_215.
57. A. Schöbel, E. Herdtweck, M. Parkinson, B. Rieger, *Chem. Eur. J.* **18** (2012) 4174–4178.
58. C. De Rosa, F. Auriemma, *Prog. Polym. Sci.* **31** (2006) 145–237.
59. D. Del Duca, D. Malucelli, G. Pellegatti, D. Romanini: "Product Mix and Properies" in N. Pasquini (ed.): *Polypropylene Handbook*, Hanser Verlag, Munich 2005, pp. 307–357.
60. M. Gahleitner, P. Jääskeläinen, E. Ratajski, C. Paulik, J. Reussner, J. Wolfschwenger, W. Neißl, *J. Appl. Polym. Sci.* **95** (2005) 1073–81.
61. P. Galli, F. Milani, T. Simonazzi, *Polym. J.* **17** (1985) 37–55.
62. C. De Rosa, F. Auriemma, P. Vollaro, L. Resconi, I. Camurati, *Macromolecules* **44** (2011) 540–549.
63. G. Cecchin, G. Morini, G. Pelliconi, *Macromol. Symp.* **173** (2001) 195–209.
64. J. Xu, L. Feng, *Eur. Polym. J.* **36** (2000) 867–878.
65. T. Macko, A. Ginzburg, K. Remerie, R. Bruell, *Macromol. Chem. Phys.* **213** (2012) 937–944.

66. D. Bouzid, T.F.L. McKenna, *Macromol. Chem. Phys.* **207** (2006) 13–19.
67. I. Kotter, W. Grellmann, T. Koch, S. Seidler, *J. Appl. Polym. Sci.* **100** (2006) 3364–3371.
68. T.C. Yu, D.K. Metzler: "Metallocene Plastomers as Polypropylene Impact Modifiers" in H.G. Karian (ed.): *Handbook of Polypropylene and Polypropylene Composites*, Marcel Dekker, New York 2003, pp. 235–295.
69. M. Gahleitner, A. Hauer, K. Bernreitner, E. Ingolic, *Intern. Polym. Proc.* **17** (2002) 318–24.
70. L.-P. Li, B. Yin, M.-B. Yang, *Polym. Eng. Sci.* **51** (2011) 2425–2433.
71. B. Jurkowski, S.S. Pesetskii: "Functionalized Polyolefins and Aliphatic Polyamide Blends: Interphase Interactions, Rheology, and High Elastic Properties of Melts" in D. Nwabunma, T. Kyu (eds.): *Polyolefin Blends*, Wiley, New York 2008, 527–555.
72. J. Li, H. Li, C. Wu, Y. Ke, D. Wang, Q. Li, L. Zhang, Y. Hua, *Eur. Polym. J.* **45** (2009) 2619–2628.
73. M. Gahleitner, C. Kock, E. Pachner, T. Pham, K. Stubenrauch, M. Tranninger, P. Popp, *Kunstst. Int.* **101** (2011) 24–30.
74. C.G. Oertel, T. Zwygers, P. Sgarzi: "Applications" in N. Pasqini (ed.): *Polypropylene Handbook*, Hanser Verlag, Munich 2005, pp. 517–546.
75. G.W.M. Peters, L. Balzano, R.J.A. Steenbakkers: "Flow-Induced Crystallization" in E. Piorkowska, G.C. Rutledge (eds.): *Handbook of Polymer Crystallization*, Wiley, New York 2013, pp. 399–431.
76. M. Fujiyama, T. Wakino, *Int. Polym. Proc.* **7** (1992) 97–105.
77. EN ISO 1873-2:2000, Polypropylene (PP) moulding and extrusion materials, Part 2: Preparation of test specimens and determination of properties.
78. J.H. Schut: "Breaking Into Medical Films", *Plast. Technol.*, Nov. 2003, 48–53.
79. D.V. Rosato, A.V. Rosato, D.P. DiMattia: *Blow Molding Handbook*, Hanser Gardner, Munich 2003, 75–119.
80. J.-C. Yu, X.-X. Chen, T.-R. Hung, *J. Intell. Manuf.* **15** (2004) 625–634.
81. M. Gahleitner, M. Kirchberger, K. Bernreitner, B.R. Kona, L. Blayac, *Kunstst. Int.* **100** (2010) 25–27.
82. A. Addeo, E.P. Moore Jr.: "Fabrication Processes" in N. Pasqini (ed.): *Polypropylene Handbook*, Hanser Verlag, Munich 2005, pp. 381–449.
83. M.M. Denn: *Polymer Melt Processing*, Cambridge University Press, Cambridge 2008, Chap. 7, pp. 83–108.
84. E. Andreassen, O.J. Myrhe, E.L. Hinrichsen, K. Grostad, *J. Appl. Polym. Sci.* **52** (1994) 1505–1517.
85. T. Matsuo, *Text. Prog.* **40** (2009) 87–121.
86. I.M. Ward, P.J. Hine, *Polymer* **45** (2004) 1423–1437.
87. Á. Kmetty, T. Bárány, J. Karger-Kocsis, *Prog. Polym. Sci.* **35** (2010) 1288–1310.
88. H. Bongaerts: "Flat Film Extrusion Using Chill-roll Casting" in F. Hensen (ed.): *Plastics Extrusion Technology*, 2nd ed., Hanser Publishers, Munich 1997, pp. 161–222.
89. K. Resch, G.M. Wallner, C. Teichert, G. Maier, M. Gahleitner, *Polym. Eng. Sci.* **46** (2006) 520–531.
90. N.J. Macauley, E.M.A. Harkin-Jones, W.R. Murphy, *Polym. Eng. Sci.* **38** (1998) 516–523.
91. R.A. Phillips, T. Nguyen, *J. Appl. Polym. Sci.* **80** (2001) 2400–2415.
92. L. Reade L.,*Film Sheet Extrusion*, June/July 2011, 23–26.
93. K. Xiao, R. Armstrong, I.-H. Lee, *Proc. SPE ANTEC* **70** (2012) 426.
94. M. Strobel, V. Jones, C.S. Lyons, M. Ulsh, M.J. Kushner, R. Dorai, M.C. Branch, *Plasma Polym.* **8** (2003) 61–95.
95. M. Gahleitner, C. Grein, R. Blell, J. Wolfschwenger, T. Koch, E. Ingolic, *eXPRESS Polym. Lett.* **5** (2011) 788–798.
96. V. Altstädt, A. Mantey: *Thermoplast-Schaumspritzgießen*, Carl Hanser Verlag, Munich 2010, 45–101.
97. C.B. Park, L.K. Cheung, *Polym. Eng. Sci.* **37** (1997) 1–10.
98. A.D. Gotsis, B.L.F. Zeevenhoven, A.H. Hogt, *Polym. Eng. Sci.* **44** (2004) 973–982.
99. J. Karger-Kocsis, *Polym. Compos.* **21** (2000) 514–522.
100. J. Markarian, *Plast. Eng.* **9** (2011) 22–29.
101. F. Garesci, S. Fliegener, *Compos. Sci. Technol.* **85** (2013) 142–147.
102. E. Ernst, J. Reußner, P. Poelt, E. Ingolic, *J. Appl. Polym. Sci.* **97** (2005) 797–805.
103. M. Soliman, F. Essers, J. Cremers, *Proc. SPE ANTEC* **63** (2005) 2146–2150.
104. T. Koch, D. Machl, *Polym. Testing* **26** (2007) 927–936.
105. V. Vasić, S. Schiesser, *Macromol. Symp.* **296** (2010) 566–574.
106. L.E. Janson: *Plastic Pipes for Water Supply and Sewage Disposal*, Borealis, Majornas CopyPrint AB, Stockholm 2003.
107. EN ISO 15874-2:2013, Plastics piping systems for hot and cold water installations–Polypropylene (PP), Part 2: Pipes.
108. S. Nestelberger, H. Herbst, C.-G. Ek, *3R Int.* 02/2007, 87–90.
109. R. Mühlhaupt, *Macromol. Chem. Phys.* **214** (2013) 159–174.
110. M.R. Yates, C.Y. Barlow, *Res. Cons. Rec.* **78** (2013) 54–66.
111. D.J. Dijkstra, G. Langstein, *Polym. Int.* **61** (2012) 6–8.
112. A.L. Andrady, M.L. Neal, *Phil. Trans. R. Soc. B* **364** (2009) 1977–1984.
113. A. Azapagic, A. Emsley, I. Hamerton: *Polymers, the Environment and Sustainable Development*, Wiley, Chichester 2003, 17–46.
114. Environmental Product Declarations of the European Plastics Manufacturers: Polypropylene (PP), Plastics Europe, Brussels 2008, download from http://www.plasticseurope.org/ (accessed 08.10.2013).
115. S.H. Patel, M. Xanthos, *Adv. Polym. Technol.* **20** (2001) 22–41.
116. G.M. Bohlmann, *Environ. Prog.* **23** (2004) 342–346.
117. S.V. Joshi, L.T. Drzal, A.K. Mohanty, S. Arora, *Compos. A* **35** (2004) 371–376.
118. M. Xanthos, *Science* **337** (2012) 700–702.
119. S.M. Al-Salem, P. Letteri, J. Baeyens, *Waste Manag.* **29** (2009) 2625–2643.
120. G. Wypych: *Handbook of Odors in Plastic Materials*, ChemTec Publishing, Toronto 2013.
121. K. Verghese, H. Lewis, L. Fitzpatrick (eds.): *Packaging for Sustainability*, Springer-Verlag, London 2012.
122. J. Scheirs, W. Kaminsky, *Feedstock Recycling and Pyrolysis of Waste Plastics: Converting Waste Plastics into Diesel and Other Fuels*, Wiley, Chichester 2006.